T0180527

Smart Innovation, Systems and Technologies

Volume 68

Series editors

Robert James Howlett, Bournemouth University and KES International,
Shoreham-by-sea, UK
e-mail: rjhowlett@kesinternational.org

Lakhmi C. Jain, University of Canberra, Canberra, Australia;
Bournemouth University, UK;
KES International, UK
e-mails: jainlc2002@yahoo.co.uk; Lakhmi.Jain@canberra.edu.au

About this Series

The Smart Innovation, Systems and Technologies book series encompasses the topics of knowledge, intelligence, innovation and sustainability. The aim of the series is to make available a platform for the publication of books on all aspects of single and multi-disciplinary research on these themes in order to make the latest results available in a readily-accessible form. Volumes on interdisciplinary research combining two or more of these areas is particularly sought.

The series covers systems and paradigms that employ knowledge and intelligence in a broad sense. Its scope is systems having embedded knowledge and intelligence, which may be applied to the solution of world problems in industry, the environment and the community. It also focusses on the knowledge-transfer methodologies and innovation strategies employed to make this happen effectively. The combination of intelligent systems tools and a broad range of applications introduces a need for a synergy of disciplines from science, technology, business and the humanities. The series will include conference proceedings, edited collections, monographs, handbooks, reference books, and other relevant types of book in areas of science and technology where smart systems and technologies can offer innovative solutions.

High quality content is an essential feature for all book proposals accepted for the series. It is expected that editors of all accepted volumes will ensure that contributions are subjected to an appropriate level of reviewing process and adhere to KES quality principles.

More information about this series at http://www.springer.com/series/8767

Giampaolo Campana · Robert J. Howlett
Rossi Setchi · Barbara Cimatti
Editors

Sustainable Design and Manufacturing 2017

Selected papers on Sustainable Design and Manufacturing

 Springer

Editors
Giampaolo Campana
Università di Bologna
Bologna
Italy

Robert J. Howlett
Bournemouth University
Poole, Dorset
UK

Rossi Setchi
Cardiff University
Cardiff
UK

Barbara Cimatti
Università di Bologna
Bologna
Italy

ISSN 2190-3018   ISSN 2190-3026   (electronic)
Smart Innovation, Systems and Technologies
ISBN 978-3-319-86071-8   ISBN 978-3-319-57078-5   (eBook)
DOI 10.1007/978-3-319-57078-5

Printed on acid-free paper

This Springer imprint is published by Springer Nature
The registered company is Springer International Publishing AG
The registered company address is: Gewerbestrasse 11, 6330 Cham, Switzerland

Foreword

For over a decade, the mission of KES International has been to provide a professional community, networking and publication opportunities for all those who work in knowledge-intensive subjects. At KES, we are passionate about the dissemination, transfer, sharing and brokerage of knowledge. The KES community consists of several thousand experts, scientists, academics, engineers, students and practitioners who participate in KES activities.

KES Conferences

For nearly 20 years, KES has run conferences in different countries of the world on leading-edge topics:

- Intelligent systems: including intelligent decision technologies, intelligent interactive multimedia systems and services, agent and multi-agent systems and smart education and e-Learning
- Sustainable Technology: including sustainability in energy and buildings, smart energy and sustainable design and manufacturing
- Innovation, knowledge transfer, enterprise and entrepreneurship: including innovation and knowledge transfer and innovation in medicine and health care
- Digital media: including archiving tomorrow and innovation in music

KES Journals

KES edits a range of journals and serials on knowledge-intensive subjects:

- International Journal of Knowledge Based and Intelligent Engineering Systems
- Intelligent Decision Technologies: an International Journal

- In Impact: the Journal of Innovation Impact
- Sustainability in Energy and Buildings: Research Advances
- Advances in Smart Systems Research

Book Series

KES edits the Springer book series on Smart Innovation, Systems and Technologies. The series accepts conference proceedings, edited books and research monographs. KES Transactions (published by Future Technology Press) is a book series containing the results of applied and theoretical research on a range of leading-edge topics. Papers contained in KES Transactions may also appear in the KES Open Access Library (KOALA), our own online gold standard open access publishing platform.

Training and Short Courses

KES can provide live and online training courses on all the topics in its portfolio. KES has good relationships with leading universities and academics around the world, and can harness these to provide excellent personal development and training courses.

Dissemination of Research Project Results

It is essential for research groups to communicate the outcomes of their research to those that can make use of them. But academics do not want to run their own conferences. KES has specialist knowledge of how to run a conference to disseminate research results. Or a research project workshop can be run alongside a conference to increase dissemination to an even wider audience.

The KES-IKT Knowledge Alliance

KES works in partnership with the Institute of Knowledge Transfer (IKT), the sole accredited body dedicated to supporting and promoting the *knowledge professional*: those individuals involved in innovation, enterprise, and the transfer, sharing and exchange of knowledge. The IKT accredits the quality of innovation and knowledge transfer processes, practices, activities, and training providers, and the professional status of its members.

About KES International

Formed in 2001, KES is an independent worldwide knowledge academy involving about 5000 professionals, engineers, academics, students and managers, operated on a not-for-profit basis, from a base in the UK. A number of universities around the world contribute to its organisation, operation and academic activities. KES International Operations Ltd is a company limited by guarantee that services the KES International organisation.

April 2017   Robert J. Howlett

Preface

Writing a book is always a cooperative work. Even in the case of a sole author, a number of colleagues, collaborators, assistants and often friends and family too are directly or indirectly involved. Writing a book is always a hard job because it requires a deep understanding and a willingness to provide useful information or concepts that will eventually help readers in developing their own projects or give inspiration for new ideas. **Writing a book means sharing ideas and sowing the seeds of new ones.**

Scientific books often gather the work of colleagues based on a specific topic, and this volume collates specifically the accepted papers of the Fourth International Conference on *Sustainable Design and Manufacturing*, SDM 2017. The event was organised and scientifically supervised by the University of Bologna, Italy, in collaboration with KES International, UK. The conference took place in Bologna on the 26–28 April 2017. The papers were submitted to a rigorous peer review before being accepted in order to guarantee the high scientific level of this publication.

The SDM conference is proposed by scientists and academics in collaboration with industrial partners and institutions with the aim of providing an occasion to share knowledge and to discuss and identify new challenges concerning the development and the application of the concept of sustainability, in particular as regards **design and manufacturing**. Industrial production is one of the main engines that drive social welfare and development, employing people and producing goods. The good design and fabrication of any industrial product is then fundamental not only to increase enterprise competitiveness, but also to improve society in general. Nowadays, neither design nor manufacturing can be considered good if they are not sustainable.

The combination between design and manufacturing proposed by this conference is quite unusual as these topics are often discussed in scientific conferences in separate sessions. This old approach considered these two activities sequential, while in the world of today, they are very much integrated and concurrent. This book echoes and embraces this important duality, and readers will find a number of ideas and directions regarding both design and manufacturing.

An **interdisciplinary approach** is necessary in the understanding of sustainability. The three well-known dimensions of sustainability are as follows: environmental, economic and social. Different competences are involved and needed to face sustainable issues and to find new and possible solutions. Any research concerning sustainability involves **different disciplines,** and the importance of **crossing competences** and professions is universally recognised. Scientific events such as the SDM conference, where scholars from different fields and countries can meet and young scientists can experience an interdisciplinary and multidisciplinary surrounding, provide an excellent platform and opportunity for the **cross-fertilisation** of ideas and solutions.

The main topics presented in this volume are as follows: sustainable design, innovation and services; sustainable manufacturing processes and technology; sustainable manufacturing systems and enterprises; decision support for sustainability.

Some articles are also dedicated to the following subjects:

– Business model innovation for sustainable design and manufacturing.
– Resource and energy efficiency for sustainability advances in process industries and business model innovation for sustainable design and manufacturing.
– Sustainability in industrial plant design & management: applications & experiences from practice.
– Sustainability of 3D printing and additive manufacturing.
– Sustainable mobility, solar vehicles and alternative solutions.
– Eco-design through systematic innovation.
– Sustainable materials such as renewable and eco-materials, bio-polymers and composites with natural fibres.
– Sustainable mobility, solar vehicles and alternative solutions.

With the increasing and pressing need for sustainability, the last two topics listed above are becoming ever more important and the attention of the scientific and industrial world towards sustainable materials is growing rapidly. The present challenge in the field of materials science also includes composites, which have introduced important innovation in industrial production but still present limits regarding recycling. At present, they cannot be considered the best solution from the point of view of sustainability, and for this reason, research is intensifying in this field.

Mobility is always an important societal challenge, and our quality of life can only be increased if the adopted solutions are sustainable. Sustainability of solar vehicles and of the numerous technical issues related to them, such as electric batteries and electric engines, is an important field of investigation that has produced relevant advancements.

This volume explores a number of these areas, and many valuable indications and practices are given.

The conference was opened by three significant keynote speakers: **Günther Seliger**, Chair of the GCSM (Global Conference on Sustainable Manufacturing) and Professor at the Department Assembly Technology and Factory Management of the *Technische Universität Berlin;* **I.S. Jawahir**, Director of the Institute for Sustainable Manufacturing at the *University of Kentucky* and Professor of Mechanical Engineering (James F. Hardymon Chair in Manufacturing Systems); **Shahin Rahimifard**, Director of the Centre for Sustainable Manufacturing and Reuse/Recycling Technologies at the *Loughborough University* and Professor of Sustainable Engineering.

The results of their studies have been published in a number of books and scientific papers and represent fundamental scientific literature for any researcher who deals with sustainable design and manufacturing. **Their seeds have been sown in a number of publications and also here**.

We would like to gratefully acknowledge all the researchers who contributed their work to realise this book. We thank the Scientific Committee and its members for their tireless revision of the papers here published and all the colleagues who chaired the sessions of this event. Heartfelt thanks are also extended to the Organising Committee for its fundamental support in making this conference possible, in particular Prof. Robert Howlett, as the Executive Chair of the SDM and as the KES International Executive Chair, and all the KES staff.

A final special thank to Springer, our publisher.

We hope that this conference proceedings book can be a useful publication, providing novel ideas and directions to develop new research concerning the topical issues of sustainability in design and manufacturing.

March 2017

Giampaolo Campana
Barbara Cimatti

Contents

Renewable Energies for Sustainable Manufacturing and Society

The Learning Supply Chain

Challenges and Opportunities of Clean Technology
in Production Engineering

Keynote Papers

Leverage of Industrial Engineering Education for Sustainable Manufacturing

Pinar Bilge$^{(\boxtimes)}$, Soner Emec, and Günther Seliger

Department of Machine Tools and Factory Management,
Technische Universität Berlin, Pascalstrasse 8, 10587 Berlin, Germany
{bilge,emec,seliger}@mf.tu-berlin.de

Abstract. The connected impact of management and technology can considerably contribute to achieve sustainability in global value creation. Industrial engineering started as an educational program at universities based on existing programs in business administration and engineering. The expanding application of this integrative approach of science and practical implementation in industry and in societal communities coins the architecture of how to implement sustainability in different societies. The practical case of setting up a joint program of universities in an early developed and an emerging country illustrates how potentials of industrial engineering can be exploited for promoting sustainable local manufacturing.

Keywords: Sustainable manufacturing · Industrial engineering · Transformative attributes · Engineering education

1 Introduction

A well-educated population is essential for local and global well-being, and sustainable development. Education and intense training play key roles in providing people with the capabilities needed to contribute to the development of sustainability. Industrial engineering (IE) can help demonstrating how sustainable manufacturing (SM) embedded in value creation becomes superior to traditional single paradigms of management and technology [1]. In traditional manufacturing engineering education, the paradigm of faster, more accurate and cheaper drives technological development. Criteria of cost and benefit drive education and practice in economics. Sustainability in its economic, environmental and social dimension is an approach to cope with the challenges of human development on earth with respect to coining wealth by business in fair partnership, by care about natural resources within the limits of their availability and by developing social life in manifold of cultural directions [2].

IE is well-established to be effective in practice and research in manufacturing. Focus on the education and implementation of IE is required to answer the question of how industrial engineers can act as change agents towards sustainability. SM is introduced in Sect. 2 in order to formulate IE objectives within educational frameworks. Section 3 presents how teaching and research in IE can contribute to sustainable value creation by innovative developments in engineering simultaneously using the dynamics of competition and cooperation in the global arena of modern logistics and

© Springer International Publishing AG 2017
G. Campana et al. (eds.), *Sustainable Design and Manufacturing 2017*, Smart Innovation, Systems and Technologies 68, DOI 10.1007/978-3-319-57078-5_1

communication [1]. A transformed IE undergraduate program, which is embedded in a newly established engineering school in cooperation between an early developed and an emerging country is presented in Sect. 4. A case integrated into practical components of the program demonstrate the leverage of IE for sustainable development.

2   Review

Potentials of methodologies in engineering and economics are explored for useful applications in industrial value creation in theory and practice. The review and analysis of the current state-of-the-art topics cover sustainable value creation architecture and with its IE attributes of education and practice. Sustainability has become an urgent requirement for both engineering and economics science, considering the limits of resources, growth and the unequal distribution of wealth.

2.1   Sustainable Manufacturing

Engineering is exploiting potentials for useful applications. Manufacturing, as a specific discipline in engineering, starts from human thinking and imagination, from knowledge about natural scientific phenomena, from physical materials and shapes value creation via processes in management and technology, objectified in tangible and intangible products, in physical artefacts and services [1].

Gaps of development between early developed and emerging countries can be closed by applying the approach of help for self-help [3]. Paradigms of manufacturing in historical development and a review of major publications related to SM and its conceptual and constituting elements concludes an architectural framework for value creation, as presented in Fig. 1. Value creation factors as products, processes, equipment, organization and people shaping value creation modules to be valuated according to sustainability impacts are rendered [4]. Perspectives of vertical and horizontal integration are pushed by cooperation and competition. This architecture can be used to model any value creation through SM.

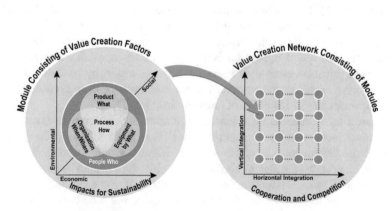

Fig. 1. Value creation architecture [4]

The solution approaches for sustainable development are specified in economic, environmental and social dimensions of value creation. Synergies of management and technology in these solution approaches are obvious:

- Economically, wealth can be achieved in the different areas of human living without increasing physical resource consumption by selling functionality rather than tangible products.
- Environmentally, non-renewable resources must not be disposed anymore but regained in product and material cycles. Chances of substituting them by renewables must be exploited, but only to the extent that renewables can be regained.
- Socially, a global village with less than one billion out of more than seven billion people consuming more than four fifths of global resources is hardly acceptable for living peacefully together. Teaching and learning for a global culture, wealth and health become vital tasks for the global human community [1].

If the lifestyles of upcoming and also developed communities will be shaped in the future by the existing, actually predominating technologies, then the resource consumption will exceed every accountable economic, environmental and social bound. Sustainable engineering represents a new scientific approach to cope with this challenge [1]. SM in its depth of technology in product development, processes and equipment and its breadth of managing human creativity, initiative and entrepreneurship has evolved a powerful leverage for organizing development and realizing physical products and services, business and wealth in the framework of regional and global markets [4]. The potentials of engineering with appropriate capabilities for leveraging SM should be exploited within this frame.

2.2 Industrial Engineering

How to adapt human living to the challenges of sustainability is considerable coined by capabilities provided by higher education. Following the European Qualifications Framework, engineering capabilities describe abilities to perform certain decisions and actions through a set of knowledge, skills and competence in various engineering disciplines [5]. IE provides capabilities to the challenges of SM. It is an area of applied science and professional activity dealing with interrelated technology and management [6]. IE is well-acknowledged with respect to engineering capabilities among others covering soft factors as awareness, motivation, application and transformation [7].

The German National Academy of Science and Engineering (acatech, in German: Deutsche Akademie der Technikwissenschaften) compared numerous courses and attributes of educational programs in IE from early developed countries in 2014 [8].

The comparison focused on identification of traditional IE education, which combines engineering and management education, as presented in Fig. 2. Around 50 curricula with around 40 modules have been analyzed to identify existing attributes by distinguishing the practices among early developed countries. Three attributes are identified as especially relevant for education and training [7]:

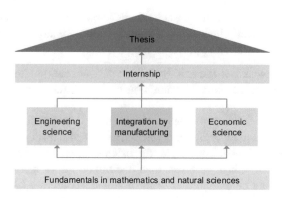

Fig. 2. Framework of traditional industrial engineering programs [8]

- Focus on an interdisciplinary profession that contains two main knowledge domains: technology and management. Interdisciplinary capabilities support engineers to understand conditions and requirements, explore technological and organizational opportunities from different perspectives, while increasing the efficiency of value creation.
- Focus on information and communication technologies, which enable the digitalization of information, data, and facts for many tasks and decisions. Digitalization of information, data, and facts, information technologies (IT) support the development and exploration of educational and practical opportunities to create value through integrated and automated solutions [9].
- Focus on problem-solving, which combines logical thinking with analysis and synthesis by applying methodological knowledge and skills. In IE, problem-solving combines logical thinking with analysis and synthesis by applying methodologies in technology and management [10].

Conflict of goals and challenges for sustainable value creation can be met by integrating different dimensions of evaluation in management and technology [11]. New attributes are needed to logically and sequentially link sustainable value creation to manufacturing.

3   Transformation

Enhancing technology and management both shaping IE by respective education and training provide knowledge, skills and motivation for initiative, creativity and hard work. The architecture of transformative shaping IE is specified as mutual relations between IE and SM [7]. Starting from traditional generic IE framework different attributes directed to competence development in IE for SM are proposed.

Transformative research according to United States National Science Foundation can overcome limits of only economic impacts by integrating environmental and social impacts in shaping manufacturing for sustainable value creation [12]. Transformation here is used as a term, which calls for a need with impactful changes without altering the fundamentals.

3.1 Requirements

Traditional attributes of IE used in practice, research, education and training are valid to create value in manufacturing. The rapid rise of IT has simplified access to information within projects and allowed for focus on interactions among stakeholders, disciplines, and regions. Simultaneously coping with the multiple challenges of SM can be achieved by IE applying transformative attributes for synthesizing solutions based on analysis of practice and research, education and training and embedded in appropriate organizational structures [13]. Discussions in round tables and workshops with different stakeholder groups, as well as interviews with individual scientists, instructors and lecturers, questionnaires with industry experts, alumni, and students, are used to determine new attributes for transformative IE. In addition to the three conventional attributes in IE, three transformative attributes are identified as especially relevant for IE education and training:

- Focus on projects that incorporate problem-solving, interdisciplinary teamwork, and project management, while emphasizing that there is hardly an individual optimal solution to any problem to increase effectiveness [14].
- Focus on sustainable solutions, which balance impacts to improve the current products and services by designing, operating and assessing value creation in manufacturing [7].
- Focus on glocalization, which is a made up word coined from a combination of the words "globalization" and "localization", which enables global environmental limitations determining the preferences for local actions and decisions [15].

3.2 Transformed Undergraduate Program

Shifting the focus of industrial engineers to SM requires the transformation of educational programs to provide future industrial engineers with the necessary capabilities for SM in terms of technical-methodological knowledge, skills and competence.

Following categories of courses in a transformed educational programs are identified and presented in Fig. 3 [7]: fundamentals in mathematics, natural sciences, engineering, economics, application in technology and management, SM, technical and international internships, soft skills, projects and thesis. All categories together aim to enable closed–loop synthesis by closing gaps of existing and by creating new solutions.

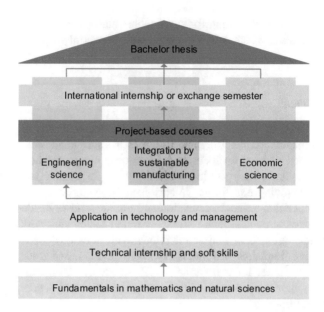

Fig. 3. Framework for transformed industrial engineering program [7]

4   Implementation

Good practices on educational programs in cooperation of early developed and emerging countries demonstrate future perspectives of transformative IE. The framework for transformed IE is exemplarily applied by a case study in education and training.

4.1   Embedding Industrial Engineering Within an Engineering School

A joint undergraduate program on IE with focus on SM is developed by adapting an existing program of an early developed country to the environment of an emerging country, also achieving feedback for improvement of courses. The educational hierarchical structure is adapted to new educational programs within an engineering school based on stakeholder interests [7]. A broader framework is required to embed a transformed IE program for education, implementation, training and research within the new university.

Yearly distribution of practical components involved in a four-year undergraduate degree program support gaining experiences in SM. Technical and international internships can leverage both awareness and motivation in companies and competence with students for sustainable value creation. Goals of the practical components, especially projects with students, regarding the enhancement of engineering capabilities were to

- train students in critical thinking through combining technological solutions and valuating them by addressing principles of SM,

- create empathy in order to increase the ability for recognizing and responding to the needs and requirements of stakeholders within the frame of regional conditions,
- raise awareness for synergies and conflict of goals within the ecosystem in order to improve the value created in local production systems,
- promote innovation and entrepreneurship to engage stakeholders, and
- demonstrate applications with competitive advantages for managerial decisions.

IE students learn and train, are enabled to synthesize solutions for SM, develop their personalities as globally thinking locally acting professionals in SM.

4.2   Sample Case – Energy and Water in Cypriot Agriculture

An undergraduate student team of six students from Germany, four from Cyprus, two from Turkey, one from Spain, and one from India with seven international researchers has executed an eight-weeks project-based course in cooperation of Germany and North Cyprus.

The production of agricultural goods in Cyprus has been mainly hampered by scarce water resources and rising energy prices. Climate changes trigger regulatory and economic counteractions as pumping ground water for irrigation and investment in new, deeper wells, including new filter technologies. Electricity prices for industry in North Cyprus have increased over 250% and annual consumer prices for fuel have increased over 300% in the last decade. Increased prices directly affect farmers, as diesel generators operate water pump systems for irrigating plants over a five-month dry period in summer from May to September. The demand of each farmer is proportional to both harvest quantity and the duration of agricultural activities. To cope with these challenges, modeling precisely the real demand of renewable resources for harvesting as well as access under variable seasonal availability is required [16].

Resource efficiency can be increased by use of regional, low priced and renewable resources for energy generation and water supply, which substitute expensive non-renewable resources. Goals include enabling the access to clean water over dry seasons, protecting of ground water level as well as applying of technologies to reuse the waste. Environmental impacts as CO_2 emissions are measured and evaluated.

Despite the current high unemployment rates, the Cypriot workforce, including farmers, has a high level of education. Workshops and meetings with the farmers how to use and share production facilities and generate renewable energy have increased awareness for sustainable value creation. Application–based software can train farmers through simple use of IT tools how to improve the capacity utilization and knowledge share to create synergies. The complete solution concept is presented in Fig. 4 and described in [16]. The hybrid generator encompasses available renewable resources in different combinations that are aligned with the energy demands of the three other value creation modules of production, atmospheric water generation, as well as water and biomass recycling.

As potential solution, the use of biomass, solar, wind, or mixed energy sources are examined to generate electricity for the power grid. Any solution including the

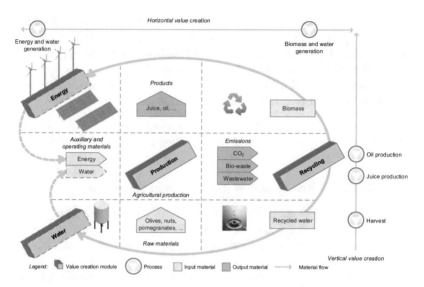

Fig. 4. Solution concept

resources and used technology for energy generation should meet the local requirements, as availability of energy resources, little space, and affordable life–cycle costs, independence of location as well as creating fewer CO_2 emissions, in contrast to currently used solutions. For example, a solution is rated positively if little or no CO_2 is emitted. The return on investment for the implementation of a new energy module is between four and six years depending on configuration and capacity [16].

A hybrid energy generator combines more than one resource to increase the availability of energy supply. It also decreases the energy expenses for the whole system with less operating costs. The hybrid generator consists of two components: clean energy and backup energy source. The clean energy component continuously generates the required energy from renewable resources and organic waste, for example, solar radiation, wind and biomass. The backup component is complemented with a diesel generator and storage batteries.

The score of a solution is built by the multiplied combination of the importance $p(c_k)$ of all categories c_k and its fulfillment $s\big(c_k(r_i), o_j\big)$ by a solution o_j. A score $S(o_j)$ presents a summed up score for the *jth* solution, as presented by Eq. (1):

$$S(o_j) = \sum_{k=1}^{u} \left[s\big(c_k(r_i), o_j\big) * p(c_k) \right]$$

$$= s\big(c_1(r_i), o_j\big) * p(c_1) + s\big(c_2(r_i), o_j\big) * p(c_2) + \ldots + s\big(c_u(r_i), o_j\big) * p(c_u)$$

(1)

The assessment methodology is illustrated for the water and energy module in Table 1. Specific evaluation criteria w_z are defined for each requirement.

This case is interdisciplinary, teamwork-based and project-oriented. Sustainability elements are inter-twinned with fundamentals of engineering science as measurement

Table 1. Evaluation of solutions for the energy module

Requirements		Importance	Solution				
			Diesel generator	Biomass generator	PV solar panel	Wind turbine	Hybrid generator
Productivity	Low area and space	1	3	1	0	0	1
	Low life-cycle costs	3	0	0	9	1	1
Resource efficiency	Temporal distribution of resources	9	9	3	1	3	9
	Low CO$_2$ emission	9	0	0	3	9	1
	High potential of resources	3	0	3	9	1	9
	Score		84	37	90	114	121

and physical tests, computer aided design, and fundamentals of economics science as business administration, marketing and accounting. Single aspects as assembly technology, energy management are integrated in a sustainable solution.

5   Conclusion

The analysis of IE education and training for contributions to SM maps engineering programs in the qualification framework of knowledge, skills and competences of awareness generation, motivation, application and transformation.

A conceptual framework coined by so-called transformative attributes of IE for implementing SM is presented within the current research. Interdisciplinarity, digitalization, problem-solving as well as focus on projects, sustainable solutions and glocalization are identified as transformative attributes.

A case of sustainability-oriented innovation in an emerging country is presented. Technological and management methodologies are applied to create sustainable local solutions with reference to global criteria of sustainable development. Thus, IE education and training can leverage SM practice.

References

1. Seliger, G.: Sustainability engineering by product-service systems. In: Hesselbach, J., Herrmann, C. (eds.) Glocalized Solutions for Sustainability in Manufacturing. Proceedings of the 18th CIRP International Conference on Life Cycle Engineering, Braunschweig, Germany, May 2–4, pp. 22–28. Springer, Heidelberg (2011)

2. UNESCO - United nations educational, scientific and cultural organization: declaration on the responsibilities of the present generations towards future generations, 12 November 1997. http://portal.unesco.org/en/ev.php-URL_ID=13178&URL_DO=DO_TOPIC&URL_SECTION=201.html. Accessed 08 Apr 2016

3. Postawa, A.B., Siewert, M., Seliger, G.: Mini factories for cocoa paste production. In: Seliger, G. (ed.) Sustainable Manufacturing. Shaping Global Value Creation, 9th Global Conference on Sustainable Manufacturing, Saint Petersburg, Russia, September 28th–30th, pp. 183–189. Universitätsverlag der TU Berlin, Berlin (2011). doi:10.1007/978-3-642-27290-5_27

4. CRC 1026 - Collaborative Research Centre 1026: Sustainable Manufacturing (2016). http://www.sustainable-manufacturing.net/en_GB/;jsessionid=9CAFE96CC87F534AA61162C732AF4441. Accessed 02 May 2016

5. EU - European Union: recommendation of the european parliament and of the council of 23 April 2008 on the establishment of the european qualifications framework for lifelong learning. In: Official Journal of the European Union, C 111/1 (2008)

6. Martin-Vega, L.A.: Chapter 1.1. The Purpose and evolution of industrial engineering. In: Maynard, H.B., Zandin, K.B. (eds.) Maynard's Industrial Engineering Handbook. McGraw-Hill standard handbooks, 5th edn., pp. 3–19. McGraw-Hill, New York (2001)

7. Bilge, P., Seliger, G., Badurdeen, F., Jawahir, I.S.: A novel framework for achieving sustainable value creation through industrial engineering principles. Procedia CIRP (2016). doi:10.1016/j.procir.2016.01.126

8. Schuh, G., Warschat, J.: Potenziale einer Forschungsdisziplin Wirtschaftsingenieurwesen. acatech DISKUSSION. acatech (2013). http://www.acatech.de/fileadmin/user_upload/Baumstruktur_nach_Website/Acatech/root/de/Publikationen/acatech_diskutiert/140130_acatech_DISKUSSION_Wirtschaftsingenieurwesen_WEB_final.pdf. Accessed 30 May 2014

9. Jovane, F., Yoshikawa, H., Alting, L., Boër, C., Westkamper, E., Williams, D., Tseng, M., Seliger, G., Paci, A.: The incoming global technological and industrial revolution towards competitive sustainable manufacturing. CIRP Ann. Manufact. Technol. (2008). doi:10.1016/j.cirp.2008.09.010

10. Bruce, B.C., Bloch, N.: Learning by doing. In: Seel, N.M. (ed.) Encyclopedia of the Sciences of Learning. Springer Reference, 1st edn., pp. 1821–1824. Springer, Berlin (2012)

11. Madni, A.M.: Transdisciplinarity. reaching beyond disciplines to find connections. J. Integr. Des. Process Sci. 11(1), 1–11 (2007)

12. NSF - National Science Foundation: Definition of Transformative Research (2015). http://www.nsf.gov/about/transformative_research/. Accessed 13 Jan 2016

13. Bilge, P., Badurdeen, F., Seliger, G., Jawahir, I.S.: A novel manufacturing architecture for sustainable value creation. CIRP Ann. Manufact. Technol. (2016). doi:10.1016/j.cirp.2016.04.114

14. Dankers, W.: Decision Making. In: Laperrière, L., Reinhart, G. (eds.) CIRP Encyclopedia of Production Engineering, 1st edn, pp. 363–367. Springer, Berlin (2014)

15. Hesselbach, J., Herrmann, C. (eds.): Glocalized Solutions for Sustainability in Manufacturing, Proceedings of the 18th CIRP International Conference on Life Cycle Engineering, Braunschweig, Germany, May 2–4. Springer, Heidelberg (2011)

16. Emec, S., Bilge, P., Seliger, G.: Design of production systems with hybrid energy and water generation for sustainable value creation. Clean. Techn. Environ. Policy (2015). doi:10.1007/s10098-015-0947-4

Forging New Frontiers in Sustainable Food Manufacturing

Shahin Rahimifard[(✉)], Elliot Woolley, D. Patrick Webb,
Guillermo Garcia-Garcia, Jamie Stone, Aicha Jellil,
Pedro Gimenez-Escalante, Sandeep Jagtap, and Hana Trollman

Centre for Sustainable Manufacturing and Recycling Technologies,
Loughborough University, Loughborough, UK
s.rahimifard@lboro.ac.uk

Abstract. One of the most prominent challenges commonly acknowledged by modern manufacturing industries is "how to produce more with fewer resources?" Nowhere is this more true than in the food sector due to the recent concerns regarding the long-term availability and security of food products. The unique attributes of food products such as the need for fresh perishable ingredients, health risks associated with inappropriate production environment, stringent storage and distributions requirements together with relatively short post-production shelf-life makes their preparation, production and supply considerably different to other manufactured goods. Furthermore, the impacts of climate change on our ability to produce food, the rapidly increasing global population, as well as changes in demand and dietary behaviours both within developed and developing countries urgently demands a need to change the way we grow, manufacture and consume our food products. This paper discusses a number of key research challenges facing modern food manufacturers, including improved productivity using fewer resources, valorisation of food waste, improving the resilience of food supply chains, localisation of food production, and utilisation of new sustainable sources of nutrition for provision of customised food products.

Keywords: Resource efficient food manufacturing · Valorisation of food waste · Resilience in food supply · Provision of customised food products

1 Introduction

Global Food Supply Chains (GFSCs) are highly complex systems developed in response to modern consumer demands for trusted food products and services, and improved choice and quality at lower prices. GFSCs within the majority of developed countries and increasingly in emerging economies are often dominated by a small number of large retailers and their supply networks. These in turn have evolved around management paradigms focussed on cost minimisation and service optimisation based on models of 'Centralised Production' (determined by economies of scale and cheap labour) and 'Just in Time' approaches to eliminate non-value adding activities. Whilst such management systems have been well suited to times of stability, they are now

© Springer International Publishing AG 2017
G. Campana et al. (eds.), *Sustainable Design and Manufacturing 2017*, Smart Innovation, Systems and Technologies 68, DOI 10.1007/978-3-319-57078-5_2

vulnerable to volatility. A number of factors including global population growth (predicted to grow to around 9.7 billion by 2050), intensification of food production, varying cost and availability of fuel, and more crucially the impacts of climate change which is projected to exacerbate challenges from drought, flooding, pests, diseases and weeds all increase volatility in the food sector [1, 2]. In addition, a wide range of factors such as the general global transition towards increasingly meat and dairy based diets, ageing populations and concerns about the health impacts of food, as well as modern social complexities associated with increased urbanisation, highlight an urgent need to improve the long-term sustainability of food manufacturing, as depicted in Fig. 1.

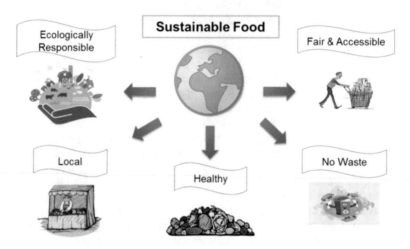

Fig. 1. New frontiers in sustainable food manufacturing

A misconception is often that 'Global Food Security' must be about growing more crops and increasing global production output. However, improving distribution, increasing productivity, and reducing waste though a range of initiatives such as enhancing food supply, better network planning of outlets and distribution to maximise efficiency and improve resilience, utilisation of new materials and biomaterial processing, multiple use of crops/waste streams and novel processes to minimise water and energy requirements, are all equally important considerations upon which the future of the food sector must be founded.

The initial section of this paper provides an overview of challenges and a number of key research questions in sustainable food manufacturing, and the latter sections of the paper briefly describe a number of relevant contemporary research areas in pursuit of improving the resource efficiency, resilience and long-term security of GFSCs.

2   Forging New Frontiers in Sustainable Food Manufacturing

It is argued that future food research activities must focus not only on identifying new sources of materials but also reducing the demand on existing resources through simultaneous considerations of innovation and development initiatives targeted at food products, processing methods, and supply networks. In this context, some of the key research questions are:

- *How* do we improve the efficiency of food production processes (e.g. through improved automation and smart technologies) to consume fewer resources (materials, energy and water)?
- *How* can we eliminate the production and post-production waste caused by inefficient supply and manufacturing activities and/or relationships?
- *How* do we use material currently discarded as waste (e.g. biomass) as a new source of raw material in food production?
- *How* do we measure, monitor and ultimately minimise the energy and water consumption per unit across the entire supply chain of food products?
- *How* can we prepare customised food products specifically tailored to the needs of consumers with restricted choice (e.g. coeliac disease)?
- *How* can supply chain resilience towards the range of aforementioned global challenges and volatility be better modelled and enhanced?

The remaining sections of this paper highlight a range of multi-disciplinary interlinked research areas in support of sustainable food manufacturing, as depicted in Fig. 2.

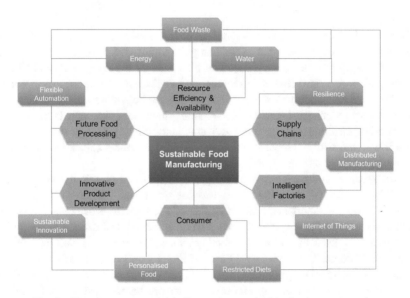

Fig. 2. Key future research challenges in sustainable food manufacturing

3 Valorisation of Food Waste

Food manufacturing is a complex process that at present is in the main linear - rather than based on circular - thinking. Globally, a staggering 1.3 billion tonnes of edible food is wasted per year both in developing and developed areas of the world [3]. Food

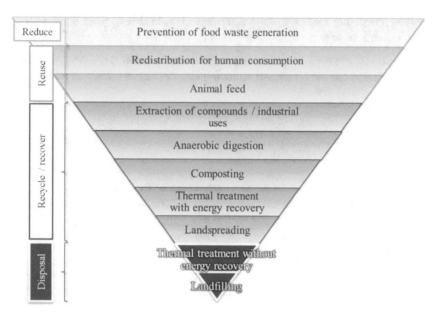

Fig. 3. The Food Waste Hierarchy [5]

waste has a dual negative environmental impact, namely undue pressure on natural resources and ecosystem services as well as pollution caused through food discards. It also entails a significant economic cost at all levels of the food supply chain, from farmers and food manufacturers to retailers and consumers. Currently, most common strategies for dealing with food waste are animal feeding, anaerobic digestion, composting, waste to energy (such as incineration), landspreading, and least preferably open burning (without energy recovery) and landfilling. Figure 3 shows the Food Waste Hierarchy, with the most sustainable waste management solutions at the top and the least sustainable options at the bottom.

Food waste is unique as a bioresource: it is rich in ingredients which may be re-used for nutritional, functional and textural properties. Hence, a deeper understanding of the characteristics and properties of food waste is necessary to identify opportunities where these under-utilised resources can be valorised. Preferably, food waste should be used to create new food products, by extracting and separating its valuable ingredients. The High Level Panel of Experts on Food Security and Nutrition from the UN 'Food and Agricultural Organisation' (FAO) concluded that valorisation approaches for food waste and by-products streams can be regarded as a key solution in global food security [4].

4 Influencing Consumer Behaviour to Reduce Food Waste

Consumer food waste is a global issue where one third of food produced never gets consumed, revealing a need to promote more sustainable consumption. Interestingly, in the developing countries, 40% of food waste is created in the field and during initial processing, whereas in developed countries, 40% of food waste is generated at the retail and consumption stages [6]. Studies have demonstrated that consumer food waste should not be conceptualised as solely a behavioural problem but rather as a symptom of an unsustainable food system that overproduces, oversupplies, and encourages consumerism [7, 8]. Manufacturers and retailers can play a crucial role in minimising consumer food waste, due to the strategic position they hold in controlling the flow of goods from producers to consumers. It is proposed that bridging the gap between food production and consumption can be achieved through a range of considerations including acquiring a deeper understanding of consumers' needs and demands, incorporating this understanding in various food production and retail activities, and using new advances in information technology to communicate more effectively with food consumers.

5 Energy Management in Food Manufacturing

The food industry is one of the largest users of energy in the world and therefore has a fundamental reliance on security of energy supply. There is some variation in energy demand between countries – those with more developed industrial systems tend to produce more highly processed foods and make them available all year round. In these countries, out-of-season agriculture, heating and drying of foodstuffs as well as transportation and storage of goods lead to large embodied energy within final food products. Therefore, the food manufacturing industry urgently needs to become resilient to changes in the energy supply grid [9]. This has been recognised and in some countries specific targets have been set; for example, in the UK the Courtauld Commitment demands a 20% reduction in greenhouse gas intensity of food and drink consumed by 2025 [10].

There are, fortunately, some major opportunities for improved energy resilience across food supply chains, which include better understanding of energy consumption in factories requiring more intelligent metering and analysis to identify energy inefficiencies, new technologies and application of technologies for food processing (e.g. microwave and infrared heating and drying), energy recovery from both process and facility levels, and finally incorporation of renewable energy technologies into food supply networks.

6 Improving Water Sustainability in Food Manufacturing

Availability and sustainable management of scarce freshwater resources is a well-recognised priority in the food sector [11]. A major barrier to effective monitoring and control of water consumption in food processing is lack of real-time and process-specific water content data [12]. Currently available data typically comes from periodic sampling conducted at the main outfall of an industrial plant, and sent to a laboratory for analysis to check for discharge consent compliance. From the point of view of tackling the fundamentals of water and effluent reduction, this approach is inadequate due to the significant lag time between generation of effluent and receipt of analysis results, and the

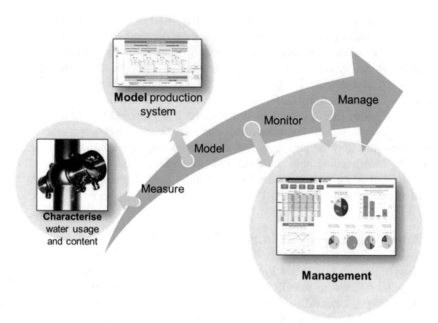

Fig. 4. Real-time monitoring and control of water consumption in food manufacturing

fact that samples are the aggregate output of multiple processes in the production chain, which makes it difficult to attribute results to specific process steps or plant operational conditions. To address these shortcomings, there is an urgent need for real-time capable instrumentation for continuous in-plant characterisation of individual water using processes and of the effluent load of the water streams they produce, as depicted in Fig. 4. Such instrumentation should be composed of two elements – a system for assessing and monitoring the contribution to water waste arising from cleaning processes [13], and another for characterising variations in water effluent magnitude and types [14]. These instrumentations can also be applied directly for real-time control of water using food processes to improve water sustainability through reuse and in-plant recycling of waste water.

7 Customised Food Products for Consumers with Restricted Choices

Food product innovation is increasingly used to enhance people's lifestyles while reducing the negative impact of manufacturing [15]. These innovations could include considerations for intelligent food packaging design that helps to increase the shelf life of a fresh product and minimizes post-production waste [16], use of alternative protein sources for nutritionally optimised foods, and product reformulation based on seasonal and locally available ingredients.

In addition, Non Communicable Diseases (NCDs) are increasingly imposing a significant burden on public health. Most cases of obesity, cardiovascular disease and type II diabetes are preventable as these diseases are directly linked to unhealthy dietary habits and sedentary lifestyles [17]. On the other hand, life expectancy is on the rise in the majority of developed countries, for example senior citizens are expected to constitute 23% of the UK population in 2035 [18]. Despite these facts, current efforts to provide the market with healthy and customised industrial foods are very limited. This highlights an urgent need for investigating various options for provision of customised and personalised food products specially tailored to the specific requirements of consumers with restricted food choices.

8   Use of Robots to Provide Flexible Automation in Food Manufacturing

The food industry is continuously being challenges to meet the demands for short-term inclinations to certain products. Many manufacturers are resorting to producing a wider variety of products in smaller batches that are suitable for a number of individual tastes. However, they face the issue of having rigid automated processes that are often designed for mass production of a small number of product types, thus limiting their flexibility in production. In addition, the substantial investment required in implementing large scale automation has often been a prohibiting factor for Small and Medium Enterprises (SMEs) to adopt automated processes within their production line. In this context, the low-cost flexible automation provided through use of robots would be an ideal solution for SMEs to improve their flexibility, productivity and product quality [19]. Furthermore, utilisation of robotic systems would enable larger manufacturers to quickly respond to market and customer changes by making the most of trends, seasonal products and frequently changing product designs, all while reducing production costs and improving quality.

Currently, use of robots in food production is mainly focused on finishing processes (e.g. packaging and palletising), but there is an inherent need for such flexible automation in processes higher up the production line to increase productivity [20]. Developing robots for food handling and processing is often challenged by irregularity in shape and non-rigidity of foodstuffs, which make them easily deformable.

9   Distributed and Localised Food Manufacturing

Traditional business models focusing on the centralisation and large scale production of food products are increasingly being challenged due to emerging demand for authentic local products and consumer concerns for the sustainability of food systems. The need for a shift towards more distributed localised food manufacturing systems (see Fig. 5) has been highlighted by a range of factors such as changes in transport and labour costs, high volumes of food waste associated with large supply chains, the availability and access to materials, energy and water, and uncertainties regarding the long-term resilience of complex global food systems [21].

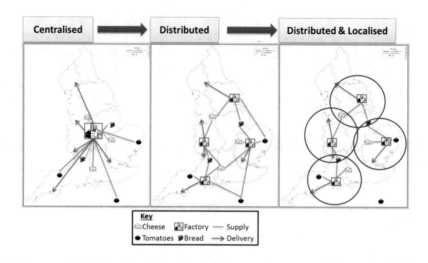

Fig. 5. Distributed and localised food manufacturing

The concept of distributed localised manufacturing has been identified as an emerging organisational theory that can support the food industry in its upcoming challenges [22]. Such distributed localised production of food is expected to support the provision of customised/personalised food products coordinated with dietary requirements, to create more agile and shorter food supply chains, and to minimise environmental impacts and costs associated with food transportation and storage.

10 Resilience in Global Food Supply Chains

Contemporary food supply chains are able to offer a huge and previously unimaginable variety of safe and competitively priced food products. A number of changes over recent decades have enabled this, including 'globalisation' of supply chain networks and 'leaning' of food processing and provision so as to remove all non-value adding features. The benefits of such an approach are substantial, but so too is the risk, when the non-value adding features which are often eliminated include traditional buffers against disruption.

Recently, we have seen vegetable shortages in Europe as a result of poor weather and we can expect the risk of further such disruptions to grow in light of global stressors such as climate change, population growth and dietary transition. Yet resilience is something of a buzzword, used interchangeably with related themes such as sustainability. Contrary to popular belief, resilience does not simply concern resisting a disruption. Rather, it is the ability to adequately anticipate potential disruptions, to react in such a way that disruption to standard operations is minimised and to learn and adapt in response to the disturbance, even if this requires fundamental changes to operational models [23]. This 'resilience cycle' is depicted in Fig. 6.

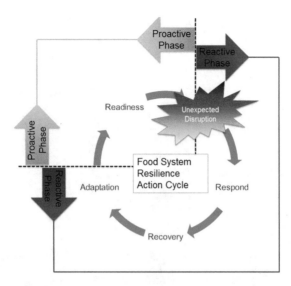

Fig. 6. The different phases of the resilience cycle in food manufacturing

Understanding resilience in this way should allow much more refined matching of a given stakeholder's management options (known as capabilities) to the specific vulnerabilities facing them [24]. However, there is a real need for further empirical research across the food supply chain (i.e. ranging from primary production to retail) to identify specific food sector capabilities and vulnerabilities, and the linkages and interactions between them.

11   Utilisation of 'Internet of Things' to Improve Food Resource Efficiency

The food sector is increasingly under pressure to improve its resource efficiency [25]. In order to achieve this, it is vital for the food supply chain actors to share and exchange knowledge and information on resource use and availability in a timely manner. The traditional methods of physically monitoring and managing resources are labour intensive and complex, and are often time consuming and costly.

The technology and tools associated with the concept of the Internet of Things (IoT) are capable of supporting numerous tasks in real-time such as tracking, locating, monitoring, measuring, analysing, planning and managing, and enhancing efficiency and transparency within food supply chains [26], as depicted in Fig. 7.

This highlights the significant potential offered through the latest IoT advancements to support innovative approaches based on an automated real-time systems for monitoring and analysing the resource usage across entire food production and supply. Such timely management and provision of appropriate knowledge and information could potentially result in more effective strategic planning and better decision making in support of resource efficient and sustainable food supply chains.

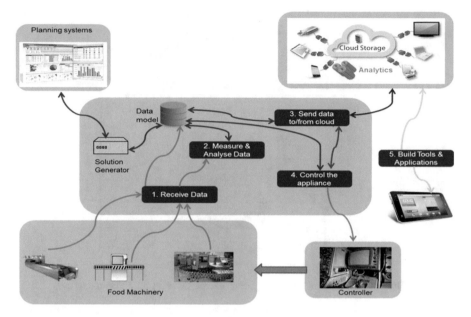

Fig. 7. Use of Internet of Things (IoT) concepts to enhance resource efficiency within food supply chains

12 Concluding Remarks

The food manufacturing industry seeks continuous improvements in most of its activities in order to not only increase profitability, but also to provide consumers with better products that can satisfy their changing needs. Competitive measures implemented by food manufacturers have focussed specifically on three key performance indicators: productivity, quality and innovation. By developing strategies which could improve any of these three business indicators, food manufacturers have been able to grow and create more valuable food products for consumers.

However, rising environmental concerns and associated pressure from governments and consumers are now increasingly forcing the food sector to strive to improve the sustainability of their products, production processes, supply chains and business strategies. Consequently, innovative approaches that can provide the industry with meaningful and significant improvements are being explored due to the ever-increasing need to adapt and change in order to remain competitive. In addition, due to the unique attributes of food products and their impact on the health and wellbeing of consumers, these improvements must go beyond environmentally conscious manufacturing and should also contribute to the shared prosperity between business and community, and between human society and natural ecosystems. The achievement of such goals clearly necessitates a complex and significant transition, which not only requires "smart" strategies to safeguard the future prosperity of food manufacturers, but also a collective and continuous drive from businesses, communities and policy makers towards long-term global food security.

References

1. United Nations Department for Economic and Social Affairs, World Population Prospects (2015). https://esa.un.org/unpd/wpp/. Accessed 23 Feb 2017
2. Allison, E.H., Perry, A.L., Badjeck, M.C., Neil Adger, W., Brown, K., Conway, D., Halls, A.S., Pilling, G.M., Reynolds, J.: Vulnerability of national economies to the impacts of climate change on fisheries. Fish Fish. **10**(2), 173–196 (2009)
3. FAO: Global food losses and food waste - extent, causes and prevention (2011). http://www.fao.org/docrep/014/mb060e/mb060e00.pdf. Accessed 3 Feb 2017
4. The High Level Panel of Experts on Food Security and Nutrition (HLPE): Food losses and waste in the context of sustainable food systems, Rome (2014). http://www.fao.org/3/a-i3901e.pdf. Accessed 23 Feb 2017
5. Garcia-Garcia, G., Woolley, E., Rahimifard, S., Colwill, J., White, R., Needham, L.: A methodology for sustainable management of food waste, waste and biomass valorization (2016). doi:10.1007/s12649-016-9720-0
6. FAO: Save food: global initiative on food loss and waste reduction (2016). http://www.fao.org/save-food/resources/keyfindings/en/. Accessed 23 Feb 2017
7. Evans, D.: Blaming the consumer – once again: the social and material contexts of everyday food waste practices in some English households. Crit. Public Health **21**(4), 429–440 (2011)
8. Aschemann-Witzel, J., de Hooge, I., Amani, P., Bech-Larsen, T., Oostindjer, M.: Consumer-related food waste: causes and potential for action. Sustainability **7**(6), 6457–6477 (2015)
9. European Commission: Energy use in the food sector: state of play and opportunities for improvement (2015). http://publications.jrc.ec.europa.eu/repository/bitstream/JRC96121/ldna27247enn.pdf. Accessed 23 Feb 2017
10. Waste & Resources Action Programme (WRAP): Household food waste in the UK (2017). http://www.wrap.org.uk/sites/files/wrap/Household_food_waste_in_the_UK_2015_Report.pdf. Accessed 23 Feb 2017
11. Ölmez, H., Kretzschmar, U.: Potential alternative disinfection methods for organic fresh-cut industry for minimizing water consumption and environmental impact. LWT Food Sci. Technol. **42**(3), 686–693 (2009)
12. Sachidananda, M., Rahimifard, S.: Reduction of water consumption within manufacturing applications. In: Proceedings of the 19th CIRP Conference on Life Cycle Engineering, Berkeley, US, pp. 455–460 (2012)
13. Simeone, A., Watson, N., Sterritt, I., Woolley, E.: A multi-sensor approach for fouling level assessment in clean-in-place. Procedia CIRP **55**, 134–139 (2016)
14. Sachidananda, M., Webb, D.P., Rahimifard, S.: A concept of water usage efficiency to support water reduction in manufacturing industry. Sustainability **8**(12), 1222 (2016)
15. Bonzanini, M., Dutra De Barcellos, M., Marques Vieira, L.: Why food companies go green? The determinant factors to adopt eco-innovations. Br. Food J. **118**(6), 1317–1333 (2016)
16. Verghese, K., Lewis, H., Lockrey, S.: The Role of Packaging in Minimising Food Waste in the Supply Chain of the Future. RMIT University, Melbourne (2013)
17. Pereira, M.A., Kartashov, A.L., Ebbeling, C.B., Van Horn, L., Slattery, M.L., Jacobs, J.R., Ludwig, D.S.: Fast-food habits, weight gain, and insulin resistance (the CARDIA study): 15-year prospective analysis. The Lancet **365**(9453), 36–42 (2005)
18. UK Parliament: Population ageing: statistics (2012). researchbriefings.files.parliament.uk/documents/SN03228/SN03228.pdf. Accessed 20 Jan 2017

19. Cederfeldt, M., Elgh, F.: Design automation in SMEs – current state, potential, and requirements. In: 15th International Conference on Engineering Design, pp. 1507–1521 (2005)
20. Nayik, G., Ahmad, K.M., Amir, G.: Robotics and food technology: a mini review. J. Nutr. Food Sci. **5**(4), 1–11 (2015)
21. Li, D., Wang, X., Chan, H.K., Manzini, R.: Sustainable food supply chain management. Int. J. Prod. Econ. **152**, 1–8 (2014)
22. Rauch, E., Dallasega, P., Matt, D.T.: Sustainable production in emerging markets through distributed manufacturing systems. J. Cleaner Prod. **135**, 127–138 (2016)
23. Christopher, M., Peck, H.: Building the resilient supply chain. Int. J. Logistics Manag. **15**(2), 1–4 (2004)
24. Hohenstein, N.O., Feisel, E., Hartmann, E., Giunipero, L.: Research on the phenomenon of supply chain resilience: a systematic review and paths for further investigation. Int. J. Phys. Distrib. Logistics Manag. **45**(1/2), 90–117 (2015)
25. Dawkins, E., Roelich, K., Barrett, J., Baiocchi, G.: Securing the future – the role of resource efficiency. WRAP, Banbury (2010). http://www.wrap.org.uk/sites/files/wrap/Securing%20the%20future%20The%20role%20of%20resource%20efficiency.pdf. Accessed 20 Feb 2017
26. Xiaorong, Z., Honghui, F., Hongjin, Z., Hanyu, F.: The design of the Internet of Things solution for food supply chain. In: 5th International Conference on Education, Management, Information and Medicine, Shenyang, China, pp. 314–318 (2015)

Metrics-based Integrated Predictive Performance Models for Optimized Sustainable Product Design

B.M. Hapuwatte, F. Badurdeen, and I.S. Jawahir[(✉)]

Institute for Sustainable Manufacturing (ISM), College of Engineering,
University of Kentucky, Lexington, KY 40506, USA
is.jawahir@uky.edu

Abstract. Implementing sustainable manufacturing principles and practices leads to innovation and sustainable value creation at product, process and system levels. In recent years, with the exponential growth in sustainable manufacturing research to meet the rapidly growing needs of industry and society, significant emphasis has been placed on designing innovative sustainable products and developing and implementing novel and advanced sustainable manufacturing processes to produce such sustainable products in automotive, aerospace, consumer products, biomedical and power industries. Sustainable manufacturing has been recognized as the driver for innovation in the manufacturing industrial sector. Achieving sustainable manufacturing targets inevitably requires a metrics-based analysis of sustainable manufacturing at product, process and systems levels.

This paper presents an overview of the 6R (Reduce, Reuse and Recycle, Recover, Redesign and Remanufacture) approach to promote sustainable manufacturing to enable closed-loop, multiple life-cycle material flow. The paper specifically focuses on sustainable product design for manufacture, with an in-depth analysis of product design and development processes by utilizing the novel 6R methodology. The transformation of conventional product design processes to sustainable product design/development is presented by expanding the recently-proposed metrics-based sustainable product evaluation method to include integrated predictive performance models for optimized sustainable product design. Designing sustainable products is presented as the most effective pathway towards promoting innovation and sustainable value creation.

Keywords: Sustainable manufacturing · 6R concept · Product development · Predictive performance models

1 Introduction

Sustainable Manufacturing: Definition, Goals and Impact. Sustainable manufacturing evolves from lean and green manufacturing concepts, and it offers a new way of designing innovative products and deploying manufacturing processes using methodologies that *minimize adverse environmental impacts, improve energy and resource efficiency, generate minimum quantity of wastes, and provide improved operational*

© Springer International Publishing AG 2017
G. Campana et al. (eds.), *Sustainable Design and Manufacturing 2017*, Smart Innovation, Systems and Technologies 68, DOI 10.1007/978-3-319-57078-5_3

safety and personnel health, while maintaining and/or improving the product and process quality with the overall life-cycle cost benefit [1].
The major goals of sustainable manufacturing are:

- Reducing *energy consumption*
- Reducing *waste*
- Reducing *material utilization*
- Enhancing *product durability*
- Increasing *operational safety*
- Reducing *toxic dispersion*
- Reducing *health hazards/Improving health* conditions
- Consistently improving *manufacturing quality*
- Improving *recycling, reuse and remanufacturing*
- Maximizing the use of *sustainable sources of renewable energy*

Sustainable manufacturing thus enables *cost-effective, environmentally-benign and societally beneficial innovative products and processes* serving as a basis for *sustainable value creation in manufacturing.*

Total Life-cycle Approach and Multi Life-cycle Products. Graedel [2] presented an extensive study of streamlined life-cycle analysis (SLCA) methods by considering five major product life-cycle stages: pre-manufacture; manufacture; product delivery; use; and recycling. Since the product delivery stage, including transportation, was considered as only one among several delivery activities involved across the life-cycle, the simplified total life-cycle of a product can be considered as consisting of only *four key stages: pre-manufacturing, manufacturing, use and post-use* [1]. To achieve multiple product life-cycles with the goal of near-perpetual product/material flow *facilitating the Circular Economy,* design and manufacturing practices for next-generation products must consider the *total life-cycle approach using innovative 6Rs* (Reduce, Reuse, Recycle, Recover, Redesign and Remanufacture). This in effect will enable sustainable value creation through innovation at all levels, in contrast to the perceived high costs of deploying sustainable manufacturing. Optimal secondary use of resources in sustainable manufacturing will lead to product/process innovation, and will provide cost-effective sustainable products.

Several other researchers have in the past attempted to quantify product sustainability. The sustainability target method (STM), developed by Dickinson and Caudill [3], correlates the economic value of a manufactured product with its environmental impacts. This method calculates resource productivity and eco-efficiency based on relevant indicators. It utilizes the estimation of earth's carrying capacity and economic information to provide a practical sustainability target and to determine if a product's end-of-life option is feasible [4]. A product sustainability index (*PSI*) method, developed by Schmidt and Butt [5] was adopted as a management tool for the sustainability assessment by Ford's product development group.

Figure 1 shows a methodology for producing sustainable products from optimized resources [1]. Developing model-based sustainable manufacturing methodologies by considering the total product life-cycle has been shown as a basis for product and process innovation in sustainable manufacturing [6].

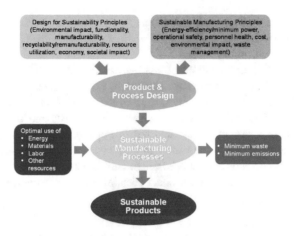

Fig. 1. Methodology for producing sustainable products from sustainable processes [1]

2 Metrics-based Product Sustainability Evaluation

It is essential to comprehensively evaluate a product's total life-cycle sustainability performance to successfully design and manufacture multi-generational sustainable products. To be effective, a metrics-based evaluation of sustainable products must integrate criteria that: (a) assess economic, environmental and societal performance, (b) consider impacts from pre-manufacturing, manufacturing, use and post-use stages of the product life-cycle, and (c) evaluate extent to which closed-loop material flow practices are implemented through the application of the 6R methodology. Beginning with early work by Fiksel et al. [7] which attempted to develop a quantitative method for product sustainability evaluation, prior studies have reported progress during the last two decades [8, 9]. Despite significant recent momentum in quantitative model development for product sustainability evaluation, most measurement schemes developed so far seem to lack in one or more of the above - (a) through (c) - required integral elements for comprehensive analysis of product sustainability.

Therefore, a more comprehensive approach to develop a framework and metrics that can help promote the design, manufacture and end-of-life (EOL) management of products to enhance the overall product sustainability becomes essential. This shortcoming has been addressed by a recent multi-year NIST-sponsored project [10] that involved an industry-university collaborative effort. The new framework developed under this project involves expansion of previously established six major product sustainability elements (environmental impact; societal impact; functionality; resource utilization and economy; manufacturability; and recyclability and remanufacturability) [11]. This effort resulted in a more comprehensive set of 13 clusters $[C_i \ (where \ i = 1, \ldots, 13)]$ developed for product sustainability evaluation. These clusters are categorized under the three triple bottom-line categories (TBL): economy, environment and society. A Product Sustainability Index $(ProdSI)$ [12] is derived for manufactured products using a five-level hierarchical structure:

- **Individual metric (M_k):** a quantifiable and measurable attribute or property related to a single parameter or indicator of product sustainability (e.g.: recovery cost, material utilization, injury rate).
- **Sub-cluster (SC_j):** aggregation of metrics to evaluate performance of specific product sustainability aspects (e.g.: labor cost, EOL product reuse, safety).
- **Cluster (C_j):** aggregation of sub-clusters to assess product sustainability directly influencing the TBL categories (e.g.: direct/indirect costs and overheads, material use and efficiency, product safety and health impact).
- **Sub-index:** combining cluster values to determine performance along each of the three TBL aspects: economy, environment and society.
- *ProdSI:* the overall aggregated product sustainability performance index.

At each level of aggregation, normalization (to address unit of measurement variations) and weighting to integrate importance of one metric (w_k), sub-cluster (w_j), cluster (w_i) or sub-index (in this case, equally weighted) relative to the others are carried out. A complete list of metrics, sub-clusters and clusters with examples and corresponding life-cycle stage(s), can be found in [12]. The identified metrics for a given manufactured product can then be used to compute its *Product Sustainability Index (ProdSI)* using the expression shown in Eq. (1) with each C_i and SC_j computed as shown in Eqs. (2) and (3):

$$ProdSI = \frac{1}{3}\left(\sum_{i=1}^{3} w_i^c c_i + \sum_{i=4}^{8} w_i^c c_i + \sum_{i=9}^{13} w_i^c c_i\right) \tag{1}$$

$$C_i = \sum SC_j w_j^{sc} \quad \forall i \tag{2}$$

$$SC_j = \sum M_k w_k^m \quad \forall j \tag{3}$$

Results from applying the metrics-based method to a manufactured product is shown in Fig. 2. This spider diagram illustrates the score (on a scale of 0–1) for all

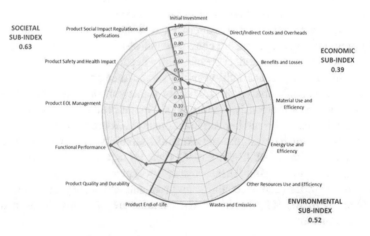

Fig. 2. Cluster-based calculated *ProdSI* for a product [12]

13 clusters, and shows the sub-indices for economy, environment and society. This ultimately leads to a *ProdSI* score of 0.51 (with equal weightage for sub-indices) indicating a slightly better than average product sustainability. This method can be applied to compare sustainability performance of a single product over a period of time or products from different generations to assess the impact of progressive design changes, and to compare similar products from competitive manufacturers [12].

Unlike most previous approaches that are biased towards only economic and environmental assessment, the *ProdSI* method provides a comprehensive evaluation of product sustainability performance by considering all relevant aspects.

3   Sustainable Product Design/Development Process

Early design considerations can significantly reduce a product's manufacturing cost. Many aspects that influence the product sustainability, eventually providing significant cost savings, improved environmental impact with societal benefits can also be established in the early product design stage. For example, the materials and resources such as energy and water used in manufacturing processes to fabricate components, as well as in all other related activities, including assembly, will all influence the overall product sustainability. Thus, while the *ProdSI* method is a comprehensive evaluation approach, that alone is not sufficient in the quest to enhancing product sustainability. It is imperative that key criteria that influence total life-cycle product sustainability performance are identified and considered during the product design process to ensure that product sustainability is enhanced during pre-manufacturing, manufacturing, use and post-use stages [13].

A new framework for such a sustainable product design/development process is illustrated in Fig. 3. As shown, initially during concept development, the design team requires better guidelines to assist selecting product/component features that meet customer requirements while also enhancing total life-cycle sustainability. Such guidelines could be developed, for example, by promoting the selection of material/ process alternatives that enhance the *ProdSI* score (by positively influencing the various metrics, sub-clusters and clusters). However, product parameters are also highly inter-dependent [14, 15] and trade-offs among them can positively or negatively affect the overall performance of products/components, including life, durability, upgradability/ maintainability, repairability, reusability, remanufacturability, etc.

With rapidly growing advanced manufacturing processes such as additive manu-facturing where a product's complex geometric requirements can be achieved with significant savings of materials/resources used, an important area that is inadequately addressed is the *functional performance of manufactured components*. This functional performance needs to be considered at the product design stage with targeted *func-tionality, product life, performance and maintenance issues*. Therefore, during the detailed design, prototype development and testing stages, product designers will require predictive performance models to develop optimal product designs considering various trade-offs. Such decision support tools can be used to maximize/minimize the product's specific objectives, incorporate constraints, and conduct sensitivity analyses to assess the influence of different product design variables.

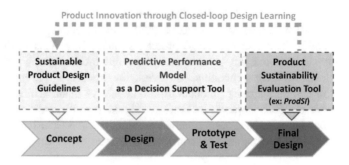

Fig. 3. Sustainable product design/development process (adapted from [13])

Once the *optimal design* is selected with support from *predictive performance models,* during the final design stage, product sustainability evaluation tools, such as *ProdSI*, can be used to evaluate the total life-cycle sustainability. The iterative application of the product development process outlined in Fig. 3 can enable designers to continuously innovate and develop successively better product designs. This approach helps identify and incorporate product sustainability drivers to enhance TBL performance, total life-cycle coverage and multi life-cycle material flow early during the design process. By incorporating predictive models for optimal product design, sustainable value is created for all stakeholders.

4   Predictive Performance Models for Sustainable Products

4.1   Significance of Predictive Product Performance Modeling

Significant progress is being made in evaluating the quality, performance and life of production equipment and machines. The need for designing and developing components used in such production equipment and machines for enhanced product performance and life is emerging as an important area of research focus.

Increasingly complex products are designed and developed to satisfy the growing functional needs. However, the anticipated functional requirements are largely feature-based, and are aimed at meeting the immediate need for functionality with quality and cost considerations and marketability, mostly with no long-term performance projections, and with very little consideration on upgradability and maintainability. Sustainability characteristics such as reusability, recyclability and remanufacturability are gaining significance in recent years with research focus on material selection, use and post-use activities. Developing lightweight designs for improved energy efficiency and performance in numerous products is another research focus area that is also closely tied to cost reduction.

Additive manufacturing provides tremendous opportunities for producing complex features in components to satisfy multiple functions with light-weight options. However, at the design stage when deciding between such alternative processes, the long-term functional performance requirements, including sustainability characteristics are important aspects to consider [13]. Predictive product performance models, therefore, are essential to help evaluate alternate product designs for their functionality, cost, quality and other sustainability characteristics over the duration over which the particular product could be in market.

4.2   Component Level Performance Needs

Product – Process integration with multi-level production systems involving component and machine level interactions is shown in Fig. 4 [16]. As seen, in a multi-level interactive production environment, the components produced become a part of an assembled, and significantly more complex, product, which in function is similar to a machine on the shop floor as it is an assembled product. Such components, whether assembled or stand-alone, are expected to perform to satisfy the functional needs. Product design for performance therefore must include predictive performance characteristics such as projected life, wear and tear rate including quality/performance deterioration rates, maintenance requirements, etc. Material and process selection for manufacturing these components is also a major responsibility of product designers.

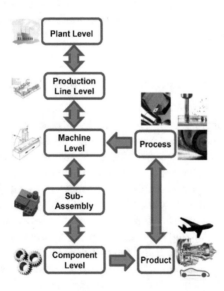

Fig. 4. Product – Process integration with interactions of multi-level production elements [16]

4.3   Product Design for Sustainability

Figure 5 shows the potential impacts of 6Rs on the previously established 13 product sustainability clusters, emphasizing tangible environmental and societal benefits and economic gains feasible. A review of the multiple sub-clusters and metrics associated with each cluster (see [12] for details) enables identifying the potential benefits of 6R

6R Elements	Economy			Environment					Society				
	Initial investment	Direct/indirect costs & overheads	Benefits & losses	Material use & efficiency	Energy use & efficiency	Other resources use & efficiency	Waste & emissions	Product EoL	Product quality & durability	Functional performance	Product EoL management	Product safety & health impact	Product societal impact regulations & certification
Reduce	x	x	x	x	x	x	x	x			x	x	x
Reuse	x		x	x	x	x	x	x			x		
Recycle		x	x	x			x	x			x		x
Recover				x			x	x			x		x
Redesign		x		x	x	x	x	x	x	x	x	x	x
Remanufacture				x	x	x	x	x			x		

Fig. 5. 6R applications in product sustainability cluster areas

implementation illustrated in Fig. 5. A more detailed analysis can be performed when a comprehensive metrics-based evaluation is performed for product design improvement for achieving greater sustainability. Analytical predictive performance models discussed in the next section, too, will enable a more quantitative assessment of 6R impacts on product sustainability.

4.4 Integrated Product Performance Models and Optimized Product Design

When designing sustainable products, multiple project objectives emerge such as energy/resource efficiency during manufacturing, use and post-use stages, projected product life/durability, product upgradability and performance, product's EOL options, etc. These aspects need to be *modeled predictively* with sufficient reliability and confidence levels, including potential risks, to estimate the total life-cycle cost and to determine the performance targets. Conflicting objectives would often require trade-offs when designing and developing such models. Integration of predictive performance models and optimizing for desired objectives, within the imposed constraints, thus become critically important at the product design stage.

Figure 6 shows the proposed comprehensive sustainable product design process through predictive performance modeling. To implement such an optimized process, different product sustainability considerations such as environmental impact, resource efficiency and economy and product functionality and their variation over the total life-cycle (covering pre-manufacturing, manufacturing, use and post-use stages) must first be quantified through predictive performance models. For example, predictive models for energy efficiency, material/resource efficiency or total life-cycle cost will enable the analytical assessment of product design performance along these different aspects. To evaluate potential trade-offs among the many conflicting aspects, the individual models must be combined to develop *Integrated Predictive Performance Models*. The optimized sustainable product design can then be developed by following

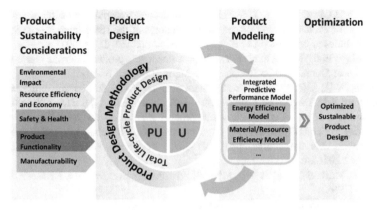

Fig. 6. Proposed optimized sustainable product design process

an iterative process. The integrated predictive performance models can be used to identify a product design through optimization, evaluate its sustainability performance using tools such as the *ProdSI,* and iteratively change product features/attributes until the final optimized sustainable product design that maximizes overall product sustainability performance is determined.

5   Summary and Outlook

Since major decisions impacting the product performance and life are made during the product development stage, it is essential to have sustainability considerations as input parameters for the product development process, along with the conventional customer and production requirements. Sustainability considerations themselves are interdependent, and also interact with the customer and production requirements, adding to the complexity of sustainable product development process. Thus, an integrated product performance model and optimization process would be required to maximize the overall product performance and sustainability. The work presented in this paper is the initial step towards developing a comprehensive optimized sustainable product design process incorporating predictive performance modeling.

References

1. Jawahir, I.S., Badurdeen, F., Rouch, K.E.: Innovation in sustainable manufacturing education. In: Seliger, G. (ed.) Proceedings of the 11th Global Conference on Sustainable Manufacturing (GCSM), Berlin, Germany, pp. 9–16, 23–25 September 2013
2. Graedel, T.E.: Streamlined Life-Cycle Assessment. Prentice Hall, Upper Saddle River (1998)

3. Dickinson, D.A., Caudill, R.J.: Sustainable product and material end-of-life management: an approach for evaluating alternatives. In: Proceedings of the IEEE International Symposium on Electronics and the Environment, Boston, MA, pp. 153–158 (2003)
4. Gao, M., Zhou, M., Wang, F.Y.: Improvement of product sustainability. In: IEEE International Conference on Robotics and Automation, pp. 3548–3553 (2003)
5. Schmidt, W.P.: Life cycle tools within Ford of Europe's product sustainability index. Case study Ford S-MAX & Ford Galaxy. Int. J. Life Cycle Assess. 11, 315–322 (2006)
6. Jawahir, I.S., Jayal, A.D.: Product and process innovation for modeling of sustainable machining processes. In: Seliger, G., Khraisheh, M., Jawahir, I.S. (eds.) Advances in Sustainable Manufacturing, pp. 301–307. Springer, Heidelberg (2011). ISBN 978-3-642-20182-0
7. Fiksel, J., McDaniel, J., Spitzley, D.: Measuring product sustainability. J. Sustain. Prod. Des. 6, 7–18 (1998)
8. Joung, C.B., Carrell, J., Sarkar, P., Feng, S.C.: Categorization of indicators for sustainable manufacturing. Ecol. Ind. 24, 148–157 (2013)
9. Singh, R.K., Murty, H.R., Gupta, S.K., Dikshit, A.K.: An overview of sustainability assessment methodologies. Ecol. Ind. 15, 281–299 (2012)
10. Jawahir, I.S., Badurdeen, F., Rouch, K.E.: Metrics-based evaluation of product and process sustainability and case studies. NIST Project Report, Institute for Sustainable Manufacturing, University of Kentucky, Lexington, KY, USA (2013)
11. De Silva, N., Jawahir, I., Dillon Jr., O., Russell, M.: A new comprehensive methodology for the evaluation of product sustainability at the design and development stage of consumer electronic products. Int. J. Sustain. Manuf. 1, 251–264 (2009)
12. Shuaib, M., Seevers, D., Zhang, X., Badurdeen, F., Rouch, K.E., Jawahir, I.S.: Product Sustainability Index (ProdSI): a metrics-based framework to evaluate the total life-cycle sustainability of manufactured products. J. Ind. Ecol. 18, 491–507 (2014)
13. Hapuwatte, B., Seevers, K.D., Badurdeen, F., Jawahir, I.S.: Total life-cycle sustainability analysis of additively manufactured products. Procedia CIRP 48, 376–381 (2016)
14. Chang, D., Lee, C., Chen, C.H.: Review of life-cycle assessment towards sustainable product development. J. Clean. Prod. 83, 48–60 (2014)
15. Chen, R.W., Navin-Chandra, D., Print, F.: A cost-benefit analysis model of product design for recyclability and its application. IEEE Trans. Compon. Packag. Manuf. Technol. Part A 17, 502–507 (1994)
16. Jawahir, I.S., Kaynak, Y., Lu, T.: The impact of novel material processing methods on component quality, life and performance. Procedia CIRP 22, 33–44 (2014)

Sustainable Design, Innovation and Services

Latent Semantic Indexing for Capitalizing Experience in Inventive Design

Pei Zhang[1,3(✉)], Cecilia Zanni-Merk[2], and Denis Cavallucci[1,3]

[1] LGéCo (Design Engineering Laboratory), Strasbourg, France
[2] INSA Rouen Normandie, LITIS, Norm@stic (FR CNRS 3638), Rouen, France
`cecilia.zanni-merk@insa-rouen.fr`
[3] INSA de Strasbourg, 24 boulevard de la Victoire, Strasbourg, France
{`pei.zhang,denis.cavallucci`}`@insa-strasbourg.fr`

Abstract. The growing complexity of the design activity in an innovation and sustainable context requires experience reuse as a means to limit unsustainable investments. It is a crucial task for both academic and industrial communities to find ways to efficiently capture and reuse past experience. Case-based reasoning (CBR) is a research paradigm that stores experience as a knowledge unit to solve a new problem from the previous design experience. A well-established method for inventive design is IDM (the Inventive Design Methodology). Its most widely used tool to solve a problem is the "Contradiction Matrix" associated with forty inventive principles. The correct use of these tools needs the mapping from freely expressed text (Specific Parameters or SPs) into a well-established set of Generic Engineering Parameters (or GEPs). This mapping requires expertise and may, if inappropriately used, lead to weak results. This paper introduces the Latent Semantic Indexing (LSI) algorithm to discover the implied semantic relations between SPs and GEPs coming from past experience. A semantic space based on the LSI results is built for guiding retrieval in case-based reasoning.

Keywords: Case-based reasoning · Inventive design · Experience capitalization · TRIZ · Latent Semantic Indexing

1 Introduction

TRIZ (the theory of inventive problem solving) [1] has been widely used for problems solving by practitioners in different industrial domains [2], to assist early stages of the innovation process. However, classic TRIZ has shown its limits when used in complex, cross domain situations [3]. A new proposal, the Inventive Design Methodology (IDM) [4] is able to represent problematic situations in Inventive Design for both moving ahead in the most promising direction for innovation and easing the use of TRIZ tools with the help of a computer and ontologies.

In the framework of IDM, one way to solve an inventive design problem is to eliminate technical contradictions. The facilitation of the resolution is achieved through

© Springer International Publishing AG 2017
G. Campana et al. (eds.), *Sustainable Design and Manufacturing 2017*, Smart Innovation, Systems and Technologies 68, DOI 10.1007/978-3-319-57078-5_4

the use of the *Contradiction Matrix*[1], *Inventive Principles* (*IP*) and semantic similarity assistance for associating the user's problem to one or several of the thirty-nine *Generic Engineering Parameters*[2] [5]. This process includes several steps. Firstly, the user needs to match the *SP*s which describe the specific problem with the 39 *GEP*s respectively (row = the *GEP*s to be improved, column = the *GEP*s that worsened). Then, the most similar *GEP*s are obtained through semantic search for each *SP*. The user has to select which one of the ranked *GEP*s he thinks is the most appropriate one. And then, based on the two selected *GEP*s, a series of *IP*s are obtained. Finally, the user has to browse through each of them to facilitate connections between the inventive direction underlying each principle and his own knowledge to build a solution concept often resulting from triggering his creativity.

For example, in a hammer, improving the *hitting efficiency (SP1)* of the surface (*action parameter*) of the hammer (*element*) consequentially leads to an undesirable conflict with the *user's well-being (SP2)*. In order to increase *user's well-being* without affecting the *hitting efficiency*, the user associates *hitting efficiency* with *Speed* (GEP-9) and *user's well-being* with *Ease of operation* (GEP-33). Then the following *IP*s are obtained in the cell at the intersection of the selected row and column of the Contradiction Matrix: *IP 32-Color change*s; *IP 28-Mechanical substitution*; *IP 13-the other way around*; *IP 12-Equipotentiality*. To understand and apply the suggested solutions one needs to identify both the desired benefits of the system and the sequence of principles. Based on their previous experience, one can design the concept solution accordingly.

The use of the *Contradiction Matrix* is simple but new users often face difficulties in using this tool, because it requires intensive experience knowledge from the user. Therefore, many works have been done with the aim of facilitating its use. We can cite, for example, the combined use of Axiomatic Design with TRIZ [6] that provides a way to systematically map design parameters with engineering parameters, or the works proposed in [7] that introduce human factors issues in the engineering parameters in the matrix.

As this task requires large amount of expertise, the new users are incapable to choose the right *GEP* rapidly the first time. As a result, the traditional "trial and error" approach is inevitable [8]. Therefore, an efficient way of reusing past experience has become extremely important to cope with the growing complexity of the design activity. Our approach addresses the need to store systematically experience knowledge contained in the inventive design activity for its reuse, to comply with the principles of sustainable design, by limiting unsustainable investment in expensive resources, such as expert time to reach breakthrough innovations. This work tries to improve this process by providing the users with the semantically similar *GEP*s to choose, minimising, at a certain extent, the abstraction task that is performed by a human. To avoid the "trial and error" approach of novice users, we propose to capitalize past experiences, implying their gathering, understanding and reuse.

---

[1] https://triz-journal.com/innovation-tools-tactics/breakthroughdisruptive-innovation-tools/resolving-contradictions-40-inventive-principles/.

[2] https://triz-journal.com/39-features-altshullers-contradiction-matrix/.

In this paper, case-based reasoning is used to store and reuse the past experience. Latent Semantic Indexing (LSI) is introduced to discover the implied semantic relations between *SP*s and *GEP*s coming from past experience and a semantic space is constructed based on the LSI result for guiding the future retrieval.

The remaining part of this paper is organized as follows: section two introduces the way to formally represent cases regarding inventive design; section three presents LSI in detail; section four describes the use of the LSI method in inventive design; section five details the retrieval process. Finally, section six provides readers with our conclusion and some perspectives of future work.

2 Case Representation

The CBR [9] approach is generally used to capture and induct experience by reproducing problem solving process of experts. Case-based reasoning adopts validated prior experience or "cases", as solutions to solve new problems [10]. The CBR process comprises four activities: Case representation; Case retrieval; Case adaptation; Case revise and retain.

A case contains different features (Table 1), they will be grouped into three sets [11] according to their use, including (a) Retrieval features: useful for computing similarities; (b) Input features: for text mining and completion rules[3] and (c) Output features: for information for the user.

Each feature set has two parts: problem features and solution features. For the retrieval features: the problem feature for retrieval are: *Specific Parameter Name* and *Semantic Distance*. The former is used to indicate whether its corresponding *Generic Engineering Parameter Type* is improving or degrading; the latter is used to determine the *Generic Engineering Parameter Classification*. The problem input feature contains the information generated from a technical contradiction: *action parameter, element, value 1, SP1, value 2* and *SP2*. The solution output feature includes the *Inventive Principle No.*, the *Inventive Principle* that has been adopted to solve the specific technical contradiction and the *concept solution* which is the solution that was designed by the experts based on the Inventive principle.

In the engineering domain, the user describes a specific problem using natural language, while experts use their knowledge to understand the problem and then to design the solution accordingly. In this context, the main difficulty in case representation is to extract features that are hidden in the texts which are used to describe the problems and its corresponding solutions [12]. This is especially the case for inventive design. When the users use the Contradiction Matrix to solve a problem, the problem is conveyed as contradictions in natural language. When the experts use Inventive Principles to solve the contradiction, one has to map one Specific Parameter to a degrading parameter and the other Specific Parameter to an improving parameter. Therefore, the experts' job can be regarded as making an abstraction effort with the aim of mapping various specific parameters to one certain generic engineering parameter.

---

[3] They adapt case descriptions, in particular, in the query.

Table 1. Features to describe a case

	Problem features	Solution features
Retrieval features	Specific Parameter Name	Generic Engineering Parameter Type
	Semantic Distance	Generic Engineering Parameter Classification
Input features	Context	
	Element	
	Action Parameter	
	Value1	
	Value2	
Output features		Inventive Principle
		Inventive Principle No.
		Concept solution

However, when a problem is conveyed by words, the formulation varies greatly based on the users' different background, knowledge, linguistic habits etc. There is no "correct" way to extract the features they share.

In order to solve this problem, we introduce the Latent Semantic Indexing (LSI) method to address the difficulty of understanding short texts because that there are many ways to refer to the same object in natural language. The goals of using LSI are:

- to find latent semantic relations between Specific Parameters and Generic Engineering Parameters;
- to identify indexes for semantic space building;
- to guide future retrieval by making the problem features comparable in the same semantic space.

3 Latent Semantic Indexing

Latent Semantic Indexing (LSI), also known as Latent Semantic Analysis (LSA), assumes that words that are close in meaning will occur in similar pieces of text. A matrix containing word counts per document is constructed from a large piece of text and a mathematical technique which is called singular value decomposition (SVD) is used to reduce the number of columns while preserving the similarity structure among rows. Words are then compared by taking the cosine of the angle between the two vectors formed by any two rows [13]. The goal is to represent the documents and terms in a unified way for exposing document-document, document-term, and term-term semantic relationships which are otherwise hidden, and in that way, discovering hidden concepts in document data. Thus, the LSI method is particularly suitable for extracting features that was originally hidden in the given corpus, to identify indexes for constructing semantic space and therefore to guide retrieval.

3.1   Main Steps of LSI

LSI is conducted following three main steps: texts pre-processing, creating the word by document matrix X and Singular Value Decomposition (SVD).

LSI represent documents as bags of words, therefore, a word segmentation is needed in the pre-process. In addition, the stop-words which has no meaning should be eliminated, such as a, of, by etc. Once the texts have been pre-processed, we can create the word by document matrix.

The word by document matrix X can be viewed as a word frequency matrix. The rows represent keywords (the words that occurs in the documents more than twice) and columns represent each document. Each cell contains the number of times one word appears in its corresponding documents. In general, the matrices built during LSI tend to be very big, but also sparse (most cells contain 0). That is because each document usually contains only a small number of keywords.

The purpose of SVD is to derive a particular latent semantic structure model. There is a mathematical proof [14] that for every $m \times n$ matrix X, there exists a singular value decomposition where

$$X = U\Sigma V^T \tag{1}$$

The U, Σ and V matrices are the outputs of SVD. U is an orthogonal $m \times r$ matrix whose columns are left singular vectors of X, while V is an orthogonal $r \times n$ matrix whose columns are right singular vectors of X. Σ is a diagonal matrix on whose diagonal are singular values of matrix X in descending order such that when the three components are matrix-multiplied, the original matrix is reconstructed.

The last step is to compare the new document vector coordinates and word vector coordinates in the target dimensional space.

The values in the Σ diagonal matrix represent the importance of the latent concept, we can set the dimension of the latent concept (i.e. the numbers of features k that we want to keep). For LSI truncated SVD is used ($k < r$), where Uk is $m \times k$ matrix whose columns are first k left singular vectors of X. $\Sigma\, k$ is $k \times k$ diagonal matrix whose diagonal is formed by k leading singular values of X. Vk is $n \times k$ matrix whose columns are first k right singular vectors of X. The rows of Uk are words and the rows of Vk are documents. Then, the word-word, document-document coordinate can be drawn (x-axes is the first dimension, y-axes is the second dimension), and the two vectors on the coordinate can be compared by the cosine similarity (Eq. (2)).

4   Latent Semantic Indexing for Inventive Design

4.1   Corpus Preparation and Texts Pre-processing

In this experiment, the corpus of is composed of 67 documents, including 28 Specific Parameters in 14 specific engineering cases and 39 Generic Engineering Parameters in the Contradiction Matrix. The aim is to capture the similar concepts that two documents share which lead to the selection of *GEPs*. The experiment corpus is presented in Fig. 1. Each parameter is considered as one document and the words in each document are separated into single words. The stop words such as for, of, or etc., are deleted.

Fig. 1. Experiment corpus

4.2   Result Analysis

This section explains the utility of LSI. As it can be seen in Fig. 2, the following results can be reported:

1. D29–D36 and D2, D15, D21, D22 are very similar. The dots and Xs which represent them are situated in the blue triangle. The concepts that they share are: ['area', 'length', 'volume', 'weight']. D2, D15, D21, D22 are specific parameters of cases and D29–D36 are Generic Engineering Parameters of Contradiction Matrix. The findings indicate that these 8 parameters (D29–D36) share the same geometrical characteristics. And specific parameters indicating geometrical problems are more similar among each other than the ones that do not, for example, D24-ease of use.
2. The finding of indexing concepts of geometrical parameters provides opportunities to find different specific parameters in the geometrical semantic space which can be constructed by synonym sets of the indexing words.
3. D50–D53 are very similar to the concept "loss" but very different from D36 and D60–D62. It suggests that concepts D50–D53 are in the degrading semantic space.
4. D60–D62 are very similar to the concept "ease" but very different from D36 and D50–D53. This observation means that concepts D60–D62 belong in the improving semantic space (see Table 2).

At this point, the cosine similarity is applied between two words to calculate their similarities before and after applying LSI. As an example, the similarity between area and length is calculated by Eq. (2).

$$\cos \theta = \frac{\vec{a} \cdot \vec{b}}{\|\vec{a}\| \|\vec{b}\|} \tag{2}$$

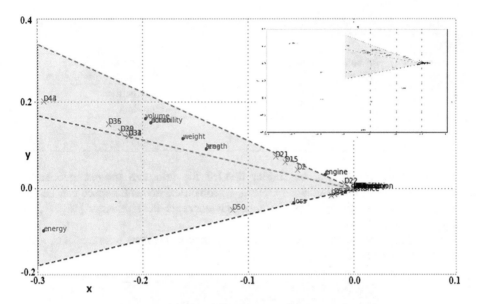

Fig. 2. LSI result with a target dimension of two

Table 2. Additional vector values of concepts

Concept	Vector
D60	(−0.076663150610301378, 0.2917930401253932)
D61	(−0.076663150610301378, −0.2917930401253932)
D62	(−0.076663150610301378, −0.2917930401253932)

In the initial matrix X, the similarity of the two vectors ($\cos \theta$) equals to 0, which means the two words area and length share no similarity. After applying LSI, the cosine similarity between the two vectors almost equals to 1, which suggests strong similarity. This result can be observed from Fig. 2, area and length are overlapped.

5   Retrieval

Originally, the *GEP*s were classified by the practice of TRIZ into three groups: the geometrical physical parameters, the generic improving parameters and the generic degrading parameters [15] (see Table 3). This classification is explicit knowledge from experts who are able to identify the underlying meaning of freely expressed specific parameters, because they possess experience over years of practice. The challenge is how we can have access to this experience of experts by means of machine learning. Our results have found hints of the semantic spaces that the experts possessed but are unaware of. Therefore, the result of LSI provides the possibility to construct semantic spaces based on identified indexes. The semantic spaces will be used to guide retrieval.

Based on the above findings, we can thus categorize retrieval features into problem features and solution features. As has been previously stated, geometrical physical

Table 3. Feature value set to determine Generic Engineering Parameter Classification

Feature value				Label
Geometrical distance	Physical distance	Improving distance	Degrading distance	Generic Engineering Parameter Classification
2	0	0	0	Geometrical
0	2	0	0	Physical
0	0	2	0	Improving
0	0	0	2	Degrading

engineering parameters can be further divided the into two groups: geometrical parameters and physical parameters. Next, a geometrical semantic space can be constructed based on synonyms of the indexing concepts in Thesaurus [16]: ['area', 'length', 'volume', 'weight'] (Fig. 3).

The other three semantic spaces can be explored as the case base expands, and it still requires further research.

```
area field operation range space breadth compass distance expanse size sphere stretch width weight burden
density gravity heft load pressure substance adiposity avoirdupois ballast gross heftiness mass measurement
net ponderosity ponderousness poundage tonnage G-factor length breadth diameter dimension duration height
limit magnitude mileage period piece portion    quantity radius range section segment space span stretch
term compass continuance elongation endlessness expanse expansion extensiveness interval lastingness
lengthiness linearity loftiness longitude longness measure orbit panorama purview reach realm remoteness
season spaciousness stride tallness unit year protractedness ranginess volume amount figure number quantity
size total aggregate body bulk compass content contents dimensions extent mass object cubic
```

Fig. 3. Geometrical semantic space

5.1 Similarity Measures

During the second step of retrieval, the central issue is to calculate the semantic distance between the new problem and the old problem. In the retrieval process, the query is inputted in the form of text, therefore, the Vector Space model (VSM) [17] is adopted in order to calculate the local similarity between the new specific parameter and the old specific parameter.

In each semantic space, the new specific parameter and the old specific parameter are represented as vectors. In order to calculate similarity between two vectors, there are numbers of algorithms, among them, Euclidian distance is chosen for its simplicity and suitability for our case.

Euclidean distance is the square root of the sum of squared differences between corresponding elements of the two vectors:

$$dist(SP_{old}, SP_{new}) = \sqrt{\sum_{i=1}^{n}(SP_{oldi} - SP_{newi})^2} \tag{3}$$

5.2 Similarity for a Specific Parameter

The Euclidean distance is used to calculate similarity for the specific parameters in each semantic space. For two different vectors, the result obtained in each space is:

$$dist\left(SP_{old}, SP_{new}\right) = \begin{cases} 0, & \text{if } SP_{old}, SP_{new} \text{ is not in the same semantic space} \\ 2, & \text{if } SP_{old}, SP_{new} \text{ is in the same semantic space} \end{cases}$$

Therefore, we can obtain a set of results to determine the classification of Generic Engineering Parameters, as can be seen in Table 3.

5.3   Decision Trees for Guiding Retrieval

A decision tree is used for retrieval, starting from the root node, intermediate nodes are generated to represent classifications of cases according to a distinctive feature. At the bottom of the tree, at final nodes called leaves, there are the source cases. Finally, in this approach, leaves represent the classification and branches represent conjunction of features that lead to these classifications, as depicted in Fig. 4.

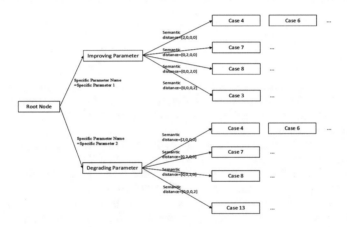

Fig. 4. Decision tree

When a user has a new problem, the semantic distance between the new Specific Parameter and the old Specific Parameter are calculated in each of the four semantic spaces, and the obtained semantic distance is used to generate the label for the new *SP*, and according to the Specific Parameter Name (whether it is *SP1* or *SP2*), the similar cases are retrieved.

6   Conclusion and Future Perspectives

This article shows the possibility of using CBR for inventive design studies. The goal is to take advantage of previous design experiences to solve new problems to cope with the difficulties in associating SPs with GEPs in the use of Contradiction Matrix [18]. The intuition of this work is simple, but the operation of CBR is difficult because of the arbitrariness of human language and the fact that TRIZ is performed by analogy based on past experience. Therefore, it brings two problems: one is that human has many ways to refer to the same object in natural language based on their different knowledge, understandings and linguistic habits. The other is the impossibility of understanding the same word that has different meanings in different contexts. By applying the LSI method, the first problem can be solved by finding the hidden concepts behind freely expressed specific problems. Based on the findings, the indexes for constructing a semantic space has been identified to guide case retrieval. Compared with the work presented in [5], which used the existing knowledge base to guide retrieval, this approach provides the possibility of a more efficient and accurate way to use actual experience.

The future perspectives of this work are to complete the other three semantic spaces through more solved cases and to validate this approach by solving new inventive design problems. It is also important to enable the system to learn by itself in the future in order to solve the polysemy problem that was addressed before. In addition, the experience and rules gained from case-based reasoning approach is going to be stored in the SOEKS [19] structure for decision making in a more flexible, creative way.

References

1. Al'tshuller, G.S., Shulyak, L., Rodman, S.: The Innovation Algorithm: TRIZ, Systematic Innovation and Technical Creativity. Technical Innovation Center Inc., Worcester (1999)
2. Abramov, O.Y.: Industry best practices and the role of TRIZ in developing new products. ResearchGate (2013)
3. Cavallucci, D., Khomenko, N.: From TRIZ to OTSM-TRIZ: addressing complexity challenges in inventive design. Int. J. Prod. Dev. 4(1–2), 4–21 (2006)
4. Cavallucci, D., Fuhlhaber, S., Riwan, A.: Assisting decisions in inventive design of complex engineering systems. Procedia Eng. 131, 975–983 (2015)
5. Yan, W., Liu, H., Zanni-Merk, C., Cavallucci, D.: IngeniousTRIZ: an automatic ontology-based system for solving inventive problems. Knowl.-Based Syst. 75, 52–65 (2015)
6. Duflou, J.R., Dewulf, W.: On the complementarity of TRIZ and axiomatic design: from decoupling objective to contradiction identification. Procedia Eng. 9, 633–639 (2011)
7. Coelho, D.A.: Matching TRIZ engineering parameters to human factors issues in manufacturing. Wseas Trans. Bus. Econ. 6(11), 547–556 (2009)
8. Campbell, B.: Brainstorming and TRIZ. TRIZ J., February 2003. http://www.triz-journal.com/archives/2003/02/index.htm
9. Aamodt, A., Plaza, E.: Case-based reasoning: Foundational issues, methodological variations, and system approaches. AI Commun. 7(1), 39–59 (1994)
10. Rousselot, F., Renaud, J.: On triz and case based reasoning synergies and oppositions. Procedia Eng. 131, 871–880 (2015)
11. Richter, M.M.: Knowledge containers. In: Readings in Case-Based Reasoning. Morgan Kaufmann Publ., San Mateo (2003)
12. Bird, S., Klein, E., Loper, E.: Natural Language Processing with Python. O'Reilly Media, Inc., Sebastopol (2009)
13. Aggarwal, C.C., Zhai, C.: A survey of text classification algorithms. In: Aggarwal, C.C., Zhai, C. (eds.) Mining Text Data, pp. 163–222. Springer, New York (2012)

14. Berry, M.W., Dumais, S.T., O'Brien, G.W.: Using linear algebra for intelligent information retrieval. SIAM Rev. **37**(4), 573–595 (1995)
15. Runhua, T.: TRIZ and Applications: The Process and Methods of Technological Innovation. Higher Education Press, Beijing (2010)
16. Synonyms and antonyms of words. www.thesaurus.com, http://www.thesaurus.com/. Accessed 07 Sept 2016
17. Salton, G., Wong, A., Yang, C.S.: A vector space model for automatic indexing. Commun. ACM **18**(11), 613–620 (1975)
18. Kosse, V.: Some limitations of TRIZ tools and possible ways of improvement. Am. Soc. Mech. Eng. Eng. Div. Publ. DE **103**, 111–115 (1999)
19. Shafiq, S.I., Sanín, C., Szczerbicki, E.: Set of Experience Knowledge Structure (SOEKS) and Decisional DNA (DDNA): past, present and future. Cybern. Syst. **45**(2), 200–215 (2014)

Optimization of Electrical Discharge Machining Parameters of Co-Cr-Mo Using Central Composite Design

Soudeh Iranmanesh[1], Alireza Esmaeilzadeh[1], and Abbas Razavykia[2(✉)]

[1] Department of Mechanical Engineering, University of Malaya, Kuala Lumpur, Malaysia
soudeh.irmanesh@gmail.com, aesmaeilzadeh2@gmail.com
[2] Department of Mechanical and Aerospace Engineering, Politecnico di Torino, Turin, Italy
abbas.razavykia@polito.it

Abstract. The optimization of electrical discharge machining (EDM) parameters of Cobalt Chromium Molybdenum (Co-Cr-Mo) is performed using central composite design to improve the process efficiency in terms of increasing material removal rate and electrode utilization time. The effects of pulse on time, pulse off time, voltage and current on electrode wear rate (EWR) and material removal rate (MRR) have been examined. The experimental results indicate that higher pulse on time, lower pulse off time, 100 v to 110 v for voltage and current at the range of 8 to 9 A are the adequate selection to achieve higher MRR and lower EWR.

Keywords: Electrical discharge machining · Efficiency · Electrode wear rate · Cobalt Chromium Molybdenum

1 Introduction

Production of high quality products at lowest cost and highest productivity are required to encourage globalization and customization market in manufacturing industry. Electrical discharge machining EDM is one of the most promising widespread manufacturing process, which is well-known to produce high verity and complex shape [1]. The main application of EDM is in die and mold manufacturing, aerospace industry, surgical and orthopedic components, due to its capability to produce complicated geometries and to cut hard material without respect to their hardness [2]. No mechanical stresses, chatter and vibration are imposed to part by EDM because of there is no direct contact between the work-piece and tool [3]. Cobalt Chromium Molybdenum (Co-Cr-Mo) is recognized as a one of the super alloys and advanced engineering materials with interesting mechanical properties such as high specific strength, high corrosion resistance, and higher biocompatibility. Co-Cr-Mo generally is used in gas turbines, orthopedics, and dental implants. It has been reported that during the electrical discharge machining of Inconel 718, material removal rate is dramatically

© Springer International Publishing AG 2017
G. Campana et al. (eds.), *Sustainable Design and Manufacturing 2017*, Smart Innovation, Systems and Technologies 68, DOI 10.1007/978-3-319-57078-5_5

affected by the voltage and current, and also most influential parameters on tool wear are pulse on time and duty cycle [4]. The EDM efficiency is influenced directly by the value of pulse on time and peak current [5]. Higher MRR could be obtained with increasing the peak current and duty factor which has been suggested for electrical discharge machining of hard material [6]. Response surface methodology is a statistical technique, which is applicable to analysis and investigation of problems that one or more responses (dependent variables) are subjective with several variables and the goal is to optimize the responses [7]. RSM calculates the relationships between the independent variables and one or more measured responses and with the help of five important steps to analyzing each response. Subsequently, multiple responses optimization will be performed, either by graphical and numerical provided tools, or with the inspection of the interpretation plots [8]. Modern machining processes such as EDM can be promising alternative to avoid negative effect of conventional machining processes, due to that the major problem associated with the machining of Co-Cr-Mo is the formation of inhomogeneous inelastic deformation on machined surface that imposes residual stresses in work-piece [9].

Cobalt Chromium Molybdenum (Co-Cr-Mo) is a unique material that there is no research addressing the adequate EDM parameters to maximize the material removal rate and electrode life. The EDM process is recognized to be not a fast manufacturing process [10]. Therefore there is a pressing need to optimize the process parameters to increase the productivity of the process and make it efficient in terms of energy consumption and machining time and cost. Thus, developing a regression model and simultaneously determining the correlation among parameters to select suitable machining parameters in order to improve machining efficiency in terms of MRR, EWR using response surface methodology (RSM) and central composite design (CCD) during the EDM of Co-Cr-Mo is the main principle aim of this study.

2 Material and Experiment Procedure

Work-piece was cut using wire cut EDM to achieve adequate dimension to use the clamp mounted on die-sinking EDM machine (Sodick-AG40L) table. Table 1 illustrates material properties and chemical composition of work-piece Co-Cr-Mo which were obtained using electron microscopy with field emission scanning (Model: Supra-35VP, Carl Zeiss, Germany) integrated with energy dispersive spectroscopy (EDS) facility. Copper electrode with dimension of 40 mm in length and 6.35 mm in diameter were applied with chemical composition, which is given by Table 2. In order to guarantee constant mechanical properties and chemical composition all electrodes come from same batch and rod. To encourage accurate and reliable report, for each observation a fresh electrode was used.

To ensure that work-piece underwent majority of erosion as opposed to the electrode, positive polarity for electrode has been considered [11]. In order to increase efficiency, oil-based dielectric fluid (PGM WHIT 3) mixed by aluminum powder was employed for all experimental trials, based on conducted observations [12, 13], Al powder additive makes discharge breakdown easier and enlarges discharge gap. Large variety of EDM parameters might be considered but the main focus of this research is on

Table 1. The material properties of the Co-Cr-Mo alloy and its chemical composition (wt.%).

Work piece	Co-Cr-Mo
Hardness	(HRC) 40–45
Density (gr/cm^3)	8.29
Yield strength (MPa)	980 ± 50
Tensile strength (MPa)	1300 ± 50
Elongation (%)	10
Co = 55.65, Cr = 25.02, Mo = 5.76, C = 10.10, O = 2.60, Si = 0.87	

Table 2. The chemical composition of Copper electrode material (wt.%).

Cu	Zn	Sn	Al	Pb	Bi
97.3	2.35	0.06	0.15	0.11	0.03

the influence of four parameters which are summarized by Table 3. It is interesting to mention that the values for influential parameter are based on the literatures referring to EDM of hard to cut material and also servo voltage kept constant at the value of 90 V.

Table 3. Machining parameters and their specifications

Factor	Unit	Low level (−1)	High level (+1)
Voltage (V)	V	60	120
Current (I_p)	A	6	10
Pulse off time (t_{off})	μs	15	23
Pulse on time (t_{on})	μs	10	30

The experimental trials were designed using face-cantered CCD by four factors with two levels, which resulted in 16 observation, as well as 3 center points and 8 axial points which resulted in 27 runs totally, with target of minimizing electrode wear rate (EWR) and maximizing material removal rate (MRR). MRR and EWR were calculated based on Eqs. (1) and (2) respectively. In order to accurate measurement of the electrode weight, precision electronic balance (Pioneer) with 0.0001 gr precision applied before and after each performed experiment.

$$MRR\left(mm^3/min\right) = \frac{Volume\ removed\ from\ work - piece\ (mm^3)}{Machining\ time\ (min)} \qquad (1)$$

$$EWR\left(mm^3/min\right) = \frac{Electrode\ weigth\ lose\ (gr)}{Machining\ time\ (min) \times Density\ \left(\frac{gr}{mm^3}\right)} \qquad (2)$$

The experimental trials and the results are demonstrated by Table 4 and each test was conducted one time. Design-Expert software version 9.0.5 was applied to analysis the results and consequently analysis of variance (ANOVA).

Table 4. Experimental design and results.

Std.	Run	t_{on}	t_{off}	V	I_p	EWR (mm³/min)	MRR (mm³/min)
1	1	10	15	60	6	0.116	1.99
2	4	30	15	60	6	0.044	2.72
3	13	10	23	60	6	0.102	1.74
4	2	30	23	60	6	0.046	2.84
5	22	10	15	120	6	1.028	4.86
6	26	30	15	120	6	0.442	6.70
7	9	10	23	120	6	1.256	4.63
8	5	30	23	120	6	0.445	6.62
9	15	10	15	60	10	0.481	3.45
10	12	30	15	60	10	0.142	4.74
11	17	10	23	60	10	0.542	2.47
12	27	30	23	60	10	0.136	4.70
13	11	10	15	120	10	1.828	6.66
14	16	30	15	120	10	0.853	8.27
15	7	10	23	120	10	2.896	5.62
16	14	30	23	120	10	1.024	8.63
17	18	10	19	90	8	1.186	5.20
18	23	30	19	90	8	0.416	6.95
19	19	20	15	90	8	0.610	6.43
20	6	20	23	90	8	0.728	6.71
21	25	20	19	60	8	0.098	2.95
22	21	20	19	120	8	1.156	7.68
23	3	20	19	90	6	0.319	5.13
24	20	20	19	90	10	0.839	6.92
25	24	20	19	90	8	0.693	6.57
26	10	20	19	90	8	0.678	6.56
27	8	20	19	90	8	0.673	6.45

3   Results and Discussion

3.1   Statistical Analysis of Factors Influencing Material Removal Rate (MRR)

From ANOVA table (Table 5), it is obviously clear that the model is significant by the F-value of 197.49. The significant model terms are those with 'Prob. > F' value of less than 0.05, it implies that A, C, D, AB, AC, AD and A2, C2, D2 are significant model terms. Factor B (Pulse off time) is considered as significant factor due to it is close to 0.05 and based on literatures it is recognized as influential factor. The Predicted R-Squared signifies how well the model predict the response values, and in this case a value of 0.99 is certainly desirable. To examine the conformity between predicted R-Squared and Adjusted R-Squared, the difference between them must be approximately less than

0.2 [14], which in this case they are in reasonable agreement with each other by difference of 0.03. Desirable ratio for Adequate Precision that shows signal to noise ratio is greater than four which in this study is 50.99.

Table 5. Analysis of variance table for MRR [Partial sum of squares - Type III]

Source	Sum of squares	df	Mean square	F value	p-value prob > F	
Model	95.54	10	9.54	197.49	<0.0001	Significant
A-Pulse on time	13.43	1	13.43	278.03	<0.0001	
B-Pulse off time	0.19	1	0.19	3.98	0.0634	
C-Voltage	54.78	1	54.78	1133.67	<0.0001	
D-Current	11.25	1	11.25	232.83	<0.0001	
AB	0.51	1	0.51	10.58	0.0050	
AC	0.60	1	0.60	12.43	0.0028	
AD	0.38	1	0.38	7.96	0.0123	
A^2	0.49	1	0.49	10.23	0.0056	
C^2	1.91	1	2.00	41.40	<0.0001	
D^2	0.61	1	0.62	12.81	0.0025	
Residual	0.66	16	0.048			
Lack of fit	0.76	14	0.055	12.31	0.0776	Not significant
Pure error	8.867E−003	2	4.433E−003			
Cor. total	96.20	26				
R-Squared	0.99		Adj. R-Squared	0.98		
Pred. R-Squared	0.97		Adeq. Precision	50.99		

MRR was affected directly by pulse on time, so it means MRR increases as pulse on time increases due to higher value of pulse on time delivers greater applied energy on work-piece and consequently higher MRR [15]. Almost all the tests when pulse on time increased from 10 to 30, MRR increased as shown in Fig. 1(a). On contrary, speed and stability of cutting are directly proportional to cycle time, on the other hand smaller pulse off time provides faster cutting. It has been observed lower MRR obtained as pulse off time increased from 15 to 23 (Fig. 1(b)). Discharge gap between electrode and work-piece is controlled by the voltage, increasing the voltage increases the discharge gap [16]. As Fig. 1(c) illustrates higher MRR achieved at higher voltage. In addition, MRR increased as current increased, MRR is directly affected by produced energy per pules and frequency of pulse. When current increase at persistent frequency results is higher value of MRR (Fig. 1(d)).

Figure 2(a) illustrate that residuals follow the normal distribution and straight line can fit between them with small departures from the straight line for MRR. This is a plot of the residuals versus the ascending predicted response values (Fig. 2(b)) shows that residuals are normally distributed with constant variance which tests the hypothesis of constant variance. Empirical model was developed in actual factors in terms of predicting MRR (Eq. 3). Empirical model based on actual factors is used to make prediction of response by using the original units for each factor.

$$MRR = -15.821 + 0.049 \times t_{on} - 0.115 \times t_{off} + 0.214 \times V + 2.121 \times I_p + 0.004 \times t_{on} \times t_{off} \\ + 0.0006 \times t_{on} \times V + 0.007 \times t_{on} \times I_p - 0.004 \times t_{on}^2 - 0.0009 \times V^2 - 0.117 \times I_p^2 \quad (3)$$

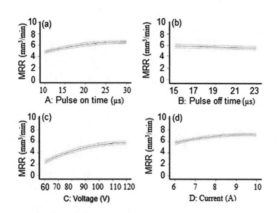

Fig. 1. Effect of individual model terms

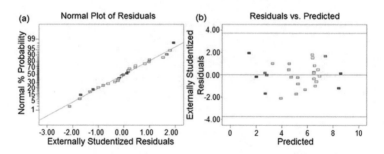

Fig. 2. (a) Normal plot of residuals, (b) residuals vs. predicted on MRR

3.2 Statistical Analysis of Factors Influencing Electrode Wear Rate (EWR)

The ANOVA analysis in Table 6 provides evidence that factors A, B, C, D and AC, AD, BC, CD interactions are significant. Predicted R-Squared and Adjusted R-Squared are reasonable in agreement with together with difference less than 0.2. Adequate Precision with value of 60.41 that shows signal to noise ratio is enough greater than four which desirable.

Pulse on time has indirect effect on EWR, therefore it means EWR decreases as pulse on time increases (Fig. 3(a)). As shown in Fig. 3(b) the pulse off time has low influence on EWR. The electrode wear increase gradually when pulse off time increase from 15 to 23. Figure 3(c, d) illustrates that lower EWR achieved at lower voltage and current. Electrode wear increase rapidly when these both parameters are increasing.

Table 6. Analysis of variance table for EWR [Partial sum of squares - Type III]

Source	Sum of squares	df	Mean square	F value	p-value Prob > F	
Model	6.54	8	0.82	259.66	<0.0001	Significant
A-Pulse on time	1.16	1	1.16	367.66	<0.0001	
B-Pulse off time	0.029	1	0.029	9.07	0.0075	
C-Voltage	3.83	1	3.83	1216.93	<0.0001	
D-Current	0.90	1	0.90	286.27	<0.0001	
AC	0.38	1	0.38	119.77	<0.0001	
AD	0.083	1	0.083	26.37	<0.0001	
BC	0.016	1	0.016	5.22	0.0347	
CD	0.14	1	0.14	46.02	<0.0001	
Residual	0.057	18	3.150E−003			
Lack of fit	0.056	16	3.514E−003	14.62	0.0658	Not significant
Pure error	4.807E−004	2	2.403E−004			
Cor. total	6.60	26				
R-Squared	0.99		Adj. R-Squared	0.98		
Pred. R-Squared	0.98		Adeq. precision	60.41		

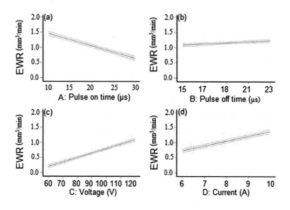

Fig. 3. Effect of individual model terms on EWR

Figure 4(a) demonstrates that residuals follow the normal distribution and straight line can fit between them with minor departures from the straight line for EWR. Figure 4(b) shows that the plot of the residuals against the arising predicted response values that residuals are normally distributed with constant variance. Empirical model based on actual factors was developed to predict EWR (Eq. 4).

$$EWR = -1.210 + 0.049 \times t_{on} - 0.014 \times t_{off} + 0.007 \times V + 0.041 \times I_p - 0.0005 \times t_{on} \times V - 0.003 \times t_{on} \times I_p + 0.0002 \times t_{off} \times V + 0.001 \times V \times I_p \tag{4}$$

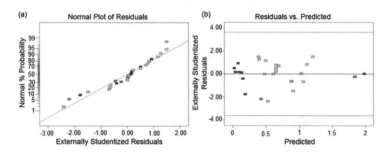

Fig. 4. (a) Normal plot of residuals, (b) residuals vs. predicted on EWR

4   Desirability, Optimization, and Confirmation

High material removal rate and lower electrode wear are desirable during rough EDM machining which directly influence the machining cost. All factors were adjusted in range and main objective is maximizing MRR and minimizing EWR. Importance of +++ and +++++ were considered for EWR and MRR respectively. Desirability is an independent function that sorts from zero outside of limits to one at the aim. In addition, numerical optimization discovers a point that maximizes the desirability function. Figure 5(a) illustrates the best setting of parameters to meet the study objective at desirability of 0.833, are at the down right corner (orange color). Based on overlay plot (Fig. 5(b)), the feasible region for high MRR and low EWR is yellow area. Based on both plot, higher Pulse on time, lower pulse off time and the 100 to 110 V the range value of voltage and Current at the range of 8 to 9 A are recommended to achieve higher MRR and lower EWR. Four confirmation tests were performed to verify the adequacy of the developed model and these are summarized in Table 7.

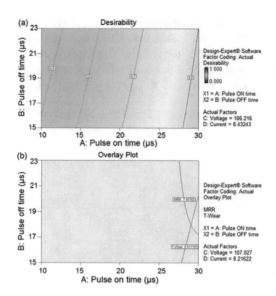

Fig. 5. (a) Desirability graph, (b) Overlay plot

Table 7. Confirmation experiments

t_{on}	t_{off}	I_p	V	MRR (mm³/min)				EWR (mm³/min)			
				Actual	Predicted	Residual	Error (%)	Actual	Predicted	Residual	Error (%)
20	25	9	110	7.37	7.52	−0.15	−2.03	0.471	0.459	0.012	2.54
30	23	9	115	8.15	8.44	−0.29	−3.55	0.139	0.141	−0.02	−1.43
25	20	8	105	7.49	7.60	−0.11	−1.46	0.149	0.145	0.04	2.68
28	18	8	108	7.89	7.83	0.06	0.76	0.061	0.062	−0.001	−1.63

5   Conclusion

An Attempt has been made to improve electrical discharge machining (EDM) efficiency in terms of reducing machining time and electrode wear rate (EWR). According
to the results and discussion during EDM of Co-Cr-Mo alloy using copper electrode in
the considered range of EDM parameters, following conclusions have been identified
throughout this observation which may direct future studies:

1. MRR is affected by significant contributions of the individual and second-order
   effects, for instance, pulse on time, voltage and current and second-order effects of
   voltage and pulse on time. The MRR enlarges with increases of the voltage, pulse
   on time and current.
2. Whiles, the significant contributions of the individual and interaction effects of
   factors affect the electrode wear rate. The EWR increases with the increases of the
   voltage and current but decreases with increases of pulse on time.

References

1. Razavykia, A., Yavari, M.R., Iranmanesh, S., Esmaeilzadeh, A.: Effect of electrode material and electrical discharge machining parameters on machining of CO-CR-MO. IJMME **16**, 53–61 (2016)
2. Tang, L., Guo, Y.F.: Electrical discharge precision machining parameters optimization investigation on S-03 special stainless steel. Int. J. Adv. Manuf. Technol. **70**, 1369–1376 (2014)
3. Puertas, I., Luis, C.J., Alvarez, L.: Analysis of the influence of EDM parameters on surface quality, MRR and EW of WC–Co. J. Mater. Process. Technol. **153**, 1026–1032 (2004)
4. Ghewade, D., Nipanikar, M.S.: Experimental study of electro discharge machining for Inconel material. J. Eng. Res. Stud. **976**, 7916 (2011)
5. Pradhan, B.B., Masanta, M., Sarkar, B.R., Bhattacharyya, B.: Investigation of electro-discharge micro-machining of titanium super alloy. Int. J. Adv. Manuf. Technol. **41**, 1094–1106 (2009)
6. Rajesha, S., Sharma, A.K., Kumar, P.: On electro discharge machining of Inconel 718 with hollow tool. J. Mater. Eng. Perform. **21**, 882–891 (2012)
7. Montgomery, D.C.: Introduction to Statistical Quality Control. Wiley, New York (2007)
8. Noordin, M.Y., Venkatesh, V.C., Sharif, S., Elting, S., Abdullah, A.: Application of response surface methodology in describing the performance of coated carbide tools when turning AISI 1045 steel. J. Mater. Process. Technol. **145**, 46–58 (2004)
9. Deshpande, A., Yang, S., Puleo, D., Pienkowski, D., Dillon, O., Outeiro, J., Jawahir, I.S.: Minimized wear and debris generation through optimized machining of Co-Cr-Mo alloys for use in metal-on-metal hip implants. In: ASME, pp. 297–305 (2012)
10. Salonitis, K., Stournaras, A., Stavropoulos, P., Chryssolouris, G.: Thermal modeling of the material removal rate and surface roughness for die-sinking EDM. Int. J. Adv. Manuf. Technol. **40**, 316–323 (2009)
11. Manjaiah, M., Narendranath, S., Basavarajappa, S.: A review on machining of Titanium based alloys using EDM and WEDM. Rev. Adv. Mater. Sci. **36**, 89–111 (2014)
12. Assarzadeh, S., Ghoreishi, M.: A dual response surface-desirability approach to process modeling and optimization of Al2O3 powder-mixed electrical discharge machining (PMEDM) parameters. Int. J. Adv. Manuf. Technol. **64**, 1459–1477 (2013)
13. Garg, R.K., Singh, K.K., Sachdeva, A., Sharma, V.S., Ojha, K., Singh, S.: Review of research work in sinking EDM and WEDM on metal matrix composite materials. Int. J. Adv. Manuf. Technol. **50**, 611–624 (2010)
14. Razavykia, A., Farahany, S., Yusof, N.M.: Evaluation of cutting force and surface roughness in the dry turning of Al–Mg 2 Si in-situ metal matrix composite inoculated with bismuth using DOE approach. Measurement **76**, 170–182 (2015)
15. Wang, C.C., Yan, B.H.: Blind-hole drilling of Al_2O_3/6061Al composite using rotary electro-discharge machining. J. Mater. Process. Technol. **102**, 90–102 (2000)
16. Kuppan, P., Rajadurai, A., Narayanan, S.: Influence of EDM process parameters in deep hole drilling of Inconel 718. Int. J. Adv. Manuf. Technol. **38**, 74–84 (2008)

Sustainable Data Collection Framework: Real-Time, Online Data Visualization

Tien-Lung Sun[✉] and Gustavo Adolfo Miranda Salgado

Department of Industrial Engineering and Management,
Yuan Ze University, Zhongli, Taiwan, ROC
tsun@saturn.yzu.edu.tw, gustavomiranda@live.com

Abstract. This paper presents a comprehensive data collection framework focused on retrieving data from an inertial sensor using a smartphone device. The aim of this study is to present a low-cost and sustainable data collection framework solution based on previous data framework model developed. The proposed framework will allow the users (doctors, physicians, patients, family members, etc.) to visualize online the real-time performance results, as the patience carries out its physical therapy. The data collection framework utilizes an in-house developed smartphone app and open source software for the server/client interaction. The data collection framework has been test by collecting information from different exercise machines and free movement exercises. The tests show that the real-time data collection framework proposed is reliable in capturing, recording and displaying the data obtained during a training exercise session, aiming toward the tracking people's quality of life.

Keywords: Data collection framework · Data visualization · Sustainable framework · Inertial sensor

1 Introduction

There has been a variety of proposed frameworks for health information systems compactible with mobile health (mHealth) devices such as Open mHealth, an open source software derived from a series of modules that allows the data flow and analysis but closed for intranet [1]. Some research papers are based on limited sensor network, sensors are been connected to a local network for data storage only, Electronic Health Record (EHR). Then the information can be view via a web-based interface, which have been prove useful by the medical staff [2]. Another study on EHR explores the option of incorporating motion sensor data to the proposed framework and its integration across different platforms [3].

Previous research had created a web-based visualization system as well, which integrates monitoring, analysis, and automated recognition it has a variety of modules such as data collection, data adapter, visualization application, etc. but limited to smart environment research only [4].

This research aims to present a low-cost and sustainable data collection framework solution based on a previous data framework prototype developed by the same team.

© Springer International Publishing AG 2017
G. Campana et al. (eds.), *Sustainable Design and Manufacturing 2017*, Smart Innovation,
Systems and Technologies 68, DOI 10.1007/978-3-319-57078-5_6

The proposed framework will allow the users (doctors, physicians, patients, family members, etc.) to visualize online the real-time performance results, as a patience carries out its physical therapy.

This paper is organize as follows: In Sect. 2, some background review of the main topics related to the research. In Sect. 3, the development of the proposed data collection framework. Section 4 presents the implementation of the proposed framework. Section 5 discusses the visual challenges of presenting the collected data. Finally, Sect. 6 describes the research conclusions and possible future work applications.

2 Background

2.1 Type of Sensors

Wearable sensors can provide information from natural body movements or from mechanical components such as exercise equipment. These wearable devices are actually micro-electro-mechanical sensors (MEMS), which are lightweight, very small and at an accessible price. They include a gyroscope and an accelerometer, and can be use conveniently in a community center [5].

There are many types of sensors that exist for health monitoring such as the passive infrared sensors for the ambient assisted living [2], a series of ambient sensors for health assessment [6], bed pressure sensor for fall risk assessment [7], and some use wearable devices which in their majority recollection angular velocity, linear acceleration and temporal data to measure health and wellness status [8].

Therefore, from the variety types of sensors, and for the purpose of this research, the inertial sensors are the best option when it comes to physical activity-monitoring systems. Moreover, their size and weight factors make them very easy to attach to any part of a person or mechanical parts of any equipment [5], e.g. arms, legs, lower back, stationary bicycle, hand weight, etc.

2.2 Sensor Data Visualization

There are many methods and techniques to visualize information depending on the needs and purpose of the user, some studies have developed their own as for example the Visualization of Time-Oriented Records (VISITORS) system. This system combines temporal analysis with different visualization techniques resulting on a successful experience to generate a variety of visualizations, but the user interface can be too complex without the proper training and practice [9]. Additionally, there is the Knowledge-based Visualization and Exploration (KNAVE-II) which allows interactive visualization through web-based architecture. It proves to be a promising tool, but it also suffers the same limitations on the user interface as VISITORS, additionally, having some complex visualizations [10]. Another visualization tool developed is Lifelines2, using a EHR system to retrieve the data it creates scatter plots, histograms, line graphs, etc. having encouraging results but still under development [11]. Furthermore, standard visualizations cannot always represent some specific data, leading to the development of new visualizations according to specific display needs [12].

While some research papers focus on developing generalized visualizations, others focus on elderly visualizations, highlighting to put attention on behavior patterns and longitudinal trends, while proposing some new visualizations styles [13]. Some studies demonstrated that to find patterns and trends it is required some long term data collection that will allow the user detect such tendencies and fluctuations [14]. Similarly, Caprani et al. approached to healthcare professionals, letting them to observe patterns and trends, not only from one visualization at a time, but also from a dashboard [15].

As the development of the data collection framework is still at an early stage, a visualization tool will not be develop, as done on previous researches [16]. For the purpose of this research, the propose system will implement a preexisting visualization tool, Tableau software. It provides predefined visualizations and filter controls that offer an easy way to explore the collected data. Moreover, it offers an online visualization platform that has the possibility to be adapted to the propose data collection system aimed on this research.

3   Framework Development

3.1   Inertial Sensor

The MEMS used for this research is the wearable device Cavy Play Band (designed by DEM and manufactured by Foxconn, CavyTech, China). It features a Bosch Sensortec BNO055 Intelligent nine-axis sensor technology; composed by a three-axis accelerometer, a three-axis gyroscope and a three-axis magnetometer, allowing a 360° synchronized human-machine interaction. Its double CPU chip allows the device to calculate the data from its sensors and send the information via Bluetooth (v2.1 + EDR, on Android) to a previously paired device in real-time and with great precision. The size of the sensor is small (40 mm × 19 mm × 7 mm) and lightweight (<26 g). These two characteristics allows the possibility to attach this sensor anywhere, enabling a variety of applications further than its original purpose of gaming entertainment.

From the MEMS, the linear acceleration data (the speed of displacement of the sensor over a fixed pre-oriented axis direction) and the Euler angles (the angle of the sensor over a fixed pre-oriented axis position) can be retrieve. By doing the calibration process, the sensor will be facing a fixed north. Having this initial position marker, it will allow the smartphone app to obtain the sensor's acceleration and rotation data independently from the smartphone position.

3.2   Prototype Data Collection Framework

The Yuan Ze University (YZU) Virtual Reality (VR) Lab Team developed a prototype model of the data collection framework, as shown on Fig. 1. The framework is compose by the following components: Exercise machine, Inertial sensor, Mobile device, Cloud storage, and Browser.

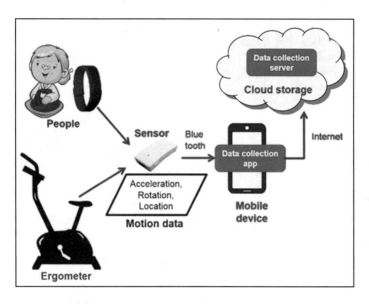

Fig. 1. Prototype data collection framework

The first component of the framework is the inertial sensor that is attached into the fitness equipment, turning them into ergometers. By using the smartphone app develop by the YZU VR Lab Team, the user will be able to track and record the linear acceleration and the Euler angles data, approximately 30 records per second, while the user is carrying out the exercise. Once the exercise session is finished, the app will automatically send a comma-separated value file containing all the collected data to a cloud server via internet connection, storing it in a database for posterior access. Each file will have an individual code name indicating the exercise equipment, the user, the date and the time of the session.

The cloud server does not only stores the data files into its database, but it also establishes a direct feed connection to an online visualization system that enables the user to access and analyze the data online, without the need of installing any additional software into its computer.

3.3 Data Collection Framework

The prototype has been proved to be stable and useful in a previous research, but it still has some major components missing, such as real-time, online data visualization. To tackle this issue, a websocket server was employ with a channel layer; this allows an open connection between the client and the server within a simultaneous, two-way communication, http request and respond.

The components of the data collection framework (Fig. 2) are still the same: Person, Inertial sensor, Mobile device, Cloud storage and the Browser. As mentioned before, the channel layer allowed us to have a more dynamic, internal communication, enabling us to have the adequate platform for the real-time data visualization component. To bring these real-time, online visualizations, HighCharts library was implement, which works perfectly in mobile and desktop browsers, supporting a variety of visualizations schemes. In addition, the new system environment require updating the database manager and its general infrastructure.

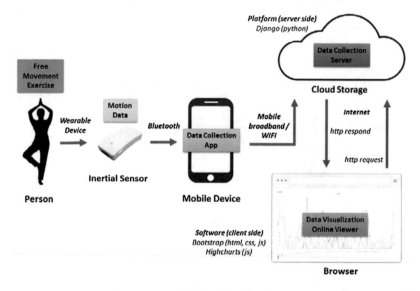

Fig. 2. Data collection framework structure

By having all these open source components and the in-house developed smartphone application it allows to have a sustainable data collection framework over time. It offers an easy, accessible and low-cost solution to doctors, physicians, patients and relatives to keep track of their physical therapy at hospitals, clinics, health community centers, home, etc.

As stated previously on Sect. 1, one part of the aim of this research is to present a more robust and stable platform with functional online services. Therefore, the finished data collection framework just presented addresses this inquire, allowing the users to access easily and to have real time visualizations from any device with internet connection as it will be shown on the next Section.

4   Implementation

4.1   Data Collection App

A crucial part of the data collection framework is the smartphone app that enables the user to record its data. The YZU VR Lab Team developed an Android App created on Unity software and based on the CavyTech SDK. The prototype design of the app features three main interfaces, as shown on Fig. 3, the Calibration screen, the User Selection screen and the Data Collection screen.

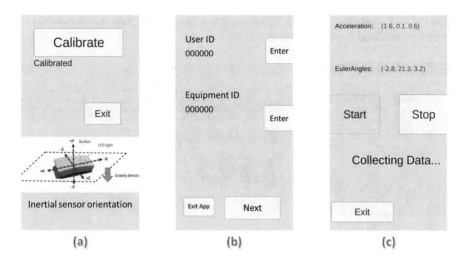

Fig. 3. Smartphone app interfaces; (a) calibration screen, (b) user selection screen, and (c) data collection screen

The Calibration screen, Fig. 3(a), helps the user to set the sensor into an initial position before starting an exercise session. Then there is a User Selection screen, Fig. 3(b), were the user inputs its ID number and the device or type of exercise it is going to perform. Finally, there is the Data Collection screen, Fig. 3(c), were the user is provide with three options (buttons): start to collect data, stop collecting the data and exit the app. While collecting the data, the user can visually observe how the acceleration and Euler angles values variate as the exercise is been perform, in real-time. When the stop button is pressed, the app automatically start to upload the file to the cloud server. Once sent, it allows to continue start another data collection session or to exit the app.

4.2 Data Query

Once the data has been upload into the cloud server, it becomes immediately available for query using any webpage browser. After entering to the website, the user can surf through the main components of the interface. Some of these options are search institution ID, user ID, device ID, date reference, equipment used for exercising, online data visualization, and download option of the data file for further data exploration using any external software.

4.3 Data Visualization

A part of the main goal of this paper is to have real-time, online visualizations from the data collected through the framework to be access easily by any user at anywhere or anytime. It has been achieve by implementing the HighCharts library and Django Channels communication platform, as shown on Fig. 4. Visualizations can be easily explore with the mouse cursor or touch screen to have some insight of the data, active or deactivate fields on the visualization, etc.

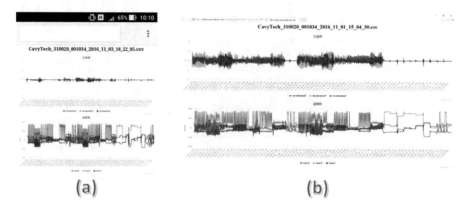

Fig. 4. Real-time, online data visualizations; (a) mobile browser view, and (b) desktop browser view

4.4 Examples

Some of the applications on which the Data Collection Framework has been test include data collection from fitness equipment. The inertial sensor is located on the pedal bar of the stationary bicycle.

In addition, it has been use to collect data from free movement exercises and routines. The user wears the inertial sensor at his right hand wrist while carryout the exercises form the training video.

The implementation of this new version of the data collection framework will be at a local health community center, as this is an early stage of development for a future public release. Currently a research experiment is been design to continue to investigate not only the potential of the framework, but to collect and analyze data from a variety of users.

5 Visualization Challenges

Nowadays, with the aid of technology, internet of things, web frameworks, mobile devices and sensors we are on the breach of Health 2.0 [17]. There is still a lot of research and development to do on big data analysis and visual analytic tools, which will have a key role for health decision makers. Under this scope, a series of challenges are proposed on how doctors, physicians, healthcare providers and individuals, by having access to electronic health records, can improve their patients quality of life through the use of interactive visualizations [18].

Based on this premise, the proposed data collection framework can address some of these challenges:

- *Offering busy clinicians' timely information in the right format:* the data collection framework is a cloud base system that only requires from the user a device with internet connection, without the need to install or give maintenance to any additional software. It also tackles this challenge by having instant access to files, tables, electronic health records, visualizations, etc. In addition, it can also upload data directly in real-time from the inertial sensor to the system for immediate processing and visualization, based on the making sense patterns.
- *Moving toward an ecosystem of visual tools:* the more information collected from an increasing variety of patients will require new ways to present data. The data collection framework addresses this challenge by having an online visualization system, allowing the exploration of the data with a series of tools included on the interface.
- *Facilitating team decision making:* by implementing the data collection framework the users will have access to an online, cloud based, information system that can be link to a patient's electronic health records. This can facilitate the process of decision making of the people involved on the patient's wellbeing, such as nurses, therapists, doctors, etc. having all interaction under a same platform.
- *Characterizing and understanding similarity:* with the use of the data collection framework, a healthcare provider can easily characterize and identify similar performance activity over an extended period. It can have an insight view of the overall progress of the patient, for example if the actual treatment or therapy is having a positive impact on the patient health condition.
- *Visualizing comparative relationships:* the data collection framework offers not only online visualizations for current data, but also offers historic data visualizations that can easily be recall for comparative analysis. Even further, users can easily identify relationships not only between the same patient data, but also between different patient groups, uncovering possible tendencies related to health problems.

6 Conclusion and Future Work

The sustainable data collection framework proved to be stable and reliable in collecting data from the inertial sensor via the in-house developed smartphone app. Sending the data by the internet to the cloud service and having real-time visualizations accessed from virtually any browser on any device, mobile or desktop.

This research work is a step into successfully implement, in the near future, a quality of life infrastructure. It is a sustainable platform solution for making sense of the data and the use of the inertial sensor to support health community centers. It also provides a low-cost solution for monitoring the elderly's quality of life while carrying out their physical therapy performance.

Although it is prove that the data collection framework is robust enough to be use at health community centers, there is always room for improvement, new ideas, latest techniques, diverse approaches and up-to-date technology and software.

On the process of achieving the presented results, some elements have been identify to be consider on future researches:

- Implementation of computer algorithms to analyze the signals from the inertial sensor, such as Decision Trees, k-Nearest Neighbor, DTW and FFT, retrieving statistical features, other exercise performance metrics, performance comparison, etc. Just to mention some outputs that could be obtain from them. It would be optimal and interesting if the algorithms could be wrote on Python and added later to the current data collection framework.
- Even furthermore, machine learning and deep learning algorithms could be implement to analyze the data. It could also be apply the approach of making sense of the data to automatize the process of analysis and visualizations, even to auto perform diagnosis and recommendations toward the user's quality of life.

References

1. Chen, C., et al.: Making sense of mobile health data: an open architecture to improve individual- and population-level health. J. Med. Internet Res. **14**(4), e112 (2012)
2. Shuang, W., Skubic, M., Yingnan, Z.: Activity density map visualization and dissimilarity comparison for eldercare monitoring. IEEE Trans. Inf. Technol. Biomed. **16**(4), 607–614 (2012)
3. Rantz, M.J., et al.: Developing a comprehensive electronic health record to enhance nursing care coordination, use of technology, and research. J. Gerontol. Nurs. **36**(1), 13–17 (2010)
4. Chen, C., Dawadi, P.: CASASviz: web-based visualization of behavior patterns in smart environments. In: International Conference on Pervasive Computing and Communications, pp. 301–303 (2011)
5. Howcroft, J., Kofman, J., Lemaire, E.D.: Review of fall risk assessment in geriatric populations using inertial sensors. J. NeuroEng. Rehabil. **10**(1), 1–12 (2013)
6. Robben, S., et al.: Longitudinal ambient sensor monitoring for functional health assessments: a case study. In: Proceedings of the 2014 ACM International Joint Conference on Pervasive and Ubiquitous Computing: Adjunct Publication, Seattle, Washington, pp. 1209–1216. ACM (2014)
7. de Folter, J., et al.: Designing effective visualizations of habits data to aid clinical decision making. BMC Med. Inform. Decis. Making **14**(1), 1–13 (2014)
8. Korhonen, I., Pärkkä, J., Van Gils, M.: Health monitoring in the home of the future. IEEE Eng. Med. Biol. Mag. **22**(3), 66–73 (2003)
9. Klimov, D., Shahar, Y., Taieb-Maimon, M.: Intelligent visualization and exploration of time-oriented data of multiple patients. Artif. Intell. Med. **49**(1), 11–31 (2010)
10. Martins, S.B., et al.: Evaluation of an architecture for intelligent query and exploration of time-oriented clinical data. Artif. Intell. Med. **43**(1), 17–34 (2008)
11. Wang, T., et al.: Extracting insights from electronic health records: case studies, a visual analytics process model, and design recommendations. J. Med. Syst. **35**(5), 1135–1152 (2011)
12. Moura, D., el-Nasr, M.S., Shaw, C.D.: Visualizing and understanding players' behavior in video games: discovering patterns and supporting aggregation and comparison. In: Proceedings of the 2011 ACM SIGGRAPH Symposium on Video Games, Vancouver, British Columbia, Canada, pp. 11–15. ACM (2011)
13. Le, T., et al.: Design of smart home sensor visualizations for older adults. Technol. Health Care **22**(4), 657–666 (2014)
14. Alexander, G.L., et al.: Generating sensor data summaries to communicate change in elder's health status. Appl. Clin. Inform. **5**(1), 73–84 (2014)

15. Caprani, N., et al.: Exploring healthcare professionals' preferences for visualising sensor data. In: Proceedings of the 2015 British HCI Conference, Lincoln, Lincolnshire, UK, pp. 26–34. ACM (2015)
16. Pham, T., et al.: Interactive visual analysis promotes exploration of long-term ecological data. Ecosphere **4**(9), 1–22 (2013). Article No. 112
17. Hesse, B.W., et al.: Social participation in health 2.0. Computer **43**(11), 45 (2010)
18. Shneiderman, B., Plaisant, C., Hesse, B.W.: Improving healthcare with interactive visualization. Computer **46**(5), 58–66 (2013)

Performance Analysis on Fitness Equipment: Application of an Inertial Sensor Toward Quality of Life

Gustavo Adolfo Miranda Salgado and Tien-Lung Sun[✉]

Department of Industrial Engineering and Management,
Yuan Ze University, Zhongli, Taiwan, ROC
gustavomiranda@live.com, tsun@saturn.yzu.edu.tw

Abstract. Exercise is important for people's quality of life, and nowadays, motion assessment has been a significant part to monitor and evaluate physical factors of people on daily basis, such as activity trackers and fit bands. Some of the issues while performing motion assessment are the use of bulky sensors, data collection restrictions, limited access to data, inability to perform further analysis of the data, etc. To address some of these problems, a data collection framework was develop with the use of a low-cost inertial sensor, and tested on three different fitness equipment. The collected data was analyze by associated to the signal to a series of performance metrics with the aid of visual interpretation. Based on the benchmark findings, the proposed approach has proven to be a viable way to analyze the data with the deployed tools, concise with the user's exercise performance while using the different fitness equipment.

Keywords: Performance analysis · Inertial sensor · Quality of life

1 Introduction

Exercise is important for people to keep and maintain their quality of life, therefore motion assessment has been an important part to monitor and evaluate physical factors while exercising. There is a variety of sensors that can track person in a room [1–5] or even with body sensors [6, 7] in which their majority were bulky and expensive. Nowadays, the use of smartphones is part of people's daily life, which has many kind of way to connect not only to internet, but also to variety of wearable devices such as external micro-electro-mechanical sensors (MEMS). These sensors can easily be integrated into any everyday life routine, effectively recording any physical activity performed. This will allow people to keep track of their own performances and wellness [8].

Many cellphone applications are used to "track" a person's activity all-day routine and have some visualizations. In the end, the user ignores these graphs due to the lack of knowledge to analyze and understand what is going on, leading them to stop using the application [8]. Most people do not use nor download health applications to keep track of their own wellness due to their unwillingness to use smartphones, lack of understanding on how it works, lack of interest, bad attitude toward technology, etc. Some researchers suggest the application ambient sensors [4] while others use Ambient Assisted Living (AAL) applying and analyzing passive infrared (PIR) motion sensors

© Springer International Publishing AG 2017

G. Campana et al. (eds.), *Sustainable Design and Manufacturing 2017*, Smart Innovation, Systems and Technologies 68, DOI 10.1007/978-3-319-57078-5_7

[9], rather than using wearable sensors due to the previous factors mention before, but lacks of the richness from direct data collection of a person, generating false data and missing important events [4].

The current visual data mining techniques and visualizations focus only on obtaining complex features, using data exploration techniques. Actually, the user requires previous knowledge and extensive analytical skills to come up with significant conclusions from a variety of visualizations [10]. Among some good practices for data interpretation through visualization, the inclusion of a time scale variable is proposed, which helps the user to understand what changes are happening over a previously determined timeframe, highlighting the importance of the user knowledge on the field of the analysis [11]. Also there were some researches on visualizations that can effectively represent spatial and temporal components of sensor data, allowing to track high detailed activity for health monitoring. [2], but those are not practical nor intuitive to use on real-life, daily basis, applications.

This paper is organize as follows: In Sect. 2, some background review of the main topics related to the research is introduced. In Sect. 3, the methods used in this study are briefly reviewed. In Sect. 4, the experimental results are shown. The Sect. 5 is the conclusion of this research. Finally, Sect. 6 describes the value of this research toward the quality of life of people.

2 Background

2.1 Sensor Data Collection Framework

There has been a variety of proposed frameworks for health information systems compactible with mobile health (mHealth) devices such as Open mHealth, an open source software derived from a series of modules that allows the data flow and analysis but closed for intranet [8]. Some works have also been made on limited sensor network, where the sensors are connected to a local network for data storage only, Electronic Health Record (EHR), then the information can be viewed via a web-based interface which have been prove useful by the medical staff [9]. Another study on EHR explores the option of incorporating motion sensor data to the proposed framework and its integration across different platforms [12].

Previous researches had created a web-based visualization system as well, such as CASASviz, which integrates monitoring, analysis, and automated recognition, having a variety of modules such as data collection, data adapter, visualization application, etc. but limited to smart environment research only [1], without any further application.

2.2 Features

Data can be analyzed and treated on many different ways. One of them is by extracting features where some of the data noise can be dispersed and hidden information discovered [13] which could be more insightful and easier to be visualized. Before extracting the features, some data preprocessing must be perform, this includes normalization of the database, admission control and projection. Then some features can be calculated such as standard deviation, magnitude, variance, energy, etc. [13–16].

3　Methods

3.1　Inertial Sensor

The MEMS used for this research is the wearable device Cavy Play Band (designed by DEM and manufactured by Foxconn, CavyTech, China). It features a Bosch Sensortec BNO055 Intelligent nine-axis sensor technology; composed by a three-axis accelerometer, a three-axis gyroscope and a three-axis magnetometer, allowing a 360° synchronized human-machine interaction. Its double CPU chip allows the device to calculate the data from its sensors and send the information via Bluetooth (v2.1 + EDR, on Android) to a previously paired device in real-time and with great precision. The size of the sensor is small (40 mm × 19 mm × 7 mm) and lightweight (<26 g). These two characteristics allows the possibility to attach this sensor anywhere, enabling a variety of applications further than its original purpose of gaming entertainment.

From the MEMS, the linear acceleration data (the speed of displacement of the sensor over a fixed pre-oriented axis direction) and the Euler angles (the angle of the sensor over a fixed pre-oriented axis position) can be retrieve. By doing the calibration process, the sensor will be facing a fixed "north"; having this initial position marker, it will allow the smartphone app to obtain the sensor's acceleration and rotation data independently from the smartphone position.

3.2　Data Collection Framework

The overall prototype for data collection framework was develop by the Yuan Ze University (YZU) Virtual Reality (VR) Lab Team. The framework is composed by the following components: an inertial sensor, exercise equipment, a smartphone app and the cloud service.

The first component of the framework is the inertial sensor that is attach into the fitness equipment, turning them into ergometers. By using the smartphone app develop by the YZU VR Lab Team, the user will be able to track and record the linear acceleration and the Euler angles data, approximately 30 records per second, while the user is carrying out the exercise. Once the exercise session is finish, the app will automatically send a comma-separated value (CSV) file containing all the collected data to a cloud server via internet connection, storing it in a database for posterior access. Each file will have an individual code name indicating the exercise equipment, the user, the date and the time of the session.

The cloud server does not only stores the data files into its database, but it also establishes a direct feed connection to an online visualization system that enables the user to access and analyze the data online, without the need of installing any additional software into its computer.

3.3   Fitness Equipment

Three fitness equipment had been use for this research, two exercise machines for the lower limbs (stepper machine and stationary bicycle) and one exercise machine for the upper limbs (shoulder wheel). Each one of this exercise machines had been equip with the inertial sensor, enabling the data collection framework to record the data from an exercise session.

3.4   Exercise Performance Metrics

From the inertial sensor, two major signals can be obtained, acceleration and Euler angles (gyroscope), as stated before. From the acceleration signal, it only registers data when it detects an acceleration force applied to the sensor, hence, the raw data can include some "noise" and irrelevant records. Furthermore, some movements are not register because the acceleration was zero due to two factors: (a) speed turns to be constant, so no acceleration is register, and (b) the movement is too slow to detect any significant record on the 2G scale. Meanwhile, the Euler angles signal records its own the movement independent of the acceleration, resulting on a clearer screening of the user's activity.

Previous researches have worked with acceleration and gyroscope raw data signals to explore and analyze activity recognition, and even elaborate many different algorithms for in-depth studies. One goal is to facilitate the use and understanding of the inertial sensor by associating the signals, without the use of elaborated algorithms, to five commonly used exercise performance metrics:

- Exercise Count (EC): number of repetitions done during the exercise session,
- Exercise Time (ET): total time of the exercise session,
- Exercise Speed (ES): indicates how fast or slow the exercise was performed,
- Exercise Range (ER): the angle arc of the movement while carrying out an exercise,
- Stretch Time (ST): indicates if the exercise is performed continuously or paused.

4   Results

For each one of the exercise machines, three different tests were conducted, lasting an average of one-minute per test, as follows:

- Test 1: the user carry out the exercise at a slow and continuous pace.
- Test 2: the user carry out the exercise at a slow and paused pace.
- Test 3: the user carry out the exercise at a fast and continuous pace.

While obtaining the inertial sensor data via the smartphone app for each one of the tests, two external observers monitored the performance of the user, recording data manually such as time, steps taken, rotation cycles, etc. The following are the results obtained:

Stepper machine results. The performance metrics of the stepper machine associated to the signal of the x-axis from the Euler angles (Fig. 1) are:

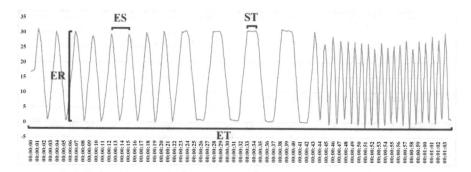

Fig. 1. Euler angles, x-axis signal, of the stepper machine

- *Exercise Count (EC):* number of repetitions done can be identified by counting the upper peaks.
- *Exercise Time (ET):* the lower axis of the visualization states the total time of the exercise session.
- *Exercise Speed (ES):* the distance between the peaks indicates how fast or slow the exercise was perform.
- *Exercise Range (ER):* the angle arc can be identify by observing the height of each peak, in this machine the average angle is of 30°.
- *Stretch Time (ST):* observing the shape of the peak can indicate if the exercise was perform continuously or paused, if its sharp it means that the user is constantly moving, but if the shape is flat it means that the user stop at that specific angle, the flat distance will indicate how long the user stop in that moment.

The benchmark results for the three tests of the stepper machine (Table 1) show that there are some gaps between the sensor and the observed records, but these differences are minimal when compared with the overall results, proving that the proposed performance analysis of the stepper machine is viable.

Table 1. Stepper machine Test 1, 2 and 3 benchmark

	Test 1		Test 2		Test 3	
	Sensor	Observed	Sensor	Observed	Sensor	Observed
EC	21 steps	20 steps	13 steps	13 steps	59 steps	58 steps
ET	00:01:03	00:01:00	00:01:01	00:01:00	00:01:01	00:01:00
ES	3 s/step	3 s/step	4.7 s/step	4.6 s/step	1 s/step	1 s/step
ER	26° to 28°	Deep steps	28° to 30°	Deep steps	15° to 20°	Short steps
ST	<1 s	No stop	2 to 3 s	2.8 s	<1 s	No stop

Stationary bicycle results. The performance metrics of the stationary bicycle associated to the signal of the x-axis from the Euler angles (Fig. 2) are:

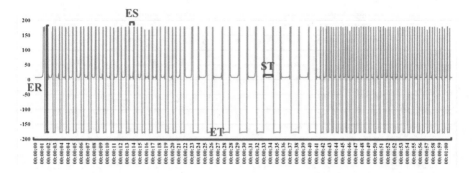

Fig. 2. Euler angles, x-axis signal, of the stationary bicycle

- *Exercise Count (EC):* number of repetitions done can be identify by counting the upper peaks.
- *Exercise Time (ET):* the lower axis of the visualization states the total time of the exercise session.
- *Exercise Speed (ES):* the distance between the peaks indicates how fast or slow the exercise was perform.
- *Exercise Range (ER):* the 360° rotatory movement of the bicycle can be identify by observing the height of each peak, going from 0° to positive 180° and from negative 180° to 0°.
- *Stretch Time (ST):* observing the intermediary peak in the center of the visualization indicates if the user stop moving while carrying out the exercise. If its sharp it means that the user is constantly moving, but if the shape is flat it means that the user stop at that specific angle, the flat distance will indicate how long it stop in that moment.

The benchmark results for the three tests of the stationary bicycle (Table 2) show that there are almost no gap between the sensor and the observed records, proving that the proposed performance analysis of the stationary bicycle is viable.

Table 2. Stationary bicycle Test 1, 2 and 3 benchmark

	Test 1		Test 2		Test 3	
	Sensor	Observed	Sensor	Observed	Sensor	Observed
EC	45 spins	44 spins	17 spins	17 spins	74 spins	68 spins
ET	00:01:03	00:01:00	00:01:02	00:01:00	00:01:04	00:01:00
ES	1.4 s/spin	1 s/spin	3.6 s/spin	3.6 s/spin	0.9 s/spin	1 s/spin
ER	360° rotation	Full rotation	360° rotation	Full rotation	360° rotation	Full rotation
ST	0.7 s/step	No stop	2 s/spin	2.1 s/spin	0.4 s/step	No stop

Shoulder wheel results. The performance metrics of the shoulder wheel associated to the signal of the x-axis from the Euler angles (Fig. 3) are:

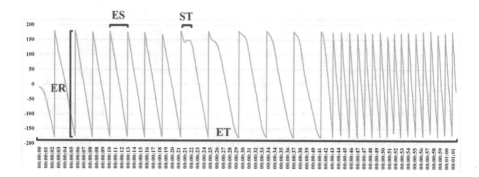

Fig. 3. Euler angles, x-axis signal, of the shoulder wheel

- *Exercise Count (EC):* number of repetitions done can be identify by counting the upper peaks.
- *Exercise Time (ET):* the lower axis of the visualization states the total time of the exercise session.
- *Exercise Speed (ES):* the distance between the peaks indicates how fast or slow the exercise was perform.
- *Exercise Range (ER):* the 360° rotatory movement can be identified by observing the height of each peak, going from 0° to positive 180° and from negative 180° to 0°.
- *Stretch Time (ST):* observing the shape of the intermediary peak indicates if the user stop moving while carrying out the exercise. If its sharp it means that the user was constantly moving, but if the shape is flat it means that it stop at that specific angle, the flat distance will indicate how long the user stop in that moment.

The benchmark results for the three tests of the shoulder wheel (Table 3) show that there are some gaps between the sensor and the observed records, but these differences are minimal when compared with the overall results, also proving that the proposed performance analysis of the shoulder wheel is viable.

Table 3. Shoulder wheel Test 1, 2 and 3 benchmark

	Test 1		Test 2		Test 3	
	Sensor	Observed	Sensor	Observed	Sensor	Observed
EC	25 spins	24 spins	14 spins	14 spins	49 spins	50 spins
ET	00:01:02	00:01:00	00:01:01	00:01:00	00:01:00	00:01:00
ES	2.4 s/spin	3 s/spin	4.3 s/spin	4 s/spin	1.2 s/spin	1 s/spin
ER	360° rotation	Full rotation	360° rotation	Full rotation	360° rotation	Full rotation
ST	None	No stop	1 s/spin	1 s/spin	None	No stop

5   Conclusion

Having in consideration the small human error while observing and recording data manually, and based on the benchmark findings, this research can confidently state that the proposed approach of performance analysis is viable, been concise with the user's exercise performance while using the different fitness equipment. Additionally, this research can also conclude the following:

- The low-cost inertial sensor used on the fitness equipment can effectively turn them into ergometers, retrieving the acceleration and gyroscope data of a person's performance during an exercise session. By using this inertial sensor, it opens new possibilities for a variety of applications such as the Internet of Things and Industry 4.0.
- It is also important to mention the high precision and sensibility of the inertial sensor, being able to track not only the acceleration data with a sense of direction, but the angle/while carrying out the exercises also, giving almost an x-ray vision of the user's performance, facilitating the process of associating of the different signals to the performance metrics.

6   Future Work: Toward Quality of Life

This research work is the first step that can led into a future implementation of the proposed method for performance analysis, data administration through the data collection framework, and the use of the inertial sensor for monitoring the progress of people's daily exercises or even physical therapies.

In addition, it is important to mention that visualizations are an important tool that enables people to navigate and understand data intuitively through an image without the need of exploring data tables, but that its usefulness depends on a series of factors such as the appropriate user, knowledge of the area, oversaturated visuals, etc.

Based on this research findings, it would be possible to make data accessible to assess the quality of life to a variety of users depending of their specific role [3], for example:

- *Doctors*: need to access large quantities of data from all his patients, analyze and compare their evolution over an extended period of time, do comparison analysis between patient groups, etc. so it can be able to take treatment decisions for the patients, it can include a variety of dense and complex visualizations.
- *Physicians:* they need to keep track of the progress of each patient's physical therapy over a timeframe. It can help to identify if the patient is performing its therapy correctly and inside the safe parameters according to its age, as some people can be more susceptible to aggravate their condition or even to create new injury by not carrying out properly the exercises. The visualizations must give insights of individual patient's performance over the therapy period, been able to compare progress between patients, such as histograms.
- *Patients*: they need to access their own performance progress in an easy, clear and understandable presentation. A colorful visualization aided with images can help to comprehend their improvement while keeping them motivated. For example the garden mapping visualizations, where according to the patient's progress the garden will have taller and bigger flowers (high performance) or have shorter and smaller flowers (low performance).

– *Family members:* they need to understand easily their relative's performance and progress while having some additional insight of their general condition, shown through a dashboard, which can include the performance metrics and overall histogram of an exercise session.

References

1. Chen, C., Dawadi, P.: CASASviz: web-based visualization of behavior patterns in smart environments. In: International Conference on Pervasive Computing and Communications, pp. 301–303 (2011)
2. Le, T., et al.: Design of smart home sensor visualizations for older adults. Technol. Health Care **22**(4), 657–666 (2014)
3. Mulvenna, M., et al.: Visualization of data for ambient assisted living services. IEEE Commun. Mag. **49**(1), 110–117 (2011)
4. Robben, S., et al.: Longitudinal ambient sensor monitoring for functional health assessments: a case study. InL Proceedings of the 2014 ACM International Joint Conference on Pervasive and Ubiquitous Computing: Adjunct Publication, Seattle, Washington, pp. 1209–1216. ACM (2014)
5. Caprani, N., et al.: Exploring healthcare professionals' preferences for visualising sensor data. In: Proceedings of the 2015 British HCI Conference, Lincoln, Lincolnshire, UK, pp. 26–34. ACM (2015)
6. Howcroft, J., Kofman, J., Lemaire, E.D.: Review of fall risk assessment in geriatric populations using inertial sensors. J. NeuroEng. Rehabil. **10**(1), 1–12 (2013)
7. Frankenberger, P., Evaluation of Visualization Techniques on Wearable Sensor Accelerometry Data: Bewertung Von Visualisierungstechniken Für Daten Tragbarer Beschleunigungssensoren (2012)
8. Chen, C., et al.: Making sense of mobile health data: an open architecture to improve individual- and population-level health. J. Med. Internet Res. **14**(4), e112 (2012)
9. Shuang, W., Skubic, M., Yingnan, Z.: Activity density map visualization and dissimilarity comparison for eldercare monitoring. IEEE Trans. Inf. Technol. Biomed. **16**(4), 607–614 (2012)
10. Klimov, D., Shahar, Y., Taieb-Maimon, M.: Intelligent interactive visual exploration of temporal associations among multiple time-oriented patient records. Methods Inf. Med. **48**(3), 254–262 (2009)
11. Aigner, W., et al.: Visual methods for analyzing time-oriented data. IEEE Trans. Visual. Comput. Graph. **14**(1), 47–60 (2008)
12. Rantz, M.J., et al.: Developing a comprehensive electronic health record to enhance nursing care coordination, use of technology, and research. J. Gerontol. Nurs. **36**(1), 13–17 (2010)
13. Yang, J., et al.: Physical activity recognition with mobile phones: challenges, methods, and applications. In: Shao, L., Shan, C., Luo, J., Etoh, M. (eds.) Multimedia Interaction and Intelligent User Interfaces: Principles, Methods and Applications, pp. 185–213. Springer, London (2010)
14. Altun, K., Barshan, B., Tunçel, O.: Comparative study on classifying human activities with miniature inertial and magnetic sensors. Pattern Recogn. **43**(10), 3605–3620 (2010)
15. bin Abdullah, M.F.A., et al.: Classification algorithms in human activity recognition using smartphones. Int. J. Comput. Inf. Eng. **6**, 77–84 (2012)
16. Su, X., Tong, H., Ji, P.: Activity recognition with smartphone sensors. Tsinghua Sci. Technol. **19**(3), 235–249 (2014)

Design Principles for Do-It-Yourself Production

Jérémy Bonvoisin[1(✉)], Jahnavi Krishna Galla[1], and Sharon Prendeville[2]

[1] Chair of Industrial Information Technology, Institute for Machine-tools and Factory Management, Technische Universität Berlin, Berlin, Germany
jeremy.bonvoisin@tu-berlin.de
[2] Institute of Design Innovation, Loughborough University London, London, UK

Abstract. The increasing access of people to fabrication capabilities has stimulated the emergence of personal fabrication settings and inspired post-industrial production scenarios. One strategy to support personal production is to increase technology literacy and access for citizens to means of production. Yet, so far, the deliberate design of products so they can be realized by individuals, an activity termed here as "design for do-it-yourself (DIY) production", has been underexplored in academia. The present article aims to formalize the know-how gained by practitioners who designed products for production in do-it-yourself settings. It provides an original definition of DIY and the formulation of 14 design principles for DIY production to support practice.

Keywords: Distributed production · Commons-based peer production · Design principles · DIY

1 Introduction

Ever-increasing accessibility of information technologies, such as cheap small-scale production tools like 3D printers, combined with today's capability to share information rapidly over the internet has stimulated the rise of the so-called "maker movement". Whether motivated to make things on a personal level or to participate in an effort to "reclaim production", the central character of this movement—the "maker"—builds his/her own products and shares publicly on digital platforms the corresponding designs, assembly manuals and best practices. While this trend is originally a matter of spare time occupation and tinkering, it also goes along with more serious and professional practices. Within the realm of "open source hardware", informal communities of interested and skilled individuals make use of the digital space to develop complex products whose design files are made publicly available. Many examples of open source hardware products are available and cover product categories such as agriculture machinery, machine-tools, means of transportation, renewable energy supply technologies or even medical equipment [1].

The maker movement and open source hardware—two distinct though interwoven phenomena—share a common characteristic termed here as "do-it-yourself production": a general willingness to diverge from standardized mass production to support distributed, informal, local and individual-scale production instead. In recent years, this

© Springer International Publishing AG 2017
G. Campana et al. (eds.), *Sustainable Design and Manufacturing 2017*, Smart Innovation, Systems and Technologies 68, DOI 10.1007/978-3-319-57078-5_8

alternative production setting has been suggested to bear great potential advantages in terms of social (democratization of production) as well as environmental sustainability (localized production, lower production volumes [2, 3]). Consequently, do-it-yourself production has been the object of growing interest, inspiring new ways of supporting the capacity of non-experts in manufacturing high quality and complex products. One of them is to promote citizen's technological awareness, production skills and accessibility to reliable means of production. In public makerspaces for example, citizens can learn how to use machine-tools such as laser-cutters or CNC lathes and doing so fabricate products by themselves. An unexplored, yet complementary strategy is to design products so their manufacturing processes fit with the constraints of low scale, low equipped and low skilled fabrication settings.

This article addresses this gap by formalizing know-how gained by practitioners who have already designed products suited to do-it-yourself production. Firstly, the article introduces the concept of do-it-yourself production and an original definition is proposed. The subsequent section introduces the methodological approach used to formalise practitioners' know-how, namely through the identification of design principles. A list of 14 design principles for do-it-yourself production are then introduced and discussed in Sect. 5.

2   Defining Do-It-Yourself Production

Commonly, the term do-it-yourself, in short DIY, refers to the "method of building, modifying, or repairing things without the direct aid of experts or professionals."[1] This approach has been the dominant production setting of pre-industrial subsistence economies and has been replaced by industrial manufacturing in industrialized economies. It was first termed "DIY" one century ago in reaction to the industrialisation and standardization of production, particularly in the context of furniture [4]. DIY may have been pushed forward by different factors—such as ideological resistance to the dominant organization of work or simply aspirations of self-development through creative activity. Nowadays this production approach shows flourishing practices, as demonstrated by the amount of DIY-manuals published over time in popular and even in scientific literature[2]. More recently, in the context of digitalisation, the increased access of individual citizens to fabrication capabilities has inspired scholars with post-industrial visions, alternatively termed "commons-based peer production" [5], "personal fabrication" [6], "direct digital manufacturing" [7], "bottom-up economy" [8], "distributed economy" [9], under the motto "design global, manufacture local" [10]. These future scenarios position individual citizens with full control over the design and fabrication of their products. How and whether this potential can be realised and what are the implications of such change are outside the scope of this article. However, to begin to distil and understand the practices that inform this trend we need to better characterize the nature of DIY, which has yet to be systematically explored in the literature. In particular, much

---

[1] Wikipedia article for the entry "Do it yourself". Accessed 03.11.2016.
[2] See for example [4].

literature tends to focus on the socio-economic and psycho-social implications of this alternative organization of production. Yet, these compelling visions tend to overlook the practical aspects of producing at an individual scale. This means that very little work has been done to accurately describe what DIY production is. In this article, we present an original definition of the concept of DIY, hence offering the necessary basis for the analysis introduced in Sect. 3.

2.1 DIY as a Production Method

As indicated by "yourself", DIY production stands for an organisation of production driven by individuals rather than organisations. In contrast with the industrial organisation of production, DIY production settings imply a voluntary limitation in access to means of production, including manpower, tools, skills and investment capacity. These limitations imply, in turn, limitations in terms of achievable product size, complexity

Table 1. Three archetypal production settings

	Home-based production setting	Makerspace-based production setting	Mass-production setting
Financing scheme	Personal, eventually com-munity-based financing	Shared funding, eventually institutional financing	Institutional and market financing
Accessible skills	Basic knowledge on how to use tools, read and understand build manual instructions. Eventually some specific knowledge about a particular technology	Training provided by technicians with extended knowledge about the tools available in the makerspace, including safety measures	Specialized and trained workers
Accessible tools	Conventional hand tools, workbench, light duty vise, drill, circular saw, jigsaw, bench grinder, jack stands, multimeter, soldering iron, eventually 3D printer	3D printer, welding torch, 3 axis CNC machine, laser cutter, plasma cutter, drill press, hand and power tools, material for PCB photoengraving, all tools of the DIY production setting	High end machinery, automated machine tools, integrated production lines, all of the tools of the makerspace based production setting
Manpower	One person, eventually family members or friends wanting to help	One person or small project team, supported by experts	Systematic division of work
Space	Typical household garage	Large workshop for collective use	Globalised supply chains

and accuracy[3]. How limited these DIY production settings are, cannot be defined universally. In the absence of a generally available definition, archetypes can be used as best available proxies.

Table 1 provides a characterization of two archetypal DIY production settings (termed as home-based and makerspace-based) compared with the current dominant production setting, i.e. industrial production. These archetypes are not strictly exclusive, but should be rather seen as coexisting singular points on a continuum between fully un-tooled production and fully industrialized production.

2.2   DIY as a Product Property

The term DIY is used to describe production methods but also to describe products (e.g. a DIY-bicycle). What does it mean for a product to be classified as DIY? Firstly, the concept of DIY is not binary, but can be described as a gradation. It needs to be considered in its vertical (along the bill of materials) and horizontal (along the supply chain) dimensions, as illustrated by Fig. 1.

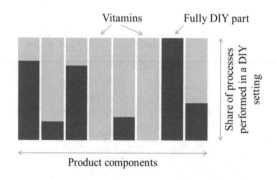

Fig. 1. Illustration of the vertical and horizontal dimensions of DIY products

Vertically, in most of the cases, not all parts of a product are suitable for a DIY production setting. For example, it may be more relevant to produce nuts and bolts in a mass production setting than in a DIY one. There is therefore a distinction between the parts that are suitable to produce within a given DIY setting and those which are not. In analogy with the vocabulary used in biological science, we term those input parts here as 'vitamins'—vitamins being compounds that are required by a given organism but that cannot be synthesized by this organism. In the context of this paper, vitamins are parts that are not suited for production in a given DIY production setting but have to be sourced in a mass production one.

Horizontally, in most cases, only the final processing steps of each part may be performed in a DIY setting. Tyl et al. [11] illustrate this principle with the label "made in France", which considers that a product is made in France as long as "the place where the

---

[3] Here, we intentionally avoid the term "quality" as it related to subjective expectations.

product acquired its principal characteristics is located in France" and "more than 50% of the product costs are acquired in France" [12].

Therefore, here we suggest that how much a product can be described as DIY is defined by the share of all processes that are performed in a DIY setting relatively to all the processes that have to be performed to build a product. Whether a process can be performed in a DIY setting is in turn defined by the capacity (skills, tools, manpower, space) of the considered DIY production setting for handling the requirements of this process.

3   Methodological Approach

The aim of this research is to identify and document design principles used by practitioners to design products so they can be reproduced in a DIY production setting. Section 3.1 introduces the concept of design principles while the subsequent Sect. 3.2 details the format chosen for the documentation of the design principles in this article. A last subsection details the method adopted for the identification of design guidelines from practice.

3.1   About Design Principles

A design principle is a form of design heuristic, i.e. a 'directive, based on intuition, tacit knowledge, or experiential understanding, which provides design process direction to increase the chance of reaching a satisfactory but not necessarily optimal solution' [13]. Design principles set procedures to orient product design towards a strategic objective [14] and are generally expressed as short instructive statements [15], that is, at least a phrase containing at least a verb and an object [16]. Bundled in compilations (e.g. frequently unordered lists), design principles are typical tools used in Design for X (DfX) approaches. Lists of principles provide pools of possible solutions approaches corresponding to a given design objective. They can be either used as learning tools to widen the perspectives of a designer or as an inspiration tools in the context of concrete problem solving.

Design principles may vary in specificity, that is, in the degree of precision of the proposed solution concept and the breadth of their applicability. Since each design principle can be stated specifically or generically depending on the context, lists of design principles are flexible tools. They can be used in generic form so to be applicable to any context (as in the design tool "information/inspiration" [17]) or be specifically adapted for the needs of a company or a product branch (as presented in [18]). They are also adaptive, since they can be enriched along the growing experience of users. Examples of design principles lists have been published by Bischoff and Blessing [19] (design of flexible products), Go et al. [20] (design for multiple lifecycles), as well as Telenko et al. [21] (design for environment).

Although design principles are typically short statements that can be collected and used as such, the documentation of contextual information specifying the meaning and the applicability of design principles may be required. Some documentation formats have been introduced in the literature recommending capturing information such as: visual illustrations, implementation examples, additional textual description, underlying generic intent, goal conflicts, applicability in the product development process (e.g. [15, 16]).

In this article, we adopt a practical approach where each principle is described by:

- Statement: short verbal phrase defining the design principle;
- Additional information: longer textual description of the principle;
- Example: illustration of the principles with the help of a concrete implementation.

3.2 Data Acquisition Method

Design principles have been gathered in a two-stage process, the first step involving desk research and the second expert interviews.

In the first stage, the design of each DIY product has been analysed to identify its specific design features that make it suitable for a DIY production setting. To be able to avail of publicly available documentation, the products under study have been selected out of a pool of open source hardware products[4] previously gathered in [1][5] according to the following criteria:

- The product is a 'discrete manufactured product', i.e. food and processing industry products are excluded.
- The product contains at least tangible and non-electronic hardware, that is, mechanical or any other type of non-electronic physical element (e.g. textile). It may eventually include electronic hardware and consequently software. Purely electronic hardware or software products such as Arduino or Linux are excluded.
- The product has at least a certain minimum complexity. Products consisting of only one part or material do not meet this criterion. Products such as business card holders or cell phone cases made of one unique 3D-printed part are out of scope.
- The product is developed for functional rather than aesthetic purposes. Jewellery, decorative items, gadgets such as personalized cell phone covers, or 3D printed rings do not fulfil this criterion and therefore were not included.
- The product is at least partly defined; undeveloped product concepts are not considered.
- The product is labelled as open source.

Products have been selected out of this pool according to following criteria:

- It is stated in the product's supporting documentation that these products are intended for a DIY production setting.
- The level of detail in each product's online documentation was sufficient to allow the analysis of the product design.

---

[4] i.e. hardware products "whose design is made publicly available so that anyone can study, modify, distribute, make, and sell the design or hardware based on that design" (Open Source Hardware Statement of Principles 1.0., http://www.oshwa.org/definition/, Accessed 22.03.2017).

[5] An updated list is maintained online at http://opensourcedesign.cc.

The selected open source hardware and DIY products have been analysed on the basis of data published online by their originators such as product descriptions including text and illustrations, CAD files, assembly instructions or bills of materials.

This analysis led to the generation of a first list of principles whose exhaustiveness and relevance has been verified in a second stage involving expert interviews. Interviews were led following a semi-structured scheme in a first part and structured scheme in the second part. First, experts were asked to explain freely on how they designed their product so it can be produced in a DIY setting (semi-structured). This allowed interviewees to potentially address unforeseen principles and hence to extend the list principles identified through desk research (exhaustiveness check). Second, experts where invited to give feedback on each of the already identified principles (structured). This allowed the interviewers to check the validity these principles and to improve their formulation (relevance check).

4   Identified Design Principles

Eight products have been found to satisfy the selection criteria defined for the first methodological step (desk research), see Fig. 2. Originators of two of these products could be interviewed in the second step (interviews).

Fig. 2. Analyzed products. From the left to the right and the top to the bottom: Open Source Ecology's LifeTrac (tractor, CC BY-SA 4.0), Opendesk's breakout table (furniture, CC BY, Joni Steiner and Josh Worley), Open PCR (PCR thermocycler, GNU GPLv3), XYZ oneseater (bicycle, CC NC-BY-SA 3.0), Multimachine (multifunctional machine-tool, unknown copyright information), RepRap Mendel (3D printer, GNU Free Documentation License 1.2, courtesy of reprap.org), rBot (CNC desktop machine-tool, CC BY-NC-SA), Precious Plastic's recycling machine (recycled plastic extruder, MIT License, copyright 2016 Dave Hakkens).

14 design principles could be identified:

1. *Use modular design.* Modularity, i.e. the distribution of sub-functions of the product among distinct functional carriers with clearly defined interfaces, can help achieve a clear separation of vitamins and DIY parts. Example: Decouple energy

source and usage, so that any type of source can be used (e.g. electric motor, hydraulic pressure) for generating a given movement.

2. *Opt for processes that can be performed with standard tools.* The more widely available the required tools are, the more probable it is that it will be accessible to anyone. Example: bolting instead of welding.

3. *Use commonplace materials.* The more widely available the required materials are, the more probable it is that they can be purchased by anyone. Example: wood plate in sizes available in hardware stores.

4. *Use discarded materials and materials that are to-hand.* Designing a product made of discarded materials that can be commonly used in households may be a solution for facilitating purchase. It should be however considered that discarded materials may have lower accuracy and require pre-processing steps. Example: contactless bicycle dynamo made of hard disk drive magnets.

5. *Use general purpose and standard components.* Standard components may be more easily available than exotic ones. Using them may therefore ease the purchasing process. Example: widely available standards nuts and bolts that can be found in hardware stores.

6. *Facilitate for tailoring.* A motivation for making a product in a DIY setting is to get a tailored product. Flexibility can be built in the product design to ease the tailoring process. Example: parameterized bicycle frame design.

7. *Facilitate for flexible construction.* Allow scope for last-minute tweaks, in case the design cannot exactly be realized as described in the product requirements. Example: high geometric and material tolerances, beams with a grid of equally separated attachment holes.

8. *Choose reversible over permanent joining features.* Using removable and adjustable joining features allows accommodating mistakes or correcting misalignments in the assembly. Examples: interlocking elements, nuts and bolts vs. welding.

9. *Reduce vitamin variation.* When several vitamins are required, using the same vitamin each time could help reduce the purchase and storage effort. Examples: one type of bolt for the assembly of all parts of the product, one type of step motors for translating the plateau of a machine-tool in all three directions or space.

10. *Reduce raw material variation.* Strive to design all DIY-components so they can be made out of the same material and using same processes. This would allow for reductions in the purchasing effort and the requirements for processing tools. Example: 3D design made of assembled laser-cut plywood.

11. *Use symmetries.* Using symmetries is another way of reducing part variation. Example: the frame of a 3D printer that is symmetrical along the slider rail of the printing head (left-right axis) can be assembled using same joining features for both left and right sides.

12. *Offer scalability through "stackability".* Building a product of a large size/power may be more difficult than building numbers of small products that can be combined to reach the desired size/power. It reduces the risk and need of handling larger and heavy products. Example: A high luminous emittance can be reached by combining several LED.

13. *Ensure ease of handling and transportation.* Bulky or fragile products may be difficult to handle and require specific handling or holding mechanisms. In those cases, use built-in handling mechanisms, alternatively provide DIY handling tools. Example: wheels and contact points built in the frame of the product, DIY building jig.
14. *Offer different depths of DIY.* Not everybody has the same skills, tools, and interest in building things. Giving the maker the possibility to choose which of the parts they consider a vitamin increases accessibility of DIY to more potential makers. Example: a product with parts that can be either 3D-printed or bought as a kit.

5   Discussion and Outlook

This article offers an initial formalization of the knowledge gained by practitioners in the development of DIY products. It first provided an original definition of the concept of DIY, which is both understood as a production environment and as a product property. It then introduced an original empirical research method whose implementation led to the identification of 14 design principles for DIY production.

The gathered design principles are useful insofar as they offer a means to share good practices learned from the open source hardware and maker communities. Broader adoption of the formalized knowledge in practice could include the development of a practitioner's guide with an extended documentation of the identified design principles including, amongst others, product examples, illustrations as well as a graphical representation of interdependencies between design principles.

From an academic viewpoint, this work can act as a basis for future research, as to-date there has been little work. Further work is however required before any claim of exhaustiveness can be made. A first approach would be to reproduce the performed study with a broader range of products as well as with increased access to product-related information and surrounding experts. A second approach would be to compare the design principles identified here with design for manufacturing and design for assembly principles. In addition to this, setting up a catalogue of technical solutions classified according to the functional requirements they solve (e.g. DIY methods for assembling two components: cable tie, interlocking parts, nuts and bolts...) could be an interesting complementary approach for practice. Finally, the process of identifying design heuristics could be improved by the development of a dedicated systematic method—no method being currently available to the knowledge of the authors.

References

1. Bonvoisin, J., Mies, R., Jochem, R., Stark, R.: Theorie Und Praxis in der Open-Source-Produktentwicklung. In: Wülfsberg, J., Redlich, T., Moritz, M., (eds.) 1. Interdisziplinäre Konferenz zur Zukunft der Wertschöpfung – Konferenzband (2016). ISBN 978-3-86818-091-6
2. Kohtala, C., Hyysalo, S.: Anticipated environmental sustainability of personal fabrication. J. Clean. Prod. **99**, 333–344 (2015)

3. Bonvoisin, J.: Implications of open source design for sustainability. In: Setchi, R., Howlett, R.J., Liu, Y., Theobald, P. (eds.) Sustainable Design and Manufacturing 2016, pp. 49–59. Springer, Cham (2016)
4. Fineder, M., Geisler, T., Hackenschmidt, S.: Nomadic Furniture 3.0 - Neues befreites Wohnen? New Liberated Living? Niggli Verlag, Zürich (2016)
5. Benkler, Y., Nissenbaum, H.: Commons-based peer production and virtue. J. Polit. Philos. **14**(4), 394–419 (2006)
6. Kohtala, C.: Addressing sustainability in research on distributed production: an integrated literature review. J. Clean. Prod. **106**, 654–668 (2014)
7. Chen, D., Heyer, S., Ibbotson, S., Salonitis, K., Steingrímsson, J.G., Thiede, S.: Direct digital manufacturing: definition, evolution, and sustainability implications. J. Clean. Prod. **107**, 615–625 (2015)
8. Redlich, T.: Open Production Gestaltungsmodell für die Wertschöpfung in der Bottom-up-Ökonomie. Universität der Bundeswehr Hamburg (2010)
9. Johansson, A., Kisch, P., Mirata, M.: Distributed economies – a new engine for innovation. J. Clean. Prod. **13**(10–11), 971–979 (2005)
10. Kostakis, V., Niaros, V., Dafermos, G., Bauwens, M.: Design global, manufacture local: exploring the contours of an emerging productive model. Futures **73**, 126–135 (2015)
11. Tyl, B., Lizarralde, I., Allais, R.: Local value creation and eco-design: a new paradigm. Procedia CIRP **30**, 155–160 (2015)
12. Association Pro France, Référentiel du label 'Origine France Garantie' V12 (2016)
13. Fu, K.K., Yang, M.C., Wood, K.L.: Design principles: literature review, analysis, and future directions. J. Mech. Des. **138**(10), 101103 (2016)
14. Vezzoli, C., Sciama, D.: Life cycle design: from general methods to product type specific guidelines and checklists: a method adopted to develop a set of guidelines/checklist handbook for the eco-efficient design of NECTA vending machines. J. Clean. Prod. **14**(15–16), 1319–1325 (2006)
15. Sarnes, J., Kloberdanz, H.: Heuristics guidelines in ecodesign. In: DS 80-1 Proceedings of the 20th International Conference on Engineering Design (ICED 2015), Design for Life, vol. 1, Milan, Italy, 27-30 July 2015 (2015)
16. Bonvoisin, J., Mathieux, F., Domingo, L., Brissaud, D.: Design for energy efficiency: proposition of a guidelines-based tool. In: Marjanovic, D., Storga, M., Pavkovic, N., Bojcetic, N., (eds.) Proceedings of DESIGN 2010, the 11th International Design Conference, pp. 629–638. The Design Society, Castle Cary (2010)
17. Lofthouse, V.: Ecodesign tools for designers: defining the requirements. J. Clean. Prod. **14**(15–16), 1386–1395 (2006)
18. Dahlström, H.: Company-specific guidelines. J. Sustainable Prod. Des. **8**, 18–24 (1999)
19. Bischof, A., Blessing, L.: Guidelines for the development of flexible products. In: DS 48: Proceedings DESIGN 2008, the 10th International Design Conference, Dubrovnik, Croatia (2008)
20. Go, T.F., Wahab, D.A., Hishamuddin, H.: Multiple generation life-cycles for product sustainability: the way forward. J. Clean. Prod. **95**, 16–29 (2015)
21. Telenko, C., O'Rourke, J.M., Seepersad, C.C., Webber, M.E.: A compilation of design for environment guidelines. J. Mech. Des. **138**(3), 031102 (2016)

Establishment of Engineering Metrics for Upgradable Design of Brake Caliper

Nurhasyimah Abd Aziz, Dzuraidah Abd Wahab$^{(\boxtimes)}$, and Rizauddin Ramli

Department of Mechanical and Materials Engineering,
Faculty of Engineering and Built Environment, Universiti Kebangsaan Malaysia,
43600 Bangi, Selangor Darul Ehsan, Malaysia
hasyimahaziz@siswa.ukm.edu.my, {dzuraidah,rizauddin}@ukm.edu.my

Abstract. Design for upgradability is one of the strategies in a remanufacturing process that can help to improve the features of a product in terms of performance and functions. This paper presents a study on the performance evaluation of a product at the engineering metric level for upgrade purpose. A brake caliper was used as a case example in which brake test was conducted to measure the brake pedal force. The brake pedal force values are used to measure the brake torque which represents the performance of the brake caliper. The upgradability of the brake caliper is then evaluated in view of improving the current design. This paper also proposed and discussed future work on the upgradability of the automotive component.

Keywords: Design for upgradability · Remanufacturing · Engineering metric · Brake caliper · Brake torque

1 Introduction

The ever increasing amount of abandoned vehicles today has affected environmental sustainability in many ways. It also reflects an improper management for End-of-Life Vehicle (ELV). It is therefore important to develop a plan at the early stage of product design and development that consider recovery options of ELV which includes reuse, remanufacture and recycle [1]. Nowadays, automotive components such as brake caliper, alternator and engine blocks have been remanufactured instead of disposal in landfill areas. Remanufacturing is referred to as the recovery process to return used products to like-new condition with a warranty to match [2, 3]. It involves several series of processes which include sorting, inspection, disassembly, cleaning, reprocessing and reassembly.

One of the important characteristics of a remanufacturable product is having a long lifetime. The introduction of upgrade strategies at the design stage can enhance the remanufacturing features of the product through the developed upgrade plan that may improve the lifetime of the product. Design for Upgradability (DfU) is a new paradigm in Design for X (DfX) that can facilitate the product design process to consider the upgrade strategies along the process [4]. The application of DfU into Multiple Lifecycle Products (MLPs) is an added advantage to the product where the upgrade plan can be

© Springer International Publishing AG 2017
G. Campana et al. (eds.), *Sustainable Design and Manufacturing 2017*, Smart Innovation, Systems and Technologies 68, DOI 10.1007/978-3-319-57078-5_9

initiated along the lifetimes of a product. Designing products for multiple lifecycles will enable the product to be preserved in a next lifecycle with proper strategies when it reaches its end-of-life at each lifecycle [5].

This paper focuses on the implementation of upgradability to improve the performances and functions of a product. A brake caliper has been chosen as a case study for this paper whereby the upgradability of the brake caliper is evaluated at three levels which are engineering metrics, component and architecture. For the purpose of this paper, the main objective is to analyse the measured parameters at the Engineering Metric Level in order to observe the relationship of each parameter with the upgrade of a product. The expected outcome of this paper is a new idea or knowledge on how the performance of a brake caliper can be improved by analyzing the functionality of a brake caliper.

2 Research Background

This study is intended to propose a new systematic approach by considering the upgradability features of a product at the design stages [6, 7]. The features of interest are those that can enhance performance and functions so that the product can be used in the next lifecycle after the remanufacturing process has been carried out. It is crucial to develop a proper upgrade plan at the initial stage of the product design process since it involves future uncertainty [8]. The next subsection will discuss in detail the development of an upgrade plan for a brake caliper.

2.1 Upgrade Plan of a Brake Caliper

Upgrade plan is the upgrade period that needs to be considered at the early stages of a design process along the generations of product [9]. The upgrade period or time is dependent on several factors which include product upgrade cycle, disposal cycle and administrative strategy [10]. Initially, the product needs to be evaluated to ensure that it is suitable for remanufacturing and thus, upgrading. A previous study by Aziz et al. [7] had discussed the remanufacturability characteristics of a brake caliper. To date, brake calipers are widely remanufactured since they are non-complex with small number of parts, easy to disassemble and reassemble and also the caliper core are generally durable and can be used for a long period of life cycle.

Previous studies [4, 11, 12] have identified the following requirements for the development of an up-grade plan, which are:

- Firstly, develop the upgrade strategies throughout the product's lifetime
- Secondly, determine the lifetime of the brake caliper
- Lastly, determine the consisting module in a brake caliper

The above requirements are the basis for initiating the upgrade plan of a brake caliper. According to Umemori et al. [8] and Inoue et al. [10], the upgrade plan is developed by specifying the required ranges of functional parameters and design parameters. Matsuda et al. [9] in his study has stated five steps in setting up the upgrade plan which is (1) choose the suitable product, (2) develop component database, (3) match the customer requirements with product structure, (4) determine the design solution and (5) evaluate the design upgradability of a product. For the case of a brake caliper, the function parameter that will be measured in this study is the braking torque at the rotor. Meanwhile the related design parameters in measuring this braking torque will be area of piston, braking line pressure, effective radius of rotor, coefficient of friction between pad and rotor disc and finally number of piston. Table 1 shows the upgrade plan initiated for a brake caliper which shows the specified range of braking torque and its related specified range of design parameters.

Table 1. Required range of brake torque for each generation

Lifecycle	1st lifecycle	2nd lifecycle	3rd lifecycle
Function parameter	Required range	Required range	Required range
Brake torque, T	1074–1170 Nm	1100–1230 Nm	1200 Nm–1350 Nm
Maximum brake pressure	4.3 MPa		

The upgrade plans describe in Tables 1 and 2 are focused on performance upgrading a brake caliper. The details of calculation and analysis on the specified range stated in Tables 1 and 2 are discussed further in a next section. A brake test was carried out to measure all the parameters involved in calculating the braking torque of a vehicle. The planning horizon of the upgrade plans need to be first specified at the early stages of the design process. In this study, the plan is set up to 10 years with three lifecycles or generations that have their own performance limit based on each design solutions as shown in Table 2. Figure 1 shows the performance limit of each generation, from the first year of a product until year 10.

Table 2. Calculated range of design parameters as a design solution

Initial value		Candidates of design solution			
		No. 1	No. 2	No. 3	No. 4
A_{piston}	0.0023 m^2	LC2:0.0022–0.0025 LC3:0.0024–0.0027	0.0023	0.0023	0.0023
R_{eff}	0.116 m	0.116	LC2:0.111–0.124 LC3:0.121–0.136	0.116	0.116
μ_{pad}	0.30	0.30	0.30	LC2:0.28–0.34 LC3:0.33–0.38	0.30
N_{piston}	1	1	1	1	1 2

LC denotes Lifecycle

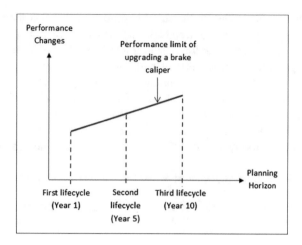

Fig. 1. Performance limit for each generation throughout upgrade plan period

3   Experimental Setup

The vehicle used in this experiment is a Malaysian car model Proton Persona 1.6L which is equipped with a Vehicle Global Positioning System (VGPS). The brake caliper used in this car is of original specifications by the manufacturer. A simulator is used to measure the desired parameters which are: (1). Brake pedal force (N), (2). Stopping time (s), (3). Stopping distance (m) and (4). Deceleration (m/s^2). For the purpose of this paper, the main parameter analysed and discussed is the brake pedal force which is the main input for a brake caliper once the brake pedal is applied. Table 3 provides a summary of the conditions for the experiment.

Table 3. Summary of the conditions for the experiment

	Description
Type of vehicle	Proton Persona 1.6L, Manual transmission, Sedan passenger car
Brake disc type	Front-ventilated disc with diameter 256 mm. Rear-brake drum
Type of equipment	Vehicle Global Positioning System (VGPS)
Range of Speed (km/h)	60, 65, 70, 75, 80, 85, 90, 95, 100 and 110
Weather condition	Clear
Road condition	Open main road, Less heavy traffic

The experiment started with the car driven at a speed of 60 km/h until it reaches the specified speed and retains the speed for a moment. As the driver was ready to press a brake pedal, a co-driver who is on stand-by position will click on the "Start" symbol on the simulator to ensure that the measured parameters are recorded throughout the braking process. Once the car is halted and the brake pedal released, the "Stop" symbol on the simulator was clicked to save the measured data. For each speed, three runs of the experiments are carried out. Figure 2 shows the test car; meanwhile Fig. 3 shows the experimental setup of the test car which is equipped with the instruments to measure the desired parameters.

Fig. 2. Test car- Proton Persona 1.6L

Fig. 3. (a) Simulator setup in a car for data recording at rear seat (b) Brake pedal force sensor installed on a brake pedal

It is important to note that the results obtained from this testing may be influenced by the following factors:

- The specified speed may be changed when a driver is about to start pressing the brake pedal since it is quite difficult to retain the exact specified speed of a car.
- The readiness of the driver to press a brake pedal
- The condition of an open road to make sure there were no vehicles on the road during the experiment was carried out
- The stability of the simulator where sometimes the GPS was unable to detect a signal that affect the data recorded during experiment

4   Results and Discussion

4.1   Performance of a Brake Caliper

In order to evaluate the performance of a brake caliper, the overall mechanism of hydraulic braking system in a vehicle needs to be first understood. Figure 4 shows the typical hydraulic braking system of a car. As the brake pedal is pressed by the driver, it will transmit the force to the power unit area where the force will be amplified by the vacuum booster. The total force will be then transmitted to the fluid pressure in the master cylinder to the piston area for the piston to actuate and thus, start the clamping action of the brake pads. The equal distribution of the fluid between front and rear is very important to ensure a stable braking action by a vehicle [13].

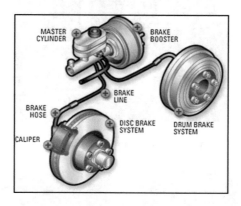

Fig. 4. Typical hydraulic brake system *Source:* https://www.pepboys.com/parts/brakes/overview/

The brake pedal force that provides the input into the braking system will be used to calculate the related parameters of the braking process which include brake line pressure (MPa), clamping force (N) and braking torque (Nm). These outputs are highly dependent on the area dimension of the component in the brake system since the fluid pressure is transmitted along them to fulfill the braking action [14, 15]. Table 4 shows the related area dimension of the components. Meanwhile, the formulas for calculating the parameters are shown in Appendix A.

Table 4. List of the dimension and coefficient

	Unit	Dimension
Brake pedal lever ratio	N/A	4:1
Vacuum booster (7″)	m^2	0.0248
Master cylinder	m^2	0.00387
Piston area	m^2	0.0023
Coefficient of friction	N/A	0.4
Effective radius of rotor	m	0.116

Braking torque is the action required to stop the motion of a vehicle when the effective radius of rotor, clamping force and the coefficient of friction of brake pads are against the rotor disc. The brake torque represents the overall results or output of a brake caliper since it comprises all technical parameters involved in a braking action [15]. Based on the following Eqs. (1) and (2), the relationship of the braking torque with its design parameters can be established through functional network diagram [8, 10].

$$Brake\ torque, T = \mu F_{clamp} R_{eff} = \mu R_{eff} \left(2 \times P_{line} \times A_{piston} \right) \tag{1}$$

Considering the number of piston as adding parameter:

$$Brake\ torque, T = \mu F_{clamp} R_{eff} = \mu R_{eff} \left(2 \times P_{line} \times A_{piston} \right) \times n_{piston} \tag{2}$$

Where μ: coefficient of friction between brake pads and disc, F_{clamp}: clamping force, R_{eff}: effective radius of rotor, P_{line}: braking line pressure, A_{piston}: area of caliper piston, and n_{piston}: number of piston.

The coefficient of friction μ and effective radius, R_{eff} are assumed to be constant along the braking process of a vehicle in this study [16]. However, in a real situation it should vary along the braking process [16, 17]. However, due to the lack of equipment facilities that can measure these two values during the brake test, it is considered as constants. The value 0.5 used in this study is the maximum value of μ during peak conditions of braking [18, 19] that isinfluenced by temperature, clamp force, pad history and sliding velocity [16]. In this study, the coefficient of friction of a brake pad is considered as the design parameters that can help to improve the braking efficiency of a vehicle by controlling the wear rate of the pad and disc. This is because the variation of μ is highly influenced by the frictional materials of brake pads and also the braking conditions [20].

Figure 5 shows the functional network diagram that represents the relationship between braking torques with its design parameters. As presented in Table 1, the specified range of brake torque in each generation will be used to calculate the required range of each design parameter by referring to Eqs. (1) and (2). Results from the acquired range as shown in Table 2 represent the candidates of design solution for upgrading of the brake caliper in terms of performance. In addition, the proposed upgrade design solutions are in-line with the guidelines provided by the manufacturers and experts in performance upgrading of brake calipers [14, 21]. Some of the guidelines are as follows:

- By increasing the effective radius, the caliper piston area, the line pressure or the coefficient of friction can improve the brake torque value [14].
- The pressure line can be improved by increasing the pedal ratio or decreasing the master cylinder diameter [14].
- Improving the area of brake pad can help in decreasing the wear rate of pad [14].
- The brake rotor disc must be bigger since it can increase the effective radius of the rotor and thus, able to reduce the clamping force. Therefore it requires less effort on the braking action to stop a vehicle [21].
- The disc thickness has an impact on quality and temperature rise when the brake disc is working therefore the optimum thickness needs to be considered [21].

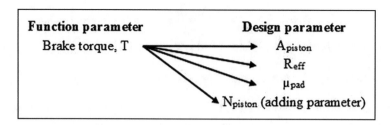

Fig. 5. Functional network diagram of a brake caliper design

The suitability of the design solutions need to be first evaluated to ensure that upgradability can be realised along the lifetime of a product as they are highly dependent on the specifications of the current design of the brake caliper and its braking system. It is important to conduct a proper evaluation or assessment in order to determine suitable candidates of the design solution that match the upgrade requirements. Therefore, this study proposed a systematic approach using a decision support model to evaluate the possible alternatives of the design solution. The approach which is based on Fuzzy-AHP is explained in the following section.

4.2   Future Work: Evaluating Upgradability of a Brake Caliper Through Fuzzy-AHP Decision Support Model

The design upgradability of a brake calliper as discussed in the earlier section is the range of design parameters that need to be acquired in each generation. The performance measure has been focused on the engineering metrics element of the brake calliper. Other than the engineering metric, the component and architecture elements have to be considered when choosing the appropriate design solutions so that a more concise evaluation and decision making can be carried out. The component level represents the reusability of components where each of the attributes or features of the brake caliper components are matched with the three upgrade plans proposed in this study. For the architecture level, the configuration of the brake caliper is examined by assessing the modules formation of the component.

A Fuzzy-AHP decision support model will be used to evaluate the alternatives of the design solutions based on the criteria that have been considered in satisfying the upgrade requirements of each element. This hybrid model was selected due to its ability to evaluate qualitative and quantitative data precisely compared to the use of the AHP technique. The rating scale used in the Fuzzy-AHP model is a linguistic variable of Triangular Fuzzy Number, while AHP only used discrete numbers of 1 to 9 as a rating scale. Preferences on the design alternatives will be measured in order to obtain the highest ranking alternative as the most suitable solution. These important criteria were selected based on the requirements and necessities in upgrading a brake caliper through in-depth interviews with experts in the field of study. Figure 6 shows the overall criteria which have been grouped into three main elements (Engineering Metric, Component and Architecture). Results validation will be accomplished through sensitivity analysis in order to ascertain their accuracy.

A recent research [10] that focused on the performance upgrading of a product, has highlighted that need for a systematic evaluation of the upgrade requirements especially when considering quantitative data which are based on opinions from the experts. Ke Xing [4] presented a systematic approach through modeling and optimization. The proposed formulation focuses on a specific product and requires detail technical information and specifications of the product. Customization is important in developing a decision support approach for design upgradability in order to facilitate the decision making in various other applications of products.

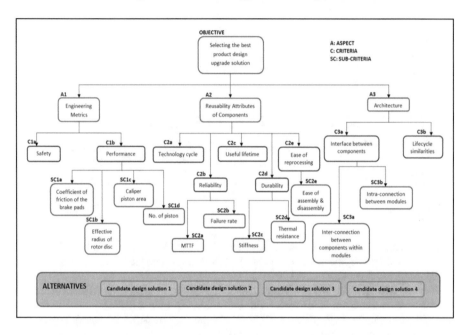

Fig. 6. Overall criteria to evaluate the design upgradability of brake caliper

5   Conclusion

As an overall conclusion, this study has focused on the evaluation of upgrade requirements that need to be considered along the upgrade plan of a brake caliper. The upgrade plan is established based on performance improvement from the analysis of brake torque and its related parameters that require upgrading. A brake test has been conducted to measure the brake torque of a vehicle through the acquired brake pedal force values. The relationship of the brake torque as the function parameter with the related design parameters is then observed in order to measure the required range or limit of the parameters for each generation. The design parameters involved is set to be the design solution that will be next assessed as a most suitable solution for upgrading. This study has also discussed a Fuzzy-AHP based decision support model for evaluating the criteria for selecting the best alternatives. Therefore, the requirements of design upgrade in terms of performance can be observed and understood throughout this study.

Acknowledgement. The authors would like to extend their appreciation to the Ministry of Higher Education, Malaysia for a student sponsorship under MyBrain15 Scholarship Program and for sponsoring this research under the GUP-2013-042 and AP-2015-008.

A Appendix

Measurement	List of formula
1. Output force to vacuum booster	$F_{output} = Brake\,pedal\,force, F_{pedal} \times Pedal\,ratio$
2. Area of diaphragm of vacuum booster, master cylinder and caliper hydraulic piston	$Area, A = \pi r^2$
3. Force at vacuum booster (MPa)	$Force, F_{booster} = Pressure\,in\,engine\,vacuum, P_{vacuum} \times A_{piston}$
4. Brake line pressure (MPa)	$Pressure, P_{line} = \dfrac{Total\,force\,at\,master\,cylinder, F_{MC}}{Area\,of\,master\,cylinder, A_{mc}}$
5. Clamping force	$Force, F_{clamp} = 2\left(P_{line} \times A_{piston}\right)$

References

1. Poulikidou, S.: Literature Review: Methods and Tools for Environmentally Friendly Product Design and Development. Identification of Their Relevance to the Vehicle Design Context. US AB, Stockholm (2012)
2. Hatcher, G.D., Ijomah, W.L., Windmill, J.F.C.: Design for remanufacture: a literature review and future research needs. J. Clean. Prod. **19**(17–18), 2004–2014 (2011)
3. Ülkü, S., Dimofte, C.V., Schmidt, G.M.: Consumer valuation of modularly upgradeable products. Manag. Sci. **58**(9), 1761–1776 (2012)
4. Xing, K.: Design for upgradability: modelling and optimisation. Ph.D. Thesis. Division of Information, Technology, Engineering and the Environment, School of Advanced Manufacturing and Mechanical Engineering, University of South Australia (2006)
5. Zhao, Y., Pandey, V., Kim, H., Thurston, D.: Varying lifecycle lengths within a product take-back portfolio. J. Mech. Design. **132**(9), 1–10 (2010)
6. Aziz, N.A., Wahab, D.A., Ramli, R., Azhari, C.H.: Modelling and optimisation of upgradability in the design of multiple life cycle products: a critical review. J. Clean. Prod. **112**, 282–290 (2015)
7. Aziz, N.A., Wahab, D.A., Ramli, R.: Evaluating design for upgradability at the conceptual design stage. JurnalTeknologi. **78**(6–9), 37–43 (2016)
8. Umemori, Y., Kondoh, S., Umeda, Y., Shimomura, Y., Yoshioka, M.: Design for upgradable products considering future uncertainty. In: Proceedings of EcoDesign 2001: Second International Symposium on Environmentally Conscious Design and Inverse Manufacturing, pp. 87–92. IEEE (2001)
9. Matsuda, A., Shimomura, Y.: Upgrade planning for upgradeable product design. In: Proceedings of EcoDesign 2003: Third International Symposium on Environmentally Conscious Design and Inverse Manufacturing, pp. 231–234. IEEE (2003)

10. Inoue, M., Yamada, S., Yamada, T., Bracke, S.: An upgradable product design method for improving performance, CO2 savings, and production cost reduction: vacuum cleaner case study. Int. J. Sup. Chain Mgt. **3**(4), 100–106 (2014)
11. Mazhar, M.I., Salman, M., Howard, I.: Assessing the reliability of system modules used in multiple life cycles. In: Kiritsis D., Emmanouilidis C., Koronios A., Mathew J. (eds) Engineering Asset Lifecycle Management, pp. 567–573. Springer, London (2010)
12. Shimomura, Y., Umeda, Y., Tomiyama, T.: A proposal of upgradable design. In: Proceeding of First International Symposium on Environmentally Conscious Design and Inverse Manufacturing, pp. 1000–1004. IEEE (1999)
13. Campbell, J.D., Jorgeson, C.M., Murphy, R.W.: Brake Force Requirement Study: Driver-vehicle Braking Performance as a Function of Brake System Design Variables. Highway Safety Research Institute, University of Michigan (1970)
14. Ruiz, S.: Brake Systems and Upgrade Selection. http://stoptech.com/technical-support/technical-white-papers/brake-system-and-upgrade-selection
15. Ho, H.P.: The influence of braking system component design parameters on pedal force and displacement characteristics. Ph.D. Thesis. School of Engineering, Design and Technology. University of Bradford (2009)
16. Neys, A.: In-vehicle brake system temperature model. Technical report No. 2012: 38. Department of Applied Mechanics, Chalmers University of Technology (2012)
17. Khaimar, H.P., Phalle, V.M., Mantha, S.S.: Comparative frictional analysis of automobile drum and disc brakes. Tribo Ind. **38**(1), 11–23 (2016)
18. Renault, A., Massa, F., Lallemand, B., Tison, T.: Experimental investigations for uncertainty quantification in brake squeal analysis. J. Sound Vib. **367**, 37–55 (2016)
19. vanIersel, S.S., Basselink, I.I., Meinders, E., van Eersel, K.J.: Measuring brake pad friction behaviour using the TR3 test bench. Technical report (DCT no. 2006.118). Eindhovan University of Technology, DCT (2006)
20. El-Tayeb, N.S.M., Liew, K.W.: On the dry and wet sliding performance of potentially new frictional brake pad materials for automotive industry. Wear **266**, 275–287 (2008)
21. Li, S., Yong-Chen, L.: The disc brake design and performance analysis. In: International Conference Consumer Electronics, Communications and Networks (CECNet), pp. 846–849. IEEE (2011)

A Manufacturing Value Modeling Methodology (MVMM): A Value Mapping and Assessment Framework for Sustainable Manufacturing

Melissa Demartini[1(✉)], Ilenia Orlandi[2], Flavio Tonelli[1], and Davide Anguitta[2]

[1] DIME - Department of Mechanical Engineering, Energetics, Management and Transportation, Polytechnic School, University of Genoa, Genoa, Italy
melissa.demartini@dime.unige.it, flavio.tonelli@unige.it
[2] DIBRIS Department of Informatics, Bioengineering, Robotics, and Systems Engineering, University of Genoa, Genoa, Italy
ilenia.orlandi@edu.unige.it, davide.anguita@unige.it

Abstract. Sustainable manufacturing is becoming increasingly important. This requires sustainable industrial system different to today's global industry with different business models, creating different products and services requiring new strategies, frameworks, and tools. The evolution towards a 'sustainable' industrial production systems requires a holistic approach, with a fundamental reassessment of the value creation. In order to achieve this target a system design approach is required. In this paper an existing and specific Manufacturing Value Modeling Methodology (MVMM) is used as a value mapping framework to help firms in creating value propositions better suited for sustainability considering economic, environmental and social perspectives. Concerning sustainability, implementing it into the MVMM requires the setting of a catalogue that presents an overview of sustainable external and internal impact factors and a mapping between them in order to translate business goals into manufacturing strategy, and allows to improve operational performance by adopting a set of sustainable industrial practices.

Keywords: Industrial sustainability · Value modeling · Value mapping

1 Introduction

Industrial sustainability is a capability that allows to achieve competitive advantage through the increase of material efficiency, energy saving, closed-loop control at industrial system level, and through the increasing competitiveness by improving economic, environmental and social performance (Demartini et al. 2016). Some companies are making progress toward the next frontier of sustainability, data from the past five years shows that many organizations are struggling to move forward (MIT Sloan Management Review). In fact, addressing significant sustainability issues has become a core strategic imperative that these companies view as a way to mitigate threats and identify powerful new opportunities. Like any business issue, addressing important sustainability issues requires specific, hard-wired organizational support, capabilities, and measurement.

© Springer International Publishing AG 2017
G. Campana et al. (eds.), *Sustainable Design and Manufacturing 2017*, Smart Innovation, Systems and Technologies 68, DOI 10.1007/978-3-319-57078-5_10

As highlighted by Smith and Ball (2012), achieving sustainability in manufacturing requires a holistic view spanning product design, manufacturing processes, manufacturing systems, and the entire supply chain. Such an approach must be taken to ensure the economic, environmental and societal goals of sustainability are achieved. Hence Authors decide to focus their attention on sustainability both as a competitive and strategic dimension in the manufacturing environment with respect to value modeling and mapping. Thus, the purpose of this paper is to present an overview regarding sustainability trends, implications and possibilities that affect the manufacturing company and supply chains, and leading a review in order to analyze the existing body of literature on value mapping tool for industrial sustainability, with the aim of creating a catalog that allow to mapping different dimension of industrial sustainability (economic, environmental and social).

Related works in the context of this paper consist of two research domains. On the one hand, there is the research regarding Industrial Sustainability and on the other hand the one on value modeling and mapping already addressed in (Taticchi et al. 2013; Taticchi et al. 2015; Tonelli et al. 2016; Taticchi et al. 2012) within the Manufacturing Value Modeling Methodology. Both domains are crucial for implementing a proper sustainability catalog; the research on industrial sustainability is important for the creation of the underlying sustainability model, while the research on manufacturing value modeling is seen as a key influencer towards constructing the framework for identifying the correct sustainability demand. Authors, report a qualitative literature review on value mapping framework for Industrial Sustainability in Sect. 2, then present the Sustainability Framework itself, with an overview on external and internal impact factors of Industrial Sustainability in Sect. 3. In Sect. 4 the mapping between the external and internal factors of the Sustainability Framework is presented. Finally, Sect. 5 shows the consequences and issues of the Authors' work and conclusions are discussed.

2   Qualitative Literature Review on Value Mapping Framework for Sustainable Manufacturing

The qualitative literature review has been performed using classic bibliometric techniques. The methodology used is a literature review based on an electronic search in "Scopus", the Authors interrogated the database searching for ("Value Mapping") AND ("Industrial Sustainability") AND ("Manufacturing"), in the titles, abstracts and keywords of papers published between 2000 and 2015. The interrogation resulted in 74 papers that constitute the base of further analysis. The earliest paper included in the dataset was published in 2002 and the most recent in 2015.

The distribution of publication per journal is made up of six journals where research has been published. Journal of Cleaner Production, IFIP Advances in Information and Communication Technology, International Journal of Operations and Production Management, International Journal of Advanced Manufacturing Technology, International Journal of Lean Six Sigma and TQM Journal lead the ranking with 5, 5, 3, 2, 2 and 2 publications, respectively. The most prolific scholars are Rana P., Badurdeen F., Bocken N., Chiarini A., Evans S., Short S. with 4, 3, 3, 2, 2, 2 publications, respectively. Instead, about the geographic diversity of scholars is relevant to note the leadership of

European academic institutions that contribute for 50% to the research field development. Moreover, there is an emerging contribution of scholars from India and Brazil. This suggests the relevance of this topic also for emerging countries. Further, the frequency of publications over time highlighting a research field that is growing very fast. The top three keywords are "Sustainability", "Value Stream mapping" and "Lean". It is apparent from the literature that most approaches for progressing towards sustainable development are generic and high level, this has been confirmed by Smith et al., that highlight a lack of guidance and tools for manufacturers to identify improvement opportunities within their own factories. Bocken et al., propose a value mapping tool that takes a multi-stakeholder perspective and considers different forms of value, such as value captured, value missed, value destroyed, and new value opportunities. Paju et al. introduced a new methodology termed sustainable manufacturing mapping (SMM) which incorporates discrete event simulation (DES) and life-cycle analysis (LCA). For Fearne and Martinez Value Chain Analysis tools need to adopt more holistic sustainable perspectives. These include addressing external factors, such as health, environmental damage and poverty, which can offer opportunities for a chain to create shared value (Porter and Kramer 2011).

For Schaltegger and Burritt, opportunities mostly originate from management decisions of the focal company. This requires both knowledge about sustainability problems, ranking of possible solutions and the assessment of consumer expectations and market strategies to make sure that the most sustainable product offering becomes a market and business success.

Other works have utilized tools such as Value Stream Mapping, discrete event simulation, and value network mapping to model the current state as well as future state maps in complex environments (McDonald et al. 2002; Lian and Van Landeghem 2002; Irani and Zhou 2003; Braglia et al. 2006). To summarize, existing tools generally tend to focus on just one dimension of sustainability, and fail to engender a holistic perspective that incorporates all three dimensions of sustainability within the business planning process (Bocken 2013). Futhermore, the literature review highlights the necessity of a tool that includes both an assessment of the sustainable external factors, and the company strategy. To identify these perspectives, we propose one approach in order to highlighting the relationships between various principles, strategies, issues, through what Authors named hierarchical Manufacturing Value Modeling Methodology (MVMM) (Tonelli 2016), it allows to assist firms in better understanding sustainable value creation within their business activities, and assist them in developing new strategies with sustainability at their core.

3   Sustainability Framework

The scope of the Authors' work is to provide a guide for manufacturing companies, in order to understand what sustainability trends and drivers are fundamental to the manufacturing environment and a set of KPIs that allows firms to control and monitor the reaching of these goals. The sustainability catalog combines these two approaches by using the MVMM, in order to include internal and external influence factors analyzing them with respect to the triple bottom line approach. The sustainability catalogue starts from the core concept of the MVMM, using the structure of external influence factors (Manufacturing Challenges), internal influence factors (Manufacturing Objectives and Sustainable Industrial Practices) and also applies the value map by using the aforementioned contents, and the concept of relationships between the value map items (Fig. 1).

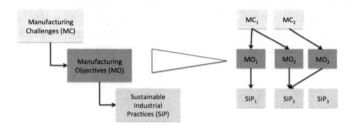

Fig. 1. Example of relationships between the value map items according to the MVMM

3.1 External Factors

The external view represents Manufacturing Challenges (MC) as shown by Table 1, where sustainable challenges that have an impact on the manufacturing environment are reported. This section gives a background on the challenges associated with embedding sustainability

Table 1. Manufacturing Challenges

External factors: MC	Sustainable dimension
Higher production flexibility and re-configurability	Economic
Highly variable and difficult to forecast market conditions	Economic
Increase in productivity	Economic
Need to enhance specific competences and skills	Economic
Need to manage dynamic and complex business networks	Economic
New models of collaboration reshoring-offshoring-nearshoring	Economic
Products to satisfy the demand for comfort, health and wellbeing of specific target groups	Economic
Pervasiveness of internet	Economic
Reduce energy consumption	Eco-Environmental
Exploitation of energy from waste and scrap	Environmental
Increase in urbanization - integration of industry in urban context	Environmental
Increase the resilience of industry to global warming and climate change (on production, procurement and markets)	Environmental
Manage environmental changes due to exploitation of farmland, deconstruction of infrastructure and urbanization	Environmental
Need to recycle components and products	Environmental
Reduce pollution in air, ground and water through improved environmental sustainability	Environmental
Use of alternative energy sources in manufacturing	Environmental
Change in the interaction with the individual (customer, worker, citizen)	Social
Growth of a new middle class at global level	Social
Increase the worker well-being in terms of high satisfaction, safety and inclusivity	Social
New forms of employment	Social
New services tailored on the people	Social
Growth of emerging countries (production and consumption)	Socio-Economic

into corporate performance management, these contents are derived from various roadmap: Factories of the Future (EFFRA), Vision 2020, Pathfinder, Cluster Fabbrica Intelligente (CFI). Examples of MC could be for instance Manage environmental changes and/or Reduce energy consumption. Due to the different markets in which companies operate, the MC might vary from scenario to scenario and related industrial context. There might be MC, which are globally valid, while there are also MC which are only true for a certain branch or industry, then it is important to study the environment of the company and the domain in which it operates in order to identify a valid set of MC. The external view is followed by the analysis of the internal process and strategies.

3.2   Internal Factors

The internal influence factors are used to represent the sustainable goals and strategies of the manufacturing company. Different internal influence factors could be identified as:

- Manufacturing Objectives (MO): describe the company strategy in terms of sustainable opportunities and issues (Table 2).

Table 2. Manufacturing Objectives

Internal factors: MO	Sustainable dimension
Client satisfaction (21)	Economic
Decarbonisation of the global energy system (22)	Environmental
Energy efficiency (23)	Economic & Environmental
Increase recycling rates (4)	Economic & Environmental
Increase usage of renewable resources (24)	Economic & Environmental
Material Efficiency (22)	Economic & Environmental
Minimize emissions to land (25)	Environmental
Minimize impact on species (21)	Environmental
Minimize water usage (28)	Economic & Environmental
Reduce usage of raw material (21)	Economic & Environmental
Reduction of air emission (27)	Environmental
Safety (21)	Social
Waste Reduction (17)	Economic & Environmental
Client satisfaction (21)	Economic
Decarbonisation of the global energy system (22)	Environmental
Energy efficiency (23)	Economic & Environmental
Increase recycling rates (4)	Economic & Environmental
Increase usage of renewable resources (24)	Economic & Environmental
Material Efficiency (22)	Economic & Environmental
Minimize emissions to land (25)	Environmental
Minimize impact on species (21)	Environmental
Minimize water usage (28)	Economic & Environmental

- Sustainable Industrial Practices (SIP): as a set of planning practices, production, purchasing and logistics aimed to incorporate a sustainable perspective in operations (Table 3).

Table 3. Sustainable Industrial Practices

Internal factors: SIP	Description
Ecodesign (17)	It is treated as the designing phase of product life cycle. It is based on Life Cycle Assessment that is a technique that summarizes the quantification of the environmental consequences of products and services
Green Supply Chain (GSC) (18)	GSC is viewed within the planning and sourcing phase of the product life cycle. GSC can be understood as sustainable operations practices together with suppliers and/or customers covering project design, selection of raw materials, selection of suppliers, green purchasing, packaging and logistics
Cleaner Production (CP) (19)	It refers to the production phase. It represents the application of an economic, environmental and technological strategy integrated with the processes and products in order to make them more efficient
Reverse Logistics (RL) (20)	RL refers to the management of waste related to the consumption of manufacturing products. Reverse logistics can be understood as the return process of moving goods in order to capture value or give the appropriate destination

3.2.1 Manufacturing Objectives

After the market related view, the MVMM suggests reviewing the MO. The goal thereby is to identify the strategy of the company and the goals that are used to achieve this strategy. Hence the aim of the sustainability catalog is to analyze the sustainability in the production process. Nonetheless when analyzing the strategy, it is also mandatory to understand the overall business strategy, since the production strategy should fit to the overall strategy of the company. The goal of this step is to set up a goal system that should identify the important areas, which have to be addressed.

Especially these two first steps of identifying the external influences, through the MC and the internal influences trough MO and SIP are mandatory for the sustainability catalog because assessing the market view as well as the internal manufacturing process related view are crucial for identifying the causes behind a certain sustainability demand. After identifying the Manufacturing Challenges and Objectives, it is important to further specify the context in which the sustainability demand occurs with the analysis of the Sustainable Industrial Practices.

3.2.2 Sustainable Industrial Practice

The identification of the context consists of selecting the correct functional areas or practices, which need a detailed analysis. Due to the focus on production, these practices are the functional areas in the manufacturing operation management domain. In this section Authors starting from this point analyze the manufacturing strategy, focusing on sustainable industrial practices in order to bring out the alignment of manufacturing strategy with business strategies. Table 3 shows the definition for each practice:

4 Contents Mapping of the Sustainability Framework

From a value modeling point-of-view, capturing the environment of the given scenario by identifying the external and internal influence factors and mapping them is necessary to find which domain specific market trends fit to which domain specific project targets. Besides the general description of the sustainability framework, it is mandatory to explain the application of the catalog itself. Since the general approach is derived from the MVMM approach it is also possible to create relationships between the different components. Generally speaking there is the possibility to create a relationship between external influence factors (Manufacturing Challenges) and the business strategy (Internal factor) that is used to tackle them. This means there is a certain set of internal influence factors that fit to a certain external factor (Table 4).

Starting with the results of these analyses, it is possible to highlight which thematic areas concerning the MC do not find a mapping with key strategic objectives. These themes, which need to be investigated more in depth, manly concern three aspects:

- New emerging markets (Highly variable and difficult to forecast market conditions, Need to manage dynamic and complex business networks, New models of collaboration reshoring-offshoring-nearshoring, Need to enhance specific competences and skills of each geographical area): increase the number of manufacturing companies strategically involved in innovation activities to cope with increasingly uncertain and unpredictable market conditions through the improvement of specific territorial skills;
- Demographic change (New services tailored on the people, Products to satisfy the demand for comfort, health and wellbeing of specific target groups): improve the social impact, making manufacturing jobs more attractive, in terms of greater safety, inclusion and personal achievement, and improve the integration of industry in an urban context that is constantly expanding to satisfy the specific demands of consumer comfort, health and well-being;
- Technological acceleration (Increase in productivity, Pervasiveness of internet, Change in the interaction with the individual (customer, worker, citizen)): increase R&D investments in the manufacturing sector exploiting the opportunities offered by technological acceleration linked mainly to the development of new technologies, the integration of advances technologies and the pervasiveness of the Internet and mechatronics.

Differently, the Manufacturing Challenges factors that are better reflected among the Manufacturing Objectives, are related to all the activities concerning the management of resources and environment: reduce the environmental impact by reducing the emission of greenhouse gases resulting from manufacturing activities, the reduction of energy consumption and of materials deriving from manufacturing activities, the reduction of waste produced by manufacturing activities and the creation of eco-products and eco-technologies.

5   Conclusions and Future Developments

Author's purpose was to examine evidence of Value mapping framework for industrial sustainability, it has been realized a structured Sustainability Catalogue within the Manufacturing Value Modeling Methodology allowing to translate sustainable trends and goals into manufacturing strategy, improving operational performance.

Companies should move away from using the traditional techniques that focused only on cost minimization and efficiency improvement to those that also take into account the environmental and societal implications of operations. It is growing ever clearer that it needs action at material, product, process, plant and system of production levels. As a result, sustainability decisions become an integral part of business decision making, the business planning cycle, and customer/supplier relationships. The sustainability catalogue is a hierarchical approach that seek to integrate consideration of the three dimensions of sustainability (economic, environmental and social) in a manner that align company and manufacturing strategy and create value for all stakeholders including the environment and society. There are some limitations to the sustainability catalogue, the model is largely qualitative, it does not allow for detailed quantitative analysis. Besides it needs to be validated with real case in order to verify and improve the catalogue. More tests are planned to further understand the applicability and suitability of the tool in different contexts. Finally, it is evident that this area still requires significant investigation at the operational and strategic levels, the framework provided will guide industry and supply chain sustainable progress and improvement.

Appendix

Table 4. Mapping between external/internal factors

External Factors: Manufacturing Challenges	Client satisfaction	Decarbonisation of the global energy system	Energy efficiency	Increase recycling rates	Increase usage of renewable resources	Material Efficiency	Minimize emissions to land	Minimize impact on species	Minimize water usage	Reduce usage of raw material	Reduction of air emission	Safety	Waste Reduction
Higher production flexibility and re-configurability			X	X	X	X	X		X	X	X		X
Highly variable and difficult to forecast market conditions	X												
Increase in productivity													
Need to enhance specific competences and skills													
Need to manage dynamic and complex business networks													
New models of collaboration reshoring-offshoring-nearshoring													
Products to satisfy the demand for comfort, health and wellbeing of specific target groups	X											X	X
Pervasiveness of internet													
Reduce energy consumption			X										
Exploitation of energy from waste and scrap		X	X	X	X	X							
Increase in urbanization - integration of industry in urban context	X				X		X	X			X		X
Increase the resilience of industry to global warming and climate change		X					X						X
Manage environmental changes due to exploitation of farmland, deconstruction of infrastructure and urbanization		X					X						X
Need to recycle components and products						X	X			X			X
Reduce pollution in air, ground and water through improved environmental sustainability		X		X			X	X			X	X	X
Use of alternative energy sources in manufacturing		X	X	X	X	X							
Change in the interaction with the individual (customer, worker, citizen)	X											X	
Growth of a new middle class at global level									X		X		X
Increase the worker well-being in terms of high satisfaction, safety and inclusivity					X		X	X			X	X	
New forms of employment													
New services tailored on the people	X												
Growth of emerging countries (production and consumption)									X		X		X

References

Demartini, M., Orlandi, I., Tonelli, F., Anguita, D.: Investigating sustainability as a performance dimension of a novel Manufacturing Value Modeling Methodology (MVMM): from sustainability business drivers to relevant metrics and performance indicators. In: XXI Summer School "Francesco Turco", pp. 262–270 (2016)

MIT Sloan Management Review: The Boston Consulting Group. Sustainability's Next Frontier: Walking the talk on the sustainability issue that matter most. Research report, available at: http://sloanreview.mit.edu/projects/sustainabilitys-next-frontier/

Smith, L., Ball, P.: Steps towards sustainable manufacturing through modelling material, energy and waste flows. Int. J. Prod. Econ. **140**(1), 227–238 (2012)

Taticchi, P., Tonelli, F., Pasqualino, R.: Performance measurement of sustainable supply chains: a literature review and a research agenda. Int. J. Prod. Perform. Manag. **62**(8), 782–804 (2013)

Taticchi, P., et al.: A review of decision-support tools and performance measurement and sustainable supply chain management. Int. J. Prod. Res. **53**(21), 6473–6494 (2015)

Tonelli, F., et al.: Approaching industrial sustainability investments in resource efficiency through agent-based simulation. In: Borangiu, T., Trentesaux, D., Thomas, A., McFarlane, D. (eds.) Service Orientation in Holonic and Multi-Agent Manufacturing, pp. 145–155. Springer International Publishing, Cham (2016a)

Taticchi, P., Balachandran, K., Tonelli, F.: Performance measurement and management systems: state of the art, guidelines for design and challenges. Measuring Bus. Excellence **16**(2), 41–54 (2012)

Bocken, N.M.P., Rana, P., Short, S.W.: Value mapping for sustainable business thinking. J. Ind. Prod. Eng. **32**(1), 67–81 (2015)

Paju, M., et al.: Framework and indicators for a sustainable manufacturing mapping methodology. In: Proceedings of the 2010 Winter Simulation Conference (WSC), pp. 3411–3422. IEEE (2010)

McDonald, T., Van Aken, E.M., Rentes, A.F.: Utilising simulation to enhance value stream mapping: a manufacturing case application. Int. J. Logistics **5**(2), 213–232 (2002)

Lian, Y.-H., Van Landeghem, H.: An application of simulation and value stream mapping in lean manufacturing. In: Proceedings 14th European Simulation Symposium, (c) SCS Europe BVBA, pp. 1–8 (2002)

Zhou, J., Irani, S.A.: A new flow diagramming scheme for mapping and analysis of multi-product flows in a facility. J. Integr. Des. Process Sci. **7**(1), 25–58 (2003)

Braglia, M., Carmignani, G., Zammori, F.: A new value stream mapping approach for complex production systems. Int. J. Prod. Res. **44**(18–19), 3929–3952 (2006)

Bocken, N., Short, S., Rana, P., Evans, S.: A value mapping tool for sustainable business modelling. Corp. Governance **13**(5), 482–497 (2013)

Tonelli, F., et al.: A novel methodology for manufacturing firms value modeling and mapping to improve operational performance in the industry 4.0 Era. Procedia CIRP **57**, 122–127 (2016b)

Tonelli, F., Evans, S., Taticchi, P.: Industrial sustainability: challenges, perspectives, actions. Int. J. Bus. Innov. Res. **7**(2), 143–163 (2013)

Fearne, A., Martinez, M.G., Dent, B.: Dimensions of sustainable value chains: implications for value chain analysis. Supply Chain Manag. Int. J. **17**(6), 575–581 (2012)

Kaiser, F.G., et al.: Ecological behavior and its environmental consequences: a life cycle assessment of a self-report measure. J. Environ. Psychol. **23**(1), 11–20 (2003)

Srivastava, S.K.: Green supply-chain management: a state-of-the-art literature review. Int. J. Manag. Rev. **9**(1), 53–80 (2007)

Fore, S., Mbohwa, C.T.: Cleaner production for environmental conscious manufacturing in the foundry industry. J. Eng. Des. Technol. **8**(3), 314–333 (2010)

Presley, A., Meade, L.: Benchmarking for sustainability: an application to the sustainable construction industry. Benchmarking Int. J. **17**(3), 435–451 (2010)

Allwood, J.M., et al.: Material efficiency: a white paper. Resour. Conserv. Recycl. **55**(3), 362–381 (2011)

Mose, M.: Analysis of the cumulative volumes, a strategy to anticipate the market (2015)

Abidin, Z.: Nazirah, C.L. Pasquire: Delivering sustainability through value management: concept and performance overview. Eng. Constr. Architectural Manag. **12**(2), 168–180 (2005)

Hutchins, M.J., Sutherland, J.W.: An exploration of measures of social sustainability and their application to supply chain decisions. J. Cleaner Prod. **16**(15), 1688–1698 (2008)

Ahi, P., Searcy, C.: An analysis of metrics used to measure performance in green and sustainable supply chains. J. Cleaner Prod. **86**, 360–377 (2015)

Vinodh, S., Ben Ruben, R., Asokan, P.: Life cycle assessment integrated value stream mapping framework to ensure sustainable manufacturing: a case study. Clean Technol. Environ. Policy **18**(1), 279–295 (2016)

Porter, M.E., Kramer, M.R.: The big idea: creating shared value. Harvard Bus. Rev. **89**(1), 2 (2011)

Schaltegger, S., Burritt, R.: Measuring and managing sustainability performance of supply chains: review and sustainability supply chain management framework. Supply Chain Manag. Int. J. **19**(3), 232–241 (2014)

Improving Sustainability in Product Development Projects

E. Lacasa$^{(\boxtimes)}$, J.L. Santolaya, and I. Millán

Department of Design and Manufacturing Engineering, EINA, University of Zaragoza,
C/María de Luna 3, 50018 Zaragoza, Spain
enrike_sena@hotmail.com

Abstract. Sustainable product development initiatives have been evolving for some time to support companies improve the efficiency of current production and the design of new products and services through supply chain management. The development of different methods and tools for considering environmental criteria in the same way as conventional design criteria through an Eco-design approach were carried out. Environmental assessment tools are generally based on a life cycle assessment (LCA) method, which can inform production and consumption choices because it assess the environmental performance of a product through accounting all the energy and material inputs and the associated emissions and waste outputs at each stage of its life cycle.

While using LCA to measure the environmental dimension of sustainability is widespread, similar approaches for the economic (LCC) and the social (S-LCA) dimensions of sustainability still have limited application worldwide and there is need for consistent and robust methods and indicators. This paper focuses on the production step and presents the redesign process of an airbrush in order to improve their sustainability performance. According to LCA evaluation methods, an approach based on the analysis of the flows exchanged by the industrial installation throughout the production step was developed. Different sustainability indicators were obtained. In particular, the environmental indicator of global warming, the economic indicator of value added and the social indicator of working hours were used to assess the sustainability performance. An improvement of the redesigned product indicators was achieved.

Keywords: Sustainability · Product design · Production · Indicators assessment

1   Introduction

Sustainability has become a requirement for competitive companies, which can also enable them to achieve higher quality products, to improve the company image and to reduce the manufacturing costs. The progress toward sustainability implies maintaining and preferably improving, both human and ecosystem well-being [1]. Achieving sustainable development in industry will require changes in organizational models and production processes in order to balance the efficiency of its operations with its responsibilities for environmental and social actions [2].

Companies are increasingly aware of environmental impact of their products. Several authors [3, 4] have contributed to the development of methods and tools considering environmental criteria in the same way as conventional design criteria through an

© Springer International Publishing AG 2017

G. Campana et al. (eds.), *Sustainable Design and Manufacturing 2017*, Smart Innovation, Systems and Technologies 68, DOI 10.1007/978-3-319-57078-5_11

Eco-design approach. Using Eco-design or Design for the Environment (DfE) all environmental impacts of a product are addressed throughout its complete life cycle, without unduly compromising other criteria and specifications like function, quality, cost and appearance. A whole product system life cycle includes five different stages: materials obtaining, production process, distribution, use and final disposition.

Nevertheless, sustainability does not only consist of the environmental impact, it consists of the three dimensions: environmental (planet), economy (profit) and social well-being (people). In order to evaluate sustainability in this triple bottom line, a new perspective is being introduced through the life cycle sustainability assessment (LCSA) framework [5, 6]. LCSA evaluate all environmental, social and economic negative impacts and benefits in decision-making processes towards more sustainable products and provide guiding principles to achieve sustainable production while stimulating innovation by identifying weakness and enabling further improvements over the product life cycle.

While using Life Cycle Assessment (LCA) to measure the environmental dimension of sustainability is widespread, similar approaches for the economic (LCC) and the social (S-LCA) dimensions of sustainability still have limited application worldwide and there is need for consistent and robust methods and indicators. Life cycle costing is a compilation and assessment of all costs associated with the life cycle of a product that are directly covered by any or more of the actors in the product life cycle [7]. Whereas, S-LCA provides information on social aspects in order to improve performance of organizations and ultimately the well-being of stakeholders. According to UNEP's guidelines [8], the socio-economic impacts, associated with the product' life, are captured in five suggested stakeholder categories: workers, local community, society, consumers and value chain actors. Many social issues are not easy to quantify, so a number of social indicators contain qualitative standards of systems and activities of the organization.

In order to apply the principles of sustainable development in practice there is a need to measure the individual sustainability dimensions and to achieving a comprehensive presentation of the results. A set of indicators for identification of more sustainable practices can be used [9]. The indicators should be developed at the appropriate level of detail to ensure proper assessment of the situation with regard to each particular challenge. Simplified indicators, able to aggregate results and weigh the most important impact categories into easily understandable and user-friendly units, are particularly useful for designers because facilitate the communication of sustainability results to the decision-makers.

In this work, different metrics and indicators are used to assess the three dimensions of sustainability in product development projects. Methodology applied and results obtained for a case study are shown in the following sections.

2   Methodology

Sustainability evaluation is focused on the production step of the product life cycle. As shows Fig. 1, the development of a more sustainable product can be achieved through a sequence of phases organized as follows: (1) Identification of inputs and outputs associated to the production process; (2) Assessment of engineering metrics and indicators for the three dimensions of sustainability; (3) Product redesign integrating sustainability criteria. Next, a new production inventory and sustainability assessment should be carried out for the redesigned product. Finally, the comparative presentation of the sustainability performance of both initial and redesigned product can be performed to detect if product was improved.

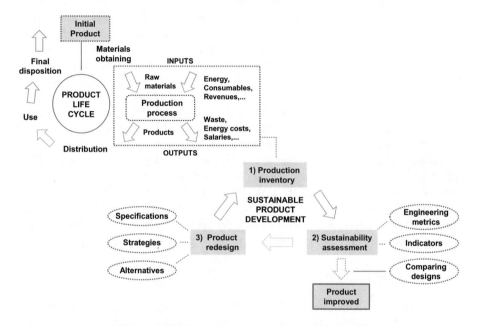

Fig. 1. Phases for a sustainable product development

2.1 Production Inventory

All existent flows associated to the production system are valued in this phase. One manufactured product is the functional unit considered within a high volume manufacturing process. The elementary flows exchanged by the industrial installation include inputs (raw materials and other components supply, energy consumption, consumables, revenues,…) and outputs (products, waste, energy costs, salaries,…).

In addition, manufacturing operations are analyzed in detail to value material transformations and resource consumptions for each part or component of the product. Particularly, the calculation of material removed, time required and power supply is carried out for each productive operation to project the manufacturing process of the redesigned product.

2.2 Sustainability Assessment

A number of engineering metrics and indicators are obtained in this phase. Metrics considered useful to assess the production activity are the mass and volume of the manufactured product, the total energy consumption, the waste percentage, the costs of raw materials and the annual production volume. To note that these metrics allow obtaining practical information for process designers and are needed to assess sustainability indicators at the production and distribution stages in the product life cycle.

Different indicators are proposed in this work to assess each of the three dimensions of sustainability. Global warming (GW) that represents total emissions of the greenhouse gases and eco-indicator 99 (E99) [10] that weighs different impact categories into a single score, are the indicators proposed to assess the environmental dimension. Midpoint categories and characterization factors from Probas [11] and MEEuP [12] databases are used in the calculus process. Moreover, the reuse-recycling potential at the final disposition phase of product life cycle is taken into account. For the economic dimension, the value added (VA) and the eco-efficiency (EE) are the indicators proposed. The value added expresses the net operating profit of the company and is obtained as the difference between sales revenues and production costs. The eco-efficiency combines and quantifies the economic and the environmental aspects because it is evaluated by the ratio of the value added and the eco-indicator E99.

Finally, the indicators selected to assess the social dimension are the working hours and the hourly wage, which are associated to the category of company workers. Metrics and indicators are expressed by manufactured product unit.

2.3 Product Redesign

Product design activities usually begin with an analytical phase where the product requirements and specifications and the diverse parameters of the problem are studied along with the anticipated market demands. Taking into account that each product consists of different parts or components and each of its components fulfils a function, the specifications of individual components should be analyzed. For each individual component, redesign alternatives, which involve the application of sustainability strategies, can be proposed.

Brezet and van Hemel [3], created the Life Cycle Design Strategies Wheel or LiDS wheel (Fig. 2), in which different strategies to achieve sustainability are identified around the product life cycle. Eight major strategies are proposed: selection of low-impact materials, reduction of materials, optimization of production techniques, selection of efficient distribution systems, reduction of the environmental impact in the use stage, optimization of initial life-time, optimization of end-of-life system and new concept development. The LiDS wheel can be used to estimate the environmental profile of an existing product or to qualify the action plan for a new and more sustainable product. Strategies listed first are considered appropriate to achieve an improvement of the product at the production phase.

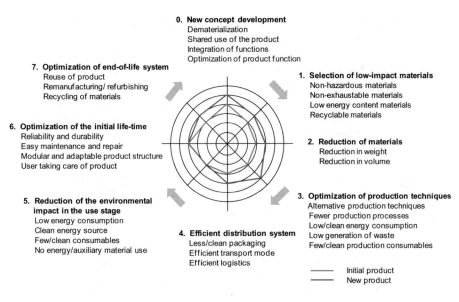

Fig. 2. The LiDS wheel. Source: Brezet and van Hemel (1997).

3 Case Study

Previous methodology was implemented in the redesign of an airbrush used in model painting jobs. This is shown in Fig. 3. The device requires a compressed air flow of high velocity to atomize liquid paint into fine droplets and to throw it over a surface. The

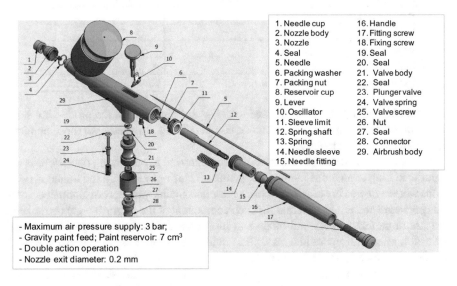

Fig. 3. Airbrush components and characteristics.

paint is supplied by gravity and different spraying effects can be achieved through the components exchange located at the airbrush exit.

Raw materials involved in the airbrush production process are stainless steel 304, CW614 N, PTFE and chromium, which is used in the surface finishing process. Material inputs and outputs associated to the production process as well as the main manufacturing operations, times required by operation and energy consumptions to manufacture one product unit are shown in Table 1. Money flows associated to the production system of one product unit are also indicated.

Table 1. Airbrush production.

Material type	Raw materials (g)	Product mass (g)
AISI 304	361	157
CW614 N	10	4.1
PTFE	0.07	0.023
Chromium	22	7.4
Total	**393.07**	**168.5**
Manufacturing operation	Time (s)	Energy (Kw·s)
Sawing	13.9	2.1
Turning	1718.7	103.3
Milling	184.7	8.8
Shearing	5	0.9
Bending	9	2.10^{-3}
Finishing	216	5.4
Assembly	610	2.3
Total	**2757.3**	**122.8**
Money flows	Production costs (€)	Revenues (€)
Raw materials	2.87	
Labour	28.7	
Energy	22.4	
Indirect costs	2.81	
Total	**56.78**	**70**

All metrics and indicators obtained to assess sustainability at the production stage are summarized in Table 3, where can be comparatively analyzed with those obtained later from the product redesigned.

It can be observed that the mass of material removed in machining operations is too high. According to previous methodology, manufacturing, assembly and finishing operations were reviewed in order to propose a more sustainable product development. Some changes in raw materials selection for each component of the airbrush were carried out. The use of calibrated bars and tubes was proposed. Thus, the waste percentage was reduced and several operations as drilling and turning processes were also avoided. Results are shown in Table 2, where the following information for both, initial design, Di, and redesign alternative, A, is shown for some components: size of raw materials, manufacturing operations that were simplified for each alternative, energy consumption and mass of material removed along the manufacturing process.

Table 2. Airbrush components. Reduction of the amount of material removed.

Airbrush component		Raw materials size (mm)	Machining operations removed (mm)	Energy (w·h)	Material rem. (g)
1. Needle Cup AISI 304	D$_i$	Ø8 × 6.2	Drilling (4)	0.39	2
	A	Ø7 × 1.5 × 6.2	Contour turning (0.75)	0.26	0.6
2. Nozzle body AISI 304	D$_i$	Ø10 × 9.8	Contour turning (0.5)	0.76	5
	A	Ø9 × 9.8		0.64	4
5. Needle AISI 304	D$_i$	Ø2 × 131.7	Facing (0.8)	0.26	2
	A1	Ø2 × 130.9		0.26	2
6. Packing washer PTFE	D$_i$	Ø4 × 2.5	Facing (0.5)	0.03	0.05
	A	Ø3 × 2.5		0.004	0.02
8. Reservoir cup AISI 304	D$_i$	Ø28 × 8	Contour turning (0.5)	3.22	30
	A	Ø27 × 8		2.94	27
9. Trigger AISI 304	D$_i$	Ø12 × 17.7	Contour turning (0.5)	5.29	15
	A	Ø11 × 17.7		4.95	13
11. Sleeve limit CW614 N	D$_i$	Ø10 × 5.7	Facing (0.3)	0.18	3
	A	Ø10 × 2.5 × 5.4	Drilling (5)	0.14	1
12. Spring shaft AISI 304	D$_i$	Ø5.5 × 42.7	Contour turning (0.5)	0.66	5
	A	Ø5 × 42.7		0.49	4
14. Needle sleeve AISI 304	D$_i$	Ø10 × 19.7	Facing (0.5)	0.5	4
	A	Ø10 × 4 × 19.2	Drilling (2)	0.42	4
15. Needle fitting AISI 304	D$_i$	Ø7 × 11.2	Facing (0.3)	0.36	2
	A	Ø7 × 10.9		0.35	2
16. Handle AISI 304	D$_i$	Ø13 × 59.7	Facing (1); Drilling (4)	6.03	44
	A	Ø12 × 4x58.7	Contour turning (0.5)	4.48	29
17. Fitting screw AISI 304	D$_i$	Ø9 × 36.7	Facing (0.8)	1.67	13
	A	Ø8 × 35.9	Contour turning (0.5)	1.25	9
21. Valve Body AISI 304	D$_i$	Ø11 × 21.7	Facing (0.8)	1.45	11
	A	Ø10 × 20.9	Contour turning (0.5)	1.1	8
23. Plunger valve CW614 N	D$_i$	Ø4 × 21.7	Facing (0.3)	0.12	1
	A	Ø4 × 21.4		0.12	1
26. Nut AISI 304	D$_i$	Ø11 × 10.7	Facing (0.5)	0.8	6
	A	Ø11 × 2 × 10.2	Drilling (7)	0.4	3
29. Body AISI 304	D$_i$	Ø13 × 82.7	Facing (0.4)	7.92	48
	A	Ø12 × 82.3	Contour turning (0.5)	6.44	36

In the case of the first component (needle cup), the operations of drilling and contour turning were eliminated by the proper selection of the raw materials size. Consequently, a significant reduction in material removed (73%) and energy consumption (18.2%) were achieved.

On the other hand, the chromed layer was proposed to be substituted by a polishing process of the stainless steel components. Product specifications were not practically modified because a high corrosion resistance was preserved.

The inventory of the production system and the subsequent sustainability assessment were carried out for the redesigned airbrush. Metrics and indicators finally obtained are summarized in Table 3. In addition to achieving a large reduction of the chromium mass and waste, a decrease of the energy consumption and costs of raw materials can be observed with respect to the initial product. The product mass and size is preserved. Consequently, production costs reduce due mainly to the use of a minor raw materials and energy consumption.

Table 3. Production metrics and indicators of the initial and redesigned products

Engineering metrics	Product mass (g)	Chromium mass (g)	Energy (Kw·s)	Waste (%)	Raw mat. costs (€)	Annual production
Initial product	168.5	7.4	122.8	57.1	2.87	26400
Redesign	168.5	0	100.4	8.2	1.45	26400
Sustainability indicators	Environmental		Economic		Social	
	GW (Kg CO_2)	E99 (pt)	VA (€)	EE (€/pt)	Working hours	Hourly wage (€/h)
Initial product	3.36	442.7	13.1	29.7	0.16	1.03
Redesign	1.84	118.7	16.8	142.2	0.16	1.03

If sustainability indicators of both initial and redesigned airbrush are compared (Fig. 4), we observed that GWP_{100} reduces 45% and EI99, which cluster different impact categories, reduces 73%. In addition, VA increases 28.2% and EE indicator greatly improves. It is assumed that sales revenues are not modified. Meanwhile, the social indicators associated to the workers group not vary.

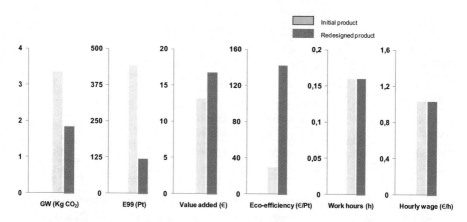

Fig. 4. Sustainability indicators comparison.

The chromed layer has a significant effect on the environmental indicators of this product, because an 82% of the total variation in global warming is produced by its removal. In addition, the vapors emission in finishing process could be avoided and the possibility of respiratory disease in workers would be reduced.

Qualitatively, the improvement of the product environmental profile can be shown through the difference between the new and initial product overall ratings in the LiDS wheel (Fig. 5). The scales on the wheel range from 0 (dead center) to 6 (outer radius). This improvement is particularly noted for the strategies involved along the initial stages of the product life cycle.

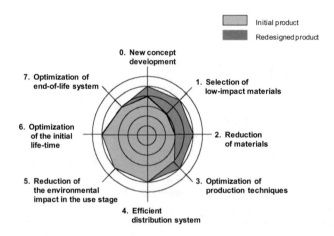

Fig. 5. Overall ratings for the initial and redesigned airbrush.

Since a lower environmental impact and an increase of the company economic profit is achieved for each manufactured product unit, a more sustainable airbrush could be developed.

4   Conclusions

This work proposes a sustainability assessment methodology aimed at design engineers to consider economic, environmental and social aspects simultaneously when developing products. The three dimensions of sustainability are quantified using a set of metrics and indicators that facilitate the communication of sustainability results during the product design decision-making process.

The redesign of an airbrush used in model painting jobs is carried out. Production inventory was obtained from the previous product development project. The fulfilment of specifications for each component of the product and the application of suitable sustainability strategies were taken into account to project an improved airbrush. As a consequence of both the optimal use of raw materials and the chromed layer removal in the finishing process, an improvement of the product sustainability indicators at the manufacturing stage was achieved.

References

1. UNCED, Agenda 21. United Nations Conference on Environment and Development, Rio de Janeiro, June 1992
2. Garner, A., Keolian, G.A.: Industrial ecology: an introduction. University of Michigan's National Pollution Prevention Center for Higher Education, Ann Arbor, MI (1995)
3. Brezet, J.C., Van Hemel, C.G.: Ecodesign: a promising approach to sustainable production and consumption. UNEP, United Nations Publications, Paris (1997)
4. Andriankaja, H., Vallet, F., Le Duigou, J., Eynard, B.: A method to ecodesign structural parts in the transport sector based on product life cycle management. J. Cleaner Prod. **94**, 165–176 (2015)
5. Kloepffer, W.: Life cycle sustainability assessment of products (with comments by Udo de Haes, H.A., p. 95). Int. J. Life Cycle Assess. **13**(2), 89–95 (2008)
6. Finkbeiner, M., Schau, E.M., Lehmann, A., Traverso, M.: Towards life cycle sustainability assessment. Sustainability **2**, 3309–3322 (2010)
7. Hunkeler, D., Rebitzer, G., Lichtenvort, K., Ciroth, A., Hunkeler, D., Huppes, G., Lichtenvort, K., Rebitzer, G., Rüdenauer, I., Steen, B.: Environmental Life Cycle Costing. SETAC Publications, Salt Lake City (2008)
8. Guidelines for Social Life Cycle Assessment of Products, UNEP/SETAC, United Nations Environment Programme, Paris (2009)
9. Azapagic, A., Perdan, S.: Indicators of sustainable development for industry: a general framework. Trans. IChemE Process. Saf. Environ. Prot. Part B **78**(4), 243–261 (2000)
10. Goedkoop, M., Spriensma, R.: The Eco-indicator 99. A damage oriented method for Life Cycle Impact Assessment (2000). (PRé Consultants B.V., Amersfoort, The Netherlands)
11. PROBAS Database. http://www.probas.umweltbundesamt.de/php/index.php
12. Kemna, R., van Elburg, M., Li, W., van Holsteijn, R.: MEEuP Methodology Report (2005)

A Living-Sphere Approach for Locally Oriented Sustainable Design

Hideki Kobayashi[✉] and Shinichi Fukushige

Department of Mechanical Engineering, Osaka University,
Yamada-oka 2-1, Suita, Osaka 565-0871, Japan
kobayashi@mech.eng.osaka-u.ac.jp

Abstract. Achieving a sustainable consumption and production pattern is one of the United Nation's sustainable development goals for 2030. To achieve this, it is necessary to consider the environmental burden from a product life cycle and the quality of life of the consumer. In this study, a systematic approach for connecting basic human needs and the product development process, called the living-sphere approach, is proposed. In this approach, value graphs, which visualize the value system of products, are connected to satisfiers fulfilling the basic needs set out by Max-Neef. A value graph links satisfiers and the traditional product development process. The significance of the proposed approach is that improving quality of daily life and traditional product development are combined in the same framework.

Keywords: Sustainable design · Environmentally conscious design (eco-design) · Local community · Human needs · Quality of life (QoL) · Sustainable consumption and production (SCP)

1 Introduction

Realizing a sustainable society is one of the most challenging problems that humanity faces. Achieving a sustainable consumption and production (SCP) pattern is a sustainable development goal that was adopted at the 2015 general meeting of the United Nations [1]. Thus, it is necessary to consider reduce environmental load caused by production and manufacturing and sufficiency or quality of life (QoL) of the consumer. It is expected that a package of policies, information services, and product development focusing on the improvement of QoL without increasing environmental load will be developed in Asian countries where high economic growth is forecast. Traditional eco-product design methods and methodologies for reducing the environmental load of a product life cycle have been developed and eco-design guidelines are being standardized [2, 3]. Although eco-products are being diffused gradually in industrialized countries, no QoL-conscious eco-product design methodology for emergent or developing countries has been established.

In this study, locally oriented sustainable design implies that site-specific conditions are reflected in the specifications or eco-design ideas of a product in the design stage. We propose a systematic approach connecting the sufficiency of the daily life of a

© Springer International Publishing AG 2017

G. Campana et al. (eds.), *Sustainable Design and Manufacturing 2017*, Smart Innovation, Systems and Technologies 68, DOI 10.1007/978-3-319-57078-5_12

consumer and the product development process, called the living-sphere approach. We describe a framework for the living-sphere approach. We focus on the part of the function or structure that is affected strongly by site-specific conditions, such as history, culture, habits, and social infrastructure. We assume that it is impossible to separate human life completely from the land that people occupy. Related studies are surveyed in Sect. 2, and the living-sphere approach is proposed in Sect. 3. After considering the significance of the work, remaining issues are discussed in Sect. 4, and concluding remarks are provided in Sect. 5.

2   Related Works

The so-called factor index describes the improvement ratio of eco-efficiency, which is the product value per unit of environmental load caused by the product life cycle [4]. Although this factor index of a product can be applied to industrial products generally, the total sufficiency of the needs or the factor index of a household cannot be quantified where various products are used in daily life. A study of the factor index calculation for a household focusing on home appliances has been reported [5]. However, the method cannot consider satisfaction and site-specific characteristics, because it estimates the factor index focusing on only the technological progress in the basic product functions. The concepts of efficiency and sufficiency are different; thus, sufficiency of needs cannot be discussed based on eco-efficiency. For instance, Cooper discussed the difference between efficiency and sufficiency with respect to extending product lifetime [6]. Eco-efficiency is improving resource and energy efficiency related to products and services, whereas sufficiency is reducing the throughput of a product or service, namely, slower consumption. Cooper claims that a solution to accomplishing eco-efficiency and sufficiency is to extend the product lifetime.

Koren suggested that the personalization and localization in the manufacturing industry after globalization is a paradigm shift [7]. Here, we discuss personalization and localization in relation to the context dependency. For example, service engineering focuses on abstracting and valuing the context of the personal user and one of the methods for this is the persona approach [8]. However, it is important to abstract product specifications depending on the site-specific context through the systematic analysis of daily life [9]. It is difficult to consider the site-specific context in full via the persona approach, because this approach is insufficient in terms of spatiotemporal considerations such as the climate, culture, and institutions. Generally, the earliest development of a consumer product is a marketing process. In the traditional marketing process, the 3 Cs, namely, the customer, the competitors, and the company, are analyzed initially [10]. However, customer analysis focuses on increasing the profit of a product or service provider; thus, it does not necessarily focus on satisfying the customers' needs. In fact, it is important to observe the sufficiency of the consumers' daily needs closely. Recently, ethnography has been used as a method of field observation for product and service marketing [11]. There is no proper procedural methodology for ethnography, and the user context is understood in detail from examining daily life. The ethnographic approach is strongly affected by the subjectivity of the observer and its reproducibility is low. For business model creation, the value proposition canvas has been proposed [12], which consists of value proposition and customer segment areas. In the value proposition area, the concepts of gain creator and pain reliever are described based on the product to be supplied. However, gains and pains are derived based on customer jobs in the customer segment area. A product is selected if it fits with the customer's gains or pains. Because the value proposition canvas is for earning sales from segmented customers, it is not sufficient to achieve SCP for all people in the target region.

There are various study fields related to QoL studies such as human needs, welfare and well-being. Various concepts of human needs have been proposed. For instance, 30 kinds of human needs have been defined in the literature [13]. Although usually people believe that human needs change with cultural progress or period, Max-Neef claimed that basic human needs are universal, but satisfiers fulfilling the basic needs depend on region, culture, and period [14]. Here, basic human needs are the opposite of high-order needs, and they are synonymous with absolute or universal needs [13]. In welfare economics, the capability approach was proposed by Sen [15]. The capability approach focuses on freedom and opportunity of choice for measuring latent ability, and it is used for the human development index (HDI) established by the United Nations Development Programme [16]. The HDI shows an integrated value consisting of the national average lifetime, literacy rate, school attendance rate, and GDP per capita. The approach for measuring subjective well-being developed by the Organisation for Economic Co-operation and Development is a typical well-being approach [17, 18]. This approach uses a questionnaire to find out whether people are satisfied with their daily life, and it is recognized as a reliable method, although there is a lack of data for emergent or developing countries. According to the traditional perspective of the linear well-being model, a consumer accomplishes well-being based on satisfaction achieved by consuming products and services [19] (Fig. 1). Although well-being and related products are connected qualitatively in this model, there is no systematic study on the relationship between well-being and the product development process.

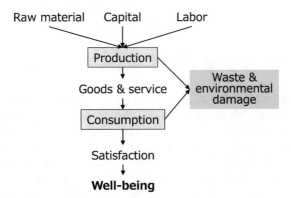

Fig. 1. Schematic of the conventional linear model.

Product design support approaches for local markets, such as developing countries, include localization of an original global standardized product, sometimes called "glocalization", and co-design, which embeds local people into the product development process [20]. In glocalization, it is difficult to modify a product for a developed country to suit a developing country just by simplifying the function and structure of the product. Moreover, the design reproducibility of both approaches is low. To accomplish high reproducibility and market-in product development, we have proposed an extended function-structure map (EFSM), in which site-specific information is mapped on a traditional function-structure map, and a visualization system for the EFSM [20, 21].

3   Framework of the Living-Sphere Approach

Figure 2 shows the framework of the living sphere approach, and we explain the concepts used in the living sphere approach.

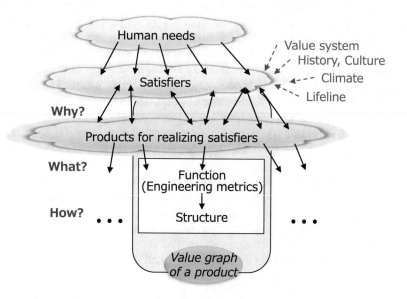

Fig. 2. Schematic of the living-sphere approach framework.

In the uppermost layer, human needs and satisfiers are set. Because this study is for SCP, the human needs we selected are basic needs, not egoistic infinite needs. Two categories of basic human needs proposed by Max-Neef are used [14]. As the axiological category, the following nine needs are introduced: Subsistence, Protection, Affection, Understanding, Participation, Idleness, Creation, Identity, and Freedom. False needs, which are not related to daily life, are beyond the scope of this work. As the existential category, the following four needs categories are introduced: Being, Having, Doing, and Interacting. On a matrix of the axiological and existential categories, satisfiers are set, which contribute to the actualization of human needs [14]. Satisfiers are not the available economic goods themselves and are selected depending on personal factors, such as religion and hobbies, and on site-specific factors, such as regional history, culture, climate, social systems, and institutions. We focus on the site-specific factors.

When we set the target product, a traditional product development methodology can be applied that focuses on required functions (What) and deploys them in the product structure (How) using a function-structure map and quality function deployment [22]. Here, in the early phase of product development, a value graph is used to confirm the reason for the existence (Why) and true value of the target product [23]. If the starting point of the thinking process is on the basic human needs side, a product is regarded as a method of realizing satisfiers. Thus, the value graph of the target product satisfying basic needs should be connected to appropriate satisfiers.

In our framework, we assume that a design engineer can develop products satisfying basic human needs and improving QoL by confirming a connection between satisfiers and value graphs. In this way, the sufficiency of human needs and product development are connected systematically. Next, we describe the procedure for connecting related concepts during the product development process based on the proposed approach.

1. The target field is investigated by a literature survey and field observations focusing on well-diffused products.
2. Based on the results, satisfiers realizing basic needs in the target area are abstracted and are placed in a nine-by-four categorized needs matrix.
3. Value graphs of well-diffused products are developed, and the value graphs and satisfiers are connected.
   (a) If the value graph of a product connects to one or more satisfiers, then the site-specific information related to average user behavior during product usage and product wear-out, and waste treatment and environmental load throughout the whole product life cycle is visualized on the EFSM to help improve the design [20].
   (b) If a value graph of a product does not connect to any satisfiers, then they are categorized as products that are not important for satisfying basic human needs. The priority for the product should be low in the proposed approach.
   (c) If a satisfier does not connect to any value graphs, then it should be confirmed whether it is realized through any tangible or intangible goods other than the product.
   (d) If goods do not realize a satisfier, then a new concept, including a new product, service, or action, should be considered for satisfying the satisfier.

As a simple example, the relationship between satisfiers and value graphs for refrigerators and rice cookers is shown in Fig. 3. The satisfiers abstracted by field studies are set in a matrix divided by needs categories. Black arrows represent the connections common to developing and developed countries, blue arrows represent connections for developed countries, and red arrows represent connections for developing countries. For example, the basic purpose of refrigerators is food storage; however, they are also used for drug and cosmetic storage in developed countries. These purposes are connected to "Physical health" or "Sensuality" as a satisfier for developed countries only. However, rice cookers in developing countries are used for cooking vegetables as well as rice. Moreover, for developed countries, refrigerators and rice cooker are connected to eating tasty food. To support eating tasty food, specific functions and parts of the products are considered. These connections are not emphasized in developing countries. In this example, two value graphs are connected to some satisfiers, allowing a design engineer to understand the relationships among the needs satisfaction, purpose, function, and structure of the two products.

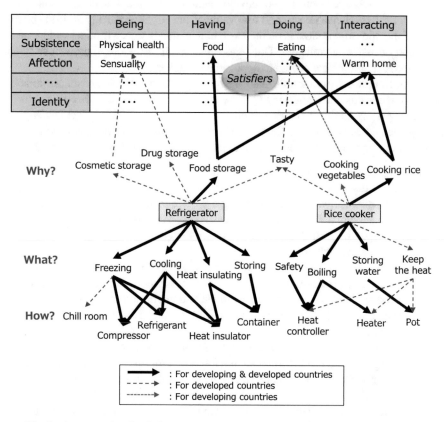

Fig. 3. An example of satisfiers and value graphs for rice cookers and refrigerators.

4   Discussion

Figure 3 shows two overlapping diagrams for different regions. A connected diagram of satisfiers and value graphs helps to understand the difference in product roles between regions. To improve understanding, each diagram should be drawn separately at first. In this case, there is a dependent relation among satisfiers. In Fig. 3, the "Physical health" satisfier depends on "Eating", and "Eating" depends on "Food" in a strict sense. A method that can treat such a case will be considered in future work.

The methods in this study and the value proposition canvas [12] are similar in terms of fitting the elements deployed from the user need side and the product side. Nevertheless, these methods are clearly different from each other. As we mention in Sect. 3, the value proposition canvas is a marketing tool to increase sales; thus, it tends to focus on "attractive qualities" [24]. However, SCP should be achieved for all people in the target regions. This study focuses on "must-be qualities" [24], which have not been understood sufficiently, especially by design engineers in other regions. Thus, universal basic human needs are the starting points in this study.

The proposed approach can be used to improve product design. For point 3(d) in Sect. 3, whether a new product or service satisfies the satisfier is examined. However, methods for fulfilling a satisfier are not limited to artificial and commercial goods; they depend on lifestyle, and thus on local or site-specific conditions. These methods require further study. There are also other important problems to overcome. For instance, balancing future environmental load reduction and improvement of current QoL is essential for sustainable design. The proposed framework focuses on social sustainability, and an eco-design approach should be added to it. Further, appropriate satisfiers must be set in the proposed framework; however, the selection process is not supported. Selection depends on the subjectivity of the observer and the engineer; thus, satisfier selection support is a difficult and important issue in future work.

5 Concluding Remarks

In this paper, the living-sphere approach and its framework were proposed as a method for locally oriented sustainable design. A simple example of using the proposed framework was presented. The main achievement is that the improvement of QoL and the traditional product development process can be connected in the proposed framework. In this study, we focus on fulfilling basic needs and SCP. By recognizing the differences in satisfiers between regions, countries, or local communities, consumption and production are considered in the context of daily life. If we propose a product-service system (PSS) for a product, then a satisfier can direct an SCP-oriented PSS. In contrast, environmental sustainability cannot be considered explicitly in this framework because the proposed framework aims for social sustainability. Both types of sustainability should be integrated in future work. In addition, case studies will be performed for different countries to evaluate the usefulness of this approach.

Acknowledgements. This research was supported by JSPS KAKENHI Grant Number 15H06347. Also, this research was supported by the Environment Research and Technology Development Fund (S-16) of the Ministry of the Environment, Japan.

References

1. UNEP: Global outlook on sustainable consumption and production policies (2012)
2. UNEP: Ecodesign - a promising approach to sustainable production and consumption (1997)
3. ISO, ISO/TR 14062: Environmental management - Integrating environmental aspects into product design and development (2002)
4. Kobayashi, Y., Kobayashi, H., Hongu, A., Sanehira, K.: A practical method for quantifying eco-efficiency using eco-design support tools. J. Ind. Ecol. **9**(4), 131–144 (2005)

5. Aoe, T., Yamamoto, R., Ikaga, T., Kondo, Y., Matsuoka, Y., Fukuda, M.: Factor X (eco-efficiency) assessment on global warming a household in Japan. J. Jpn. Inst. Energy **89**, 1070–1087 (2010). in Japanese

6. Cooper, T.: Slower consumption. J. Ind. Ecol. **9**(1–2), 51–67 (2005)

7. Koren, Y.: The Global Manufacturing Revolution. Wiley, New York (2010)

8. Kimita, K., Shimomura, Y., Arai, T.: Evaluation of customer satisfaction for PSS design. J. Manuf. Technol. Manag. **20**(5), 654–673 (2009)

9. Spencer, J., Lilley, D., Porter, S.: The opportunities that different cultural contexts create for sustainable design: a laundry care example. J. Clean. Prod. **107**, 279–290 (2015)

10. Ohmae, K.: The Mind of Strategist: The Art of Japanese Business. McGraw-Hill, New York (1991)

11. Oda, H.: Doing Ethnography (2010). Syun-Jyuu-sya, in Japanese

12. Osterwolder, A., Pigneur, Y., Smith, A., Bernarda, G., Papadakos, T.: Value Proposition Design. Wiley, Hoboken (2014)

13. Dean, H.: Understanding Human Need. The Policy Press, Bristol (2010)

14. Max-Neef, M.: Human Scale Development. The Apex Press, New York (1991)

15. Sen, A.: Capability and well-being. In: Nussbaum, M., Sen, A. (eds.) The Quality of Life, pp. 30–52. Oxford University Press, Oxford (1993)

16. UNDP: Human development report (2015)

17. Stiglitz, J.E., Sen, A., Fitoussi, J.: Mismeasuring Our Lives: Why GDP Doesn't Add Up. The New Press, New York (2010)

18. OECD: OECD guidelines on measuring subjective well-being (2013)

19. Jackson, T.: Motivating Sustainable Consumption, p. 10. Sustainable Development Research Network, London (2005)

20. Kobayashi, H.: Perspectives on sustainable product design methodology focused on local communities. In: Matsumoto, M., Masui, K., Fukushige, S., Kondoh, S. (eds.) Sustainability Through Innovation in Product Life Cycle Design, pp. 79–92. Springer, Singapore (2016)

21. Sugita, Y., Fukushige, S., Kobayashi, H.: A visualization system of design information for locally-oriented sustainable product. In: Proceedings of the 24th CIRP Conference on Life Cycle Engineering, Kamakura (2017). (Proceedings CD-ROM)

22. Closing, D.: Total Quality Development. ASME, New York (1993)

23. Ishii, K.: Textbook of ME217 Design for Manufacture: Product Definition, Stanford University (1998)

24. Kano, N., Seraku, N., Takahashi, F., Tsuji, S.: Attractive quality and must-be quality. J. Jpn. Soc. Qual. Control **14**(2), 39–48 (1984). in Japanese

What Stops Designers from Designing Sustainable Packaging?—A Review of Eco-design Tools with Regard to Packaging Design

Xuezi Ma and James Moultrie[(⊠)]

Department of Engineering, Institute for Manufacturing,
University of Cambridge, Alan Reece Building,
17 Charles Babbage Road, Cambridge CB3 0FS, UK
jm329@eng.cam.ac.uk

Abstract. Packaging has caused much waste and its sustainability has received much attention in the past decades. Designers have made efforts to mitigate environmental impacts of packaging. However, many packaging designs are still far from achieving their sustainability goals. The purpose of this study is to perform a literature review of the principal design methods and tools for sustainable packaging published over the last twenty years. The objective is to understand the main obstacles that limit their effective implementation in the packaging design process. This study develops a sustainable packaging design and development model and proposes criteria for accessing packaging tools and methods. This study has found that to achieve sustainable design, many tools have limitations in demonstrating usage and balancing trade-off situations. Most of the tools focus on defining problems rather than suggesting possible solutions.

Keywords: Packaging · Eco-design · Tools · Sustainable production

1 Introduction

Packaging is not deemed useful after it has fulfilled its common purpose which is to protect and promote its product. As a result, packaging is considered a burden for the environment and a disrupting waste to the consumers (Verghese 2005). Therefore in recent decades, packaging sustainability has received a lot of attention due to its major impact on the environment (Beitzen-heineke 2015; Byggeth and Hochschorner 2006). Consumers have increasingly taken products' sustainable performance into consideration while purchasing (Magnier and Schoormans 2015; Hoogland et al. 2007). Due to increasing environmental consciousness, governments have launched standards and regulations to regulate green packaging.

Research into the environmental and economic impacts of packaging sustainability has been stimulated by regulations and market pressure. This research has produced a number of packaging sustainability guidelines, theories, strategies and tools. These have been made available to various stakeholders, including designers, engineers,

© Springer International Publishing AG 2017
G. Campana et al. (eds.), *Sustainable Design and Manufacturing 2017*, Smart Innovation, Systems and Technologies 68, DOI 10.1007/978-3-319-57078-5_13

technologists, marketers and environmental managers in the production, transportation and distribution areas of packaging production. However, it is argued that this proliferation of sustainability assessment methods and regulations has created confusion for packaging designers and other stakeholders (Slavin and Coordinator 2016; Navarro et al. 2005). The available methods and tools may be inaccessible for designers to use in the design process. For example, designers may not know which tool applies better for different design stages, or they may get confused about what to expect from results of the tools and where to apply them. Furthermore, although organizations such as The Sustainable Packaging Alliance (SPA) in Australia, the Sustainable Packaging Coalition (SPC) in the US and The Industry Council for Research on Packaging and The Environment (INCPEN) in the UK have worked with brand owners, packaging companies and retailers to promote responsible packaging production, these research activities have established issues in packaging design, they have not managed to integrate packaging sustainability into packaging design and manufacturing decisions.

Table 1. Literature reviews.

Subject	Author/Year	Description
Product design and development	Bovea and Pérez-Belis (2012)	Reviewed and classified the eco-design tools according to environmental assessment methods, multi-criteria product development approaches, life cycle perspectives, qualitative or quantitative tools, the stages of the conceptual design process where the tool can be applied and the related methodology.
	Casamayor and Su (2013)	Analyzed prescriptive and analytical eco-design tools in product design and development process and used them to facilitate the development of a prototype of a LED.
	Vallet et al. (2013)	Compared three eco-design tools by redesigning a disposable razor.
Eco-design function	Navarro et al. (2005)	Presented 65 eco-design tools according to their functional criteria (i.e. design stage, life cycle stage and problem level).
	Svanes et al. (2010)	Compared four methodologies that are frequently used for packaging sustainability design by listing their characterization into resources indicators, economy, social elements, whole life cycle considered, product loss considered, product protection, user friendliness and market acceptance.
Trade-off situations	Byggeth and Hochschorner (2006)	Compared 15 eco-design tools, especially their contributions in trade off situations
End-of-life strategies	Rose (2000)	Examined the existing end-of-life treatments, and presented an end-of-life model.
Customer behavior	Niedderer et al. (2014)	Reviewed the key theories, models and approaches for behavior change in the sustainable design field.

To access the usability of the eco-design tools, various literature reviews have been carried out. These are summarized in Table 1. The listed literature reviews are mainly focused on summarizing the generic methods in sustainable product design. Two of these reviews have looked explicitly at packaging as a critical issue, but none of them have discussed the usability of tools in packaging design. The packaging design process is a process balancing requirements such as protecting the content, promoting the product and fulfilling the transportation needs. Packaging design tools should help designers make choices between these requirements. Therefore tradeoff, as an important process during packaging design and development, should be specifically reviewed.

To address the above mentioned issues, this study will review the existing sustainable packaging design tools and assess them from designers' perspectives, so that their usability can be evaluated in a structured manner to facilitate further packaging design research and activities.

This paper is structured as follows. Firstly, existing sustainable packaging design tools are reviewed and clustered according to the packaging design and development phases to which they can be applied. Next, results for these tools and the expected use for these results are discussed. This is followed by a discussion of whether these tools deal with trade off situations in the packaging design process. Lastly, the tools are analysed by their utility in facilitating packaging design. The paper concludes with opportunities for further research in this area.

2 Method

The review is covered from the perspective of three disciplines: packaging engineering, design and policy. Search terms differed slightly for the three perspectives owing to the different use of words among engineers, designers and policy researchers. For example, the term 'sustainable' gives a relatively large number of hits when searching a policy database, whereas the term 'environmental' is more effective when searching an engineering database (Baumann et al. 2002).

Sustainable packaging design literature was analyzed in the three steps described below:

- Search of six main databases (i.e. ScienceDirect, Google Scholar, Springer, ResearchGate, Wiley, Europe PMC) and screening the main references covering a time span of 20 years by scanning the abstract. Checking of references in the most important publications for additional references.
- Overview of the existing techniques for evaluating the environmental requirements of packaging in the references from the first step. Each method is briefly summarized.
- Review of tools that have been developed for improving packaging sustainability. Each tool is classified according to its methodology, measure and measurement.

After analysing the literature, it is clear that there are few tools and methods focused on sustainable packaging design and development in the public domain.

To analyse these tools from a designer's perspective, we selected tools and methods that are publicly accessible. Twelve tools and methods were found in the relevant packaging design and development field (Table 2). In the next section, the tools will be analysed according to their implementation phases in packaging design and development, their results and their usability of dealing with trade-off situations.

Table 2. Analyzed tools for sustainable packaging design and development.

Tools	Author/Year	Description
Design guidelines for sustainable packaging	Sustainable Packaging Coalition (2006)	Provides a framework for sustainable packaging design while outlining various design strategies and reference materials
Australian Sustainable Packaging Guideline	Australian Packaging Covenant (2010)	Assists Covenant signatories and others to review and optimize consumer packaging to make efficient use of resources and reduce environmental impac.
Sustainable Packaging Framework	Sustainable Packaging Alliance (2010)	Defines sustainable packaging in four dimensions: effective, efficient, cyclic and safe. Gives the strategies for packaging design, manufacture, logistic and marketing in each dimension
Envirowise: Packaging design for the environment	Envirowise and INCPEN (2008)	Details the considerations needed to develop pack systems that optimize use of materials, energy and water and minimize waste, and looks at the trade-offs between different goals with the minimum environmental impact
Sustainable packaging indicators and metrics framework	Sustainable Packaging Coalition (2009)	Develops a common set of indicators and metrics for companies to use to measure progress toward the vision of sustainability articulated in the SPC Definition of Sustainable Packaging
Maturity grid	James Moultrie, Sutcliffe Laura and Anja Maier (2016)	Develops a metric and grids to access the sustainability of packaging design
A guide to packaging material flows and terminology	GreenBlue (2009)	Provides a framework for communication along the supply chain
Packaging Impact Quick Evaluation Tool (PIQET)	Verghese et al. (2010)	An on-line tool allows for assessments of incoming raw materials packaging systems and outgoing product packaging systems of an organization
Eco-Costs/Value Ratio Model (EVR)	Renee Wever and Joost Vogtländer (2013)	Deals with the environmental assessment of packaging design alternatives with functionalities such as environmental burden and marketing

(continued)

Table 2. (*continued*)

Tools	Author/Year	Description
Packaging scorecard	Carl Olsmats and Chris Dominic (2003)	Evaluates different criteria from supplier, transportation, retailer and consumer's perspectives
COMPASS	SPC (2009)	Assess packages on resource consumption, emissions and packaging attributes
SimaPro	Pre Sustainability	Uses metrics to collect, analyze and monitor the sustainability performance of products
GaBi	Thinkstep	Evaluates product scenario by calculating the environmental impacts

3 Analysis of Available Tools in the Packaging System

Companies typically share the following steps in the product design and development phases: understanding the issue, exploring the possible solutions, defining and refining the solution, implementing the idea, manufacturing, distribution and sales (Waage 2007; Poulikidou et al. 2014; Biju et al. 2015; Joore and Brezet 2015). Compared to the normal product design and development phases, packaging design and development is proved to change due to its characteristics. Based on several packing development methodologies (Gordon 1994; Griffin Jr. 1985; Brody 1999; Paine 1990; Buccia and Forcellini 2007; Boylston 2009), the research in this paper proposes a Sustainable Packaging Design and Development Model. In general, it presents the packaging development process in three main phases: pre-development, development and post development. These three phases are broken into the following sub-processes: initial research, concept design, detail design, testing, packaging launch and packaging review. Note that the design stage begins after a period of comprehensive research. Along with the packaging development process, the applied stage of available sustainable packaging design tools for sustainable purpose is classified as follows:

1. Pre-development

   - Initial research. Identifying the objectives of the product-packaging system that aligns with the company's policy and governments' regulations. Collecting information from internals (Production, Quality, Logistics, Retail environment, Marketing etc.) and externals (Competitive companies, Raw material and Packaging equipment suppliers).

2. Development

   - Concept design. Proposing and selecting the most feasible ideas, economically, technically and environmentally (Design for X).
   - Detail design. All levels of packaging are detailed. In this phase most of the activities run for packaging as well as product.

Table 3. Overview of sustainable packaging design tool.

	Pre-development	Development	Post-development
Qualitative research	Design guidelines for sustainable packaging Australian sustainable packaging guideline sustainable packaging framework Envirowise: packaging design for the environment Sustainable packaging indicators and metrics framework Maturity grid	A Guide to Packaging Material Flows and Terminology	
Quantitative research	Packaging Impact Quick Evaluation Tool (PIQET) COMPASS/SimaPro/GaBi Eco-Costs/Value Ratio Model (EVR)		Packaging Scorecard Life cycle Inventory

- Testing. Test packaging function as well as the market acceptance.
- Packaging launch. Including the planning, production and packing of the product and delivery to the sales point, and promoting packaging's environmental impacts.

3. Post-development

- Packaging review. Keeping a record of the environmental performance of packaging, including its energy and water consumption, waste indicators, consumers' satisfaction and recycling process.

According to the packaging development process described above, the identified 12 methods and tools were clustered according to the three main applied stages described above for packaging design and development, with the corresponding qualitative and quantitative results (Table 3). It should be noted that only aids in sustainable packaging design have been included, although several tools exist to evaluate "sustainability". It is interesting that no quantitative tools have been found in the development phase. No qualitative tools have been found in the post-development phase. The following sections will look at how each tool could achieve its aim by analysing its functions along the packaging design and development process.

3.1 Pre-development

Most of the listed tools can be applied at the pre-development stage. They include guidelines that are mainly developed by NGOs and governmental institutions such as SPC, SPA, Wrap and INCEPN. Guidelines like "Design guidelines for sustainable packaging" aim to help designers to understand the life cycle of sustainable packaging. During the life cycle of the packaging, these guidelines remind designers of the imperative design questions related to sustainable design during each packaging phase and also list the regulations and resources to refer to during the design of the corresponding stage. Similarly, "The Australian Sustainable Packaging Guideline" helps designers to reflect on their design by raising questions when they review the whole supply chain of designed packaging. Also, the Sustainable Packaging Framework defines sustainable packaging in four dimensions: effective, efficient, cyclic and safe. It gives generic strategies for packaging design, manufacture, logistics and marketing for each dimension, as well as the key performance indicators for these strategies.

Metrics and maturity grids are specified for sustainable packaging design. The Sustainable Packaging Indicators and Metrics Framework introduces indicators and metrics to help stakeholders to measure the sustainability of packaging (resource usage and waste produced) during the supply chain before production. It declares the different terms by giving definitions and explaining how to measure these items as well as why to measure them. (Moultrie et al. 2016) develops a maturity grid of sustainable packaging through interviews with practitioners from companies of different sizes. It selects different criteria to judge the sustainable performance of a package in different stages of a packaging's life cycle.

Based on Streamlined life cycle assessment, the Packaging Impact Quick Evaluation Tool (PIQET) has been developed to assess the sustainability of packaging by calculating the product/packaging ratio, environmental impact indicator, as well as analysing the inventory in each life cycle stage of packaging. Software such as COMPASS, SimaPro and GaBi make information accessible to non-LCA professionals to manage data and compare design concepts. Based on the LCA theory, (Wever and Vogtländer 2013) developed the Eco-Costs/Value Ratio Model (EVR Model), which can be used to compare the eco-burden of a packaging with the value created.

3.2 Development

A Guide to Packaging Material Flows and Terminology creates a close loop material system for nine major packaging materials. These unified terms are used across stakeholders.

3.3 Post-development

The packaging scorecard method is used to evaluate different criteria like handleability, flow information, product protection, volume and weight efficiency from supplier, transportation, retailer and consumer perspectives. It scores each criterion so that companies can work towards their own improvement.

3.4   Looking at the Results for Sustainable Design Tools

In qualitative tools, the results may simply be some yes or no answers of some sustainable design questions along the supply chain. The outcomes of qualitative tools, however, may not be sufficient because they could evaluate very different products with similar results. For example, in design guidelines for sustainable packaging, it is hard to make choices between two design concepts if they both satisfy the same conditions such as eliminating all necessary packaging components, optimizing a package's dimensions to best fit the product and considering the effect of using recycled material on a package's technical performance.

For quantitative tools, LCA based software such as PIQET, COMPASS, SimaPro and GaBi provides detailed life cycle environmental impacts of packaging life cycle phases such as the accurate amount of carbon dioxide. The Eco-Costs/Value Ratio Model defines the sustainability of packaging by calculating its eco-costs/value and comparing the relative location in the eco-costs and value diagram. The Packaging scorecard defines the problems in the packaging supply chain by providing scores for each phase. The Life Cycle Inventory qualifies the material use, energy use, environmental discharges and wastes associated with packaging life cycle phases, from raw material extraction to material processing, packaging fabrication, use, reuse or recycling, and ultimate disposal.

3.5   Looking at How Trade Offs Are Incorporated

Current research indicates that consumers' attention on environmentally friendly packaging have steadily increased during the past decades (Nordin and Selke 2010; Martinho et al. 2015; Magnier and Schoormans 2015; Lofthouse et al. 2009). The pressure of balancing trade-offs also comes from regulations. Nearly 200 European Union directives have been released to regulate the sustainability of packaging (Giancristofaro and Bordignon 2016). Packaging in the whole system (primary, secondary and tertiary packaging) needs to be minimized without compromising its function of safety, protection and promotion. However, minimizing the packaging usage does not mean minimizing the environmental impacts. For example, if manufacturers reduce the packaging materials in primary packaging to save the raw materials and decrease the transport weight, the content may be damaged during the transportation, which causes a lager waste to the environment. To balance this situation, a valuation (e.g. rating of the importance of criteria or strategies within each tool) has to be included in the tool (Byggeth and Hochschorner 2006). Based on this criterion, tools were classified in Table 4.

Guidelines such as Design Guidelines, Australian Sustainable Packaging Guidelines, Sustainable Packaging Framework and Envirowise offer strategies to facilitate designers' balance of possible consequences and sensible choices. LCA based tools such as PIQET, COMPASS, SimaPro and GaBi rate the importance of each environmental impact that the design concept may have by quantifying the facts (i.e. carbon dioxide emission, water waste, land waste, recyclate etc.). Tools with valuations are

Table 4. Sustainable packaging tools classified according to whether they contain valuation.

	Pre-development	Development	Post-development
Valuation in the tools	Design guidelines for sustainable packaging Australian sustainable packaging guideline Sustainable packaging framework Envirowise: packaging design for the environment Packaging Impact Quick Evaluation Tool (PIQET) COMPASS/SimaPro/GaBi		Packaging Scorecard Life cycle Inventory
No valuation in the tools	Eco-Costs/Value Ratio Model (EVR) Sustainable packaging indicators and metrics framework Maturity grid	A guide to packaging material flows and terminology	

feasible for sustainable packaging design. Based on these analysis results, tools with valuation are picked and classified by criteria in the following sub-section.

3.5.1 Looking at How Trade Offs Are Incorporated

To make the tools accessible and feasible for packaging designers to use in trade-off situations as well as to assist in picking a suitable design concept, they have to satisfy certain criteria. For example, in order to choose the suitable packaging concept tools designers need a list of important requirements for sustainable packaging solutions or the results of tools have to be meaningful with regard to developing sustainable packaging solutions. Also, some of the tools have other purpose, but we will not regard it as satisfying certain criteria if that is not its main purpose. Through reviewing the literature, criteria for aiding design sustainable packaging have been proposed as follows:

1. Whether the tool gives specific directions or generic guidance?

   - Design decisions vary due to different types of results for different tools. Specific directions give designers specific strategies to achieve certain sustainable goals, whereas generic guidance facilitates designers to identify crucial design issues.

2. Whether the tool takes the total packaging/product system into consideration?

   - The tool should take a holistic overview of the packaging/product system from the sustainable perspective. It should include indicators such as the mass of material that has been used, whether it has been recycled or not, the energy use along the supply chain, product waste and the degree filling for primary, secondary and tertiary packaging.

3. Whether the tool provides design alternatives?

   - Alternatives should be provided to demonstrate the "right" direction. Compared to abstract indicators, concrete examples are an easier reference to designers.

Table 5. Sustainable packaging tools analyzed according to whether tools fulfill the criteria.

Tools	1	2	3	4	5	6	7	8
Design guidelines for sustainable packaging	Generic	✓		✓		✓		
Australian sustainable packaging guideline	Generic	✓		✓		✓		
Sustainable packaging framework	Generic	✓		✓				
Envirowise: packaging design for the environment	Generic	✓		✓		✓		
Packaging Impact Quick Evaluation Tool (PIQET)	N.m.	✓				✓		
Packaging scorecard	Specific	✓						
COMPASS	N.m.	✓					✓	✓
SimaPro	N.m.	✓					✓	✓
GaBi	N.m.	✓					✓	✓

4. Whether the tool includes examples to illustrate its guidance?

   – Guidance should give examples to show how to fulfil the requirements in the instructed way.

5. Whether the tool demonstrates hierarchy for sustainable decisions in different aspects?

   – Packaging design involves complicated trade-off situations. Designers have to balance requirements from companies, manufacturers and consumers as well as raw material extraction, distribution, marketing and recycling. It is impossible for designers to satisfy all requests. Tools should help designers to prioritise requirements in different situations.

6. Whether the tool considers the preservation of product quality?

   – The original aim for packaging is to protect its content from physical and chemical damages such as stacking pressure, moist, oxidization and toxicity. This helps designers avoid potential harm to the product.

7. Whether the tool calculates the distribution cost?

   – Distribution costs include packaging material cost, packaging process cost, transportation cost, handling cost and product loss.

8. Whether the tool requires designers to have pre-knowledge or experience to use the tool?

   – Tools should be accessible for designers to an effectively and efficiently check their design innovations. Requesting pre-knowledge may cause barriers to designers with limited sustainable knowledge and training.

For the purpose of this reason, tools were screened based on the above criteria and the results are shown in Table 5. It is clear that none of the tools demonstrates hierarchy for sustainable decisions. This may cause confusion for designers when making design decisions. After defining the problems and issues, it would be difficult for designers to balance the environmental issues and decide which "less worse" decision to make. Also, none of the tools calculate the costs, which should be prioritized during the decision make process.

4 Conclusion

The design of packaging is not a priority in many companies. The complexity of packaging with regard to its varied life cycle and a relatively small market size limit the professionalism within this field. However, due to the growing environmental issues that packaging has caused, the sustainability of packaging has been pushed up the agenda. It seems that knowledge of the subject of packaging design is still fragmented and there is a need to tackle packaging design problems more structurally. The thorough analysis of literature on sustainable design tools and methods in the sustainable packaging design field confirms that despite the great number of approaches proposed by researchers in this field, designers still have difficulties in their practical and effective implementation and use. Tools are mainly focused on defining problems rather than giving solutions. For example, tools like PIQET and Design guidelines for sustainable packaging are keen on reminding designers of the environmental impacts by qualifying the waste or encouraging reflection on potential issues. Concrete suggestions and possible ways to refine the design, however, are missing. Together, the review results amount to saying that researchers in the packaging design and development field need to research existing tools in more depth to make them more usable as well as develop new tools to better address designers' real needs.

This paper reviewed existing eco-design tools in the packaging domain from a designer's perspective. Future work requires discussion with designers about the practical use of these tools as well as testing of the packaging framework and proposed criteria in a real life context to better understand how sustainable packaging design tools can best be implemented. It would be beneficial to collect designers' real needs for improving packaging sustainability through case studies.

Related to this, it is evident that in different firms there are complex trade-offs to be made between different elements through packaging design. Due to the complicated nature of packaging, how firms handle the trade-offs during sustainable packaging design might provide fruitful opportunities for research.

Finally, assessing the current eco-design methods is only part of the story. To be effective in the long term, changes to design processes and practices need to be more formally institutionalised. There is thus work to be done in better understanding how such changes can be implemented and good practices anchored as part of a company's design activity.

References

Australian Packaging Covenant: Australian sustainable packaging guideline (2010). http://www.packagingcovenant.org.au/data/Resources/Sustainable_Packaging_Guidelines.pdf. Accessed 3rd Mar 2017

Baumann, H., Boons, F., Bragd, A.: Mapping the green product development field: engineering, policy and business perspectives, vol. 10, pp. 409–425 (2002)

Beitzen-heineke, E.F.: The prospects of zero-packaging grocery stores to improve the social and environmental impacts of the food supply chain. J. Cleaner Prod. **140**(3), 1528–1541 (2015). Elsevier Ltd., September 2014–2015

Biju, P.L., Shalij, P.R., Prabhushankar, G.V.: Evaluation of customer requirements and sustainability requirements through the application of fuzzy analytic hierarchy process. J. Cleaner Prod. **108**, 808–817 (2015). Elsevier Ltd.

Bovea, M., Pérez-Belis, V.: A taxonomy of ecodesign tools for integrating environmental requirements into the product design process. J. Clean. Prod. **20**(1), 61–71 (2012)

Boylston, S.: Designing Sustainable Packaging. Laurence King Publishing, London (2009)

Brody, A.L.: Development of packaging for food products. In: Brody, A.L., Lord, J.B. (eds.) Developing New Food Products for a Changing Marketplace, pp. 313–351. CRC Press, Boca Raton (1999)

Buccia, D.Z., Forcellini, F.A.: Sustainable packaging design model. In: Loureiro, G., Curran, R. (eds.) Complex Systems Concurrent Engineering, pp. 363–370. Springer, London (2007)

Byggeth, S., Hochschorner, E.: Handling trade-offs in ecodesign tools for sustainable product development and procurement. J. Clean. Prod. **14**(15–16), 1420–1430 (2006)

Casamayor, J.L., Su, D.: Integration of eco-design tools into the development of eco-lighting products. J. Clean. Prod. **47**, 32–42 (2013)

Envirowise and INCPEN: Envirowise: packaging design for the environment (2008). http://www.packagingfedn.co.uk/images/reports/Incpen&Envirowise%20Guide%20to%20Packaging%20Eco%20Design.pdf. Accessed 3rd Mar 2017

Gordon, W.F.: New Food Product Development: From Concept to Marketplace. CRC Press, Boca Raton (1994)

Giancristofaro, R.A., Bordignon, P.: Consumer preferences in food packaging: CUB models and conjoint analysis. Br. Food J. **118**(3), 527–540 (2016)

GreenBlue: A guide to packaging material flows and terminology (2009). http://www.sustainablepackaging.org/Uploads/Resources/guide-to-packaging-materials.pdf. Accessed 3rd Mar 2017

Griffin Jr., R.: Materials and package testing. In: Principles of Package Development, pp. 130–167. AVI Publishing Company Inc., Connecticut (1985)

Hoogland, C.T., de Boer, J., Boersema, J.J.: Food and sustainability: do consumers recognize, understand and value on-package information on production standards? Appetite **49**(1), 47–57 (2007)

Joore, P., Brezet, H.: A multilevel design model: the mutual relationship between product-service system development and societal change processes. J. Clean. Prod. **97**, 92–105 (2015)

Lofthouse, V.A., Bhamra, T.A., Trimingham, R.L.: Investigating customer perceptions of refillable packaging and assessing business drivers and barriers to their use. Packag. Technol. Sci. **22**(6), 335–348 (2009)

Magnier, L., Schoormans, J.: Consumer reactions to sustainable packaging: the interplay of visual appearance, verbal claim and environmental concern. J. Environ. Psychol. **44**, 53–62 (2015). Elsevier Ltd.

Martinho, G., Pires, A., Portela, G., Fonseca, M.: Factors affecting consumers' choices concerning sustainable packaging during product purchase and recycling. Resour. Conserv. Recycl. **103**, 58–68 (2015)

Moultrie, J., Sutcliffe, L., Maier, A.: A maturity grid assessment tool for environmentally conscious design in the medical device industry. J. Clean. Prod. **122**, 252–265 (2016)

Navarro, G., Rizo, T.C., Ceca, S.B., Ruiz, M.J.C.: Ecodesign function and form. classification of ecodesign tools according to their functional aspects. In: ICED 2005, 15th International Conference on Engineering Design: Engineering Design and the Global Economy, vol. 12, no. 5, p. 3839 (2005)

Niedderer, K., Mackrill, J., Clune, S., Lockton, D., Ludden, G., Morris, A., Cain, R., Gardiner, E., Evans, M., Gutteridge, R., Hekkert, P.: Creating sustainable innovation through design for behaviour change: full project report (2014)

Nordin, N., Selke, S.: Social aspect of sustainable packaging. Packag. Technol. Sci. 23(6), 317–326 (2010)

Olsmats, C., Dominic, C.: Packaging scorecard–a packaging performance evaluation method. Packag. Technol. Sci. 16(1), 9–14 (2003)

Paine, F.A.: Packaging Design and Performance. Pira, Leatherhead (1990)

Poulikidou, S., Björklund, A., Tyskeng, S.: Empirical study on integration of environmental aspects into product development: processes, requirements and the use of tools in vehicle manufacturing companies in Sweden. J. Clean. Prod. 81, 34–45 (2014)

Rose, C.M.: Design for environment: a method for formulating product end-of-life strategies. Doctoral dissertation, Stanford University (2000)

Slavin, B.C., Coordinator, S.: How to assess sustainable packaging: an overview of the tools and resources available (2016). http://leadwise.mediadroit.com/files/18826HowtoAssessSusPkg fordistribution.pdf. Accessed 26 Nov 2016

Sustainable packaging alliance: Sustainable packaging framework (2010). http://www. sustainablepack.org/database/files/filestorage/sustainable%20packaging%20definition% 20july%202010.pdf. Accessed 3rd Mar 2017

Sustainable packaging coalition: Design guidelines for sustainable packaging (2006). Sl: sn

Sustainable packaging coalition: Sustainable packaging indicators and metrics framework (2009). http://www.sustainablepackaging.org/Uploads/Resources/spc_indicator_metrics_ framework.pdf. Accessed 3rd Mar 2017

Svanes, E., Vold, M., Møller, H., Pettersen, M.K., Larsen, H., Hanssen, O.J.: Sustainable packaging design: a holistic methodology for packaging design. Packag. Technol. Sci. 23(3), 161–175 (2010)

Vallet, F., Eynard, B., Millet, D., Mahut, S.G., Tyl, B., Bertoluci, G.: Using eco-design tools: an overview of experts' practices. Des. Stud. 34(3), 345–377 (2013)

Verghese, K.L., Horne, R., Carre, A.: PIQET: the design and development of an online 'streamlined' LCA tool for sustainable packaging design decision support. Int. J. Life Cycle Assess. 15(6), 608–620 (2010)

Verghese, K.L.: Sustainable packaging: how do we define and measure it. In: 22nd IAPRI Symposium, ResearchGate, Australia (2005)

Waage, S.A.: Re-considering product design: a practical 'road-map' for integration of sustainability issues. J. Clean. Prod. 15(7), 638–649 (2007)

Wever, R., Vogtländer, J.: Eco-efficient value creation: an alternative perspective on packaging and sustainability. Packag. Technol. Sci. 26(4), 229–248 (2013)

Impact of a Sustainable Manufacturing-Related Learning Game on Basic Knowledge and Network Thinking

A Study with High School Students

Ina Roeder[✉], Mustafa Severengiz, Rainer Stark, and Günther Seliger

Technische Universität Berlin, Berlin, Germany
{ina.roeder,rainer.stark}@tu-berlin.de,
{severengiz,seliger}@mf.tu-berlin.de

Abstract. Modern challenges require modern thinking, which again calls for modern teaching methods. The complex field of sustainable manufacturing cannot be taught out of school books but must be experienced. Learning games have a great potential for that. Therefore, to meet the gap in educational resources in the field of sustainable manufacturing, a learning game for high school students has been developed and its effect tested with 76 players.

Keywords: Sustainable manufacturing · Learning game · Serious game

1 Sustainability Stakeholders of Tomorrow – Teaching the Youth

The United Nations considers the mobilization of the broad public to be the essential requirement for achieving a shift towards a more sustainable development, especially with regard to Education for Sustainable Development targeting school children as stakeholders of tomorrow. Surveys have shown that knowledge of sustainability and sustainable manufacturing in particular is underrepresented in German children and teenagers [1]. Moreover, a market analysis found a severe lack of high quality teaching materials when it comes to manufacturing-related sustainability [2].

Goods-Loop is an educational board game which has been designed as an introductory teaching tool for school students, covering the product life cycle, end of life (EOL) scenarios and the three pillars of sustainability as well as their interdependencies in manufacturing. It is a result of a joint activity of the Department of Industrial Information Technology and the Department of Machine Tools and Factory Management of Technische Universität Berlin.

The learning game's design was chosen due to its ability to simulate complex situations such as the three-dimensionality of sustainable development and thus making them available for personal experience – something e.g. written text cannot accomplish.

2 Playing for Fun, Learning for Life – Serious Games

Over the last decades a shift in education from instruction-based to learner-centered teaching methods can be noticed (cf. [3, 4]). This trend questions the traditional "learning

G. Campana et al. (eds.), *Sustainable Design and Manufacturing 2017*, Smart Innovation, Systems and Technologies 68, DOI 10.1007/978-3-319-57078-5_14

by listening" approach and focuses rather on the ability to find and use information in a meaningful way [3]. One element to diversify teaching is the implementation of so-called learning games [5]. Games are widely used in education ([6, 7]) and there is an increasing interest in how games may influence learning (cf. [8–10]). One traditional game type is the board game. Board games are being used in various fields (e.g. English teaching [11], medicine [6], artificial intelligence [12]) and are generally seen as interesting, enjoyable and collaborative [13]. As board games are most often associated with the context of leisure time with family or friends, the players experience it as a relaxing learning environment. Moreover, a board game can have additional educational potential compared to IT-based solutions by being socially more interactive [14]. Furthermore, board games also offer the potential of experiencing greater levels of flow [15], so that players may be driven by pure pleasure rather than external incentives [16], thus increasing the overall motivation for the learning action.

Serious games, gamification and game-based learning all apply the game idea or elements of it but are to be distinguished from simply entertainment-oriented games [17, 19] describe a game as "a set of activities involving one or more players. It has goals, constraints, payoffs, and consequences. A game is rule-guided and artificial in some respects. Finally, a game involves some aspect of competition, even if that competition is with oneself." The term "serious games" was in contrast first described as having an "explicit and carefully thought-out educational purpose and not [being] intended to be played primarily for amusement" [20]. Nowadays the definition of a *serious game* or *serious gaming* is often directed towards the usage of a computer (e.g. [21]). In this context, game-based learning is often understood as a subcategory of serious games or even as a synonym [22]. [23, 24] widened the meaning of serious games by considering context or situations rather than artifacts. But even beyond that, there are discussions were to draw the line between serious games and games for pure pleasure [25]. This is more so since children themselves do not seem to distinguish between fun and learning when playing [26].

Proponents of game-based learning highlight several advantages of the approach. E.g. complex subjects can be taught and tested within a gaming environment without taking risks [27]. Therefore, games can reduce the anxiety of making mistakes [28]. Also games can promote teamwork, team building and further social skills just as game-based learning can provide a link between theory and practice [29]. Several researchers also share the opinion that fun improves the learning productivity (cf. [30]). Still, the benefits of game-based learning are an object of discussion (cf. [31]). In a large number of cases in which the intended effects could be measured for certain learning games (cf. [32–35]), many were said to be limited in their generalizability [36].

3   *Goods-Loop* – A Sustainable Manufacturing Learning Game

The game *Goods-Loop* is an open source board game for three to four players which aims at teaching the complexity of sustainability in a production environment for high school students. The players are set in the position of producers who want to maximize their capital throughout their product's life cycle. However, capital does not only exist as economic capital, but also as ecological and social capital – so that there are three currencies to be handled at the same time. In order to teach the necessity of respecting all three dimensions of sustainability, the player with the lowest imbalance between all three types of capital and yet the overall highest performance wins the game. The game ends after the first player has sold his or her third product.

3.1 Let's Play – Game Mechanics

The outer circle of the playing field reflects the life cycle of a product (compare Fig. 1). Within the game the phases *product design*, *raw material extraction*, *production*, *assembly*, *quality control*, *selling*, *usage*, and *collection* are being distinguished. This simple representation has been chosen with respect to the little previous knowledge the players are expected to possess. Within the inner circle there are several EOL scenarios considered: When the player reaches the *collection* field is when he or she can decide whether the product gets disassembled and recycled, remanufactured, repaired, reused or put the product to the landfill. Depending on this decision, the user starts the next round on his or her journey through the product life cycle in the phase of production, assembly, quality control, selling, or raw material extraction respectively. The earlier his or her starting phase is located within the product life cycle, the more economic capital can be gained by selling his product when he reaches the *selling* field again, but also the longer does it take to reach that phase of selling – a process continually costing at least economic and environmental resources.

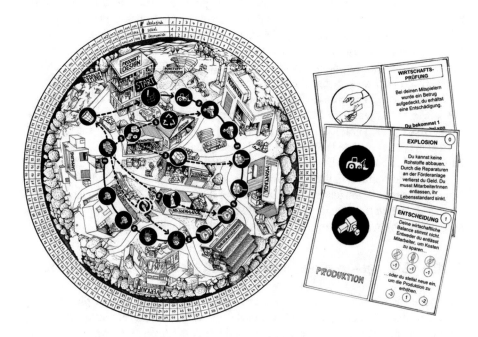

Fig. 1. Game board (left), interaction card (top right) and process cards

Within the game players are drawing process as well as interaction cards. Both kinds of cards are meant to reflect real issues of companies. *Process cards* are drawn whenever a player reaches a certain phase, represented by the respective field. They contain information which raise or lower one or more of the capitals (economic, ecological or social). They are either passively performed (production stop due to a machine breakdown; loss of economic capital) or do contain active choices (either fire five percent of your staff or start a new production line; lose social or economic capital). To raise interactivity between the players, they pick up *interaction cards* every time they enter a new life cycle phase. Interaction cards can either have a positive impact (your product outperforms the ones of your competitor, so you gain their customer loyalty) or a negative one (you lose a highly valuable employee to the competitor to your right). Interaction cards which have a negative effect need to be played directly. The positive ones can be played at any strategically convenient time.

3.2   EOL Strategies – Heart of the Game

The basis for the choice of displayed EOL scenarios and the calculation of their impact in the game was the EU directive 2008/98/EG [37]. Out of this list, the following have been chosen as exemplary EOL scenarios: *Re-use*, *Remanufacturing*, *Repair*, *Recycling*, and *Landfill*.

Reconditioning has been ignored with regard to the learning game since it resembles remanufacturing and the game should display only strongly differing EOL strategies for an easy introduction. *Prevention* is not displayed as an EOL strategy in the game because following the logic of handling ones product (or raw materials) in a second and third life cycle, non-production does not fit the game idea. However, it is included in the process cards that lead the player through the single life cycle phases. *Energy Recovery* has been subsumed under *Landfill*, regarding the materials as "lost" for further life cycles in the logic of the game.

4   Into the Wild – Field Test

[25] emphasizes the fact that the intended learning effects of game developers do not necessarily need to occur and therefore need to be proven within studies. This should obviously be done during the development process. Therefore, field tests have been conducted in three different development stages of *Goods-Loop* in order to advance the game's playability, increase its fun factor and of course its learning effects. The latest prototype, which is presented in this paper, has been played by n = 76 teenagers (45 boys and 31 girls, 8th to 11th grade) whose learning outcome was measured using questionnaires in a controlled test environment at Technische Universität Berlin.

4.1   Design of Experiment

The survey was designed as starting element of three project days that were offered for teenagers from various Berlin high schools. The project day character was chosen to reduce the intimidating atmosphere that a research environment can have especially on young probands. Furthermore, the test was not to be conducted in a classroom setting with an already elaborated social situation with its difficult to control socio-psychological effects on the learning outcome, e.g. (assumed) teacher's preference, negative learning experiences, or group dynamics.

A pre-post-test was designed using almost identical questionnaires before and after the treatment "game play". The questionnaire was designed to operate the hypotheses according to the teaching goals as were:

H1: Playing the learning game "Goods-Loop" improves the knowledge of general manufacturing processes.

H2: Playing the learning game "Goods-Loop" improves the knowledge of manufacturing-related aspects within the three dimensions of sustainability.

H3: Playing the learning game "Goods-Loop" improves the ability of network thinking with regard to the topic of sustainable manufacturing.

A control group design had been dismissed since the impact of the game was not to be compared to other teaching or learning methods, but according to the individual knowledge gain.

In the first part of the survey, designed to measure H1, the probands were asked to explain the term "product life cycle" and to bring six phases of the life cycle in the right order. They were further asked to assess their own knowledge on different manufacturing processes and explain a variety of EOL scenarios. The code for the answers was specifically tailored for each task. When explaining "product life cycle" in an open format the code distinguished between wrong/no answer – no knowledge (0), simple answer (e.g. "what happens to a product") – little former knowledge (1), and elaborated answer (e.g. "the idea of saving resources through circular economy") – good knowledge (2). When arranging the life cycle phases the coding was no answer/two mistakes or more – no knowledge (0), one mistake – little knowledge (1), all correct – good knowledge (2). Regarding the knowledge on EOL, the probands were asked to check a list of EOL scenarios including two non-existent control items. In a first step they should state whether they had heard the term before and in a second give an explanation for it. The coding for each item went as follows: not heard before/heard before while also stating this for the control scenarios/heard before but completely wrong explanation – no knowledge (0), heard before while excluding the control scenarios – little knowledge (1), correct explanation – good knowledge (2). The arithmetic mean was calculated for all scenarios to receive the overall EOL knowledge classification of each proband.

In the second part, designed to measure H2, the probands took on the role of consultants and stated general possibilities for an enterprise to act socially acceptable, environmental-friendly and economically sound. The coding differentiated between no knowledge (0) – no/wrong answer, little knowledge (1) – simple answer that points in the correct direction (e.g. "pay good salary"), and good knowledge (2) – complex answer that links or weighs different aspects or shows detailed knowledge by using technical terms (e.g. "allowing for labor unions and sick leave"). Again, the arithmetic mean was calculated for all three items.

In the third part, designed to measure H3, the probands consulted fictional manufacturing enterprises with regard to more complex decisions in which topics of sustainability lay hidden, e.g. "What could be the effects of a higher level of automation on the living standard of the employees?" Apart from one such question for each dimension of sustainable development there was a fourth question, asking the probands to weigh the three dimensions against each other in a specific scenario. The coding was conducted as for the items on H2.

The probands were welcomed to the Production Technology Center of Berlin and handed out the questionnaires right away. Afterwards the rules of the game were introduced without explaining the content. The probands were divided into groups of four and played the game for 90 min. Afterwards they were handed out the second questionnaire with identical questions on the hypotheses but arranged in a different order. For each item the score change was calculated and the arithmetic mean of all changes targeting the same hypothesis was calculated. Consequently the score difference was decoded as knowledge loss for <0, unaffected for 0, slight knowledge gain for 0.1–0.5, regular gain for 0.6–1.0, and strong gain for 1.1–2.0. Regular gain was therefore attributed when at least for some items the quality of answers had changed from no to little or from little to good knowledge. A strong knowledge gain implies a leap in quality for at least several answers from no to good knowledge within the classification framework.

4.2   Results

For each question, the change of knowledge or ability was only calculated for those probands who did not earn full score for the respective question in the pre-test already, e.g. who had the mathematical chance of improvement. This was especially necessary since a great number of probands did not fill in sections of the second questionnaire, which they had already easily answered the first time. Mathematically this would have indicated a great knowledge loss where there was simply an understandable lack of motivation. Furthermore, probands were only partly included who had not completed

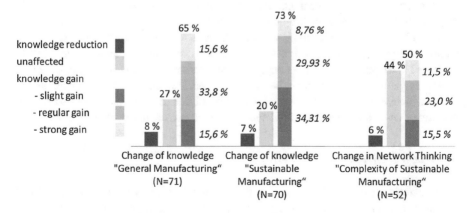

Fig. 2. Effects of playing the learning game

the questionnaire. The consideration of their answers was restricted to the sections in which they had answered at least some questions.

Looking at the results of the first survey section, H1, the assumption that playing the game may improve the knowledge of general manufacturing processes, can be confirmed. Out of those players who did not already have good knowledge on general manufacturing processes, 65% showed improved knowledge after playing the game as shown in Fig. 2. Of those, 24% showed a slight knowledge gain, 52% a clear knowledge gain and another 24% even showed a strong knowledge gain.

When evaluating the second section of the survey it was found that H2, the assumption that knowledge of sustainable manufacturing can be improved, can be confirmed even more clearly, showing a knowledge gain rate of 73%. However, the intensity of knowledge gain was less distinct than in Sect. 1 with 47% of the knowledge gainers showing only a slight improvement and only 12% a strong improvement.

The most ambitious teaching goal, formulated as H3, improving the ability of network thinking by understanding the complexity of sustainable manufacturing, could also be confirmed by a 50% improvement rate, although 44% of the sample remained unaffected. In this section, the improvement intensity was higher than in Sect. 2 although lower than in Sect. 1 with 69% of the sample showing clear or even strong improvement.

In each section, there is a sample that shows a reduction of knowledge or ability after the treatment. The most likely explanation is the tool design itself with the repetitiveness of the questionnaire, having an effect on the probands' will to cooperate. As mentioned before, there were a number of probands who did not fill in the second questionnaire completely, especially if they had given elaborated answers in the first survey. It must be assumed that the same de-motivational effect has also occurred in probands who still answered the questionnaire and whose answers were therefore taken into account but who might have done so less enthusiastically, leading to less elaborated answers and a seeming decrease of their performance. The possibility of this effect had been considered in advance. However, the repetition of the same questions had proven the least susceptible to measuring errors in the pre-tests.

All in all, the game has been proven to have the potential to improve knowledge on manufacturing in general as well as on its sustainability aspects. Moreover it can convey the complexity of sustainable manufacturing to at least half of the players who enter the game with fragmentary or no previous knowledge.

It can be seen that the teaching success increases with the simplicity of the topic, as could be expected. Hence, weighing sustainability factors in manufacturing can be conveyed to a smaller part of the sample than the steps of a product life cycle. However, a learning game can never in itself be a complete teaching unit. It needs to be embedded within a broader educational context. Given the fact that the test was conducted without briefing or de-briefing, the measured success can be considered very high.

5   Keep Learning – Next Steps

To focus more strongly on player action, the *Goods-Loop* cards are currently digitalized, so that every event that occurs during the life cycle of a certain product can be connected to choices the player has made before. By this, a product chronic can be created for each player in the end of the game, showing the sustainability path of his or her product. The learning game is offered as Open Educational Resource under a free commons license.

References

1. Roeder, I., Scheibleger, M., Stark, R.: How to make people make a change – using social labelling for raising awareness on sustainable manufacturing. Procedia CIRP **40**(2016), 359–364 (2015)
2. Roeder, I., Severengiz, M., Stark, R., Seliger, G.: Open educational resources as a driver for manufacturing-related education for learning of sustainable development. Procedia Manuf. **8**, 81–88 (2017)
3. Garris, R., Ahlers, R., Driskell, J.E.: Games, motivation, and learning: a research and practice model. Simul. Gaming **33**(4), 441–467 (2002). SAGE Publications
4. Hainey, T.: Using games-based learning to teach requirements collection and analysis at tertiary education level. Doctor of Philosophy. University of the West of Scotland (2010)
5. Ravenscroft, A.: Promoting thinking and conceptual change with digital dialogue games. J. Comput. Assist. Learn. **23**(6), 453–465 (2007). Wiley
6. Bochennek, K., Wittekindt, B., Zimmermann, S.Y., Klingebiel, T.: More than mere games: a review of card and board games for medical education. Med. Teach. **29**(9), 941–948 (2007)
7. Blakely, G., Skirton, H., Cooper, S., Allum, P., Nelmes, P.: Educational gaming in health sciences: a systematic review. J. Adv. Nurs. **65**(2), 259–269 (2009)
8. Ke, F.: A qualitative meta-analysis of computer games as learning tools. In: Ferdig, R.E. (ed.) Handbook of Research on Effective Electronic Gaming in Education, vol. 1, pp. 1–32. IGI Global, Hershey (2009)
9. Kebritchi, M., Hirumi, A., Bai, H.: The effects of modern math computer games on learners' math achievement and math course motivation in a public high school setting. Br. J. Educ. Technol. **38**(2), 49–259 (2008). BERA
10. Wu, W.H., Chiou, W.B., Kao, H.Y., Hu, C.H.A., Huang, S.H.: Re-exploring game-assisted learning research: the perspective of learning theoretical bases. Comput. Educ. **59**(4), 1153–1161 (2012). Elsevier
11. Gaudart, H.: Games as teaching tools for teaching english to speakers of other languages. Simul. Gaming **30**(3), 283–291 (1999). SAGE Publications
12. Tesauro, G.: Temporal difference learning and TD-Gammon. Commun. ACM **38**(3), 58–68 (1995). Association for Computing Machinery
13. Zagal, J.P., Rick, J., Hsi, I.: Collaborative games: lessons learned from board games. Simul. Gaming **37**(1), 24–40 (2006). SAGE Publishing
14. Rossiter, K., Reeve, K.: It's your turn!: exploring the benefits of a traditional board game for the development of learning communities. In: Edvardsen, F., Kulle, H. (eds.) Educational Games. Design, Learning and Applications, pp. 331–337. Nova Science Publishers, New York (2010)
15. Khan, A., Pearce, G.: A study into the effects of a board game on flow in undergraduate business students. Int. J. Manag. Educ. **13**(3), 193–201 (2015). Elsevier
16. Csikszentmihalyi, M.: Flow: The Psychology of Optimal Experience. Harper Perennial, New York (1990)
17. Davidson, D.: Beyond Fun: Serious Games and Media. ETC Press, Pittsburgh (2008)
18. Caillois, R.: Man, Play, and Games. Schocken Books, New York (1961)

19. Dempsey, J.V., Haynes, L.L., Lucassen, B.A., Casey, M.S.: Forty simple computer games and what they could mean to educators. Simul. Gaming **33**(2), 157–168 (2002). SAGE Publications

20. Abt, C.C.: Serious Games. University Press of America, Lanham (1987)

21. Zyda, M.: From visual simulation to virtual reality to games. Computer **38**(9), 25–32 (2005). IEEE Computer Society Press

22. Corti, K.: Games-based learning: a serious business application. PIXE Learning Limited (2006). www.pixelearning.com/docs/games_basedlearning_pixelearning.pdf. Accessed 29 Nov 2009

23. Steinkuehler, C., Duncan, S.: Scientific habits of mind in virtual worlds. J. Sci. Educ. Technol. **17**(6), 530–543 (2008)

24. Sanchez, E., Jouneau-Sion, C.: Les jeux, des espaces de reflexivite permettant la mise en cewre de demarches d'imestigation. In: Paper presented at the Ressources et trauail collectif dans la mise en place des demarches d'investigation dans l'enseignement des sciences - Actes des joumees scientifiques DIES 2010., Lyon, 24 et 25 Novembre 2010

25. Haring, P., Chakinska, D., Ritterfeld, U.: Understanding serious gaming: a psychological perspective. In: Felicia, P. (ed.) Handbook of Research on Improving Learning and Motivation through Educational Games: Multidisciplinary Approaches, vol. 1, pp. 413–430. Information Science Reference (2011)

26. Ritterfeld, U.: Serious gaming: assumptions and realities. In: Invited Address Meaningful Play Conference, Michigan State University (2008)

27. De Tornyay, R., Thompson, M.: Strategies for Teaching Nursing. Wiley, Hoboken (1987)

28. Deck, M., Silva, J.: Getting Adults Motivated Enthusiastic and Satisfied. Resources for Organisation, Edina (1990)

29. Kuhn, M.A.: Gaming: a technique that adds spice to learning. J. Continuing Educ. Nurs. **26**(1), 35–39 (1995)

30. Lepper, M.R., Greene, D.: Overjustification research and beyond: toward a means-end analysis of intrinsic and extrinsic motivation. In: Lepper, M.R., Greene, D. (eds.) The Hidden Costs of Reward, pp. 109–148. Lawrence Erlbaum, Hillsdale (1978)

31. Connolly, T.M., Boyle, E.A., MacArthur, E., Hainey, T., Boyle, J.M.: A systematic literature review of empirical evidence on computer games and serious games. Comput. Educ. **59**, 661–686 (2012)

32. Durkin, K.: Game playing and adolescents' development. In: Vorderer, P., Bryant, J. (eds.) Playing Video Games: Motives, Responses, and Consequences, pp. 415–428. Lawrence Erlbaum, Mahwah (2006)

33. Lee, K.M., Peng, W.: What do we know about social and psychological effects of computer games? A comprehensive review of the current literature. In: Vorderer, P., Bryant, J. (eds.) Playing Video Games: Motives, Responses and Consequences, pp. 325–346. Lawrence Erlbaum Associates Inc., Mahwah (2006)

34. Lieberman, D.A.: What can we learn from playing interactive games? In: Vorderer, P., Bryant, J. (eds.) Playing Video Games: Motives, Responses, and Consequences, pp. 379–397. Lawrence Erlbaum Associates, Mahwah (2006)

35. Ritterfeld, U., Weber, R.: Video games for entertainment and education. In: Vorderer, P., Bryant, J. (eds.) Playing Video Games: Motives, Responses, and Consequences, pp. 399–413. Lawrence Erlbaum, Mahwah (2006)

36. Watt, J.H.: Improving methodology in serious games research with elaborated theory. In: Ritterfeld, U., Cody, M.J., Vorderer, P. (eds.) Serious Games: Mechanisms and Effects, pp. 374–388. Routledge/LEA, New York (2009)

37. Eriksen, M., Lebreton, L., Carson, H., Thiel, M., et al.: Plastic pollution in the World's Oceans: more than 5 trillion plastic pieces weighing over 250,000 tons afloat at sea. PLoS ONE **9**(12), e111913 (2014)

Sustainable Manufacturing Processes and Technology

Improvement of Sustainability Through the Application of Topology Optimization in the Additive Manufacturing of a Brake Mount

Stefan Junk[1(✉)], Claus Fleig[2], and Björn Fink[1]

[1] Laboratory for Rapid Prototyping, Department of Business
and Industrial Engineering, University of Applied Sciences Offenburg,
Klosterstr. 14, 77723 Gengenbach, Germany
stefan.junk@hs-offenburg.de,
b.fink@stud.hs-offenburg.de
[2] Department of Mechanical and Process Engineering,
University of Applied Sciences Offenburg,
Badstr. 24, 77653 Offenburg, Germany
claus.fleig@hs-offenburg.de

Abstract. In recent years, the additive manufacturing processes have rapidly developed. The additive manufacturing processes currently present a high-performance alternative to conventional manufacturing methods. In particular, they offer the opportunity of previously hardly imaginable design freedom, i.e. the implementation of complex forms and geometries. This capability can, for example, be applied in the development of especially light but still loadable components in automotive engineering. In addition, waste material is seldom produced in additive manufacturing which benefits a sustainable production of building components. Until now, this design freedom was barely used in the construction of technical components and products because, in doing so, both specific design guidelines for additive manufacturing and complex strength calculations must be simultaneously observed. Yet in order to fully take advantage of the additive manufacturing potential, the method of topology optimization, based on FEM simulation, suggests itself. It is with this method that components that are precisely matched and are especially light, thereby also resource-saving, can be produced. Current literature research indicates that this method is used in automotive manufacturing for reducing weight and improving the stability of both individual parts and assembly units. This contribution will study how this development method can be applied in the example of a brake mount from an experimental vehicle. In this, the conventional design is improved by means of a simulation tool for topology optimization in various steps. In an additional processing step, the smoothing of the thus developed component occurs. Finally, the component is generatively manufactured by means of selective laser melting technology. Models are manufactured using binder jetting for the demonstration of the process. It will also be determined how this weight reduction affects the CO_2 emissions of a vehicle in use.

Keywords: Additive manufacturing · Topology optimization · Selective laser melting · Sustainable design · Binder jetting · Carbon footprint

© Springer International Publishing AG 2017
G. Campana et al. (eds.), *Sustainable Design and Manufacturing 2017*, Smart Innovation, Systems and Technologies 68, DOI 10.1007/978-3-319-57078-5_15

1 Introduction

In recent years, Additive Manufacturing (AM) processes have established themselves alongside the conventional, i.e. subtractive and formative, manufacturing processes [1]. While these processes were initially employed for the manufacturing of prototypes (Rapid Prototyping, RP) and small series, they are in the position to economically manufacture medium quantities, e.g. in automotive and aircraft manufacturing, today. Overall, the number of components additively manufactured has increased greatly over recent years [2]. Tools for thermoforming, along with prototypes and components, are produced using additive manufacturing (Rapid Tooling, RT) [3].

It is common in all additive processes that the components are built layer by layer. The manufacturing also occurs directly; that is to say that complex tools or devices or programming are not required. In this way, it is possible to, on one hand, create complex forms such as, e.g. arched cooling channels at forming tools. An additional advantage in terms of sustainability is that hardly any material is wasted in additive manufacturing [4]. While, for example, up to 90% of the row material is transformed to shavings in conventional machine cutting, only the support material is considered waste in additive manufacturing.

Today there is a variety of different processes on the market that use a variety of materials and methods to structure the layers. The materials range from plaster powder, paper and photopolymer to plastics and metals. Selective laser sintering is especially suitable here for automotive and aircraft manufacturing because components can be produced with complex geometries and various metal powders which provide sufficient stability for the use of the products. Lately, new processing centers are also being developed where additive and subtractive manufacturing processes are combined (hybrid additive manufacturing) [5]. In this, complex forms can be produced with laser sintering, and a high surface quality is subsequently achieved with the aid of a machining during post processing. Even the manufacturing of tools for cooling channels for hot forming has been successfully tested [6].

2 State of the Art in Topology Optimization

The limiting conditions for the application of additive manufacturing processes set high standards for the design engineer. On one hand, the construction guidelines for additive manufacturing must be observed. These rely on the method of structuring the layers selected and the designated material [7]. On the other hand, light and thereby sustainable products, in particular, should be primarily developed in automotive and aircraft manufacturing in order to reduce resource and fuel consumption and with that CO_2 emissions. Generally, this leads to a reduced wall thickness in an effort to reduce the weight. However, this reduction should not cause a weakness in the stability or lead to the failure of the component. In order to simultaneously fulfill these complex demands, analytical calculations are, as a rule, no longer sufficient [8]. Rather, the methods of topology optimization based on FEM simulation offers the possibility of combining complex design and sufficient stability.

Individual parts are usually first optimized for weight reduction in automotive engineering. Steel structural components lend themselves to topology optimization as well because weight reduction is easily realized here [9]. Additional reductions in weight can be implemented by combining constructive lightweight engineering and the application of tailored blanks, i.e. the reduction in sheet thickness using welded steel sections with various thicknesses or material properties [10]. Even in the manufacture itself, additive manufacturing can lead to a resource and energy saving in comparison with conventional methods under certain conditions (e.g., small numbers of pieces) [11].

Topology optimization can also be applied in the automotive manufacturing of trucks. Here, the weight as well as the rigidity and the strength can be considerably improved through topology optimization [12]. Entire assembly kits can also be optimized along with individual parts. It is evident in the example of a multi-part cross beam from automotive production that a weight reduction of approximately 35% in comparison to conventional design can be achieved in converting the material used from steel to aluminum in combination with a re-design and a topology optimization [13]. However, the integration of a crash simulation is not yet possible due to the complex correlations among the individual parts of the cross beam which, in addition, necessitate a non-linear simulation.

In order to exploit the possibilities of the design for additive manufacturing new design methods were developed recently. Thus, the design space can be significantly expanded by the application of axiomatic design theory in combination with experience of practitioners and a subsequent topology optimization [14]. In addition, topology optimization can also be used effectively in order to optimize and evaluate different variants during product development for additive manufacturing [15].

3 Processing Steps in Topology Optimization

A range of processing steps are conducted in topology optimization that exceed the conventional new construction of components or the re-design of existing components (see Fig. 1). In order to do so, the requirements and the constraints (e.g. materials, loads, construction space) must be known ahead of time. The basis for topology optimization is a virtual model of the component developed with Computer Aided Design (CAD). This is transferred via an interface to the topology optimization software.

The result of topology optimization is based on a simulation with finite elements whereby lesser loaded areas will be spared. In the course of this, a new component surface originates, composed of the often irregularly distributed and "jagged" surface of the finite elements. Since this irregular surface is generally difficult to manufacture, and, moreover, there is also a risk of notching, it is smoothed by specialized software in the next step. The optimized and smoothed model can then be transferred to data processing where the further processing for the additive manufacturing begins (also see Sect. 4).

Virtual Model

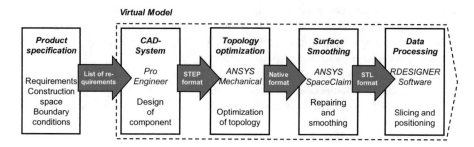

Fig. 1. Processing chain in topology optimization.

3.1   Starting Point and Requirements for the Component

In order to study the application of additive manufacturing processes in automotive manufacturing, a representative demonstrator from the chassis of an experimental vehicle is studied. The vehicle itself was developed by students as a study project at the University of Applied Sciences Offenburg. It is the "Schluckspecht" (boozer) vehicle, version V, with which the objective, among other things, pursued was to travel a long distance consuming as little fuel as possible (see Fig. 2a).

A brake mount was chosen as the demonstrator component for this contribution; it serves as the mount for the disc brake fastening (Fig. 2b). The brake mount is shaped in the I-form and is currently processed conventionally out of aluminum by milling. It weighs 74.7 g. The current assembly situation of the brake mount is pictured in Fig. 2c. The requirements for the component are, first, to be of a sufficient stability because the focus is on safety-relevant components. In addition, the component, in having as little weight as possible, should have a positive influence on the dynamics of vehicle movement and fuel consumption. In order to perform the assembly of the topology-optimized version as simply as possible, the connection compartments, in particular beds, drill holes and threads, should not be altered in their form.

Fig. 2. Experimental vehicle (a), conventional brake mount (b) and assembly situation (c).

3.2   Re-design of the Brake Mount

The requirements are considered in re-designing the component. The re-design is accomplished with the aid of the CAD system Pro-Engineer from PTC. As a first step

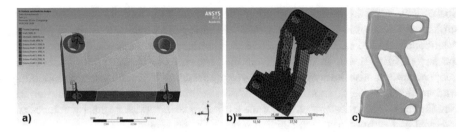

Fig. 3. Boundary conditions (a), result of topology optimization (b) and after smoothing (c).

towards lightweight construction and for the conversion of the component to additive manufacturing, the construct is altered so that it is optimized for manufacturing through the aid of selective laser melting (SLM). A Z-form is selected for this as shown in Fig. 3; this form is difficult to apply to machining but is suited very well to SLM. Furthermore, in comparison to the conventional variant in I-form, a significant weight reduction in the amount of 62.3 g can be expected with this new design. The requirements are taken into account with respect to the construction of the connection points.

3.3   Implementation of Topology Optimization

The prepared CAD model is transferred to topology optimization in the software using, in this example, ANSYS Mechanical-Academic. The software ANSYS Mechanical together with the software tool ANSYS SpaceClaim for the needed smoothing offers an integrated data flow from the simulation to the optimization up to the post processing of the results. This integrated data flow prevents problems with data transfer via interfaces and ensures a reliable data quality for the later additive manufacturing process.

First, the material and element size of the mesh were determined. Yet more precise results would be achieved with a finer mesh with a smaller element size [16]. The minimum element size is set at 2.1 mm (the smallest possible element size in this version) which provides sufficient precision in the finite element mesh. In the next step, a fixed bearing is selected as bearing (see Fig. 3a). A moment of 39.7 Nm is applied on the component and a load of 9000 N is assumed as bolt force. Subsequently, a target value for the intended mass reduction (in this case: 65%) must still be created.

The calculation algorithm of the software does not precisely convert the mass reduction intended by the user but balances the result with the component features. Here the target function is minimum volume with maximum rigidity [17]. The result is represented in Fig. 3, middle. By reducing the material in the center of the component, two bars arise which lead to a significant weight reduction in the component (34.9 g).

3.4   Repairing and Smoothing the Geometry

The result of the topology optimization reveals a pitted, uneven and severely jagged surface (see Fig. 3b). Therefore, repairs must follow in the next step using the ANSYS

SpaceClaim software. After transferring the data in STL format, the holes on the surface will be closed because the virtual model for 3D printing must be watertight. A smoothing of the surface is also necessary in order to prevent a notching effect with the jagged FEM mesh. In doing so, the element size of approx. 2 mm is set in an iterative process also known as "shrink-wrap packaging". Then a second smoothing with an element size of approx. 1 mm follows in order to further improve the surface (see Fig. 3c). Due to the smoothing of the surface, possible weak points by notches are removed and errors during the additive process are prevented.

4 Processing Steps in Additive Manufacturing

The conversion of the optimized component into a physical model should occur using additive manufacturing. A number of processing steps must be conducted for this at the conclusion of the topology optimization (see Fig. 4). The result of the processing steps repairing and smoothing must thus first be transferred to the software for pre-processing. Various tasks for data preparation, in particular the division of layers and the creation of the support structure, are executed at this point. Following this, the layer-by-layer construction of the component occurs using selective laser sintering. A subsequent processing step will still be required that consists of, in particular, the removal of the support structure and a surface treatment by sandblasting.

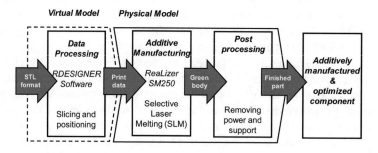

Fig. 4. Processing steps in additive manufacturing.

4.1 Data Processing

The complete CAD model is usually imported in STL format to the software for the SLM device (RDESIGNER) after the smoothing and sliced in horizontal layers with defined layer thickness here. If need be, a support structure is required dependent on the geometry of the component. This serves to stabilize the component and also protects against the deformation of the component which could occur during the cooling process due to its own proper weight or warpage. Additionally, increased internal tensions can be reduced with better heat dissipation in using such a support structure [18]. The support structure (see Fig. 5a and b) is automatically generated during data processing. Lastly, the SLM device for the model is transferred to a manufacturer-specific print data file.

Fig. 5. Support construction in data preparation (a), additively manufactured brake mount on the processing platform (b) and re-processed component (c).

4.2 Manufacturing the Brake Mount Using Selective Laser Sintering

The manufacturing equipment for the newly generated brake mount is the SLM device ReaLizer SLM250. The device consists of a sieving station, processing area and main tank as well as a system control. In the processing area of the device, parts with dimensions with a maximum of 248 × 248 × 240 mm can be manufactured. Using a microwelding process, the component emerges, layer by layer, almost with its final contours on a bed of powder in the processing area with the aid of the laser at an output of 400 W. The material is applied as metal powder. The layer thickness depends on the requirements for the component with respect to the production speed and surface quality. The brake mount is produced with a 50 μm thick layer. The processed components are formed with a very high thickness of over 99% whereby it is guaranteed that the mechanical features of the generated component nearly correspond to those of the base material. Unmelted metal powder is up to 100% re-useable.

AlSi10Mg is used as material for the brake mount. This alloy is typically used in the automotive and aeronautic industry because it is very lightweight. Thus, the material is suitable for lightweight engineering and also this project. The combination of silicon and magnesium should lead the alloy to a marked increase in rigidity and stability wherewith it can be used for complex and thin-walled geometries. Because the material also bears highly dynamic loads, it is often used for components with high loads.

4.3 Re-processing the Brake Mount

When the SLM machine has completed the additive manufacturing process, the generated component can be uncovered from the unmelted powder after a brief cooling period. This is conducted, depending on the device, using small and large brushes, a suction device or with compressed air. The powder removed can be re-used for the next component. Once the components have been uncovered, the base plate on which the components were produced can be removed from the processing area and the parts can be freed from the support constructions.

Pliers are used for the rough removal of the support construction. Then the components are filed, and, in the last step, the surface is treated with a sandblaster. The surface of the brake mount, in particular, on which the support construction was bonded to the component must be reworked. The surface here is distinctly rougher where it was bonded to the support construction than the parts that had no contact with the support construction (see Fig. 5c).

5   Discussion of the Results

Already through the use of additive manufacturing the material used can be significantly reduced. While a raw volume of the bulk material of 40.8 cm^3 is necessary for the machining operations, only material volume of 12.7 cm^3 is required for the additive manufacturing process. This represents a reduction of material consumption of approximately 68%.

But also the generation of a component plays a significant role for sustainability and thereby also the CO_2 emissions [18]. Components with a high ratio of blank to pre-manufactured capacities exhibit a markedly better result for additive manufacturing than massive components. Likewise, the material used is important as the energy necessary for the manufacturing has an essential influence on the result. A thin-walled hollow part can emit less CO_2 in additive manufacturing than in machining manufacturing. On the other hand, the CO_2 emissions for a massive steel shaft are considerably lower in conventional manufacturing than in additive manufacturing.

Particular value is placed on CO_2 emissions during the utilization period of the component in this study. For this purpose, the alteration in weight is first analyzed during the various processing steps. This is illustrated in the preliminary results from the CAD model, topology optimization and smoothing in Fig. 6. The preliminary results are prepared as a presentation model in gypsum using the Binder Jetting (BJ) process. Here it becomes especially clear that repairs and smoothing are necessary for the virtual model after the topology optimization.

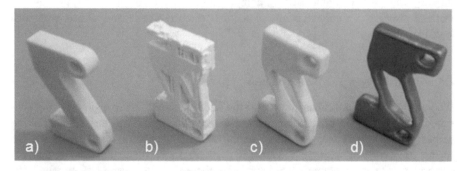

Fig. 6. Preliminary results as BJ model from the CAD model (a), topology optimization (b) and smoothing (c) and end result from the SLM process (d).

Table 1. Comparison of the brake mount weight after the different processing steps

	Mass [g]	Deviation [g]	Deviation [%]
Conventional design (I-shape)	74,7	+10,8	+19,9
Adapted design (Z-shape)	63,9	–	–
Topology optimization	34,9	−29,0	−44,0
Additively manufactured part (SLM)	34,2	−29,7	−45,2

A comparison of the brake mount weight after the different processing steps (see Table 1) first indicates that, by adapting the production to additive manufacturing in using the Z-form, a reduction of weight by 10.8 g is possible. With topology optimization, the weight could be reduced by 29.0 g. A slight weight reduction to 34.2 g occurred with the smoothing and re-processing of the additive manufactured component. Thus, the intended weight reduction by 65% could not be achieved. The reasons for this results are, among other things, that an additional reduction of the material would lead to a no longer tolerable decrease in the stability and strength of the component.

To determine the effect of these actions on the CO_2 emissions, the fuel consumption must first be observed. It is assumed that the brake mount is installed twice in the vehicle; that is, the entire weight reduction is 81.0 g. One can suppose that a reduction of the vehicle weight of 100 kg causes a reduction in the fuel consumption of approx. 0.4l/100 km [19]. This signifies a potential reduction of 0.324 ml/100 km for both brake mounts. Under the assumption that a carbon footprint of 2.640 g CO_2 occurs per liter [20], this would lead to a CO_2 emissions reduction of approx. 855 mg/100 km.

6 Conclusion

This contribution studies how, in implementing the methods of topology optimization in combination with additive manufacturing processes, the CO_2 emissions of vehicles can be reduced and thereby improve the sustainability. A brake mount is chosen here as an example; it is conventionally manufactured using machining manufacturing. It was exhibited that a customized construct of the CAD model was necessary to be able to use the advantages of selective laser melting. It also exhibited that, in applying topology optimization and the essential repairing and smoothing of the model, it was possible to reduce the mass by approximately 45%. In the subsequent SLM process several components could be successfully produced with re-processing.

It must be noted that this method amounts to a great effort as it is necessary to first adapt the design to the additive manufacturing process. Along with this, the application of topology optimization presumes knowledge in FEM simulation. The analysis of the processing chain also indicated that different software packages were utilized whereby the operation of which the users needed to learn. The data transfer between the software packages with the support of various data formats had to be mastered by the operator. It could be shown in this that binder jetting is well-suited to illustrate the varied stages of this method through presentation models.

References

1. Gibson, I., Rosen, D.W., Stucker, B.: Additive Manufacturing Technologies: 3D Printing, Rapid Prototyping, and Direct Digital Manufacturing. Springer, New York (2014)
2. Wohlers, T.: Wohlers Report - Additive Manufacturing and 3D Printing State of the Industry Annual Worldwide Progress Report. Wohlers Associates Inc., Fort Collins (2014)
3. Junk, S., Sämann-Sun, J., Niederhöfer, M.: Application of 3D printing for the rapid tooling of thermoforming moulds. In: Srichand, H., Lin, L. (eds.) Proceedings of the 36th International MATADOR Conference, pp. 369–372, Springer, London (2010)
4. Yoon, H.-S., Lee, J.-Y., Kim, H.-S., Kim, M.-S., Kim, E.-S., Shin, Y.-J., Chu, W.-S., Ahn, S.-H.: A comparison of energy consumption in bulk forming, subtractive, and additive processes review and case study. Int. J. Precis. Eng. Manuf. Green Technol. 1(3), 261–279 (2014)
5. Campbell, I., Bourell, D., Gibson, I.: Additive manufacturing: rapid prototyping comes of age. Rapid Prototyping J. 18(4), 255–258 (2012)
6. Mueller, B., Hund, R., Malek, R., Gebauer, M., Polster, S., Kotzian, M., Neugebauer, R.: Added value in tooling for sheet metal forming through additive manufacturing. In: International Conference on Competitive Manufacturing, pp. 1–7 (2013)
7. Frisch, M., Glenk, C., Dörnhöfer, A., Rieg, F.: Topology optimization for small and medium-sized enterprises - from experience-based construction to the use of topology optimization for the product development process. ZWF Zeitschrift für wirtschaftlichen Fabrikbetrieb 111(5), 243–246 (2016)
8. Pacurar, R., Pacurar, A.T.: Topology optimization of an airplane component to be made by selective laser melting technology, Modern Technologies in Manufacturing, MTeM. In: International Conference on Modern Technologies in Manufacturing (2015). Appl. Mech. Mater. 808, 181–186. Trans Tech Publication, Zürich
9. Oliveira, J., Teixeira, P., Lobo, G., Duarte, J., Reis, A.: Topology optimization of a car seat frame, the current State-of-the-Art on material forming. In: 16th International ESAFORM Conference on Material Forming, ESAFORM (2013)
10. Guangyao, L., Fengxiang, X., Xiaodong, H., Guangyong, S.: Topology optimization of an automotive tailor-welded blank door. Trans. ASME J. Mech. Des. 137(5), 1–8 (2015)
11. Chen, D., Heyer, S., Ibbotson, S., Salonitis, K., Steingrímsson, J.G., Thiede, S.: Direct digital manufacturing: definition. Evol. Sustain. Implications J. Cleaner Prod. 107, 615–625 (2015)
12. Stroobants, J., Campestrini, P.: Component optimisation through simulation driven design. Topology & topography optimization. In: Proceedings of 3rd Commercial Vehicle Technology Symposium on Commercial Vehicle Technology, CVT (2014)
13. Li, C., Kim, I.Y.: Topology, size and shape optimization of an automotive cross car beam. Proc. Inst. Mech. Eng. Part D J. Automobile Eng. 229(10), 1361–1378 (2015)
14. Salonitis, K.: Design for additive manufacturing based on the axiomatic design method. Int. J. Adv. Manuf. Technol. 87, 989–996 (2016)
15. Salonitis, K., Al Zarban, S.: Redesign optimization for manufacturing using additive layer techniques. Procedia CIRP 36, 193–198 (2015)
16. Madenci, E., Guven, I.: The Finite Element Method and Applications in Engineering Using ANSYS®. Springer, New York (2016)
17. Rozvany, G., Lewiński, T. (eds.): Topology Optimization in Structural and Continuum Mechanics. Springer, Heidelberg (2013)

18. Lachmayer, R., Lippert, R.B., Fahlbusch, T. (eds.): 3D-Druck beleuchtet - Additive Manufacturing auf dem Weg in die Anwendung. Springer, Heidelberg (2016)

19. Cheah, L., Heywood, J., Kirchain, R.: The Energy of Impact US Passenger Vehicle Fuel Economy Standards. MIT, Cambridge (2008)

20. Kenny, T., Gray, N.F.: Comparative performance of six carbon footprint models for use in Ireland. Environ. Impact Assess. Rev. **29**(1), 1–6 (2009)

Sustainability of Die-Assisted Quenching Technology and Comparison with Traditional Processes

Giampaolo Campana[1(✉)], Fabio Lenzi[1], Francesco Melosi[1], and Andrea Zanotti[2]

[1] Department of Industrial Engineering, University of Bologna, Viale del Risorgimento 2, 40136, Bologna, Italy
{giampaolo.campana,fabio.lenzi5, francesco.melosi2}@unibo.it
[2] Proterm S.p.A., Via Piretti 4, 40012 Calderara di Reno, Bologna, Italy
dir.quality@proterm.it

Abstract. The open tank oil quenching process is a traditional heat treatment that gives the final microstructures and performance to high quality mechanical parts, typically made by alloyed steels. Due to the microstructure transformations that occur during a heat treatment, the heat treated mechanical part is subjected to shape modifications and distortions. In order to match design tolerances, a machining allowance must be planned and re-machining operations must be realised. The die-assisted oil quenching process utilises a hydraulic press in order to apply a high pressure through a die to the mechanical part during the cooling stage of the heat treatment. The force exerted by the die on the part determines a reduction of distortions along with the control of the shape and of certain dimensions depending on the part geometry and the die design. In the present paper, the technical sustainability of the die-assisted oil quenching process is discussed and compared with the traditional heat treatment in terms of distortion control and reduction of machining operations.

Keywords: Process sustainability · Manufacturing · Heat treatment · Quenching

1 Introduction

In general, sustainability refers to the avoidance of the depletion of natural resources in order to maintain an ecological balance [1]. The *Brundtland Report* of United Nations World Commission on Environment and Development defines sustainability as the "development that meets the needs of the present without compromising the ability of future generations to meet their own needs" [2]. One of the most important domains, where sustainability has to be considered, is industrial production. Any object or production process should respect some sustainable criteria in terms of respect for the environment; a trade-off between performance and cost, social attention and care. Therefore sustainable manufacturing aims to create and distribute innovative goods that optimize the usage of resources in input - in terms of design, manufacturing process,

G. Campana et al. (eds.), *Sustainable Design and Manufacturing 2017*, Smart Innovation, Systems and Technologies 68, DOI 10.1007/978-3-319-57078-5_16

energy, material, packaging and transportation - and remove unnecessary process outputs, including waste, toxic materials, CO_2 emissions over the entire product life-cycle [3].

The dimensions of sustainability have been introduced through the concept of the Triple Bottom Line or the following three pillars: Environment; Economy; Society [4].

Achieving sustainable manufacturing means meeting the requirement of the triple bottom line of environmental, social and economic factors while the green manufacturing targets matches only the environmental and social factors. Dornfeld et al. [5] made a clear distinction between the different types of manufacturing paradigms in terms of the three pillars of sustainability (Table 1).

Table 1. Manufacturing categories based on the three pillars of sustainability (summarised from [5]).

Manufacturing categories	Sustainability aspect covered		
	Environmental	Social	Economic
Green	X	X	
Lean	X		X
Mass		X	X
Sustainable	X	X	X

A failure to realize any one of the three pillars will lead to the system becoming not sustainable. Sustainability implies systematic thinking or a multi-stakeholder approach. Sustainable manufacturing can be defined as the production of manufactured products that use processes which minimize negative environmental impacts, conserve energy and natural resources, are safe for employees, communities and consumers and are economically sound [6–9]. In other words, in order to match the sustainability concepts, products, processes and services should meet not only the challenges concerning their functions, performance and cost, but also the environmental and social issues. Therefore, it is required that the manufacturing industry continues to be proactive in the development of cleaner manufacturing products and processes.

This paper aims to highlight the sustainable manufacturing aspects related to a Die Assisted Oil Quenching (DAOQ) process, also in comparison with a more tradition Open Tank Oil Quenching (OTOQ) process, as far as it is a cleaner and more efficient production method. The DAOQ will be compared with the OTOQ also in terms of sustainability with the help of an experimental activity.

The societal aspects related to the health and the safety of the workers will not be deepened here because they are not directly influenced by the present technical analysis of the production method. The direct benefits or lack thereof as related to the production process falls outside the scope of this work.

2   Comparison Between Quenching and Die Assisted Quenching in Terms of Sustainability

The quenching treatments of steels is an important heat treatment that is performed in order to achieve the highest performance of the material. The OTOQ is today the most common process even if it involves uncontrolled volume variations and distortions of the treated parts due to the specific volume increase caused by the martensitic transformation of the material microstructures. These geometric variations must be corrected by means of the cold straightening or the grinding process to keep the required dimensions [10]. These manufacturing processes must be realized at the end of the production cycle, only after the final heat treatment, with additional cost and with the aim to achieve the dimensional or shape tolerances according to design requirements [10, 11]. The cold straightening process is critical because it can induce undesired defects in the component. The grinding process is expensive and it takes time [10, 11].

A definition of DAOQ is as follows: "Heating a part to the austenitic condition and then immersing the same into a liquid to achieve metallurgical transformation, while at the same time forcing the part to hold size and shape" [12]. The pressure, which has been generated by a hydraulic press and transferred through a die to the treated component, permits the contrast of the forces induced by phase transformations of the material. This technology guarantees the achievement of precise and accurate tolerances. Gears, bearing rings and axisymmetric parts are typically treated with the DAOQ due to the critical aspect related to the need to control the dimensional variations.

Changes in the shape and dimensions that are induced by the quenching process in the mechanical part are mainly distortions of the geometry with respect to the initial shape and dimensions. Types of possible induced deformations are: increase of planarity errors, roundness errors of holes, enlargement or reduction of the main geometrical characteristics such as inner or outer diameters.

Distortions are a natural unavoidable consequence of the quenching process and depend on a huge number of process parameters: the thermal cycle; the quality of the material; the technological and thermal history of the component; residual stresses that are generated by every single step of the production cycle; cooling rate and methods of cooling; critical temperatures for phase transformations. All these parameters must be controlled during the heat treatment in order to reduce distortions.

With the aim to minimize distortions, the most efficient process route is to set a slow cooling rate and to maintain a uniform temperature throughout the part during the whole process without compromising phase transformations.

As already mentioned, dimensional variations of the part generally generate a high cost of rework. Typical operations after the OTOQ process are the grinding or the cold straightening processes if the treated parts are gears, bearing rings or shafts. This type of quenching can treat a single piece or a batch.

The DAOQ process allows highly controlled processing conditions through a hydraulic press and a die. The die has typically three main parts (Fig. 1):

- **A centre expander cone** that controls the inner hole of the components.
- **An inner upper die** that reduces the flatness of the part.

Fig. 1. A scheme of the Die Assisted Quenching process

- **An outer upper die** that controls the external diameter of the components and it can act on the planarity with a different pressure of internal pressor.

An independent hydraulic circuit controls each section of the mould. It is possible to set a precise level of the die-exerted pressure for different areas of the part geometry during the cooling phase. At the furnace exit, the workpiece is placed on the lower die. The upper die closes the mould and then the worker starts the cooling cycle. The cooling rate should be higher from the austenitizing temperature (T_{aus}) to the temperature at which the martensitic transformation (M_s) begins. Then, it should be lower from M_s to the end of the transformation (M_f) to minimize thermal gradient. Finally, any cooling rate can be set from M_f to the room temperature (T_{room}).

The DAOQ ensures the best cooling condition in comparison with the OTOQ and it means that deformations are minimised due to the heat treatment and thus rework operations can be reduced. The DAOQ is an effectively controlled heat treatment and it implies that it is also possible to decrease the machining allowance because it determines less deformation. For this reason, rework operations can be reduced or even avoided.

On the other hand, this heat treatment process has some complications: the management of the process control system and the mould design. The mould must be designed to control specific dimensions and it must be precise and accurate. DAOQ requires a higher initial investment than the traditional hardening process due to the press, the die and the automation system employed in order to improve working conditions. This hardening process involves skilled workers in every single step: from the design of the die and the dimensional control of each component to the tuning of an increased number of process parameters.

Open tank oil quenching permits the treatment of any geometry; otherwise die assisted hardening is linked to the press dimensions.

In Table 2, the properties of both the analysed quenching processes are compared in terms of the three pillars of sustainability for a conical crown that undergoes DAOQ or OTOQ treatments. This part must be coupled with a pinion by a run-in stage that was also considered. Advantages and disadvantages of DAOQ and OTOQ have been illustrated taking into account an industrial case study.

Table 2. Differences between the OTOQ and the DAOQ process by analysing a technological cycle of a conical crown in term of manufacturing and sustainable issues

		Open Tank Oil Quenching (OTOQ)	Die Assisted Oil Quenching (DAOQ)
Manufacturing issues	Technological cycle of a conical crown	– Machining; – Carburizing; – OTOQ; – Relieving tempering; – Grinding operation; – Run-in stage	– Machining; – Carburizing; – DAOQ; – Relieving tempering; – Grinding operation; – Run-in stage One more cooling and heating. The DAOQ is indirect quenching.
Sustainable pillars	Economic	– Cost of the OTOQ furnace: 50 euro/h; – OTOQ cost: 5 euro/pz; – Grinding operation; – Run-in stage: 20 min; – Machining allowance: 0,2–0,4 mm;	– Cost of the DAOQ equipment: 110 euro/h; – DAOQ cost: 6,5 euro/pz; – No grinding operation; – Run-in stage: 10 min; – Machining allowance: 0,08–0,15 mm;
	Environmental	– Machining allowance: 0,2–0,4 mm; – Grinding is unavoidable; – High raw material consumption.	– Machining allowance: 0,08–0,15 mm; – No grinding process; – Low raw material consumption.
	Social	– The contact area between crown and pinion is not ideal. It means decrease of reliability and increase of early failure. – Low-level manpower; – High manual work content.	– The best contact between crown and pinion is obtained. – Skilled and specialised workers; – Low manual work content (due to automation).

DAOQ extends the technological cycle because it is an indirect quenching; it costs more than OTOQ due to the more expensive equipment and tools but machining allowances are significantly reduced and grinding operations are completely avoided. DAOQ also permits better quenching conditions so as it ensures less distortions and improved microstructures with better performances.

As a conclusion, the DAOQ is a final heat treatment that reduces machining allowance by avoiding the grinding machining and, as a consequence, a smaller amount of raw material is utilized by means of a simplification of the part design.

3   Experimental Activity

The DAOQ technology helps the achievement of a precise and accurate shape and dimensions by controlling microstructural evolution and distortions according to Table 2. The experimental activity was engineered as a Randomized Block Design.

The first control factor was the sample geometry. Five different geometrical series were realized. The geometries were characterized by: the same outer diameter (De), same thickness (t) and a variable inner diameter (Di), as it showed in Fig. 2. The realised diameter ratio, Di/De, is in the range between 0 and 0.84 and the sample sets were composed of a disk and different rings with a variable transversal section (Fig. 2a).

The second design factor was the technological cycle. Three replications for each geometry were submitted to two subsequent phases that included six different operations. The first one included: *Machining (Op01)*, *OTOQ (Op02)* and *Tempering (Op03)*. The second stage contained: *Spheroidizing Annealing (Op04)*, *DAOQ (Op05)* and *Tempering (Op06)*.

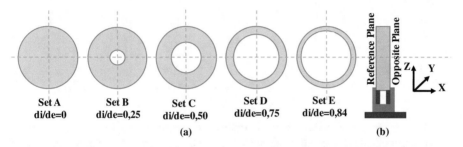

Fig. 2. Geometries (a) and position (b) of specimens in the Coordinate Measuring Machine

The material of the sample is a high carbon alloyed steel, 100CrMo7. In order to ensure the same chemical composition and the same microstructure only a billet of material was used for machining all the samples.

The experimental outcomes were the sample measurements. A Coordinate Measurement Machine (CMM) made six scans for each sample between consecutive operations in order to describe the dimensional evolution. The definition of the measurement standard was ensured to limit the errors during detections. Each sample had a signature and an orientation thus it was possible to align the pieces on the plane of the CMM in the same position for all the measurement operations, as Fig. 2b shows.

The geometric tolerance of flatness was considered in order to evaluate sustainability because this geometric measurement is directly in relationship with the eventual need of a grinding. The measurements consisted of multiple scans of both the planes of the sample.

In particular, the die that was used in order to realize the DAOQ process was simply composed of the outer upper die in order to control planar tolerances. The sustainability of the DAOQ process was then evidenced through the analysis of the tolerances of the planar surfaces of samples.

4 Data Analysis and Discussion

Table 3 shows the experimental data for the series C (di/de = 0,50). It points out flatness of both planes, obtaining from CMM scans without standard deviation after: Machining (Op01), OTOQ (Op02), Annealing (Op04) and DAOQ (Op05). Tempering was not considered because it had less influence on the planarity tolerance.

Table 3. Planarity tolerances for the series C

Planarity tolerances						
Operations	Sample C.1		Sample C.2		Sample C.3	
	Ref. Plane [mm]	Opp. Plane [mm]	Ref. Plane [mm]	Opp. Plane [mm]	Ref. Plane [mm]	Opp. Plane [mm]
Machining (Op01)	0,011	0,014	0,007	0,012	0,024	0,013
Open quench (Op02)	0,103	0,114	0,057	0,072	0,088	0,117
Annealing (Op04)	0,234	0,237	0,176	0,161	0,137	0,068
Press quenching (Op05)	0,070	0,157	0,131	0,138	0,188	0,138

These three single values have permitted to calculate the mean value and the standard deviation of geometric tolerance. The data were processed for a whole set of geometries and operations so as to generate, for the reference plane and the opposite plane (Tables 4 and 5), a statistical interpretation of the data collected.

Table 4 refers to the OTOQ and Table 5 to the DAOQ. The OTOQ was compared with the machining operation and the DAOQ was matched with the spheroidizing annealing.

Figures 3 and 4 show the planar tolerance variations for the reference plane and the opposite plane, respectively. The variations were calculated in comparison with the previous operation of the technological cycle: in the case of the OTOQ the planar tolerance was compared with the same measure for the machined sample; for the DAOQ samples, the comparison was with the previous spheroidizing annealing treatment. When the variations are positive it means that the planarity tolerances increased due to the quenching because of induced distortions. At the opposite, when the variations are negative it means that the planarity tolerance decreased due to the quenching. For the whole set of geometries, the reference plane and the opposite plane show a similar behavior and, in particular, an improvement of planarity tolerance after DAOQ and a worsening for the OTOQ process.

Table 4. Mean value and standard deviation of planarity tolerance after Machining (Op01) and OTOQ (Op02) for whole the series: A, B, C, D, E

Planarity tolerances				
Name of the series	Machining (Op01)		Open Tank Oil Quenching OTOQ (Op02)	
	Ref. plane: mean value ± std. deviation [mm]	Opp. plane: mean value ± std. deviation [mm]	Ref. plane: mean value ± std. deviation [mm]	Opp. plane: mean value ± std. deviation [mm]
A $di/de = 0$	0,035 ± 0,015	0,034 ± 0,011	0,117 ± 0,063	0,116 ± 0,050
B $di/de = 0,25$	0,021 ± 0,016	0,025 ± 0,021	0,087 ± 0,015	0,098 ± 0,028
C $di/de = 0,50$	0,014 ± 0,009	0,013 ± 0,001	0,083 ± 0,023	0,101 ± 0,025
D $di/de = 0,75$	0,048 ± 0,023	0,018 ± 0,003	0,108 ± 0,018	0,058 ± 0,010
E $di/de = 0,84$	0,053 ± 0,028	0,027 ± 0,018	0,175 ± 0,017	0,096 ± 0,010

Table 5. Mean value and standard deviation of planarity tolerance after Spheroidizing Annealing (Op04) and DAOQ (Op05) for whole the series: A, B, C, D, E

Planarity tolerances				
Name of the series	Spheroidizing Annealing (Op04)		Die Assisted Quenching DAOQ (Op05)	
	Ref. plane: mean value ± std. deviation [mm]	Opp. plane: mean value ± std. deviation [mm]	Ref. plane: mean value ± std. deviation [mm]	Opp. plane: mean value ± std. deviation [mm]
A $di/de = 0$	0,283 ± 0,070	0,321 ± 0,088	0,101 ± 0,009	0,135 ± 0,019
B $di/de = 0,25$	0,240 ± 0,088	0,296 ± 0,094	0,096 ± 0,005	0,206 ± 0,015
C $di/de = 0,50$	0,182 ± 0,049	0,195 ± 0,039	0,090 ± 0,036	0,144 ± 0,011
D $di/de = 0,75$	0,353 ± 0,031	0,319 ± 0,037	0,114 ± 0,012	0,129 ± 0,006
E $di/de = 0,84$	0,379 ± 0,085	0,309 ± 0,084	0,124 ± 0,015	0,136 ± 0,008

Starting from a distorted shape, the DAOQ allows a recovery of the tolerances thanks of the pressurized die-assistance. This process controls diameters and holes too but this experimental campaign was realized in order to put in evidence the control of planarity tolerances.

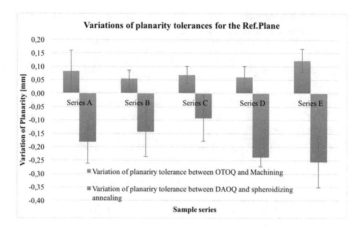

Fig. 3. Comparison of planarity tolerances for Reference Plane

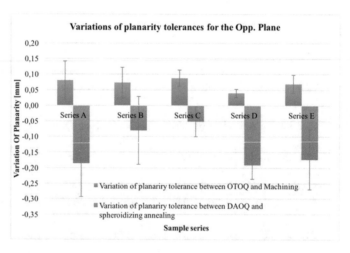

Fig. 4. Comparison of planarity tolerances for Opposite Plane

5   Conclusions

According to Table 2, the DAOQ ensures the most homogeneous treatment conditions in order to effectively control the material phase transformation and to achieve satisfactorily mechanical proprieties together with strict tolerances. The DAOQ reduces distortions that means a reduction of machining allowances during the product design and thus minimising the amount of employed material. Besides, it means a reduction or usually an avoidance of rework operations after quenching processes. DAOQ achieves the customer requirements by saving money, the environment and enhancing worker operations.

Through an experimental activity, the DAOQ process was demonstrated as capable of a recovery of flatness due to an important reduction of distortions. The comparison between OTOQ and DAOQ showed clearly that the second one controls efficiently the volumetric distortion that is strictly connected with the microstructure change due to quenching.

Acknowledgements. The authors would like to thank the MiUR and the Proterm S.p.A. enterprise for financial support and the cooperation during the whole project development. In particular, we would like to thank Andrea Zanotti for enriching technical discussions.

References

1. Stevenson, A.: Oxford Dictionary of English, 3rd edn. Oxford University Press, New York (2010)
2. United Nations General Assembly, 11 December 1987: Report of the World Commission on Environment and Development, A/RES/42/187, 96th Plenary Meeting
3. Rachuri, S., Sriram, R.D., Narayanan, A., Sarkar, P., Lee, J.H., Lyons, K.W., Srinivasan, V., Kemmerer, S.J.: Summary of the NIST workshop on sustainable manufacturing: metrics, standards, and infrastructure. Int. J. Sustain. Manuf. (IJSM) **2**(2/3) (2011)
4. Elkington, J.: Cannibals with Forks: Triple Bottom Line of 21st Century Business. Capstone Publishing Ltd., Oxford (1997)
5. Dornfeld, D.A.: Green Manufacturing: Fundamentals and Applications. Springer, New York (2013)
6. Davim, J.P.: Sustainable Manufacturing, Control Systems, Robotics and Manufacturing Series. ISTE Ltd., Wiley, London, Hoboken (2010)
7. Garetti, M., Taisch, M.: Sustainable manufacturing: trends and research challenges. Prod. Plan. Control **23**, 83–104 (2012)
8. Pusavec, F., Krajnik, P., Kopac, J.: Transitioning to sustainable production – Part I: application on machining technologies. J. Cleaner Prod. **18**, 174–184 (2010)
9. Seliger, G.: Sustainability in Manufacturing. Recovery of Resources in Product and Material Cycles. Springer, Heidelberg (2007)
10. ASM Handbook, vol. 4B: Steel Heat Treating Technologies, Basics of Distortion and Stress Generation During Heat Treatment, pp. 339–354. ASM International (2014)
11. Jones, L.E.: Fundamentals of Gear Press Quenching. Lindberg Technical & Management Services Group, Charlotte (1994)
12. ASM Handbook, vol. 4A: Steel Heat Treating Fundamentals and Processes, Press Quenching, pp. 252–256. ASM International (2013)

A Tool to Promote Sustainability in Casting Processes: Development Highlights

Emanuele Pagone$^{(\boxtimes)}$, Mark Jolly, and Konstantinos Salonitis

Manufacturing Department, Cranfield University, Cranfield MK43 0AL, UK
{e.pagone, m.r.jolly, k.salonitis}@cranfield.ac.uk

Abstract. The validity of traditional manufacturing decision variables (i.e. cost, quality, flexibility and time) is questioned by some important challenges of our time: the scarcity of natural resources and environmental pollution. Increasing energy cost to extract and process natural resources, alongside regulatory pressures against pollution, pushes very mature and competitive processes like casting towards a holistic approach where sustainability contributes to strategic decisions together with the mentioned traditional manufacturing variables. As a contribution to this industrial necessity, a modular tool able to analyse material and energy flows in casting processes is under development. In particular, the ability to represent automatically Sankey diagrams of the flows recently implemented is described and validated.

Keywords: Sustainable manufacturing systems · Energy efficiency · Casting

1 Introduction

Traditional manufacturing decision variables comprise cost, quality, flexibility and time [1]. However, in recent times scarcity of resources and pollution problems (e.g. the Paris UN Climate Change Conference agreement in December 2015) are imposing a revision or integration of these standards that are not able to take these aspect into consideration.

Moreover, the increasing competition derived by the globalised economy, depicts a challenging scenario for the whole manufacturing industry and, in particular, for a mature technology like metal casting.

One promising solution to these challenges is to implement sustainability alongside traditional manufacturing decision variables [1] (Fig. 1). Sustainability has been defined rather broadly as meeting the present needs of mankind without hindering the same opportunity to future generations [2]. In more pragmatic terms, a "triple bottom line" is generally accepted where environmental, economic and societal aspects are considered together [3].

The tool presented in this paper has been designed to support this change although the current effort is more focussed on energy resilience and environmental aspects. However, its modular architecture allows future integrations to potentially encompass the entire sustainability spectrum.

© Springer International Publishing AG 2017
G. Campana et al. (eds.), *Sustainable Design and Manufacturing 2017*, Smart Innovation, Systems and Technologies 68, DOI 10.1007/978-3-319-57078-5_17

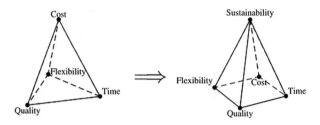

Fig. 1. Shift of decision-making attributes in modern manufacturing to include sustainability.

2 Aims and Rationale

To discuss the challenges presented in the introduction, Cranfield University, with the collaboration of the UK Cast Metals Federation (CMF), organised a workshop that was attended by several stake-holders of the sector (i.e. foundries, suppliers, consultants and academics) [4, 5]. Among the key outcomes of the event, it was highlighted the need for a tool able to rapidly analyse foundry measurements. The computer program presented in this work implements this tool in a user friendly fashion. In particular, the focus is on the effective visualisation of material and energy flows encompassing the entire chain of processes from charge to waste. The program is designed to be used as a stand-alone tool or to be integrated with existing manufacturing tools (see Sect. 5). Its main goal is to support decision-making, offering a range of capabilities such as explore improvements, identify synergies or opportunities for energy scavenging, benchmark practices or train personnel to a correct behaviour. These features intend to promote a more energy resilient casting practice with the long-term potential to incorporate also the three areas of sustainability. Moreover, attention has been devoted to allow flexibility in selecting the type and level of detail of the input data required to adapt to different types of foundries (different processes) and different data sets available (very detailed versus broad information).

3 Energy and Material Flows

As briefly mentioned, the energy intensive nature of foundries make them particularly sensitive to the cost of energy because it affects significantly their competitiveness. The on-going depletion of natural resources (including fossil fuels) clearly appears one important motivation to minimise energy inputs in foundries.

However, efforts aimed at the direct reduction of the energy required by casting processes must be supported by equally important endeavours related to material flows. In fact, a significant amount of energy is spent to melt metal that does not eventually remain part of the finished product. Relevant examples are the removal of the gating and oxidised or deteriorated scrap metal. One effective metric to represent the overall performance of the process with respect to material flows of the cast metal is the Operational Material Efficiency (OME), i.e. the ratio of the amount of shipped casting over the metal melt measured in a representative amount of time [6].

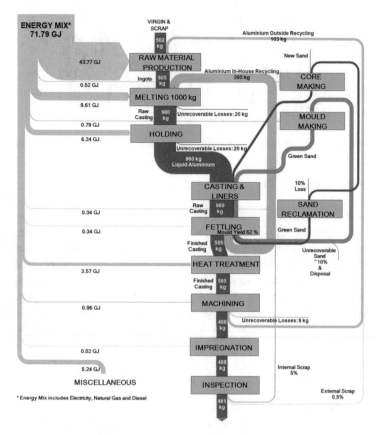

Fig. 2. Example of material and energy flows in a foundry represented by a Sankey diagram [20].

However, also ancillary material flows can impact negatively (and sometimes significantly) on the energetic performance of the plant. Examples of this type of material losses are the release in the environment of flue gasses obtained by the combustion of fossil fuels and the replacement or reclamation of sand in casting processes that make use of it.

Valid strategies applicable to the improvement of casting practices include audits and lean philosophy tools [7]. However, both show shortcomings for their implementation. Audits usually suggest improvements in the equipment [8, 9] that require capital investments often not viable for Small and Medium Enterprises (SMEs) like foundries [10–12]. Lean philosophy tools have the advantage of not requiring large investments but are usually less implemented in businesses like foundries where large stocks of raw materials are present [7].

Sankey diagrams can be an effective tool to better understand the process (as exemplified by Fig. 2) and aid decision-making [13]. Hence, it appears appropriate the decision to implement the ability to automatically generate Sankey diagrams in the computer program presented.

Previous research efforts have been devoted to the modelling and visualisation of processes in the form of Sankey diagrams. Viere *et al.* implemented a modelling methodology based on Petri networks with the inclusion of economic and environmental aspects in the energy and material flows [14]. Other examples are more focussed on the visualisation in the energy sector (e.g. flows related to primary and secondary fuels) [15]. Interestingly, Sankey diagrams have been considered unsuitable for real-time visualisation because "inherently lumped over a certain time period" and they "lack dynamic" [16]. However, the ability of the computer program presented in this work to automatically generate diagrams allows its encapsulation in real-time manufacturing systems (as explained in Sect. 5). Another key aspect of originality of the tool is its open source nature in contrast with the commercial nature of mature implementations present in the literature (e.g. Umberto, e!Sankey pro). Finally, it should be noted that Sankey diagrams are only one visualisation option among others that can be flexibly implemented and selected in the program.

4 Methods

A general overview of the current state of development of the computer program is initially provided, whereas a more detailed analysis of the Sankey diagram module is provided in a separated section.

4.1 Computer Program Overview

A broad representation of the tool work-flow shows three main stages (Fig. 3).

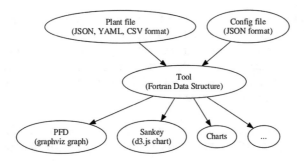

Fig. 3. General work-flow of the presented data visualisation tool.

Firstly, textual input files are prepared in specific formats to describe the foundry specifications ("Plant file") and provide configuration settings ("Config file"). It is mandatory for the user to provide information about the plant, whereas the inclusion of non-default configuration specifications is entirely optional. Work is ongoing to create a Graphical User Interface (GUI) that will collect the information from the user in a more user-friendly manner and will generate the relevant files automatically.

Secondly, after having parsed the input files, a data structure is generated (specifically, a derived data type according to the Fortran computer language parlance). Finally, an automatic conversion of this information to a graphical representation is executed. In a modular fashion, different graphical outputs are available and currently the program can generate Process Flow Diagrams (PFDs) by means of a graphviz [17] directed graph and Sankey diagrams using the javascript library d3.js [18, 19].

The data structure that contains the information about the foundry comprises an ordered series of process phases ("Components" in the "Plant file"). Each phase may include a variable number of inward or outward flows (including none) that can be categorised as of material or energy type. At least one process phase must be entered, as well as the name of the first phase of the entire process.

Foundry data can be collected with great flexibility adapting to the specific measurement system already available because there is no strong requirement on the level of accuracy required. Thus, either aggregated or instantaneous data can be provided and a sanity check based on the consistency of the unit of measures is performed by the program. However, the number of checks and restrictions imposed to the input data (concerning, for example, the sequence of different phases) has been kept to a minimum to avoid hindering flexibility. Potential gaps in the measurements of the inputs are not currently addressed directly by the program: it is the user that has to decide to change the level of accuracy that describes the plant or calculate or assume the missing datum or data. In fact, the tool is conceived primarily as an effective visualisation program whereas the simulation capabilities are kept to a minimum (basic conservation laws). However, it is not excluded that future development will also implement more extended simulation capabilities.

Moreover, the standardised format of the textual input files provides the advantage to link with relatively small effort the visualisation tool to external programs or databases (see Sect. 5).

The general design of the tool implements object-oriented concepts. Among the advantages of this approach, a notable example is the implementation of the routine to perform an action on all the elements of the plant, exploring the entire data structure automatically (e.g. calculate values at each "component" according to conservation laws). Such routine is a polymorphic object that applies the same algorithm (implemented only once) for every post-processing task required (more details will be provided in Sect. 4.2).

In a previous paper, more implementation details of the overall program and the Process Flow Diagram (PFD) module (including its validation) were presented [20].

4.2 Sankey Diagram Module

Compared to the general work-flow of Fig. 3, a more detailed look at the process to convert the internal data structure into a Sankey diagram (an object generated by the open source javascript library d3.js), reveals a few intermediate steps (Fig. 4).

Initially, the derived data type is converted into a CSV format file, filtering and structuring the information to be subsequently processed by a script in the R programming language [21], automatically generated in parallel. The R script uses the

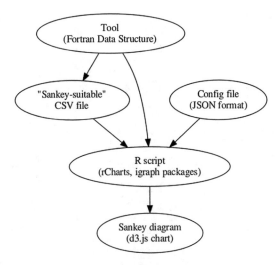

Fig. 4. Sankey diagram generation work-flow of the proposed tool.

packages rCharts [22] and igraph [23] to convert the mentioned CSV file into the final Sankey diagram object using the specifications present in the configuration file.

The original implementation of the Sankey diagram d3.js plug-in [24] did not support the representation of "cycles", i.e. flows of energy or material that are fed back to a previous step of the overall process (e.g. re-melting scrap metal after fettling). The name "cycles", adopted from the terminology of graph theory, is also used in the implementation to name the relevant derived data types and type bound procedures. A modified version of the Sankey diagram d3.js plug-in [25] that allows the presence of "cycles" has been interfaced to the presented work-flow.

A more detailed description of the implementation is provided analysing a minimal UML class diagram that completely describes the Sankey module (Fig. 5). The UML chart (obtained using the ForUML tool [26]) comprises the involved derived data types, type bound procedures and their connections. The naming of type bound procedures follows the well-established convention of "set", "get" and "predicate" methods [27].

In the lower part of Fig. 5 (from right to left) are shown the essential derived data types that implement the model of the manufacturing plant. At the highest level, type plant_t contains the information about the entire plant and encapsulates a number of comp_t types, i.e. the phases that describe the overall process (e.g. melting, pouring, machining). Type comp_t encapsulates a variable number of flow_t elements that represent the inward or outward, material or energy flows of that specific phase of the process. In order to model cycles, a cycle_t derived data type is encapsulated in the flow_t type. A more accurate description of these data types and their methods is presented in a previous paper [20].

In the upper part of Fig. 5, the data types and associated methods of the R script file (R_script_t), the CSV file (csv_t) and the Sankey diagram (sankey_t) are shown. All these data structures are identified by a name (fname for the two files and simply name for the sankey_t type) and an initialisation status flag (initialised).

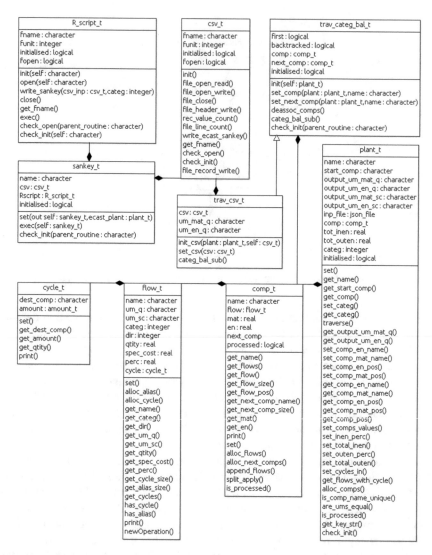

Fig. 5. Minimal UML class diagram describing the Sankey module implementation (obtained by ForUML [26]).

Type `sankey_t` (the simplest among them) encapsulates the necessary `csv_t` and `R_script_t` types alongside methods to set the Sankey diagram from an instance of type `plant_t` (set), to create the final output (`exec`), and to check the initialisation status of the object (`check_init`). Types `R_script_t` and `csv_t` represent files and then include a unit number (`funit`) and logical flags necessary to check if the file is initialised (`initialised`) or open (`fopen`). Among the relevant type bound procedures (mostly with simple functionalities and self-explanatory names) there are two notable ones. The `R_script_twrite_sankey` routine that generates the R script based on a CSV file (already created) and a specified category (i.e. material or energy) of the flows. More complex is the implementation of procedure `write_ecast_sankey` that is bound to type `csv_t`. This routine traverses the entire data structure of type `plant_t` to generate a CSV file with the following structure: source, target, value, category, unit of measure.

To accomplish elegantly the goal to traverse the entire data structure contained in an instance of `plant_t`, the `traverse` type bound procedure is invoked. This routine has the advantage to implement only once the algorithm necessary to explore the data structure, independently of the specific task required that will be executed using a polymorphic mechanism.

In this specific case, while traversing the data structure, it is necessary to operate on an instance of type `csv_t` that is not present in the scoping unit of `plant_t` and its `traverse` procedure. However, the `traverse` method uses also the `trav_cat_bal_t` data type (implemented in the same scoping unit) that contains the minimal information to explore the entire `plant_t` data structure. Extending type `trav_cat_bal_t` with the specific information necessary to generate the CSV file, it is possible to generate the super-class `trav_csv_t` that can be invoked with a polymorphism by the `traverse` routine algorithm. Figure 5 contains information also about this rather complex (but very effective and elegant) mechanism available in modern Fortran.

5   Industrial Implementation Scenarios

A number of options to integrate the computer program with traditional, existing manufacturing system tools are available. In this context, additional advantages that go beyond the visual representation of data become accessible. Six main areas are identified among these options (Table 1).

Table 1. Main features of the scenarios for the integration of the tool with existing manufacturing systems.

Scenario	Input	Additional benefit
Production improvement	Audited data	Accurate specifications (via interfaced tools, e.g. CFD)
Product design	Manufacturing processes database	Accurate specifications (via interfaced tools, e.g. CFD)
Benchmarking	Reference plants database	Basic Pareto analysis (i.e. find "low-hanging fruits")
Process monitoring	Real-time data	Process monitoring tool (via Internet of Things)
Training	Real-time data	Personnel didactic tool (via Internet of Things)
Life cycle assessment	Materials life cycle database	Product life cycle analysis

An interface to specialised Computational Fluid Dynamics (CFD) codes used to model the casting phase can provide an additional level of detail to the "overall picture" represented by the tool. Hence, it becomes possible to explore different design options or improve an existing product if enough data is exchanged in a standardised fashion. Work is currently undergoing to achieve these capabilities. One step further in this direction would be the implementation of automatic optimisation functions. To assist in the improvement of an existing product, the tool requires access to audited data. Alternatively, the design of a new product would require to access a database of manufacturing processes.

In a similar way, using a database of reference foundries, it is possible to benchmark the existing performance identifying the improvements with the highest return on investment (a basic Pareto analysis).

Feeding the software with real-time data from the production equipment, it becomes possible to monitor the process and also to control it, if a two-way communication channel is established. This last case can be easily imagined in the context of a "smart foundry" where the concept of "Internet of Things" is applied. Networking and suitable protocol communication capabilities would need to be implemented to achieve this goal. Furthermore, the availability of real-time data, transforms the computer program also into an effective training tool able to educate the workforce towards a virtuous behaviour including, in this way, also the "human" factor and completing its spectrum of action to the entire factory.

Finally, thanks to the modular design of the computer program, it is possible to implement embodied energy or CO_2-footprint flows that would allow the use of data provided by a materials life cycle database. In this way, a complete life cycle analysis of the product becomes possible.

6   Validation

The capabilities of the computer program (including the automatic generation of Sankey diagrams) are validated considering a complete, generic casting process that reproduces most of the key features that can be found in a foundry and it is based on industrial data producing aluminium alloy products.

The PFD (Fig. 6) shows the process steps with light blue rectangular boxes, the material flows in yellow septagons and the energy flows in red elliptical shapes. Every node of the diagram reports its content and, in particular, the process step boxes show the calculated material content at the *end* of the phase. If the calculated material or energy content of any process step is negative (the obvious result of a wrong input data set), execution halts and an error message is issued. The user-defined unit of measure of the flows are kg for materials and GJ for energy. As mentioned in Sect. 4.1, an error is issued if an inconsistency in the units of measure is identified. Other errors related to missing data in the input file are identified at parsing-time and a relevant error is returned. The program calculates and shows also the contribution in percentage points of each input and output energy flow compared to the total (respectively, total input and output) to highlight quickly the main flows.

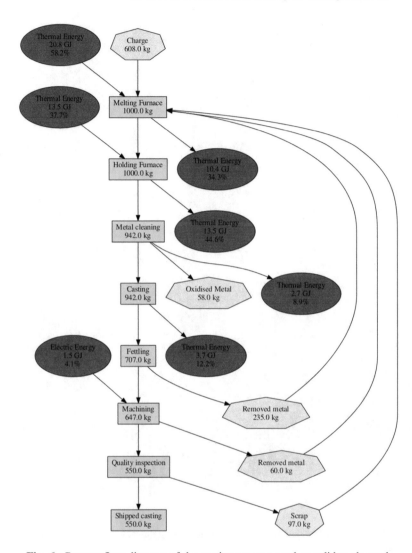

Fig. 6. Process flow diagram of the casting process used to validate the tool.

In this example, values are normalised for 1000 kg of melt material. Metal is initially heated until it changes phase in a furnace that is the main contributor to the energy input in the entire process and also is the place where material removed at later stages is collected to be re-melt. This step is also a major contributor of the energy losses (accounting for about 34% of the total energy output).

The liquid metal is then transferred to another furnace where it is put on hold to accommodate different production rates and let the metal oxides float to the surface. During this step another significant amount of energy is spent (more than a third of the total) to maintain the temperature constant, with significant losses in sensible heat (about 45% of the total).

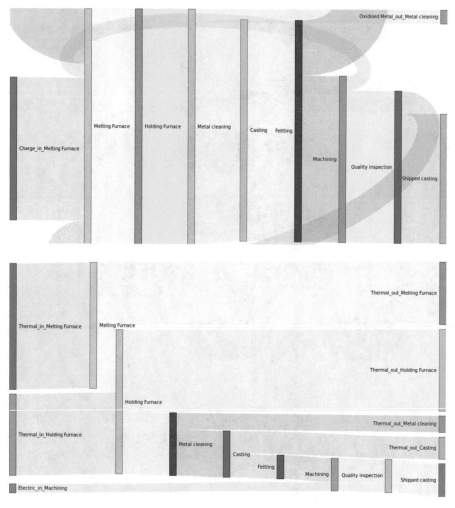

Fig. 7. Material (top) and energy (bottom) Sankey diagrams of the casting process used to validate the tool.

Before pouring, metal is cleaned (in this case, to remove oxides that inevitably form at high temperature) to improve the quality of the castings. Hence, the metal cleaning phase is associated with both material and energy (again, in the form of sensible heat) irrecoverable losses of medium-small proportions.

During the casting phase metal is poured into the mould cavity and cools down, generating another outgoing flow of energy (about 12% of the total outgoings).

After solidification, the product is then fettled with the removal of the gating and runners that are internally re-cycled in the melting furnace and account for a significant amount of metal (235 kg).

To achieve the required surface finish and dimensional accuracy, machining operations are required and generate additional metal scrap (60 kg) that is re-melt in the furnace. The energy input required during the machining operations is small compared to the other phases (about 4% of the total).

A quality inspection (that involves negligible flows of energy) is finally performed before shipping and the defective products (97 kg) are scrapped and returned to the melting furnace.

The relevant material and energy Sankey diagrams are presented in Fig. 7. The figure is a screen-shot because the standard output is an interactive chart accessible by any web-browser where nodes can be dragged within the canvas and a mouse-over action highlights the flows and reports its numerical value in the relevant tooltip.

Finally, it should be noted that the program offers wide flexibility about the degree of accuracy to describe the foundry process and it is dependent on the information included in the input file describing the foundry. It is possible to span from a broad, simplified, lumped approach to keep the diagram and description of the plant simple, up to a fine-grained visualisation based on very detailed information (an example of a more accurate description of a foundry plant can be found in [20]). The option to build-up progressively a more and more accurate model when additional information become available, is another advantage of the design of this tool.

7 Conclusion

A program designed to understand more accurately foundry processes and to promote the development of more energy efficient casting practices has been presented. In particular, the recently introduced ability to automatically generate Sankey diagrams has been described and validated. This effective representation helps identify main areas of improvement, discover synergies or opportunities to recover waste energy, benchmark processes or improve personnel behaviour.

Several promising options are available for the integration of this tool with legacy manufacturing systems, bringing additional benefits besides data visualisation to reach the mentioned aims. Such options have been briefly outlined and, in particular, the integration of CFD codes is being actively explored.

Efforts to make the tool more user friendly to set-up are focussed to creating a Graphical User Interface (GUI) that can generate a suitable input file.

Acknowledgements. The authors would like to acknowledge the UK EPSRC project "Small is Beautiful" for funding this work under grant EP/M013863/1.

References

1. Salonitis, K., Stavropoulos, P.: On the integration of the CAx systems towards sustainable production. Procedia CIRP **9**, 115–120 (2013)
2. UN General Assembly: Report of the World Commission on Environment and Development: Our Common Future, United Nations (1987)
3. Elkington, J.: Cannibals with Forks: The Triple Bottom Line of 21st Century Business. Capstone, Oxford (1997)

4. Mehrabi, H., Jolly, M., Salonitis, K.: Road-mapping towards a sustainable lower energy foundry. Foundry Trade J. Int. **190**(3732), 52–53 (2016)
5. Jolly, M.: Reducing energy in manufacturing – and how it affects the foundry industry. Foundry Trade J. Int. **190**(3735), 144–145 (2016)
6. Zeng, B., Jolly, M., Salonitis, K.: Manufacturing cost modelling of castings produced with CRIMSON process. In: TMS Annual Meeting, pp. 201–208 (2014)
7. Salonitis, K., Zeng, B., Mehrabi, H.A., Jolly, M.: The challenges for energy efficient casting processes. Procedia CIRP **40**, 24–29 (2016)
8. Klugman, S., Karlsson, M., Moshfegh, B.: A, Scandinavian chemical wood pulp mill. Part 1. Energy audit aiming at efficiency measures. Appl. Energy **84**(3), 326–339 (2007)
9. Salonitis, K.: Energy efficiency assessment of grinding strategy. Int. J. Energy Sect. Manag. **9**(1), 20–37 (2015)
10. Cagno, E., Trucco, P., Trianni, A., Sala, G.: Quick-E-scan: a methodology for the energy scan of SMEs. Energy **35**(5), 1916–1926 (2010)
11. Anderson, S.T., Newell, R.G.: Information programs for technology adoption: the case of energy-efficiency audits. Resour. Energy Econ. **26**(1), 27–50 (2004)
12. Thollander, P., Ottosson, M.: Energy management practices in Swedish energy-intensive industries. J. Cleaner Prod. **18**(12), 1125–1133 (2010)
13. Schmidt, M.: The Sankey diagram in energy and material flow management. Part II: methodology and current applications. J. Ind. Ecol. **12**(2), 173–185 (2008)
14. Viere, T., Stock, M., Genest, A.: How to achieve energy and resource efficiency: material and energy flow modeling and costing for a small and medium-sized company. In: Energy-Related and Economic Balancing and Evaluation of Technical Systems – Insights of the Cluster of Excellence eniPROD, Proceedings of Workshop on Cross-Sectional Group 1 "Energy Related Technologic and Economic Evaluation" of the Cluster of Excellence eniPROD. Wissenschaftliche Scripten, Auerbach (2013)
15. Subramanyam, V., Paramshivan, D., Kumar, A., Mondal, M.: Using Sankey diagrams to map energy flow from primary fuel to end use. Energy Convers. Manag. **91**, 342–352 (2015)
16. Herrmann, C., Zein, A., Wits, W.W., van Houten, F.J.A.M.: Visualization of environmental impacts for manufacturing processes using virtual reality. In: 44th CIRP Conference on Manufacturing Systems. University of Wisconsin, Madison (2011)
17. Gansner, E.R., North, S.C.: An open graph visualization system and its applications to software engineering. Softw. Pract. Exp. **30**(11), 1203–1233 (2000). www.graphviz.org
18. Bostock, M., Ogievetsky, V., Heer, J.: D3: data-driven documents. IEEE Trans. Vis. Comput. Graph. **17**, 2301–2309 (2011). (Proc InfoVis)
19. D3: Data-Driven Documents. https://github.com/d3/d3. Accessed 2 June 2016
20. Pagone, E., Jolly, M., Salonitis, K.: The development of a tool to promote sustainability in casting processes. Procedia CIRP **55**, 53–58 (2016)
21. R Core Team: R: A Language and Environment for Statistical Computing. R Foundation for Statistical Computing, Vienna, Austria (2016). https://www.R-project.org/
22. rCharts – Interactive Charts from R. http://ramnathv.github.io/rCharts/. Accessed 14 Nov 2016
23. igraph – The network analysis package. http://igraph.org/. Accessed 14 Nov 2016
24. D3 Plugins. https://github.com/d3/d3-sankey. Accessed 14 Nov 2016
25. D3-plugin-captain-sankey – Friendly (subtree) fork of the "sankey" plugin. https://github.com/soxofaan/d3-plugin-captain-sankey. Accessed 14 Nov 2016
26. ForUML – Extracting UML Class Diagrams from Object-Oriented Fortran. https://github.com/t2time/ForUML. Accessed 14 Nov 2016
27. Chapman, S.J.: Fortran 95/2003 for Scientists and Engineers. McGraw-Hill Education, New York (2007)

Supply Chain Major Disruptions and Sustainability Metrics: A Case Study

Luisa Huaccho Huatuco[1(✉)], Guljana Shakir Ullah[2], and Thomas F. Burgess[2]

[1] The York Management School, University of York,
Freboys Lane, York, YO10 5GD, UK
luisa.huatuco@york.ac.uk
[2] Leeds University Business School, University of Leeds,
Woodhouse Lane, Leeds, LS2 9JT, UK
gul.su1992@gmail.com, tfb@lubs.leeds.ac.uk

Abstract. Major Disruptions and Sustainability metrics in supply chains (SCs) are presented by means of a case study involving a large manufacturing organisation. The main findings point to four main strategies which organisations could use in the face of disruptions, namely: maintaining stock, sharing information, disaster management planning, and pursuing initiatives with suppliers (e.g. dual sourcing, outsourcing, SC visibility and risk modelling). The sustainability metrics indicate that being successful at managing disruptions in SCs does not preclude manufacturing organisations from also being successful in the sustainability dimensions of the triple bottom line.

Keywords: Supply chain · Disruptions · Sustainability · Case study · Metrics

1 Introduction

The nature of SC major disruptions is that they are usually unexpected and have a low frequency of occurrence [1]. They range from natural disasters e.g. flooding, to man-made disasters, e.g. wars, how a company can deal with these SC major disruptions will determine their survival and growth [2] and sustainability. The research question addressed in this paper is: *How does an organisation's SC disruption preparedness relate to its sustainability in the supply chain?*

2 Literature Review

2.1 Supply Chain Disruption (SCD)

Early literature regarding disruption emphasises the need to prevent and protect one's company against SCD [3–5]. However, this emphasis has now shifted to a longer-term approach, which is to recognise SCDs and strengthen the company's preparedness in order to build resilience towards disruption risks [6–10]. Researchers have recognised that SCs have become increasingly interconnected so effects of disruptions can surpass

© Springer International Publishing AG 2017
G. Campana et al. (eds.), *Sustainable Design and Manufacturing 2017*, Smart Innovation, Systems and Technologies 68, DOI 10.1007/978-3-319-57078-5_18

the actual point of disruption, potentially, across entire SCs thereby having far-reaching effects [11, 8]. A number of researchers [11–13] believe that the phenomenon of just-in-time (JIT) has worsened the effects of SCDs. The use of JIT to reduce cost and improve efficiency may be effective in a stable environment, but can be destructive if a disaster strikes, due to the JIT system being less flexible [9]. Barker and Santos [6] evidenced that having inventory available can decrease the burden which the SC disruption has caused; whereas this option would not be available if a JIT approach was being practiced.

2.2 Sustainable Supply Chains (SSC)

Due to organisations looking to become more environmentally-friendly and to use their materials more economically, a trend in SSC literature is reverse SC, [14]. Kusumastuti et al. [15] extended previous studies by Krikke et al. [14] who provided models for reverse SCs, by incorporating location and other complexities which are present in a SC. Their study investigates the difficulties of reverse SCs due to them being dispersed as organisations seek to manufacture in low-cost countries, such as China.

2.3 Combining SCDs and SSCs

Recently, research has combined SC disruptions and sustainability [16–20]. Some of them appear in academic journals whereas some papers were published at conferences, which indicate the early phases of the current topic. Rush et al. [19] provides recommendations for communities in case of disruptions, however, it does not provide quantifiable evidence. Hofmann et al. [20] study how SC disruptions can arise from sustainability issues. These studies emphasise the need for further research on the combined SC disruption and sustainability with quantifiable results. Furthermore, the latest Global risk report [21] shows how changing climate with extreme weather events have become a source of substantial risk for individuals and businesses. The constructs of SCDs and SSCs emerged from the literature review and were explored further through semi-structured interviews in the case study.

3 Methodology

The methodology consisted of a case study comprising semi-structured interviews, historical data and observations. Company A is a Danish Multinational engineering and electronics company with its headquarters in Denmark. It was established in 1943 and currently has a network of sales offices in 55 countries worldwide. Its markets are in different sectors, such as: aerospace, airport environment, automotive/ground vehicles, defence, environmental noise and vibration, space, wind energy and telecom and audio. This case study was chosen because it was expected that a large organisation would have the resources to dedicate to responding to SCD and to enhancing sustainability metrics. The data collection phase took place at two different locations. First, a day data collection in the UK followed by a two-day data collection in Denmark (HQ). The interviewees were: SC manager, group sourcing manager, global logistics manager, product engineering manager and manufacturing manager. The SC manager and the group sourcing manager facilitated one day in the UK, whereas the group sourcing manager additionally facilitated two days in Denmark. The rest of the interviewees granted an interview (with some follow ups) lasting for an average of one hour. The questions used in the semi-structured interviews were developed prior to the visits allowing for some open questions that could emerge during the course of the interview. The interviews covered two parts. First, questions in relation to SCDs, e.g. rating the frequency and impact of SCDs, the role of information sharing in tackling SCDs. Second, questions in relation to SC sustainability metrics, e.g. importance of sustainability targets and assessing their achievement of those, and rating their sustainability metrics and SSC performance. The qualitative data analysis comprised of three different stages: transcribing the semi-structured interviews taken at both the UK and Denmark sites, using NVIVO coding and classifying direct quotations to analyse the semi-structured interviews and writing the report including production of tables, charts, graphs and detailed results. The results were presented in the form of a written report which was sent via email to the key facilitators (SC Manager and Group Sourcing Manager) so they could assess confidentiality and accuracy of the report, as well as provide feedback comments.

4   Results

4.1   Safety Stock

To guard against disruption, Company A has two months' safety stock of most components and this is extended to three years' worth of stock for one core product with a single supplier for (despite searching for an alternative). It believes this safety stock strategy will provide them with sufficient robustness. If a disaster were to occur, it would still be able to operate with the stock levels they have and do not see severe disruptions to their organisations as the Product Engineering Manager states *"we would be up and running three to four months in worst case scenario after a disaster"*. Therefore:

P1: Firms with high levels of safety stock in their SC are less likely to be negatively affected by SCDs.

4.2   Suppliers

Dual Sourcing. Sourcing from a single supplier will enable a firm to reduce some costs, but can create problems on occasions where a major disruption occurs. Dual sourcing is not always available for every one of Company A's critical products. As indicated above for one critical product, they have not yet found an alternative source and therefore they source from a single supplier and have safety stock of up to three years. However, it has recognised the need for dual sourcing and the strategy they have used to select suitable products is as follows: *"Firstly, it was which product has the highest turnover or highest value for us and our customers and where this comes from in the SC and where we have the highest risk in the SC. Suppliers that make small bits and pieces – we can get elsewhere so it was depending on criticality"* – Group Sourcing Manager.

Outsourcing. Although in the last 20 years Company A has outsourced all its lower level components, which are not the main value-adding components, it has still retained the value-added production stage at its own site. Outsourcing could increase the risk of a disruption occurring; however, it has taken steps to ensure their robustness against such disruption. *"I think we have a very complex SC, but [...] we went from buying raw materials to buying semi-finalised goods. So we have strategically moved off the SC that makes it easier for us."* – Group Sourcing Manager.

SC Visibility. To ensure Company A is not affected by a disaster in the SC, they obtain the exact information they require from the suppliers including they require to know exactly where their components are. They therefore have full visibility of their products, both incoming and outgoing. Added to this, the suppliers are checked annually on their financial position to ensure there is no risk of bankruptcy. *"If for example, there is a risk of bankruptcy from one of our suppliers, we follow the supplier every year, are they in good shape, etc."* – SC Manager.

Risk Modelling. Another way in which Company A has actively sought to reduce the risk of disaster is by carrying out a simulation with a supplier in a region which is prone to disasters; flooding in particular. They simulated that a real-life flood had occurred in the area and that, despite the disaster, the supplier had to deliver the goods to it within eight weeks. This target was achieved by the supplier and the simulation was declared successful. It A has also ensured that its suppliers have a contingent supplier in case of floods. However, it needs to guard against becoming complacent; the nature of disasters is that they are unpredictable, which means a disaster could take place in a region where its suppliers are located and no disaster plan had been devised. Therefore, it seemed wise to carry out simulations with other suppliers too, starting from the ones most prone to disaster. Therefore:

P2: Firms with strong supplier management practices are less likely to be negatively affected by SCDs.

4.3   Disaster Management Planning

Company A has a disaster management plan in place which covers events ranging from the likely to the very unlikely. This plan is audited regularly by a group that owns it. Within this disaster management plan, many areas are covered. Each step of recovery is presented along with who is responsible for each stage. Along with these are contact numbers for the relevant staff members. The fact that it has not yet had to use this disaster management plan could be an indication of their robustness. Having a plan stands in contrast with some organisations which take disasters on a 'as they come' basis and do not have a set-up procedure to re-start the system to normal operations if a disaster were to occur. *"We have different kinds of contingency plans: We have specific insurance if something happens to specific suppliers for us – it's only for the biggest one. There is insurance for loss of turnover. Then we have another plan – divided into financial/office suppliers/transport – we have all the major suppliers – if something happens, we know what to do. Then, we have a third one –*

how often does the disaster happen, then we have different plan to take action and then they choose from a high risk to a low risk." – SC Manager. Therefore:

P3: *Firms with strong disaster management planning are less likely to be negatively affected by SCDs.*

4.4 Information Sharing

Information sharing is key to preventing disruption occurring and responding appropriately when a disruption does occur. Company A shares information both internally and externally. Each department gets regular feedback from questionnaires completed by employees, which is then discussed within the department to see whether targets have been met; comparisons are made with previous targets and plans within each department are produced following this. This information sharing also happens between departments, via shared files in the Information System they work with internally; these are typically in the form of MS Excel files, in which they have traffic light colour coding to represent progress on solving a problem. It also shares information externally with two other companies that are also owned by the same group; this sharing occurs approximately once a year in order to *"exchange best practices"* – Group Sourcing Manager. Therefore:

P4: *Firms with strong information sharing within their SC are less likely to be negatively affected by SCDs.*

4.5 Sustainability Metrics

Economic. The interviewed managers claim economic sustainability is built into Company A's ethos of providing a quality product. This means *"not just quality in terms of touch and feel but quality in terms of surrounding"* – Group Sourcing Manager. Due to this desire for high quality, it has not increased price because they are already operating in a high quality area, i.e. Its approach is not based on low cost. It is known that some multinational organisations operate in countries where unethical issues, such as child labour, are very common and go unnoticed.

Environmental. In addition to having a high quality product, Company A has a goal to be 3% more cost efficient than the previous year. The researchers recommended that it investigates eco-efficiency and how it can help them, as the survey responses of how eco-efficient it is varied significantly. In theory, being more eco-efficient should also yield economic benefits. It has the resources, such as highly skilled staff, to look more in to the area of eco-efficiency which could yield long-term benefits. Company A is certified by ISO14001 for which they are audited every year. Other aspects such as reduction of waste and energy consumption are also measured. In order to be more environmentally friendly, it sends out consolidated shipments rather than shipping out every day. The waste ratio is medium while one employee believed it to be very low and one employee believed it to be very high. These results may suggest that more progress is needed in terms of looking at environmental factors, such as the waste ratio and also making employees aware of progress. This point is backed up by a statement made by Product Engineering Manager regarding sustainability targets "It is available so everybody can access them, but that's not the same as being aware of it – more than half would say they don't know. All employees can look at our KPI's, but not many do it."

Social. Company A has many initiatives in place, for example, they run an ethics programme where they train all their employees in relevant areas. When selecting suppliers, it looks into the social characteristics of the suppliers and have dismissed some suppliers due to unethical behaviour, e.g. child labour practices: *"When we are selecting suppliers, we look into the social dimension from the suppliers – we wouldn't use child labour etc. They do challenge us that we do have low cost suppliers but not for any cost – it has to be ok in an ethic and social correct way. We have had suppliers that have been dismissed because they were not fulfilling the standards we required by [owner Company]"* - Group Sourcing Manager. Company A's HR department looks at how employees feel they are treated. At the UK location, employees are free to give suggestions about production improvements. It is certified to ISO14001, but they may also want to consider the ISO18001 certificate in this situation. All responding employees believed that it is a good employer and *"employees here have a huge degree of flexibility"* – Group Sourcing Manager. The average number of years an employee works there is 17 years and retired employees also return to keep in contact at their 'senior club', which indicates a degree of satisfaction. H&S is extremely important and they have occupational health advice which they can call on, as required per case. An audit is also carried out to ensure employees are not working more than the hours they are allowed. From the employees surveyed, half believed the standard of H&S is high whereas half deemed it to be very high. Therefore:

P5: Firms with strong SCDs preparation are more likely to exhibit high levels of SC sustainability metrics.

5   Conclusions

This paper has addressed the research question: *How does an organisation's SC disruption preparedness relate to its sustainability in the supply chain?* Company A has made great efforts to address the different SCDs possibilities by adopting several of the reported strategies in the literature, e.g. information sharing, disaster management planning, hence making it strongly prepared to SC disruptions, i.e. robust. It has also taken into account its sustainability metrics achieving a strong performance overall. However, it needs to pay more attention to the long lead times, the waste ratio and the resource utilisation rate. On the positive side, labour equity shows that all employees interviewed are extremely satisfied to be working there. The limitations of this paper are mainly related to the drawbacks of single case studies, namely the lack of generalisation of results, which other research methodologies could overcome, e.g. survey. Thus, the results presented here apply to Company A and its SC without further implications for other organisations. Nevertheless, the in-depth understanding gained through the relatively close contact with the organisation is valuable to understand the phenomenon under scrutiny, i.e. SC disruption and sustainability. Future research avenues could include more case studies, especially at the other end of the spectrum of expected resilience, e.g. SMEs which do not have the resources to dedicate to major disruptions with "low probability and high impact" occurrence. Furthermore, empirically testing the five propositions stated in this paper can shed some light on the topic.

References

1. Oke, A., Gopalakrishnan, M.: Managing disruptions in supply chains: a case study of a retail supply chain. Int. J. Prod. Econ. **118**, 168–174 (2009)
2. Knemeyer, A.M., Zinn, W., Eroglu, C.: Proactive planning for catastrophic events in supply chains. J. Oper. Manag. **27**(2), 141–153 (2009)
3. Chopra, S., Sodhi, M.: Managing risk to avoid supply-chain breakdown. MIT Sloan Manag. Rev. **46**(1), 53–61 (2004)
4. Christopher, M., Peck, H.: Building the resilient supply chain. Int. J. Logistics Manag. **15**(2), 1–13 (2004)
5. Sheffi, Y., Rice Jr., J.B.: A supply chain view of the resilient enterprise. MIT Sloan Manag. Rev. **47**(1), 45–47 (2005)
6. Barker, K., Santos, J.R.: Measuring the efficacy of inventory with a dynamic input–output model. Int. J. Prod. Econ. **126**(1), 130–143 (2010)
7. Ergun, O., HeierStamm, J.L., Keskinocak, P., Swann, J.L.: Waffle house restaurants hurricane response: a case study. Int. J. Prod. Econ. **126**(1), 111–120 (2010)
8. Kleindorfer, P.R., Saad, G.H.: Managing disruption risks in supply chains. Prod. Oper. Manag. **14**(1), 53–68 (2005)
9. Schmitt, A.J., Singh, M.: A quantitative analysis of disruption risk in a multi-echelon supply chain. Int. J. Prod. Econ. **139**(1), 22–32 (2012)
10. Chopra, S., Sodhi, M.: Reducing the risk of supply chain disruptions. MIT Sloan Manag. Rev. **55**(3), 72–80 (2014)
11. Bakshi, N., Kleindorfer, P.: Co-opetition and investment for supply chain resilience. Prod. Oper. Manag. **18**(6), 583–603 (2009)
12. Thevenaz, C., Resodihardjo, S.: All the best laid plans conditions impeding proper emergency response. Int. J. Prod. Econ. **126**(1), 7–21 (2010)
13. Vachon, S., Klassen, R.D.: Environmental management and manufacturing performance: the role of collaboration in the supply chain. Int. J. Prod. Econ. **111**(2), 299–315 (2008)
14. Krikke, H., Blanc, L.L., Krieken, M.V., Fleuren, H.: Low-frequency collection of materials disassembled from end-of-life vehicles on the value of on-line monitoring in optimizing route planning. Int. J. Prod. Econ. **111**(2), 209–228 (2008)
15. Kusumastuti, R.D., Piplani, R., Hian Lim, G.: Redesigning closed-loop service network at a computer manufacturer: a case study. Int. J. Prod. Econ. **111**(2), 244–260 (2008)
16. Shakir Ullah, G., Huaccho Huatuco, L., Burgess, T.F.: A literature review of disruption and sustainability in supply chains. In: 2014 Conference Sustainable Design and Manufacturing, Cardiff, 28th–30th April 2014 (2014a)
17. Shakir Ullah, G., Huaccho Huatuco, L., Burgess, T.F.: Disruption and sustainability in supply chains. In: White Rose Business and Management Doctoral Conference, York Management School, York, 14th–15th July 2014 (2014b)

18. Gonzalez, E., Sarkis, J., Huisingh, D., Huatuco, L.H., Maculan, N., Montoya-Torres, J., de Almeida, C.M.V.B.: Making real progress toward more sustainable societies using decision support models and tools: Introduction to the special volume. J. Cleaner Prod. **105**, 1–13 (2015)
19. Rush, C., Houser, R., Partridge, A.: Rebuilding sustainable communities for children and families after disaster: recommendations from symposium participants in response to the April 27th, 2011 Tornadoes. Community Ment. Health J. **51**, 132–138 (2015)
20. Hofmann, H., Busse, C., Bode, C., Henke, M.: Sustainability-related supply chain risks: conceptualization and management. Bus. Strategy Environ. **23**(3), 160–172 (2014)
21. World Economic Forum: The Global Risk Report 2017, 12th edn., Insight report (2017)

Multi-Layer Stream Mapping: Application to an Injection Moulding Production System

M.N. Gomes[1], A.J. Baptista[2], A.P. Guedes[2,3], I. Ribeiro[1], E.J. Lourenço[2], and P. Peças[1(✉)]

[1] IDMEC, Instituto Superior Técnico, Universidade de Lisboa,
Av. Rovisco Pais, 1049-001 Lisbon, Portugal
{manuel.gomes,ines.ribeiro,ppecas}@tecnico.ulisboa.pt
[2] INEGI, Instituto de Ciência e Inovação em Engenharia Mecânica e Engenharia Industrial,
Rua Dr. Roberto Frias, Campus da FEUP, 400, 4200-465 Porto, Portugal
{abaptista,apguedes,elourenco}@inegi.up.pt
[3] FEUP, Faculdade de Engenharia da Universidade do Porto,
Rua Dr. Roberto Frias, Campus da FEUP, 4200-465 Porto, Portugal
apguedes@fe.up.pt

Abstract. The Multi-Layer Stream Mapping (MSM) methodology addresses current challenges regarding the applicability of Lean Thinking concepts in the domain of sustainability assessment tools. Therefore, MSM aims to assess the overall performance of a production system, while evaluating the productivity and efficiency of resource utilization as well as evaluate the costs related to missuses and inefficiencies and other process and domains variables. This paper highlights the benefits arising from the application of the MSM methodology in a real industrial case regarding the injection moulding process, namely fostering the quantification of the efficiency of different resources streams, for its improvement, for the several production processes involved. So, it is explained how MSM can contribute for a more sustainable production system with a continuously increasing productivity.

Keywords: Multi-Layer Stream Mapping · Resources efficiency · Operational efficiency · Production system

1 Introduction

The continuous population growth as led to increased concerns regarding the protection of the environment and resource scarcity by our society and economy. The concept of sustainability was defined in 1987 in a Report of the World Commission on Environment and Development as "development that meets the needs of the present without compromising the ability of future generations to meet their own needs" [1]. Since then, the interest and concern about such subjects has aroused growing interest and the sustainability concept was never as meaningful and important as it is nowadays [2]. Sustainability is in the agenda of many organizations that wish to integrate the ecological and social issues in their business strategy [3]. One consequence is the exploration and use

© Springer International Publishing AG 2017
G. Campana et al. (eds.), *Sustainable Design and Manufacturing 2017*, Smart Innovation, Systems and Technologies 68, DOI 10.1007/978-3-319-57078-5_19

of management systems adapted to the current global challenge of sustainability. The lean manufacturing, which is based on the identification and elimination of waste from the value stream, is in line with these demanding challenges [4]. In fact, some studies point out that the concepts of lean and sustainability are deeply related and synergies do exist being in already observed cases where organizations familiar with lean will easily grasp sustainability and vice-versa [4, 5].

In order to face the current challenge of sustainability it is expected that organizations adopt, not just an environmental conscience and practical tools and methods to apply this attitude to reduce the environmental impact of their processes and products, but also a managerial approach that allows to decouple the environmental burdens from economic growth, while retaining or improving the quality of the products and competitiveness. Lean Manufacturing has proved in the last decades to be a powerful approach, containing a wide set of tools, that allow gains in terms of productivity, efficiency and cost reductions, while improving product quality. In recent years, the MSM - Multi-Layer Stream Mapping methodology was introduced to address current challenges regarding the applicability of Lean Thinking concepts in the domain of sustainability assessing tools.

MSM aims to assess the overall performance of a production system, while evaluating the productivity and efficiency of resource utilization (e.g. energy, raw materials, various consumables, etc.) as well as evaluate the costs related to missuses and inefficiencies and other process and domain variables (e.g. quality aspects, specification metrics, bottlenecks, etc.). The MSM contains an intrinsic link with the lean tool Value Stream Mapping (VSM). However, this new approach, introduces disruptive innovations related with its applicability and wide assessment solutions for complex systems analysis.

The MSM is intended to be used, not only for analytical evaluation of production system efficiency, but also to support the decision making process, namely for greenfield design or online systems monitoring.

This paper highlights the benefits arising from the application of the MSM methodology in a real industrial case with an injection moulding process, namely fostering the quantification of the efficiency of different resources streams, enabling to support decision-making process for its improvement, for the several production processes involved. In particular it is explained how MSM can contribute for a more sustainable production system with a continuously increasing productivity.

2   From VSM to MSM

VSM is a lean tool that can be applied to visualize the product or service flow in a value based analysis. The tool consists on a representative flow diagram that enables the identification of inefficient processes, regarding time, that have room for improvement in order to increase the final value delivered to the client [6].

The analysis promoted by the VSM allows a broader comprehension of all the processes involved and its boundaries in a continuous view and is not just focused on individual processes. The outputs from VSM allow both the added-value time, which is the time required to perform a specific process, and the non-added-value time, which refers to the remaining time (waiting time, transport, etc.) to be considered.

VSM is a particularly useful tool for mapping the production systems and identifying critical situations for improvement actions. In addition, it is well suited, due to its highly visual component contribution, to a more intuitive and broad understanding about the need to reduce non-added-value time (wastes). Several authors point out that these positive characteristics of VSM could be further improved by including (i) a dynamic perspective allowing an understanding of the system behaviour over time; (ii) uncertainty aspects allowing to include the influence of variability in performance and (iii) resources-efficiency related indicators towards introducing sustainability aspects in daily decisions in companies [7–10].

Based on these recognized gaps of the VSM, the MSM was developed. MSM takes into account, not only the time, but also other streams: energy, materials, consumables, etc. Keeping a similar graphic layout to VSM, MSM offers a multi-layers (multi-variables) mapping, which explains the Multi-Layer Stream Mapping name given to the methodology. Following VSM logics, MSM is characterized by distinguishing the "add-value" and "non-add-value" events/operations in each step of the production system.

So, MSM allows the detection of different forms of wastes and inefficiencies, as well as the related costs that are currently affecting the system performance, therefore this proposed tool can be integrated in the lean methodologies and tools universe. By having more variables involved in the assessment, it also enables to perform a more detailed and wider analysis on the system's performance.

3 The MSM Methodology

The MSM is a lean tool originally developed in INEGI by Lourenço et al. (2013) in the framework of studying new methodologies used to evaluate the eco-efficiency performance trough resource efficiency in an industrial environment.

The MSM approach suggests the mapping of more variables than just the time for the system's performance assessment, so it was important to find a way to integrate/compare them. For that, the MSM suggests the non-dimensionalization of the different variables in a generic based ratio between the portion of the variable that "adds value" to the product and the "total amount" of the variable that enters (is used by) the unit process (see Eq. (1)).

$$\Phi = \frac{Value\,Added\,Fraction}{(Value\,Added\;Fraction + Non\text{-}Value\,Added\,Fraction)} \tag{1}$$

This ratio can range from 0–100% thus allowing the aggregation of efficiency ratios along a production system, sectors or even plants in a bottom-up analysis. Depending on the value obtained for the ratio, the MSM approach suggests the use of a 4 colour scheme to ease the results interpretation and to instill the tool with a highly visual management component (efficiency 100%–90%: Green; 89%–70%: Yellow; 69%–40%: Orange; <40% Red). This way, the tool can be used as an "alarm", allowing an easy and direct way of detecting critical situations with inferior efficiency results.

In order to apply the MSM and interpret the assessment results, the following steps should be performed:

1. Identification of the system boundaries;
2. Identifications of the process steps;
3. Identification of all relevant process variables and parameters;
4. Definition of the associated KPI to each variable, always to be maximized and with figures always comprehended between [0–100%];

By guiding the analysis to be performed closer to the real characteristics of the system, and by not functioning as a rigid model, the MSM provides a high flexibility capacity to adapt to different productive system types.

After defining the variables of the system, the next step of the MSM methodology consists on monitoring those variables. The monitoring, which can be performed in-line, will allow the understanding of the current behaviour of the different variables that were previously selected to characterize the system (Fig. 1) .

Fig. 1. MSM approach on the assessment of a productive system

The outcome of the MSM methodology is presented in the form of MSM "scorecards" (dashboards) that clearly display all the information regarding resources and operational aspects. The multiple variables used on the efficiency assessment end up creating multiple value streams that, ultimately, allow a quick identification and quantification of the system's current inefficiencies, as well as the related costs [11].

The critical situations identified and marked as red in the final scorecard, must then be further studied so that they can be a target of future improvement actions. After the implementation of the improvement actions the system is expected to evolve towards a better scenario, where waste is minimized thus enhancing a culture of continuous improvement within the organization.

3.1 Types of Process Parameters

The MSM approach, by not having any particular pre-defined variables in the assessment, allows an organization to select variables according to the process and type of analysis it is intended to perform. In an industrial environment, where high productivity is often the measure of performance of the system, it is important to select variables that allow the understanding of the conditions on how the system is currently operating compared to its potential. However, with the increased concern about sustainability, industry should not only be focused on the performance of production systems but also on improving the efficiency level, namely the use of resources. The challenge is then to be able to create more with less, and through increased resource efficiency reduce raw material demand and emissions.

The variables for the efficiency assessment, of a given production system, can be grouped in two categories: resource and operational. The first category includes variables such as: energy, water, gas and other consumables. On the other hand, the operation variables are often: machine speed loss, machine availability, process temperature, quality, product dimensions, etc. (Fig. 2).

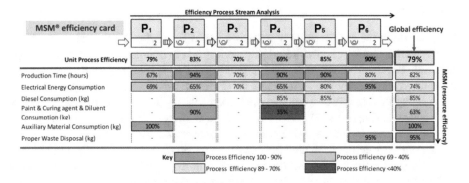

Fig. 2. Outcome of the MSM methodology - scorecard example

3.2   Potentialities

Despite having its roots in the lean tool VSM, the MSM tool introduces innovations, namely; a wider variety of variables on the assessment of the overall system performance, and the possibility to be used not only as evaluation tool (for diagnosis) but rather as in-line efficiency control tool, by receiving in streaming production data.

According to Lourenço et al. 2016, the application of the MSM tool to a production system, must allow:

- The identification of the most critical resources or process parameters;
- The identification and quantification of inefficiencies of a given production system and unit process;
- The quantification of resources and operational efficiency, and overall production system performance;
- The prioritization for implementation of improvement actions and optimization actions;
- The evaluation of efficiency regarding progress and incite for continuous improvement sustainability within organizations.

The MSM, by decomposing a variable into just two portions: "add-value" and "non-added-value", can also be used to perform other kind of cost analysis. It allows the conversion of those portions into monetary values, making it possible, for example, to estimate how much capital is currently being used but does not add value, i.e. cost of the missuses/inefficiencies. This particular feature is highly appreciated by the industry, which often does not understand the real dimension and impact of inefficient processes until it is translated into currency. That way, the MSM can also be used as a support for the decision making process, since the process improvements actions may be studied with more precision for its payback time (return of investment analysis).

4   Case Study

4.1   Scope

To understand the real benefits of applying the MSM tool to a production system, a particular case study was performed in an industrial environment. The company where the study occurred belong to the plastic industry, being specialized in the creation and assembling of different plastic injection parts used in water conveyance systems.

Although the referenced company is responsible for the creation of a wide variety of products and applications, this particular case study will consider the study of a production flow associated to one particular, but representative, product.

The process steps, considered for this study, are presented and described in Table 1. All data were collected on the production site during a period of 3 months, where different measurements were performed to assess the system's current efficiency level. According to each specific process steps, the variables were chosen and categorized into two distinct groups: resources and operational. The different variables used in this case study can be found in Table 2.

Table 1. Description of the process steps that takes place along the production system

Process steps	Description
Plastic injection	Creation of the basic components for later assembly
Assembly	Assembly of the different components into the final product
Film packing	Application of a protective film on the final product already packed on a card box
Shipping	Process of moving the finished product from the company's facilities towards the customer

Table 2. Variables used in the case study and respective classification

Variable type	Variables	
Resources	Time [min.]	Raw material [kg]
	Electric energy [kWh]	Consumables [%Roll/pallet]
	Worker availability [h]	Area [m^2]
Operational	Availability [h]	
	Performance [h]	
	Quality [units]	

The data regarding processing time were collected by measuring the time that a specific plastic component, later used as an assembly part, spent through the different process steps. The time measured includes both the process time, which is considered the absolute time required by an operation, and the lead time that is the total time that the component spent on each process steps from its arrival. The electrical energy consumption was measured only on the plastic injection unit since it is the process steps with the most representative electrical energy consumption. The electrical consumption was measured using metering devices during a total period of time of 86 h. The amount of raw material used to create the desired plastic injection component was also measured in the same process. Since some of the plastic material is retained in the sprue and runners of the injection mold during the solidification phase, it will not be present in the final form of product, and, for that reason, is classified as waste. In this particular case, the consumable used in the process consists of plastic film during the packing of the finished product. The measurements included an analysis on the total number of pallets that a single plastic film roll could be used to wrap. The last resource variable measured was the building area. The current shop-floor area occupied by the machines and work stations were measured for the different process steps and compared to the total available area. The operational variables, that includes the availably, performance and quality parameters, were measured in a period of 8-hour, since the company is currently operating during 3-shifts (8 h each) and the production is often defined by shift. The availability allowed the detection of problems that are affecting the equipment and, consequently, causing breaks on the production rate. The performance parameter measures the current performance of a process step compared to the expected (ideal cycle time), allowing the understanding of how much room for improvement still exists in the production system. The quality parameter includes an analysis of the number of rejected units (out of quality specification) compared to the total number of units produced.

4.2 MSM Efficiency Results

The results achieved regarding the resource efficiency of the system obtained by the application of the MSM approach are presented in Fig. 3. The process unit efficiency results regarding the resources variables reveal that the injection and film packing are currently the least efficient operating units. Within the different resource parameters used for the efficiency assessment, the parameter "time" is the most critical one and the production system overall efficiency regarding resource efficiency is currently 80%.

In terms of operational efficiency (Fig. 4) the results indicate that the injection unit has the worst operational performance and the least efficient parameter identified is the machine and cell performance.

In Fig. 5 is presented an overall analysis of the system and the correspondent Overall Equipment Effectiveness results, integrating both the performance of resources and the operational variables. The overall system performance is the result of the product between the resources and overall operation performance. The injection and film packing process units presents the lowest performance level. The overall production system's performance, which results from the average value of all unit process performance values, is, for this specific case, 65%.

Fig. 3. Resource efficiency scorecard

The overall analysis (Fig. 5) also included a performance indicator, namely the OEE (Overall Equipment Effectiveness). The OEE is calculated from the multiplication between the availability, performance and quality parameters which were previously taken into account in the selected operational variables. Analyzing the results achieved for the OEE on this production system, it is clear that the most critical process stage is the injection process. The low result obtained for this process (38%) is however expected since the performance of the injection unit was also low.

Fig. 4. Operational production efficiency scorecard

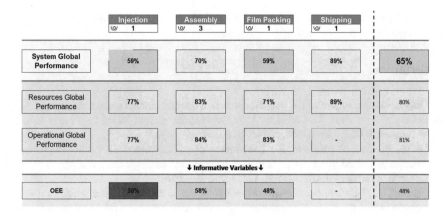

Fig. 5. Resumed analysis scorecard

5   Conclusions

The MSM framework has been applied in different case studies and sectors to the assessment of overall efficiency and performance of multi-domain complex system analysis. Thus, supporting the decision making process and allowing continuous improvement of production systems through systematic and standardized analysis of the efficiency of a given production line, production sector, process step, or even to assess the efficiency of several factories (by a complete bottom-up analysis). The methodology and supporting tool can be applied in three types of analysis: green-field analysis (new production systems design), in situ analysis (static analysis on running systems), monitoring and continuous analysis (dynamic analysis upon streaming of data feeding).

As the sustainability theme gains importance, new methodologies and industry-oriented tools to integrate this component have been developed. The industry, based on the high consumption of natural resources on the productive activities, can play a relevant role in the search for a sustainable future for the planet. However, besides environmental concerns, the main focus of industry is still the production performance. Based on the needs of the planet and industry goals, an innovative tool that can integrate both these components has been developed. MSM tool is a novel approach to support the control of production systems, enabling the visualization of performance linked to the monitoring systems available nowadays. It is based on the lean principles, seeking to improve the system's current performance while accounting for the efficiency level regarding the resources used in the different production activities. The MSM approach is based on the quantification of value and waste in each individual activity of the productive flow, using different type of variables to evaluate the system performance. The variables used in this case study are related to the resources consumed in the operations and the general performance of the system. The MSM Scorecards, that are a key output of the MSM tool, allow the identification of different types of inefficiencies that are currently affecting the system. The highly visual component of the tool, allied with the translation of the variable's units measured in a simple percentage form, allows any worker within the organization to interpret the results, promoting the participation of all in the improving/success of the company.

References

1. United Nations: Report of the world commission on environment and development (1987)
2. Fercoq, A., Lamouri, S., Carbone, V.: Lean/green integration focused on waste reduction techniques. J. Clean. Prod. **137**, 567–578 (2016)
3. Luthra, S., Garg, D., Haleem, A.: The impacts of critical success factors for implementing green supply chain management towards sustainability: an empirical investigation of Indian automobile industry. J. Clean. Prod. **121**, 142–158 (2016)
4. Cherrafi, A., Elfezazi, S., Chiarini, A., Mokhlis, A., Benhida, K.: The integration of lean manufacturing, six sigma and sustainability: a literature review and future research directions for developing a specific model. J. Clean. Prod. **139**, 828–846 (2016)
5. Ng, R., Low, J.S.C., Song, B.: Integrating and implementing lean and green practices based on proposition of carbon-value efficiency metric. J. Clean. Prod. **95**, 242–255 (2015)
6. Wilson, L.: How to Implement Lean Manufacturing. Mc Graw, New York (2013)
7. Womack, James P.: Value Stream Mapping. The Manufacturing Engineering, Brookline (2006)
8. Braglia, M., Frosolini, M., Zammori, F.: Uncertainty in value stream mapping analysis. Int. J. Logist. Res. Appl. **126**(6), 435–453 (2009)
9. Rother, M., Shook, J.: Learning to See: Value Stream Mapping to Add Value and Eliminate Muda. Lean Enterprise Institute, Brookline (1999)
10. Brown, A., Amundson, J., Badurdeen, F.: Sustainable value stream mapping (Sus-VSM) in different manufacturing system configurations: application case studies. J. Clean. Prod. **85**, 164–179 (2014)
11. Lourenço, E.J.J., Pereira, J.P.P., Barbosa, R., Baptista, A.J.J.: Using multi-layer stream mapping to assess the overall efficiency and waste of a production system: a case study from the plywood industry. Procedia CIRP **48**, 128–133 (2016)

Sustainability of Micro Electrochemical Machining: Discussion

Mina Mortazavi[✉] and Atanas Ivanov

Department of Mechanical, Aerospace and Civil Engineering, Brunel University London,
Kingston Lane, Uxbridge, UB8 3PH, UK
{Mina.mortazavi,Atanas.ivanov}@brunel.ac.uk

Abstract. Micro electrochemical machining is one of the promising non-conventional machining methods which has created new horizon in Micro and Nano product technologies including MEMS, defense, medical and automobile industries. An existing challenge in manufacturing has been known as the lack of identified methodology and measurement science to evaluate the sustainability of the process performance. This challenge would be more critical when it comes to Micro and Nano manufacturing process. This paper presents a review on challenges encountered in micro electrochemical machining considering it as a sustainable manufacturing process.

Keywords: Sustainability · Micro-electrochemical machining (micro-ECM) · Sustainable micro manufacturing

1 Introduction

U.S. Department of Commerce defines the Sustainable manufacturing as the creation of manufactured products which use processes that minimize negative environmental impacts, conserve energy and natural resources, are safe for employees, communities, and consumers and are economically sound [1].

Deficiency of measurement science and methodologies to compare the performance of manufacturing processes with respect to sustainability has resulted in inaccurate and uncertain comparisons. However different efforts have been made to suggest indicators and measurements for sustainability of manufacturing processes. Also a few organizations made efforts to introduce a comprehensive framework for sustainable manufacturing indicators. As an example, National Institute of Standards and Technology (NIST) addressed five dimensions of sustainability including environmental effects management, economic growth, social well-being, technological development and performance management. [1]

In addition to that manufacturing industries need to remain globally competitive, be able to improve productivity and reduce environmental impact and energy usage.

All these would emphasis the need for a transformation/improvement from manufacturing process based on operator experience towards automated, advanced and innovated technologies considering the need for high quality products, competitive process and sustainable system. It seems that the industrial trend from traditional manufacturing

G. Campana et al. (eds.), *Sustainable Design and Manufacturing 2017*, Smart Innovation, Systems and Technologies 68, DOI 10.1007/978-3-319-57078-5_20

process to advanced technology based manufacturing process is a response to this demand.

Current manufacturing industries experience the lack of effective methodologies and measurement science with respect to sustainability and it is worse when it comes to micro manufacturing as there is still huge knowledge gap in selection and utilization of the manufacturing methods and technologies. Lack of knowledge and standards, manufacturing and production guidelines will affect the selection of appropriate technology and their competitiveness which will hugely influence the assessment of processes sustainability.

Therefore, it requires defining approved measurement methods, methodologies and sustainability assessment technologies based on a scientific, computable and comparable model covering all manufacturing process and methods. It should not be forgotten that any sustainability assessments have to consider three different levels including system, process and product. Although these three levels would be considered in order to make the manufacturing a sustainable process but each level would have its own criteria and indicators [2].

In this paper sustainability of micro electrochemical machining (micro ECM) has been reviewed as a young non-conventional manufacturing process which has promising future in industry.

2   Micro ECM

Non-conventional manufacturing- known as advanced manufacturing method as well- can involve mechanical, thermal, electrical or chemical energies or a combination of these energies. Also it is applicable for variety of materials including hard machining, brittle and conductive materials. By applying suitable control methods and selecting optimum physical parameters for process, they can be used successfully for different type of workpiece material regardless material hardness, toughness and brittleness.

Micro-ECM is one of these non-conventional manufacturing processes which presented substantial advantages in compare with other manufacturing methods. Micro ECM has broadly been subject of the academic and industrial research activities as a result of its compatibility in machining chemically resistant materials which are extensively used in biomedical, MEMS and electronics applications [3].

Micro ECM in particular has been used for chemically resistance materials, high strength materials and hard to machine materials.

Machined workpiece is burrs free with no thermal or physical strain. There is no tool wear due to the nature of the chemical reaction which takes place at the anode and the cathode with no contact between tools. Capability of machining different materials, simpler machine set up, no tool wear, high MRR rate, dimensional accuracy and high surface integrity have made the micro-ECM more interesting for manufacturing industry.

Micro-ECM is an electrochemical process which works on the basis of anodic dissolution of workpiece based on Faraday's laws. This process is established on anodic dissolution of workpiece (anode) using a desired shaped tool (cathode) based on Faraday's laws of electrolysis. A continuous flow of electrolyte through the Inter Electrode Gap (IEG- the gap between two electrodes), is necessary to dissipate removed materials from anode and prevent it to be deposited on cathode (Fig. 1).

By placing the two conductive electrodes in a suitable electrolyte and Appling voltage, due to pass of current and surface oxidation, electrons (ions or group of ions) will be transferred between electrodes; therefore workpiece surface will face physical changes and will be machined. Machining accuracy can be improved by maintain very small IEG during operation and setting optimized initial values for process parameters such as voltage and current levels, pulse parameters and electrolyte features.

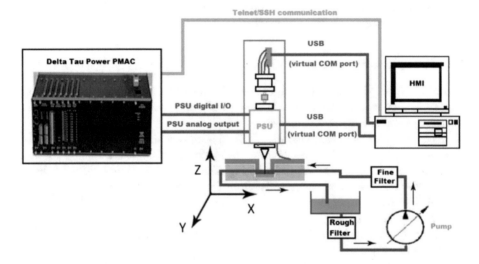

Fig. 1. Schematic diagram of the micro ECM machine [4]

General advantages and disadvantages of micro-ECM can be summarized as below:

- Capability of processing various material
- Free of burrs products
- No upper layer deformation
- No thermal or physical strain in product (no contact between tool and workpiece)
- Multiple use of tool (No tool wear)
- High dimensional accuracy
- High surface quality
- Efficient complex geometries
- High MRR
- Cost effectiveness

  Limitations:

- Environmental issue (possible harmful chemical products)
- Poor fatigue properties
- Lack of capability to create sharp corners
- Requires high knowledge base operation

- Requires widespread research for different materials and products

The critical condition in micro ECM is the gap between two electrodes which should be maintained in stable small size; it is important to keep the IEG as small as possible as it dictates the resolution of the machining. Machining process in general and machined workpiece are very much under the effect of different parameters.

Considering the complex nature of micro-ECM machining and lack of documentation and standards with regard to the sustainability assessment creates more challenge and difficulties for establishing fundamentals to assess and analyze the impact of the process with regard to energy, environment, tooling, cost and other effective criteria in evaluation and assessment of sustainability.

2.1   Example of Application of Micro ECM

Brunel University London has designed bespoke micro ECM machine which presents next generation machine for targeted industry including automotive industry, semiconductor, medical and metrology sectors.

It has three axes of linear motion(X, Y, Z) using air bearings with linear DC brushless motors and 2-nm resolution encoders for ultra-precise motion and a spindle allowing the tool-electrode to rotate during machining. The innovative control system is based on the Power PMAC motion controller from Delta Tau which applying two process control algorithms: fuzzy logic control and adaptive feed rate. Also an in-house-developed pulse generator capable of ±10 V, 5 A and down to 50 ns pulse on-time has been used to apply various short pulses. Electrolyte circuitry of the machine consists of Tank, filters and pipes. [4]

Capability to preparing the tool cathode by reverse ECM is another advantage of this machine. Therefore, a fully automated on-line tool-preparation can take place and followed by workpiece machining which would be cost saving on tool manufacturing.

This machine has been used as a research platform to create comprehensive data base for different material and electrolytes which have been used for various proposes. In addition to that, in partnership with different industries, high value manufacturing and machined samples been tested to verify and validate the technology [5].

This machine has been used to test a new technology in sharpening medical needles for German needle manufacturer which does not create defective layer, do not burn the material and can create surface roughness down to Ra10 nm. The results from quality of the surface and freedom to shape the needle according to the biophysics recommendation and machining time of only 10 s per needle (in the preliminary tests), is a proof for the employment of this technology in production lines. [5]

Although this is very successful implementation of micro-ECM for industrial applications but there is still need to have a comprehensive methodology to compare and analyze the results of both micro ECM process and current manufacturing process (grinding).

Figures 2 and 3 present the SEM image of needle using grinding and ECM technologies respectively.

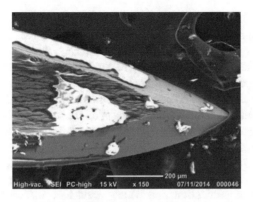

Fig. 2. SEM image of medical needle

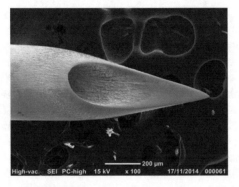

Fig. 3. SEM image of micro ECM 3D sharpened needle

In order to bring this method to practice, assessing the success of the method based on sustainability will be very helpful as it may justify the initial cost of the setting the technology.

3   Sustainability of Micro ECM

Although the common belief would suggest that micro scale manufacturing implies more sustainability by reducing raw material usage, less energy consumption and less environmental impact but it may not be always the case. Recent researches show that some factors can prevent achieving the expectation in sustainable micro-manufacturing [6].

In general and in terms of energy, although there is possibility to use less energy at production level for micro-scale machining but it is important to consider the need for ventilation, filtering and maintaining the clean room which would increase the cost of energy.

In terms of materials, micro manufacturing process usually produce very value added products for specific applications; and in most cases required materials have significant commercial value. Therefore, any defeat or waste can lead in significant raw material costs.

In addition to that the impact of chemical substances or gases would increase the environmental impact and costs.

Despite the importance of sustainable manufacturing, there is not much publication in area of sustainable micro manufacturing technologies. K. Kellans et al. discussed the environmental impact of non-conventional processes [7], G. Tristo et al. has presented and analyzed the online energy consumption in micro EDM [8] and F. Modica et al. has discussed the sustainable micro manufacturing of micro components for micro EDM [9]; and author is not aware of any specific publications related to micro ECM sustainability assessment if there is any.

Micro-ECM as mentioned before is known for burrs free products with no thermal and physical effects as a result of the nature of process. This is a positive sign towards decreasing defeats in production line which leads less waste and therefore would increase the sustainability of method in compare with other non-conventional manufacturing methods.

In addition to that, micro ECM proved to have no or minimum tool wear as there is no direct contact between work-piece and tool. Therefore, this would be additional capacity in saving materials by less or no wear and tear in tool electrode.

In terms of chemical impacts, it is clear that micro ECM requires electrolyte to activate the process and create the current pathway between tool electrode and workpiece. According to Bhattacharyya, two main categories of electrolytes are being used in micro ECM. "Passive electrolytes" which contain oxidizing anions and they are known for better machining precision. "Non-passive electrolytes" which contain aggressive anions and would have less effect on electrode due to formation of soluble products as they can be completely swept from IEG area [10].

Regardless this classification, micro ECM uses nontoxic and less aggressive electrolyte. This is an advantage in measuring the sustainability of micro ECM but one should consider that the performance of machining would be affected by remaining sludge from removing materials during pulse on time as it would create sparks. Therefore it is very important to assure that any sludge and gases would be flushed away from IEG and also filtered or clean electrolyte would be used during process.

Based on available research, some materials would be machined better with passive electrolyte and some would be machined better with acidic electrolyte. Although a various range of materials and electrolytes have been the subject of researches in micro ECM machining but there is still no comprehensive data base available to create shortcut in selection of materials and relevant electrolytes. But what is obvious is the need for clean, non-affected electrolyte to run the operation and its value in creating sustainable process as the electrolyte features can affect accuracy, surface finishing and material removal rate.

The other important area is the tool components and its preparation. By recognizing most suitable tooling material and tooling shape, the energy and material waste would be definitely decreased; as it would affect material removal rate, accuracy and efficiency of process.

In addition to that, tool isolation has proved significant change in micro ECM machining, specifically in milling.

Table 1 has summarized the areas and factors which can influence the sustainability of micro ECM process. The productivity, accuracy and reliability of the process depend on different factors and there is a complex relation between these factors such as voltage, current, pulse on-time, pulse width, frequency, electrolyte conductivity and tool material and shape.

Table 1. Influencing area and factors of micro ECM process on Sustainability

Dimensions of sustainability	Influencing areas and factors of process
Environment	Non-toxic electrolyte
	Non-hazardous waste generated
	Decreasing noise pollution
	Energy and material resources
Economy	Energy usage
	No defeat, No tool wear
	Tool preparation
	Maintenance cost
Society and Technology	Worker health
	Worker safety
	Product pricing
	Product quality
	Technology advancement

Therefore to define sustainability criteria and indicators, it should be intended to introduce critical parameters in relation to effective and important factors in a way to provide more efficient comparability and evaluation such as energy consumption in relation to material removal rate.

However, as a result of current uncertainties in electrochemical process, it would not be easy to analyze the results very straightforward as they can be affected by other parameters.

4   Conclusion

There is increasing demands for precision micro- manufacturing for MEMS, biomedical applications, automotive industry and IT applications which will lead research to utilization of micro ECM technology widely.

As highlighted in this paper and previous research work, micro ECM method can be used as one of the main alternatives in precision manufacturing. The results of recent activities have proved that micro ECM has valuable potential to be used in different areas while its full capacity has not been explored. But further research needs to utilize this method effectively in industry and be able to assess and evaluate its advancement and sustainability.

Sustainability of manufacturing is appeared as one of the main industrial concerns, also the manufacturing processes have shifted to non-conventional methods and considering the increased demand for micro and nano products, but there is not much effort or at least documentations and publications to establish fundamental approaches toward assessment and evaluation of sustainability of micro manufacturing. It seems that the speed of these two activities do not match at this stage and sustainability assessment and evaluation need to move forward fast enough to fulfill the current requirements.

In spite of lack of valid measures and indicators to present sustainability of manufacturing process and based on current available knowledge and experiences, it would be clear that micro ECM can be recognized as economically, environmentally and energy consumption very beneficial to industry based on its advantages in process features, productivity and quality.

However, by maximizing the metal removal rate, minimizing error and improving energy consumption and electrolyte selection, the optimized and sustainable process can be achieved.

But it is urgently needed to create a comprehensive database for this process as it is very much knowledge based operation and there is still uncertainty in process level as various materials in combination with different process parameters' setting and electrolyte selection can lead to different results and end product quality.

References

1. Mani, M., Maden, J., Lee, J.H., Lyons, K.W., Gupta, S.K.: NISTIR 7913 review on sustainability characterization for manufacturing processes, February 2013. http://nvlpubs.nist.gov/nistpubs/ir/2013/nist.ir.7913
2. Jayal, A.D., Badurdeen, F., Dillon Jr., O.W., Jawahir I.S.: Sustainable manufacturing: modeling and optimization challenges at the product, process and system levels. CIRP J. Manuf. Sci. Technol. 2(3), 144–152 (2010)
3. Datta, M., Harris, D.: Electrochemical micromachining: an environmentally friendly, high speed processing technology. Electrochimica Acta 42, 3007–3013 (1997)
4. Spieser, A., Ivanov, A.: Design of an electrochemical micromachining machine. Intl. J. Adv. Man. Tech. 78(5), 737–752 (2015)
5. Ivanov, A., Mortazavi, M.: Advanced applications of micro ECM technology. In: 5th International Conference on Nanomanufacturing (NanoMan2016), Macau
6. De Grave, A., Olsen, S.: Challenging the sustainability of micro products development. In: Proceeding of 2nd International Conference on Multi-Material Micro Manufacturing, pp. 285–288 (2006)
7. Kellens, K., Dewulf, W., Lauwers, B., Kruth, J.P., Duflou, J.R.: Environmental impact reduction in discrete manufacturing: examples for non-conventional processes. In: 17th CIRP Conference on Electro Physical and Chemical Machining (ISEM), pp. 27–34, (2013)
8. Tristo, G., Bissacco, G., Lebar, A., Valentincic, J.: Real time power consumption monitoring for energy efficiency analysis in micro EDM milling. Intl. J. Adv. Man. Tech. 78, 1511–1521 (2015)
9. Modica, F., Marrocco, V., Copani, G., Fassi, I.: Sustainable micro-manufacturing of micro-components via micro electrical discharge machining. Sustain. Open Access J. 3, 2456–2469 (2011)
10. Bhattacharyya, B., Malapati, M., Munda, J.: Experimental study on electrochemical micromachining. J. Mater. Process. Technol. 169, 485–492 (2005)

Application of Design for Environment Principles Combined with LCA Methodology on Automotive Product Process Development: The Case Study of a Crossmember

S. Maltese[1,2], M. Delogu[3], L. Zanchi[3(✉)], and A. Bonoli[2]

[1] Department of Civil, Chemical, Environmental and Materials Engineering,
University of Bologna, Bologna, Italy
[2] Magneti Marelli S.P.A. – Powertrain Division, Bologna, Italy
[3] Department of Industrial Engineering, University of Florence, Florence, Italy
laura.zanchi@unifi.it

Abstract. The existing Community regulation pushes the carmakers to design eco-sustainability of the vehicle over its life cycle to limit the consequences of the current state and the expected growth of the sector. In this sense, one of the primary aim is reducing raw materials consumption and emissions through the adoption of innovative materials and technologies. This implies the need for the carmakers to integrate Design for Environment (DfE) principles at the early Research and Development (R&D) stage. The article presents a concrete example of integration of DfE and LCA methodology application in the R&D process of a vehicle component produced by Magneti Marelli. The study allowed drawing a balance between the advantages of a lightweight solution with respect to the standard one both from performance and environmental point of view.

Keywords: Automotive sector · Sustainable manufacturing · Design for environment · Lightweighting · Life cycle assessment

1 Introduction

The motivators that have pushed the Governments to focus their attention on sustainability activities are mainly due to the alarming data recorded on non-renewable resources depletion and global climate change. The transportation sector accounts for two-thirds of total crude oil consumption, and, one third for GHG emissions [1, 2]. Vehicles are extremely resource intensive products, especially during their use phase (particularly for internal combustion engines vehicles), causing a relevant amount of fuel consumption and CO_2 emissions generation. Another matter is originated by vehicle disposal; every year, in Europe, End-of-Life Vehicles (ELVs) constitute about 8-9 million tonnes of waste [3]; hence European Directive 2000/53/EC fixed new targets for vehicle recovery and specific standards (i.e. ISO 22628) exist for calculating the recyclability and recoverability rate of a vehicle [4, 5]. To tackle these problems, the automotive sector has started to focus the attention on environmental impact reduction initiatives,

© Springer International Publishing AG 2017
G. Campana et al. (eds.), *Sustainable Design and Manufacturing 2017*, Smart Innovation,
Systems and Technologies 68, DOI 10.1007/978-3-319-57078-5_21

by getting involved into sustainability programs, incorporating policy regulations from the organization to the product level.

In order to meet environmental impact reduction, it is important to integrate environmental friendly principles and solutions in the R&D process, paving the way for the introduction of Design for Environment (DfE). The overall characteristics of DfE approach are the application at the early design phase and the perspective of the whole product life-cycle [6]. The Life Cycle Assessment (LCA) represents one of the most spread methodology providing useful set of environmental indicators for the DfE process and a clear procedure to compare and select the most favorable scenario; few case study application regards the automotive sector [7–16].On the one hand, the use of LCA to drive DfE choice could provide in the LCA result interpretation since the environmental consequences are addressed by means of indicators which are distant from the designer comprehension (i.e. Abiotic Depletion Potential) [17, 18]. Moreover, an intensive data collection is required to obtain reliable results and, in most cases, the LCA analysis necessary relies on a number of assumptions [19]. On the other hand, LCA is a comprehensive methodology that, despite its inherent challenges in terms of data availability and impact indicators improvements [11, 20], is still considered a useful and practicable approach for designers.

This article presents a concrete example in which these challenges are addressed in the context of vehicle component lightweight design. First, an overview of the R&D workflow integrating DfE principles and LCA is shown. Then, a case study is presented concerning a part of the suspension system designed by Magneti Marelli©. In this study, two important environmental issues have been focused, as relevant for the automotive sector: the mitigation of GHG emissions and the raw material reduction. Particular attention is given to the relationship between the lightweight strategy and these specific environmental objectives along the product life cycle, furthermore outcomes are discussed in terms of DfE strategies.

2   Method

Figure 1 shows the R&D workflow in which DfE principles are integrated with the LCA. First, the *DfE Conceptual Study* is defined, in particular the DfE approaches guiding the procedure [6] and the design strategy (i.e. lightweighting, power efficiency).

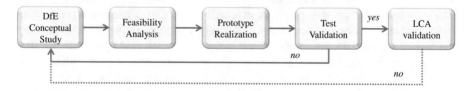

Fig. 1. R&D workflow which integrates Design for Environment principles and LCA

Then, the *feasibility analysis* compels the product functional requirements definition (i.e. corrosion resistance) which is preparatory for the design phase. This step is followed by the *prototype realization*. Finally, *Test Validation* is carried out to check if the innovative design satisfies the technical performance, accordingly to the specific component function. If the prototype is validated then its environmental performances are evaluated and compared to the standard design by means of LCA. Developing the environmental assessment after the prototype step would guarantee more reliable LCA results since they could be based on detailed data collection about geometry, materials, technologies etc.

3 Case Study

3.1 Component Description

The method has been applied for a new design concept of an existing automotive component of the suspension system: a front crossmember, hereafter CM, made of stainless steel. The main reason guiding the CM selection is its mass (19 kg), so significant results could be expected through the implementation of a lightweighting strategy. Among the several DfE approaches [6], four of them can be considered particularly relevant for the sector: (i) design to minimize material use; (ii) design for manufacturing; (iii) design for energy efficiency; (iv) design for recycling. Lightweighting is obtained through the raw material substitution using an aluminum alloy. Moreover, additional changes are expected also in the production technology, the supply chain management and the recovery process during End-of-Life cycle.

The CM is a structural component that takes part either in the suspension system and transmissions connected to the body at different points and linked to the lower arms through elastic bushings. CMs design solutions are depicted in Fig. 2; the standard design is stainless steel (19 kg) constituted by several sub-components (highlighted in difference colors), whereas the innovative design is one-piece unitary structure of aluminum (15.65 kg) allowing an overall reduction of about 22%.

Fig. 2. CM standard design (left) and innovative design proposal (right)

Previous works have addressed the economic and environmental convenience of using cast aluminum in the design of an automotive crossmember [21, 22].

The substitution of the material did not lead to a geometry variation of the reference crossmember design; nevertheless specific tests were carried out in order to verify the aluminium conformity for crossmember functionality. Based on crossmember technological requirements, structural, static and dynamic stiffness, frequencies and corrosion resistance (CM is located under vehicle chassis and so exposed to a corrosive environment) have been accomplished. Results demonstrated that the innovative aluminum-component satisfy the functional requirements; in addition better corrosion-resistance can be provided.

3.2   LCA Goal and Scope, Inventory and Impact Categories

The LCA analysis is performed according to ISO 14040 standard, following a "cradle-to-grave" approach. The functional unit of the assessment is the CM mounted on an Alfa Romeo Giulietta 1.4 Turbo 105 CV gasoline for an operational lifetime mileage of 150000 km. Figure 3 shows CM life cycle phases for the two alternatives, which differ for the initial phases of production and manufacturing, affected by the employment of different materials and therefore manufacturing. For the End-of-Life (EoL) phase it was assumed the worst case scenario, where the component is not previously disassembled from the vehicle and the material recovery occurs during downstream treatments. To environmentally characterize all processes an analytical model reproducing real CM life cycle for each scenario has been developed by means of the software GaBi 6.5.

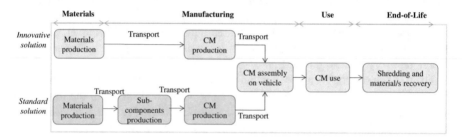

Fig. 3. CM life cycle breakdown for the two solutions of analysis according to the system boundaries considered

Primary data, concerning energy consumption of the machineries, auxiliary materials (i.e. process water) and scraps rate, have been collected for manufacturing and EoL phase, whereas raw materials and energy production eco-profile have been retrieved from the GaBi database. The raw material production category includes all stages from their extraction to the final state. The standard CM is produced with two different raw materials: austenitic stainless steel FEE 430 and ferritic stainless steel FEE 316, instead the innovative design with secondary aluminum 6061 - T6. In the modelling of the raw materials and auxiliary materials, GaBi 6.5 database has been used.

The production technology of the standard design compels sub-components production, welding and painting process (pre-treatment and cataphoresis). Such processes are critical since involve several auxiliary materials and the production of sludge and waste water that need to be properly treated. The innovative CM includes one-piece unitary structure of secondary aluminum through die casting and machining processes. The scraps generated during steel-based sub-components and aluminum machining process were considered as steel and aluminum recycling credit of 22% and 32%[1] respectively. Process parameters (materials and energy flows) were obtained by direct measurements on industrial processes on site during the production period of one week dividing the energy expenditure and auxiliary materials consumption for the relative productivity; an extract is provided in Table 1. The logistic analysis takes into account all supply-chain actors from the raw materials extraction to the final phase of component assembly on vehicle suspension system. Logistic data on itinerary, transport typology and distance travel were collected (Table 1); the differences of the two supply-chain scenarios are in the up-streaming phase; in fact, for the steel based solution three different suppliers are involved whereas for the innovative solution only one supplier is present, while the means of transport is equal (Trucks 28 - 34t gross weight, 22t payload capacity, diesel driven, Euro 5 - cargo consumption mix, Gabi software modelling parameter). To calculate the environmental impact imputable to the CM mass during the use phase it was used an analytical model based on the approach of Koffler [23], taking into account the technical data referred to the specific vehicle (i.e. vehicle mass, type, fuel consumption, driving cycle) (see Eq. 1). The amount of CO_2 and SO_2 are directly dependent from the fuel consumed during the operation life time of the vehicle equipped with the CM (see Eq. 2).

Table 1. Manufacturing processes, EoL treatment electricity consumptions (CM standard with ferrous recovery and CM innovative with Al non-ferrous recovery) and logistic data (distance for the delivery of goods across CMs supply chain within manufacturing gate)

	Standard	Innovative
Manufacturing		
The electricity consumption manufacturing phase	Sub-component production = 9.5 MJ/FU	Machining = 7.42 MJ/FU
	Welding = 2000 MJ/FU	
	Painting = 88 MJ/FU	
Electricity consumption EoL treatments	Shredding → Aeraulic separation → Magnetic separation = 2.76 MJ/FU	Shredding → Aeraulic separation → Magnetic separation → Eddy current → Inductive resonance → Ballistic separation = 3.66 MJ/FU
Logistic		
Total distance travelled (km)	2700	1129

$$Fuel\,component = Fuel\,vehicle(Mcomponent/Mvehicle) * FRV \qquad (1)$$

- $Fuel_{component}$, $Fuel_{vehicle}$ = fuel consumption of the reference component and the reference vehicle, (l/100 km);
- $M_{component}$, $M_{vehicle}$ = mass of the reference component and the reference vehicle, (kg);
- FRV_{NEDC} = Fuel Reduction Value for the NEDC driving cycle (0.12) [12]

---

[1] Based on GaBi 6.5 steel rebar world steel data and aluminium credit data.

$$emiss\,i = emiss\,i\,km * use\,km * (100 * fuel\,component/fuel\,vehicle * use\,km) \qquad (2)$$

- emiss i = emissions of pollutant i during the entire component life-time (g);
- emiss i km = vehicle per-kilometre emission of pollutant i (g/km);
- fuel vehicle = vehicle per- kilometre fuel consumption (l/100 km).

During the EoL phase, it is assumed that the CM remains on the vehicle which is shredded and then material flows are sorted and recycled. Overall, the EoL management system is characterized by a high heterogeneity since different technologies and processes exist, moreover they are frequently developed in different plants [24]. In this study, it has been assumed a typical Italian craft-type Authorized Treatment Facilities where two main stages targeted to ferrous and non-ferrous metals sorting. The first stage includes shredding, aeraulic separation and magnetic separation for ferrous metals separation. The remaining waste flows are then treated by means of magnetic separation, eddy current, inductive resonance and ballistic separation for the non-ferrous metals separation. To model the initial phase of vehicle shredding was considered the *car drained* process within Gabi 6.5 database, then for the further processes and energy consumption, primary data were collected from an EoL plant during one day of operation (Table 1). To perform the recovery process, it has been considered the *sorted automotive casting scrap credit* process for steel solution and *aluminum auto fragments scrap credit* process for aluminum, already modeled in Gabi software.

The impact categories selected aim at evaluating the aforementioned environmental issues (GHG emissions and resource depletion) which the new design strategy intends to decrease. In particular the Global Warming Potential (100 years) (GWP), the Abiotic Depletion Potential Elements ($ADP_{elements}$) and Primary Energy Demand (PED) from the CML method are calculated.

4   Results and Discussions

The material and technology variation mainly influenced the emissions generated during vehicle operation and the recoverability portion of the material at the component disposing stage, in Table 2 are summed up the results.

LCA results demonstrate that the new design solution entails a significant impact decrease up to a 70% for all the impact categories with the most outstanding linked to the $ADP_{elements}$ with a sharp reduction of more than 90%. It can be observed that the highest discrepancy between the two solutions is given by raw material and manufacturing life cycle phases. However, further benefit are achieved also in the use phase, in fact the lower density of the new material allowed an overall mass reduction of about 22%. It can be observed that the three impact categories are affected by different component life cycle phases. The $ADP_{elements}$ is mainly related to the raw material phase, whereas the GWP and PED are affected by use phase and manufacturing (Fig. 4).

Table 2. Overview of standard and innovative component general feature over their life cycle

	Standard	Innovative
CO_2 emissions (kg/FU 150 000 km)	92	75
SO_2 emissions (kg/FU 150 000 km)	5.07E–04	4.16E–04
Fuel consumption (l/FU 150 000 km)	34	28
Metals recovery ratio	0.023[a]	0.32[b]

[a]Data from Gabi process data set "*Steel mill scales - scrap credit*" based on average price ratio between Steel Benchmarker GLO plate and EU scrap price 2007–2010.

[b]Data from Gabi process data set "*Aluminium auto rads - scrap credit*" based on average price ratio between LME Al99.7 and EU scrap price 2007–2010.

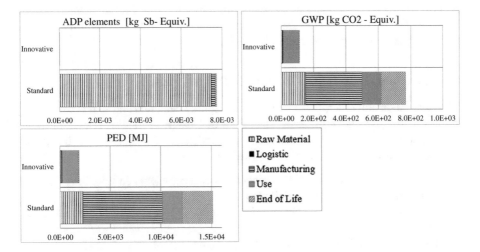

Fig. 4. LCIA results for the comparison of the standard and the innovative solutions for PED, GWP_{100} and $ADP_{elements}$ indicators

Concerning the raw material phase, the standard solution involves the usage of heterogeneous-heavyweight raw materials (austenitic and ferritic stainless steel). The production process of stainless steel involves a numerous step and a consumption of hazardous materials as nitric and hydrofluoric acid during pickling process, ester and paraffinic mineral oil for surface quality control and so forth. The innovative solution leads to a significant reduction of resource consumption due to the use of recycled material, which also involves a lower energy consumption and less auxiliary materials consumption decreases (i.e. O_2, CH_4, $Ca_2O_4Si_{powder}$).

The reduction of the number of the parts in the innovative design concept lead to a simplification of the supply chain. In the standard case, the total distance covered by all the transportation is more than 3000 km with five suppliers involved. Instead, the innovative solution requires only one supplier with a total distance travelled of 780 km. As far as manufacturing phase is concerned, the CM standard solution involves processes that are more critical. In particular, welding and painting process are responsible for a great quantity of auxiliary materials (i.e. water and chemicals), take more time and consume more energy if compared to the die-casting production process. During the standard production process, the semi-finished product is transferred across difference lines dislocated along the production plant via huge-tape transport conveyors thus increasing electricity consumption. In contrast, the production process of the aluminum component consists of only two stages: die casting and machining. The energy expenditure for the production of the innovative CM is noticeably less and this influence the GWP and PED figures considerably. Overall, the advantages in terms of manufacturing phase are multiple: (i) savings of energy expenditure due to the convey system simplification and production process substitution; (ii) time cycle shortened; (iii) more stable process and greater control through make- process strategy; (iv) reduction of auxiliary materials; (v) decrease of process waste generation. Vehicle use phase accounts for the greatest part of the generation of CO_2 emissions and fuel consumption whose decrease is 18.5% (Table 2). Concerning the EoL phase the innovative solution is still preferable since a higher recovery ratio can be obtained.

4.1 Combining LCA Impact Results and DfE Approaches

$ADP_{elements}$ indicator quantifies the impact on resource depletion and so it is particularly influenced by raw material phase. The effect of Design to minimize material usage strategy could be measured through $ADP_{elements}$ indicator. The lightweight strategy has revealed to be effective for the following reason:

- The production process of stainless steel requires more energy and auxiliary materials compared to the secondary aluminum [25];
- The production process of standard CM requires a great amount of auxiliary materials;
- The recycling credit of aluminum is higher than the steel.

GWP quantifies the GHG emissions, whose great generation is related to vehicle operation and manufacturing process, could be representative of the Design for Energy Efficiency effect, but also answers for Design for Manufacturing objectives in the measurement of process energy-expenditure and Design for End-of-Life in terms of GHG saving for the material recovered. The PED indicator measures the total energy demand in terms of renewable and non-renewable energy resources. According to the LCA results, PED is particularly related to Design for Energy Efficiency approach but also to Design for Manufacturing and Design of End of Life.

Finally, the relationship between the selected DfE approaches and the CM life cycle phase is presented in Fig. 5. The results revealed that the selection of Design to Minimize Material Consumption and Design for Energy Efficiency, through the substitution of aluminum as a lightweighting-strategy driver, revolutionized the whole life cycle of the crossmember. This is a clear example demonstrating that life cycle phases could be interlinked by specific driver selection in design decision: for the case in question the raw material substitution. The following life-cycle framework (Fig. 5), offers a visible explanation of such concept; the lightweight strategy through the material substitution lead to consequences along the whole component life cycle and visible responses to all the DfE approach. The choice of aluminum has affected the manufacturing, therefore, a reformulation of a new production technology was necessary; this was occurred through the derivative of Design for Manufacturing: the Design to Minimize Energy Consumption and Design for Assembly, resulting in the diminution of production process energy expenditure and time cycle. From EoL point of view, the aluminum offers a good potential of recovery from the Automotive-Shredded Residue (ASR). In addition, the existing technology allows for a higher recovery-ration and less energy expenditure compared to the steel. Aluminum solution perfectly matches Design for End-of-Life objectives.

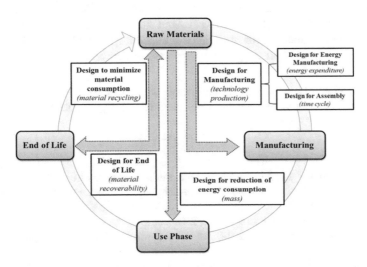

Fig. 5. Life cycle framework of design decision for environment application strategy for crossmember case study

5   Conclusions

Although the LCA is considered a powerful method for the environmental impacts evaluation of new product, there is a general skepticism bore out from some practical difficulties that could emerge during its implementation in the R&D process. This study tries to get through this by means of a concreate example of integration of DfE and LCA methodology application in the R&D process of a vehicle component produced by Magneti Marelli. In order to enhance the LCA results interpretation, an effort was dedicated to the selection of those impact categories able to evaluate the selected environmental issues - GHG emissions and resource depletion - and their analysis in relation to the product life cycle phases and DfE approaches. In this sense the GWP, $ADP_{elements}$ and PED were found relevant for evaluating the environmental issues GHG emissions and resource depletion, which the new design strategy intends to address. The lightweight strategy was found to produce several benefits in addition to the mass reduction: the production process of stainless steel requires more energy and auxiliary and hazardous materials compared to the secondary aluminum, moreover the recycling credits of aluminum is higher than the steel ones, further benefits regard the supply chain simplification and manufacturing time cycle shortening. LCA results demonstrate that replacing steel with secondary aluminum entails an overall impact decrease up to a 70%. Moreover, besides the significant mass reduction (−22%) and thus the use phase impact decrease, it was observed that the most benefits regard raw material and manufacturing life cycle phases. Such results confirmed the importance of primary data gathering as a way to obtain analysis that is more precise and reliable. Thought the collection of data directly at the manufacturing plant could be time intensive, nevertheless this could provide useful information that could be further systematically structured to build a company database more detailed that the current database. Future research directions should necessary regard further discussion about the implementation of a set of relevant and suitable environmental indicators, trade-off handling and multidisciplinary analysis as key elements from improving the R&D process in a sustainability perspective.

Acknowledgment. The authors would like to thank Magneti Marelli S.P.A. for the cooperation. In particular, the authors are very grateful to Mrs. Rubina Riccomagno for her fruitful contribution.

References

1. IEA (International Energy Agency): Key world energy statistics 2016 – total final consumption by fuel (2016). www.iea.org/publications/freepublications/publication/KeyWorld2016.pdf

2. EPA (Environmental Protection Agency): Global greenhouse gas emissions data - global emissions by economic sector (2014). www.epa.gov/ghgemissions/sources-greenhouse-gas-emissions

3. Eurostat: Environmental data centre on waste, key waste streams, end of life vehicles (ELVs) (2015). http://ec.europa.eu/eurostat/web/waste/key-waste-streams/elvs

4. Berzi, L., Delogu, M., Pierini, M., Romoli, F.: Evaluation of the end-of-life performance of a hybrid scooter with the application of recyclability and recoverability assessment methods. Resour. Conserv. Recycl. **108**, 140–155 (2016). doi:10.1016/j.resconrec.2016.01.013

5. Delogu, M., Del Pero, F., Berzi, L., Pierini, M., Bonaffini, D.: End-of-life in the railway sector: analysis of recyclability and recoverability for different vehicle case studies. Waste Management (2016, in Press). doi:10.1016/j.wasman.2016.09.034

6. Mayyas, A., Qattawi, A., Omar, M., Shan, D.: Design for sustainability in automotive industry: a comprehensive review. Renew. Sustain. Energy Rev. **16**, 1845–1862 (2012)

7. Arena, M., Azzone, G., Conte, A.: A streamlined LCA framework to support early decision making in vehicle development. J. Cleaner Prod. **41**, 105–113 (2013)

8. Bevilacqua, M., Ciarapica, F.E., Giacchetta, G.: Development of a sustainable product lifecycle in manufacturing firms: a case study. Int. J. Prod. Res. **45**(18–19), 4073–4098 (2007)

9. Le Duigou, A., Baley, C.: Coupled micromechanical analysis and life cycle assessment as an integrated tool for natural fibre composites development. J. Cleaner Prod. **88**, 61–69 (2014)

10. Corona, A., Madsen, B., Hauschil, M.Z., Birkved, M.: Natural fibre selection for composit eco-design. CIRP Ann. Manufact. Technol. **65**, 13–16 (2016)

11. Zanchi, L., Delogu, M., Ierides, M., Vasiliadis, H.: Life cycle assessment and life cycle costing as supporting tools for EVs lightweight design. Smart Innov. Syst. Technol. **52**, 335–348 (2016). doi:10.1007/978-3-319-32098-4_29. Cited 3 times

12. Delogu, M., Del Pero, F., Romoli, F., Pierini, M.: Life cycle assessment of a plastic air intake manifold. Int. J. Life Cycle Assess. **20**(10), 1429–1443 (2015). doi:10.1007/s11367-015-0946-z. Cited 6 times
13. Del Pero, F., Delogu, M., Pierini, M., Bonaffini, D.: Life Cycle Assessment of a heavy metro train. J. Clean. Prod. **87**(1), 787–799 (2015). doi:10.1016/j.jclepro.2014.09.023
14. Raugei, M., Morrey, D., Hutchinson, A., Winfield, P.: A coherent life cycle assessment of a range of lightweighting strategies for compact vehicles. J. Clean. Prod. (2015). doi:10.1016/j.jclepro.2015.05.100
15. Kim, H.C., Wallington, T.J.: Life cycle assessment of vehicle lightweighting: a physics-based model of mass-induced fuel consumption. Environ. Sci. Technol. **47**, 14358–14366 (2013). doi:10.1021/es402954w
16. Kelly, J.C., Sullivan, J.L., Burnham, A., Elgowainy, A.: Impacts of vehicle weight reduction via material substitution on life-cycle greenhouse gas emissions. Environ. Sci. Technol. **49**, 12535–12542 (2015). doi:10.1021/acs.est.5b03192
17. Baumann, H., Boons, F., Bragd, A.: Mapping the green product development field: engineering, policy and business perspectives. J. Cleaner Prod. **10**, 409–425 (2002)
18. Millet, D., Bistagnino, L., Lanzavecchia, C., Camous, R., Poldma, T.: Does the potential of the use of LCA match the design team needs? J. Cleaner Prod. **15**, 335–346 (2005)
19. Klocke, F., Kampker, A., Döbbeler, B., Maue, A., Schmieder, M.: Simplified life cycle assessment of a hybrid car body part. Procedia CIRP **15**, 484–489 (2014)
20. Delogu, M., Zanchi, L., Maltese, S., Bonoli, A., Pierini, M.: Environmental and economic life cycle assessment of a lightweight solution for an automotive component: a comparison between talc-filled and hollow glass microspheres-reinforced polymer composites. J. Cleaner Prod. **139**, 548–560 (2016)
21. Brown, K., Juras, P.: The 1997 Chevrolet Corvette suspension CMs. SAE Technical Papers, February (1997). www.worldstainless.org/process_and_production/production-process. Accessed May 2012
22. Randon, V., Lee, N.: Design of a lightweight aluminium cast crossmember. Paper presented at SAE 2002 World Congress and Exhibition, 4 March 2002
23. Koffler, C.: On the calculation of fuel savings through lightweight design in automotive life cycle assessments. Int. J. Life Cycle Assess. **15**(1), 128–135 (2010)
24. Berzi, L., Delogu, M., Giorgetti, A., Pierini, M.: On-field investigation and process modelling of end-of-life vehicles treatment in the context of Italian craft-type authorized treatment facilities. Waste Manag. **33**(4), 892–906 (2013)
25. Center for sustainable systems: Update material production modules in the GREET 2 model. University of Michigan (2011). css.snre.umich.edu/project/update-material-production-modules-greet-2–model

A Conceptual Framework to Support Decision-Making in Remanufacturing Engineering Processes

Awn Alghamdi[1,2(⊠)], Paul Prickett[1], and Rossitza Setchi[1]

[1] School of Engineering, Cardiff University, Queen's Buildings,
The Parade, Cardiff CF24 3AA, UK
{AlghamdiAH1,Prickett,Setchi}@cardiff.ac.uk
[2] Mechanical Engineering Department, Engineering College, Albaha University,
P.O Box 1988, Albaha, Kingdom of Saudi Arabia

Abstract. Remanufacturing is a promising industrial activity where products and materials are upgraded and considered for at least another life cycle. In addition to being an environmentally conscious action, remanufacturing has the potential to support circular economy within which significant profit opportunities exist. However, high levels of uncertainty can be experienced during, before and after remanufacturing. This makes its planning stochastic and hard to control. As each component or product is different, with for example high levels of geometrical variation; they may require a unique strategy and process planning. To aid this process, a conceptual decision making framework to support process planning of remanufacturing engineering processes (REP) is proposed. Quality Function Deployment (QFD) method is employed to support the proposed framework (hereafter referred to as REP-QFD). The application of the QFD based methods rely heavily on inputs from experts, in the form of their experience and knowledge. The paper considers how the proposed framework can be engineered with the aim to substantially reduce this reliance on experts and their expertise. The term "Engineering" here reflects the study's focus on technical decisions at the reconditioning stage. To further support the framework a taxonomy of metal manufacturing/remanufacturing processes is also developed.

Keywords: Remanufacturing · Manufacturing · Decision making · House of quality · QFD · Repair processes · Uncertainty · Additive manufacturing (AM)

1 Introduction

1.1 Remanufacturing

The main focus of this research is remanufacturing process planning (RPP), and the importance of selecting the most feasible and practical operations from among many alternatives. Remanufacturing involves a series of industrial operations and processes aimed at engineering a faulty product that is at the end of its life into "like new" or "better" condition. Rebuilding and refurbishing operations such as conventional machining, additive manufacturing or surface engineering can be applied to restore the

G. Campana et al. (eds.), *Sustainable Design and Manufacturing 2017*, Smart Innovation, Systems and Technologies 68, DOI 10.1007/978-3-319-57078-5_22

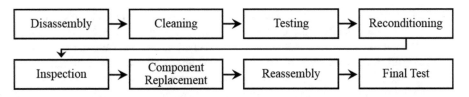

Fig. 1. Generic remanufacturing process

specifications and structure of a given component. Several stages can be considered to accomplish any remanufacturing activity; these stages and their possible sequence are shown in Fig. 1.

Many of the existing remanufacturing methodologies are applicable at the product and/or component level. These approaches typically reflect the high variation, dimensional deviation and uniqueness of each worn component. They usually aim to inform the use of different repair and remanufacturing strategies. However, the opportunity to share and more widely apply these approaches may be limited by their scope. For instance, electronic remanufacturing employs technologies and techniques that differ from those used in machinery remanufacturing. The framework proposed here is intended to be more widely applicable, whilst addressing initially considerations of relevance to the remanufacturing of high-value mechanical components. The intention is to consider how the framework outlined below can be populated with expertise in this context, whilst demonstrating the opportunities that are offered on a wider basis.

1.2 Quality Function Deployment

Quality Function Deployment (QFD) provides a visual representation of the most important parameters and attributes from a customer perspective and a means for converting them into technical parameters.

Four Houses of Quality (HoQ) need to be built during the product development life cycle. In the first HoQ, Product Planning, the "voice of the customer" (VoC) forms the inputs; its function is to translate these demands into engineering characteristics (outputs). These outputs feed into the second HoQ, Product Design, to assist designers in generating the parts characteristics. These are then carried forward into the process and production planning stages, forming the third and fourth HoQs.

QFD is usually focused on the first two design stages but some work has considered applying it in the context of remanufacturing [1–4]. The use of QFD has been proved very successful, reducing development time by half and engineering costs by thirty per cent [5]. Although the QFD tool can help to simplify product development in many sectors, the tool itself is not an easy process to establish. It requires very complex information to prepare the technical charts required to move from the product development planning stage to the production planning stage. Among the downsides of QFD are the ambiguity associated with required data, (such as customer preferences) and the expertise required to translate such non-measurable data into measurable and technical form. As QFD is primarily an expertise-assisted process, there can be a lack of more quantitative data to help experts and engineers to select the most feasible and ideal process plan. To overcome these disadvantages, several publications have combined the QFD method with other mathematical, statistical or decision-making tools; these are discussed in the next section. The fuzzy QFD approach plays an important role in taking account of different judgments from decision-makers and one main contribution from the current study is to propose a framework to reduce or eliminate the reliance on experts. It also considers the information (such as design requirements) from all stages of product development to create adequate process plan.

The next section of the paper briefly outlines research within the remanufacturing field aimed at utilising QFD to better support product design and process and production planning. Section (3) will introduce the different elements of the framework and how the QFD has been modified to support it.

2   Remanufacturing Processes and QFD

A methodology combining fuzzy set theory and QFD has been proposed [1] to incorporate several remanufacturing design attributes into the initial stage of product design. The modified QFD formulated design requirements as engineering characteristics to support product remanufacturability (Remty). This approach introduced three fundamental modifications to standard QFD. First, the established range of VoC was extended to involve users, the environment, remanufacturers and cost. Remanufacturers are considered to be customers of the original equipment manufacturers (OEM), as any successful DfRem activity depends on good links between them. The approach then builds a hierarchical requirements structure and determines weights to support the prioritization of the requirements. Finally, the approach aims to reduce process ambiguity by employing the fuzzy-sets approach to assign an interval weight rather than a single number weight to categorize importance.

Other work created a novel reuse-oriented redesign method for used products using axiomatic design theory and QFD [2]. It presented the QFD customer as the user of a machine tool whose features and processing were characterized as QFD customer requirements. Comparing these studies [1] and [2], it becomes clear that the two employed different methodologies and inputs within the QFD approach. Although sharing the same research aim of supporting design for remanufacturing, they also chose entirely different kinds of customer. Nevertheless, despite minor variations of accuracy in comparing a remanufactured machine tool with the quality standard of the same machine, these studies confirmed the power of the QFD method. Similarly, an integrated eco-design decision making (IEDM) methodology created using QFD as one stage of three key stages of the method development [6].

Another study, integrated the QFD and fuzzy linear regression to relate remanufacturing performance to process quality characteristics [7]. This involved three phases: a QFD selection framework, a fuzzy-linear regression stage to estimate the QFD parameters, and a selection stage to choose the optimal process plan. In the case study, the corporation's key workers recommended maximum and minimum values for the process quality characteristics of a lathe machine guide. In a similar study, analytical

Table 1. Process Planning selected literature and potential relationship to QFD

Ref.	Research focus	Methods	Case study	HoQ
[9]	The impact of uncertainty of used parts refurbishability (Refty) and its profit function	Mathematical analysis	None	4
[10]	RPP involving inputs with multiple and uncertain qualities with capacity constraints	Stochastic programming	None	4
[11]	Analysis of the impact of the uncertainty of used part quality and the remanufacturing process performance	Analytical algorithm	Testing	3
[12]	Near optimum buffer allocation plan for remanufacturing	Algorithm/network/decomposition principle	Testing	4
[13]	Analysis of a remanufacturing system with multiple process routings	Graphical Evaluation and Review Technique (GERT)	Lathe spindle	3 & 4
[14]	Analysis of remanufacturing performance with uncertain reliability of capacity, processing time, and demand	Algorithm/Linear optimization approach/discrete-event simulation	Mobile phones	4
[15]	Design, evaluation and implementation of Remanufacturing processes	Mixed integer program	Flat screen monitors	4
[16]	Examination of all possible remanufacturing processes for used parts	Graphical Evaluation and Review Technique (GERT)	Lathe spindle	4
[17]	Decision making in selecting remanufacturing technology regarding cost, quality, time, service, resource consumption and environmental impact	Multi-criteria decision making model (MCDM)/Analytical Hierarchy Process (AHP)	Valve stems remanufacturer	3

(continued)

Table 1. (*continued*)

Ref.	Research focus	Methods	Case study	HoQ
[18]	Improving final quality components using additive manufacturing processes for finishing and to automate derivation of the repair section	STEP-based numerical control (STEP-NC). using ISO 14649 standard	Remanufacturing of a pocket part	3
[19]	Production schedule and process plan for remanufacturing	Monte Carlo simulation/Pareto-based opt	Experimental Testing	4
[20]	Determination of best reconditioning process sequence by analysing the condition of core components	FMEA/Conceptual framework	Camshaft remanufacturing	2 & 3
[21]	Using current parts remanufacturing information to generate a sound process plan	Hybrid method of rough set (RS) and case-based reasoning (CBR)	Machine tool saddle remanufacturing	3 & 4

hierarchy process (AHP) and linear regression techniques were combined with QFD to select an optimal process alternative [8].

Table 1 lists other significant studies, encompassing uncertainty management, processes routings, demand, technology portfolio, surface engineering quality, and production scheduling. In this summary, the last column cross-references each with the corresponding HoQ in the normal QFD process. The gap noticed when reviewing the literature is that different stages of product life-cycle should be considered using HoQs 1 to 4 with more engineering specifications and attributes. The proposed framework is designed to achieve this goal.

3 QFD Framework for Remanufacturing Engineering Processes (REP-QFD)

This section describes the proposed modification to traditional QFD method to support remanufacturing processes. The focus is on evaluating the technical issues of the reconditioning stage, as this controls the most significant operations of remanufacturing. For that reason, the term "process" is used here to refer to the technologies and industrial operations intended to accomplish the reconditioning of worn components. There is no limit to which components and products can be classified as

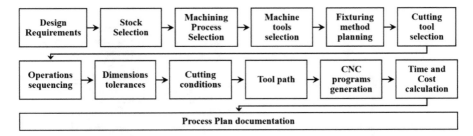

Fig. 2. Process plan for reconditioning stages. adapted from [22]

remanufacturable, and the number of different possible combinations of component materials and conditions is infinite. As an example, the process plan for the reconditioning stage of a rough-cut process includes the steps as shown in Fig. 2.

The REP-QFD framework, uses a modified HoQ to support the decisions that are needed to formulate and apply the most appropriate remanufacturing strategies. In this case, the HoQ was limited to investigate the application of metal-based remanufacturing processes. At this early stage of development, the process will start with the consideration of the three main options of component reconditioning processes as process alternatives. These are termed additive, subtractive and surface enhancing. The first step towards the design of the proposed framework was to produce a hierarchical representation of all possible operations to bring the worn component to "like new" condition. This representation was the starting point of this current research. The aim was to bring together and represent existing knowledge [23–29] in such a way as to support future considerations of the most applicable remanufacturing process. Doing so resulted in Figs. 3, 4 and 5 which show the developed taxonomy of remanufacturing using additive, subtractive and surface finish processes, respectively. The classification encompasses modern and traditional manufacturing processes as well as processes and techniques developed specifically to support remanufacturing engineering. It forms an important research contribution upon which future work can build.

The second step is listing all possible failure modes for the component undergoing remanufacturing. The main divisions and subdivisions of component condition and damage mechanisms must be reviewed to help in matching each required process against damage. For example, surface damage might require operations to repair micro-cracks.

The intention here is to assess the most appropriate way by which the existing state of the component can be re-engineered to meet the demanded quality(s) listed in Sect. (1) of the REP-QFD (Fig. 6). Here, failure mode and effect analysis (FMEA) can support current and future operations by identifying the degree of damage in each category (wear, fracture, corrosion or deformation).

The next element, remanufacturing process planning, is undertaken in Sect. 3. This requires the linking three important criteria: defect, remanufacturing process or processes required and surface feature. In this stage, different alternatives arise, and rankings need to be specified as in the next element of the framework. This stage

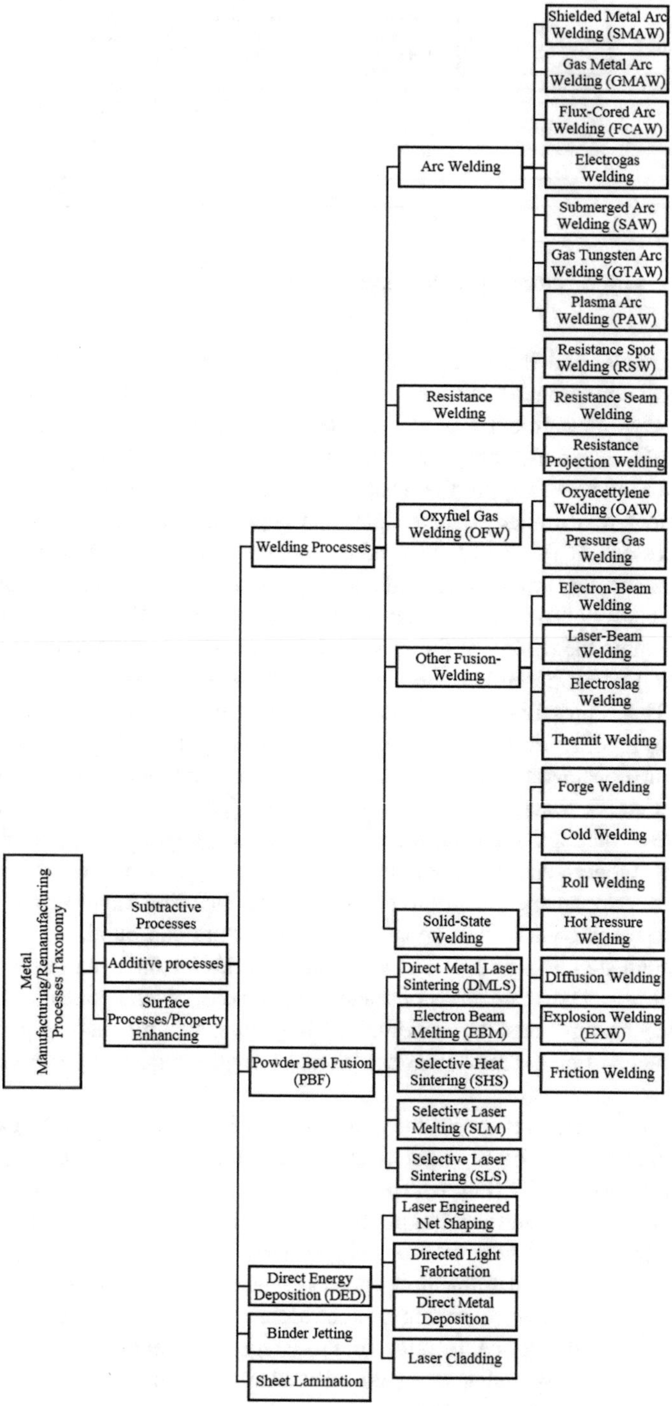

Fig. 3. Taxonomy of metal manufacturing/remanufacturing processes (Additive)

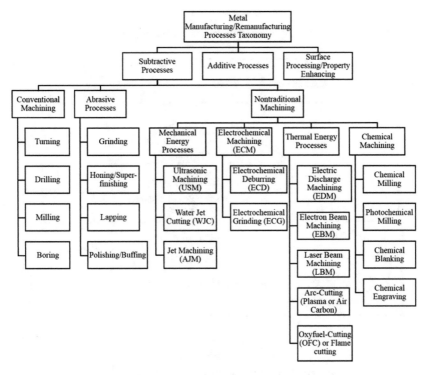

Fig. 4. Taxonomy of metal manufacturing/remanufacturing processes (Subtractive)

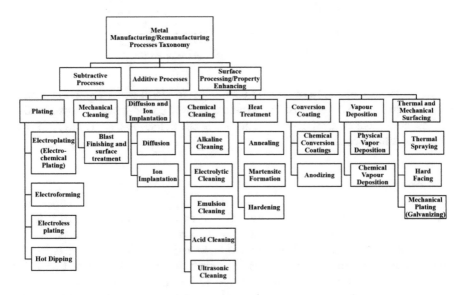

Fig. 5. Taxonomy of metal manufacturing/remanufacturing processes (surface enhancing)

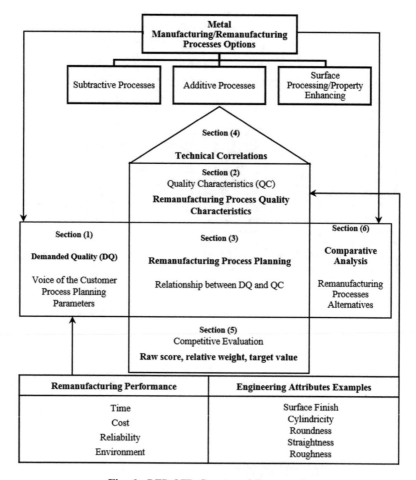

Fig. 6. REP-QFD Conceptual Framework

includes analysis of (1) process instruction, (2) performance (resources, service, emission, quality, cost, time), and (3) required technology (in-house or outsourced).

The next element uses QFD to rank each alternative and to identify the best alternative among competing processes.

4   Conclusion

This paper introduces an initial conceptual framework to reduce the stochastic and fuzzy nature of engineering attributes during remanufacturing process planning. The key contribution of the proposed model is to reduce dependence on subjective judgement and consider different stages of product life-cycle using HoQs 1 to 4 with more engineering specifications and attributes included. It also provides a schematic representation of the important elements of the process plan and suggested QFD modification. Future work will introduce the complete REP-QFD method and an illustrative case study will be used to evaluate the practicality of the framework.

Acknowledgements. The author acknowledges financial support and scholarship from Albaha University.

References

1. Yang, S., Ong, S.K., Nee, A.Y.C.: Design for remanufacturing-a Fuzzy-QFD approach. In: Re-engineering Manufacturing for Sustainability, pp. 655–661. Springer (2013)
2. Du, Y., Cao, H., Chen, X., Wang, B.: Reuse-oriented redesign method of used products based on axiomatic design theory and QFD. J. Clean. Prod. **39**, 79–86 (2013)
3. Armacost, R., Balakrishnan, D., Pet-Armacost, J.: Design for Remanufacturability Using QFD (2005)
4. Yüksel, H.: Design of automobile engines for remanufacture with quality function deployment. Int. J. Sustain. Eng. **3**, 170–180 (2010)
5. Clausing, D., Pugh, S.: Enhanced quality function deployment (1991)
6. Romli, A., Prickett, P., Setchi, R., Soe, S.: Integrated eco-design decision-making for sustainable product development. Int. J. Prod. Res. **7543**, 1–23 (2016)
7. Jiang, Z., Fan, Z., Sutherland, J.W., Zhang, H., Zhang, X.: Development of an optimal method for remanufacturing process plan selection. Int. J. Adv. Manuf. Technol. **72**, 1551–1558 (2014)
8. Ullah, S.M.S., Muhammad, I., Ko, T.J.: Optimal strategy to deal with decision making problems in machine tools remanufacturing. Intl. J. Precis. Eng. Manuf. Green Technol. **3**, 19–26 (2016)
9. Zikopoulos, C., Tagaras, G.: Impact of uncertainty in the quality of returns on the profitability of a single-period refurbishing operation. Eur. J. Oper. Res. **182**, 205–225 (2007)
10. Denizel, M., Ferguson, M., Souza, G.G.C.: Multiperiod remanufacturing planning with uncertain quality of inputs. IEEE Trans. Eng. Manag. **57**, 394–404 (2010)
11. Tang, X., Mao, H., Li, X.H.: Effect of quality uncertainty of parts on performance of reprocessing system in remanufacturing environment. J. Southeast Univ. **27**, 92–95 (2011)
12. Aksoy, H.K., Gupta, S.M.: Optimal management of remanufacturing systems with server vacations. Int. J. Adv. Manuf. Technol. **54**, 1199–1218 (2011)
13. Li, C., Tang, Y., Li, C.: A GERT-based analytical method for remanufacturing process routing. In: IEEE International Conference on Automation Science and Engineering, pp. 462–467 (2011)
14. Seliger, G., Franke, C., Ciupek, M., Başdere, B.: Process and facility planning for mobile phone remanufacturing. CIRP Ann. Manuf. Technol. **53**, 9–12 (2004)
15. Kernbaum, S., Heyer, S., Chiotellis, S.: CIRP J. Manuf. Sci. Technol. Process Plann. IT-equipment Remanuf. **2**, 13–20 (2009)
16. Li, C., Tang, Y., Member, S., Li, C., Li, L.: A modeling approach to analyze variability of remanufacturing process routing. TASE **10**, 86–98 (2013)
17. Jiang, Z., Zhang, H., Sutherland, J.W.: Development of multi-criteria decision making model for remanufacturing technology portfolio selection. J. Clean. Prod. **19**, 1939–1945 (2011)
18. Um, J., Rauch, M., Hascoët, J., Stroud, I.: STEP-NC compliant process planning of additive manufacturing : remanufacturing. Int. J. Adv. Manuf. Technol. **88**, 1215–1230 (2016)

19. Zhang, R., Ong, S.K., Nee, A.Y.C.: A simulation-based genetic algorithm approach for remanufacturing process planning and scheduling. Appl. Soft Comput. J. **37**, 521–532 (2015)
20. Kin, S.T.M., Ong, S.K., Nee, A.Y.C.: Remanufacturing process planning. In: Procedia CIRP, vol. 15, pp. 189–194 (2014)
21. Jiang, Z., Jiang, Y., Wang, Y. et al.: A hybrid approach of rough set and case-based reasoning to remanufacturing process planning. J. Intell. Manuf. (2016). doi:10.1007/s10845-016-1231-0
22. Chin, K.S., Zheng, L.Y., Wei, L.: A hybrid rough-cut process planning for quality. Int. J. Adv. Manuf. Technol. **22**, 733–743 (2003)
23. Groover, M.P.: Principles of modern manufacturing (2013)
24. Kalpakjian, S., Schmid, S.R.: Manufacturing Engineering and Technology (2010)
25. Unune, D.R., Mali, H.S.: Current status and applications of hybrid micro-machining processes: a review. Proc. Inst. Mech. Eng. Part B J. Eng. Manuf. **229**, 1681–1693 (2015)
26. Bras, B.: Design for Remanufacturing Processes (2007)
27. Parkinson, H.J., Thompson, G.: Analysis and taxonomy of remanufacturing industry practice. Proc. Inst. Mech. Eng. Part E-J. Process Mech. Eng. **217**, 243–256 (2003)
28. Herderick, E.: Additive manufacturing of metals: a review. In: Materials Science and Technology 2011 Conference and Exhibition, MS&T 2011, vol. 2, pp. 1413–1425 (2011)
29. Zhang, H., Liu, S., Lu, H., Zhang, Y., Hu, Y.: Remanufacturing and remaining useful life assessment. In: Handbook of Manufacturing Engineering and Technology, pp. 3137–3193. Springer (2015)

Optimized Production Process of a Supporting Plate as an Improvement of the Product Sustainability

G. Bertuzzi[1], S. Di Rosa[2(✉)], and G. Scarpa[2]

[1] Sacmi Imola S. C., Via Selice Provinciale 17/a, 40026 Imola, BO, Italy
giacomo.bertuzzi@sacmi.it
[2] EnginSoft S.p.A., Via Della Stazione 27, Frazione Mattarello, 38123 Trento, Italy
s.dirosa@enginsoft.com, g.scarpa@enginsoft.it

Abstract. The main goal of the described activity is the verification of an innovative production process of a steel supporting plate by using numerical simulation. The normal production process consists of different phases and processes: the laser cutting of blanks, welding, stress relieving, heat treatment and machining. The studied alternative is a sand casting process in order to directly obtain the final supporting plate. Numerical simulations were used to investigate the impact behavior of complex products in order to define the best production solution. The new design chain represents an important simplification in order to reach a simpler, less expensive and more sustainable manufacturing process and these aspects were confirmed performing a Life Cycle Assessment.

Keywords: Casting process · Numerical simulation · Multi-objective optimization · Life Cycle Assessment · Design chain

1 Introduction

Nowadays, the development of a new product is a multi-disciplinary work that involves structural and functional design, ergonomics, reliability, manufacturing process, quality control and assembly. According to literature, all manufacturing processes have two outputs: geometry - the macroscopic shape of the product - and properties - all intrinsic material properties. These two outputs completely define the performance of the product and the design specifications that it must meet. All the manufacturing processes also involve the transformation of material from an initial condition, in term of geometry and properties, to the final needed outputs. This transformation is accomplished through the application (or removal) of energy and material. While industry was historically focused only on outputs and the economic impact of the manufacturing process, today, it is also concerned about product sustainability in relation to its production processes and, in particular, to its environmental impact.

In this project an industrial example allows to describe the critical analysis of the Current Production of a steel supporting plate [7], hereinafter referred to as CP, the definition of an alternative production process, the simulation and optimization of the innovative process, the evaluation of the improvements in term of environmental impacts by Life Cycle Assessment (LCA) of the proposed production process in

G. Campana et al. (eds.), *Sustainable Design and Manufacturing 2017*, Smart Innovation, Systems and Technologies 68, DOI 10.1007/978-3-319-57078-5_23

comparison with the CP. Optimized component designs and casting processes using new engineering tools are achieved in concert with casting engineers and designers. This integration and human collaboration is critical for the successful speed-up of the design-process chain. Designers need to be supported by casting experts to be able to take full advantage of casting performance, concerning its design and properties [14].

Quantitative results about casting performance provided by casting process simulation, help designers to understand the process impact on the performance of the castings in use [10, 11, 12].

2   Current Geometry and Current Production Description

The supporting plate is made by steel S235 JR as reported in Table 1 (reference standard EN 10025-2:2014). The production process is achieved with 5 different steps:

- Cutting of the blanks [7].
- Squaring and laser cutting.
- Welding [7].
- Stress relieving heat treatment.
- Machining.

Table 1. Nominal chemical composition of S235JR.

C% max	Si% max	Mn% max	P% max	S% max	N% max	Cu% max
0,17[c]	–	1,40	0,035	0,035	0,012[a]	0,40
0,19[c]	–	1,50	0,045	0,045	0,014[b]	0,45

The final component is shown in the Fig. 1. The length of the welding perimeter is 30.25 m, while the component total weight is 515 kg.

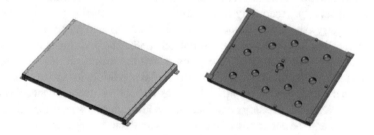

Fig. 1. Steel supporting plate.

3   The Proposed New Geometry and the New Production Process for the Supporting Plate

The designers developed a new geometry of the plate, see in Fig. 2, constituted by a honeycomb structure that was optimized both for castability and final machining. Concerning the material, steel was replaced with a castable material with similar properties (gray iron) [9]. The casting process typically used for such a cast iron component is sand casting. The weight of this component is around 625 kg.

Fig. 2. New geometry of the supporting plate.

The optimization of the casting process, in order to get the best component performances, was run using MAGMA5 and its Optimisation tool [16]. Casting simulation is nowadays mandatory to get a reliable and robust process layout: gating, risering and chilling system. It also helps to predict metallurgy and melting practice, and offers predictions concerning shrinkage, porosity, microstructure and mechanical properties [17].

4   Casting Process Analysis and Optimization

To simulate the casting process, the geometry symmetry is assumed, so that only one fourth of the component is simulated, thus speeding up the optimization process. The high numbers of variables which define the foundry process definitely make it one of the most difficult problems to optimize since, generally speaking, there is no single solution which will satisfy the stated aims in the best possible way. In order to select the best solution, it is necessary to evaluate and compare a large number of potentially possible solutions. In the case of foundry, it is not possible to analyze the entire range of solutions with a high number of variables and goals by means of a single - objective tool. Despite dividing the problem into sub-problems of a lesser entity (filling, solidification, thermal aspects, residual stress, etc.), the goals are frequently in contrast or interlinked with each other and must therefore be pursued separately without attributing individual degrees of incidence a priori. The optimization process of a system takes into account the synergetic interaction among various parameters. A multi-objective optimization procedure returns trade-off solutions (good compromises in relation to various objectives). The optimization approach is based on automatized and parametric analysis and can reduce time-effort when many configurations and responses of a system need to be considered [11].

It is necessary to define parameters, objectives, constraints and responses to be monitored. The casting optimization was led by the following goals:

1. The reduction of component porosity under 10% in critical areas as the machined surface or where threated holes will be machined.
2. The minimization of the allowances, feeders and chillers used in the process, so to reach the maximum process efficiency.

Objectives and constraints have to be defined as follows (Fig. 3):

- Objective 1: to minimize feeders and allowances volumes.
- Objective 2: to minimize chiller volumes.
- Constraint 1: maximum accepted porosity in the evaluation area 1 = 10%.
- Constraint 2: maximum accepted porosity in the evaluation area 2 = 10%.
- Constraint 3: maximum accepted porosity in the evaluation area 3 = 10%.

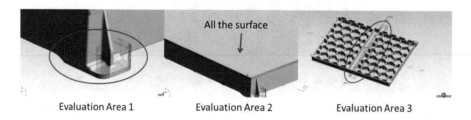

Evaluation Area 1   Evaluation Area 2   Evaluation Area 3

Fig. 3. Definition of the evaluation areas.

In order to define the DOE plan, necessary to identify the optimized conditions, the following geometry input variables are considered (Fig. 4):

- Diameter and height of feeders.
- Presence or absence of feeders.
- Diameter and height of chillers.
- Presence or absence of chillers.
- Height of allowances.

The optimization phase is based on an initial population of configurations to simulate (the DOE, Design of Experiment) selected through the use of the algorithm "Reduced Factorial" (this algorithm grounds on two distinct levels of "full factorial" algorithm to cover the extreme of the intervals considered) in combination with the algorithm "Sobol" (this algorithm creates design "quasi random" ensuring that all the factorial design space is covered more uniform as possible) [8]. With the combination of these algorithms, it is possible to completely cover the space vector of input variables guaranteeing the representation of the complete design space. Subsequent generations are created by the, above mentioned, genetic algorithm called MOGA (Multi Objective Genetic Algorithm), which allows the user to define new additional designs based on elitism and mutation attributes

[13, 14]. Despite the complexity of the topic, the user define only the number of configurations of the first population and the number of successive generations as a function of the available time. A total of 42 independent variables where introduced in the optimization loop. The casting optimization required to run 1250 simulations in total with the aim of defining the optimum casting layout generated by the DOE plan and the optimization algorithm allowing to achieve the objective 1 and 2 and taking into account the constraints 1, 2 and 3.

The diagram in Fig. 5 considers the trend of the objectives versus the optimization run designs and shows the optimization convergence towards the reduction of both the objectives, guarantying low porosity (under 10%) in the evaluation areas.

The 2D Scatter Chart (Fig. 6) shows the distribution of the designs versus the defined objectives:

- Objective 1: to minimize feeders and extra metals volumes
- Objective 2: to minimize chiller volumes

The optimum design is the n. 1106.

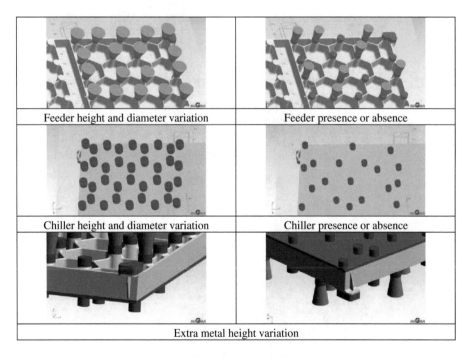

Fig. 4. Geometry input variations.

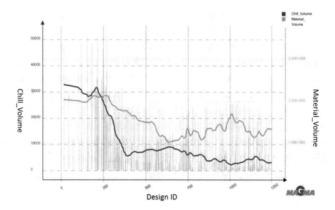

Fig. 5. Optimization history chart.

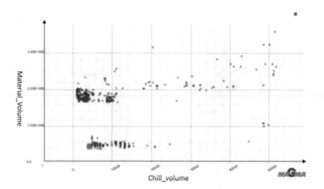

Fig. 6. 2D scatter chart.

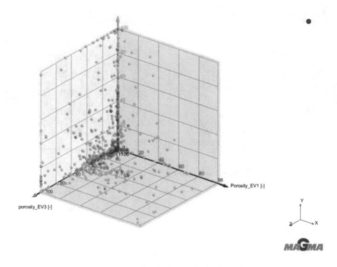

Fig. 7. 3D scatter chart.

The 3D scatter chart (Fig. 7) shows the calculated designs versus the maximum porosities in the three different evaluation areas. The design 1106 (the optimum one) respects the constraints (porosity under 10%).

The optimum design solution has the following distribution and dimensions of feeders, chillers and allowances (Fig. 8).

Fig. 8. Optimum design solution: feeders, allowances and chillers geometry definition.

The map of porosity is analyzed in order to control the respect of the constraints (1, 2 and 3): porosity value less than 10% in the evaluated areas for the optimum design (Fig. 9).

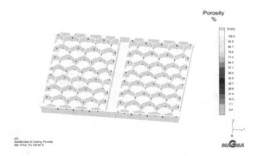

Fig. 9. Map of porosity for the optimum design component.

5   Life Cycle Assessment: Comparison Between Normal Production and the Casting Process

Life Cycle Assessment is such a system analysis tool that can be used to assess the environmental aspects and potential impacts associated to a production system [1]. LCA gives a holistic view of the impacts associated to a product system during:

- the whole span of life (from cradle to grave),
- from the raw material extraction through the production phase (cradle to gate),
- from the use and end-of-life phases (gate to grave),
- includes the processes from the production phase only (gate to gate).

The main phases of an LCA, as commonly assumed, are the following ones [2–4]:

- Goal & Scope.

- Inventory Analysis.
- Impact Assessment.
- Interpretation.

The main environmental impact categories are [5, 6]:

- Global Warming Potential (GWP). It is the index that measures the warming grade of earth that is produced by the greenhouse gas and the unit of measure is kg CO_2 eq.
- Acidification Potential (AP). The emission of NOx, SOx (from combustion of fuel) in atmosphere produces the reduction of pH in different ecosystem, with heavy consequences on organism. The unit of measure is kg SO_2 eq.
- Stratospheric Ozone Depletion Potential (ODP). Ozone is produced in the stratosphere and it is destroyed by different chemical reactions that transform it in molecular oxygen. ODP is the index that measures the ozone depletion and it is expressed in kg CFC-11 eq.

GaBi software is the LCA modeling program produced by Thinkstep that EnginSoft has used to perform the impact evaluation. The production process is constituted by 5 different steps that were mentioned in the previous paragraph 2. A great simplification of the process was obtained by applying the casting process because the final component is obtained only with two steps:

- Casting.
- Machining.

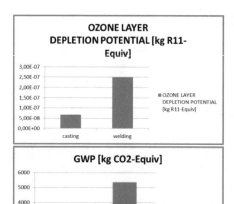
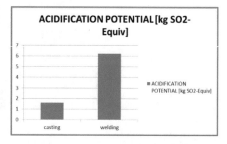

Fig. 10. Environmental impacts: comparison between normal production and casting process.

In order to make a comparison in term of environmental impacts it is possible to measure the consumed energy used for the CP against the one required for the proposed casting process. The length of the weld bead of the steel plate is 30.25 m and the energy necessary for the production of the welded and assembled plate is 10890 MJ. To cast the steel plate, the necessary energy is 2800 MJ. The most significate calculated environmental impacts are proposed in the graph below (Fig. 10). It is clear from those diagrams that the casting solution produces less environmental impacts than the CP, so from an environmental point of view it is better to produce the supporting plate using casting process and machining instead of complex welding operations.

6   Conclusions

The production of the steel supporting plate requires a complex process consisting of five different phases. The definition of a new geometry, that can be obtained by a casting process, generates a simplification of the process, thus only 2 steps are needed: casting and machining.

Thanks to casting simulation and optimisation, the production costs are reduced by 30% and the LCA comparison of CP production and casting process shows that the casting process is convenient also from an environmental point of view [6].

Acknowledge.   The Authors want to thank Sacmi Imola S. C. and EnginSoft S.p.A. for their support during the development of the project.

References

1. Baldo, G.L.: LCA - Life Cycle Assessment, IPASERVIZI (2000)
2. Fontana, M.: Diffusione ed uso dei risultati del programma di ricerca ExternE, Inquinamento nr. 11, vol. 199, pp. 18–19 (1999)
3. ISO 14040: Environmental Management - Life Cycle Assessment - Principles and Framework (1997)
4. ISO 14041: Environmental Management - Life Cycle Assessment - Goal and Scope Definition and Inventory Analysis (1998)
5. ISO 14042: Environmental Management - Life Cycle Assessment - Life Cycle Impact Assessment (2000)
6. ISO 14043: Environmental Management - Life Cycle Assessment - Life Cycle Interpretation (2000)
7. Rinaldi, E.: Saldatura e taglio dei metallic. HOEPLI (2001)
8. modeFRONTIER 4.0: User's Manual. ESTECO, Trieste (2009)
9. Karsay, S.I.: Production, Ductile Iron, vol. 1. QIT, Fer et Titane Inc. (1981)
10. Sobol', I.M.: On the systematic search in a hypercube. SIAM J. Numer. Anal. **16**(5), 790–793 (1979)
11. Poles, S., Vassileva, M., Sasaki, D.: Multiobjective optimization software. In: Branke, J., Deb, K., Miettinen, K., Słowiński, R. (eds.) Multiobjective Optimization. LNCS, vol. 5252, pp. 329–348. Springer, Heidelberg (2008). doi:10.1007/978-3-540-88908-3_12
12. Poles, S., Geremia, P., Campos, F., Weston, S., Islam, M.: MOGA-II for an automotive cooling duct optimization on distributed resources. In: Obayashi, S., Deb, K., Poloni, C., Hiroyasu, T., Murata, T. (eds.) EMO 2007. LNCS, vol. 4403, pp. 633–644. Springer, Heidelberg (2007). doi:10.1007/978-3-540-70928-2_48

13. Pettersson, F., Chakraborti, N., Saxen, H.: A genetic algorithms based multi-objective neural net applied to noisy blast furnace data Source. Appl. Soft Comput. **7**(1), 387–397 (2007)
14. Pettersson, F., Biswas, A., Sen, P.K., Saxen, H., Chakraborti, N.: Analyzing leaching data for low-grade manganese ore using neural nets and multiobjective genetic algorithms. Mater. Manuf. Processes **24**(3), 320–330 (2009)
15. Egner-Walter, A., Hartmann, G., Kothen, M.: Integration of manufacturing process simulation in to the process chain. In: 20th CAD-FEM User Meeting, Kultur-und Congress Centrum Graf Zeppelin Haus, Friedrichshafen, Lake Costance, Germany, 9–11 October 2002
16. Bonollo, F., Odorizzi, S., Numerical Simulation of Foundry Processes. SGEditoriali (2001)
17. Gramegna, N., Bucchieri, L., Furlan, L.: Integrated CAE development of innovative grey iron heat exchanger. In: NAFEMS World Congress 2005, Malta, 17–20 May 2005

Sustainable Manufacturing Systems and Enterprises

Sustainable Manufacturing for Thai Firms: A Case Study of Remanufactured Photocopiers

Jirapan Chaowanapong[1(✉)] and Juthathip Jongwanich[2]

[1] School of Management, Asian Institute of Technology, Pathumthani, Thailand
Jirapan.chaowanapong@ait.asia
[2] Faculty of Economics, Thammasat University, Bangkok, Thailand
Juthathip@econ.tu.ac.th

Abstract. Remanufacturing represents a significant mean encouraging sustainability. This paper aims to investigate the critical factors influencing the decisions of firms to engage in remanufacturing through conducting a case study of Thai remanufactured photocopiers employing qualitative and quantitative approaches. The results show that business feasibility is the prominent determinant driving firms' decisions, followed by firm's strategic factors, and policy factors. Totally derived from the area of business feasibility, as the top four individual factors, financial aspects is ranked first as the most critical factor influencing remanufacturing, followed by availability of skilled workers, product maturity, and technical aspects. Firm-level characteristics matter significantly in ranking the factors. Concrete support by government towards implementing comprehensive policies is needed to strengthen remanufacturing development in Thailand.

Keywords: Remanufacturing · Sustainability · Photocopiers

1 Introduction

Remanufacturing represents a critical strategy within sustainability [1]. Sustainability is explained in terms of three pillars: economy, society, and environment [2]. Remanufacturing generates environmental and business benefits by extending product longevity, whilst minimizing raw material consumption [3]. Remanufacturing is defined as "The process of returning a used product to at least OEM original performance specification from the customers' perspective and giving the resultant product a warranty that is at least equal to that of a newly manufactured equivalent" [4, 5]. Remanufacturing is, therefore, considered as constituting an End of Life Strategy (EOL).

Although remanufacturing is still concentrated in developed countries, for example the U.S. and U.K., overconsumption of natural resources has also prompted remanufacturing practices to become increasingly important in developing countries. In developing countries remanufacturing research has mostly been conducted in China, India, and Brazil and is still in the nascent stage. Due to the limited handful of empirical studies in developing countries, this study aims to examine the key factors influencing the decisions of firms to engage in remanufacturing through conducting a case study of remanufactured photocopiers in Thailand. The photocopier industry in Thailand is

© Springer International Publishing AG 2017
G. Campana et al. (eds.), *Sustainable Design and Manufacturing 2017*, Smart Innovation, Systems and Technologies 68, DOI 10.1007/978-3-319-57078-5_24

chosen because it is suitable for conducting remanufacturing activities due to the high market value, suitable specifications, and long useful lifespan of photocopiers [6]. The market value estimate of used photocopiers in Thailand amounts to approximately US$ 60 million, or 50% of the value of new products [6]. This has prompted photocopiers to represent an attractive remanufacturing option in Thailand. In addition, Thailand has the potential to become the remanufacturing center of the Association of South East Asian Nations (ASIAN) countries. The Thai workforce has accumulated the manufacturing knowledge required for undertaking remanufacturing since being integrated into the global manufacturing network of large multinational firms [6, 7].

The rest of the paper is organized as follows. The next section presents the analytical framework. The third section elaborates on the research methodology. Findings and discussion are outlined in Section four, while our conclusions and policy inferences comprise the final section.

2 Analytical Framework

The determinants driving a firm's decision to undertake remanufacturing are categorized into three areas: business feasibility, a firm's strategic factors, and policy factors, subdivided into thirteen individual factors (Fig. 1).

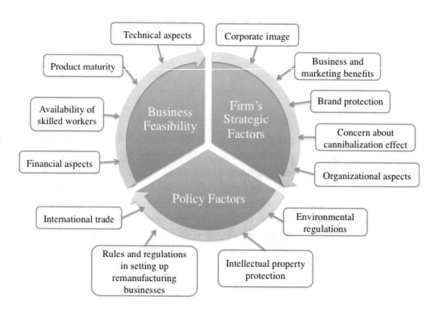

Fig. 1. A summary of the factors influencing a firm's decision to conduct remanufacturing

2.1 Business Feasibility

First, technical aspects represent product designs and their capability to be disassembled and re-assembled. In making products more remanufacturability, the development of product design should take into consideration the processes involved in performing remanufacturing and the costs incurred in the product life cycle [8–10]. To enable remanufacturing feasibility, design for remanufacturing must have the potential to be easily disassembled, cleaned, replaced, and re-assembled in order to prolong the remanufactured product's lifespan [11, 12]. Second, product maturity comprises the product's physical lifespan and the speed of technological change [13]. If a new product is subject to rapid technological change, product upgrades are recommended, rather than remanufacturing. If a product lifespan is quite short, the product is unsuitable for remanufacturing. The next key driver concerns the availability of skilled workers. Remanufacturing is labor intensive because its processes are relatively complex and must be handled by skilled workers [14]. A lack of such expertise can constitute a major hurdle for remanufacturing operations, as outlined in a Chinese case study [15]. Finally, financial aspects play an important role in remanufacturing [13, 16–19], through costs and demand. The majority of remanufacturing costs are incurred from the additional resources required to resume a product to its original performance standards. An effective network encompassing reverse logistics constitutes the remanufacturing backbone [3, 10, 17, 20–23]. This reduces uncertainties in balancing core supply and remanufacturing demand [14]. Customers may have a negative perception of remanufactured goods as they believe they represent inferior quality [13, 15]. Such negative perceptions are even more prevalent in developing countries where remanufacturing knowledge and environmental concerns are limited.

2.2 Firm's Strategic Factors

The growing concerns of environment and corporate social responsibility (CSR) have encouraged many businesses to offer green products to their consumers to enhance corporate image [9, 10]. Remanufacturing represents such a green solution since it consumes comparatively less resources as opposed to traditional manufacturing. Moreover, remanufacturing comprises business and marketing benefits through building new business strategies and creating new product sales [16, 24, 25]. This enables a firm to utilize feedback information in addressing product design, market solutions and customer demand. Securing a supply of spare parts represents another business benefit minimizing the dependency on suppliers after terminating a production line [16, 24, 26]. Brand protection is the third factor determining whether firms remanufacture. Original Equipment Remanufacturers (OEMs) basically hold full information on product design, product specifications, and material sourcing. OEMs prefer to operate in-house remanufacturing as they need to protect their brand from Independent Remanufacturers (IRs) [26]. They also try to prevent their cores running to IRs by controlling product quality. Conversely, this causes difficulty for IRs due to the lack of OEM cooperation [16, 24]. Concerns about cannibalization effects become an additional factor influencing firms' decisions. Remanufactured products may cannibalize new product sales because of preferences transferring from new to remanufactured goods within some price ranges [13, 15, 27]. The belief that remanufacturing may destroy the prospects of new products could reduce remanufacturing development [25, 28]. Nevertheless, some remanufactured and new products target different market segments. Thus, cannibalization effects are minimized [16]. Finally, organizational aspects through effective management, alignment and communication among remanufacturing and other departments represent another critical factor in remanufacturing [15]. Collaboration and flexibility between original and remanufacturing activities can help generate effective cost management and optimum production planning [13, 29].

2.3 Policy Factors

The factor to consider first in conducting remanufacturing concerns the rules and regulations affecting setting up remanufacturing businesses. A lack of relevant governmental rules and regulations represent barriers to operating remanufacturing, as revealed in the cases of India [30] and Brazil [25, 28]. Moreover, international trade policy, in some countries, for example Brazil, China and Malaysia, remanufactured products are banned or limited through either higher tariffs and fees, or overly stringent regulations, certification, and inspection requirements [31]. These increase the costs of remanufacturing production. Third, Intellectual Property (IP) protection represents an incentive for OEMs to become involved in remanufacturing to protect their IP from IRs [32]. Breaking IP potentially occurs when patents are not protected due to the lack of governmental regulations found predominately in emerging countries [9]. Finally, under the banner of environmental regulations, the U.S. enacted take-back laws strengthening remanufacturing since they force firms to take responsibility for their products once they reach their end of life [33]. The legislation driving remanufactured toner cartridges is regulated under WEEE[1] directives that require companies to be responsible for post-consumption recycling [16, 24]. Environmental regulations such as Extended Producer Responsibility (EPR) in the U.S. and Europe constitute key drivers for undertaking remanufacturing [10].

3 Research Methodology

This research applied both qualitative and quantitative methods through employing industrial surveys administered to respondents of photocopier remanufacturing firms. A qualitative approach was conducted through semi-structured interviews. Quantitative methodologies involved the collection of questionnaires and descriptive statistics, t-test and One-way ANOVA (Analysis of Variance) were performed. A purposive sampling strategy was applied for informant selection to gain insight and identify information-rich participants, rather than probability sampling techniques [34].

---

[1] WEEE DIRECTIVE (2002/96/EC): Waste Electrical & Electronic Equipment requires producers to manage post-consumer recycling and the disposal of electronic products effective August 13, 2005.

Business owners and executive managers with the authority to make remanufacturing decisions constitute the key informants. As there is no current registration of photocopier remanufacturing firms in Thailand, our estimation of firms comprised approximately 80 firms[2]. The thirteen sampled firms constituted 16% of the overall total. Firms were selected bearing in mind their relevance to this research and their experiences, rather than randomly [35].

The data collection consumed three months, during January–March 2015. The length of interviews varied from half an hour to two hours. The interview questions were first tested on a firm owner with over ten years experience of remanufacturing activities. The questionnaires were sent to six experts for their suggestions. The respondents were asked by a structured questionnaire to indicate on a 5-point Likert scale [36]. The questionnaire was composed of three sections: firm information, influencing factors for undertaking remanufacturing, and personnel information. The influential levels, derived from the mean scores, standard deviation (S.D.), and median, were categorized into five levels: (1) not at all = 1.00–1.79, (2) slightly = 1.80–2.59, (3) somewhat = 2.60–3.39, (4) very influential = 3.40–4.19, and (5) extremely influential = 4.20–5.00. Comparing group means was performed through t-test and One-way ANOVA [37].

4 Findings and Discussion

Using questionnaires, the mean scores of factors were converted into five influential levels. The results show that business feasibility is the most crucial determinant driving firms' decisions, followed by firm's strategic factors, and policy factors. Figure 2 presents the mean scores of the thirteen influencing factors. The top four mean scores were totally derived from the area of business feasibility. Only financial aspects gained a highest rating score at the extremely influential level. Availability of skilled workers and product maturity were determined as being very influential. Technical aspects were categorized as being somewhat influential.

One-way ANOVA[3] was calculated for firm sizes[4]: small, medium, and large. It revealed that corporate image is the only factor differing significantly among firm sizes (Table 1).T-test[5]revealed that concerns about cannibalization effects is the only factor that presented a significant difference between the firms selling and not selling new products along with remanufactured goods (Table 1). A detailed representation of each factor revealed through both qualitative and quantitative methods is as follows:

---

[2] Remanufacturing firms in Thailand are defined using the broader term, comprising both remanufacturing and semi-remanufacturing [6, 7, 38].

[3] Note that a post hoc analysis could not be performed to identify which couple of firm sizes is statistically significantly different because among the three groups the large-sized group comprised only one firm.

[4] Small-scale enterprise has fewer than 50 employees and medium-scale enterprise has fewer than 200 employees and over of these is large-scale [39].

[5] It is crucial to note that t-test could not performed for actors and ownerships because there is one group that is OEM and only one foreign-owned firm.

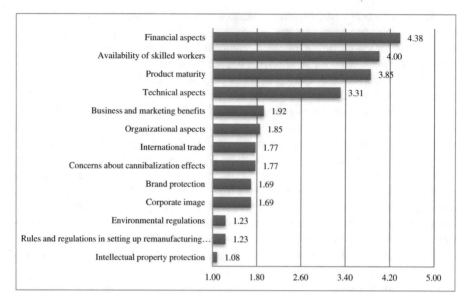

Fig. 2. Mean scores of thirteen influencing factors for photocopiers

Table 1. Oneway ANOVA categorized by firm sizes and T-test categorized by new product sales

Factor	ANOVA		T-test	
	F	Sig	T-test	Sig
Technical aspects	0.224	0.803	−1.745	0.109
Product maturity	0.405	0.678	−1.556	0.148
Availability of skilled workers	0.625	0.555	−1.045	0.319
Financial aspects	0.325	0.730	0.622	0.546
Corporate image	8.623	0.007[*]	−0.575	0.577
Business and marketing benefits	1.058	0.383	−0.508	0.621
Brand protection	3.385	0.075	−1.300	0.274
Concerns about cannibalization effects	1.012	0.398	−2.886	0.015[*]
Organizational aspects	0.192	0.828	−1.749	0.108
Rules and regulations in setting up remanufacturing businesses	–	–	−1.000	0.391
International trade	2.102	0.173	−2.190	0.051
Intellectual property protection	–	–	−1.000	0.391
Environmental regulations	–	–	−1.000	0.391

Note: [*] P-value $\leq$ 0.05

4.1 Business Feasibility

(1) **Financial Aspects.** Financial aspects were the only factor that reached the extremely influential benchmark in firms' decisions to undertake remanufacturing. Remanufactured photocopiers potentially generate high profitability for firms. The price difference between new and remanufactured photocopier is around four times in high-speed series. This attracts firms to undertake remanufacturing due to high profits and also attracts customers in view of lower prices. Core management constitutes the most important ingredient when decreasing remanufacturing production costs. The majority of respondents revealed that the core condition before remanufacturing is considerably important when making decisions to conduct remanufacturing because it impacts production costs. Core acquisition comprises two approaches determined by the remanufacturing actors: IRs or OEM. With IRs, the majority of core acquisitions are imported as used photocopiers, mostly from European countries and the U.S., which are of high quality with warrantee contracts of around two years. With OEM, the eco-manufacturing center of Asia Pacific, core acquisition is mostly obtained from both import and domestic sources. Domestic cores are acquired through the "close loop supply chain" to secure cores derived from the expiration of five years tenancy agreements. At the end of their useful life, they are recycled. The majority of firms simultaneously operate both selling and rental business concerning remanufactured photocopiers. With IRs, key customers such as copy centers currently face competitive business through the low price of copies per page. This drives them to demand low cost photocopiers coupled with prompt service and maintenance. Therefore, rental remanufactured photocopiers are needed to meet demand. After the expiration of the tenancy agreements of remanufactured photocopiers, firms will also resell them as the third hand goods to CLMV, i.e. Cambodia, Lao, Myanmar and Vietnam. Moreover, the key customer of remanufactured products for OEM is their own headquarters in Japan through export under the Basel Convention on the Control of Transboundary Movements of Hazardous Wastes and Their Disposal. Such remanufactured products comprise two categories as consumable parts, including cartridges and fusing units.

(2) **Availability of Skilled Workers.** At the very influential level in the second ranking, skilled and knowledgeable workers are needed when remanufacturing photocopiers [14, 15, 31]. Workers are required to be constantly trained how to read manual code concerning maintenance, remanufacturing, providing solutions to customers, and updating to new technology. One photocopier unit necessitates five to seven days labor, using up to three workers to be remanufactured, depending on particular product specifications.

(3) **Product Maturity.** As the third rank, product maturity was considered as very influential. Interviewees revealed that technological change in photocopiers is quite slow and relatively less complex, particularly concerning hardware. This potentially supports performing remanufacturing activities [13]. It is also worth

remanufacturing photocopiers because they retain high value and are able to work post-remanufacturing as new.

(4) **Technical Aspects.** This factor was determined as being somewhat influential. Photocopier technology is obviously important because it allows multi-functionality in response to various customer demands concerning copying, printing, faxing, scanning, file online sending, memory database systems, connectivity, and security. OEM perceived that design issues are very important in remanufacturing. They must be considered at the early stage of production through collaboration between designers and engineers. Specifically, remanufacturing design has an impact on production costs, particularly labor. This was also suggested in previous study [9].

4.2   Firm Strategic Factors

(1) **Business and Marketing Benefits.** This was rated as the highest factor in this area, though its absolute score lies at the slightly influential level. From our interviews, business and marketing benefits represent another reason for photocopier OEM to operate remanufacturing. This represents green marketing initiatives promoting corporate considerations concerning the environment. Undertaking remanufacturing potentially creates new product sales and market segments [16, 24]. However, this factor is not valid for IRs due to the fact that they focus particularly on business profitability. Thus, overall this factor was rated as only slightly influential.

(2) **Organizational Aspects.** Participants identified organizational aspects as slightly influential. Two firms mentioned that cooperation between sales and technical divisions for providing service to customers being important regarding effective organizational management. Another pair stressed that honesty in using appropriate replacement parts needs to be closely mornitored throughout the organization. However, many respondents argued that without sophisticated organizational demands, remanufacturing operations could occur more smoothly. The cost of production becomes higher when sophisticated organizational restrictions are applied.

(3) **Concerns about Cannibalization Effects.** Cannibalization effects were confirmed as not at all influential. The majority of firms revealed that there has been no conflict between new and remanufactured products through cannibalization due to differences in market segments [16, 17]. In particular, OEM separately targets new and remanufactured products as reflecting different customer needs. New goods are provided for end-users, whereas remanufactured goods target rental businesses. However, t-test analysis shows that firm selling new products along with remanufactured products perceived this variable as more critical than firms selling only remanufactured products (Table 1). Firms selling both products rated this factor higher because their markets for new and remanufactured

products are close substitutes. They, thus, need to be concerned more about cannibalization effects between them.

(4) **Brand Protection.** This factor was regarded as not at all influential. Most firms, particular IRs, are relatively unconcerned about protecting their brand because they do not have their own brand identities. However, one OEM addressed brand protection as one reason for undertaking remanufacturing. Remanufacturing helps protect their brand reputation and prevents cores moving to IRs. They also need to make sure that remanufactured products qualify to their OEM's standards and prevent inferior remanufactured products entering the market possibly produced by IRs. This is in line with many previous studies [9, 16, 24–26].

(5) **Corporate Image.** This variable was on average rated as not at all influential, reflecting the fact that the majority of firms do not pay attention to their corporate image when conducting remanufacturing. Interestingly, one-way ANOVA (Table 1) shows that OEM firm tended to rate this factor more crucial than smaller firms because of their strong environmental policies, including EPR and Corporate Social Responsibility (CSR) initiatives.

4.3   Policy Factors

(1) **International Trade.** Although international trade was regarded as not at all influential, this factor turned out to be ranked first within policy factors. However, Thai regulations regarding international trade still allow import of used photocopiers as cores which have a useful lifespan of lower than five years, unlike Malaysia and India [31]. In addition, international trade policy enables OEM to run remanufacturing businesses through the import of cores and export remanufactured products under the Basel Convention. Thus, some firms argued that this factor, to some extent, could help facilitate remanufacturing processes in Thailand.

(2) **Environmental Regulations.** Environmental regulations were also rated as being not at all influential, being in line with other emerging countries like China [12, 15]. Environmental regulations are not a driving factor influencing IRs to implement remanufacturing because Thailand has currently no specific environmental regulations concerning remanufacturing. Meanwhile, OEM tends to rely on their strong environmental policies. Firms in Thailand are forced to be concerned about the environment through factory-directed laws, for instance ISO9001: 2008 and ISO14001: 2004.

(3) **Rules and Regulations in Setting up Remanufacturing Businesses.** This factor was considered as not at all influential by IR respondents. Nevertheless, OEM remanufacturing was established under the Basel Convention mainly facilitated by Thai government. OEM argued that government policy, starting from the accurate definition of remanufacturing and clear rules and regulations in setting up the business, could help to promote remanufacturing activities in the country.

(4) **Intellectual Property (IP) Protection.** As the least important factor, IP protection is not particularly relevant for IRs because they are not brand owners. OEM also mentioned that IP protection is not an obvious driver affecting the decision to become involved in remanufacturing. This is in contrast with developed countries where IP encourages remanufacturing due to the effective enforcement of regulations [32].

5   Conclusions and Policy Inferences

This study aims to examine the significant factors driving the decisions of firms to perform remanufacturing, using the Thai photocopier industry as a case study. The results show that business feasibility is the most important factor influencing firms' decisions, followed by particular firm's strategic factors, and policy factors. Occupying the top four individual factors, financial aspects is ranked first as most critical in remanufacturing, followed by availability of skilled workers, product maturity, and technical aspects. Firm characteristics matter significantly in ranking key factors. Large-sized firm tends to place a higher weight on corporate image than smaller-sized operations. Concerns about cannibalization effects tend to be more crucial in influencing firms who sell both remanufactured and new products than firms who process only remanufactured products.

In order to maximize opportunities, whilst minimizing obstacles in remanufacturing implementation, three missions need to be strengthened. These comprise policies and regulations related to remanufacturing activities, human resource development (HRD) and infrastructure. The establishment of comprehensive policies and standards is needed in this initial stage of development. The strategic direction of HRD should be determined by the requirements of photocopier firms as key customers. The foundation of remanufacturing centers and industry standards, in terms of infrastructure, is required. Research and Development (R&D) is also needed to improve remanufacturing activities. With respect to potential contributions, this study will potentially stimulate remanufacturing research and development in Thailand, or other countries, where remanufacturing lies in the infancy stage. This manuscript focused only on one industry with limited factors investigated. Further empirical research should extend to deepening any knowledge and generalizability derived to other products, industries, and countries.

References

1. Matsumoto, M., Ijomah, W.: Remanufacturing. In: Kauffman, J., Lee, K.-M. (eds.) Handbook of Sustainable Engineering, pp. 389–408. Springer, Netherlands (2013)
2. Morrison-Saunders, A., Therivel, R.: Sustainability integration and assessment. J. Environ. Assess. Policy Manag. **8**, 281–298 (2006)
3. Lund, R.T.: Remanufacturing: the experience of the United States and implications for developing countries. World Bank, Washington, DC (1984)

4. Ijomah, W.: A model-based definition of the generic remanufacturing business process (2002)
5. Ijomah, W.L., Childe, S., McMahon, C.: Remanufacturing: a key strategy for sustainable development. In: The 3rd International Conference on Design and Manufacture for Sustainable Development, pp. 99–102. Cambridge University Press, Loughborough (2004)
6. Kohpaiboon, A., Jongwanich, J., Tanasritanyakul, A., Jhanhorm, K., Vatinpongphan, V.: Study on strategies of the preparation and development of environmental products under international trade framework (phase 2). Thammasart University, Bangkok (2011)
7. Kohpaiboon, A., Jongwanich, J., Jhanhorm, K., Tanasritanyakul, A., Rojanakanoksak, K., Wongwat, K., Suradej, W.: Study on strategies of the preparation and development of environmental products under international trade framework (phase 3). Thammasart University, Bangkok (2012)
8. Ijomah, W.L.: Addressing decision making for remanufacturing operations and design-for-remanufacture. Int. J. Sustain. Eng. **2**, 91–202 (2009)
9. Subramoniam, R., Huisingh, D., Chinnam, R.B.: Remanufacturing for the automotive aftermarket-strategic factors: literature review and future research needs. J. Clean. Prod. **17**, 1163–1174 (2009)
10. Subramoniam, R., Huisingh, D., Chinnam, R.B.: Aftermarket remanufacturing strategic planning decision-making framework: theory & practice. J. Clean. Prod. **18**, 1575–1586 (2010)
11. Bras, B., Hammond, R.: Towards design for remanufacturing – metrics for assessing remanufacturability. In: Flapper, S.D., de Ron, A.J. (eds.) Proceedings of the 1st International Workshop on Reuse, pp. 5–22 (1996)
12. Zhang, T., Chu, J., Wang, X., Liu, X., Cui, P.: Development pattern and enhancing system of automotive components remanufacturing industry in China. Resour. Conserv. Recycl. **55**, 613–622 (2011)
13. Matsumoto, M.: Development of a simulation model for reuse businesses and case studies in Japan. J. Clean. Prod. **18**, 1284–1299 (2010)
14. Lundmark, P., Sundin, E., Björkman, M.: Industrial challenges within the remanufacturing system. In: Swedish Production Symposium, Stockholm, pp. 132–138 (2009)
15. Abdulrahman, M.D.-A., Subramanian, N., Liu, C., Shu, C.: Viability of remanufacturing practice: a strategic decision making framework for Chinese auto-parts companies. J. Clean. Prod. **105**, 311–323 (2015)
16. Östlin, J., Sundin, E., Bjorkman, M.: Business drivers for remanufacturing. In: 15th CIRP International Conference on Life Cycle Engineering, pp. 581–586 (2008)
17. Hammond, R., Amezquita, T., Bras, B.: Issues in the automotive parts remanufacturing industry: discussion of results from surveys performed among remanufacturers. Int. J. Eng. Des. Autom. **4**, 27–46 (1998). Spec. Issue Environ. Conscious Des. Manuf.
18. Hatcher, G.D., Ijomah, W.L., Windmill, J.F.C.: Design for remanufacturing in China: a case study of electrical and electronic equipment. J. Remanufacturing **3**, 3 (2013)
19. Toffel, M.W.: Strategic management of product recovery. Calif. Manage. Rev. **46**, 120–141 (2004)
20. Haynsworth, H.C., Lyons, T.: Remanufacturing by design, the missing link. Prod. Invent. Manag. J. **28**, 24–29 (1987)
21. Rahman, S., Subramanian, N.: Factors for implementing end-of-life computer recycling operations in reverse supply chains. Int. J. Prod. Econ. **140**, 239–248 (2012)
22. Steinhilper, R.: Remanufacturing: The Ultimate form of Recycling. Fraunhofer IRB Verlag, Stuttgart (1998)

23. Steinhilper, R.: Recent trends and benefits of remanufacturing: from closed loop businesses to synergetic networks. In: Proceedings Second International Symposium on Environmentally Conscious Design and Inverse Manufacturing, pp. 481–488 (2001)
24. Östlin, J.: On Remanufacturing Systems Analysing and Managing Material Flows and Remanufacturing Processes (2008)
25. Saavedra, Y.M.B., Barquet, A.P.B., Rozenfeld, H., Forcellini, F.A., Ometto, A.R.: Remanufacturing in Brazil: case studies on the automotive sector. J. Clean. Prod. **53**, 267–276 (2013)
26. Seitz, M.A.: A critical assessment of motives for product recovery: the case of engine remanufacturing. J. Clean. Prod. **15**, 1147–1157 (2007)
27. Linton, J.D.: Assessing the economic rationality of remanufacturing products. J. Prod. Innov. Manag. **25**, 287–302 (2008)
28. Oiko, O.T., Barquet, A.P.B., Aldo, O.R.: Business issues in remanufacturing: two Brazilian cases in the automotive industry. In: Proceedings of the 18th CIRP International Conference on Life Cycle Engineering, Braunschweig, pp. 470–475 (2011)
29. Lund, R.T., Denney, W.M.: Opportunities and Implications of Extending Product Life. Center for Policy Alternatives, Massachusetts Institute of Technology, Cambridge (1977)
30. Rathore, P., Kota, S., Chakrabarti, A.: Sustainability through remanufacturing in India: a case study on mobile handsets (2011). http://linkinghub.elsevier.com/retrieve/pii/S0959652611002277
31. USITC: Remanufactured goods : an overview of the U.S. and global industries, markets, and trade. http://www.usitc.gov/publications/332/pub4356.pdf. Accessed 22 Aug 2014
32. Pagell, M., Wu, Z., Murthy, N.N.: The supply chain implications of recycling. Bus. Horiz. **50**, 133–143 (2007)
33. Amezquita, T., Hammond, R., Salazar, M., Bras, B.: Characterizing the remanufacturability of engineering systems. In: Proceedings of ASME Advances in Design Automation Conference, DE, 17–20 September, vol. 82, pp. 271–278 (1995)
34. Patton, M.Q.: Qualitative Evaluation and Research Methods, 2nd edn. Sage, Newbury Park (1990)
35. Stuart, I., McCutcheon, D., Handfield, R., McLachlin, R., Samson, D.: Effective case research in operations management: a process perspective. J. Oper. Manag. **20**, 419–433 (2002)
36. Sekaran, U.: Research Methods for Business: A Skill-Building Approach, 3rd edn. Wiley, New York (2000)
37. Park, H.M.: Comparing group means: t-tests and one-way ANOVA using Stata, SAS, R, and SPSS (2009)
38. OIE: Remanufacturing development in Thailand. http://www.oie.go.th/sites/default/files/attachments/km_oie/km_oie_06.pdf
39. OSMEP: Thai definition of small and medium enterprises. http://www.sme.go.th/Lists/EditorInput/DispF.aspx?List=15dca7fb-bf2e-464e-97e5-440321040570&ID=1781. Accessed 02 May 15

Steps in Organisational Environmental Change: Similarities Across Manufacturing Sectors

Peter Ball[(✉)]

University of York, York, UK
peter.ball@york.ac.uk

Abstract. With increasing expectations on manufacturers to show leadership in reducing the negative environmental and social consequences of their operations, there are many examples of successful journeys towards sustainability. Potentially there are common steps in how organizations change and these could be common across different sectors. This research seeks to uncover what companies do to instigate and create momentum in their organization and what commonality exists. Seven steps were identified through analyzing case data across companies operating in different sectors. The steps of vision, leadership, education, simplicity, pilot, momentum and broadcast are presented and the implications for further research assessed.

Keywords: Sustainable manufacturing · Organizational change · Process steps

1 Background

The desire for socially responsible business and reduced environmental impact has received significant attention [1, 2]. For sustainable progress [3] companies need to initiate and advance their programs [4]. The lean philosophy [5] is one path to follow to relentlessly reduce by aggressively removing waste and limiting impact. Another is to do more and flourish [6] by positively impacting the surroundings.

So how should companies embark on their sustainability journeys? A valid approach would be to learn from the leaders in the field. Operationally, the technical achievements and implemented practices are well documented [7–9]. Additionally, strategically, much has been published at the firm level [4, 10, 11]. However, little has been published on the paths that companies take from establishing their vision to gaining momentum across the company.

The organizational change literature [12] offers potential, as does the work on processes [13] and barriers [14], to understand how companies guide change within operations and enable success. In support, the sustainable leadership body of knowledge [15–18] captures how change is directed. There is, however, a gap in the literature how companies build from initial pilots to ramping up activity both within their own operations and beyond. There is need to better understand how companies cautiously extend their early work to a level that they then have confidence and credibility to broadcast externally.

© Springer International Publishing AG 2017
G. Campana et al. (eds.), *Sustainable Design and Manufacturing 2017*, Smart Innovation,
Systems and Technologies 68, DOI 10.1007/978-3-319-57078-5_25

This paper details work on understanding the steps that manufacturing companies, known for their sustainability leadership, start and create momentum for their environmental journeys. A grounded research methodology is used to provide four detailed company cases. It shows how the data was coded and the codes clustered to create seven significant steps along with associated aspects for each step. To close, issues are discussed and further work identified.

2   Methodology

Conversations with manufacturing companies recognized to be leaders in advancing their sustainability agenda revealed common traits in the way they approached the challenge. Staff spoke of senior support, gave anecdotes about intensifying activity and related caution communicating achievements. The literature in this area lacks insight into how companies lower their environment impact to become less unsustainable. The impetus here was therefore to capture the steps companies were taking on their sustainability journey and uncover any significant differences between sectors.

The research used a grounded strategy [19] to develop a rich picture of the company pathways that could be condensed into generic steps. The steps uncovered could then be compared to the sustainable manufacturing and change literature. The approach allowed companies to relate their stories rather than restrictively interviewing them for case studies against a proposed theoretical model.

Individuals from leading companies were contacted out of convenience and reputation in the sustainability community for significant progress (anecdotally and through company reports). Hence selection was purpose-led [20] rather than random. This ensured that the journeys captured were 'mature' with numerous projects completed.

Prompt questions were: What was the program called? What was the trigger to start? What was the motivation to pursue long term? Who was involved, how was involvement achieved? Who supported? Where does the inspiration come from? What are the top tips to help others? The questions were not leading on process steps or demanding specific examples. In part this allowed focus on how they progressed rather than what they changed and in part revealed whether they emphasized process or technology. The interview stopped when the prompt questions were answered and the interviewee volunteered no further insight. Interviews were by telephone and face-to-face lasting one hour or more over one or more meetings. Initial guidance was given to the interviewees to prompt the accounts of their company's approach and continuous notes were taken of their accounts without audio recording. Later, clarifications were sought on the accounts and feedback was gained on initial write-ups

In line with the grounded approach, open, axial and selective coding [20, 21] was used progressively on the original notes for the individual cases, the cross-case analysis and finally to condense the aspects of each of the emergent steps. The codes were assigned to each distinct feature of the company's journey. Given that the literature was not used in advance, the coding was open and identified distinct features. The coded write-ups of each company's account were compared and commonalities were identified. Some codes were refined to use common language across cases. Through iteration, clusters of codes formed intermediate categories and finally characterized key 'steps' in the companies' collective journeys. The codes for each account are shown in italics in the next section with the steps shown in bold. Note that these steps emerged after iteration, hence are shown within each case for efficiency of presentation. From the clusters of codes that formed the steps, the codes were condensed into more general 'aspects' (shown in Table 1). Aspects are the potential features of a company's journey rather than an essential checklist or necessary sequence. Likewise, companies will iterate through the steps in parallel and not proceed sequentially.

3 Case Studies: How Companies Lead Change

This section details the approaches used by four companies with the initial codes used to create key steps (in bold). The subsequent section clusters the codes to create aspects that are generic across the cases from the different sectors.

3.1 Automotive

The company is global. It developed its **Vision** for sustainability using a *journey metaphor* of seeing the end point without necessarily knowing the intermediate steps. The *sustainability vision* was general to allow everyone to *connect to* and create *ambition*. Specifically, it sought to *minimize impact* and to *live in harmony with nature*. **Passion in leadership** was shown by *deeply held beliefs* of its *leaders* who communicated a vision that staff could *believe* was achievable but distant on how to reach the end point. These *champions* worked to *win hearts and minds* across the company.

The **education of people** was important to *understand the basics* and why the current process was the way it was. The company used its *overall philosophy of zero* waste for the journey and its leaders *shared knowledge* on the 'what' and particularly on the *'how'*. Their focus on *educating staff on why* is a key lesson here. The company had a **simple approach** and the need to always *go and see* the challenges at first hand in the workplace. Staff focused on what was behind the *fundamental impacts* on the environment for the *underpinning principles*. *Rules for simple processes* were developed that in turn became *standard processes*.

Successful pilots were carried out in a small number of *pilot plants*. *Trials were performed* to understand the best way forward. They *proceeded cautiously*, taking many *small, incremental steps* and ensured they could *return to known condition* and start again if needed. By *encouraging trying* then *mistakes could be permitted* in a controlled way. After the pilots, to **develop momentum** the *few full time staff* then *persistently engaged others* to *increase activity* whilst being *sensitive to local conditions*. The company valued consistent *long term evolution* of practice and used *central funding* to remove risk from operations and *avoid barriers* to progress.

The company **broadcast success** more and more widely as they progressed on their journey giving *reward and recognition* where due. They *communicated* with *consistent messages* to *cascade awareness and advances* throughout the organizational hierarchy but always ensured *local relevance*.

3.2   Aerospace

This company has a number of plants across Europe. It **created a vision** that was *ambitious* to *reduce energy and carbon*. **Passion in leadership** was demonstrated through a *network* of *local champions* who were *nominated by each plant* and *supported by seniors* to *foster enthusiasm in other staff* and in turn *change behaviors*.

In the **education of people** step *many staff were trained* in with materials that contained tools and *inspiration from others* to give confidence the changes could be achieved. Emphasis was placed on *questioning the fundamentals* by constantly *asking why* to avoid jumping to conclusions on solutions. They used a **simple approach** that was *adopted as a standard process* from another company *based on principles* of the waste hierarchy and *tailored it to their needs*. The *standard way of working* balanced *promoting quick wins* with *supporting disruptive projects*.

Successful pilots for their energy efficiency program were developed by a *pilot plant*. It needed to *understand what works locally* having recognized its lean production success was achieved by avoiding 'copy and paste' of standard methods. *Successful initial projects* helped *generate confidence* for more projects. The company used *external funding* to start the energy reduction activity that involved *industrial and academic collaboration*. To **develop momentum** they used their initial success to create an energy efficiency *network to cascade* the work for *wider adoption*. Whilst *support from the seniors* was important, the *motivation for change* came from the shop floor *improvement groups*. Earlier training *provided tools* to *help others carry out their own projects* but *avoid dictating detail*. They recognized that *'mundane' changes* that were *widely adopted* could be more effective than 'shiny' technology projects and *avoided local inertia* from a 'not invented here attitude'. The local *successes enabled expansion* as groups were keen to *learn from one another* and *understand why* some areas had better performance. The *internal benchmarking* developed into *external benchmarking* as well as *working with suppliers*.

To **broadcast success**, and *communicate clear practices* it was realized that the approach had to *fit with the existing performance* measurement system. Significantly the *network of champions* wanted to *celebrate success*, and find *time to have fun*.

3.3   Food

This case company has a small number of plants in the UK. Its **vision** contained their *mantra of practical sustainability* driven by the *desire to reduce* their energy consumption. Later work showed *staff were aligned* to the *sustainability vision*. The **leadership passion** came from the *initial motivation* for the sustainability vision but it was not seen as *the responsibility of one lone leader*.

Through **educating people** they created *wider understanding* of their carbon footprint and a *focus on the hotspots*. The company sees its *inspiration* coming from staff *discussion and understanding* on what will *help others*. By *helping others* they *helped themselves* and in doing so *removed fear* in calculations. They worked to *avoid lack of interest* and *avoid defensive practices*. Their overall **simple approach** was *simple and effective engagement* with a *practical mind-set* and a *desire to demystify*. The view was to do the *basics first* and get *clever afterwards* by *adopting standard methods* for assessment and *fast-tracking analysis* to *identify hotspots* in the business. They were *not interested in offsets*, etc. There was *emphasis on the practical and basic aspects* and *using the information available* and *not demanding accuracy*.

The company had many **successful pilots** but *initial progress was challenging* to achieve *initial successes* due to the *lack of information*. The company persisted and **developed momentum**, by *engaging others* and seeking *advances from all staff*. Their *initial successes* led to them *working with suppliers*. Their offer to *share tools and data* with suppliers was *initially resisted* as it was *perceived as a threat* but moved to *collaboration with suppliers* within a couple of years so that the *benefits were mutual*. The *emphasis on simplicity* meant that assessment of carbon footprint *across the supply chain* was *matched to the data suppliers had available*, recognizing their *suppliers did not have the time and skills to obtain data* demanded by sophisticated tools. The company later *worked with customers* to *help their customers calculate the carbon impact*. Having worked on the basics for many years their next big thing was *investing in technology* for renewable energy.

Through the process of **broadcasting success,** the company has *attracted interest from others* including new and existing customers. Their work with others was seen as *triple win* through driving lower cost, premium pricing and lower carbon footprint. Within the company *staff were surveyed* and results showed a majority of *staff were aligned* to the *sustainability vision*.

3.4   Drink

This company has sites within a region of the UK. Presenting the **vision creation**, it had long-standing *deep-rooted values* of *working with the local community* and its environmental progress has been built around its *long-term* strategy. There is *pride in acting responsibly* and being *purpose led*, not simply profit led. **Passion in leadership** is evidenced from staff *championing* the environmental and resilience agenda. Building on its heritage, its *enthusiasm* is guided by *doing what is right*. Leaders focus and communicate the *ambition* through *high level environmental metrics*.

Education of people is both *internal* to the company as well as with the *wider community* whether through education establishments or other manufacturers. The **approach of simplicity** is demonstrated through simple presentation of high level environmental and social *metrics*. *Justification for change* is built around *long-term business case*.

Successful pilots are related through the *story of their progress* and traced to three *bold investments* that supported *long-term* growth and significantly lowered the *environmental impact* and *exposure to shortages*, thereby building *resilience*. Regardless, most projects have *fast payback* and *build credibility for further change*. **Momentum has developed** through recognising the need for a *diverse spectrum of activities* of low effort/high gain through to high effort/low gain developments. The latter showing their commitment to the *long term*. Many advances are a consequence of *shaping wider business investment* and *not solely focused on environmental impact*. The sustainability values within the company are used to *engage the local community* to improve the local environment and health of those living within it.

Finally, the step of **broadcast success** is shown through events and graphical portrayal of change. *Extensive communication internally* and *appropriate communication externally* conveys significant messages *readily backed up* by data. Communication is through multiple channels including web, blogs, newsletters and reports. The communication has established the company's *reputation* and has led to *trust* within and collaboration with the *local community*.

4   Analysis

The accounts of the four case companies presented show significant similarities despite operating in different manufacturing sectors. Each company seeks to progress along their journey towards sustainability with (*coded*) activities that have potential to be clustered. For efficiency of presentation, the clusters of codes that form the **steps** were inserted into the case write ups, although it must be emphasized that these steps emerged from clustering the codes across the cases later. This section considers each of those steps and corresponding aspects that draw directly from the codes (Table 1). Whilst differences were shown in detail, the overall approaches used share the common steps presented here.

The **vision** step was immediately evident from all companies with most companies explicitly using that word. The companies were selected for the research because they were known in the sustainability community to be leaders and, in part, from awareness of their visions. All cited environmental motivations within their vision, some presented themselves more widely.

Passion in leadership was demonstrated by all companies. The leaders of sustainability showed enthusiasm in the progress of their companies and talked at length about engaging others either directly or indirectly through other local leaders. They all showed consistency and obsession (not in any meant in a rude way!) in their style.

Education of people featured strongly in the companies. It varied from informal internal knowledge sharing through to engaging external providers to formally deliver in-house materials. Some educated other companies' staff as well as their own.

Simplicity of approach was emphasized by all. Both the formal process as well as the ease of communication of the process itself. This is presented as a single step as, it was challenging to separate the technical approach from the simplicity its use. The companies described the technical steps, the use of any tools to support those steps and how the wider adoption was considered.

Successful pilots captures the first projects that the company described. The early projects in most companies were low cost and focused on the particular technical problems in hot-spots. Although it is notable that one company used its long term perspective to make early gains through combining the environmental objectives with other large scale business investment.

Develop momentum captures the volume of projects that used the pilots as a foundation and established the companies' reputation in the business community and sustainability community more broadly. The technical activities in both this and the pilot step were the same but the mind-set of the pilots was to be cautious, allow mistakes and build credibility with small groups of people whilst the momentum step engaged many others to follow.

Table 1. Steps and aspects to support organizational change towards sustainable manufacturing

Steps and aspects to support
1. Vision creation
Creation of an ambitious sustainability vision
Detail of (long term) vision gives staff ability to connect and align
Vision provides focus on minimizing impact
2. Passion in leadership
Leaders foster enthusiasm in others to engage
Leaders provide the inspiration and storylines for changing behaviors
Seniors create local champions to ensure the responsibility and activity is shared
3. Education of people
Staff are educated in an overall philosophy
The focus is on understanding the basics and questioning why
Knowledge is shared as staff discuss and learn from one another
Ways to allay threats and fears are promoted
4. Approach with simplicity
Standard, simple, practical "go and see" approach is adopted and developed
Approach has core principles that focus on identifying fundamental impacts
Support for combination of short/incremental projects and disruptive projects
Analysis is based on available data, recognizing some data is hard to obtain
5. Successful pilots
Early projects demonstrate success (in pilot plants) and builds confidence
Progress is challenging in both how to analyze and how to change
Work proceeds cautiously with small steps and ability to return to previous state
Staff encouraged to try. Mistakes are permitted in controlled way to learn more
Internal and external collaboration and funding helps initiation
6. Develop momentum
From successful pilot activity is increased by engaging others over the long term
Few full-time staff engage network/community to motivate change and adoption
Persistence develops motivation for change and gain momentum
Activity within same plant, suppliers or customers benchmarked for awareness
In helping others and sharing, mutual benefits can be realized and avoid barriers
Central funding can help with acceleration of adoption of new methods
Promote 'mundane' process change and appropriate technology change
Leaders are sensitive to local conditions and avoid dictating detail/allow tailoring
7. Broadcast success
Consistent messages are communicated
Success recognized within existing performance structure
Success is celebrated, often emphasizing multiple benefits
As advances are cascaded local relevance is ensured
Engagement of staff is measured and communicated
Communication is local as well as (cautiously) outside plant/company boundary
Important to set aside time to have fun

Broadcasting success was placed as the last step but it was clear that, like other steps, the sequence was not linear, especially for the communication that started off as local limited reporting and eventually was external once sufficient momentum and credibility had been gained.

Table 1 consolidates the approaches taken by the companies into steps and aspects. The steps are evident in all companies; the aspects were common to many companies but companies did not engage in all aspects. The aspects are potential ways of achieving each step rather than a strict recipe. Accordingly, they have not been presented as a numbered checklist. It must be emphasized that steps will take place in parallel, for example education is given in multiple waves as more and more staff internal and external to the company are drawn into the vision. Finally, the developing momentum step had a significant cluster of aspects and further research could justify subdivision of this step. For example, some aspects relate to initial expansion whereas others relate to later maturity when sustaining the change is emphasized.

5   Discussion

There has been significant research in what achievements have been made, what technical changes have been implemented and what objectives companies are establishing. The research presented here addresses a weakness in the literature on the organizational change companies have taken to reduce impact across the breadth of the organization. In response, this research captures the steps and associated aspects of how companies undertake their journey towards sustainable manufacturing from initiation to wider adoption.

A grounded method was used to collect and interpret accounts of companies recognized to be leaders in the field of sustainability and establish common characteristics. Individual cases were coded and the codes clustered across the cases to establish major steps that companies take and common aspects to support those steps. The decision made to collect data prior to the creation of a model, framework or set of guidelines allowed a practice-led enquiry without the influence of published theoretical models. This allowed distinctive features to be drawn out: education (as opposed to training), the distinct ramp-up to create the momentum, community/networks (as distinct from top down) and caution in how success was publicized.

There are similarities in the sequence of steps that would be expected of any project management process [22] in which: a project is initiated with a charter (vision); planning with the creation of a team (leadership); executing by setting up a team (education, approach) and then carrying out the work (pilots, momentum); and finally closure (broadcast). Considering the steps as an overall innovation process, the steps share similarities with the more generalized process of diffusion of innovation [23]. The initial step has characteristics of the understanding of potential changes, the potential benefit and what and why they should be pursued. The next steps are about persuading others that the change is necessary and the leaders ensuring compatibility with values and past experience. This is followed by a decision stage of piloting and accepting. Then the implementation of change is adopted more widely, possibly with continued experimentation, before the confirmation stage of communicating, evaluating and gaining support. In effect change is viewed as a journey [24] rather than an imitative in which staff are empowered [18] to progress.

6   Conclusions

This paper presents common steps taken by four companies with a reputation for sustainability progress. Whilst the detailed activity for each company was different, the companies have common journeys overall.

The research reveals that companies have clear vision and passionate leadership to drive the change. Companies have a process for improvement that staff are educated as well as trained in. Companies pilot before ramping up their activity to create momentum across the company. Throughout their journeys the companies communicate their successes and the reach and strength of the messages build as they achieve more.

Many opportunities for further work exist to address shortcomings in the research. More cases would allow any subtleties in the way companies operate between sectors, at different supply chain tiers, different supply chain power and different parts of the world. Additionally, the companies used for this research were known leaders and therefore have established programs of change guided by well-shaped visions. It is likely that companies early in their journeys towards sustainability would have different priorities and emphasis. Contextual factors such as the maturity of their lean programs, the ownership model, the level of regulation and company size could also be revealing. Further, who was operationally leading the change within the companies would be of interest as some were led from an environmental function, others from operations. Finally, the level of success of each company approach was not measured, simply a judgement was made on the basis of the reputation and therefore more quantification could be introduced.

Acknowledgements.   The author wishes to thank companies interviewed for this research for their time and enthusiasm and notes some data collection whilst employed by Cranfield University.

References

1. Kleindorfer, P.R., Singhal, K., Wassenhove, L.N.: Sustainable operations management. Prod. Oper. Manag. **14**(4), 482–492 (2005)
2. Garetti, M., Taisch, M.: Sustainable manufacturing: trends and research challenges. J. Prod. Plan. Control **23**(2–3), 83–104 (2012)
3. World Commission on Environment and Development: Our Common Future. Oxford University Press, Oxford (1987)
4. Montabon, F., Pagell, M., Wu, Z.H.: Making sustainability sustainable. J. Supply Chain Manag. **52**(2), 11–27 (2016)
5. Womack, J.P., Jones, D.T.: Lean Solutions: How Companies and Customers Can Create Value and Wealth Together. Simon & Schuster, London (2005)
6. Ehrenfeld, J.R.: Sustainability by Design: A Subversive Strategy for Transforming Our Consumer Culture. Yale University Press, London (2008)

7. Hajmohammed, S., Vachon, S., Klassen, R.D., Gavronski, I.: Lean management and supply management: their role in green practices and performance. J. Clean. Prod. **39**, 312–320 (2013)
8. King, A.A., Lenox, M.J.: Lean & green? An empirical examination of relationship between lean production & environmental performance. Prod. Oper. Manag. **10**, 244–256 (2001)
9. Zhu, Q., Sarkis, J., Lai, K.: Confirmation of a measurement model for green supply chain management practices implementation. Int. J. Prod. Econ. **111**(2), 261–273 (2004)
10. Bansal, P., Roth, K.: Why companies go green: a model of ecological responsiveness. Acad. Manag. J. **43**, 717–736 (2000)
11. Moretto, A., Lion, A., Macchion, L., Da Giau, A., Caniato, F., Caridi, M., Danese, P., Rinaldi, R., Vinelli, A.: A roadmap towards a sustainable supply chain: steps and impacts on performance. In: Proceedings of the European Operations Management Association (EurOMA) Conference, Trondheim (2016)
12. Senge, P., Smith, B., Kruschwitz, N., Laur, J., Schley, S.: The Necessary Revolution: How Individuals and Organisations are Working Together to Create a Sustainable World. Nicholas Brealey Publishing, London (2010)
13. Smith, L., Ball, P.D.: Steps towards sustainable manufacturing through modelling material, energy and waste flows. Int. J. Prod. Econ. **140**(1), 227–238 (2012)
14. Lunt, P., Ball, P., Levers, A.: Barriers to industrial energy efficiency. Int. J. Energy Sect. Manag. **8**(3), 380–394 (2014)
15. Avery, G.C., Bergsteiner, H.: Sustainable Leadership: Honeybee and Locust Approaches. Routledge, Abingdon (2011)
16. Fullan, M.: Leadership & Sustainability: System Thinkers in Action. Corwin Press, Thousand Oaks (2004)
17. Waldman, D.A., Siegel, S.S., Javidan, M.: Components of CEO transformational leadership and corporate social responsibility. J. Manag. Stud. **43**(8), 1703–1725 (2006)
18. Visser, W., Courtice, P.: Sustainability Leadership: Linking Theory & Practice, University of Cambridge Institute for Sustainability Leadership, SSRN Working Paper Series (2011)
19. Glaser, B., Strauss, A.L.: Discovery of Grounded Theory: Strategies for Qualitative Research. AldineTransaction, London (1999)
20. Robson, C.: Real World Research: A Resource for Social Scientists and Practitioner-Researchers. Blackwell Publishers Ltd., Oxford (2011)
21. Easterby-Smith, M., Thorpe, R., Jackson, P.: Management Research. Sage Publication, London (2012)
22. PMBOK: A Guide to the Project Management Body of Knowledge, 5th edn. Project Management Institute, Pennsylvania (2013)
23. Rogers, E.M.: Diffusion of Innovations, 5th edn. Free Press, New York (2003)
24. Esty, D.C., Simmons, P.J.: The Green to Gold Business Play Book. Wiley, New Jersey (2011)

From the Treatment of Olive Mills Wastewater to Its Valorisation: Towards a Bio-economic Industrial Symbiosis

Yannis Mouzakitis, Roxani Aminalragia-Giamini, and Emmanuel D. Adamides[✉]

Department of Mechanical Engineering and Aeronautics, University of Patras, Patras, Greece
{ymouzakitis,adamides}@upatras.gr,
roxannero.1989@gmail.com

Abstract. Although there is a significant progress in treatment technologies, Olive Mills Wastewater (OMWW) remains a source of environmental degradation to olive-oil producing regions. In this paper, the management of OMWW is examined as a trigger for a bio-economic industrial symbiosis, based on the valorisation of OMWW for biopolymers and bioenergy production. In addition, the valorisation of OMWW is considered in an eco-industrial context, as a node of a wider network of material, energy and information exchanges. The aim of the paper is to discuss the benefits and the feasibility of such a venture from a technical, economic, as well as social perspective using the context of a specific prefecture in Greece as a reference implementation environment.

Keywords: Bioeconomy · Industrial ecology · Industrial symbiosis (IS) · Waste valorisation · Olive mill wastewater (OMWW) · Biopolymers · Bioplastics · Polyhydroxyalkanoates (PHAs)

1 Introduction

Olive-oil production is an ever-growing activity in olive producing countries. As a result, olive mill wastewater (OMWW), which is a basic by-product of the olive oil extraction process, constitutes a major source of pollution in many regions. Although environmental technology offers a wide spectrum of treatment techniques and methods to olive-mills [1–3], in most cases, the 'safe' environmental-friendly disposal of OMWW still remains an important (if not unresolved) problem [3, 4].

An eco-industrial approach to manage waste and extract value from it, thus becoming a more attractive activity compared to uncontrolled disposal, is to use it as a raw material or energy source in (different) production processes [5, 6]. For the case of OMWW, various uses have been proposed, including biopolymers, bioenergy, fertilizers etc. [2, 7]. In this paper, we examine the use of OMWW as a basic material for the production of biopolymers (PHAs), thus triggering a symbiotic scheme consisting of oil mills, PHAs producers and PHAs consumers, which are usually plastics manufacturers. The creation of such as network of production facilities necessitates, beyond the consideration of the technological challenges involved, a thorough understanding of the social and economic issues involved

© Springer International Publishing AG 2017

G. Campana et al. (eds.), *Sustainable Design and Manufacturing 2017*, Smart Innovation, Systems and Technologies 68, DOI 10.1007/978-3-319-57078-5_26

in forming, as well as in maintaining this symbiosis. This paper, using primary and secondary level research, explores these issues in the context of the prefecture of Achaia in Western Greece, which, in this way, acts as a reference implementation environment.

The rest of the paper is organised as follows: In Sect. 2, we provide a short review of the environmental impacts and the treatment/valorising methods of OMWW. Based on documented technologies and theoretical frameworks stemming from the related literature, in Sect. 3 we describe the PHAs-based valorisation approach, at both *facility* (production of biopolymers) and *symbiotic* (partners and exchanges of the network) levels, while in Sect. 4 we discuss the proposed framework in the context of Achaia, emphasizing its social embeddedness. The paper ends with the conclusions and the suggestions for future work (Sect. 5).

2   The Problem of OMWW

Neolithic people collect wild olives as early as the 8th millennium BC, while, according to archaeological evidence, olives were turned into olive oil by 6000 BC in Middle East. Today, there are more than 750 million productive olive trees worldwide, and the three quarters of the annual olive oil production comes from Mediterranean Countries (Spain is by far the largest producer of olive oil, followed by Italy and Greece) [3]. The production of olive oil (which is commonly used in cooking, cosmetics and pharmaceuticals) is rapidly increasing (see Table 1), while the pollution caused by the OMWW constitute an important environmental problem, especially in Mediterranean Basin.

Table 1. Olive oil production (10^3 tns) (based on [8]) (World/EU/Spain, Italy & Greece)

Year	90/91	95/96	00/11	05/06	10/11	13/14
World	1453	1735	2565	2573	3075	3252
EU	993	1371	1878	2357	2224	2482
(% of World Pr.)	(68)	(79)	(73)	92	(72)	(76)
Sp–It–Gr	970	1335	1824	2311	2148	2378
(% of EU Pr.)	(98)	(97)	(97)	(98)	(97)	(96)

The quantity and quality (chemical synthesis and polluting ingredients) of OMWW varies significantly according to factors such as: the extraction method, the type of olive trees, the type of soil and irrigation water, the climatic conditions, the harvest time and the stage of maturity of the olive, and finally, the use of pesticides & fertilizers [1–3]. This variation results in important divergence in both measurements and estimations which can be found in the related literature. As for the extraction method, one may distinguish between the traditional press extraction process, the continuous 3-phase centrifugation process which was introduced during the 1970s, and the 2-phase process, which can be found mainly in Spain and uses no process but only washing water. The problem with OMWW is present only in the 3-phase process, which is still dominant in most olive productive countries. The fact that the 2-phase process produces no serious wastewater should not lead us to the conclusion that it is more environmental friendly, since the wet sludge produced presents similar (if not more important) treatment and disposal difficulties. In addition, the 2-phase

Table 2. Quantity and environmental impacts of OMWW (based on [1–4])

Quantity	Impacts
7–30 millions m^3 (world production per year)	– *on water bodies (surface & ground water, sea(shores:* discolouration, eutrophication, intoxication
1:100–1:200 (equivalency to m^3 domestic sewage)	– *on the ground (soil):* affection of cation exchange capacity, changes in the fertility, soil porosity, immobilization of available nitrogen, decrease of available magnesium
1–1.6 m^3 (per 1tn of processed olives)	
4.7–7.6 m^3 (per 1 tn of olive-oil)	– *on the crops (plants):* germinations of seeds, early growing stage, leaf & fruit abscission

process results in a 7% reduction in olive oil production which is also of lower quality (for details concerning the energy consumption and the mass balance of inputs/products/wastes between the three methods, see [1–3, 10]).

Independent of their constitution, OMWW have the following general properties: an intensive dark brown to black colour, a strong offensive smell, an acid pH, a high degree of organic load & solid matter, and significant concentration of pollutants such as polyphenols, flavonoids, phosphorus, potassium, tanins, reduced sugars and (acetic, formic and oleanolic) acids [1–3]. They are considered as 'strong' industrial wastes, with antimicrobial, phytotoxic & biotoxic action, while their disposal disposal causes major environmental impacts on the ground, water bodies and crops (Table 2).

Table 3. A typology of valorisation of OMWW

Reference	Category	Specific products
[2]	Use after treatment	Fertilizer/soil conditioner, herbicide/pesticide, animal feeding, human consumption
	Recovery	Residual oil, organic compounds (pectins, antioxidants, enzymes)
	New products	Alcohols, biosurfacants
[3]	Production of fertilizers	Biofertilization/bioremediation, composting
	Recovery of antioxidants	Polyphenols, flavonoids. anthocyanins, tannins, oleanolic acid, maslinic acid
	Production of biopolymers	Exopolysaccharides (EPSs), Polyhydroxyalkanoates (PHAs)
	Production of biogas	CH_4, CO_2
	Production of animal feed	
[7]	Agriculture	Edible fungi, microbial biomass (animal feed & agronomic use), compost
	Energy	Biogas (CH_4, H_2), biofuel (ethanol)
	Industry	Biopolymers (polysaccharides & bioplastics), enzymes (lipases, laccases, pestinaces)

Almost all of treatment processes developed for domestic and industrial wastewaters have been tested on OMWW but none of them appeared suitable to be generally adopted [3]. However, given that all these approaches are technically feasible, but they are lacking of economic viability, the future lies in valorisation of OMWW which have the ability to be transformed into products for use in agriculture, biotechnology, pharmaceutics and food industry (Table 3 includes a review of existing categories and products for OMWW valorisation, while an evaluation of different treatment methods is presented in Table 4.

Concluding, coping with OMWW is a complex issue, depending on several factors such as the variability in both quantity and quality of the wastes, the intense & seasonal

Table 4. Evaluation of treatment methods (based on [2])

Method		Evaluation
Biological	Aerobic	– operate efficiently in specific feed concentrations – in higher concentrations it is uneconomical – large quantities of waste sludge – unable to removal efficiently certain pollutants – unsuitable for direct/efficient treatment – used in combination to increase the efficiency of main process
	Anaerobic	– more suitable than aerobic – pretreatment or posttreatment is also needed – feasibility to treat high organic load – low energy requirements, produce less waste sludge – possibility for valorisation (natural gas) – ability to restart easily after months of shut down
Physico-Chemical	Neutralization, Precipitation/ Flocculation	– requires use of additional chemicals – important reductions in odour remissions – large quantities of waste sludge, – cheap & simple method, suitable for pretreatment – unsuitable for animal feed valorisation
	Oxidation	– lack effectiveness due to high cost of antioxidants & availability of the plant (low interval of COD for which the system is suitable) – suitable alternative when biological degradation is not applicable
Thermal	Evaporation, Distillation	– already used in desalination & food industry – reduction of waste volume – great divergence in bibliography for their effectiveness – post-treatment is necessary – high energy consumption & equipment cost
	Lagooning	– low energy cost (natural evaporation through solar energy) – simple process, applicable in all olive producing areas – possibility of ground water contamination – insect & odour nuisances
	Combustion, Pyrolisis	– destructive technique, eliminate possible valorisation – necessary pretreatment and posttreatment (gaseous emissions) – high energy & equipment cost, suitable for centralised plants – possible use of OMWW as source of energy
Physical	Membrane	– effective without adding solvents – extremely high capital cost/complicated procedure – pre-treatment is necessary/not suitable for strong OMWW – valorisation: can be used as pre-treatment steps in processes aiming at recovery (polyphenols, flavoring agents)

production, the variability in treatment & valorisation methods, the high regional scattering & small size of olive mills, the high investment, operation and the high transportation costs (in the case of a centralised facility), the high level of technological know-how, the necessity of large storage facilities and the proximity to human settlements [1–4, 9, 10].

3   Towards a Bioeconomic Industrial Symbiosis

Our approach for managing OMWW is based on two ideas: the concept of *bioeconomy*, which is based on the sustainable production and conversion of renewable biomass (such as OMWW) into a range of bio-based products, chemicals and energy [10, 11], and the concept of *industrial ecology,* where separate entities (such as an olive mill, an OMWW treatment/valorising facility etc.) achieve environmental and economic benefits through collaboration and exchanges of materials, energy and by-information. [5, 6]. As a result, each one of the concept, defines a different level of analysis. More specifically, bioeconomy refers to the *facility level*, where the transformation of the OMWW into valuable products takes place, while industrial ecology refers to the *symbiosis level,* where the inputs/outputs of each facility level are exchanged in a context of a wider system of participating facilities.

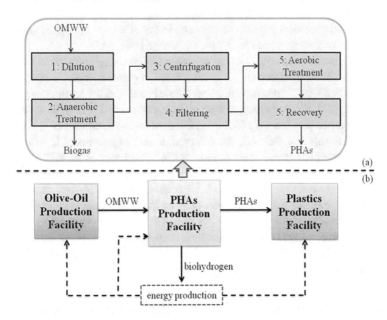

Fig. 1. The proposed bioeconomic industrial symbiosis (a) flow chart of PHAs & biogas production (facility level), (b) partners & exchanges of the network (symbiosis level)

Starting with the *bioeconomic level*, our proposed valorisation scheme is based on the production of two valuable outputs: biopolymers (PHAs) and bioenergy (H_2). (PHAs) are microbiologically produced polyesters which combine high functionality (tunable mechanical and physical properties) with low environmental impact (biodegradability and non-toxicity) and may substitute the three main products (polypropylene, polyethylene and polystyrene of the global polymer market [12, 13]. PHAs can be produced from various industrial residues from different branches and although they have gained high interest in industry and academia, their manufacturing costs are still high to compete with conventional petroleum-based polymers [14]. While evidence for PHAs production from OMWW [15], and biohydrogen production from wastewater [16] can be traced back in the 1990s, this paper proposes a five-step process (see Fig. 1a), based on recent advances in the specific valorisation techniques and methods [17–19].

At the *symbiosis level*, we may distinguish three major partners (see Fig. 1b): the olive-oil production facility, which provides its OMWW to the PHAs production facility, which in turn provides its outputs to the plastics production facility. In addition, biohydrogene produced during the PHAs production process can be valorised in term of bioenergy (mainly from the PHAs production facility. Clearly, the enabler of the proposed scheme is the PHAs producer, which has the role of *biorefinery* (in the jargon of bioeconomy) or *decomposer* (in the jargon of industrial ecology). The PHA unit enables the "3–2" criterion [6] (three different production units exchanging at least two different resources) for a "Category 4" [5] industrial ecosystem (exchanges among nearby firms, (usually) not collocated).

In the proposed industrial symbiosis, the olive-processing units form the anchor [5] of the system as they provide the relatively constant (though seasonal) flows of waste [20]. The PHAs unit whose operation depends on the availability of waste constitutes the scavenger that collects and processes waste closing the environmental loop.

Industrial symbiosis theory, as it was developed by studying cases of IS implementation, identifies collocation/proximity, governmental regulation, clear anchor-scavenger roles, common strategic visions as well as economic benefits as the main antecedents of IS implementation [21]. Similarly, power asymmetries, excessive diversity, exits of player(s), cost/risk and relation problems (lack of trust, different interests) can act as limiters. On the other hand, lubricants that facilitate the accomplishment of successful IS efforts include intermediaries/coordinators and socio-psychological factors, such as trust, knowledge creation, embeddedness and culture. Main benefits/consequences of IS are innovation and environmental and economic co-benefits [21]. Especially, for the initiation phase of and IS effort that is not imposed and/or coordinated by an external agency, economic efficiency [22], resource security, rising costs of waste disposal and regulatory pressure are decisive events [23]. Solutions/implementations (or their proposals of) need be assessed in terms of economic efficiency, compliance to regulatory frameworks and the degree of social embeddedness of local community(-ies) [24].

4   Socio-Economic Evidence from a Case-Study in Greece

In this section, we briefly describe the uncovering of an Industrial Symbiosis case in the region of Achaia in Western Greece, stimulated by the (potential) investment in PHAs production facility with the technical characteristics presented in Sect. 3. The case was based on primary and secondary research using various sources of information (telephone interviews, visits and face-to-face interviews, as well as secondary sources/

documents) from all three parties involved in the symbiosis, as well as from governmental organizations. Due to the large number of small olive-processing units (mills), the necessary information was obtained from three of the largest units. Similarly, the main source of the data for the plastics production was obtained from the largest firm of the sector in the region. The discussion is of explanatory nature aiming at surfacing the opportunities and challenges of the innovative industrial symbiosis, primarily through a socio-economic lens. To facilitate the overall assessment of the PHAs-based IS effort, Table 5 summarizes the attributes with regard to the benefits, challenges and risks, as well as the similarities and discrepancies of the three players of the symbiosis.

Table 5. Attributes of participants in the symbiosis

Characteristic	Olive Oil Mill	PHAs producer	Plastics producer
Business model	Produce to supply/order	Produce to supply/order	Produce to stock
Strategy	Cost	Cost	Cost/differentiation
Production process	Simple, rigid	Simple, may use different inputs	Some flexibility in inputs Flexibility in outputs
Culture/ Embeddedness	Agrarian, relations with suppliers	Agrarian/industrial	Industrial
Operation	Seasonal	Continuous (may be seasonal)	Continuous
Supply/suppliers	Constant base	Variable, diverse	Shopping May need incentives to use PHAs
Importance of a single company in IS	Not critical (other suppliers exist)	Critical	Not critical (other customers may be found)
Benefits(business)	Free from the burden of waste treatment	Business opportunity	Indifferent May have interest in CSR
Benefits (environmental)	Environmental-friendly waste processing	Production of PHAs and biogas	Production of biogradable plastics
Risks (individual)	Natural disaster limits supply	Technical or business failure	Technical or business failure
Risks (IS)	PHAs producer fails or asks for higher prices	Competition for waste collection, or even demand for payment for the waste Pressure for lower pricing by plastics producer(s)	Damage to eco-friendly image if supply is disrupted Alternative sourcing may be more expensive
Possible risk (IS) mitigation strategies	Own- or shared investment in PHAs production (forward integration)	Own- or shared investment in PHAs production (integration)	Own- or shared investment in PHAs production (backward integration)

The Prefecture of Achaia with an area of 3,271 Km^2 and population of about 310,000 inhabitants is the largest prefecture in Western Greece with the third largest city of Greece (Patras). The economy of the region is based on services (70% of GDP). Manufacturing and agriculture amount to about 20% and 10% of GDP, respectively. Due to the mountainous morphology of the area, agricultural activities are mainly tree-growing. There are almost 3.5 million of olive trees in the area producing around 21,000 tns of olive oil every year in 59 olive pressing facilities, most of them acting as independent firms (SMEs). Most of them have a local character, as far as supply of olives is concerned. All the 59 facilities produce

129,675,000 lt of OMWW per year. This waste is processed either using government approved methods or it is disposed uncontrollably to nearby rivers vastly contributing to the environmental degradation of the area. At the receiving end of the PHAs-based symbiotic scheme, there are three major plastics producers in the area, the largest requiring 4.4 tns of plastic as input each year (data for 2014)

Detailed input/output economic analysis of different scenarios (see [25]), taking into account all streams of revenue (provision of waste management service and valorisation of PHAs and H_2) and costs (operational and labour costs, equipment depreciation), over a time horizon of 10 years, has indicated that a PHAs unit serving 1–10 olive processing facilities (which are located at a distance of at most 60 km from it) is economically feasible. The best economic performance is when the PHAs unit serves ten olive-oil production facilities (an average facility produces around 1,400 m^3 of OMWW per season), valorising 6,000 kgr/year of PHAs and 17,000 m^3/year of H_2. In this case, the equipment costs 84,000 €, the annual operational and labor cost is 35,000 €, while the total annual revenues may reach 150,000 €.

As far as the plastics production is concerned, the production of plastics from PHAs is more expensive compared to conventional plastics. However, premium pricing to account for the environmental characteristics in products for niche markets (e.g. medical prosthetics) may compensate the additional cost. In addition, such products in the product portfolio of the firm augment its environmental image.

Regarding the social context of the symbiosis under consideration and the related mechanisms of embeddedness [22] the dominant attitude of the inhabitants of the area towards environmental protection stems from an instrumental logic of "cleanness", not to inhibit tourism activities. This frequently leads to a Not-In-My-Back-Yard behavior and competition between local communities. Sometimes this attitude is present in the behavior of olive processors concerning their waste.

Environmental regulation frameworks exist for all three parties in the IS. The symbiosis facilitates cost-effective compliance, especially for the olive mills, which are embedded in the social processes of the communities of olive tree growers (the majority of units are located in small villages or in the outskirts of small towns). The plastics production factories are also located in the outskirts of small towns and of the city of Patras, mostly within industrial zones. This is also the most appropriate place for the PHAs production unit. Although situated near towns and cities and having their personnel from urban environments, they are closely linked to agrarian life due to the links that their management and personnel has with villages: urbanization in the area is a relatively recent phenomenon. This is particularly important for the specific context as, in general, in the institutional environment of Greece, usually personal relationships make business relationships [26].

Finally, the research for the development of the case has revealed the risks for sustainability of the symbiosis. Unless trust is built among the participants, or integration of activities along the supply chain takes place, the symbiosis is very venerable to the individualistic rational profit maximizing behavior of economic agents. A step towards the sustainability of the IS would be to exert pressure towards making compulsory the treatment of waste for the substitution of conventional raw materials for plastics. Alternatively, the cooperative ownership and operation of the PHAs unit by the olive processors may guarantee its sustainability, at least, as far as the supply side is concerned.

5   Conclusions

The inappropriate disposal of OMWW results in severe environmental impacts and choosing a treatment method constitute a mutli-parameter decision, which should take into account technical, organisational, economic, and social issues. Nowadays, the trend is to turn the problem into an economic opportunity through the valorisation of waste. In this paper, we proposed a bio-economic valorisation scheme (building a facility for the production of biopolymers (PHAs) and bioenergy (H_2) based on OMWW), which was examined in an eco-industrial context in the geographic, economic and social environment of the Achaia region in Greece exposing its benefits and implementation challenges.

References

1. Azbar, N., Bayram, A., Filibeli, A., et al.: A review of waste management options in olive oil production. Critic. Rev. Environ. Sci. Technol. **34**, 209–246 (2004)
2. Niaounakis, M., Halvadakis, C.P.: Olive Processing Waste Management: Litterature Review and Patent Survey. Waste Management Series 5. Elsevier (2006)
3. Tsagaraki, E., Lazarides, H., Petrotos, K.B.: Olive mill waste water treatment. In: Oreopoulou, V., Russ, W. (eds.) Utilization of By-products and Treatment of Waste in the Food Industry, pp. 133–157. Springer, Boston (2007)
4. Justino, C.I.L., Pereira, R., Freitas, A.C., et al.: Olive oil mill wastewaters before and after treatment: a critical review from the ecotoxicological point of view. Ecotoxic. **21**, 615–629 (2012)
5. Chertow, M.R.: Industrial symbiosis: Literature and taxonomy. Annu. Rev. Energy Environ. **25**, 313–337 (2000)
6. Chertow, M.R.: "Uncovering" industrial symbiosis. J. Ind. Ecol. **11**, 11–30 (2007)
7. Morillo, J.A., Zntizar-Ladislao, B., Monteoliva-Sancez, M., et al.: Bioremediation and biobalorisation of olive-mill wastes. Appl. Microbiol. Biotechnol. **82**, 25–39 (2009)
8. Olive Oil World Council. http://www.internationaloliveoil.org/
9. Vlyssides, A.G., Loizides, M., Karlis, P.K.: Integrated strategic approach for reusing olive oil extraction by-products. J. Clean. Prod. **12**, 603–611 (2004)
10. Organisation for Economic Co-Operation and Development: The Bioeconomy to 2030. OECD, Paris (2009)
11. de Besi, M., McCormick, K.: Towards a bioeconomy in Europe: national, regional and industrial strategies. Sustainability **7**, 10461–10478 (2015)
12. Dietrich, K., Dumont, M.-J., Del Rio, L.F., et al.: Producing PHAs in the bioeconomy – towards a sustainable bioplastic. Sustain. Prod. Consum. **9**, 58–70 (2017)
13. Koller, M., Marsalek, L., de Sousa Dias, M.M., et al.: Producing microbial polyhydroxyalkanoate (PHA) biopolyesters in a sustainable manner. New Biotech. **37**, 24–38 (2017)
14. Anjum, A., Mohammad, Z., Zia, K.M., et al.: Microbial production of polyhydroxyalkanoates (PHAs) and its copolymers: a review of recent advancements. Int. J. Biolog. Macromol. **89**, 161–174 (2016)

15. Gonzalez-Lopez, J., Pozo, C., Martinez-Toledo, M.V., Rodelas, B., Salmeron, V.: Production of polyhydroxyalkanoates by Azotobacter chroococcum H23 in wastewater from olive oil mills (alpechín). Int. Biodeter. Biodeg. **38**, 271–276 (1996)
16. Sunita, M., Mitra, C.K.: Photoproduction of hydrogen by photosynthetic bacteria from sewage and wastewater. J. Biosci. **18**, 155–160 (1993)
17. Ntaikou, I., Kourmentza, C., Koutrouli, E.C., et al.: Exploitation of olive oil mill wastewater for combined biohydrogen and biopolymers production. Biores. Technol. **100**, 3724–3737 (2009)
18. Beccari, M., Bertin, L., Dionisi, D., et al.: Exploiting olive oil mill effluentsas a renewable resource for production of biodegradable polymers through a compines anaerobic-aerobic process. J. Chem. Technol. Biotechnol. **84**, 901–908 (2009)
19. Ntaikou, I., Valencia Peroni, C., Kourmentza, C., et al.: Microbial bio-based plastics from olive-mill wastewater: Generation and properties of polyhydroxyalkanoates from mixed cultures in a two-stages pilot scale system. J. Biotech. **188**, 138–143 (2014)
20. Wang, C., Zhang, G., Wang, W.: Research on the industrial symbiosis supporting system of eco-industrial park. Chinese J. Pop. Res. Env. **7**, 61–66 (2009)
21. Walls, J.L., Paquin, R.L.: Organizational perspectives of industrial symbiosis: a review and synthesis. Organiz. Environ. **28**, 32–53 (2015)
22. Baas, L.W., Boons, F.A.: An industrial ecology project in practice: exploring the boundaries of decision-making levels in regional industrial systems. J. Clean. Prod. **12**, 1073–1085 (2004)
23. Chertow, M., Ehrenfeld, J.: Organizing self-organizing systems: towards a theory of industrial symbiosis. J. Ind. Ecol. **16**, 13–27 (2012)
24. Boons, F., Janssen, M.A.: The myth of Kalundborg: social dilemmas in stimulating eco-industrial parks. In: Van den Bergh, J.C., Janssen, M. (eds.) Economics of Industrial Ecology: Materials, structural change, and spatial, pp. 337–355. MIT Press, Cambridge (2005)
25. Aminalragia-Giamini, R.: Production of bio-polymers from olive mill wastewater (in Greek). Diploma Thesis, Department of Mechanical Engineering & Aeronautics, University of Patras, Greece (2016)
26. Psychogios, A.G., Szamosi, L.T.: Exploring the Greek national business system. EuroMed J. Bus. **2**, 7–22 (2007)

A Case Study of Sustainable Manufacturing Practice: End-of-Life Photovoltaic Recycling

Jun-Ki Choi[✉]

Renewable and Clean Energy Program, Department of Mechanical Engineering,
University of Dayton, 300 College Park, Dayton, OH 45469-0238, USA
jchoi1@udayton.edu

Abstract. The usage of valuable resources and the potential for waste generation at the end of the life cycle of photovoltaic (PV) technologies necessitate a proactive planning for a PV recycling infrastructure. To ensure the sustainability of PV in large scales of deployment, it is vital to develop and institute low-cost recycling technologies and infrastructure for the emerging PV industry in parallel with the rapid commercialization of these new technologies. There are various issues involved in the economics of PV recycling and this research examine those at macro and micro levels, developing a holistic interpretation of the economic viability of the PV recycling systems. This study will present mathematical models developed to analyze the profitability of recycling technologies and to guide tactical decisions for allocating optimal location of PV take-back centers (PVTBC), necessary for the collection of end of life products. The economic decision is usually based on the level of the marginal capital cost of each PVTBC, cost of reverse logistics, distance traveled, and the amount of PV waste collected from various locations. Results illustrated that the reverse logistics costs comprise a major portion of the cost of PVTBC; PV recycling centers can be constructed in the optimally selected locations to minimize the total reverse logistics cost for transporting the PV wastes from various collection facilities to the recycling center. In the micro- process level, automated recycling processes should be developed to handle the large amount of growing PV wastes economically. The market price of the reclaimed materials are important factors for deciding the profitability of the recycling process and this illustrates the importance of the recovering the glass and expensive metals from PV modules.

PV manufacturing has been growing over the past 10 years and further annual growth of 15% is expected until 2025 [1]. Studies on positioning a grand plan for solar power shows how vast PV arrays and other renewable energies can provide significant amount of electricity and total energy needs by 2050 [2]. Various new PV technologies have been introduced in the market and existing technologies have undergone further development. How all these developments will affect the fate of the end-of-life PV modules is uncertain. In addition, the market price of some rare earth materials utilized in the manufacturing of the various PV technologies has exponentially increased in the past five years [3]. Therefore, it is necessary to set a proactive strategic recycling plan for the treatment of the disposed PV wastes.

© Springer International Publishing AG 2017
G. Campana et al. (eds.), *Sustainable Design and Manufacturing 2017*, Smart Innovation, Systems and Technologies 68, DOI 10.1007/978-3-319-57078-5_27

There are three different types of PV waste; end-of-life modules, manufacturing scraps, and defect form packaging and transportation. Among these, end-of-life PV modules are the major source for the recycling process and our waste prognosis showed that the future amount of PV waste will grow exponentially. For example, c-Si recycling processes consists of five major steps. First the unloaded modules transported from the collection sites will be loaded to the automatic conveyor system to enter into the recycling process. Then the junction boxes are removed manually. Thermal treatment burns off the laminates to facilitate the separation processes. From the separation steps, copper wire, aluminum frame, glass, and waste are separated. During the next step the solar cells are treated chemically. Surface and diffusion layers are removed subsequently by cleaning steps. Cells and wafer breakage are cleaned by etching techniques. Regarding to the reclaimed materials and waste, the following outlet parameters are considered. Junction box is processed by an electronic scrap waste treatment company (collection cost paid by PVTBC). Plastic is burned off after the thermal treatment (i.e. incineration cost paid by PVTBC). Waste goes to land fill and PVTBC pays landfill tipping fees. Aluminum can be reused while glass, copper, and silicon can be sold to recycling companies. The thermal process could be improved with regards to its throughput, cycle time and yield. The yield of recovered cells depends largely on type, design and state of the modules to be processed. Design dependent factors that affect results of the thermal process are the type of laminate and crystal, the dimensions of the embedded cells, and the material and dimensions of bonds and soldering.

There are various issues involved in the economics of PV recycling in the macro and micro level. In the macro-level, strategies are needed for allocating the centralized/decentralized collection and recycling facilities in the optimal locations to minimize the total recycling system costs. This includes issues such as the optimal level of marginal capital costs to open up a PV take-back center (PVTBC), costs associated with the reverse logistics services for the collection of PV modules and transporting them to the recycling facilities. Various stakeholders (e.g., dismantlers, recyclers, smelters) must be taken into account in the recycling infrastructure. In the micro-level, optimized process planning is required to ensure the profitability of the PVTBC. Potential PVTBC will face some challenging decisions in the following issues; material separation, revenue structures of current and future recycling processes with regard to the volatility of the market price of materials/components, cost associated with processing, reverse logistics costs, and external social costs, such as landfill-tipping fees.

Therefore, this study developed a generic mathematical modeling framework to evaluate the economic feasibility of the macro-level reverse logistics planning and the micro-level recycling process of the PV waste by considering the complex issues of the PV recycling planning listed above. A mixed integer programming and a linear programming are applied to the macro logistics and micro process planning models respectively. A case study of the crystalline silicon PV waste recycling in Germany is presented to illustrate the applicability of the models.

First, the macro-level reverse logistics model is designed to allocate the optimized locations of PVTBC by considering the amount of PV wastes to be collected, distance traveled (routing schemes) to PVTBC, and capital cost of opening the facility. The base model solves the optimization problem of the location of the capacitated facility by minimizing the objective function subject to the various constraints. The objective function is the sum of the transportation costs (i.e., fuel price, fuel-efficiency of lorry, and distance traveled), and the costs of logistics services provided by the registered logistics company. With the variation of the marginal capital cost for opening up a PVTBC, the model suggests the best candidate locations to open up PVTBCs by considering the amount of waste from each collection locations and the cost associated with the reverse logistics to transport the waste from each location to the designated PVTBCs. In the micro-level recycling process level, the main objective of each PVTBC is to maximize the revenues from selling the materials recovered from the collected PV modules to the price varying markets for reclaimed materials while minimizing the cost associated with processes, transportation, capital, and inventories. The base optimization model decision set determines how much material to process by which equipment, in what period to process it, and if applicable, how much inventory should be held each period. Various experimental designs provide sensitivity analysis on key parameters.

Based on the current study, following general conclusions can be claimed from German case study. In order to ensure the economics of the PV end-of-life management systems, PVTBC should be constructed in an optimally decentralized location to minimize the total reverse logistics cost to transport PV wastes from various collection facilities to the PVTBC. In the recycling process level, advanced and automated energy efficient recycling processes should be integrated to handle the large amount of growing PV wastes economically. Market price of the reclaimed materials is important factor for deciding the profitability of the recycling process. Therefore, it is important to recycle thin-film PV modules (i.e., CdTe, CIGS) where some rare earth materials can be reclaimed. This study focused on the short term planning which currently accounts for the available PV waste. However further study will adopt strategies to consider the complex waste flows generated from different spatial (i.e. US), temporal (i.e. future), and technical (i.e. various technologies) aspects. Lastly, further study will experiment the life cycle environmental implication of the PV recycling along with the economic feasibility.

References

1. Choi, J.-K., Fthenakis, V.M.: Design and optimization of photovoltaics recycling infrastructure. Environ. Sci. Technol. **44**(22), 8678–8683 (2010)
2. Zweibel, K., Mason, J., Fthenakis, V.M.: A solar grand plan. Sci. Am. **298**, 48–57 (2008)
3. Choi, J.-K., Fthenakis, V.M.: Economic feasibility of recycling photovoltaic modules: survey and model. J. Ind. Ecol. **14**(6), 947–964 (2010)

Supply Chain Risk Management for Sustainable Additive Manufacturing

Daniel R. Eyers[(✉)]

Cardiff Business School, Cardiff University, Aberconway Building, Colum Drive,
Cardiff, CF10 3EU, UK
eyersDR@cf.ac.uk

Abstract. Additive Manufacturing technologies are becoming increasingly prevalent in commercial practice, and as a result attention is gradually being devoted to their implications for sustainability and supply chain management. One pertinent topic for these technologies remaining largely unexplored is that of Supply Chain Risk Management. This paper serves to provide an initial contribution in the form of a conceptual framework to identify risk sources, consequences, responses, and controls from a sustainability perspective, and provides directions for future research on this topic.

Keywords: Additive Manufacturing · Sustainable manufacturing · Supply chain · Risk management

1 Introduction

The unique characteristics of Additive Manufacturing technologies that enable the production of a wide range of different products are now well-established in literature [1], with success particularly evidenced in medical, aerospace, and automotive applications [2]. There is much enthusiasm for Additive Manufacturing to enable future Digital Manufacturing initiatives [3, 4], to support customer-integrated manufacturing [5], and to achieve sustainability objectives [6]. In short, there is great expectation that the technologies will have a profound impact on the way future production is conducted.

If Additive Manufacturing is to be a significant contributor to a future world of sustainable manufacturing, researchers need to think very carefully about how this will be achieved. Much of the current Additive Manufacturing research has focused on individual machine technologies, but these alone do not achieve 'manufacturing', and there has been a call for increased attention to the wider manufacturing and supply chain systems [7]. The notion that an individual customer may 'just press print' to fabricate their product is (unfortunately) a fallacy. A whole raft of entities in the supply chain are involved in the achievement of a customer product, including designers, material suppliers, manufacturers, post-processors, distributors, warehouses, and retailers. To support a future of sustainable digital production, all contributors need to be considered – not just specific Additive Manufacturing machine processes.

In recognition of this need to focus on the supply chain, this paper provides an initial exploration on the nature of risk within supply chains that employ Additive

© Springer International Publishing AG 2017
G. Campana et al. (eds.), *Sustainable Design and Manufacturing 2017*, Smart Innovation,
Systems and Technologies 68, DOI 10.1007/978-3-319-57078-5_28

Manufacturing technologies, and identifies some approaches to its management. At present Additive Manufacturing's share of total manufacturing output is tiny: for all types of manufacturing, the combined total output of the top-20 manufacturing countries in 2014 exceeded £8.2tr [8], whilst Additive Manufacturing's worldwide market was estimated to be £4.103bn [2]. However, for some products (e.g. ITE Hearing Aids) Additive Manufacturing is now standard practice, and as companies venture into these technologies, they need to have confidence in the way risks within the supply chain are managed. The purpose of this paper is therefore to provide a framework with which to characterise risks for Additive Manufacturing supply chains, and to understand potential responses that can be made to these.

2   Literature Review

2.1   Supply Chain Risk Management (SCRM)

The management of supply chains became of increasing interest to practitioners as a result of decreasing vertical integration, increasing requirements for competitiveness, and the realization that optimization strategies need to take a holistic approach to ensure overall system improvement [9]. The 1980's and 90's saw much emphasis on the practical challenges of co-ordinating and controlling supply chains, and more recently research has extended to consider issues of sustainability. Whilst there have been multiple interpretations of what a 'sustainable' supply chain comprises, there is increasing consensus of its alignment with the 'triple bottom line' concept in terms of social, economic, and environmental components [10].

Supply chain research has typically focused on the decisions that can be made to best configure and manage a supply chain to achieve its objectives. Researchers have long recognized that supply chains are complex and plagued with uncertainties [e.g. 11], requiring effective strategies to manage these challenges. More recently, attention has been extended to formal risk management practices for supply chains. Risk is an inherent part of any supply chain, and the concept of SCRM has gained much traction since year 2000, particularly in the wake of serious disruptions to increasingly global supply chains [12]. Today, SCRM is recognized as an important element of supply chain management [13]. Much literature has been published on SCRM, and synthesis of the knowledge base has recently been provided by several literature review papers [e.g. 14–17].

The term 'risk' in a traditional supply chain context concerns disruption to flows (of material, information, products, and money) between organizations [18] and such disruption has the potential to affect the efficient management of the supply chain [14]. Hofmann et al. [19] identify that in addition to these disruptions to flow, consideration needs also to be given to the potential risks arising from sustainability issues. Whilst emphasis in SCRM is often on risks with an economic effect, Freise and Seuring [20] further summarize a multitude of different social and environmental risks that also may be relevant to supply chains.

Definitions of risk typically imply negative connotations for the supply chain, but this need not be the case, as risks can also lead to positive outcomes. Risks are therefore uncertain events: they may happen, or they may not, and their impact can be different depending on a multitude of factors. As a result, Harland et al. [21] identify that risk assessment in a supply chain context should examine the likelihood (probability) and significance (consequence) of risks. Strategies for managing risk are often divided into those which are proactive (planned in advance of risks materialising) or reactive (dealing with the consequences of risks that have materialised). For example, Dani (2009) proposed a predictive-proactive methodology that would provide adequate quantitative data (gained through a variety of mechanisms) to help organizations understand risks and form proactive plans.

2.2 SCRM for Sustainable Additive Manufacturing

Additive Manufacturing's potential contribution to the improvement of supply chains has gained considerable research attention in recent years [22, 23]. Whilst this is a positive development for the technologies, it is notable that such attention has not been extended to SCRM in this context. This is particularly notable since SCRM in general is a rapidly growing research topic [17], and so is Additive Manufacturing [5] – but as yet the two have not been unified.

A structured review was conducted of four databases (EBSCO Business Source Premier, Emerald, Science Direct, and Scopus) to identify the extent of existing research. The selection of this combination of databases was motivated by their inclusion of the principal scholarly journals relevant to this work, together with access to appropriate conference publications and trade articles. Full-text searches of the databases were conducted using keywords "Additive Manufacturing" or the more commonly used term "3D Printing", together with the focal "Supply Chain Risk Management" topic.

This search of the literature yielded just 9 papers (5 journal articles, 3 conference papers, and 1 practitioner article). Each paper was read fully by the author, and its relevance to the topic evaluated. Three papers were only included in the search result because of the keyword being found in their references, and were therefore identified as having no relevance to the study. A further five papers mentioned the topics of Additive Manufacturing/3D Printing and risk within their text, but their treatment of the concepts was disparate (with the keywords typically found in different sections of the paper).

As a result of this review, only one practitioner article [24] could be identified as offering *some* relevance to this study. Rather than the management of risks in Additive Manufacturing supply chains that are the focus of the current study, it considered the benefits the technologies could bring to risk in conventional supply chains. These are summarized in Table 1, and focus on the radical changes that may come about through the adoption of Additive Manufacturing technologies. Whilst these can be identified as offering an optimistic perspective on the technologies, notably these claims are made without empirical evidence to substantiate the identified possibilities. Furthermore, this assessment does not attempt to explore the new risks that arise in the Additive Manufactured supply chain, nor approaches to risk management, and it is important that potential adopters consider risk when considering the implementation of these technologies.

Table 1. Envisaged contributions 3D Printing (Additive Manufacturing) could make to supply chain risks (Adapted from: Caccamo 2016)

Benefit	Contribution of Additive Manufacturing
Simplification of the supply chain	Reshoring of production Simplification of products On-Demand Production
Lessening business continuity risk	Removal of suppliers and sites that may pose risks Removal of need to identify single-supplier risks
Lessening brand and compliance risk	Elimination of unmonitored geographically remote sub-tier suppliers through reshoring Improvement of brand reputation by reducing waste
Eviscerating capacity risk	On-demand production eliminates the need for inventory or safety stock
Changing security risks	Elimination of most physical supply chain risks as supply chain is digital Transformed cyber risks: Supply chain simplification eliminates risks from entities upstream in the supply chain, but shifts points of failure downstream

These findings highlight the paucity of research available for SCRM and Additive Manufacturing, and also an absence of literature concerning sustainability issues within SCRM.

3   Research Method

The lack of research concerning SCRM for Sustainable Additive Manufacturing motivated the adoption of an exploratory, conceptual study which serves as a precursor to further empirical investigation. Such conceptual approaches provide a firm grounding for field research, helping to focus pertinent topics, develop theories, contextualize work within existing bodies of knowledge, and support the selection and implementation of research methods in future studies [25]. The study was conducted by the author following extensive prior work that explored the implications of Additive Manufacturing systems in terms of flexibility for Operations and Supply Chain Management [26], and whilst data from this research is not included in this study, its findings served as important contributors to the development of this work.

4   A Proposed Supply Chain Risk Management Framework for Sustainable Additive Manufacturing

The wealth of literature pertaining to SCRM (in general) offers a wide variety of different implementation frameworks. These were reviewed, and the work of Jüttner et al. [27] identified as being particularly relevant as a result of its conceptual simplicity, and opportunity for fruitful extension in the context of sustainable Additive Manufacturing.

This is illustrated in Fig. 1, and supported by explanations for each component in the next four sub-sections.

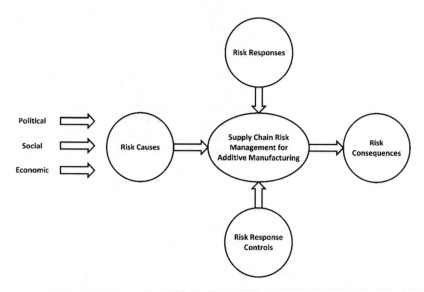

Fig. 1. A framework for sustainable Supply Chain Risk Management in Additive Manufacturing supply chains

4.1 Risk Causes

Risks are uncertain events (or sets of events) that (should they occur) will have an impact on the way in which Additive Manufacturing supply chains operate. Risk causes can come from any source, and this can include political, social, and economic sources. As noted by Harland et al. [21] it is important to understand the likelihood (or probability) of these events occurring. These events will either affect the supply chain positively (i.e. they provide opportunities), or negatively (i.e. they are a threat). They may be within the control of supply chain entities, or be completely external and cannot be influenced by them. For example, a change in government could have a positive effect (if this led to increased economic prosperity, and subsequently increased sales), or a negative effect (if this led to an economic downturn and reduced customer demand). A recent example in 2015 is the Federal Aviation Administration's assessment of an Additive Manufactured sensor for GE commercial jet engines. The risk event here is whether the sensor would be approved: the positive outcome being approval, whilst potential negative outcomes being either a complete rejection, or a requirement for revision.

4.2 Risk Consequences

A risk consequence is the effect (good or bad) of the risk event occurring, and is typically the focus of SCRM strategies that are enacted by risk responses and controls. These consequences may affect just a single firm, or multiple entities within the whole supply chain. Many different consequences are possible, and they may also have political, social, or economic impacts. A common approach to evaluating consequences is the use of 'risk impact scales', which provide a relative measure of the impact of the risk on defined objectives.

Extending the GE example from Sect. 4.1, the positive consequence is the ability to use these new parts in real aircraft engines, offering time and money saving potential for the future. The negative outcome of rejection (whether in full, or partial) would be increased costs on further development and testing, together with delays and lost opportunities for the integration of the technologies in future aerospace applications. From a supply chain perspective, it is necessary to understand how different entities may be affected by these consequences, and the magnitude of the impact.

4.3   Risk Responses

A major development of this work over that of Jüttner et al. [27] is in the way that risks were handled. In this earlier study, the objective was to *mitigate* the risk, and general strategies were offered. In the current paper a broader perspective is taken: recognizing that risks can be both positive or negative, it is necessary to *respond* to the risk, which can have many more outcomes than mitigation. Nine generic risk responses are well-established in risk management literature, and it is suggested that these can be applied for Additive Manufacturing supply chains (Table 2).

Table 2. Potential risk response actions for Additive Manufacturing supply chains

Response action	Description
Avoid	Take action to prevent the risk event from occurring
Reduce	Take action to lessen the impact of the risk event
Fallback	Have a backup position ready in event of the risk event occurring, so that the impact is lessened
Transfer	Identify ways to transfer negative risk effects to other parties, so that the effect is lessened for the focal supply chain
Accept	Acknowledge that the risk event may occur, but do nothing in preparation for it
Share	Identify ways to share the positive or negative effects of a risk event with others
Exploit	For risk events that will have a positive outcome, identify ways to fully exploit this
Enhance	For risk events that will have a positive outcome, identify ways to further increase its magnitude
Reject	For risk events that will have a positive outcome, do not capitalise on it

4.4   Risk Response Controls

The fourth (and often overlooked) aspect of risk management in a supply chain context is the need to maintain overall control of the supply chain. Consistent with general theories about the management of any system or supply chain, risk response controls are identified as a further extension of the Jüttner et al. [27] framework. In practice, such controls concern the continual monitoring of potential events that will lead to risks occurring, instigating the appropriate responses, monitoring the outcome of these responses, and looking out for new (as yet undefined) risks that may arise and forming plans to deal with them.

5   Discussion

The achievement of sustainability within manufacturing is a challenging and multifaceted activity. Technologies such as Additive Manufacturing have been identified to offer many potential opportunities to improve sustainability, but as yet their significance and impact is not well understood [6]. However, what is clear from sustainability research in general is that a single technology is unlikely to offer a panacea, and sustainability will arise from a multitude of different contributors. To be sustainable, manufacturing firms will need to focus both on their internal production competencies, but also on the achievement of sustainability within their supply chains [28], and it is the management of risk within these that is the focus of this study.

Managing risk in supply chains is often a complex endeavour, and the rapid growth of research in this area underlines the range potential approaches that can be undertaken [17]. The lack of Additive Manufacturing related SCRM literature evidenced in this study has motivated the development of a conceptual framework which serves to unify the manufacturing, sustainability, and supply chain risk management themes (Fig. 2). It has provided an initial understanding of how knowledge from these currently disparate bodies of research can be intertwined, and contributes to the overall discussion on the management of sustainability within manufacturing supply chains.

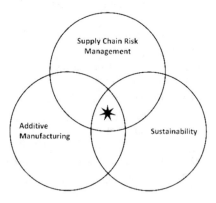

Fig. 2. Unification of sustainable Supply Chain Risk Management in Additive Manufacturing supply chains

Further work is required to validate and extend this conceptual contribution for practical implementation. It is suggested that initial qualitative studies are needed to better understand the nature of each of the components of the framework, particularly in terms of the nature of risk and means of evaluating its likelihood and impact within a supply chain context. Given that Additive Manufacturing potentially allows for radically different supply chain models [29], will they be subject to the same types of risks as their conventional counterparts? Will existing risks be eliminated, or will they be replaced by new ones? Will these risks affect overall sustainability objectives, and if so, what will be the outcome? Or will sustainability requirements introduce risks of their own? Questions such as these need thorough exploration with empirical data to fully understand the potential impact of Additive Manufacturing, which may then in turn support the extension of the work for quantitative evaluation.

6   Conclusion

This exploratory study has provided an initial contribution to the concept of SCRM for sustainable Additive Manufacturing. It has overviewed the SCRM concept, and highlighted its current lack of discussion in an Additive Manufacturing context. By extensively developing an existing SCRM framework with an explicit focus on opportunities for supply chains employing Additive Manufacturing, this paper has presented a feasible framework that is suitable for empirical assessment, and some pertinent questions for further research.

References

1. Gibson, I., Rosen, D.W., Stucker, B.: Additive Manufacturing Technologies: 3D Printing, Rapid Prototyping, and Direct Digital Manufacturing. Springer, New York (2015)
2. Caffrey, T., Wohlers, T.T.: Wohlers Report 2015. Wohlers Associates Inc., Fort Collins (2015)
3. BMBF: The new High-Tech Strategy Innovations for Germany, Rostock, Germany (2014)
4. KPMG: The Digitalisation of the UK Automotive Industry. KPMG (2016)
5. Ryan, M.J., Eyers, D.R.: Sustainable scenarios for engaged manufacturing: a literature review and research directions. In: 4th International Conference on Sustainable Design and Manufacturing, Bologna, Italy (2017)
6. Ford, S., Despeisse, M.: Additive manufacturing and sustainability: an exploratory study of the advantages and challenges. J. Cleaner Prod. **137**, 1573–1587 (2016)
7. Eyers. D.R., Potter, A.T.: The concept of an industrial Additive Manufacturing System. In: 26th Annual POMS Conference, Washington D.C. (2015)
8. Rhodes, C.: Manufacturing: International Comparisons (Briefing Paper 5809). House of Commons Library, London (2016)
9. Lummus, R.R., Vokurka, R.J.: Defining supply chain management: a historical perspective and practical guidelines. Ind. Manag. Data Syst. **99**, 11–17 (1999)
10. Carter, C.R., Rogers, D.S.: A framework of sustainable supply chain management: moving toward new theory. Int. J. Phys. Distrib. Logistics Manag. **38**, 360–387 (2008)
11. Childerhouse, P., Towill, D.R.: Reducing uncertainty in European supply chains. J. Manuf. Technol. Manag. **15**, 585–598 (2004)

12. Dani, S.: Predicting and managing supply chain risks. In: Zsidisin, G.A., Ritchie, B. (eds.) Supply Chain Risk: A Handbook of Assessment, Management, and Performance, pp. 53–66. Springer, New York (2009)
13. Ritchie, B., Brindley, C.: Supply chain risk management and performance: a guiding framework for future development. Int. J. Oper. Prod. Manag. **27**, 303–322 (2007)
14. Ghadge, A., Dani, S., Kalawsky, R.: Supply chain risk management: present and future scope. Int. J. Logistics Manag. **23**, 313–339 (2012)
15. Ho, W., Zheng, T., Yildiz, H., et al.: Supply chain risk management: a literature review. Int. J. Prod. Res. **53**, 5031–5069 (2015)
16. Rao, S., Goldsby, T.J.: Supply chain risks: a review and typology. Int. J. Logistics Manag. **20**, 97–123 (2009)
17. Fahimnia, B., Tang, C.S., Davarzani, H., et al.: Quantitative models for managing supply chain risks: a review. Eur. J. Oper. Res. **247**, 1–15 (2015)
18. Jüttner, U.: Supply chain risk management. Int. J. Logistics Manag. **16**, 120–141 (2005)
19. Hofmann, H., Busse, C., Bode, C., et al.: Sustainability-related supply chain risks: conceptualization and management. Bus. Strategy Environ. **23**, 160–172 (2014)
20. Freise, M., Seuring, S.: Social and environmental risk management in supply chains: a survey in the clothing industry. Logistics Res. **8**, 1–12 (2015)
21. Harland, C., Brenchley, R., Walker, H.: Risk in supply networks. J. Purchasing Supply Manag. **9**, 51–62 (2003)
22. Maccarthy, B., Blome, C., Olhager, J., et al.: Supply chain evolution – theory, concepts and science. Int. J. Oper. Prod. Manag. **36**, 1696–1718 (2016)
23. Tuck, C.J., Hague, R.J.M., Burns, N.: Rapid manufacturing: impact on supply chain methodologies and practice. Int. J. Serv. Oper. Manag. **3**, 1–22 (2007)
24. Caccamo, W.: 3D printing blows up supply chain risk management. Supply Demand Chain Executive **17**, 34–37 (2016)
25. Ravitch, S.M., Riggan, M.: Reason & Rigour. SAGE Publications, Thousand Oaks (2012)
26. Eyers, D.R.: The flexibility of industrial Additive Manufacturing Systems. Cardiff Business School. Cardiff University, Cardiff (2015)
27. Jüttner, U., Peck, H., Christopher, M.: Supply Chain Risk Management: outlining an agenda for future research. Int. J. Logistics Res. Appl. **6**, 197–210 (2003)
28. Paulraj, A.: Understanding the relationships between internal resources and capabilities, sustainable supply management and organizational stability. J. Supply Chain Manag. **47**, 19–37 (2011)
29. Eyers, D.R., Potter, A.T.: E-commerce channels for Additive Manufacturing: an exploratory study. J. Manuf. Technol. Manag. **26**, 390–411 (2015)

Decision Support for Sustainability

Sustainable Design: An Integrated Approach for Lightweighting Components in the Automotive Sector

C.A. Dattilo[(✉)], L. Zanchi, F. Del Pero, and M. Delogu

Department of Industrial Engineering of Florence (DIEF), Florence, Italy
{caterinaantonia.dattilo,laura.zanchi,francesco.delpero,
massimo.delogu}@unifi.it

Abstract. In past years the European Union (EU) set targets to reduce emissions in order to encourage and develop a more sustainable society. As a consequence of this, the carmakers began to study new materials and innovative technologies in order to lightweight their vehicles, thus reducing use stage fuel consumption and environmental impact. A promising strategy for this is replacing steel with composites although the adoption of these materials often involves negative effects on production and End-of-Life (EoL) stages. For this reason, a comprehensive assessment of the entire component Life Cycle (LC) is needed, not only in terms of environmental issues but also economic and social ones. This paper presents a sustainable design approach based on TOPSIS methodology functional to compare different design solutions in the automotive sector; the approach is also validated by an application to a real case study.

Keywords: Lightweighting · Sustainability · Integrated approach · TOPSIS

1 Introduction

The automotive is considered a sector on the rise. In 2010 global vehicle registrations were estimated around 1,015 billion of units and this number is expected to grow up to 2,5 billion by 2050 [1]. Much of this growth is foreseen to occur in emerging markets such as China and India, and it will result in significant increases in air emissions, global fuel demand and material requirements, and a corresponding increase of waste produced during the End-of-Life (EoL) is expected [2]. More specifically light-duty vehicles account for approximately 10% of total energy use and greenhouse gases [3] and they could increase from roughly 700 million to 2 billion over the coming decades. For this reason car manufacturers have been implementing several technical solutions to meet legislation requirements and satisfy consumer expectations. In this regard, Electric Vehicles (EVs) represent a great opportunity as they are responsible for less impact than conventional vehicles. Additionally to electric vehicles, another viable solution for road transport decarbonisation is lightweighting. In this regard lightweight materials and innovative design solutions offer great potentialities to reduce energy and impact of use stage [4]. In the EVs context the lightweight design is particularly relevant since directly influences driving range and batteries size. On the other hand, although lightweighting enables lowering the use stage impact through a reduction of energy consumed during

© Springer International Publishing AG 2017

G. Campana et al. (eds.), *Sustainable Design and Manufacturing 2017*, Smart Innovation, Systems and Technologies 68, DOI 10.1007/978-3-319-57078-5_29

operation [5], it often involves negative effects on production and EoL stages [6, 7]. Indeed, several lightweight polymeric materials, such as carbon fibre reinforced polymers, are energy-intensive to produce and involve higher CO_2 emissions prior to the use stage [8]; furthermore, carbon fibre and composites are more difficult to be recycled if compared with metals [9]. As a consequence, the effect of lightweighting on production/EoL and use stages could be responsible for a thorny balance of benefits and disadvantages over the entire Life Cycle (LC). In this context the comparative Life Cycle Assessment (LCA) is the most appropriate methodology for establishing the effective environmental convenience of innovative lightweight design solutions with respect to the reference ones [10]. Meanwhile, pressure from stakeholders and corporate strategies towards sustainability are the main drivers for the performances improvements of products within a wider sustainability approach, by taking into account environment, economy and society [11]. The aim of this paper is presenting an integrated approach for sustainable design of vehicle components lightweighting and describing its application to a case study in order to validate it. The proposed methodology is carried out according to an integrated Design-For-X (DFX) approach [12] and it combines environmental (LCA), cost (LCC) and social aspects (S-LCA) by the direct application of the TOPSIS (Technique for Order Preference by Similarity to Ideal Solution). The TOPSIS method is widely used to provide effective results for the ranking of alternatives that have absolute data for different indicators; however, few studies exist about its use in the integration of LCA, LCC and S-LCA [13–15]. Finally, the research provides an original example where TOPSIS is applied to determine the best sustainable solution between two different lightweight designs for EVs components, during the early design phase.

2   Method

Life Cycle Thinking (LCT) is the key concept for sustainable design and it deal with several sustainability assessments such as environmental (LCA), economic (LCC) and social (S-LCA). The need for an interdisciplinary method leads to a holistic approach for sustainable design, where environmental, economic and social performances are combined [16]. In this section, the proposed methodology for sustainable analysis of components for EVs is described in three main steps (Fig. 1):

- **Step 1:** Life Cycle Inventory (LCI), that is a detailed gathering of all data associated to entire LC (from-cradle-to-grave) of components, in terms of environmental, economic and social questions;
- **Step 2:** LCA, LCC and S-LCA, which deal with environmental, economic and social assessment, respectively;
- **Step 3:** Integrated assessment means by TOPSIS.

The quality of data collected in LCI (Step 1) is very important in order to guarantee the accuracy of the results; data are classified as *primary data* (i.e. direct measurement), *secondary data* (i.e. databases, studies) or *assumptions* in case primary or secondary data are not available.

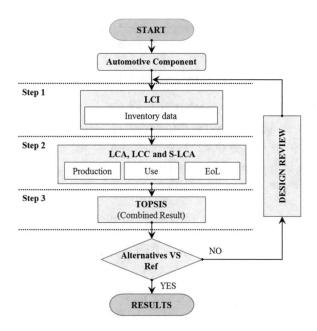

Fig. 1. Flow chart of sustainability method for automotive components

In the second step the LCA is conducted according to ISO 14040:2006 and ISO 14044:2006 standards and impact assessment methods (i.e. CML2001). For LCC analysis the Code of Practice [17] and standards for other sectors (i.e. ISO 15686-5:2008 for buildings) currently represent the only methodology references; as far as the automotive sector is concerned, only voluntary approaches and LCC case studies can be found [12, 18, 19]. For this reason a tailored costs model for vehicle components is proposed (Eq. 1):

$$LCC = C_{Production} + C_{Use} + C_{EoL} \tag{1}$$

Where: (a) $C_{Production} = \sum_{j=1}^{n} C_{Material_j} + \sum_{i=1}^{m} C_{Manufacturing_i}$, includes the cost of main machines and tools, manufacturing energy consumption, cost of consumables, labor. (b) $C_{Use} = C_{Propulsion} + C_{CO2}$, compels the cost for electricity and CO_2 emissions by assuming the Emission Trading System values [20]; (c) $C_{EoL} = C_{Treatment} - C_{Recycling}$ includes the wrecking operations, treatment for recycling and disposal for fluff materials and the gain obtained from the recycling of component parts made of recyclable materials. The S-LCA is the youngest technique of analysis; it is a methodology aimed to assess the potential social and economic impacts of products/services throughout the entire LC and it has been conducted according to the Guidelines for Social Life Cycle Assessment of Products [21].

Step 3 consists in the application of the TOPSIS method for the definition of sustainable result, starting from the previous analysis (LCA, LCC, S-LCA) which represent the criteria. TOPSIS method is based on the assumption that the best alternative should have the shortest distance from an ideal solution and the farthest distance of the negative ideal solution. The procedure of TOPSIS can be expressed in several steps:

1. Create a decision matrix $[D] = \left[r_{ij}\right]_{mxn}$ consisting of m alternatives and n criteria.

2. Calculate the normalized decision matrix $[R] = \left[r_{ij}\right]_{mxn}$ with $r_{ij} = a_{ij}/\left(\sqrt{\sum_{j=1}^{m} a_{ij}^2}\right)$;

3. Calculate the weighted normalized decision matrix: $[V] = \left[v_{ij}\right]_{mxn}$ with $v_{ij} = w_i \cdot r_{ij}$. It is generated by multiplying the columns of matrix $[R]$ with weights of criteria $\left(w_1, w_2, \ldots, w_n\right)$ estimated by the pairwise matrix where each entry $a_{i,j}$ represents the relative importance of the criterion C_i when it is compared with the criterion C_j. [22].

4. Determine the positive ideal (A^*) and negative ideal solutions (A^-):

$$A^* = \left\{\left(\max_i v_{ij} | j \in J^b\right), \left(\min_i v_{ij} | j \in J^c\right), i = 1, \ldots, m \right\} = \left\{v_{1^*}, \ldots, v_{n^*}\right\}_{ij} \qquad (2)$$

$$A^- = \left\{\left(\min_i v_{ij} | j \in J^b\right), \left(\max_i v_{ij} | j \in J^c\right), i = 1, \ldots, m \right\} = \left\{v_{1^-}, \ldots, v_{n^-}\right\}_{ij} \qquad (3)$$

where the J^b index is associated with benefit criteria and J^c is associated with cost criteria.

5. Calculate the separation measures from the positive ideal (S_{i^*}) and negative solution (S_{i^-}), using the n-dimensional Euclidean distance:

$$S_{i^*} = \sqrt{\sum_{j=1}^{n} \left(v_{ij} - v_{j^*}\right)^2} \text{ and } S_{i^-} = \sqrt{\sum_{j=1}^{n} \left(v_{ij} - v_{j^-}\right)^2} \text{ for } i = 1, \ldots, m \qquad (4)$$

6. Calculate the index, C_{i^*} to the ideal solution by the formula:

$$C_i^* = S_i^- / (S_i^* + S_i^-) \qquad (5)$$

In particular, if $A_i \equiv A^- \Rightarrow S_i^-$ then: $C_i^* = 0$; if $A_i \equiv A^* \Rightarrow S_i^*$ then: $C_i^* = 1$.

7. Rank the alternatives in decreasing order, using C_{i^*} index.

The proposed methodology can be used especially to compare an innovative alternative (or more than one) with the reference one (that represents the current solution) in order to define the best vehicle component in terms of sustainable issues. For this reason, if the reference solution results better than the new one then there is the need to improve it (*design review*).

3   Case Study

3.1   Description

In order to validate the proposed methodology, an innovative suspension arm is chosen as case study (Fig. 2). The suspension arm, developed by Magneti Marelli (MM), weights 1,8 kg and it is manufactured using an innovative technology called Advanced Sheet Compression Molding (ASCM) [23]. One of the ASMC process advantages is the possibility to create a strong cohesion between metallic inserts and composite material during the molding process. This arm is composed by four parts (Fig. 2-a): (1) *Front bushing attachment* made of aluminum (0,121 kg); (2) *Rear bushing attachment* made of aluminum (0,292 kg); (3) *Ball joint attachment* made of aluminum (0,257 kg); *Laminate* made of Carbon Fiber (CF) reinforced Vinyl Ester (VE) (1,14 kg). The metallic inserts allow connecting the arm to the rest of the vehicle. In particular, it is connected to the wheel knuckle through the ball joint.

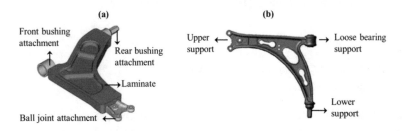

Fig. 2. Suspension module: MM concept (a); reference component (b)

3.2   Goal and Scope

The *goal* of this study is evaluating the environmental, economic and social performances along the whole LC (production, use, EoL) of the proposed suspension arm. The *scope* is to verify and demonstrate through an integrated approach the benefits of adopting innovative lightweight materials, in comparison with those used on a reference component.

Functional Unit. The Functional Unit (FU) is a suspension arm, connecting the wheel with the suspension system of an electric passenger vehicle, with a life-distance of 150000 km for 10 years. FU has been set in order to allow a further comparison with the reference module.

System Boundaries. System boundaries include all stages of components LC (Fig. 3); *cut-off criteria* are applied in order to exclude those processes that have a negligible influence on the overall sustainability analysis (i.e. transportation). Concerning social data, the analysis is focused on raw material phase as it is generally perceived relevant.

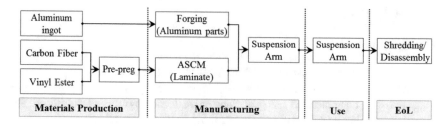

Fig. 3. System boundaries suspension arm (MM)

Modeling. The suspension arm module is modeled by using a breakdown approach; data collection is performed by considering the different sub-modules and mono-material parts involved.

Reference Component. The reference component is composed by an unique metallic part made of forged steel with a total mass of 4 kg (Fig. 2-b).

3.3   Step 1: Life Cycle Inventory

In this paragraph all the data associated to the component LC stages are described for each one of the three pillars of sustainability.

Environmental Data. Data gathering (in terms of materials, energy, waste) is carried out distinguishing *primary data* (direct data, measurements) and *secondary data* (database, literature) (Table 1). The datasets used in the analysis are included in GaBi software.

Table 1. Sources and quality of data for materials and manufacturing

Data	Source	Quality of data	
		Primary	Secondary
Aluminum ingot	GaBi		X
Thermal energy production	GaBi		X
Electricity consumption for ingot casting	[24]		X
Electricity production	GaBi		X
Forging energy consumption	MM	X	
Forging scraps (per 1 kg of forged Al)	Estimated		X
Percentage of Carbon Fiber	MM	X	
Carbon fiber production	GaBi		X
Percentage of Vinyl Ester	MM	X	
Vinyl Ester production	[23]		X
Electricity consumption for pre-preg	[25]		X
Electricity consumption for ASCM process	MM	X	
ASCM scraps	MM	X	X
Aluminum recycled	Estimated		X
Carbon fiber recycled	Estimated		X

In the modelling of use stage the environmental impact due to each module weight is estimated by correlating the variation in car mass between the scenarios of presence and absence of the module to the corresponding variation in energy consumption [12]. The EoL stage is modelled considering: (a) shredding and post-shredding technologies for material recycling or Automotive Shredding Residue (ASR) energy recovery, for innovative solution; (b) shredding and sorting technologies for material recycling, in case of reference suspension arm. Since the commercial database does not provide specific information about technologies involved in vehicle EoL, literature data about energy consumption are assumed.

Economic Data. Data about materials and manufacturing are collected from literature and industry. For the use stage, the EU average cost of energy is assumed.

Social Data. A database called PSILCA (Product Social Impact Life Cycle Assessment) is used for the social analysis [26]. In particular, social indicators and structure are mainly inspired by UNEP/SETAC guidance book. Currently, this database includes 88 qualitative and quantitative indicators classified in 23 subcategories (topics) and 5 stakeholder groups (workers; value chain actors; society; local community; consumers). The indicator assessment is performed according to an ordinal risk scale of 6 different risk levels (no risk; very low risk; low risk; medium risk; high risk; very high risk).

3.4   Step 2: LCA

The evaluation of the environmental impacts is performed by the CML 2001 method and the Primary Energy Demand (PED) indicator is assumed; GaBi software is used for modeling and implementing the LCA analysis. In particular, the three following impact categories are taken into account: (a) Global Warming Potential (GWP), expressed in kg CO_2-eq.; (b) Abiotic Depletion Potential (ADP), expressed in kg Sb-eq.; (c) PED, expressed in MJ. In the following Table 2 and Fig. 4 the LCA results produced by the innovative suspension arm (Inn) in comparison with the reference one (Ref) are reported.

Table 2. Environmental impacts: results of "Inn" in comparison with "Ref"

	GWP (kg CO_2-eq.)		ADP (kg Sb-eq.)		PED (MJ)	
	Ref	Inn	Ref	Inn	Refe	Inn
LCA	3,08E+01	3,22E+01	3,37E–05	1,21E–05	6,98E+02	6,52E+02
Raw mat.	–	2,07E+01	–	1,09E–05	–	4,46E+02
Manufact.	–	6,31E–01	–	–3,70E–08	–	1,52E+01
Product.	5,86E+00	2,13E+01	2,89E–05	1,09E–05	9,16E+01	4,62E+02
Use	2,84E+01	1,10E+01	4,97E–06	1,92E–06	6,41E+02	2,48E+02
EoL	–3,42E+00	–4,49E–02	–1,98E–07	–7,25E–07	–3,46E+01	–5,76E+01

Concerning the global LCA results, the bar graphs in Fig. 4 show that the innovative solution is better than the reference one for only ADP (–64,2%) and PED (–6,6%); on the other hand it produces a GWP increase by 4,7% mainly due to the Production and EoL.

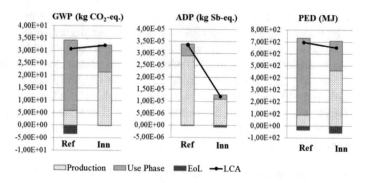

Fig. 4. Impacts assessment results in comparison

However, a separate analysis of all LC stage reveals that the innovative solution involves a smaller impact for the only use stage.

3.5 Step 2: LCC

The results of LCC analysis, carried out by the Eq. 1 (see Sect. 2), is summarized in the following table (Table 3).

Table 3. LCC results for the two alternative solutions

	$C_{Production}$ (€)		C_{Use}(€)	C_{EoL} (€)	LCC (€)
	$C_{Material}$	$C_{Manufacturing}$			
Innovative	23,92	15,13	3,61	−0,22	42,44
Reference	10,68	5,95	8,04	−2,16	22,51

3.6 Step 2: S-LCA

The social assessment is carried out by the PSILCA database selecting the "manufacture of rubber and plastic production" and "manufacture of basic metal" respectively associated to the innovative solution and the reference one. In particular, the values described in Table 4 are calculated by allocating the social results to the two alternative solutions, choosing three indicators among those included in the list of "workers" stakeholder group.

Table 4. Social data for each alternative

	Fair Salary, FS (med risk hours)	Gender Wage gap, GW (med risk hours)	Weekly Hours of work per employee, WH (med risk hours)
Innovative	1,561	0,282	0,535
Reference	0,23	0,09	0,08

3.7 Step 3: Integrated Assessment by TOPSIS

According to TOPSIS method (see Sect. 2), the more sustainable solution is determined considering seven criteria (C1: Total Cost; C2: GWP; C3: ADP; C4: PED; C5: FS; C6: GW; C7: WH) and two alternatives (A1: Innovative solution; A2: Reference solution) as the decision matrix shows in Table 5.

Table 5. The decision matrix [D]

	C1	C2	C3	C4	C5	C6	C7
	(€)	(kg CO_2-eq.)	(kg Sb-eq.)	(MJ)	(med risk h)	(med risk h)	(med risk h)
A1: Inn	42,44	3,22E+01	1,21E–05	6,52E+02	1,5609	0,282	0,535
A2: Ref	22,51	3,08E+01	3,37E–05	6,98E+02	0,23	0,09	0,08

The next steps involve the construction of the normalized decision matrix (Table 6) and the generation of the weighted normalized decision matrix (Table 9). In particular, the weights of criteria $(w_1, \dots, w_7)$ (Table 8) are calculated on the basis of the pairwise matrix (Table 7). It is worth to note that a more detailed analysis shall report the opinion of largest number of experts, in order to reflect a more objective judgment on the relative importance of criteria. For the purpose of this study it is assumed that the forum of experts is composed by the group of authors.

Table 6. The normalized decision matrix [R]

	C1	C2	C3	C4	C5	C6	C7
A1	0,88	0,72	0,18	0,68	0,99	0,96	0,99
A2	0,47	0,69	0,98	0,73	0,14	0,29	0,14

Table 7. The pairwise matrix

	C1	C2	C3	C4	C5	C6	C7
C1	1	2	2	2	2	2	2
C2	1/2	1	2	2	2	2	2
C3	1/2	1/2	1	2	2	2	2
C4	1/2	1/2	1/2	1	2	2	2
C5	1/2	1/2	1/2	1/2	1	2	2
C6	1/2	1/2	1/2	1/2	1/2	1	2
C7	1/2	1/2	1/2	1/2	1/2	1/2	1

"1" represents an equal relative importance, "2" a slight favour of C_i compared to C_j

Table 8. Weights of criteria

	C1	C2	C3	C4	C5	C6	C7
w_i	0,24	0,20	0,16	0,13	0,11	0,09	0,07

Table 9. The weighted normalized decision matrix [V]

	C1	C2	C3	C4	C5	C6	C7
A1	0,21	0,14	0,03	0,09	0,11	0,09	0,07
A2	0,11	0,14	0,16	0,10	0,02	0,03	0,01

Assuming that all the v_{ij} are cost criteria ($j \in J_c$, see Sect. 2), the positive and negative ideal solutions are in Table 10, while the separation measures are in Table 11.

Table 10. The positive (A^*) and negative (A^-) ideal solutions

	C1	C2	C3	C4	C5	C6	C7
A^*	0,11	0,14	0,03	0,09	0,02	0,03	0,01
A^-	0,21	0,14	0,16	0,10	0,11	0,09	0,07

Table 11. Separation measures (S_i^* and S_i^-) of each alternatives

	S_i^*	S_i^-
A1	0,160	0,129
A2	0,129	0,160

Therefore, after calculating the relative closeness index C_i^*, the best sustainable solution is the reference one (Table 12).

Table 12. The relative closeness index (C_i^*) and the preference order

Alternatives	C_i^*	Preference order
A1: Innovative	0,45	2
A2: Reference	0,55	1

4   Discussions and Conclusions

Besides the diffusion of LCA as a supporting tool during design phase in the automotive sector, there is a need of creating an approach to integrate environmental with economic and social aspects to evaluate sustainability of alternative design solutions. Starting from the current literature and available standards and guidelines for LCA, LCC and S-LCA, this paper would contribute proposing a concrete application of the TOPSIS method for the integration of LCA, LCC and S-LCA results. Strengths and weakness of such method are evaluated by means of a case study concerning two different lightweight designs for EV component, in particular for the suspension arm. The results demonstrate that the reference solution is preferable with respect to the innovative one. In particular, LCA results suggest that a delicate trade-off between production and use stage impacts exists and it needs to be carefully handled. Moreover, the final result is affected by the EoL scenario; this means that more advanced technologies mainly for post-shredding treatments and materials recycling could provide better environmental performances for the innovative solution. Concerning economic issue, nowadays the composite solution is more expensive to produce than the reference one, even if innovative materials are lighter and comparable in terms of performances. The social issues show that the innovative material provides higher risk of impacts and this is mainly ascribable to the supply chain as well as the high unit cost of the material. TOPSIS was found a viable method to integrate heterogeneous results but actually it presents critical issues due to the weighted criteria matrix that is built in a subjective manner. In particular, it was observed that increasing the weight for each criteria and the gap between the criteria values for each one of the alternatives, the final result significantly changes. Further research would necessarily regard improvements in terms of results integration and interpretation; yet, stakeholder engagement techniques (i.e. surveys) could be used to evaluate weight of criteria thus improving the consistency of indicators relevance.

Acknowledgements. The presented work was funded by the European Commission within the project ENLIGHT (Grant agreement No: 314567): www.project-enlight.eu. The authors, as partners of the project, wish to thank all ENLIGHT partners for their contribution, particularly Fabio Pulina from Magneti Marelli.

References

1. ITF Executive summary, in ITF Transport Outlook 2015. OECD Publishing, Paris (2015). doi:10.1787/9789282107782-3-en
2. Berzi, L., Delogu, M., Giorgetti, A., Pierini, M.: On-field investigation and process modelling of End-of-Life Vehicles treatment in the context of Italian craft-type Authorized Treatment Facilities. Waste Manag. **33**(4), 892–906 (2013)
3. Solomon, S., Qin, D., Manning, M., Chen, Z., Marquis, M., Averyt, K.B., Tignor, M., Miller, H.L.: Climate Change 2007: The Physical Science Basis. Contribution of Working Group I to the Fourth Assessment Report of the Intergovernmental Panel on Climate Change. Cambridge University Press, Cambridge (2007)
4. Delogu, M., Del Pero, F., Pierini, M.: Lightweight design solutions in the automotive field: environmental modelling based on fuel reduction value applied to diesel turbocharged vehicles. Sustainability **8**, 1167 (2016). doi:10.3390/su8111167
5. Kelly, J.C., Sullivan, J.L., Burnham, A., Elgowainy, A.: Impacts of vehicle weight reduction via material substitution on life-cycle greenhouse gas emissions. Environ. Sci. Technol. **49**, 12535–12542 (2015)
6. Delogu, M., Del Pero, F., Romoli, F., Pierini, M.: Life cycle assessment of a plastic air intake manifold. Int. J. Life Cycle Assess. **20**, 1429–1443 (2015)
7. Delogu, M., Zanchi, L., Maltese, S., Bonoli, A., Pierini, M.: Environmental and economic life cycle assessment of a lightweight solution for an automotive component: a comparison between talc-filled and hollow glass microspheres-reinforced polymer composites. J. Clean. Prod. **139**, 548–560 (2016)
8. Modaresi, R., Pauliuk, S., Løvik, A.N., Muller, D.B.: Global carbon benefits of material substitution in passenger cars until 2050 and the impact on the Steel and Aluminum industries. Environ. Sci. Technol. **48**, 10776–10784 (2014)

9. Delogu, M., Del Pero, F., Pierini, M., Bonaffini, D.: End-of-Life in the railway sector: analysis of recyclability and recoverability for different vehicle case studies. Waste Manage. (2016) http://dx.doi.org/10.1016/j.wasman.2016.09.034

10. Dhingra, R., Das, S.: Life Cycle energy and environmental evaluation of downsized vs. lightweight material automotive engines. J. Cleaner Prod. **85**, 347–358 (2014)

11. Zanchi, L., Delogu, M., Zamagni, A., Pierini, M.: Analysis of the main elements affecting social LCA applications: challenges for the automotive sector. Int. J. Life Cycle Assess. (2016). doi:10.1007/s11367-016-1176-8

12. Zanchi, L., Delogu, M., Ierides, M., Vasiliadis, H.: Life cycle assessment and life cycle costing as supporting tools for EVs lightweight design. Smart Innov. Syst. Technol. **52**, 335–348 (2016)

13. Onat, N.C., Gumus, S., Kucukvar, M., Tatari, O.: Application of the TOPSIS and intuitionistic fuzzy set approaches for ranking the life cycle sustainability performance of alternative vehicle technologies. Sustain. Prod. Consum. **6**, 12–25 (2016)

14. Doukas, H., Karakosta, C., Psarras, J.: Computing with words to assess the sustainability of renewable energy options. Expert Syst. Appl. **37**, 5491–5497 (2010)

15. Streimikiene, D., Balezentis, T., Krisciukaitiené, I., Balezentis, A.: Prioritizing sustainable electricity production technologies: MCDM approach. Renew. Sustain. Energy Rev. **16**, 3302–3311 (2012)

16. Dattilo, C.A., Delogu, M., Berzi, L., Pierini, M.: A sustainability analysis for Electric Vehicles Batteries including ageing phenomena. In: 16th International Conference on Environment and Electrical Engineering (EEEIC). IEEE (2016)

17. Swarr, T.E., Hunkeler, D., Klöpffer, W., et al.: Environmental life-cycle costing: a code of practice. Int. J. Life Cycle Assess. **16**, 389–391 (2011). doi:10.1007/s11367-011-0287-5

18. Witik, R.A., Payet, J., Michaud, V., Ludwig, C., Manson, J.A.E.: Assessing the life cycle costs and environmental performance of lightweight materials in automobile applications. Composites **42**, 1694–1709 (2011)

19. Kim, H.J., Keoleian, G.A., Skerlos, S.J.: Economic assessment of greenhouse gas emissions reduction by vehicle lightweighting using aluminium and high strenght steel. J. Ind. Ecol. **5**, 64–80 (2010)

20. Koch, N., Fuss, S., Grosjean, G., Edenhofer, O.: Causes of the EU ETS price drop: Recession, CDM, renewable policies or a bit of everything?—New evidence. Energ. Policy **73**, 676–685 (2014). doi:10.1016/j.enpol.2014.06.024

21. UNEP/SETAC: Guidelines for Social Life Cycle Assessment of Products (2009). http://www.unep.org/pdf/DTIE_PDFS/DTIx1164xPAguidelines_sLCA.pdf. Accessed 9 Nov 2015

22. Saaty, T.L.: Decision making with the analytic hierarchy process. Int. J. Serv. Sci. (2008). doi:10.1504/IJSSci.2008.01759

23. Roos, S., Szpieg, M.: Life cycle assessment of Z-Bee. Project report, Swerea IVF (2012)

24. U.S. Department of Energy: Energy Efficiency and Renewable energy U.S. Energy Requirements for Aluminum Production – Historical Prospective, Theoretical Limits and Current Practices (2007)

25. Suzuki, T., Takahashi, J.: Prediction of energy intensity of carbon fiber reinforced plastics for mass-produced passenger cars. In: 9th International SAMPE Symposium (2005)

26. Ciroth, A., Eisfeldt, F.: PSILCA – A Product Social Impact Life Cycle. Assessment database. Database version 1.0. Documentation Version 1.1 (2016)

A Monitoring and Data Analysis System to Achieve Zero-Defects Manufacturing in Highly Regulated Industries

Theocharis Alexopoulos[✉] and Michael Packianather

School of Engineering, Cardiff University, Queen's Buildings, The Parade, Cardiff CF24 3AA, UK
alexopoulost@cardiff.ac.uk, packianatherms@cf.ac.uk

Abstract. In order to become more competitive, manufacturing companies exploit new technologies and practices that can improve their production efficiency, and reduce the number of rejected products. This work is about a Monitoring and Data Analysis System (MDAS), a software system that combines data mining, neural networks modelling and graphical data analysis to assist the company in identifying patterns, trends or problems that increase the risk of rejected products. A pilot version of the proposed system is tested on two production lines of a pharmaceutical company and has identified previously unknown patterns and trends that were hindering the quality of the end product. Since the operation of the proposed system does not affect the production it is suitable for industries bound by strict regulation. In general, the proposed system could be adopted for other products and industries.

Keywords: Data mining · Big Data analytics · Zero-defects manufacturing · Neural networks · Business intelligence · Knowledge engineering

1 Introduction

The idea of zero-defects manufacturing is a natural outcome of the costs that a defective product induces to the manufacturer. Since the 1960s, guidelines and methodologies are being published to reduce the defective parts and therefore to make the manufacturing process more efficient [1]. On the other hand, manufacturing facilities are often the source of many environmental and ecological impacts, making them the focus of sustainability-related research, reports, and legislation [2]. As a result, zero-defects manufacturing is still a topic with increased interest and many new technologies are being utilised in order to minimise rejected products.

The strategic initiative called "Industry 4.0" [3] is promoting the digitalisation of traditional industries in order to create intelligent factories. This initiative has intensified the trend of monitoring the manufacturing processes and therefore huge amounts of data are becoming available. Big Data concept is also growing because of the numerous sensors that are being used [4] so with advanced data analytics the manufacturing sector can achieve zero-defects productions.

© Springer International Publishing AG 2017
G. Campana et al. (eds.), *Sustainable Design and Manufacturing 2017*, Smart Innovation, Systems and Technologies 68, DOI 10.1007/978-3-319-57078-5_30

At the moment, the companies are analysing only part of the available manufacturing process related data, typically to safeguard quality and to prevent major disruptions in production. This ensures that the products comply with the required standards but does not ensure a zero-defect production. Every rejected product is a defect waste as identified by Taiichi Ohno in the Toyota Production System and therefore contributes to unnecessary costs [5]. From a sustainability point of view, defect products consume raw materials, resources and finally create unnecessary environmental impact.

This work reported in this publication is about combining modern concepts and data mining techniques in order to develop a data management system that assists in spotting patterns or disturbances in the production of a highly-regulated manufacturing facility. The aim was to keep end-products' parameter values within the limits set by the regulations while increasing the level of knowledge about the manufacturing process itself and increasing the level of accuracy of the production control.

2 Literature Review

Big Data is characterised by the 3Vs theory: Volume, Variety, and Velocity [6]. In modern manufacturing facilities, a high number of sensors are used to collect large amounts of real-time data related to the state of the resources, the state of the product, and other factors that affect the product, the process or the reliability of resources. This trend supported by Industry 4.0 initiative and Internet of Things [4] has accelerated the process of integrating Big Data methodologies into business intelligence. Consultancy companies like Gartner have prepared guidelines for the implementation of business intelligence process [7]. Auschitzky et al. [8] present an in-depth analysis of how to utilize Big Data mining and advanced analytics to make rational manufacturing decisions. Chen et al. [9] on their review present the commonly accepted methodology to process data and provide a detailed roadmap for data mining techniques and methodologies applicable to Internet of Things (IoT) applications. Che et al. [10] are examining the challenges that the big amounts of data present and discuss cloud based solutions that can deliver high processing power. Recently, there have been many publications that discuss the changes that data availability and data mining bring to industry. Senthilkumaran et al. [11] examine applications for pharmaceutical industry, Amiri M. et al. [12] for construction industry, Wang et al. [13] for sustainable cement production, Ariyawansa et al. [14] for airports and flight industry, Rostami et al. [15] for manufacturing quality control and many more examples of publications that analyse how IoT, Big Data and data mining can increase business intelligence, autonomy and therefore efficiency and sustainability in every sector.

The exploitation of data mining tools in manufacturing sector has led to the development of many decision support systems, architectures and methodologies that rely on this practice. Benane and Yacout [16] used classification techniques to improve the accuracy of condition based maintenance. Munoz [17] is proposing two methods to analyse the data available from testing the raw materials so variation in the quality of the products could be reduced. Zhang et al. [18] developed an architecture for data analysis within and outside the shopfloor covering the whole product lifecycle while

Fysikopoulos et al. [19] used monitoring data to adapt the production line simulation models and propose the optimum production setup. Khan et al. [20] propose a methodology based on comparison of probabilities and genetic algorithm to increase the efficiency of quality control in a manufacturing facility.

In highly regulated industries, embedding new practices is significantly more difficult as the legislation has to be amended by the regulators before new technologies are utilised. Therefore, the effort in the current work is to apply the knowledge available in literature without introducing risks related to production line compliance requirements.

3 A Monitoring and Data Analysis System

The proposed Monitoring and Data Analysis System (MDAS) has been developed with the following principles in mind:

- Minimum user input
  The system has been developed for industrial users in mind so what is key for its success is to minimise the additional man-hours needed to train the users. This is achieved by creating interfaces that automatically detect the format of data and clean raw data so it can be used in the next processing steps. In addition, many graphs are produced without the need of specific parameter selection or manual manipulation of the dataset.
- Modularity
  Although there is one graphical user interface to control the system, different modules are responsible for the different stages of data processing. This adds flexibility to the system as it can quickly include new modules or interfaces to adapt to the unique characteristics of a company. Moreover, maintenance and updating the system is easier and there is low risk of major failures as problems will affect only the module concerned and not the whole system.
- Expandability
  The proposed system is currently in pilot usage and can produce a series of intuitive graphs and information. However, depending on the needs of the user, new types of graphs can be added just by adding a micro-module, responsible for the graph. In addition, new data input interfaces can be embedded to the system so it can directly read data by new types of sensors.

The proposed system was developed using MATLAB programming language. This language has been selected for its popularity among the engineers who contributed in the development of the modules and because of the interfaces that MATLAB offers for connectivity with various data sources. The input data used for testing and verification purposes was supplied in .csv and text files as the system of this work is not initially developed with signal processing capabilities. If such capability is needed, then it can be implemented as an additional data input module. The current version of the system does not support dynamic changes in data input so real-time changes in the already received data are discarded.

3.1   Data Preparation

Input data should be correlated to an identification number of a product or batch of products. In highly regulated industries, it is common practice to have tracking systems in place for each product or batch so that when a production fault occurs, the affected products can be quickly identified and be examined or immediately rejected. This work requires that the company has a tracking system in place. The pilot version of the tool depends on the tracking provided by RFID tags placed on each batch of products.

Step 1 - Data loading and cleaning. Initially, the sources of data should be specified. Each source may have some corrupted data and for this reason a first check for data format issues or missing data should be performed. Once the errors are identified the user would be informed. The module that reads the data would be "expecting" that each data file includes the product ID at the same line with the examined parameter. Suppose if a problem occurs with the values of this line, it discards the line and moves to the next one.

Step 2 - Verification. There are plotting functionalities provided to the user in order to ensure that there are no problems with the input data. These include simple plots of input data, plots of average, maximum, minimum values and raw input data plotting on MATLAB's or operating system's command window.

As soon as a new source of input data is provided, the system will generate a scatter plot depicting how well the input parameters are describing the output ones. Each point of the graph would show the linear difference between the output values of two product IDs d_{output} versus the linear difference of their input parameters d_{input}. Similarity would be specified by the user, who should set the weight of each input parameter. The following example is for 2 input parameters x_1, x_2 and one output y_1. The user would also define weights of importance for these parameters, w_1, w_2 respectively. For product IDs A and B the differences will be calculated as follows:

$$d_{input} = w_1\left(x_{1A} - x_{1B}\right) + w_2\left(x_{2A} - x_{2B}\right) \tag{1}$$

$$d_{output} = y_{1A} - y_{1B} \tag{2}$$

Initially, an exhaustive algorithm was developed identifying the 100 closest input data sets, but in a Big Data scenario, this algorithm would be too slow. Application of heuristics or genetic algorithms can solve this problem but such functionality has not been implemented in the proposed version.

In Fig. 1 left, the graph shows that the smaller the difference between the input parameters (abscissa) the smaller the difference of output (ordinate). This shows that the input parameters sufficiently describe the output. If a key parameter has not been taken into consideration, then 2 product IDs with the same input would have different output. Figure 1 right is an example of missing key input parameters as product IDs with small differences in input have a completely random output. In the latter case, the user of the system should identify more sources of data until the graph resembles the one shown on the left.

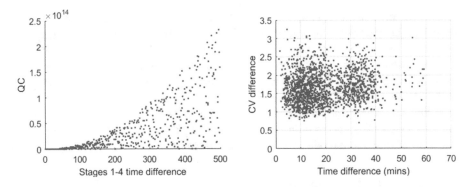

Fig. 1. Difference of output parameters vs. difference of input parameters (On the left input parameters can describe output. On the right, similar input parameters produce random output).

3.2   Data Analysis

After cleaning the input data, the system can process this data to produce more graphs or create the input to feed models that will identify trends or patterns. The graphs that the user can display are based on traditional statistics, including the differential of a parameter and pseudo-colour checkerboard plots that provide an excellent way of understanding changes in values without the need to read the value.

In cases where data is considered as Big Data, the traditional plotting methods would not be sufficient to create the previous plots. The amount of memory and time that MATLAB needs to produce such graphs would cost the company many man-hours as the user would wait minutes or hours in order to view each graph. To solve this problem, the input data is used to train a neural network and every new batch of data that becomes available would be used to retrain this network. The previous graphs could be produced by curve fitting of the neural network and in this case there would be low requirements for memory and computing power. In the proposed version, a predefined type of neural network has been used. The training algorithm is Bayesian Regularization with 25 neurons and 1 hidden layer. The number of input parameters depend on the case but the output is always one. For a different output parameter, a second neural network should be trained.

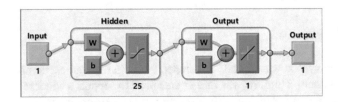

Fig. 2.   The default neural network for 1 input parameter (example)

The data analysis part of the system is a toolkit offering an extensive list of graphs. As this system is not a commercialised piece of software, the list of graphs have been customised in each case to provide the company with the needed graphs and not with an endless list that would be practically difficult to use. In the two cases that the system was tested, the pre-installed data processing functions used the same engine but different representations. As an example, on the first case, graphs showing the behaviour of the manufacturing process over time were more important. So, the custom graphs were based on how similar input was producing different outputs over time. In the second test case, the focus was on the patterns of output parameter based on characteristics that were not affected by time. As the proposed system is at a pilot state, it is expected that more graph producing micro modules will be added in the future.

3.3 Support for Production Autonomy

The trained neural network can be used as a model for self-adaptation of the production depending on the characteristics of the already produced products. The neural network in this case is used as an objective function for an optimization algorithm. This function could be updated every time when new monitoring data from the production line becomes available and it could be used by the production management system to set the optimum parameters on each resource.

In principle, the neural network should be retrained every time the production line uses new raw materials or consumables or after maintenance of machinery. This will ensure that the curves fitted accurately represent the current production and not continue showing results heavily based on past batches with different raw materials or resource characteristics. Finally, the neural network would be an accurate simulation model of the production itself so it can be used in production virtualization and cloud manufacturing services.

In future, as an expansion of the proposed system, a self-organisation map will be created from the provided input data. The map will be used to further automate the process of understanding the data by the system itself. Identification of correlations that currently depend on the studied case, will be performed automatically as parts of the map will be directly correlated to types of input data. In practice, this will assist the company in quickly identifying the characteristics of raw materials related to the product or categorising the semi-manufactured products based on their probability of being rejected down the line and avoiding unnecessary costs. How the proposed system could benefit a company in practice is demonstrated in Sect. 4.

4 Findings

The proposed system was initially developed for a pharmaceutical company to look at a specific production line and was further enhanced to support a wider range of production lines. Because of confidentiality restrictions the labels on diagrams' axis have been changed.

In the first pilot case, simple graphs of product quality versus product ID for a batch of products, ordered by manufacturing date, showed that the output parameter was proportional to the length of time since the beginning of the process. The same pattern was repeated after a new batch entered the line. The system prints the differential of the same graph and this revealed an additional pattern showing that the value increased rapidly every 9 products as shown in Fig. 3. The company was suspecting the time dependence pattern but was unaware of the 9 products pattern. Before the introduction of the proposed system, data was cleaned manually and the whole process of data mining was dependant on the judgement/intuition of the person doing it.

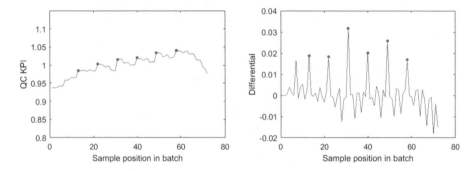

Fig. 3. Left: Output parameter value average vs. position of product in batch. Right: Differential vs. position of product in batch. The dots show the "non-obvious" pattern occurring every 9 products.

After identifying the trends, the analysis focused on identifying the parameters that cause or amplify these trends. Traditional methods of statistics (least squares method,

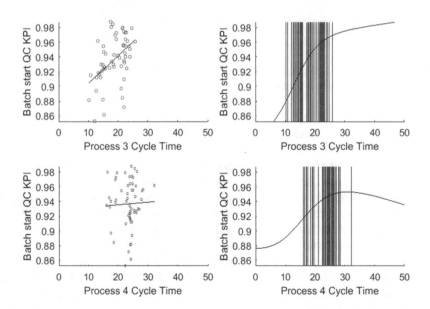

Fig. 4. Left: Least squares line of samples (unreliable). Right: Signals extracted by the neural networks.

standard deviation etc.) were used initially but they were unable to produce reliable results. The non-linear curve fitting by neural network that is provided by the proposed system, was used to calculate the output parameter value while keeping all but one input parameters constant. The constants' values are the average of all samples. After identifying the trends that the neural network model indicated, minor changes to the production were made and a verification using the new production data was performed.

Figure 4 shows the two most significant correlations that were identified between each input parameter and the output parameter. The system, produced the diagrams and turned the focus on the correct parameters of interest. The parameter presented on the top two graphs was verified to strongly affect the uptrend of the examined output parameter of Fig. 4, left. The scatter plots of Fig. 4 show that a simple least squares fitting curve was not reliable enough to be used when compared to the trends produced by the neural network.

In Fig. 5 the result of the new setup based on the neural network model findings is presented.

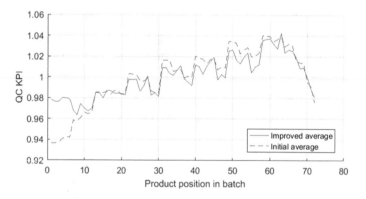

Fig. 5. Output parameter vs. position of product in batch. The dashed line is the initial average. The solid line is the improved average.

The new graph shows that the output parameter of the amended production setup has been significantly increased for the first products of the batch and it is much closer to value 1 that is the target value for the company. After this change in the output parameter the risk of rejected products was significantly reduced as the quality of the end product was within the regulation limits and significantly closer to the optimal.

The second parameter that was suggested by the neural network signal (Fig. 4 bottom graphs) was also examined. The least squares curve gives a first indication that there is not a strong correlation but as in the first finding, the deviation of output parameter values for this parameter is high and makes the least squares curve unreliable. After changing the production setup, there was a minor improvement of the end-product quality but this improvement was within the statistical error limits and cannot be considered as success due to changing the production setup. However, the achievement in this case is that the setup proposed by the system did not have a negative impact on the production which is significant as the system suggestions did not put the product quality into risk. The reliability factor in systems that feed autonomous production lines is very important as high reliability ensures that the production line will remain stable with constant product characteristics.

Finally, apart from the direct impact to the product quality, the developed system provided one more important graph that assists in understanding the production line characteristics and in simulating production changes. Figure 6 shows the range of values that an output parameter can take based on a defined range of values for each input parameter. In order to produce this graph, the system relies on the user to specify the range of permitted values for each input parameter. Then, by using the trained neural network, all combinations of input values are calculated and the output is shown in a graph. If all input parameters would be shown graphically, then a graph of n+1 dimensions would be needed where n is the number of input parameters. As this is not possible for most of applications, the graphs are typically 2 dimensional with the most significant input parameter on the abscissa.

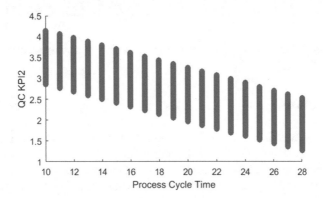

Fig. 6. Diagram showing the range of possible values of an output parameter vs. one input parameter (typically the one with highest impact)

The proposed system was used to analyse data belonging to another production line of the same company. In the second case, the products were packed together and were going through the quality control. In this case, the checkboard graphs were used as this type of graph resembled how the products were packed together. Although the quality control was based on the average of all the measured value for each product comprising the pack the system demonstrated that the value of each product had a strong correlation to the position in the pack.

This finding was an outcome of a good visualization technique that the proposed system supports but in the current version there is no automated method for identification of such patterns. However, visualization assists the production manager to quickly understand and interpret the presented data and in the case of Fig. 7 the company quickly identified that the product was manufactured with no variation but sample handling and measuring methods used in quality control, were sensitive to the position of the product in the pack.

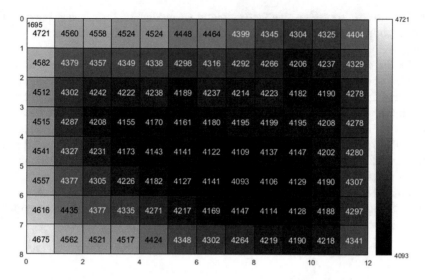

Fig. 7. Measured values for each product in the pack represented with a checkboard graph (bright high, dark low)

5   Conclusions

The exponential growth of data availability in manufacturing lines has created opportunities and challenges. The opportunity that every company has is to increase their whole knowledge about its manufacturing process and therefore become more efficient and competitive. The challenge is to invest in technologies and practices that will not disrupt the company's production but will produce high return on investment. The proposed Monitoring and Data Analysis System (MDAS) is providing a manufacturer with quick access to new production knowledge and with its modular architecture can quickly adapt to the needs of a manufacturing facility. The system provides the managers of highly regulated facilities a combination of graphical representations and neural network models to assist them in spotting trends, patterns and problems in order to achieve near zero-defect situation before the products exceed the acceptable tolerances set by the regulations and get rejected. Application of the proposed system has been applied in two different lines of a pharmaceutical company which demonstrated its capabilities in assisting the company in identifying problems in the production that were the cause for their rejected products. Because of the versatility and the expandability of the system's architecture, applications in other industries with more complex data are feasible and will contribute into enhancing the functionality of the proposed system and its usability in industry at large.

References

1. Assistant Secretary of Defense (Manpower Installations and Logistics) Washington DC: Guide To Zero Defects. Quality and Reliability Assurance Handbook (1965)

2. Garretson, I.C., Mani, M., Leong, S., Lyons, K.W., Haapala, K.R.: Terminology to support manufacturing process characterization and assessment for sustainable production. J. Clean. Prod. **139**, 986–1000 (2016)
3. National Academy of Science and Engineering: Recommendations for implementing the strategic initiative "INDUSTRIE 4.0" (2013)
4. Mourtzis, D., Vlachou, E., Milas, N.: Industrial Big Data as a result of IoT adoption in manufacturing. In: 5th CIRP Global Web Conference Research and Innovation for Future Production (2016)
5. Ohno, T.: Toyota Production System: Beyond Large-Scale Production. Productivity Press, Cambridge (1988)
6. Laney, D.: 3D Data Management: Controlling Data Volume, Velocity and Variety. Application Delivery Strategies. METAGroup (2001)
7. Buytendijk, F., Oestreich, T.W.: Organizing for Big Data Through Better Process and Governance. Gartner Publications, Stamford (2016)
8. Auschitzky, E., Hammer, M., Rajagopaul, A.: How Big Data can improve manufacturing (2014). http://www.mckinsey.com/business-functions/operations/our-insights/how-big-data-can-improve-manufacturing
9. Chen, F., Deng, P., Wan, J., Zhang, D., Vasilakos, A.V., Rong, X.: Data mining for the internet of things: literature review and challenges. Int. J. Distrib. Sens. Netw. **11**(8) (2015). doi: 10.1155/2015/431047
10. Che, D., Safran, M., Peng, Z.: From Big Data to Big Data Mining: challenges, issues, and opportunities. In: International Conference on Database Systems for Advanced Applications (2013)
11. Senthilkumaran, U., Manikandan, N., Senthilkumar, M.: Role of data mining on pharmaceutical industry-a survey. Int. J. Pharm. Technol. **8**(3), 16100–16106 (2016)
12. Amiri, M., Ardeshir, A., Zarandi, M.H.F., Soltanaghaei, E.: Pattern extraction for high-risk accidents in the construction industry: a data-mining approach. Int. J. Inj. Contr. Saf. Promot. **23**(3), 264–276 (2016)
13. Wang, Y., Shao, Y., Matovic, M.D., Whalen, J.K.: Recycling combustion ash for sustainable cement production: a critical review with data-mining and time-series predictive models. Constr. Build. Mater. **123**, 673–689 (2016)
14. Ariyawansa, C.M., Aponso, A.C.: Review on state of art data mining and machine learning techniques for intelligent Airport systems. In: International Conference on Information Management (2016)
15. Rostami, H., Dantan, J.-Y., Homri, L.: Review of data mining applications for quality assessment in manufacturing industry: support vector machines. Int. J. Metrol. Qual. Eng. **6**(4), 1–59 (2015). Article 401. doi:10.1051/ijmqe/2015023
16. Bennane, A., Yacout, S.: LAD-CBM; new data processing tool for diagnosis and prognosis in condition-based maintenance. J. Intell. Manuf. **23**(2), 265–275 (2012)
17. Garcia-Munoz, S.: Two novel methods to analyze the combined effect of multiple raw-materials and processing conditions on the product's final attributes: JRPLS and TPLS. Chemometrics and Intelligent Laboratory Systems (2014)
18. Zhang, Y., Ren, S., Liub, Y., Si, S.: A Big Data analytics architecture for cleaner manufacturing and maintenance processes of complex products. J. Cleaner Prod. **142**, 626–641 (2016). Part 2. doi:10.1016/j.jclepro.2016.07.123
19. Fysikopoulos, A., Alexopoulos, T., Pastras, G., Stavropoulos, P., Chryssolouris, G.: On the design of a sustainable production line: the MetaCAM tool. In: ASME International Mechanical Engineering Congress and Exposition (2015)
20. Khan, A.R., Schiøler, H., Knudsen, T., Kulahci, M.: Statistical data mining for efficient quality control in manufacturing. In: 20th IEEE International Conference on Emerging Technologies and Factory Automation (2015)

A Combination of Life Cycle Assessment and Knowledge Based Engineering to Evaluate the Sustainability of Industrial Products

Giampaolo Campana[1(✉)], Mattia Mele[1], and Barbara Cimatti[2]

[1] Department of Industrial Engineering, University of Bologna, Viale del Risorgimento 2, 40136 Bologna, Italy
{giampaolo.campana,mattia.mele3}@unibo.it
[2] Institute of Advanced Studies, University of Bologna, Via P. Miliani 7/3, 40132 Bologna, Italy
barbara.cimatti2@unibo.it

Abstract. The Life Cycle Thinking and the Knowledge Based Engineering approaches can be integrated in order to allow a preliminary-rough but effective Life Cycle Assessment, since the first phases of the design of an industrial product. In the present paper, we propose a general framework considering a number of different aspects, mainly concerning the manufacturing choices, but also related to the design of the product. The aim is to overcome the *eco design paradox* and to provide a tool supporting designer during the product concept to increase its sustainability.

Keywords: Life Cycle Assessment · Knowledge Based Engineering · Manufacturing · Product design

1 Introduction

One of the most important issues concerning design and manufacturing of industrial products is sustainability. Efforts to make manufacturing more sustainable must consider issues at all relevant levels – product, process, and system and not just one or more of these in isolation [1].

Sustainable manufacturing implies the creation of products that use processes that minimize negative environmental impacts, conserve energy and natural resources, are safe for employees, communities, and consumers and are economically sound [2]. *Concurrent Engineering* can be considered the present most popular approach to effectively design and fabricate a product. Appropriate sustainable production processes must be defined since the first design phase in order to optimize the whole development of the product and to achieve the sustainability goal.

The concept of the Triple Bottom Line has defined three dimensions of sustainability: Environment, Economy, Society [3] and several methods and techniques to increase sustainability in industrial design and production have been introduced.

© Springer International Publishing AG 2017
G. Campana et al. (eds.), *Sustainable Design and Manufacturing 2017*, Smart Innovation, Systems and Technologies 68, DOI 10.1007/978-3-319-57078-5_31

Life Cycle Assessment (LCA) has been identified as the most efficient technique to assess the impact of industrial products in particular from the environmental point of view [4]. Other tools such as the Life Cycle Costing (LCC) and the Social Life Cycle Assessment (SLCA) respectively focus the Economic and Social dimensions of the Sustainability of industrial products.

Nowadays a Life Cycle Thinking (LCT) approach is essential to become aware of how everyday life (consuming products, engaging in activities and so on) affects the environment. LCT implies a holistic view of products and activities systems. For designers, it is a fundamental to adopt this perspective in order to design new sustainable effective products

LCA can be realized through different methods and techniques but the International Organization for Standardization [5] has established four main phases, which are: goal and the scope definition, Life Cycle Inventory (LCI), Lifecycle Impact Assessment (LCIA) and interpretation of the result. In particular, LCI and LCIA are activities requiring a high amount of accurately documented information [6]; therefore, specific datasets and software are often employed [7]. Many of the required information are acquired during the development of the product and LCA is commonly employed as a support tool to make decisions at the end of the design phase [8] because it allows considering all the useful data. As data are needed before and to carry out the LCA, the so-called *eco-design paradox* occurs [9]. This paradox is due to the relation between the product knowledge and the possible environmental improvement of the product itself and of its manufacturing processes. In fact, when the knowledge of the product is sufficient to define the huge amount of information necessary for a reliable analysis of its environmental impact, the design and the production processes have already been defined and the implementation of changes is not convenient anymore.

The possibility for designers and engineers to carry out a preliminary-rough LCA in the first steps of the design process allows them to increase the sustainability of the product. A number of studies have been developed to forecast the LCIA result for a new product when only uncertain data are available. In most cases, the proposed solutions employ heuristic methods based on previous datasets to compute a forecast of the environmental impact. Examples of this kind using Artificial Neural Networks (ANNs) are shown in [10, 11].

A Knowledge Based Engineering (KBE) system enables the extraction and the representation of several characteristics of the product, also related to design, manufacturing and other phases of the product life cycle, e.g. [12]. The use of KBE to provide an approximate LCA has been proposed in [13] by the development of a collaborative environment: product features are used as input for ANNs in order to assess the environmental impact of new products.

A general framework for the integration of a preliminary LCA analysis into a Knowledge Based Engineering System (KBES) to design industrial products is proposed in the present paper. The theoretical approach will be described in the next section. The proposed method has been applied developing a software tool that can assist designers and engineers in the industrialisation phase of industrial products. In particular, the application has been tested for plastic bottles.

2   A Knowledge Based Approach to Forecast Sustainability

2.1   General Operations Sequence

The definition of a Knowledge Based Engineering System (KBES) is not unambiguous, in fact a number of methodological approaches have been purposed in the scientific literature [14]. A general sequence of macro operations, which in many cases are subdivided in internal steps, can be summarised and it is described in Fig. 1. The first step is a proper definition of the aim, which is affected by several factors such as the required details, the time and the money necessary for coding, the number of planned future variants in the design and so on.

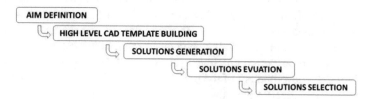

Fig. 1. General sequence of operations in the proposed KBE framework

In this context, a KBES will be described by the use of a High Level Computer Aided Design Template (HLCADT) [15]. A HLCADT can ensure that the purposed solutions fulfil all the constraints that are related to design and manufacturing issues for the product feasibility. In order to reach this aim, the system has to integrate some knowledge based data concerning the specificities of the environment in which the design process is developed. Nevertheless, the main aspects here exposed can be applied to KBESs for a general approach and thus the generation of a CAD model is not necessary if other representations of the product are provided.

Each solution generated by the KBES is defined by a set of parameters describing its properties from different points of view such as the design, the manufacturing technology and processes. The highest the number of parameters is the more precise the result and the more helpful the tool for the decision-making process will be. Therefore, KBES architecture must be designed to provide the needed data, in terms of quality and quantity, for a reliable LCA that requires a description of the target solution as much complete as possible in order to produce significant results. According with this consideration, the definition of the scope within the LCA analysis [5] should be part of the first phase of the system development, Fig. 1.

The goal of the present study is a comparison among different possible solutions in terms of design and manufacturing process; a first choice regards the boundaries of the life cycle study, which, for example, can be cradle-to-grave or cradle-to-gate.

2.2   The LCIA Framework

The definition of boundaries directly affects the architecture of the lifecycle framework that has to be implemented in the KBES. Even if a LCA should consider all the different steps of the product life cycle [5], just a part of the productive system can be considered to compare solutions if the rest of the chain is assumed to be independent from design variables. Therefore, the proposed approach can be divided in two steps: a global description of the life cycle of the product and an individuation of the aspects that are affected by design and manufacturing parameters, which are managed by the KBES.

This framework requires a high flexibility in order to be adaptable to every choice made by the user-designer-engineer. Independently, some variables must be provided as "*entry points*" for the parameters of the KBES. One possible solution is the generation of a production framework that includes all the possible flows and processes by assigning to each of them a specific weight in the LCI.

In the proposed framework, let k be the index for one of the m flows of the production system (e.g. energy), which can be supplied by using n_k alternative processes indexed i_k (e.g. electrical power). It is possible to assume the generic j-th flow (input or output) of a production system as q_j and to introduce Eq. (1):

$$q_j = \sum_{k=1}^{m} q_k \left(\sum_{ik=1}^{nk} a_{j,ik} w_{ik} \right) \tag{1}$$

where q_k is the output flow for the process k (normalized to its reference flow); $a_{j,ik}$ is the reference amount of the flow j-th for the process i_k-th, as available in reference literature (e.g. [7]). The process i_k-th satisfies only a fraction w_{ik} of the task; a combination of processes is used under the condition in Eq. (2):

$$\sum_{ik=1}^{nk} w_{ik} = 1; w_{ik} \in [0, 1] \tag{2}$$

The w_{ik} factors were adopted to describe the combination of the manufacturing processes involved in a specific production system; the necessary amount of flow q_k depends on a specific product. Both w_{ik} and q_k can be extracted by the representation of the production environment through the KBES.

2.3 Parameter Translation

As mentioned, different architectures of KBES can be proposed depending on the specific aims of the application. According to the purpose of the KBES, the input parameters, as they were inserted by the user-designer-engineer, can be of different kind. As a consequence, they have to be converted into a knowledge based representation to be efficiently managed by an inference engine. The inference engine provides the descriptive parameters of the actual solution. These parameters are processed a second together with the information coming from the LC framework described in Sect. 2.2. This phase provides necessary LCI parameters.

The output must be descriptive and generally has to provide a quick overview about the main features of the considered set of solutions and thus it facilitates the decision-making process. In most cases, a translation of the output parameters coming out from the first steps is required to perform the LCI, as schematically displayed in Fig. 2.

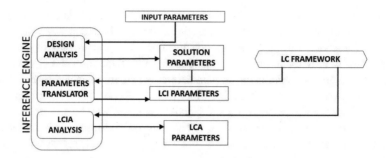

Fig. 2. Parameter management in the integrated LCT-KBES

The translation phase consists of several steps, including the use of formulas to derive data, unit conversion and others. The KBES has to include any possible scenario, applying all the correct inferences to the analysed cases. Once the LCI inventory has been completed, the LCIA can be carried out using one of the methods proposed in literature [16]. The specificity of each methodology can be transferred into a KBE database, allowing the user-designer-engineer to apply the most suitable one for the considered application. The output parameters can be used to compare in the same moment a number of possible solutions.

2.4   Comparison Among Solutions

In general, the comparison of solutions within the KBES is a multi-objective problem by the optimization of a general set of functions, as for example shown in Eq. (3):

$$S = \{f_1(X), f_2(X) \dots f_r(X)\} \quad with\, X = \{x_1, x_2 \dots x_p\} \tag{3}$$

Where p is the number of input parameters that are managed by the user, which are used by the KBES to produce the r output parameters $y_i(X)$, which describe a specific solution, including the results of the LCIA.

In most cases, the relation between the input and the output parameters is not explicitly expressed and it derives from a set of operations carried out from the inference engine, which makes the problem non-linear.

To reduce the task to the minimization of a set of functions, even when the target is the maximization or the research of a predefined value (which can be obtained by calculations in the KBES), the r objective functions in Eq. (3) representing the solution S can be obtained from the output parameters $y_i(X)$ as follows in Eq. (4):

$$f_i(X) = \begin{cases} y_i(X), & if\ y_i(X)\ should\ be\ minimized. \\ -y_i(X), & if\ y_i(X)\ should\ be\ maximized \\ |y_i(X) - t_i|, & if\ y_i(X)\ should\ reach\ the\ target\ value\ t_i \end{cases} \quad ;\ i \in [1, r] \tag{4}$$

Once target functions have been defined, various multi criteria decision-making and multi-objective optimization techniques can be applied [4].

Indicating as S_u and S_v the generic u-th and v-th solutions belonging to the collection of solutions C^s produced by the KBES and defined by the input sets X_u and X_v, we can consider the following Eq. (5):

$$S_u < S_v \leftrightarrow \{f_i(X_u) \leq f_i(X_v) \forall i \in [1,r]\} \wedge \{\exists i^* \in [1,r] | f_{i^*}(X_u) \leq f_{i^*}(X_v)\} \tag{5}$$

Based on this definition, the Pareto Front, PF^*, of non-dominated solutions can be built through the condition in Eq. (6):

$$S_u \in PF^* \leftrightarrow \nexists v \in [1,r] | S_v < S_u \tag{6}$$

This operation allows the user to get a first screening of the solutions generated through the KBES, focusing his energies on more creative activity. This approach also allows realizing iterative procedures, taking the non-dominated solution as a starting point for the research of possible new solutions (even using multi-objective optimization techniques [17]).

The weighted sum method can be adopted to get a mono dimensional problem when the relevance of each aim of the design can be certainly quantified. The function to be optimized can therefore be expressed as in Eq. (7):

$$F(X_u) = \sum_{i=1}^r f_i(X_u) w_i, \tag{7}$$

where w_i is the weight expressing the importance assigned to the function $f_i(X)$ by the designer. When the results of an LCIA analysis are considered, an accurate choice of the weights must be done.

Product properties can have different orders of magnitude and units. A solution is the normalization of the target functions Eq. (8):

$$f_i^*(X_u) = \frac{f_i(X_u) - f_i(X_{min,i})}{f_i(X_{max,i}) - f_i(X_{min,i})} \tag{8}$$

where $X_{min,i}$ and $X_{max,i}$ are the sets of parameters, which correspond to the solutions $S_{min,i} \wedge S_{max,i} \in PF^*$ that make the function minimum and maximum respectively (with reference to the Pareto front). So, the function to be minimized is Eq. (9):

$$F^*(X_u) = \sum_{i=1}^r f_i^*(X_u) w_i \tag{9}$$

This kind of representation allows the user-designer-engineer a more intuitively setting of the relative importance of each requirement, reducing the necessary time for the selection of the final solution.

2.5 Further Considerations on the Purposed Method

The proposed method consists of a simplified procedure to carry out an analysis of sustainability for the concept of products, since the preliminary phase of the design process development. The accuracy of the results and the consistency of them is strongly affected by the quality of the LC framework and of the reference flows data, as well as for any LCA study.

The integration with a KBES requires the analysis of several data and elements. As an example, it is necessary that the database includes information about the materials and the manufacturing processes. The more information is available the better the system will be able to accurately describe the effects of the production environment on the sustainability of the product. An exhaustive LCA should efficiently describe all the environmental aspects related to the product, nonetheless, in case the adopted LC model does not cover all the phases of the product life, a qualitative interpretation of other KBES output parameters should be realized to understand their effects on remaining aspects.

3 A Case Study: Design and Manufacturing of a Plastic Bottle

3.1 Introduction

In order to provide an application of the exposed method, the results obtained by the integration of a LCIA tool in a KBES for the design of a plastic bottle are described. The tool is used to generate possible solutions for bottles design, combining geometrical, manufacturing and functional parameters. The LCIA tool has been used within the proposed methodology and applied to the case of a plastic bottle.

The described tool is focused on the evaluation of environmental impact issues related to materials and energy consumption in case of a plastic bottle production.

3.2 LCIA Framework

The LCIA included in the KBES is a cradle-to-gate study based on simplified hypotheses. The transportation, in terms of ways and distances, of the raw materials are assumed the same for each material selected. Therefore, the impact of transportation is not influent and can be excluded by the LCI computation.

The KBES allows the user-designer-engineer to consider data imported from a number of sources, as well as data coming from customers. A default dataset extracted from the European Life Cycle Database (ELCD) [18] is integrated into the application. The inference engine automatically recognizes the kind of plastic and can define the flow for the production of the raw material. The analysis refers to the unity of product and for this reason the weight of a bottle, which is one of the descriptive parameters provided by the KBES, is used as an input.

The KBES, through combining information concerning machine, material and shape, provides an estimation of the production rate (p_R, in item/hour) and the energy consumption of the production system (E_c, in kW). Therefore, a transformation of the introduced parameters is necessary to obtain the energy required for the production of each bottle (e_s, in MJ/item), as shown in Eq. (10):

$$e_s = 3.6\, E_c p_R \tag{10}$$

This energy amount is taken as the reference for the LCI analysis. A reference process can be defined by the user and be adapted to every production environments.

3.3 Analysis of the Results

The descriptive parameters outputs of the LCIA have to be managed together with the ones produced from the KBES. This fact means that the analysis is a "black-box" for the user-designer-engineer who can examine only the results of the elaboration and can focus on activities related to creativity: the evaluation of the analysed solutions based on proposed design and manufacturing issues.

In order to compare different design solutions, a first possible approach is monodimensional and based on the values assumed by a single property. The user can make a choice among a number of analyses for each generated solution. It is not necessary that all the solutions of a certain group have the same number of descriptive parameters. Therefore, before the comparison among properties $f_{i^*}(X)$, the KBES has to perform a preliminary screening to select only the solutions S_k that satisfy the condition $f_i^*(X_k) \in S_k$.

The Graphical User Interface (GUI) of the KBES provides a graphic and numerical representation of the selected properties using the interactive window shown in Fig. 3, in which the eco footprint indicator of different solutions is displayed. The indicator was obtained using the EU 27 normalizing factors per person [19].

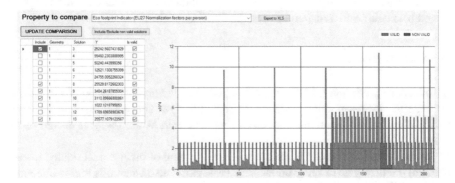

Fig. 3. Mono dimensional comparison of eco footprints indicator for different solutions.

The most efficient way to graphically display the results is mapping the different solutions on a scattered plot by a bi-dimensional comparison. This kind of representation also allows showing the Pareto front of non-dominated solution: elements belonging to this set are also effectively listed for the next design activities of the user. A possible interface for this kind of analysis is showed in Fig. 4.

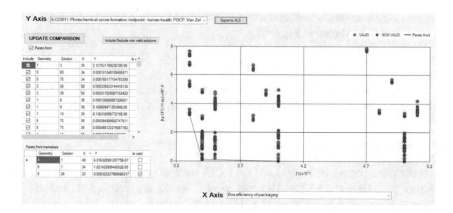

Fig. 4. Bi-dimensional comparison of solutions.

When a more general n-dimensions analysis is carried out, a graphic representation of the comparison is not possible. In this case the user interface only allows setting the requirements and exporting the list of non-dominated solutions to the group of properties indicated by the user. Nonetheless, the weighted sum method can support the designer in identifying the best solution in terms of the best combination for the product of design and manufacturing processes.

As not all the aspects of product life cycle can be considered by the LCIA, a critical analysis of some other descriptive parameters is necessary to assess the behaviour of the product during the successive steps of the life cycle. As an example, the X-Axis property showed in Fig. 4 is the "box efficiency of the bottle" or the ratio between the filling volume of the bottle and the volume of the minimum rectangular parallelepiped, which could enclose the product. This parameter is an indicator of the effects that the shape has on packaging and transportation of the finished product.

4   Conclusions

Introducing a preliminary LCIA in the first phases of the product design process can be very useful for a proper and sustainable product design. This early assessment allows anticipating the choice of the manufacturing processes while the best design of the product is defined.

An accurate LCIA calculation needs a large amount of precise data related to each phase of the life cycle of the product. This issue exists also when this tool is integrated with a KBES for the preliminary design, when the output solutions can be defined only from some uncertain data. Nevertheless, by the presented application, it is possible to produce some qualitative significant information, useful to compare different solutions and to make the best choice.

The architecture of the presented tool allows a direct interaction between the user and the KBES at different levels of accuracy, according to the quality of the available information. It constitutes an efficient aid to assess the environmental impact of the product; therefore, since the very beginning of the product development, it can be an effective guide for the decision making in order to select the best and most sustainable solutions. Furthermore, this tool permits a time and a cost reduction. Allowing saving time, which should have been dedicated to define technologies and manufacturing processes, gives designers the possibility to focus on more creative activities. Considering several aspects related to environmental sustainability, it helps designing sustainable products choosing the most efficient and effective combination between the design and the fabrication aspects of the commodity.

References

1. Jayal, A.D., Badurdeen, F., Dillon Jr., O.W., Jawahir, I.S.: Sustainable manufacturing: modeling and optimization challenges at the product, process and system levels. CIRP J. Manufact. Sci. Technol. **2**, 144–152 (2010). Elsevier
2. International Trade Administration 2007: How Does Commerce define Sustainable Manufacturing? U.S. Department of Commerce
3. Elkington, J.: Cannibals with forks: triple bottom line of 21st century business. Capstone Publishing Ltd., Oxford (1997)
4. Curran, M.A. (ed.): Life Cycle Assessment Handbook - A Guide for Environmentally Sustainable Products. Wiley, Hoboken (2012)
5. ISO 14040:2006, Environmental management - Life cycle assessment - Principles and framework
6. Vigon, B.W., et al.: Life-Cycle Assessment: Inventory Guidelines and Principles. CRC Press, Boca Raton (1993)
7. Office of the European Union: International Reference Life Cycle Data System (ILCD) Handbook - General guide for Life Cycle Assessment - Detailed guidance, Joint Research Centre - Institute for Environment and Sustainability, EUR 24708 EN, Luxembourg (2010)
8. Scipioni, A., et al.: LCA to choose among alternative design solutions: the case study of a new Italian incineration line. Waste Manag. **29**, 2462–2474 (2009). Elsevier
9. Poudelet, V., et al.: A process-based approach to operationalize life cycle assessment through the development of an eco-design decision-support system. J. Clean. Prod. **33**, 192–201 (2012)
10. Wallace, D.R. et al.: Approximate life-cycle assessment in conceptual product design. In: Proceedings of DETC ASME 2000 (Design Engineering Technical Conferences and Computers and Information in Engineering) Conference, Baltimore, Maryland (2000)
11. Sousa, I. et al.: A learning surrogate LCA model for integrated product design. In: Proceedings of 6th CIRP International Seminar on Life Cycle Engineering, Kingston, Canada, pp. 209–219 (1999)
12. Sandberg, M.: Knowledge based engineering. In: Product Development, Lulea University of Technology (2003)
13. Park, J.-H., Seo, K.-K.: A knowledge-based approximate life cycle assessment system for evaluating environmental impacts of product design alternatives in a collaborative design environment. Adv. Eng. Inform. **20**, 147–154 (2006)
14. Wim, J., Verhagen, C., Bermell-Garcia, P., van Dijk, R.E.C., Curran, R.: A critical review of Knowledge-Based Engineering: an identification of research challenges. Adv. Eng. Inform. **26**, 5–15 (2012)
15. Amadori, K., Tarkian, M., Ölvander, J., Krus, P.: Flexible and robust CAD models for design automation. Adv. Eng. Inform. **26**, 180–195 (2012)

16. ILCD Handbook: Analysing of existing Environmental Impact Assessment methodologies for use in Life Cycle Assessment. Joint Research Centre - European commission (2010)
17. Miettinen, K.: Nonlinear Multiobjective Optimization. Springer, New York (1998)
18. Recchioni, M., Mathieux, F., Goralczyk, M., Schau, E.M.: ILCD Data Network and ELCD Database: current use and further needs for supporting Environmental Footprint and Life Cycle Indicator Projects. Joint Research Centre - European commission (2013)
19. Office of the European Union: Normalisation method and data for Environmental Footprints, European Commission - Joint Research Centre - Institute for Environment and Sustainability, Luxembourg (2014)

Eco-Intelligent Factories: Timescales for Environmental Decision Support

Elliot Woolley[✉], Alessandro Simeone, and Shahin Rahimifard

Wolfson School of Mechanical, Electrical and Manufacturing Engineering,
Centre for Sustainable Manufacturing and Recycling Technologies,
Loughborough University, Loughborough, UK
{e.b.woolley,a.simeone,s.rahimifard}@lboro.ac.uk

Abstract. Manufacturing decisions are currently made based on considerations of cost, time and quality. However there is increasing pressure to also routinely incorporate environmental considerations into the decision making processes. Despite the existence of a number of tools for environmental analysis of manufacturing activities, there does not appear to be a structured approach for generating relevant environmental information that can be fed into manufacturing decision making. This research proposes an overarching structure that leads to three approaches, pertaining to different timescales that enable the generation of environmental information, suitable for consideration during decision making. The approaches are demonstrated through three industrial case studies.

Keywords: Manufacturing · Environmental impact · Decision support · Artificial intelligence

1 Introduction

Globally, factories account for roughly one third of energy use [1], and one third of energy related CO_2 production [2]. This is in addition to other air, land and water emissions, chemical use and demand for materials. The world's factories are a hotspot of human induced environmental impacts and therefore require effective environmental management programmes.

In contrast to this need, current manufacturing management systems and related decision making are optimised for cost effectiveness, time efficiency (productivity) and quality control [3], but not environmental impacts. These complex networks of data and information systems enable manufacturers to remain competitive by making informed short-term decisions and by forecasting over longer time scales. However, despite legislative developments in this area (e.g. [4]), environmental considerations are not routinely included in this planning (Fig. 1), and it is becoming clear that their inclusion could lead to a significant reduction in environmental impacts [5]. In this work, an analysis of industrial decision making is contrasted with modern approaches for generation of consideration of environmental data in manufacturing decisions.

© Springer International Publishing AG 2017
G. Campana et al. (eds.), *Sustainable Design and Manufacturing 2017*, Smart Innovation, Systems and Technologies 68, DOI 10.1007/978-3-319-57078-5_32

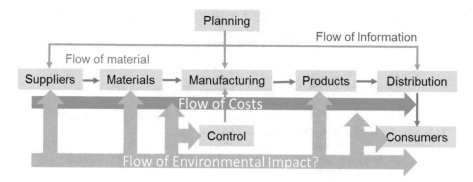

Fig. 1. Opportunities for the inclusion of environmentally related data in industrial decision making.

A gap in the availability of environmental data is identified, with a three timescale (short, medium and long) approach proposed to enable more systematic generation of environmental information to feed into decision making within manufacturing.

Correspondingly a number of case studies are presented demonstrating the generation of eco-intelligent information across these timescales. The paper concludes with a discussion of how the approaches described in this work can be applied more routinely to the wider industry.

2   Literature Review

2.1   Manufacturing Decision Making

Enterprise Resource Planning (ERP) is the generic term used for modern manufacturing management infrastructures (not just the software component) whereby there is one set of rules for balancing supply and demand, linking customers and suppliers in one chain, employing proven business processes for decision making, and providing cross-functional integration across departments and activities [6]. Because of its effectiveness, some form of ERP can be found in almost every manufacturing company worldwide, and the provision of the software and support for ERP (e.g SAP, Oracle) is in itself a global multi-billion dollar business [7]. Due to the perceived low economic value of environmental performance, none of these systems are configured to allow comprehensive consideration of environmental impacts in decision making. In order to be able to consider environmental impacts in production planning and control, there is a need to measure and compare environmental metrics but these are difficult to define and vary widely depending upon specific manufacturing activities.

Clearly within manufacturing companies there is a substantial amount of information that is created on a daily basis which is used across many departments to enable efficient, profitable operation of their production plants [8]. Despite much of this data being used to support manufacturing activities across production, logistics, customer promising, etc., there are other, environmentally focused, decisions that could be supported using this data. For example, by comparing production cell energy consumption against product throughput would yield an indication of product energy embedded by that cell – which can be used as a benchmark to highlight and investigate periods of over-consumption. By utilising this latent capability in environmentally related data, there are often opportunities for reducing environmental impact [9].

2.2 Environmental Information

The idea of incorporating environmental considerations into manufacturing activities is not new. The establishment of Environmental Management Systems (EMS) have allowed manufacturers to make decisions on their activities with respect to environmental performance [10]. In addition there are a number of both complex and simplified LCA type tools (SimaPro [PRé Consultants], CES [Granta Design], etc.) which can be and are used by manufacturing companies to assess the environmental impact of the products they make, and thus allow them to improve product design ('design for X' approaches) to reduce resource consumption and avoid other negative environmental impacts [11]. However LCA tools still only support slow, progressive improvements to manufacturing activities rather than optimising in the shorter term.

Not only is there a complexity in understanding and implementing the soft (information) side of incorporating environmental considerations into decision making, but there are issues regarding collecting or accessing sufficient information (both real-time and longer term) to understand existing performance and thus influence decision in planning and control. Such a network of information within a manufacturing enterprise is typically only partially present. In particular the problems manufacturers face with real-time energy metering, management and optimization has been addressed [12, 13], and its relative low importance (in management agendas) coupled with a range of technical and economic implications. In reality, there is a lot of usable data generated within factories, but without the infrastructure to interpret and communicate eco-performance metrics, it is not possible to influence operation and planning decisions.

As a sign of progress within industry, smart metering has been used to help inform decisions by tracking not only the total electrical work and consumed energy, but also the characteristics of specific power consumption over time [14]. Subsequently, research has focused on smart metering systems which involve the use of sensors, processors and analysers to capture, transfer, identify and resolve energy and resource flows in manufacturing systems [15]. Unfortunately however, although some of this information could be used for the assessment of environmental impact, it currently is not due to the lack a suitable infrastructure.

This highlights a need for a methodical approach to information gathering within factories, and for decision makers to have access to appropriate data, and be equipped with the ability to process this data such that it can be fed into decision making processes. Only through these capabilities can manufacturers have the opportunity to improve their environmental performance through improved decision making.

3   Eco-Intelligent Information

Manufacturing decisions require the varied approaches to information processing depending upon the level at which they are taken [8] – clearly manufacturing decision at the machine level will have different requirements to those taken at the enterprise level as has been reported for manufacturing energy management [16]. Therefore when considering the generation of environmental data to feed into decision-making, a range of timescales (roughly corresponding to different manufacturing levels) with appropriate methodologies must be defined. In the current research, manufacturing decisions are segregated into short, medium, and long timescales pertaining to seconds-hours, hours-months and months-years respectively.

The fundamental thesis being that short-term decisions (such as machine optimisation) require the availability of near real-time data for increased autonomy, medium term decisions (such as maintenance scheduling) require suitable modelling approaches based on appropriate key performance indicators (KPI), whilst long term decisions (such as heavy investments in capital equipment) require forecasting of future impacts. The proposed structure for eco-intelligent information generation is shown in Fig. 2.

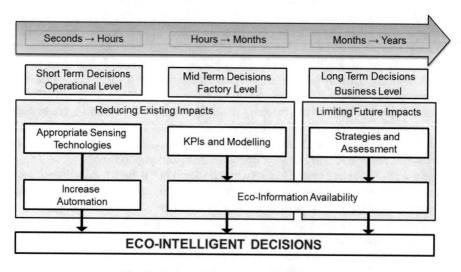

Fig. 2. Eco-intelligent information generation

For the three different timescales the data requirements and processing into information are quite different. There will be variations in the type of data, amount of data, speed of acquisition and processing required, accuracy and complexity, repetition rate, use of intelligence (natural or artificial), importance to a company amongst many others. It is therefore not suitable to consider all decisions using same approach and consequently three approaches relating to the different timescales are presented and described in the remainder of this section.

3.1 Short-Term Decision Making

Within the short term decision making timescale the research scope seeks to minimise a set of environmental factors in a manufacturing process by monitoring specific variables using the most appropriate sensing units, with the possibility of using intelligent decision making support systems.

The first step of the approach (Fig. 3) is the analysis of the manufacturing process under investigation. This phase is aimed at identifying the aspects of the process and highlighting the related environmental impacts. The problem definition phase results in the identification of the environmental factors to be minimised.

At this point, further consideration about the process needs to be undertaken, classifying which variables can be actively controlled and which variables can be monitored. The sensing unit selection considers the physical and chemical aspects of the process, taking into account commercial availability. Prior to any industrial implementation, the eco-intelligent process monitoring approach requires extensive experimental work in order to calibrate the system and to obtain reliable, repeatable results.

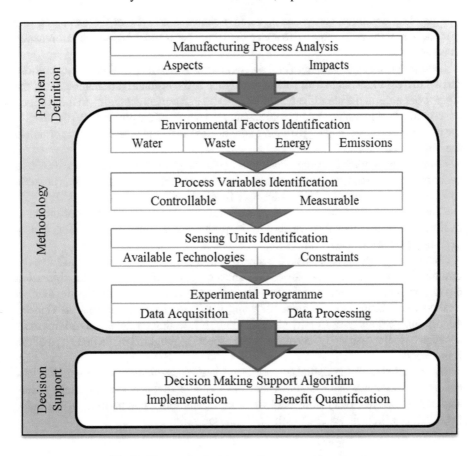

Fig. 3. Short-term decision making support framework

Once the sensing units have been validated, the experimental setup can be defined, along with a comprehensive experimental plan [17]. The data processing procedure is finalised to extract important features to transform data into useful information [18]. The most common methodologies involve time domain analysis, such as statistical features [19] and Principal Components Analysis (PCA). Whenever time domain features are not suitable, an alternative approach is the frequency domain analysis, such as Wavelet Transform [20].

In terms of decision making support systems and paradigms, neural networks are mainly used for pattern recognition, time series prediction and data fitting [21, 22]. Other DM support systems include Fuzzy Logic paradigms [23], Genetic Algorithms (GA) [24], and Ant Colony Optimisation (ACO) [25].

The decision support algorithm will be implemented using the processed data and will generate a result.

3.2   Medium-Term Decision Making

The problem definition comprises of a comprehensive analysis of the manufacturing process or system and is required to understand the aspects and the related environmental impacts.

The proposed methodology (Fig. 4) starts with a characterisation of the environmental drivers to take into account. The next step is the problem formulation: here, the framework aims at the identification of boundaries and targets according to the problem description. In this respect, taking into account the nature of the problem, firstly identify the decision variables, paying particular attention to units, and utilise them to formulate the objective function.

Analogously, formulate the constraints, either logical or explicit to the problem description by expressing them in terms of decision variables. At this point it is possible to identify the data needed for the objective function and constraints.

The model identification is crucial phase of information generation for medium-term decisions as it describes the structure of the problem and allows the definition of key performance indicators (KPIs).

According to the specific task, the identification of a suitable algorithm to solve the optimisation problem must be carried out. The most common categories of optimisation algorithms, are the finitely terminating algorithms, such as Simplex [26], the iterative methods, e.g. Conjugate Gradient [27], and the heuristic methods, such as Genetic Algorithms [28] and Ant Colony Optimisation [29]. In this phase, the algorithm must be adapted to the case study, considering the aspects highlighted in the first steps of the framework.

A critical step in the optimisation process is the presentation of the solution in a concise and comprehensible summary for stakeholders. In this phase, the results generated need to be effectively comparable in terms of environmental performance in order to quantify the benefit obtained with the optimisation.

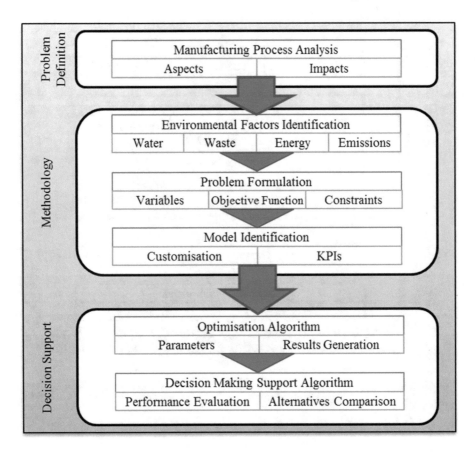

Fig. 4. Medium-term decision making support framework

3.3 Long Term Decision Making

There is a need to ensure that the outcomes of environmentally focussed strategic decisions, made over long timescales are in alignment with the greater business strategy [30].
However, in contrast to short and medium term decisions, rather than attributing environmental impact to existing processes or activities of an enterprise, it is more appropriate to forecast and attribute environmental impact to activities required to support
and deliver a new business activities. Therefore the first and most difficult phase of the
process is planning, which incorporates the definition of the scope of the analysis (the
decision in question) and sets the boundaries of consideration (e.g. timescales, areas of
business, lifecycle stages) as shown in Fig. 5.

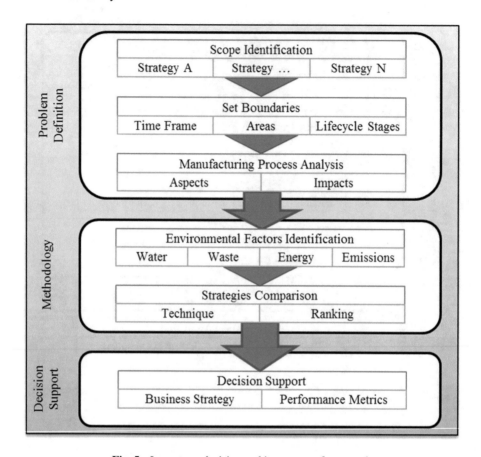

Fig. 5. Long-term decision making support framework

Defining the scope, allows a comparison of different strategies that fulfil the same business need and thus allows a certain level of creativity within long-term decision making. Devising the potential strategies to overcome the problem can only be undertaken by an appropriate team from within a company and will be highly problem specific. Such guidance is outside the scope of this work.

Once potential strategies have been established, the boundary conditions, as described above, allow the identification and quantification of aspects and impacts associated with those solutions. Analysis of the aspects across the different manufacturing levels and lifecycle stages can be undertaken in a systematic manner (see [31] for example). There are a number of tools that assist in the evaluation of environmental impacts, such as LCA and EMS, as described in Sect. 2.2.

Depending upon information generated and the particular impacts considered and appropriate technique for comparison of results [32] should be used and ranking of the potential strategies carried out. The strategies with the best environmental performances should be compared against the broader business strategy before being considered along-side economic and social performance metrics. Outcomes from the implementation of the strategy can be used to feed back into future long-term decision making analysis.

4   Case Studies

The following brief case examples demonstrate the application of the eco-intelligent methodologies presented in Sect. 3.

4.1   Clean-in-Place Monitoring – Short Term

Clean-in-place (CIP) is a widely used technique applied to clean industrial equipment without disassembly [33]. Cleaning food deposits, which contain both proteins and minerals, is a complex process that involves interactions between surface, deposits and detergent. It requires a multistage process, having many steps that may be controlled by shear stress, mass transfer, and chemical reaction [33].

Existing CIP processes are time intensive and waste large amounts of energy, water, and chemicals [34, 35]. Furthermore, it is estimated that on average, a food and beverage plant will spend 20% of each day on cleaning equipment, which represents significant downtime for a plant [35].

The purpose of this case study is to reduce the cleaning time, so it is necessary to monitor the food traces left within the process tank (Fig. 6). Due to the chemical compo-sition of the food deposit to be monitored, (in this case) milk proteins, the sensing unit selected was a digital camera endowed with UV light set [36].

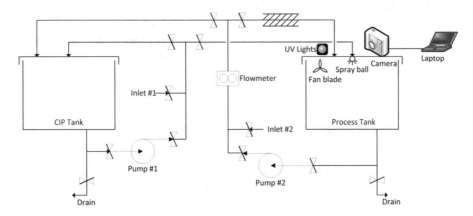

Fig. 6. Clean-in-place experimental rig scheme

Data acquisition for this case study consists in the acquisition of a series of digital images, through a time-lapse technology. Data processing is needed to assess the surface fouling level within each digital image and allow the monitoring of the cleaning process.

Image processing algorithms based on image segmentation and thresholding [37] were employed for the quantification of remaining fouling.

The decision making associated to this research aims at optimising the cleaning process and is specifically required to understand when each phase of the cleaning process becomes redundant and automatically switch to the next phase.

4.2   Production Scheduling – Medium Term

In a food manufacturing plant, an inventory of 50 products was considered. An important qualitative feature of many of the materials in the inventory was that they were a potentially hazardous contaminant if carried over between different production runs. These materials were categorised into multiple different types of Potentially Cross Contaminating Materials (PCCM). The content of each PCCM in each product was designated according to content levels 0–3. The changeover cleaning protocols, defined by the PCCM content of the former and latter product in a scheduled sequence, is defined in [24]. In this case study the identification of the environmental factor is straight forward, hence the cumulative change over time required (which includes water, energy and other overheads) was used as a proxy environmental impact for overall resource consumption. The production sequence is to be optimized for minimal cumulative resource consumption during changeovers.

Finding the optimal sequence of products with minimised resource consumption was determined to be analogous to the asymmetric travelling salesman problem (ATSP) [24], where each product was represented by a node and the 'distance travelled' between nodes was represented by the changeover cleaning time. With the model identified, expedient solving of the ATSP in this context was approached using a genetic algorithm (GA) [24], which enables the determination of a near optimal solution of complex problems using feasible computing resources.

The GA generated an optimal sequence for 50 products with the minimum changeover cleaning time requirement. Repeat implementation of the GA provides alternative product sequences with equivalent total cleaning time. In this way, a selection of optimum sequences may be performed.

4.3   Energy Efficient Business Modelling – Long Term

One of the long-term decisions faced by modern manufacturing companies is how best to deliver value into the market. An increasing number of companies are moving towards the delivery of product service systems (PSS) in place of the more traditional make-sell business model [38]. PSSs have many potential economic, social and environmental advantages. However, it is not always clear as to how beneficial a PSS may be, if at all, in comparison to the make-sell alternative.

In this example a comparison of different business strategies for the provision of steel roofing is made, with a particular focus on lifecycle energy requirements. In one strategy the company supplies steel roofing panels via a traditional make-sell business model, and in the other, supplies identical roof panels via a PSS business model. In the latter instance, the manufacturer is responsible for the panels' maintenance throughout their lifetime plus their end-of-life (EoL) recovery.

For fair comparison, the performance metric is set as energy per square meter per year (MJ/m^2yr) and the scope includes manufacture of the panels, use (maintenance) and end-of-life recovery. In addition, the lifetime of the steel roofing for the make-sell and PSS business strategies has been assumed to be 15 and 25 years respectively; the PSS roofing having, on average, an extended lifetime due to a regular maintenance schedule.

The energy requirement per square-metre per year of the roofing for the two business strategies is shown in Table 1.

Table 1. Energy requirements considered through product life cycle for comparison between make-sell and PSS business strategies for steel roofing

Energy contributor	Make-Sell	PSS
Manufacture		
Production Energy	$33 \ MJ/m^{2}$ *	$33 \ MJ/m^{2}$ *
Σ(Process Energy + Plant Energy)	$145 \ MJ/m^{2}$ #	$145 \ MJ/m^{2}$ #
ΣCorporation Energy	$2 \ MJ/m^{2}$ #	$4 \ MJ/m^{2}$ &
Use (maintenance)	N/A	
ΣCorporation Energy		$2 \ MJ/m^{2}yr$ &
End of Life		
Production Energy		$-48 \ MJ/m^{2}$ #
ΣCorporation Energy		$4 \ MJ/m^{2}$ &
Lifetime of panel	15 yr	25 yr
Performance metric	$12 \ MJ/m^{2}yr$	$7.5 \ MJ/m^{2}yr$

* = data calculated from physical material properties, # = data taken or inferred from [39], & = data simulated from company/customer location

The energy demand for the manufacturing stage of the product represents the largest energy outlay for the company, and so preserving this investment in energy (through the use of additional energy during use and EoL) by adopting a PSS business strategy becomes beneficial from an energy consumption standpoint.

Based on the analysis of energy requirements for each strategy, a decision is likely to be made to proceed with the PSS business model. In this exemplifying study, only one performance metric was considered: in a more detailed application it is likely that a greater number of indicators would need to be calculated, considered and compared with the wider business strategy.

5   Concluding Discussion

There is a need to more routinely incorporate eco-intelligent information into manufacturing decision making if the industry is to reduce its environmental impacts whilst still meeting the need of consumers. Three approaches for the generation of eco-intelligent information have been described that related to different types of manufacturing decision (based on varying timescales). For short-term decisions (seconds to hours) understanding the requirements for sensing and automation are important task. For medium term decisions (hours to months) developing KPIs and associated data models is of primary importance. In contrast, for longer term decisions (months to years) the key challenge is in the problem definition and setting of system boundaries. Each approach has been presented and demonstrated using three industrial case examples.

In summary the possibility of routinely incorporating environmental information into manufacturing decision making across all timescales is possible, but requires markedly different approaches. Precisely how to compare eco-intelligent information with economic and social considerations, remains an active topic of global research.

References

1. International Energy Agency. Key world energy statistics (2016). http://www.iea.org/statistics/statisticssearch/
2. Allwood, J., Cullen, J.: Sustainable Materials – With Both Eyes Open. UIT Cambridge, Cambridge (2012)
3. Öker, F., Adıgüzel, H.: Time-driven activity-based costing: an implementation in a manufacturing company. J. Corp. Account Financ. **27**(3), 39–56 (2016). http://doi.wiley.com/10.1002/jcaf.22144. Accessed 25 Nov 2016
4. U.N.: United Nations Framework Convention on Climate Change, vol. 21, Paris Agreement C.N.63.2016.TREATIES-XXVII.7.d (2016). https://treaties.un.org/doc/Publication/MTDSG/VolumeII/ChapterXXVII/XXVII-7-d.en.pdf
5. Epstein, M.J., Buhovac, A.R., Yuthas, K.: Managing social, environmental and financial performance simultaneously. Long Range Plann. **48**(1), 35–45 (2015)
6. Wallace, T.F., Kremzar, M.H.: AM ERP: Making It Happen Resource Planning, 385 p. Wiley, New York (2001). http://cyberrvr.weebly.com/uploads/5/9/7/8/5978545/wiley_-_erp_making_it_happen.pdf
7. Holsapple, C.W., Sena, M.P.: ERP plans and decision-support benefits. Decis. Support Syst. **38**, 575–590 (2005)
8. Vijayaraghavan, A., Dornfeld, D.: Automated energy monitoring of machine tools. CIRP Ann. Manuf. Technol. **59**(1), 21–24 (2010)
9. Rahimifard, S., Seow, Y., Childs, T.: Minimising embodied product energy to support energy efficient manufacturing. CIRP Ann. Manuf. Technol. **59**(1), 25–28 (2010)
10. Melnyk, S.: Assessing the impact of environmental management systems on corporate and environmental performance. J. Oper. Manag. **21**(3), 329–351 (2003). https://scholars.opb.msu.edu/en/publications/assessing-the-impact-of-environmental-management-systems-on-corpo-3. Accessed 14 Nov 2016
11. Fiksel, J., Wapman, K.: How to design for environment and minimize life cycle cost. In: Proceedings of 1994 IEEE International Symposium on Electronics and the Environment, pp. 75–80 (1994). http://ieeexplore.ieee.org/articleDetails.jsp?arnumber=337290
12. Wenzel, K., Riegel, J., Schlegel, A., Putz, M.: Semantic web based dynamic energy analysis and forecasts in manufacturing engineering. In: Hesselbach, J., Herrmann, C. (eds.) Glocalized Solutions for Sustainability in Manufacturing, pp. 507–512. Springer, Heidelberg (2011)

13. Kara, S., Bogdanski, G., Li, W.: Electricity metering and monitoring in manufacturing systems. In: Hesselbach, J., Herrmann, C. (eds.) Glocalized Solutions for Sustainability in Manufacturing - Proceedings of the 18th CIRP International Conference on Life Cycle Engineering, pp. 1–10. Springer, Heidelberg (2011)

14. Kara, S., Bogdanski, G., Li, W.: Electricity metering and monitoring in manufacturing systems. In: Hesselbach, J., Herrmann, C. (eds.) Glocalized Solutions for Sustainability in Manufacturing - Proceedings of the 18th CIRP International Conference on Life Cycle Engineering, pp. 1–10. Springer, Heidelberg (2011). http://link.springer.com/10.1007/978-3-540-89644-9. Accessed 14 Nov 2016

15. Karnouskos, S., Izmaylova, A.: Simulation of web service enabled smart meters in an event-based infrastructure. In: IEEE International Conference on Industrial Informatics (INDIN), pp. 125–130 (2009)

16. Woolley, E., Seow, Y., Arinez, J., Rahimifard, S.: The changing landscape of energy management in manufacturing. In: Llamas Moya, B., Pous, J. (eds.) Greenhouse Gases (2016)

17. Byrne, D.M., Taguchi, S.: The Taguchi approach to parameter design. Qual Prog. 20(12), 19–26 (1987)

18. Guyon, I., Elisseeff, A.: An introduction to feature extraction. In: Feature Extraction - Foundations and Applications, vol. 207, 740 p. Springer, Heidelberg (2006). http://www.springerlink.com/content/j847w74269401u31/%5Cn, http://link.springer.com/10.1007/978-3-540-35488-8

19. Segreto, T., Simeone, A., Teti, R.: Sensor fusion for tool state classification in nickel superalloy high performance cutting. Procedia CIRP 1, 593–598 (2012)

20. Karam, S., Teti, R.: Wavelet transform feature extraction for chip form recognition during carbon steel turning. Procedia CIRP 12, 97–102 (2013)

21. Simeone, A., Watson, N., Sterritt, I., Woolley, E.: A multi-sensor approach for fouling level assessment in clean-in-place processes. Procedia CIRP 55, 134–139 (2016). http://www.sciencedirect.com/science/article/pii/S2212827116307685. Accessed 7 Nov 2016

22. Simeone, A., Segreto, T., Teti, R.: Residual stress condition monitoring via sensor fusion in turning of Inconel 718. Procedia CIRP 12, 67–72 (2013)

23. Simeone, A., Woolley, E.B., Rahimifard, S.: Tool state assessment for reduction of life cycle environmental impacts of aluminium machining processes via infrared temperature monitoring. Procedia CIRP 29, 526–531 (2015). http://www.sciencedirect.com/science/article/pii/S2212827115001134. Accessed 10 July 2015

24. Gould, O., Simeone, A., Colwill, J., Willey, R., Rahimifard, S.: A material flow modelling tool for resource efficient production planning in multi-product manufacturing systems. Procedia CIRP 41, 21–26 (2016). http://www.sciencedirect.com/science/article/pii/S2212827116000111. Accessed 19 Feb 2016

25. Dorigo, M., Gambardella, L.M.: Ant colony system: a cooperative learning approach to the traveling salesman problem. IEEE Trans. Evol. Comput. 1(1), 53–66 (1997)

26. Nash, J.C.: The (Dantzig) simplex method for linear programming. Comput. Sci. Eng. 2(1), 29–31 (2000). http://ieeexplore.ieee.org/stamp/stamp.jsp?arnumber=814654

27. Hestenes, M.R., Stiefel, E.: Methods of conjugate gradients for solving linear systems. J. Res. Natl. Bur. Stand. (1934) 49(6), 409 (1952). http://nvlpubs.nist.gov/nistpubs/jres/049/jresv49n6p409_A1b.pdf

28. Goldberg, D.E.: Genetic Algorithms in Search, Optimization, and Machine Learning, 432 p. Addison Wesley, Boston (1989). http://www.mendeley.com/research/genetic-algorithms-in-search-optimization-and-machine-learning/

29. Dorigo, M., Maniezzo, V., Colorni, A.: Ant system: optimization by a colony of cooperating agents. IEEE Trans. Syst. Man Cybern. Part B Cybern. 26(1), 29–41 (1996)

30. Andrews, K.R.: The concept of corporate strategy. In: Resources, Firms, and Strategies: A Reader in the Resource-based Perspective, pp. 52–59 (1997). http://books.google.at/books/about/Resources_Firms_and_Strategies.html?id=Zj7JwMcMoY0C&redir_esc=y

31. Woolley, E., Sheldrick, L., Arinez, J., Rahimifard, S.: Extending the Boundaries of Energy Management for Assessing Manufacturing Business Strategies (2013)

32. Simeone, A., Luo, Y., Woolley, E., Rahimifard, S., Boër, C.: A decision support system for waste heat recovery in manufacturing. CIRP Ann. Manuf. Technol. **65**(1), 21–24 (2016)

33. Palabiyik, I., Yilmaz, M.T., Fryer, P.J., Robbins, P.T., Toker, O.S.: Minimising the environmental footprint of industrial-scaled cleaning processes by optimisation of a novel clean-in-place system protocol. J. Clean. Prod. **108**(Part A), 1009–1018 (2015). http://www.sciencedirect.com/science/article/pii/S0959652615010458. Accessed 26 Apr 2016

34. Thomas, A., Sathian, C.T.: Cleaning-in-place (CIP) system in dairy plant-review. IOSR J. Environ. Sci. Ver. III **8**(6), 2319–2399 (2014). http://www.iosrjournals.org

35. Jude, B., Lemaire, E.: How to Optimize Clean-in-Place (CIP) Processes in Food and Beverage Operations. Schneider Electric White Paper (2013). http://www2.schneider-electric.com/documents/support/white-papers/energy-efficiency/how-to-optimize-clear-in-place-CIP-processes.pdf

36. Whitehead, K.A., Smith, L.A., Verran, J.: The detection of food soils and cells on stainless steel using industrial methods: UV illumination and ATP bioluminescence. Int. J. Food Microbiol. **127**(1–2), 121–128 (2008)

37. Otsu, N.: Threshold selection method from grey-level histograms. IEEE Trans. Syst. Man Cybern. **9**, 62–66 (1979). http://www.scopus.com/inward/record.url?eid=2-s2.0-0018306059&partnerID=tZOtx3y1

38. Tukker, A.: Eight types of product-service system: Eight ways to sustainability? Experiences from suspronet. Bus. Strateg. Environ. **13**(4), 246–260 (2004)

39. Kara, S., Manmek, S.: Impact of manufacturing supply chain on the embodied energy of products. In: Proceedings of the 43rd International Conference on Manufacturing Systems, Vienna, Austria, pp. 187–94 (2010)

Assessing Sustainability Within Organizations: The Sustainability Measurement and Management Lab (SuMM)

Mariolina Longo[✉] and Matteo Mura

Department of Management, University of Bologna, Bologna, Italy
{mariolina.longo,matteo.mura}@unibo.it

Abstract. Sustainability measurement represents a key element if companies aim to translate sustainability from compliance with standards to core organizational asset. This paper presents the development of a structured dataset of European companies – named as Sustainability Measurement and Management (SuMM) Lab– that aims to assess company's performance accordingly to sustainability metrics. Data have been collected on a pilot sample of 400 Italian companies and the results provide insights on how SuMM Lab could be exploited to assess the diffusion of sustainability practices in organizations.

Keywords: Sustainability · Measurement and management · SuMM Lab · Multivariate statistical analysis

1 Introduction

Despite a thirty-year-long, often inconclusive, debate over the effect of sustainability on company performance, the conversation on whether sustainability pays off has finally moved forward and now revolves around the fact that "there is no alternative to sustainability" (Nidumolu et al. 2009: 57). Indeed, our society is currently consuming the equivalent of 1.5 planet Hearths to support human activities (Global Footprint Network 2014), and this certainly represent an unsustainable rate even at today's level of production (Randers 2012). Also, recent economic downturns have radically challenged the dominant economic model of development, which is based on the primacy of shareholders over stakeholders and linear value chains over network structures. These elements, have started to force individuals, companies and society to re-consider the whole process of value creation, therefore companies that challenge current assumptions in terms of value creation and value capture and embrace a sustainability-oriented approach towards business (Adams et al. 2016), gain a significant advantage over their competitors (SustainAbility 2014). These companies succeed thanks to their transition towards sustainability-oriented business models that trigger innovation in products, and services, but also due to innovative measurement tools that help companies to measure, monitor, and assess their performance according to both an economic, social, and environmental bottom line (Searcy 2012).

© Springer International Publishing AG 2017

G. Campana et al. (eds.), *Sustainable Design and Manufacturing 2017*, Smart Innovation, Systems and Technologies 68, DOI 10.1007/978-3-319-57078-5_33

The performance measurement literature as well as the engineering management community (Leite et al. 2013) has devoted considerable attention to the development of measurement tools that take sustainability into account (Keeble et al. 2003). Since the early 1990s, the shortcomings of traditional financial and cost-related metrics clearly emerged and new approaches for measuring performance were introduced. These systems included both financial and non-financial measures, and focused on multiple performance dimensions to comply with different stakeholder's wants and needs (Neely et al. 2002; Semenova and Hassel 2015). Despite these huge efforts, numerous gaps still emerge in both theory and practice of sustainability measurement. Firstly, a comprehensive and integrated framework for sustainability assessment at company level is still lacking (Singh et al. 2007). Also, there is no agreement over which indicators should be disclosed by companies, and also whether the type of indicators should vary accordingly to different industries. As a consequence, reporting frameworks show great variations in content and scope (Searcy 2012). Finally, although existing frameworks (i.e. GRI, CDP, Asset4) have provided interesting insights on the measurement of company's sustainability, they mostly focus on public companies or large organizations, and leave SMEs behind. However, SMEs represent 99% of all European businesses, 67% for number of FTEs and 58% for value added (EU Commission 2012), and could represent a fundamental driver of sustainable development.

Consequently this paper presents the development of a structured dataset of European companies – named as Sustainability Measurement and Management (SuMM) Lab – that aims to assess company's performance accordingly to sustainability metrics.

Data for this study were collected through secondary sources – i.e. companies' websites and sustainability reports. Overall, a pilot sample of 400 companies was identified and assessed, and data were analyzed through multivariate statistical techniques. The results presented in this paper are preliminary and exemplify how SuMM database could be exploited to assess the diffusion of sustainability practices among EU companies.

2   Theoretical Background

Sustainability metrics and scorecards support organizations in assessing how well they are doing in meeting their sustainability priorities (Searcy 2012). Research on sustainability indicators and indices has focused on both the individual organization and the sector in which the organization acts. The most well-known set of corporate sustainability indicators are the 79 measures included in the Global Reporting Initiative's (GRI) G4 reporting guidelines (GRI 2015). Despite GRI's acceptance, several sustainability measurement frameworks have been developed and propose indicators that move beyond the ones suggested by the GRI, for example by linking together different sustainability measures in cause and effect relationships (Searcy 2012), by proposing process together with outcome metrics (Delmas et al. 2013), or to standardize the different sustainability indicators accordingly to different organizations (Semenova and Hassel 2015). Case studies have been deployed to demonstrate the importance of sustainability measurement within organizations, to explore which indicators need to be developed, to highlight the importance of involving external stakeholders in the development of the indicators, and the use of existing standards as a reference point (Keeble et al. 2003). Additionally, a number of publications have also focused on sustainability Balanced Scorecard (Dias-Sardinha and Reijnders 2005). Together with research at the company-level that has focused on individual indicators, the development of composite sustainability indices that focused on the sustainability of companies and sectors (Veleva and Ellenbecker 2001; Singh et al. 2007) has also been investigated. Therefore composite indices on sustainable production systems, energy and material usage, emissions to the natural environment, economic performance, products, workers, and community development and social justice have been developed (Fan et al. 2010).

Despite many contributions on this topic, corporations still struggle to develop, implement, use, and improve sustainability metrics that address the needs of both internal and external stakeholders.

3   Method

3.1   Metrics

In order to collect data for this study we developed a set of 56 indicators selected from the Global Reporting Initiative G4 Guidelines. In order to define which indicators to include, the two authors developed a preliminary list of indicators which characterize the field of interest. The list was then refined through the interaction with a panel of ten managers belonging to five companies that are very proactive in the sustainability field (for a similar procedure). The 56 metrics are dummy variables and represents the presence (code = 1) or the absence (code = 0) of the sustainable practice within the company. These indicators were subsequently grouped into 11 sustainability processes based on homogeneous metrics. All indicators focused on process rather than outcome measures, as detailed figures of outcome measures are rarely disclosed by organizations (see Delmas et al. 2013). Table 1 details the 11 sustainability processes and the related number of KPIs.

Table 1.   Indicators and sustainability processes

Sustainability process	Metric
Environmental certifications	8
Social certifications	6
Energy use	7
Use of raw materials	9
Waste management	10
Environmental impact	3
Sustainability reporting	3
Employees welfare	2
CSR	4
Sustainable supply chain	1
Sustainable labeling	3

3.2 Sample and Data Collection

A sample of 400 companies was selected in order proceed with the assessment process. Companies were selected based on their size. Specifically, we employed data on 2014 turnover provided by Aida, a Bureau Van DijK database containing the balance sheets of all Italian companies. We identified the first four industries in terms of total turnover 2014 in the Emilia Romagna region - i.e. food (Ateco 10), machineries (Ateco 28), ceramics (Ateco 23), and mechanics (Ateco 25) – and then selected the first top 100 companies for each industry. Overall these companies represented about 35 billion euros and about 70,000 full time employees.

Data for this study were then collected through secondary sources – i.e. company's website and sustainability report. In order to collect the data we explored and went through all 400 companies' websites and sustainability reports searching for the 56 identified sustainability indicators. Some companies didn't have neither their website available nor the sustainable report, therefore were omitted from the data collection process. Overall, 336 companies were assessed, providing a response rate of 84%.

3.3 Methodology

Data have been analysed using univariate and multivariate techniques. Descriptive analysis and boxplot detailing sample distribution have been proposed, together with correlation analyses exploring the relations among sustainability processes.

4 Results

Three sets of results are presented. Firstly, a descriptive analysis details the level of diffusion of sustainability practices among the selected companies. Then, we present a benchmarking analysis that compares the sample average diffusion of sustainability processes in our sample, with companies representing leaders in the adoption of such practices. Finally, a correlation analysis details the relationship among the different sustainability processes.

It is worth noting that the aim of this paper is to present the development of the SuMM Lab dataset and not to discuss the results obtained by means of the pilot study. Therefore the results shown in the following sections have been used to exemplify possible uses and implementation of the SuMM Lab.

4.1 Descriptive Analysis

Figure 1 reports a descriptive analysis of the environmental certifications adopted by companies. The figure is based on a split sample analysis comparing small, medium and large organizations based on turnover values (Euro).

Besides reporting results on the specific indicators, we also computed an aggregated index for each sustainability process analyzed. The index was developed based on the ratio of the number of indicators that scored 1 compared with the total number of

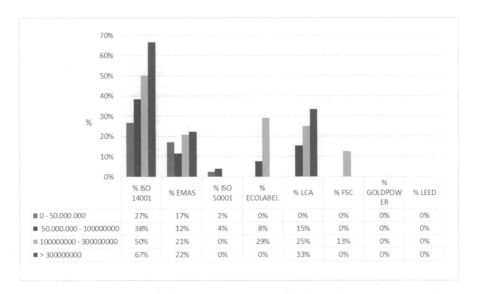

Fig. 1. Environmental certifications

indicators within a specific process. For example, given that the process on "environmental certifications" contains 8 metrics, a company reporting 4 certifications was assign a value of 50% in the "environmental certifications index" Fig. 2 reports the scores for all sustainability processes divided by *company* size, while Fig. 3 details information on the distribution of the sustainability process indices based on a boxplot analysis.

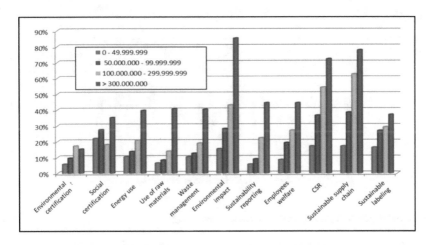

Fig. 2. Descriptive analysis of sustainability processes

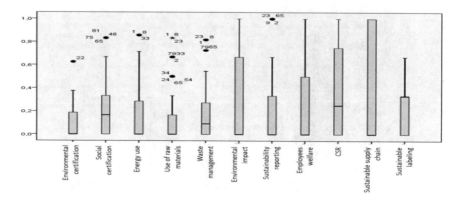

Fig. 3. Boxplot analysis: distribution of sustainability processes

4.2 Benchmarking Analysis

Figure 4 details the benchmarking analysis, that compares sample average with the companies that are leaders in the specific sustainable process analyzed. Companies can use this analysis in order to assess how they perform compared to the "best-in-class" among industry peers.

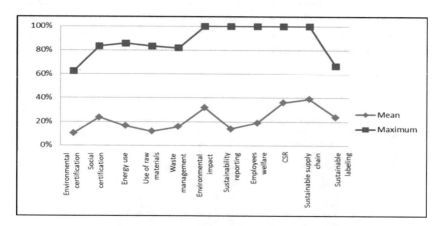

Fig. 4. Benchmarking analysis

4.3 Correlation Among Sustainability Processes

Table 2 reports the correlations among aggregated indexes of sustainability processes, together with turnover and FTEs. The use of raw materials emerge as significantly correlated with energy use, waste management, and environmental impact. Additionally, environmental certifications emerge as clearly related to sustainability in the supply chain, while waste management is related to sustainability practices towards consumers (i.e. sustainable labeling).

Table 2. Correlation analysis among sustainability processes

	1	2	3	4	5	6	7	8	9	10	11	12	13
1. Turnover	1.00												
2. Full Time Employees	.69	1.00											
3. Environmental certify.	.25	.27	1.00										
4. Social certifications	.13	.23	.39	1.00									
5. Energy use	.39	.28	.47	.39	1.00								
6. Use of raw materials	.49	.36	.56	.38	.68	1.00							
7. Waste management	.40	.28	.56	.38	.71	.85	1.00						
8. Environmental impact	.44	.34	.55	.27	.63	.71	.75	1.00					
9. Sustainability reporting	.45	.39	.68	.28	.60	.72	.72	.57	1.00				
10. Employees welfare	.42	.43	.48	.22	.45	.51	.50	.49	.55	1.00			
11. CSR	.36	.32	.55	.22	.42	.43	.49	.52	.59	.78	1.00		
12. Sustain. supply chain	.31	.26	.68	.31	.36	.40	.47	.41	.58	.45	.68	1.00	
13. Sustainable labeling	.27	.28	.42	.37	.52	.47	.58	.46	.38	.50	.43	.40	1.00

Note: Pearson's correlation coefficients higher than .60 are significant at $p < .05$

5   Conclusions and Further Research

This paper has presented the development of a structured dataset of sustainable practices that have been developed and implemented by European companies. Although the paper reports the pilot phase of the research project, data collected so far provide interesting insights over the opportunities offered by the SuMM Lab dataset. By collecting companies' sustainability data based on standards metrics it is possible to map and compare sustainable performance of EU companies with details on different industries, company size, and countries; to perform benchmarking analyses in order to identify who are the leaders and what they do; to detect the main drivers of the adoption of sustainability practices and explore the relations among different sustainability processes.

There are numerous possibilities for future research in the sustainability measurement and management field. Considering the type of KPIs that could be developed, industry-specific risk needs to be taken into account through the use, for example, of GES risk evaluation scores for each industry. Also, the metrics used should be statistically validated, therefore factor analyses based on polychoric correlations could be employed to assess whether the metrics belong to the proposed sustainability processes. Additionally, considering the data retrieval process. Collaborations with computer science scholars are already in place in order to develop parsing algorithms for automatic retrieval of secondary data throughout the web. Preliminary results show that data on environmental and social certifications could be effectively collected through automated parsing algorithms as the reliability of such algorithms ranges from 94% to 97% according to different company size. Furthermore, the data collected process could be extended beyond Italy to EU organizations.

References

Adams, R., Jeanrenaud, S., Bessant, J., Denyer, D., Overy, P.: Sustainability-oriented innovation: a sistematic review. Intl. J. Manage. Rev. **18**, 180–205 (2016)

Delmas, M.A., Etzion, D., Nairn-Birch, N.: Triangulating environmental performance: what do corporate social responsibility ratings really capture? Acad. Manage. Perspect. **27**(3), 255–267 (2013)

Dias-Sardinha, I., Reijnders, L.: Evaluating environmental and social performance of large Portuguese companies: a balanced scorecard approach. Bus. Strategy Environ. **14**, 73–91 (2005)

European Commission (2012). http://ec.europa.eu/eurostat/statisticsexplained/index.php/File: Key_size_class_indicators,_nonfinancial_business_economy,_EU-28,_2012.png

Fan, C., Carrell, J.D., Zhang, H.-C.: An investigation of indicators for measuring sustainable manufacturing. In: 2010 IEEE International Symposium on Sustainable Systems and Technology, Arlington, May 2010

Global Footprint Network (2014)

Global Reporting Initiative: G4 Reporting Guidelines (2015)

Keeble, J., Topiol, S., Berkeley, S.: Using indicators to measure sustainability performance at a corporate and project level. J. Bus. Ethics **44**, 149–158 (2003)

Leite, L.R., Van Aken, E., Martins, R.A.: Case study on the effect of sustainability on performance measurement systems. In: Krishnamurthy, A., Chan, W.K.V. (eds.) Proceedings of the 2013 Industrial and Systems Engineering Research Conference (2013)

Neely, A.D., Adams, C., Kennerley, M.: The Performance Prism: The Scorecard for Measuring and Managing Business Success. Prentice Hall Financial Times, London (2002)

Nidomolu, R., Prahalad, C.K., Rangaswami, M.R.: Why sustainability is now the key driver of innovation. Harvard Bus. Rev. **87**(9), 56–64 (2009)

Randers, J.: 2052: A Global Forecast for the Next Forty Years. Chelsea Green Publishing (2012)

Searcy, C.: Corporate sustainability performance measurement systems: a review and research agenda. J. Bus. Ethics **107**, 239–253 (2012)

Semenova, N., Hassel, L.G.: On the validity of environmental performance metrics. J. Bus. Ethics **132**, 249–258 (2015)

Singh, R.K., Murty, H.R., Gupta, S.K., Dikshit, A.K.: Development of composite sustainability performance index for steel industry. Ecol. Ind. **7**, 565–588 (2007)

SustainAbility: Model Behavior. 20 Business Model Innovations for Sustainability (2014)

Veleva, V., Ellenbecker, M.: Indicators of sustainable production: framework and methodology. J. Clean. Prod. **9**, 519–549 (2001)

A Multi-Criteria Decision-Making Model to Evaluate Sustainable Product Designs Based on the Principles of Design for Sustainability and Fuzzy Analytic Hierarchy Process

Chanjief Chandrakumar[1], Asela K. Kulatunga[2(✉)],
and Senthan Mathavan[3]

[1] School of Engineering and Advanced Technology, Massey University,
Palmerston North, New Zealand
C.Chandrakumar@massey.ac.nz
[2] Department of Production Engineering, University of Peradeniya,
Peradeniya, Sri Lanka
aselakk@pdn.ac.lk
[3] Department of Civil and Structural Engineering, Nottingham Trent University,
Nottingham, UK
s.mathavan@ieee.org

Abstract. The exponential and adverse increase in global conditions like climate change, ocean acidification and aerosol loading stresses the urgency to achieve sustainable development in all global activities, including manufacturing. Despite the attempts of developing disparate approaches and concepts at different scales of economies, the global issues are still worsening. As a consequence, United Nations Environment Programme (UNEP) has introduced the concept of Design for Sustainability (DfS) to evaluate the sustainability aspects of product designs, acknowledging the importance of developing sustainable products (and services). Even though the DfS approach has been recognised as an appropriate base to establish a sustainable global economy, the scope of the assessment still requires further research, specifically with the social impacts assessment. Hence, in this study, we propose a multi-criteria decision support system based on the DfS principles, including an exhaustive set of criteria referred in the Social Life Cycle Assessment framework for the purpose of evaluating three chosen sanitation system designs. We applied the Fuzzy Analytic Hierarchy Process to solve the proposed model. Additionally, a sensitivity analysis was performed to understand the effects of the chosen priority weights. The analysis shows that the sanitation system design choice is heavily influenced depending on whether the high importance the packaging and logistics phase or the usage phase, emphasising the need for a comprehensive life-cycle analysis. The research reported in this paper is a step toward realising the targets of DfS at a practice level by simultaneously building on the original proposals and using them to solve real word design selection problems.

Keywords: Multi-Criteria Decision-Making · Design for Sustainability · Sanitation system design · Fuzzy Analytic Hierarchy Process

© Springer International Publishing AG 2017
G. Campana et al. (eds.), *Sustainable Design and Manufacturing 2017*, Smart Innovation,
Systems and Technologies 68, DOI 10.1007/978-3-319-57078-5_34

1   Introduction

Increasing awareness on the sustainable development issues has compelled the scientific and political communities to adopt the ideology of Sustainable Consumption and Production (SCP) in all global activities [1]. Consequently, manufacturers have started adopting the principles of sustainable manufacturing through developing eco-friendly products which conserve energy and natural resources, while maintaining minimal social impacts. In sum, the concept aims at incorporating the Triple Bottom Line (TBL) dimensions of sustainability into manufacturing: economic, environmental and social [2]. Nevertheless, achieving sustainability in all three dimensions is not quite straightforward, and it can only be achieved through improved product design and development [3]. In this sense, researchers have come up with disparate approaches and frameworks, including the concepts of eco-design [4, 5] and Design for Sustainability (DfS) [1]. The concept of eco-design benefits the manufacturers economically and environmentally; however, it lacks comprehensibility as it fails to assess a product's social impacts [3], and this limitation has been dealt through developing a comprehensive approach called DfS [1]. DfS extends its scope of assessment by embracing social impacts while linking product innovation with sustainability [1]. Interestingly, both concepts of eco-design and DfS adopt a life cycle thinking approach in determining the environmental hotspots. Although the scientific community has acknowledged the DfS concept as a base for assessing a product's overall impacts, the concept still requires further research, specifically with the criteria and indicators relevant to the social sustainability assessment. As a consequence of this limitation, until today, a very few papers have been published based on DfS [3, 6, 7]. Meanwhile, the UNEP launched a concept called Social Life Cycle Assessment (SLCA) to estimate the social impacts related to a product's life cycle [8]. Realising that the DfS framework as an appropriate base to assess the sustainability of a product's life-cycle and the SLCA methodology as a solution addressing the limitation of the DfS framework, we developed a comprehensive Multi-Criteria Decision-Making (MCDM) model. However, our intention was not to perform an SLCA, instead, we selected a set of relevant criteria and indicators (hereafter referred as Social Sustainability Indicators (SSI)) from the SLCA framework. Furthermore, we applied the Fuzzy Analytic Hierarchy process (FAHP) to solve the developed model.

2   Methodology

This section explains the five-fold MCDM model (hereafter called as "DfS+SSI"), as shown in Fig. 1. The complexity involved in the process of the selecting a sustainable design was broken down to integrate the economic, environmental and social impacts assessment approaches. Within this context, we determined a set of criteria and attributes to evaluate a product design based on the principles of DfS and SLCA. Considering the DfS+SSI criteria and assessment attributes, a five-level decision hierarchy model was developed, as illustrated in Fig. 2. Level 1 indicates the overall objective of selecting the best design alternative, while Level 2 aims at the entire life cycle from material extraction to end of life disposal (phases 1 to 5 in Fig. 2). The TBL

dimensions of sustainability are evaluated in Level 3, whereas, Level 4
focuses on the assessment attributes
based on the principles of DfS and
SLCA, and finally, Level 5 represents
the chosen alternatives. Chosen
designs were compared in terms of
both quantitative and qualitative attributes, and the comparisons were
translated into pairwise comparison
matrices (PCMs) based on experts'
views. Subsequently, the PCMs were
solved using the FAHP process (for
theories refer to [9]). Afterwards,

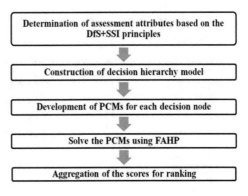

Fig. 1. Proposed framework

using the experts' knowledge, weights were estimated for all sustainability dimensions
and attributes, starting from the life-cycle (LC) phases. Lastly, an aggregation step was
performed to rank the designs based on their sustainability performances. The next
section explains how the proposed model works in a practical scenario.

3 An Illustrative Case-Study

Ensuring availability and managing water and sanitation sustainably has become a
global priority. As a result, the scientific community has launched a new research area
called sustainable sanitation - that is, closing the nutrient loops and minimising the use
of freshwater. The ideology of developing sustainable sanitation systems has gained
significance in many conferences. In this sense, a sustainable sanitation system design
competition for students was organised at a recent conference held in Malaysia, 2014:
the 12th Global Conference on Sustainable Manufacturing, and three top designs were
chosen.

In this analysis, we
developed a set assess-
ment criteria and attri-
butes pertaining to a
sustainable sanitation
system design to
re-evaluate the deci-
sions of the organisers
of the competition. For
this purpose, we selec-
ted three of the four
finalists [10–12]. The
first two were chosen as
the second runner-up
and the winner, and
the latter one was the

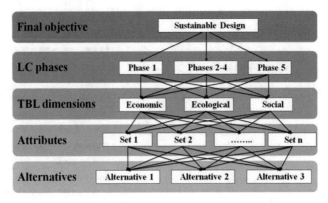

Fig. 2. Decision hierarchy model

Table 1. The selected assessment attributes for sustainable design selection process

LC phase (L_i)	TBL (T_i)	Assessment attributes ($C_{i,j}$) i-LC phase, j-attribute		Designs (X_i)
Material extraction or acquisition (L_1)	Economical (T_1)	Amount of virgin & packaging materials	$C_{1,1}$	Design 1 (X_1)
	Environmental (T_2)	Amount of consumption of recovered materials	$C_{1,2}$	Design 2 (X_2)
	Social (T_3)	Lifetime	$C_{1,3}$	Design 3 (X_3)
		Cost of assembly	$C_{1,4}$	
Manufacturing or assembly (L_2)		Time consumption to complete	$C_{2,1}$	
		Space required	$C_{2,2}$	
		Process improvements	$C_{2,3}$	
		Technology	$C_{2,4}$	
		Energy consumption	$C_{2,5}$	
Packaging and logistics (L_3)		Transport costs	$C_{3,1}$	
		Optimized methods: packaging important parts	$C_{3,2}$	
		Amount of consumption of packaging materials	$C_{3,3}$	
		Emissions	$C_{3,4}$	
		Energy consumption	$C_{3,5}$	
		Green materials	$C_{3,6}$	
Usage (L_4)		Water efficiency	$C_{4,1}$	
		Frequency of cleaning	$C_{4,2}$	
		Cleaning costs	$C_{4,3}$	
		Cost of repairing or replacements	$C_{4,4}$	
		Number of subassemblies	$C_{4,5}$	
		Ease of handling	$C_{4,6}$	
		Complexity of the model	$C_{4,7}$	
		Waste utilization-fertilizers	$C_{4,8}$	
		Urinary reuse	$C_{4,9}$	
		Odor or gas emissions	$C_{4,10}$	
End of life disposal (L_5)		Amount of reused packaging materials	$C_{5,1}$	
		Recovery of sub-assemblies for reuse	$C_{5,2}$	
		Cost of recovery or recycle	$C_{5,3}$	
		Number of recovered components	$C_{5,4}$	
		Maintenance	$C_{5,5}$	
		Amount of dumped or not utilized materials	$C_{5,6}$	
		Reparability	$C_{5,7}$	
		Easiness for recovery	$C_{5,8}$	

fourth semi-finalist (the data relevant to second runners-up was not accessible). These are referred as X1, X2 and X3, respectively, from now on. The three designs were then compared based on the DfS+SSI principles. Furthermore, an LC approach was adopted for evaluating the entire LC of each design. Afterwards, the assessment attributes were chosen based on the design specifications. Table 1 illustrates the selected criteria and assessment attributes.

One of the challenges we faced during the model development process was that certain attributes were common to different LC phases, for instance, the amount of consumption of packaging materials (C1,1) was common for phases like materials extraction or acquisition (L1), packaging and logistics (L3) and end of life disposal (L5). However, this challenge was addressed by assigning different priority weights for the common attributes in different LC phases. Next, the decision hierarchy model was developed and the linguistic pairwise comparisons were proposed for each TBL dimension, attribute and corresponding design alternatives. We then transformed the linguistic comparisons into triangular fuzzy numbers (refer to [9] for more details), and established the PCMs. Finally, the PCMs were evaluated using the FAHP process and the choices were ranked following a weighting step. For the purpose of comparing the design alternatives and setting priority weights, a group of six experts from the Environmental Science and Engineering field were consulted. The paragraphs below explain how the process was followed for the case study in consideration.

Initially, the TBL aspects under the usage phase (L4) as shown in Table 2 was evaluated using FAHP, and the weight for each TBL dimension was estimated. Afterwards, the PCM constructed for comparing the criteria under the each TBL dimension (Table 3) was evaluated to determine the priority weights for each criterion. In this paper, we present the social dimension (T3) of the usage phase (L4). Later, the chosen design alternatives were compared corresponding to each criterion (Table 4), and the corresponding weights were obtained. Likewise, the aforementioned approach was iterated for other phases. Ultimately, the weight for each design alternative was calculated with regard to the weights estimated for each TBL dimension, LC phase and assessment attributes. Table 5 presents the aggregated weights for the design alternatives against the LC phases. Calculations and the details of other phases, criteria and attributes were left out here for space considerations. Now, the importance of each LC phase must be considered in order to aggregate the weights shown in Table 5.

We used a reasonable set of normalised values proposed by the experts in the absence of a full-LCA: L1-0.1, L2-0.35, L3-0.1, L4-0.35, and L5-0.1. According to [12], most of the sustainability-related impacts of a sanitation system can be identified in the phases of manufacturing and usage. Therefore, the proposed weights could be considered as a good proxy. We obtained the mean value of the weights proposed by the experts; however, this distribution of numbers could be still subjective. After choosing the priority weights for the LC phases the aggregate weights for design alternatives were estimated. Table 6 presents the aggregated weights of the alternatives. Hence, in real life sanitation system design evaluation problems, performing a full-LCA would be highly recommended; although there are limitations in the LCA methodology (performing a full-LCA is out of the scope of this study).

Table 2. PCM for sustainability dimensions: Usage phase (L4)

TBL	T_1	T_2	T_3
Economic (T_1)	(1, 1, 1)	(1, 3/2, 2)	(1, 3/2, 2)
Environmental (T_2)	(1/2, 2/3, 1)	(1, 1, 1)	(1, 3/2, 2)
Social (T_3)	(1/2, 2/3, 1)	(1/2, 2/3, 1)	(1, 1, 1)

Table 3. PCM for social dimension (T3)

Social	$C_{4,10}$	$C_{4,3}$	$C_{4,8}$	$C_{4,6}$
Odor/gas emissions	(1, 1, 1)	(3/2, 2, 5/2)	(3/2, 2, 5/2)	(1, 3/2, 2)
Cleaning costs	(2/5, 1/2, 2/3)	(1, 1, 1)	(1/2, 2/3, 1)	(1/2, 2/3, 1)
Waste utilization	(2/5, 1/2, 2/3)	(1, 3/2, 2)	(1, 1, 1)	(2/3, 1, 2)
Ease of handling	(1/2, 2/3, 1)	(1, 3/2, 2)	(1/2, 1, 3/2)	(1, 1, 1)

Table 4. PCM for cleaning costs (C4,3)

Cleaning costs	X_1	X_2	X_3
Design X_1	(1, 1, 1)	(3/2, 2, 5/2)	(2, 5/2, 3)
Design X_2	(2/5, 1/2, 2/3)	(1, 1, 1)	(1, 3/2, 2)
Design X_3	(1/3, 2/5, 1/2)	(1/2, 2/3, 1)	(1, 1, 1)

Table 5. Priority weights of the alternatives: all phases

		LC phases				
		L_1	L_2	L_3	L_4	L_5
Weight	X_1	0.319	0.329	0.347	0.336	0.329
	X_2	0.356	0.377	0.327	0.425	0.369
	X_3	0.325	0.294	0.326	0.239	0.303

Table 6. Aggregate weights of the alternatives

Design	X_1	X_2	X_3
Aggregated weight	0.332	0.386	0.282

4   Results

According to Table 6, the ranking order of the three sustainable sanitation system designs is as follows: X2 > X1 > X3. Thus, based on our analysis, we conclude that X2 is the most preferred design. It is interesting to note that these design choices derived by our methodology concur with the award rankings for the three designs at the conference. Furthermore, as shown in Table 5, different designs show better performance in individual phases. For instance, X2 is preferred in the phases of material extraction, manufacturing, usage and end-of-life. Likewise, X1 performs well in the phase of packaging and logistics. Hence, to understand the effects of incorporating DfS+SSI principles (S0) in the product design and development stage we re-iterated the model for the following scenarios as well: only economic (SI), only environmental (SII), only

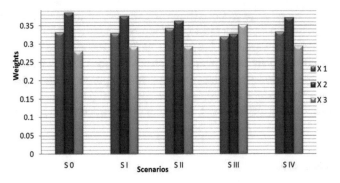

Fig. 3. Comparison of the five scenarios

social dimensions (SIII) and eco-design (SIV). The obtained results are illustrated in Fig. 3. According to Fig. 3, X2 shows better performance in all scenarios, except for the scenario SIII. Meanwhile, X3 demonstrates better social performance, compared to X1 and X2. We, therefore, recommend adopting an LC approach in evaluating the sustainable design choices to avoid burden shifting. Even though the sensitivity analysis was performed for 79 combinations of different weights, we do not include all the results due to space limitations.

5   Conclusions

The sustainable designs evaluation process is an MCDM problem as it involves with complex and critical decisions. Until today, many attempts have been made to assess sustainability; however, the scope of the assessment was only limited to economic and environmental dimensions. Lately, the UNEP introduced the DfS framework to develop sustainable products. Nevertheless, the DfS framework was criticised for its inability to comprehensively assess social impacts. Hence, we developed an MCDM model to evaluate designs by integrating the UNEP's DfS framework with the criteria and attributes recommended in the S-LCA methodology. The proposed five-fold approach aimed at developing a decision hierarchy model and PCMs, and solving them using FAHP. The leverage of FAHP was used to assess the vague and complex criteria and attributes, specifically with environmental and social dimensions, using fuzzy numbers instead of crisp values. Afterwards, the proposed MCDM model was applied to select the best sustainable sanitation system design, and the outcomes were compared with the already published results. Besides, a sensitivity analysis was carried out to understand the impacts as a result of varying priority weights. The robustness of the decision for small changes in relative weights around the chosen set of weights showed that as long as major numbers were captured in an LCA for sanitation systems, the design choice outcome can be predicted reasonably. Furthermore, the sensitivity analysis provided us with an insight on how different combinations from the economic, environmental and social influence the decision on best design. This consideration would be beneficial when evaluating any other general products. However, there are few challenges involved with this approach. The availability of data required for the application of the proposed approach was one of them, and also the case-study was not carried out based on a full-LCA. But, conducting a full-LCA using reliable data sources would increase the precision of the analysis. Besides, we recommend that the proposed framework can be combined with a number of evaluation criteria to assign more accurate priority weights. In sum, the proposed novel approach of evaluating sustainable product designs will be beneficial from sustainable product design and development perspectives, as the study provided a base to integrate key frameworks like DfS and SLCA.

References

1. UNEP Division of Technology Industry and Economics: Design for Sustainability – A Practical Approach for Developing Economies (2007)
2. Jayal, A.D., Badurdeen, F., Dillon, O.W., Jawahir, I.S.: Sustainable manufacturing: modeling and optimization challenges at the product, process and system levels. CIRP J. Manuf. Sci. Technol. 2(3), 144–152 (2010)
3. Spangenberg, J.H., Fuad-Luke, A., Blincoe, K.: Design for Sustainability (DfS): the interface of sustainable production and consumption. J. Clean. Prod. 18(15), 1485–1493 (2010)
4. Chandrakumar, C., Kulatunga, A.K., Mathavan, S.: Fuzzy AHP based multi-criteria decisions support system for eco-design. In: Proceedings of 6th International Conference on Industrial Engineering and Operations Management, pp. 322–328 (2016)
5. Chandrasegaran, S.K., Ramani, K., Sriram, R.D., Horváth, I., Bernard, A., Harik, R.F., Gao, W.: The evolution, challenges, and future of knowledge representation in product design systems. Comput. Aided Des. 45(2), 204–228 (2013)
6. Küçüksayraç, E.: Design for sustainability in companies: strategies, drivers and needs of Turkey's best performing businesses. J. Clean. Prod. 106, 455–465 (2015)
7. Küçüksayraç, E., Keskin, D., Brezet, H.: Intermediaries and innovation support in the design for sustainability field: cases from the Netherlands, Turkey and the United Kingdom. J. Clean. Prod. 101, 38–48 (2015)
8. UNEP-Life Cycle Initiative: Guidance on organizational life cycle assessment.http://www. lifecycleinitiative.org/wp-content/uploads/2015/04/olca_24.4.15-web.pdf. Accessed 20 Dec 2015
9. Büyüközkan, G., Kahraman, C., Ruan, D.: A fuzzy multi-criteria decision approach for software development strategy selection. Int. J. Gen. Syst. 33(2–3), 259–280 (2004)
10. Mainguy, G., Wernitz, C.: Manufacturing for a sustainable terra preta sanitation system. Procedia CIRP 26 (2015). http://www.gcsm.eu/files/contributions/GCSM2014_Gabrielle% 20Mainguy.pdf
11. Xavier University: Terra Preta sanitation system for post disaster transitional communities in the Philippines. Procedia CIRP 26 (2015). http://www.gcsm.eu/files/contributions/GCSM 2014_Xavier%20University.pdf
12. Ihalawatta, R.K., Kuruppuarachchi, K.A.B.N., Kulatunga, A.K.: Eco-friendly, water saving sanitation system. Procedia CIRP 26, 786–791 (2015)

Renewable Energies for Sustainable Manufacturing and Society

Implementation of an Advanced Automated Management System for the Optimization of Energy and Power Terms in a Water Purification Plant (WPP) with a Photovoltaic Plant (PP)

Jesús Chazarra Zapata[1(✉)], Imene Yahyaoui[2], Javier Castellote Martínez[3],
José Miguel Molina-Martínez[4], Manuel Estrems Amestoy[3],
and Antonio Ruiz Canales[1]

[1] Departamento de Ingeniería, EPSO, Universidad Miguel Hernández de Elche, Elche, Spain
jesuschazarra@gmail.com, acanales@umh.es
[2] Federal University of Espírito Santo, Espírito Santo, Brazil
imene.yahyaoui@ufes.br
[3] Departamento de Materiales y Fabricación, Universidad Politécnica de Cartagena,
Cartagena, Spain
javiercastellotemartinez@gmail.com, manuel.estrems@upct.es
[4] Departamento de Ingeniería Agromótica y del Mar, Universidad Politécnica de Cartagena,
Cartagena, Spain
josemiguel.molina.martinez@gmail.com

Abstract. Currently, the use of software systems for optimizing the sustainable use of water and energy resources has been implemented in a wide range of productive sectors. In the sector of Water Purification Plants (WPP) the sustainable control of water and energy consumption is carried out by means of advanced Supervisory Control And Data Acquisition (SCADA) systems. The main objective of these SCADA systems is controlling the consumption of these resources in real time. In the majority of the occasions, optimization systems for the sustainability consumption of these resources are not integrated in the SCADA system. With the integration of systems for optimization, the determination in real time of the most viable option for the sustainable management of resources is expected. In this paper, a case study in the Southeast Spain of an advanced automated management system for the optimization of energy and power terms in a WPP with renewable energy based in a Photovoltaic Power Plant (PVPP) is presented. Firstly, the introduction and some additional details for the entire problem are presented. Moreover, the main parts of the software integrated in the SCADA system of the WPP will be detailed.

Keywords: Environmental sustainability software · Water Purification Plant · Photovoltaic plant · Optimization · Energy consumption

© Springer International Publishing AG 2017
G. Campana et al. (eds.), *Sustainable Design and Manufacturing 2017*, Smart Innovation,
Systems and Technologies 68, DOI 10.1007/978-3-319-57078-5_35

1   Introduction

In a Water Purification Plant (WPP), the energy and water costs generally represent an important part of the total costs. This is a function on the energy dependence of the plant system. However, these costs have increased in an alarming way in the last years. This is due to the increasing of the energy price. In the case of Spain, the prices of the access tolls, the tendency of the power billing term (PBT) was increasing from 2008 to nowadays. From 2008 and 2009 the increase was a 30%. During 2009 and 2010 they increased a 20%. From 2010 to nowadays the increment has been reached to 2%. The billing term of active energy (BTAE) has been maintained along the same period [1–3]. Currently the market is free and it is possible to meter measure to the tariffs depending on the type of contract. For this reason it is possible to manage the tariffs with the commercializing company.

The use of renewal energies has been an alternative for saving energy costs in this kind of systems. One of the common renewal technologies in WPP are the photovoltaic plants. This renewal energy provides the power to the main pumps of this system and it allows to the auto consumption. This is the maintenance of energy of these plants without the dependence of the energy network. Improving reliability and reducing maintenance and operating costs have become important factors in increasing the competitiveness of photovoltaic (PV) systems. A plant's operator can only adopt prompt measures to eliminate operational faults when these are immediately signaled. For this reason, continuous, absolute and comparative measurements are necessary to ensure the highest efficiency and availability of photovoltaic plants. In this context, a modeling software for taking data from energy and predict the operation in real time has been developed [4].

The quality of the measured data has a significant influence on the estimation accuracy of the software and thus on the SCADA system of the plant. In this case, models for irradiance and PV module temperature are included to check the measured data for inconsistency and other anomalies [5, 6].

All this has been developed not only seeking to reduce economic costs, efficiency and reduction of water and energy consumption, but to protect against climate change, reducing CO_2 emissions and trying to comply with the main COP21 agreements. This conference had as main objective to avoid that the increase of the global average temperature exceeds 2 °C with respect to the pre-industrial levels and also seeks to promote additional efforts that allow global warming not to exceed 1.5 °C. Small steps such as the introduction and management of this type of software, will revert in a future great progress towards the sustainable conservation of the planet. [7].

Specific software based in the previous items has been applied to a case study PVPP included in a WPP, where a Supervisory Control and Data Acquisition (SCADA) system has been installed; operational data are saved on a remote server and available on a dedicated website. The use of this software shows the effectiveness of the proposed approach. In this study a case study in the Southeast Spain of an advanced automated management system for the optimization of energy and power terms in a WPP with a PVPP is presented. Firstly, the introduction and some additional details for the entire problem will be presented. Moreover, the main parts of the software integrated in the SCADA system of the WPP will be detailed. This software is based in genetic algorithms.

2 Materials and Methods

The specific software is a decision support system (DSS) that is programmed in PC computing environment. Specifically, is programmed in Visual Basic. This is due to the necessary power and flexibility in the calculations. Nevertheless, because of the operation of this WPP for a big population, the SCADA system is independent of the DSS. Another reason for this difference is the configuration of security of water supply and security protocols of operation in the WPP. For this reasons, software is not operating in real time. The information is sent to the SCADA system once per day and previously to the approval of the supervisor of the plant.

The specific software for energy optimizing (SOE, with the initial letters of Software de Optimización Energética, in Spanish) consists of a main window with five information tabs. First tab shows a general synoptic panel of all the system of the WPP. In the next tabs the detailed information of every part of the WPP is shown. In this specific system these parts are: (a) Elevation to the Security Swamp/Security Swamp, (b) Elevation from pumping station 1/Reserve tank, (c) Elevation from pumping station 2 and pumping station 3/Dam. Moreover, the information of the PP is included.

The input data of the SOE comes from several sources.

The data of the PVPP are obtained from a file with *.XML format. This file is generated by the control machine of the PVPP. In this file, the information of every 15 min along one is included. This file is processed by the SOE and it is considered as the production of the next day.

The price and CO_2 production linked to the generated power (kW) is directly introduced daily in a window of the DSS software. Along the day and hourly both data are inserted. The obtaining of these data is difficult to forecast and to calculate. This is due to the volatility of the electrical demand and the complexity of the energy mix generation in the Iberian Peninsula. This is very varied because the nuclear and the hydraulic energy are operating in continuous. Wind energy is operating continuously but it is important during the night. Solar energy is important only during the warm months. All of them have an obligatory consumption and they get into the electric tender from zero price. Only when the consumption is important, this is working days and very warm or very cold days, the production of combined cycles and coal is included. This generates the highest prices and CO_2 production and the hourly variation of them. Moreover, the energy price is low when the consumption is low. In this case, the energies with a low quantity of CO_2 emissions are used. These data are obtained from www.omie.es [8] or www.ree.es [9].

Data of water production in the WPP are programmed by the supervisor in the PC where the SOE is installed. This is an hourly flow data. Instead of using the data from the SCADA the introduced data are supervised by the responsible and the supervisor of the plant. This method has been included in order to adapt the procedures of the WPP to the SOE.

The data from the accumulated water volume in every reservoir is obtained from a TXT extension file. This file generates the control order of the plant. This date is sent hourly in order to prepare the daily scheduling to the SOE or sending an advertisement to the supervisor. This advertisement will be sent if some significant deviation between the programmed flow by the SOE and the real flow in the reservoirs are produced.

Security swamp volume. This data is obtained and sent by the automaton that it is controlling the pumping system of the swamp. A TXT file is generated and it is saved in the same ftp where are saved all the programs.

The supplied data by the PLC-SCADA of the WPP to the SOE are the next (Fig. 1):

Functioning regime of Apolonia pumping system.
Functioning regime of Swamp pumping system.
Functioning regime of elevation IV of Lorca.
Functioning regime of elevation Águilas to Puerto Lumbreras.
Functioning regime of Blowing system of WPP.
Functioning regime of water recirculation pumps in the plant.

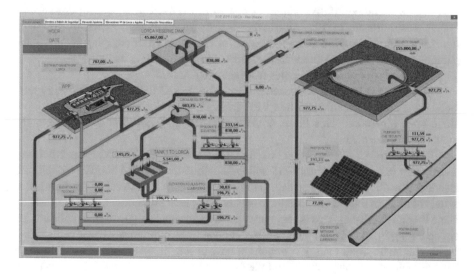

Fig. 1. General synoptic tab of WPP.

All the data are supplied, as the previous, by means of interchange data files. There is no direct communication between SOE system and the SCADA system. This is ought to allow the fiability in the plant functioning.

In summary, the data that are exported by the SOE system to the SCADA system are sent once per day and previously to the approval by the plant's supervisor. Moreover, the SOE system sends an email to the included addresses in the configuration system. This is presented with a PDF format file that includes the functioning format for the plant in the next day. The data sending is once per day, excepting the case where the supervisor changes the order. This has been configured in order to guarantee the knowledge of plant functioning one week in advance.

2.1 General Synoptic Tab

In this tab a general map of the WPP system is presented. This is included in Fig. 1. In this figure, the hydraulic system with the main pumping systems and reservoirs connected by several pipelines are represented.

In this tab, the optimized operation of the system in real time is shown. By means of graphical figures, the active pumps for every pump system in every elevation are represented in green. The active scorings of the entire system are presented in dark blue. The direction of the water circulation is presented with arrows and in green color. Moreover, the water level in the dams is presented.

Additionally, textual information from information panels is presented. These frames are showing information about entering and outgoing flows, volume, demanding power, produced energy and avoided CO_2 emissions to the atmosphere.

The information panels follow the next color pattern in order to help to identify what kind of information is presented:

- Square with blue letters: Entering flow to elevation, dam or WPP.
- Square with red letters: outgoing flow of elevation, dam or WPP.
- Square with black letters on the right of an elevation: Demanding power by the elevation.
- Square with black letters on a dam: Current volume of a dam.
- Square with green letters: Expected produced energy in the photovoltaic system.

2.2 Elevation to the Security Swamp/Swamp Tab

In this tab (Fig. 2) the graphical and textual information as the previous tap is shown. Moreover, a detailed information frame with a daily resume of the relevant operation parameters is shown. This frame is located on the left of the screen and pumping elevation data and security swamp data are reflected.

The data that are shown are:

- Energy consumption.
- Energy per elevated volume ratio.
- Economic cost.
- CO_2 emissions.
- Avoided CO_2 emissions.
- Incoming flow to the Security swamp.
- Outgoing flow to the Security swamp.
- Initial volume of the Security swamp.
- Final volume of the Security swamp.

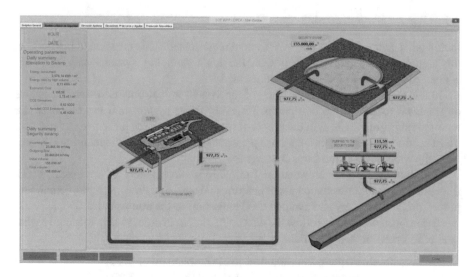

Fig. 2. Elevation to the Security Swamp/Swamp tab

2.3 Elevation from Pump Station 1/Reserve Tank Tab

In the Elevation from pump station 1/Reserve tank tab (Fig. 3) the initial general information tab is presented.

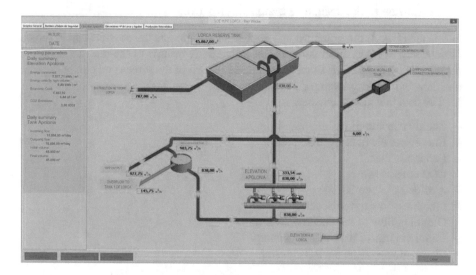

Fig. 3. Elevation from pump station 1/Reserve tank.

Moreover, a frame of detailed information with a daily resume of the more relevant parameters is shown. This frame is located on the left of the frame and these are the included main data:

– Consumed energy.
– Energy per elevated volume ratio.
– Economic cost.
– CO2 emissions.
– Incoming flow to the reserve tank.
– Outgoing flow of the reserve tank.
– Initial volume of the reserve tank.
– Final volume of the reserve tank.

2.4   Elevation from Pumping Station 2 and Pumping Station 3/Dam

This tap (Fig. 4) includes graphical information and textual as the general synoptic frame. A detailed frame of the daily resume of the main functioning parameters is integrated. This is located on the left of the screen and these is the main data:

– Consumed energy for elevation from pumping station 2.
– Energy per elevated volume for pumping station 2 ratio.
– Energy per elevated volume for dam ratio.
– Economic cost for elevation of pumping station 2.
– Economic cost for elevation of pumping station 3.
– Emissions of CO2 for elevation of pumping station 2.
– Emissions of CO2 for elevation of pumping station 3.
– Avoided emissions of CO2 for elevation of pumping station 2.
– Incoming flow for dam.
– Outgoing flow for dam.
– Initial volume of dam.
– Final volume of dam.

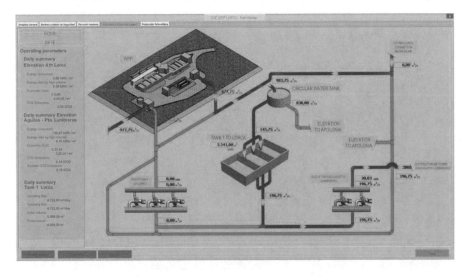

Fig. 4. Elevation from pump station 2/Dam.

2.5 Photovoltaic Production Tab

In this tab (Fig. 5), the graphical information of the expected photovoltaic production for the current data is presented. Moreover, graphical information about the expected energy consumption of the system against the produced energy by the photovoltaic system along the day is provided.

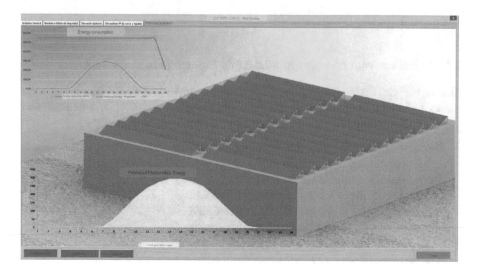

Fig. 5. Photovoltaic production tab.

For all the system, is possible to increase the information about elevations with an emergent window (Fig. 6). This window shows an image of a pumping system. In this

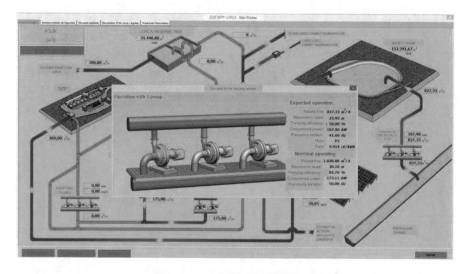

Fig. 6. Pumping information tab.

image, the number of active pumps (Green), textual information of the expected operation for this moment and nominal operation is presented.

This is the presented information:

– Raised flow.
– Manometric head.
– Pumping efficiency.
– Consumed power.
– Frecuency of the variator.

The SOE software has been designed as information software. For this reason some basic functionalities has been included:

– Access to the hourly prediction.
– Introduction of entering data.
– Configuration of dam volumes.
– Exportation of the result informs.

2.6   Acces to the Hourly Prediction

Using the daily prediction button (Fig. 7) the prevision per hour of the operation in the current data can be per hour can be consulted hourly. In this way it is possible to check the predicted functioning for the WPP system along the 24 h of the day.

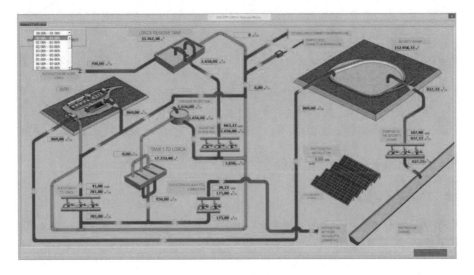

Fig. 7. Hourly prediction access.

For this purpose, it is only necessary to select the hour with the folder located at the top on the left. Then the data will be updated in the screen.

Some of the restrictions of this software are us follows: (a) the data cannot be negative; (b) the maximum limit of the distribution network demand is established in

$1{,}300 \text{ m}^3 \cdot \text{h}^{-1}$. This limit is over the worse demand scenario of $20{,}000 \text{ m}^3 \cdot \text{day}^{-1}$ that was registered in the last years. As initial point, an input database has been generated from the obtained information of the users.

In order to calculate the prediction of the PPW system two operation modes can be selected. Firstly, manual introduction of data can be selected. Moreover, it is possible working with an integrated database in the software.

In order to generate a new prediction it is necessary to push the button "Apply" and wait to finish the simulation.

In order to use the introduced manual data it is necessary to use the option "New data". Then, the data will be introduced in the activated zone of the lower part of the screen on the left and push the arrow in order to conform the data.

From this emerging window (Fig. 8) the timetable and monthly energy prices depending on the current tariff (Spanish company Iberdrola España S.A.U.) can be observed.

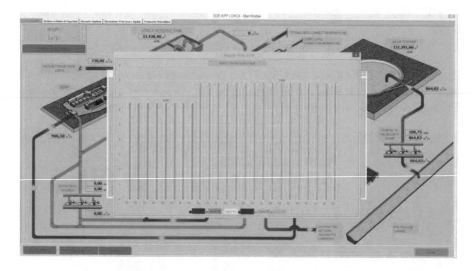

Fig. 8. Hourly and monthly energy prices screen.

2.7 Configuration

With this button (Fig. 9) it is possible to access to the emergence window where the configuration of the parameters of the tanks can be done. The configurable parameters are:

- Maximum volume.
- Minimal volume.
- Initial volume.

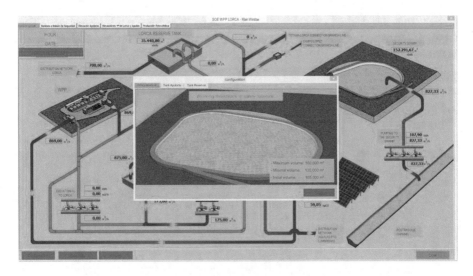

Fig. 9. Configuration screen.

2.8 Export of the Result Informs

Using this button (Fig. 10) it is possible to access the emergent window of report generation. From this function is possible to create daily informs and/or monthly reports with the obtained results after the simulation.

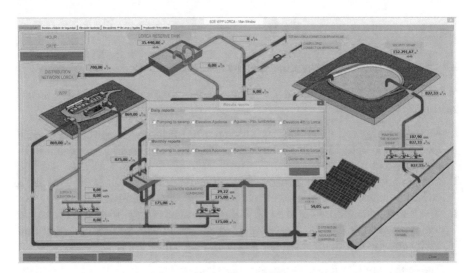

Fig. 10. Export of results.

Selecting the desired checkbox (Fig. 11) some PDF files can be generated. They supply information about daily functioning parameters of elevations and dams. The generated reports are:

- Elevation A.
- Pumping to the Security dam.
- Elevation B.
- Elevation C.

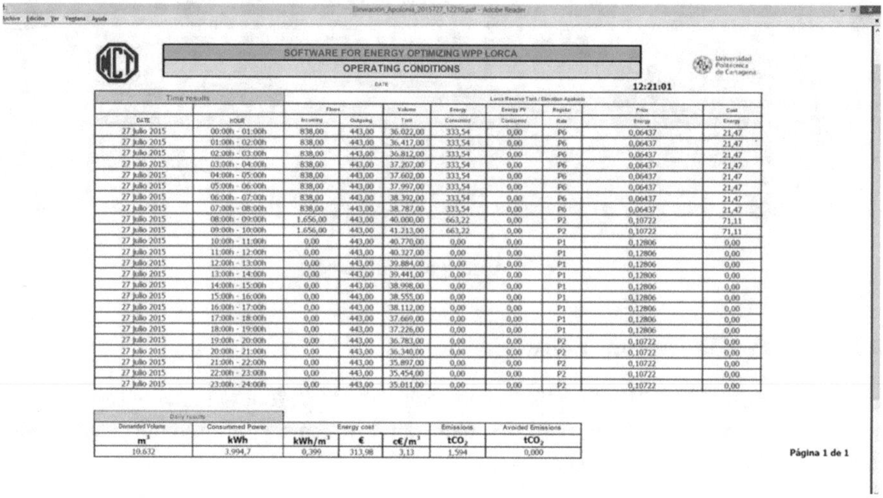

SOFTWARE FOR ENERGY OPTIMIZING WPP LORCA

OPERATING CONDITIONS

DATE 12:21:01

Time results		Lorca Reserve Tank / Elevation Apolonia							
		Flows		Volume	Energy	Energy PV	Regular	Price	Cost
DATE	HOUR	Incoming	Outgoing	Tank	Consumed	Consumed	Rate	Energy	Energy
27 julio 2015	00:00h - 01:00h	838,00	443,00	36.022,00	333,54	0,00	P6	0,06437	21,47
27 julio 2015	01:00h - 02:00h	838,00	443,00	36.417,00	333,54	0,00	P6	0,06437	21,47
27 julio 2015	02:00h - 03:00h	838,00	443,00	36.812,00	333,54	0,00	P6	0,06437	21,47
27 julio 2015	03:00h - 04:00h	838,00	443,00	37.207,00	333,54	0,00	P6	0,06437	21,47
27 julio 2015	04:00h - 05:00h	838,00	443,00	37.602,00	333,54	0,00	P6	0,06437	21,47
27 julio 2015	05:00h - 06:00h	838,00	443,00	37.997,00	333,54	0,00	P6	0,06437	21,47
27 julio 2015	06:00h - 07:00h	838,00	443,00	38.392,00	333,54	0,00	P6	0,06437	21,47
27 julio 2015	07:00h - 08:00h	838,00	443,00	38.787,00	333,54	0,00	P6	0,06437	21,47
27 julio 2015	08:00h - 09:00h	1.656,00	443,00	40.000,00	663,22	0,00	P2	0,10722	71,11
27 julio 2015	09:00h - 10:00h	1.656,00	443,00	41.213,00	663,22	0,00	P2	0,10722	71,11
27 julio 2015	10:00h - 11:00h	0,00	443,00	40.770,00	0,00	0,00	P1	0,12806	0,00
27 julio 2015	11:00h - 12:00h	0,00	443,00	40.327,00	0,00	0,00	P1	0,12806	0,00
27 julio 2015	12:00h - 13:00h	0,00	443,00	39.884,00	0,00	0,00	P1	0,12806	0,00
27 julio 2015	13:00h - 14:00h	0,00	443,00	39.441,00	0,00	0,00	P1	0,12806	0,00
27 julio 2015	14:00h - 15:00h	0,00	443,00	38.998,00	0,00	0,00	P1	0,12806	0,00
27 julio 2015	15:00h - 16:00h	0,00	443,00	38.555,00	0,00	0,00	P1	0,12806	0,00
27 julio 2015	16:00h - 17:00h	0,00	443,00	38.112,00	0,00	0,00	P1	0,12806	0,00
27 julio 2015	17:00h - 18:00h	0,00	443,00	37.669,00	0,00	0,00	P1	0,12806	0,00
27 julio 2015	18:00h - 19:00h	0,00	443,00	37.226,00	0,00	0,00	P1	0,12806	0,00
27 julio 2015	19:00h - 20:00h	0,00	443,00	36.783,00	0,00	0,00	P2	0,10722	0,00
27 julio 2015	20:00h - 21:00h	0,00	443,00	36.340,00	0,00	0,00	P2	0,10722	0,00
27 julio 2015	21:00h - 22:00h	0,00	443,00	35.897,00	0,00	0,00	P2	0,10722	0,00
27 julio 2015	22:00h - 23:00h	0,00	443,00	35.454,00	0,00	0,00	P2	0,10722	0,00
27 julio 2015	23:00h - 24:00h	0,00	443,00	35.011,00	0,00	0,00	P2	0,10722	0,00

Daily results						
Demanded Volume	Consummed Power	Energy cost			Emissions	Avoided Emissions
m^3	kWh	kWh/m^3	€	$c€/m^3$	tCO_2	tCO_2
10.632	3.994,7	0,399	313,98	3,13	1,594	0,000

Página 1 de 1

Fig. 11. Generated inform.

2.9 Looking for the Energy Efficiency by the SOE

In order to looking for the optimal value in the energy consumption, the SOE system is looking for two objectives.

1. Minimal energy consumption. This is trying so the functioning range of every system locates in the optimal point of the curves. This has to be taken into account in the pipelines too.
2. Consumption of the entire generated energy in the PVPP that it is installed in the cover of the reservoir for the water outflow of the plant (nominal 330 kW).
3. The non-supplied energy by the PVPP and non-saved energy in the same season of the program. It will be tried to consume in the period without CO2 linked to electricity energy or with the minimal.

The CO2 emissions linked to the plant after the starting of the SOE are reduced in a percentage over a 35%.

The restrictions that have been taken into account are as follows:

1. The functioning flow of the PPW is constant and it is not possible to stop it. Change in a maximum of once per day. The usual change is a maximum of once or twice

per week. The PPW has been designed in order to consume a low level of energy during the purification process.

2. Any change in the functioning mode in the PPW has to be supervised by the supervisor. For this reason, during the weekend, it is not possible to change the protocol.

3. The water consumption is not possible to manage. This can be predicted but not managed. E.g. the water consumption during the weekend in summer is low because there is no industrial activity.

4. The water capacity of the reservoirs is constant.

5. The water that remains in the reservoirs is limited. The maximum time for water in the reservoir is three days.

6. The Águilas-Puerto Lumbreras elevation has to supply water continuously in order to maintain the network pressure. The rest of the pumping systems can be managed because they move the water between them.

7. The maximum of supplied power by the electric network is limited to the contracted power in the six specific periods of the Spanish law.

8. The pipeline network has very important losses of pressure with the use of the highest value of pressure against the use of several pumps in a pumping system.

9. The maintenance has to be developed in the morning because the majority of the staff is working during these hours.

The output data of the SOE system are only the next: Apolonia pumping system, Elevation IV of Lorca and Swamp pumping system. These pumping system are the 85% of the electric consumption of the WPP.

2.10 Optimization Calculation

The calculation formula is double. Firstly, the system is modelled according to the functioning normative. The main aspects are the restriction for the functioning. E.g. minimal volume of the reservoirs, maximum of pressure in the pipelines, among others. In this phase the parameters related to the calculation as photovoltaic production in the previous days, flow to generate the next day, needs of filter washed, among others, are included. Has to be taken into account that the volume in the reservoirs cannot be low when the weekend is near. During the weekend and according to the Spanish energy system the linked production of CO_2 is low.

Moreover, once the restrictions are introduced to the daily operations, the programming of a macro is activated using the theory of genetic evolution. This algorithm tries successfully looking for the best combinations in order to finish when the differences between the combinations reach a fixed threshold. In this case it is the third decimal. When the solutions of the iteration are similar to over three decimals, the date is considered as the correct for the optimal functioning in the next day. The result is the functioning of every pumping system and the cumulated volume in every reservoir during the 24 h of the day. In this case, more than twenty million combinations are calculated. With an Intel I5 CPU, the calculation is obtained after more than a minute.

Once the calculation has been finished, the generation of the screens of results is obtained. These have to be validated by the supervisor in order to be sent to the SCADA system for the PPW management.

3   Results and Discussion

Some examples of simulation have been developed in order to estimate the possibilities of SOE software. The case study is located in an extended zone from the South of the Región de Murcia to the South of the city of Alicante (Comunidad Valenciana), Spain. This plant provides potable water to several cities and villages along this extended zone (Orihuela, Elche, Crevillente, Alicante, among others).

In the majority of the cases this is the common procedure of the software. Automatically, when SOE is started, a simulation of operation is generated for the current moth and day. The input data (demanded flow by the pipe networks of the PPW system and elevations 1, 2 and 3 and additional flows of the system) is obtained from the internal database of the software. In this way, when the program begins, the first simulation is realized. All the operation data of the system can be visualized and all the data can be seen.

The obtained results of the simulations are related to graphics of volume balance of the dams and prediction per hours for the selected day. The introduction of varied flows and volume conditions can generate different scenarios.

Based in the daily and monthly serial data from ten years, the consumption flow for several days can be estimated. Moreover, with the data of stored volumes the energy consumption can be reduced using the developed software by means of simulation scenarios. With this reduction, the footprint of CO_2 can be reduced. In the same way, with the optimization of the exploitation management can be reduced the pressures in the pipe network of the system. This is related to the reduction escapes in the pipeline and a major control of it. A flow balance in the entire system has to be developed (inflows and outflows) to control the points where there is an excess of pressure.

The estimation of the escapes in the entire pipe network is around the 20% of the total volume for the entire system. The use of this software can avoid these escapes. Moreover, an energy saving can be obtained avoiding the escapes. In this case, the daily water quantity that is in this system is around $22,000 \, m^3 \cdot h^{-1}$. The maximum of the water saving is $4,000 \, m^3 \cdot day^{-1}$ (20%). In addition, with a night coefficient of 12 h ($C_n = 0{,}5$) the maximum potential water saving is about $2,000 \, m^3 \cdot day^{-1}$. The consequence of it is that water saving can be obtained as well as an energy saving because of the reduction of the pressures. The variation of pressures in this system ranges from 41 to 50 m. This case study is a complex system with a very high level of energy value, with several pumps in the previous cited pumping stations and several interconnected dams and tanks.

The management of this system with this software can save a quantity of 1,100 or 1,200 €/day.

4   Conclusions

In this case study, the use of SOE software gives a powerful tool that can save water and energy in a WPP.

The real reductions obtained by applying this software as a management system are to reduce energy consumption by 30% in the case of WPP, considering that the energy cost of the WPP is 50% of the total cost of the plant (approximately 1,000 .000 €/year) our system gets a saving of 1000 €/day. In the case of pumping, the savings will be as high as 40%.

For environmental impact indicators, it should be noted that the plant currently consumes an average of 7.35 Gw/year and that with the implementation of the system a CO_2 reduction has been obtained, derived from the use of energy during normal hours with a value of 227 $grCO_2$/Kw, compared to 150 $grCO_2$/Kw, resulting in an average reduction of 532,000 kg of CO_2 per year.

With the SOE software, several scenarios can be generated to obtain the optimal purpose by means of genetic algorithms.

With the implementation of the SOE system, energy savings over 20% has been reached. About the saving of non-emitted CO_2, the values are over the 30%. In parallel, economical savings of the electric bill over the 30% have been reached. With an investment under 20,000 €, savings over 100,000 €/year are obtained.

The integration of a PP in the entire system of the WPP allows an additional energy supply that increases the global energy efficiency. Moreover, the entire system has the highest isolation degree against the energy network.

This optimization system can contribute to the reduction of the CO_2 emissions to the atmosphere. Another improvement is the reduction of the exposed surface of the water level in the dams of the entire system. This can reduce the evaporation of the water films in the dams. In this way, the loss of water is reduced.

References

1. Orden IET/3586/2011 peajes acceso, tarifas y primas en régimen especial. Orden IET/3586/2011 que establece peajes de acceso, tarifas y primas del Régimen Especial
2. Orden IET/1491/2013, de 1 de agosto, por la que se revisan los peajes de acceso de energía eléctrica para su aplicación a partir de agosto de 2013 y por la que se revisan determinadas tarifas y primas de las instalaciones del régimen especial
3. Orden IET/2735/2015, de 17 de diciembre, por la que se establecen los peajes de acceso de energía eléctrica para 2016 y se aprueban determinadas instalaciones tipo y parámetros retributivos de instalaciones de producción de energía eléctrica a partir de fuentes de energía renovables, cogeneración y residuos
4. Ventura, C., Tina, G.M.: Utility scale photovoltaic plant indices and models for on-line monitoring and fault detection purposes. Electr. Power Syst. Res. **136**, 43–56 (2016)
5. Etherden, N., Vyatkin, V., Bollen, M.H.J.: Virtual power plant for grid services using IEC 61850. IEEE Trans. Ind. Inform. **12**(1), 437–447 (2016)

6. Filali-Yachou, S., González-González, C.S., Lecuona-Rebollo, C.: HMI/SCADA standards in the design of data center interfaces: a network operations center case study. Estándares (HMI/SCADA para el diseño de interfaces en los centros de datos: El centro de control y operaciones como caso de estudio). DYNA **82**(193), 180–186 (2016)
7. Outcomes of the U.N. climate change conference in Paris. In: 21st Session of the Conference of the Parties to the United Nations Framework Convention on Climate Change (COP21), Paris. FCCC/CP/2015/10/Add.1 (2015)
8. www.omie.es
9. www.ree.es

The Learning Supply Chain

Barriers and Enablers to Supply Chain Knowledge Sharing and Learning Using Social Media

Susan B. Grant[(✉)]

College of Engineering, Design and Physical Sciences, Brunel University,
Kingston Lane, UB8 3PH, London, UK
Susan.Grant@brunel.ac.uk

Abstract. This research looks at the idea of interactive supplier social networks (SSN's), a novel and comparatively unexplored area in the field of supply chain management. The paper aims to understand the motivations prompting suppliers in a horizontal supply chain to share knowledge within a supplier network. A social constructionist perspective is employed to explore the factors that motivate/prevent engagement in knowledge sharing using social media tools from a customer and supplier's perspective across an insurance supply chain. The findings reveal corporate and industrial culture, work routines, technology, and a high regulatory environment can have a limiting effect on the generation of voluntary engagement in knowledge sharing between organizations and their supply chains in this sector.

Keywords: Knowledge sharing · Supply chains · Knowledge networks · Peer to peer communities · Insurance industry

1 Introduction

Knowledge is the foundation of a firm's competitive advantage and ultimately the driver of a firm's value [1]. Organizations therefore need to recognise it as being a valuable asset and develop a mechanism for tapping into the collective intelligence and skills of employees and supplier partners in order to create added value to a customer offering [2]. Most companies' today, are struggling with interconnecting knowledge, talent, ideas and relationships in their organisational environment and across their supply chains. Vital corporate knowledge is often trapped in information silos like email inboxes and information systems like ERP, and CRM. Recent evidence shows companies are beginning to consider web based 'social networking' as a community-building platform to sharing knowledge, [3]. Indeed, the development and evolution of social networking sites such as Facebook, Linked In. etc., is fuelling the appeal of social networking for companies, where achieving close communities with employees, customers, and suppliers is difficult to accomplish [4].

Given the importance of sharing knowledge, it becomes necessary to understand what factors lead to or hinder engagement and successful knowledge sharing amongst groups and individuals in virtual communities.

Business processes in the insurance industry traditionally requires the input, participation and decisions of many stakeholders. In motor vehicle claims processing,

G. Campana et al. (eds.), *Sustainable Design and Manufacturing 2017*, Smart Innovation, Systems and Technologies 68, DOI 10.1007/978-3-319-57078-5_36

repairers, assessors' claims staff, policy holders and legal representatives need to provide inputs and make decisions at different stages of the claims process. Indeed, for sectors such as insurance which depend on complex processes of multiple individuals exchanging information, and knowledge, interaction via social networks could potentially deliver a huge set of efficiencies and opportunities for rethinking core processes. Despite this need, up to now, firms associated with insurance are not seen as conducive to fostering knowledge sharing and generating collaborations across their supply chains in a pro-active way [5].

Social media is one of the key technology areas that insurers have had to quickly adapt to in order to remain competitive and visible to their consumers, brokers and employees.

A key challenge for insurers in online communities, concerns motivating a critical mass of supply chain members to engage in knowledge sharing activities in the first place [6]. Despite the proliferation of virtual communities of practice in businesses today, there is still have a lot to learn about what factors motivate active participate in community knowledge-sharing activities.

2 Knowledge Sharing

Knowledge sharing is "the act of making knowledge available to others" [7]. It is a voluntary, conscious act between two or more individuals resulting in joint ownership of the knowledge between the sender and the receiver [7]. Nonaka and Takeuchi [8] drew a clear distinction between tacit and explicit knowledge, with tacit knowledge (constructed by people) being highly personal and hard to formalize, making it difficult to communicate or to share with others. Subjective insights, intuitions, and hunches typically fall into this category of knowledge. Furthermore, tacit knowledge is deeply rooted in an individual's action and experience, as well as in the ideals, values, or emotions he or she embraces.

Whatever approach is adopted for sharing knowledge, whether face to face or on line, the willingness of individuals to share knowledge and engage in the process is key. Engaging in Knowledge sharing cannot be forced, but can be encouraged and facilitated [9]. While the literature on facilitators and obstacles to participant engagement in online communities has been growing, a disproportionate amount tends to be based on intra-organizational based studies, rather than across a supply chain. This seems at odds given that organizations increasingly operate within a connected and often global environment, where supply chain networks are typically the norm. Several early studies have emphasised the significance and benefits of supply chain networking for knowledge transfer, via supplier associations and open innovation networks [10].

2.1 Social Media Defined

There has been much debate on the definition of social media [11]. However, despite this, the literature seems to generally agree that social media software is represented by a range of emerging tools (wikis, blogs etc.) and platforms e.g. (Yammer) where users are able to share information and importantly collaborate and create networks of communities [12]. Given this, it appears that community driven and information-centric social media tools have tremendous potential for organisations to facilitate communities for information and knowledge exchange.

Indeed, while the literature points to some early cases of the knowledge transfer potential of social tools in industrial contexts (predominantly intranets), little is still known about drivers of use, the forms of use and likely potential of such platforms as a technology to group communication, knowledge sharing and information exchange, especially when extended across organisational boundaries to include supply chains.

3 Factors that May Hinder Engagement in Virtual Social Networks

3.1 Role of Organizational Culture in Knowledge Sharing Communities

Organizational culture plays a primary role in the likelihood that employees will be willing to engage in knowledge sharing and working together. If the culture is not supportive, or the reward system favours only individual effort, it may be difficult to get people to work together [13]. Ho [14] found that an organizational knowledge sharing culture will affect the attitudes of employees towards knowledge sharing in general, and indeed, organizational culture and inter-unit competition (for example encouraging knowledge hoarding behaviour amongst sales employees) was the second biggest barrier to knowledge sharing in numerous business organizations [15]. Closely related to organizational culture is the significance of leadership in facilitating knowledge sharing behaviours [16]. Management support for knowledge sharing has been positively associated with employee's perception of a knowledge sharing culture (e.g. employee trust, willingness of experts to assist others). Top management support was shown to affect both the level and quality of knowledge sharing through its influence on the employee [16]. Leadership, and in particular the employees perceptions of management's support for knowledge sharing, were seen to have a significant impact on participation in knowledge sharing. Leadership is particularly important for organizations willing to 'evolve' their culture to a knowledge-supporting culture, with suggestions that leaders should model the proper behaviours causing culture to evolve in a way that enables and motivates knowledge workers to create, codify, transfer, and use and leverage knowledge [17]. Vuori [18] case study revealed the organizational culture or general attitude didn't set particular challenges for intra-organizational knowledge sharing if it was supported and considered worthwhile by all employees, suggesting buy in from the top plays a crucial role. Management support is not the only facilitator of employee attitude towards knowledge sharing, support from co-workers, mentors and supervisors, had a similar effect on employees to share knowledge and encourage the habituation of such behaviour patterns [13, 19]. In addition to leadership support, habitual behaviours, including old habits of doing things within job roles were shown to affect knowledge sharing in online platforms [20].

Other cultural dimensions are likely to influence knowledge sharing also. Chiu et al [21] found social interaction ties, trust, norms of reciprocity, identification, shared vision and shared language (facets of social capital) and community related outcome expectations and personal outcome expectations can enable knowledge sharing in virtual communities. Wasko and Faraj [22], in contrast found a negative relationship between reciprocity (one dimension of social capital) and knowledge sharing. The inconsistent results suggest the relationship may be based on other factors such as participant's personality and the perceived usefulness of the community [23]. Kankallei et al [24] found perceived reciprocity to be positively related to the likelihood that individuals would contribute to knowledge sharing under weak rather than strong pro sharing norms, which suggests such norms may compensate for the weak levels of reciprocity in the community. Trust in particular is regarded as an essential factor for tacit knowledge sharing, and a necessary condition to knowledge sharing [23], with an absence of trust potentially affecting the flow of knowledge over time.

3.2 Role of Technology in Knowledge Sharing Communities

Knowledge management technology can provide the network of links between geographically dispersed groups and individuals that enables effective knowledge sharing. Research which considers knowledge management within an information processing view, sees the flow and transformation of information through a network of processing nodes (individuals and groups) of the network [25] where knowledge sharing inputs can be transferred and processed using technological networks to produce certain outputs. This view assumes that building networks that provide structural links between different groups will somehow automatically produce knowledge creation and sharing [25]. Given this, it is not surprising IT tools/platforms have been sold as pure knowledge sharing solutions. In contrast, a community model emphasising dialogue through active and systematic networking which may be IT-enabled, [26] appears a more realistic approach. Swan [25], contrasts two cases whereby the focus on IT for knowledge sharing had the effect of reinforcing organizational and social 'electronic barriers between individuals, thereby limiting the extent of knowledge sharing'. This demonstrates the importance of understanding the softer drivers to engagement right from the outset to any knowledge sharing implementation initiative. Technology can also be seen as an obstacle to organizational knowledge sharing as it may not allow the flexibility and functionality that users would like, leading to the need for workarounds or alterations in the way that people work in order to accommodate the technology [27].

3.3 Individual/Personal Obstacles to Engagement in Knowledge Sharing

Among the reasons why individuals may not want to share their knowledge, are organizational incentives and rewards, including boosting self-esteem, altruism and conforming to organizational practices. Osterloh and Frey [28] demonstrated that intrinsic factors such as viewing and sharing are more powerful than extrinsic (e.g. monetary or administrative) stimuli. Lin [29] highlights expected organizational rewards and reciprocal benefits as key sources of extrinsic motivation to share knowledge, whereas in other cases, both extrinsic and intrinsic motivational factors have a positive effect on knowledge sharing attitudes in communities of practice, with factors such as enjoyment in helping others and need for affiliation as being more significant [30]. In recent research, factors affecting knowledge sharing in social media platforms were found to be to partially related to an individual's expected benefits and rewards [20]. Bock and Kim [31], show that rewards only work in the short term, as long as they are offered, but do not permanently change the attitude to knowledge sharing. In contrast, rewards may even impede knowledge sharing, as the effect of a reward can boost knowledge sharing temporarily, but once it is withdrawn, employees can revert to other activities with a higher expectation of utility. In some environments, the incentive to share knowledge occurred when individuals viewed their knowledge as a public good, belonging to the organization, rather than themselves [26].

Some individual reasons given for poor engagement in online communities include a fear of criticism, fear of misleading others through the wrong information, a lack of clarity as to the best ways to share knowledge and cultural assumptions about appropriate and inappropriate ways to communicate and share knowledge/information [26].

While generally, incentives have had a positive effect on knowledge sharing, the empirical results of studies examining the effects of extrinsic rewards has been mixed. Bock and Kim [31], found that anticipated extrinsic rewards had a negative effect on attitudes towards knowledge sharing. Vuori [18], also found financial rewards to be one of the least motivating factors to engaging in sharing knowledge across a social media platform, although rewards were highlighted as important to changing people's current practices (e.g. excessive e-mailing, sorting information in several databases). In contrast, they found intrinsic rewards were identified as focal to motivating knowledge sharing across colleagues - knowledge is shared to help the organization or colleagues.

3.4   Competition Versus Collaboration

The literature has identified a key barrier is the handling of confidential information and lack of trust [32]. Supply chain members are often reluctant to share information because of fear of opportunistic behaviour, i.e. partners exploiting information for self-interest. Companies may, therefore, refrain from sharing information unless prevention of leakage to competitors is guaranteed. In a similar vein, there is a risk that shared information may negatively affect the competitive position of the buyer or supplier in relation to their competitors. This issue could negatively impact on firms' commitment to relationships and their willingness to share information with supply chain partners without concern that this might be misused.

4   Research Methodology

The objectives of the study were to explore the attitudes to knowledge sharing to via a private networking platform across an insurer and its supply chain. The research was exploratory in nature and took a three stage approach. As such, the qualitative case study methods were deemed to be the most suitable to empirically investigate the real life context of the on line knowledge network across a supply chain [33]. Of key importance to this research, was information such as attitudes and opinions which can best be obtained with the open ended questions, allowing the researcher to gain spontaneous information about attitudes and actions, rather than a rehearsed position.

Stage one addressed the main challenges in the implementation of knowledge networks from an insurer's perspective.

A series of interviews lasting in the region of 1.5 h long with the insurer were conducted over a 3-month period. The initial interview with the Head of the claims Executive was intended to gain a general overview of the industry and culture, as well as management attitudes to sharing knowledge internally and externally; and rewards to sharing information. A second interview involving middle and senior management in Motor insurance claims, was conducted 3 weeks later. The aim of this discussion was to develop the concept of social networking across the supply chain and to identify areas where an SSN may apply in the insurer's supply chain. Further in-depth interviews were conducted 3 weeks later with motor insurance managers.

Stage two involved carrying out in-depth interviews with the CEO's of eight body-shop networks and independents at their premises. Suppliers were located in different regions of the UK. Participants were asked a series of questions around specific themes. Whilst some questions asked for factual data, the majority of questions were attitude and opinion based.

Interviews lasted in the region of 1–2 h long, to be followed some days later by a list of follow-up questions. Eliciting self-reports on attitudes orally via interviews etc. is a useful way to collect attitudinal data in people who have the self-awareness to recognise their own beliefs and feelings and the ability to articulate them [34]. Indeed, self-report procedures represent the most direct type of attitude assessment [34].

The data collection procedure lasted for a period of 10 months from October 2012–July 2013. A total of 38 h was recorded across all interactions over the period of the study. The findings from the interviews are presented in the following section. All interviews, and audience discussions were recorded and transcribed verbatim.

Data from interviews, were coded and analysed using qualitative analysis methods [35]. The qualitative analysis involved an iterative process. Different categories of barriers were highlighted by the coding process. The first step in the analysis of the data involved coding each of the interview transcripts using Nvivo 10 software to speed up the process. And follows the coding process of grounded theory [35]. Grounded theory techniques can capture the interpretive experiences of individuals and develop theoretical propositions from them.

Analyses on the texts of respondents to ascertain how many respondents mentioned a particular barrier to engagement was used. The accuracy of the research findings was validated by the use of participant checks. A random selection of participants from both the insurance and supplier side were shown the interview summaries to validate. All participants in this sample agreed the interview summaries accurately reflected their opinions.

5   Findings (Enablers and Barriers)

5.1   Insurer's Perspective

On Communication, Information Exchange and Knowledge Sharing Internally and Externally: Currently communication and information exchange with suppliers is both planned and adhoc, via telephone, e-mail and in person. The degree of communication between the garages however varies. Current knowledge sharing practices across the supply chain tends to be via quarterly executive and yearly meetings (not peer to peer). Information on costs of parts for example are shared within the supply chain, as visibility is required for best practice.

Manager C *'A plethora of software tools exist connecting body shop suppliers, customer relationship teams, and other interested suppliers in the claims process, but as for a forum where they can exchange ideas, innovations, experiences etc – no this doesn't exist'.*

Perceived Risks, Barriers and Tensions to Social Networking via an SSN: Insurers key concerns included Trust, IPR, time (staff have priorities), and organizational culture. Security and Data protection in particular was seen as a big issue. As such a major barrier in setting up a system was *'data protection and safeguarding IP rights'.* In addition, the insurer's perception of *'the level of sophistication of some of the individual companies in the motor supply chain, was not high in using technology. This could prove to be a barrier to using social networking technology'.*

Work Practices, Culture, Competitive Positioning: The insurer was concerned about the value of sharing across horizontal suppliers.

Manager A… *'I think we have to be really carefully about when we start getting body shop suppliers to chat to each other-there are some forums where they can do that today and some of it well it's a bit like news of the world stuff-they could make a drama out of a real crisis and vice versa-you have to be very careful what these people say. I think in terms of sharing innovation and data about each other supplier, I think we need to be careful we don't lose that competitive edge, we don't do anything that's sitting inside our contracts which are heavily written in terms of confidential data',*

IPR and Awareness of IPR: Manager B … *'There are those contractual terms and commitments which will include IPR and competitive advantage - which we as an insurance company would want to retain over our competitors and over their supply chains-and that certainly would be a factor that would need to be considered within a knowledge networking system such as the one you outline-we don't want to lose any of what we may have in that regard'.*

Perceived Benefits: *Manager D… In terms of the principles, we would like a facility where we could pull/put information there for all of our suppliers to see at the same time-like an SSN, and where we could get suppliers to share ideas. Suppliers are likely to be concerned about the use of commercial data within such a system.*

Management Attitudes Towards Sharing Knowledge Internally and Externally:
From the insurer's perspective, while knowledge sharing is encouraged across the organization, it doesn't extend in a formalised structure to the supply chain. Knowledge sharing internally is encouraged via competitions, staff ideas boards and via internal news publications, and tends to occur with across individuals within departments silos rather than cross functionally.

Rewards and Investments into Knowledge Sharing: The insurer doesn't directly offer incentives or rewards for knowledge sharing both internally or externally to the supply chain, and no official incentive policy exists.

5.2 Suppliers Perspective

On Collaboration and Sharing Knowledge: Supplier B *'Collaboration in the industry has been slow; sharing knowledge is likely to be about customer service, improving the customer experience. Standardising the customer journey, so the insurer could measure it once that was established as a norm, the industry could look at repair technologies, estimated practice etc. but suppliers are worried about giving away their IP'.*

Perceived Benefits: Acommunity around a claim:
Supplier A *Everyone wants to put themselves within the claims process, because that is where we can add value. I think a he community around a claim is excellent because there are so many people interacting in the claims process, and that drives so much the cost into the process both in terms of indemnity and handling cost, that you could actually strip out the cost by having that interaction and ultimately, that will have a positive impact on customer premium cost, and that will have an impact on retention.*

Incentives: Suppliers expressed an interest in engaging in SSN's if there were appropriate incentives in place, in particularly using league tables.

6 Discussion

The interviews revealed useful insights about potential facilitators and inhibitors to engagement in a knowledge sharing platform across an insurer and a horizontal supply chain of motor body shops. These included:

Knowledge Sharing Cultural Inertia: Suppliers perceived the insurer's contribution to knowledge sharing as being too little. While there wasn't a formalised policy on knowledge sharing across suppliers, a champion did organize knowledge exchange workshops on an ad hoc basis. This resulted in infrequent meetings (face to face) and limited peer to peer sharing of 'tacit' knowledge. Additionally, management support for knowledge sharing was inconsistent, with some insurance managers keener to promote the idea of virtual communities over others.

Loss of Intellectual Property: A key fear for suppliers was the potential loss of intellectual property.

Trust Issues: The evidence revealed limited trust between suppliers, which was a key stumbling block into voluntarily entering into knowledge sharing behaviours across the supply chain. Lack of trust was typical in this highly competitive industrial setting where body shops are competing against each other on insurance claims contracts. In addition to this, trust is formed as individuals get to know each other.

Regulatory Environment: The regulatory environment within insurance concerns data protection. This high regulatory environment has had the effect of reinforcing a closed and tightly controlled information and knowledge sharing environment.

IT Overlap and Overload of Systems: Within the insurance supply chain, a number of systems were installed to support different tasks but which also generated some overlap in activity. Implementing additional systems and the resultant information systems overload, was perceived as potentially capable of 'turning off individuals from using it, even though its role as a tool to support knowledge exchange was unique'.

Economic Rewards: Extrinsic rewards in the form of positive reputational effects and the concomitant financial rewards that go with higher customer volume were cited as a motivator to engagement by many suppliers. Intrinsic rewards including helping others or contributing towards the community were mainly absent. This runs counter to research which highlights intrinsic rewards as being more significant over extrinsic incentives in facilitating knowledge sharing [30]. The culture of the sector (financial, insurance) is likely to play a part in this departure from the literature.

7   Conclusions and Limitations

With the growing rise in adoption of collaborative social media tools, such as Yammer, and pressures on businesses to adopt these new technologies, this research sought to understand the potential role social media tools can have as a knowledge and information sharing conduit across a knowledge intensive supply chain, in automotive insurance claims market. The paper seeks to explore whether information sharing and knowledge exchange can take place across weakly integrated non-linear extended supplier relationships, using social media tools?

Extending the study by conducting a large-scale survey of motor insurance body shops and other insurers/areas of insurance would be a useful follow-up. Equally, the study is restricted to the insurance industry, and other industrial contexts (e.g. manufacturing) including a focus on SME's would be a useful way forward.

Acknowledgments. This research was supported by a grant from the British Academy, Ref: SG101426 'Exploring supplier attitudes to knowledge networking: a pilot study in the UK insurance market' 2012.

References

1. Teece, D.J.: Strategies for managing knowledge assets: the role of firm structure and industrial context. Long Range Plan. **33**(1), 35–54 (2000)
2. Bellinger, A.S., Smith, R.: Managing organizational knowledge as a strategic asset. J. Knowl. Manag. **5**, 8–18 (2001)
3. Bredl, K., Grob, A., Hunniger, J., Fleischer, J.: The avartar as a knowledge worker. Electron. J. Knowl. Manag. **10**(1), l5–25 (2012)
4. Khan, A., Khan, R.: Embracing new media in Fiji: the way forward for social network marketing and communication strategies. Strateg. Dir. **28**(4), 3–5 (2012)
5. Casemore, S.: Social Media and the Coming Supply-Chain Revolution Feb 29, 2012, CFO.com (2012)
6. Grant, S.B.: Exploring attitudes to knowledge networking in UK insurance supply chains. In: 21st Euroma Conference 2014, Palermo, Italy (2014)
7. Ipe: Knowledge sharing in organisations: a conceptual framework HRD review vol. 2 (2003)
8. Nonaka, I., Takeuchi, H.: The Knowledge-Creating Company. Oxford University Press, New York (1995)
9. Kaser, P., Miles, R.: Understanding Knowledge Activists. Long Range Plan. **35**(1), 9–28 (2002)
10. Chesborough, H.: Open Innovation: The New Imperative for Creating and Profiting from Technology. Harvard Business School Press, Cambridge (2003)
11. Constantinides, E., Fountain, S.: Web 2.0: conceptual foundations and marketing issues. J. Direct Data Digital Mark. Pract. **9**, 231–244 (2008)
12. McAfee: The effects of culture and human resource management policies on supply chain management strategy. J. Bus. Logistics **23**(1), 1–18 (2009)
13. Cabrera, E., Cabrera, A.: Fostering knowledge sharing through people management practices. Int. J. Hum. Resour. Manag. **16**(5), 720–735 (2005)
14. Ho, C.-H.: The relationship between knowledge management enablers and performance. Ind. Manag. Data Syst. **109**(1), 98–117 (2009)
15. Wah, L.: Making knowledge stick. In: Cortada, J.W., Woods, J.A. (eds.) The Knowledge Management Yearbook 2001–2001. Butterworth-Heinemann, Boston (2000)
16. Lee, W.B., Cheung, C.F., Tsui, E., Kwok, S.: Collaborative environment and technologies for building knowledge work teams in network enterprises. Int. J. Inf. Technol. Manag. **6**(1), 5–22 (2007)
17. Ribière, V.M., Sitar, A.S.: Critical role of leadership in nurturing a knowledge-supporting culture. Knowl. Manag. Res. Pract. **1**(1), 39–48 (2003). (10)
18. Vuori, V., Okkonen, J.: Knowledge sharing motivational factors of using an intra-organizational social media platform. J. Knowl. Manag. **2012**, 592–603 (2012)
19. Cameron, P.D.: Managing knowledge assets: the cure for an ailing structure. CMA Manag. **76**(3), 20–23 (2002)
20. Paroutis, S., Al Saleh, A.: Determinants of knowledge sharing using web 2.0 technologies. J. Knowl. Manag. **13**, 52–63 (2009)
21. Chiu, C., Hsu, M., Wang, E.: Understanding knowledge sharing in virtual communities: an integer of social capital and social cognitive theories. Decis. Support Syst. **42**, 1872–1888 (2006)
22. Wasko, M.M., Faraj, S.: Why should i share? Examining knowledge contribution in networks of practice. MIS Q. **29**(1), 35–57 (2005)
23. Wang, S., Noe, R.A.: Knowledge sharing: review and directions for future research. HRM Rev. **20**(2), 115–131 (2010)

24. Kankanhalli, A., Tan, B.C.Y., Wei, K.-K.: Contributing knowledge to electronic knowledge repositories: an empirical investigation. MIS Q. **29**(1), 113–145 (2005)
25. Swan, J., Newell, S., Scarbrough, H., Hislop, D.: Knowledge management and innovation: networks and networking. J. Knowl. Manag. **3**(4), 262–275 (1999)
26. Ardichvili, A., Page, V., Wentling, T.: Motivation and barriers to participation in virtual knowledge sharing communities of practice. J. Knowl. Manag. **7**(1), 64–77 (2003)
27. Dotsika, F., Patrick, K.: Implementing a social intranet in a professional services environment through web2.0 technologies. The Learning Organization (2013)
28. Osterloh, M., Frey, B.S.: Motivation, knowledge transfers and organizational forms. Organ. Sci. **11**(5), 538–550 (2000)
29. Lin, H.F.: Effects of extrinsic and intrinsic motivation on employee knowledge sharing intentions. J. Inf. Sci. **33**(2), 135–149 (2007)
30. Jeon, S., Kim, C.M., Koh, J.Y.: Individual, social and organizational contexts for active knowledge sharing in communities of practice. Expert Syst. Appl. **38**(10), 12423–12431 (2011)
31. Bock, G.W., Kim, Y.G.: Breaking the myths of rewards: an exploratory study of attitudes about knowledge sharing. Inf. Res. Manag. J. **15**(2), 14–21 (2002)
32. Spekman, R.: Supply chain competency: learning as a key component. Supply Chain Manag. Int. J. **7**(1), 41–55 (2002)
33. Yin, R.K.: Case study Research: Design and Methods (1994)
34. Henerson, M., Morris, L., Fitz-Gibbons, C.: How to Measure Attitudes. SAGE Publications, Newbury Park (1987)
35. Miles, M.B., Huberman, A.M.: Qualitative Data Analysis: An Expanded Sourcebook. Sage, Beverly Hills (1992)

Supply Chain Learning Using a 3D Virtual World Environment

Olinkha Gustafson-Pearce[(✉)] and Susan B. Grant

College of Engineering, Design and Physical Sciences, Brunel University,
Kingston Lane, London, UB8 3PH, UK
{Olinkha.Gustafson-Pearce,Susan.Grant}@brunel.ac.uk

Abstract. This paper discusses the use of virtual world technology in relation to the unsustainability of the current levels of greenhouse gas emissions, related to business travel. If it can be demonstrated that the use of virtual worlds enables users to participate in meetings and other events in a manner that benefits the individual and the organisation, without the need for the individuals to meet 'face to face', then overall, business travel can be reduced. However creating the virtual environment that engages the user in 'meaningful' discourse, requires testing the environment against specific targets. This paper discusses that in the context of Supply Chain Management within the Insurance business, knowledge transfer is a key factor, that is currently conducted through 'standard' channels, primarily emails and the telephone. A number of team meetings are also organised, since it is felt that 'face to face' contact between members is necessary. Business travel for participants contributes to greenhouse gas emissions. Therefore, for this study, the use of 3D Virtual World (VW) tools to discover if knowledge sharing and learning within a horizontal supply chain managed by a principal insurer, was effective and reduced the need for 'face to face' meetings. A set of web based tools, applications and exercises supporting the formation of communities of inquiry and promoting knowledge transfer and learning, through social interaction is presented. These results are from a pilot study that was run over a four month period across an insurance supply chain, to explore how suppliers and the principal insurer shared knowledge, using these tools. With the IoT (Internet of Things) generating multiple sources of 'streamed' data, the potential for using this type of data in a format that allows users to access data that is 'understandable' to them, is expanding. Within the insurance industry, and specifically home claims, a key priority is to have current and meaningful data on physical events and conditions available to their stakeholders and members of the supply chain. This is to enable them to make correct and timely decisions on claims, for example, weather related claims. Therefore an environment was designed and created, which used live streaming data from the United States Geological Survey, and a variety of VW tools and techniques to illustrate this data, and to orient it to make it relevant to the home claims teams.

Keywords: Knowledge sharing · Virtual world environment · IoT · Virtual reality · Greenhouse gasses · Business travel

© Springer International Publishing AG 2017
G. Campana et al. (eds.), *Sustainable Design and Manufacturing 2017*, Smart Innovation, Systems and Technologies 68, DOI 10.1007/978-3-319-57078-5_37

1 Introduction

A great deal of time and cost is associated with human interaction, related to contact both within and between organisations. This is especially prevalent when non co-located individuals or teams have to meet face to face. These costs include travel, hotels and 'away from office' time. The costs can be monetary for the organisation, but in addition a significant amount of environmental damage, through increase in greenhouse gas emissions, is caused by 'business' travel. As business becomes more global, over time, the environmental impact of increased business travel, is not sustainable. The UK Government and the World Wildlife Fund (WWF) have released a paper highlighting the environmental costs of business travel, in which they state that 'eliminating unnecessary meetings would … make a major impact on the travel foot print' [1]. In a WWF report conducted by Pamlin and Szomolanyi, they stated that 'if all European companies cut their business travel by 20% it would save 22 million tonnes of CO_2, equivalent to taking one third of UK cars off the road [2]. Complex supply chains are often global and the need for effective communication between the individual nodes (individuals, teams or organisations) is vital.

This study focusses on knowledge sharing and demonstrates that social networking, through the use of virtual world engagement, can engage the individual in social networking, leading to enhanced engagement with the 'team', and reduce the need for physical 'face to face' meetings.

2 Background

The KNOWNET project examines the potential of current social networking technologies, to support sustained knowledge sharing and generation across a horizontal supply chain, in the insurance market. The concept of collaborative networking is particularly timely in an industry that seeks to strengthen the inter-organizational ties between insurers, suppliers and external agencies, for improving processes, accelerating innovation, fostering creativity, sharing experiences, and generating ideas, amongst the supplier network [3]. Knowledge and learning is the foundation of a firm's competitive advantage and ultimately the driver of a firm's value [4]. Organizations therefore recognise it as being a valuable asset and develop a mechanism for tapping into the collective intelligence and skills of employees and supplier partners in order to create a greater organizational knowledge base [4].

Supply chain management research increasingly is expected to make a significant contribution to the knowledge transfer and productivity debate, and indeed there is increasing recognition that supply chains are beginning to prioritise knowledge creation and exchange [5]. Current literature suggests that the adaptation and implementation of successful ideas and practices can enable the development of innovative mechanisms, which in turn may result in productivity improvements [6]. Successful management of a supplier network in particular, can potentially enhance the productivity of the supply chain through diffusion of knowledge. Despite this, however, there remains a generally adopted view that the potential of SCM synergies for the creation and transfer of useful knowledge has not yet been materialised [7]. Indeed, the findings of a recent study for the creation of value in organizations for example, suggests that although firms in the UK, assign great importance to their suppliers as sources of new knowledge creation, their involvement in the generation of knowledge is low [6]. There are a number of reasons and challenges associated with this. A key challenge concerns motivating supply chain members to engage in knowledge sharing and generating activities in the first place. A potentially contributing factor to this is that, within complex supply chains many supply chain partners or nodes within the system, may be geographically distant. This presents difficulties with communication and, for the companies, has associated costs with increasing contact between the nodes. Simple contact in the form of telephone calls or emails, may be possible, but the intricate nature of face to face contact, may be lost. In addition, many current methods of contact can create difficulty in generating and transforming knowledge into organizational action, and subsequently it is even more difficult to transfer good ideas, insights and knowledge to supply chain partners. Connecting the partners in a supply chain in face to face contact, can be costly and therefore may only happen sporadically, and over time may lead to a lack of sustainability in regard to the veracity of the Supply Chain. In a broader context, environmental sustainability related to the travel involved in bringing the individuals or companies together, (fuel use, hotels, etc.) means that the planets resources are depleted and, business travel is noted as a key factor in the increase in greenhouse gasses [1].

This paper reports on a customised socially interactive virtual environment which was designed to examine whether knowledge sharing across supply chain members can be facilitated for this pilot study. Social interaction encourages the sharing of ideas, discoveries, successes and failures and provides general social support [8, 9]. These elements are often missing from traditional information portals. Individuals who are removed from a social interactive experience may feel isolated, or start to lose motivation, which may lead to dropping out of the knowledge sharing and learning process. In contrast, a virtual 3D environment requires an 'avatar' (human or other 'shape'), which can travel inside the virtual space and communicate with others in real time. Using the web 3D virtual world as a business application, allows an 'immersive' experience, determined by the degree to which the user's senses are engaged, and the desirability and meaningfulness of the activity in which the user is participating [10]. Within the environment participants can communicate with each other via public or private voice chat, local or group or private text chat, messaging, document and object sharing, screen sharing, etc. The applications and information the user needs to complete a task, for example have a meeting, deliver a presentation or collaborate on a model, are accessible from and can be displayed within the virtual environment. This reduces the need for participants to physically travel to a central location and means that 'face to face' interactions, more akin to 'real world' interactions, can be sustained and enhanced. In addition, if the use of virtual world technology is shown to be effective, reduction in costs for travel and 'out of office' time, can be reduced.

The social interactive environment considers the interpretational process *of knowledge* through four distinctive consecutive stages [11] through which knowledge is transferred. Cognitive IT led approaches to knowledge transfer typically fail to take into account such factors that lead different groups to have divergent, possibly even irreconcilable, interpretations of knowledge. The community view [12] recognises that knowledge has to be continuously negotiated through interactive social networking processes. The community model emphasises dialogue occurring through active and systematic networking (which might be IT enabled), rather than linear information flows.

Virtual worlds [13] are digital worlds accessed through a 'viewer'. These viewers are a form of 'web browser' but they allow the user to engage with the virtual 3D environment. The use of VW environments has been gaining importance in the commercial world. For example, IBM's Global Innovation Outlook, hosted business orientated discussions around Smart cities. The meetings were designed to gain insights from influencers to smarter cities, explore the effectiveness of virtual worlds for collaboration, and extend relationships to build on an existing IBM partnership. The meetings yielded new insights and discussions, brainstorming and best practices to consider during planning of the Smart cities activity [14]. Operations management is also an area where virtual world environments have played a role in training. The use of different VW/VR technologies are becoming a useful method to improve the understanding of the plans and to support interdisciplinary discussions. Indeed, virtual world based training is the world's most advanced method of teaching manufacturing skills and processes to employees. Using cutting-edge VR/VW technology, training takes place in a realistic, simulated version of the actual facility, complete with the actions, sights, and sounds of the plant floor [15].

There are a large number of virtual worlds, from the 'youth friendly' 'Minecraft' [16], through to the 'photorealistic' complex worlds created with Unity [17]. One of the more popular platforms for training, education and business is the open source platform Open Simulator [18]. This platform allows 'users' to design and create virtual worlds that fulfil their needs. In this paper the authors report on a use for a virtual environment, which was created using the Open Simulator platform, specifically designed and developed to enhance knowledge sharing for a complex supply chain within the insurance industry.

2.1 Methodology

The pilot took a two phase approach. In the first instance, the researchers met with the company to understand what the user knowledge sharing needs were. Phase two involved the build of a customised virtual world environment to trial knowledge sharing, using streamed data and the testing of that environment with employees of the company. The methodology used on this trial was qualitative in nature. Data was collected via taped conversations between users to the VW. Qualitative analyses software: Nvivo 10 was used to generate themes and analyse the data. The following section details the methodology used in detail.

2.2 Initial Exploration into Stakeholder and User Needs

In 2013, the KNOWNET project was launched under the leadership of Dr. Susan Grant (PI). The Project (a Marie Curie funded IAPP under FP7) was carried out in collaboration with Royal Sun Alliance Insurers (RSA). A key objective was to develop and build a web based interactive environment – a supplier social network (SSN) to support and facilitate exchange of good ideas, insights, tacit and explicit knowledge, across a diverse group of stakeholders in a multi-level supply chain, within the Insurance sector. For this purpose two components were designed, a social network and a dedicated VW environment. The KNOWNET project designed and developed a 'social interaction' Yammer [19] platform to enable conversations between stakeholders and researchers for knowledge sharing. This platform was used for initial research in the discovery phase. Users were asked what information they would find useful to enhance their role in the insurance industry. In the conversations that followed, contributed to by a range of employees and other stakeholders, it was found that a 'conversation' forum which enabled users to view 'real time' data on weather or other geophysical events would be considered useful. It was also stated that current methods of data viewing were considered hard to understand. Users had to search, possibly multiple sources, to discover information useful to them and forums, whilst useful, tended to have a range of 'less useful for their needs' data.

The authors had previously designed, developed and used a virtual world environment for teaching and training in a university environment. This experience showed that engaging stakeholders and users in the early processes of design and development was crucial to success, especially the discovery phase of interests and needs of what would engage the user, and enable them to process information in a comprehensive manner. In addition it was also shown that setting clear targets for outcomes by the stakeholders and developers, was essential to success.

2.3 Creating the Virtual World and the Use of 'Streamed' Data

To this end a virtual world environment was created by Dr. Gustafson-Pearce on the Brunel University virtual world platform, which enabled users to 'log in' to the world as avatars. An avatar in a virtual world has a 'real' presence in the world and is the graphical representation of the individual [20]. It was discovered that a range of 'streamed data', available through various sources could be used to create artefacts and effects within the virtual world. This use has implications that relate to the growing expansion of the IoT (Internet of Things). The IoT comprises of the exchange of M2M (machine to machine) related data and can be leveraged by appropriately implemented artefacts and systems. 'The IoT is a global infrastructure for the information society, enabling advanced services by interconnecting (physical and virtual) things based on existing and evolving interoperable information and communication technologies [21].

In the created virtual world, real time streamed data from the United States Geological Survey (USGS) Earthquake Hazards Program was utilized by the virtual world and represented by a variety of designed artefacts. The main 'area' had a large (10 m sphere) which was a representation of the Earth (Fig. 1). This Earthglobe had a number of functions; objects were 'scripted' to 'appear' when earthquakes events happened. These artefacts had a number of 'states'. They showed location on the Earthglobe by physical presence; appearing at the location of the earthquake, colour defined Magnitude (M on the Richter scale [22] from red = M7+, green = M4 to M7 and yellow = M2.5 to M4.

Depth was defined by a series of 'circles' where if the earthquake was below 80 km then the circles appeared slow moving, to fast moving 'spiral' effects which meant that the earthquake was close to the surface. The closer to the earths' surface of the earthquake, the faster the circles appeared to move. When the earthquake first occurs the colours are bright, as time passes the colour slowly fades (using opacity) over a three day period. Since this data was updated every 3 min, real time information was available to the viewer. Therefore the user could, at a glance, see that a representation which appeared over Tokyo showing bright red, fast moving circles would be a major earthquake that had just occurred, close to the surface and a potential 'disaster' situation. There were several additions to the data. If the user wanted further information about the earthquake they could click on the artefact and a text description of the event would appear, they could also click on this description and go to the web page from the USGS where considerable more detailed information was available. Also a representation of the moon and night/day on the planet showed the Terminator zone [23] and light/dark on the Earth in real time. This was felt to be important information for rescue and other services, since an earthquake that happens at night could be more difficult for access and might require different rescue equipment.

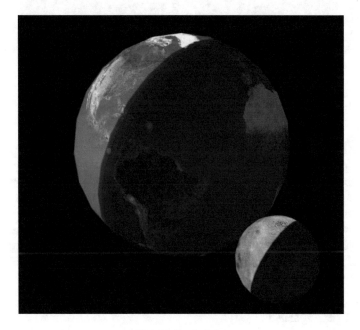

Fig. 1. Earthglobe

Additional areas were also created. These include an exhibition which showed various aspects related to earthquakes including a 'what to do in an earthquake' with an interactive Q&A panel that tested participants knowledge about how to stay safe in the event of an earthquake. There was also a 'discussion' area with seating where partici- pants could sit and chat (Fig. 2). Both these areas were created to encourage and enable conversation and knowledge sharing. 'Voice' was enabled in the environment, which enabled the engagers to discuss topics in real time.

Fig. 2. Knownet social area

3   RSA Participants

An Instant Message was sent out on the RSA Yammer platform to ask for participants for a VW platform pilot study. A number of people responded, which resulted in 10 people who were the initial group of volunteers for the first VW pilot study. None of the volunteers had previously used any virtual world technology. Therefore, since this would be an entirely new experience for the volunteers, user guides for the viewer and the virtual world environment were developed, to enhance the user experience and these were first distributed to them. In addition, the first time that the volunteers engaged with the virtual world platform an 'orientation' session lasting on average 30 min was conducted.

3.1   The User Experience

The initial users had not previously met each other in the physical world, although some contact had been established through the yammer platform. This was primarily organisational – convenient times to meet, etc. and the group mainly consisted of 'home workers'. All sessions were supervised by Dr. Gustafson-Pearce and Dr. Grant.

The first user engagement in the VR environment session was orientation and exploration and primarily focussed on the individual. This session lasted on average 30 min. The second session, on average a week later, had groups (3–4 people) and in this session they engaged with the Earthglobe and discussed the display and information provided (Fig. 3). This session was not time constrained and on average lasted 90 min.

Fig. 3. USGS data and members inworld

The researchers found participants (through the medium of the avatar) reacted in a manner that suggested presence [24] after the second session. Establishing 'presence' in the virtual world where all users felt comfortable and felt that they fully engaged with their colleagues and the environment, happened very quickly in the second session. Presence was defined and demonstrated by individuals apologising if their avatar 'bumped' into another avatar and generally behaving as if the avatar was 'them'. Statements such as 'I will follow you ...' and 'over here' also indicated immersion in the created environment. Interest in the displays was high, with many questions about what they were viewing. They were keen to understand the various forms of the displays and, in further discussions in the 'breakout' areas, they had extensive conversations about how a platform of this kind could be used for the RSA. In addition to the focussed discussions which related to knowledge exchange, specific to the RSA, it was also found that participants engaged in 'water cooler' conversations. These included exchanges about the weather and their individual locations etc.

During the second session it was proposed by the participants that a useful 'tool' would be an area where they could discuss current 'claims'. These primarily focussed on the difficulties of insurance assessments related to claims. Therefore a new area was created which had photographs and artefacts relating to a hypothetical 'domestic fire damage' claim was designed and built (Fig. 4). Although this area would not use the 'streamed data' previously discussed for the example, it would enable the participants to explore the potentials for functions and opportunities such an area might provide.

Fig. 4. Fire damage discussion area

4   Potential Areas for Collaboration, Knowledge Sharing and Learning Identified by Participants

Participants were invited to engage with the virtual world for a third session where use case modelling was broadly utilized to "allow description of sequences of events that, taken together, lead to a system doing something useful." [25]. Again groups of 3–4 participants met inworld, and much as in the physical world, greetings were exchanged and 'water cooler' chat happened whilst all participants gathered. Again demonstrating that the individuals who engaged with this project, displayed clear engagement and presence when inworld. During the more formal discussions participants listed a number of functions that they felt would be useful if the system displayed similarities in connectivity that the Earthglobe example employed. A brief overview of the main discussion areas are listed in Table 1.

Table 1. Overview of main discussion areas

Suggestion	Opportunities and potential benefits identified by participants	Current
Creating relevant virtual artefacts and a 'display' area for current 'in process' claims	Early career loss adjusters meeting more experienced loss adjusters inworld to discuss the case. In addition the often many people involved in a claim can meet inworld from any location globally, to discuss the claim	Many email exchanges Hard to know who is doing or has done what Site meetings have to be organised, which can take considerable time, both to organise and to travel to
Linking relevant virtual artefacts to sensors and/or other monitoring systems (video etc.) on site to monitor progress	Insurance companies often handle many facets of a claim. These might include builders, hotels, local authorities, furniture and other suppliers. If sensors/bar code scanners/video/etc. were employed the site could be monitored and progress updated in real time Additionally it was also discussed that the site of the claim was often 'at risk' from burglary and such a system may reduce this risk	Many email exchanges Hard to know who is doing or has done what Site meetings have to be organised, which can take considerable time both to organise and to travel to
Linking forensic data to the created model and artefacts	The participants felt that if this was possible it would enable communication between the forensic teams and the loss adjusters. Working inworld with voice and an actual model and artefacts would enable and enhance communication	Forensic reports are sent to the loss adjusters and although it was felt communication was generally good, sometimes aspects of the report needed further clarification

5   Discussion and Conclusion

The findings suggest that the virtual world environment is an effective platform for supply chain social interaction and learning, and it was stated by the participants that the platform was 'almost as good as meeting in the real world'. Participants agreed that such physical meetings took up a considerable amount of 'unprofitable' time – time taken getting to and from the meeting, overnight stays, etc. In documents relating to other companies experience with virtual worlds, a key factor for the company was the reduction of travel costs. IBM who have been building their virtual world assets, states that '… the Annual Meeting was executed beautifully at one-fifth the cost of a real world event. [26] Many of these costs relate to travel of the participants, which in turn creates greenhouse gasses, clearly linked to unstainable environmental impact.

   To ensure useful user participation it is essential that the virtual world environment is 'fit for purpose'. In this study the existence of streamed data which could be made relevant to an insurance context, was used as the basis for further discussion, and for generating new information and knowledge between users, as well as learning from the information. The use of the virtual world by the participants was found to be easy to engage with and provided a medium to explore data and exchange knowledge, concepts and ideas, that was not possible through other platforms (emails, Yammer, WebX, etc.). The focus on the preparation for engagement with the VW; making it 'useful' for both the users and the management team, was demonstrated to be crucial in developing 'meaningful' outcomes. Table 1 describes a range of current processes which mainly involve communication channels. The participants felt that these current channels lacked 'presence', which meant, for example, that they were unsure if emails had been received, if someone was 'at their desk' and able to respond, or that the message had been 'fully understood'. It was also felt that in relation to more extensive incidents (flooding, weather related, etc.) that often they found out about the incident through news channels, and then monitoring the situation was reliant on them searching a range of media outlets. All participants related that they had felt a sense of 'being there' and that the virtual world experience had 'felt real'. It is felt that for this group in this pilot study, a clear sense of presence experienced by the participants, had been demonstrated. After engaging with the virtual world they felt that a range of the issues they encountered in their jobs, could be addressed, through the use of virtual world technology as presented to them in this pilot study.

Further work, which would relate the use of virtual worlds in the business context to specific cost parameters should be considered. If virtual world meeting are shown to be effective in a range of areas, then costs, specifically travel costs, can be reduced. This in turn would reduce the significant amount of greenhouse gasses caused by business travel.

References

1. Wreford, L., Leston J.: WWF-UK policy position statement on business travel (2013). http://assets.wwf.org.uk/downloads/business_travel_ps_0709.pdf
2. Pamlin, D., Szomolanyi, K.: Saving the Climate @ the Speed of Light. WWF and Enso, Brussels (2006)
3. Grant, S.B.: KNOWNET: exploring interactive knowledge networking across insurance supply chains. Int. J. Prod. Manag. Eng. 2(1), 7–14 (2014)
4. Bollinger, A.S., Smith, R.: Managing organizational knowledge as a strategic asset. J. Knowl. Manag. 5(Iss1), 8–18 (2001)
5. Wu, C.: Knowledge creation in a supply chain. Supply Chain Manag. Int. J. 13(3), 241–250 (2008)
6. Edwards, T., Battisti, G., Neely, A.: Value creation and the UK economy: a review of strategic options. Int. J. Manag. Rev. 5/6(3/4), 191–213 (2004)
7. Giannakis, M.: Facilitating learning and knowledge transfer through supplier development. Supply Chain Manag. Int. J. 13(1), 62–72 (2008)
8. Chiu, C., Hsu, M., Wang, E.: Understanding knowledge sharing in virtual communities: an integration of social capital and social cognitive theories. Decis. Support Syst. 42, 1872–1888 (2006)

9. Leug, C.: Knowledge sharing in online communities and its relevance to knowledge management in the e-business era. Int. J. Electron. Bus. **1**(2), 140 (2003)
10. Nevo, S., Nevo D., Carmel, E.: Unlocking the Business Potential of Virtual Worlds. MIT Sloan Magazine: Spring 2011, 23 March 2011
11. Gilbert, M., Cordey-Hayes, M.: Understand the process of knowledge transfer to achieve successful technological innovation. Technovation **16**(6), 301–312 (1996)
12. Swan, J., Newell, S., Scarbrough, H., Hislop, D.: Knowledge management and innovation: networks and networking. J. Knowl. Manag. **3**(4), 262–275 (1999)
13. Wikipedia definition of virtual worlds. https://en.wikipedia.org/wiki/Virtual_world. Accessed 09 Apr 2016
14. Gandi, S.: IBM dices into Second Life (2010). www.ibm.com/developerworks/library/os-social-secondlife. Accessed 01 May 2016
15. Mujber, T.S., Szecsi, T., Hashmi, M.S.J.: Virtual reality applications in manufacturing process simulation. J. Mater. Process. Technol. **155–156**, 1834–1838 (2004)
16. Wikipedia definition of Minecraft. https://en.wikipedia.org/wiki/Minecraft. Accessed 04 Apr 2016
17. http://unity3d.com/unity. Accessed 04 Apr 2016
18. Wikipedia definition of Open Simulator. http://opensimulator.org/wiki/Main_Page. Accessed 04 Apr 2016
19. https://www.yammer.com/. Accessed 04 Apr 2016
20. Wikipedia definition of avatar. https://en.wikipedia.org/wiki/Avatar_%28computing%29. Accessed 04 Apr 2016
21. Internet of Things Global Standards Initiative. http://www.itu.int/en/ITU-T/gsi/iot/Pages/default.aspx. Accessed 04 Apr 2016
22. Wikipedia definition of the Richter magnitude scale. https://en.wikipedia.org/wiki/Richter_magnitude_scale. Accessed 04 Apr 2016
23. Wikipedia definition of the Terminator (solar) zone. https://en.wikipedia.org/wiki/Terminator_(solar) Accessed 04 Apr 2016
24. Schuemie, M.J., Van der Straaten, P., Krijn, M., Van der Mast, C.A.P.G.: Research on presence in virtual reality: a survey. Cyberpsychol. Behav. **4**(2), 183–201 (2001). Mary Ann Liebert, Inc.
25. Use Case Modelling by Kurt Bittner and Ian Spence. https://tvolodi.files.wordpress.com/.../use-case-modelling-by-kurt-bittner Accessed 06 Apr 2016
26. http://www.virtualpublichealth.com/docs/Second_Life_Case_IBM.pdf. Accessed 28 Jan 2017

Manufacturing Lead Time Reduction
and Its Effect on Internal Supply Chain

Atanas Ivanov[✉] and Twana Jaff

Brunel University, Uxbridge, London UB8 3PH, UK
atanas.ivanov@brunel.ac.uk

Abstract. Companies seek to reduce manufacturing lead-time in order to reduce the cost of the production; short lead-times are a major source of potential competitive advantage and also can help achieve internal supply chain optimization and better sustainability. This paper proposes a study on reducing manufacturing lead time. The research methodology based on survey questionnaire and cased study in order to find potential methodologies that can reduce lead-time and its effect on internal supply chain. This research study will present a conceptual framework of the causes of excessive lead-time. The aim is to provide simple strategies for reducing manufacturing lead-time also to provide internal supply chain more efficiency.

Keywords: Manufacturing lead time reduction · Throughput time reduction · Quick response manufacturing · Operation management · Internal supply chain

1 Introduction

This research presents a study on reducing lead-time in manufacturing industry and aims to provide guidance to the industry practitioner on how this will affect the internal supply chain. Initial idea is to reduce manufacturing throughput time and to discuss how this will influence the flexibility of the internal supply chain and improve the sustainability of a company. At present theses links are not always well understood [1, 5, 6, 8]. Many benefits and competitive advantages can be obtained from a shorter manufacturing lead-time (MLT) and the benefits go far beyond improving profitability and increasing the process of quick response to customer [2, 3]. This work will propose some key factors which should be considered in the MLT reduction studies and the links with the supply chain. In the discrete manufacturing production nearly 80–85% of the time is a non-productive time [4, 10].

The manufacturing practices and processes have faced increasing pressure from global competition and the need for a shorter lead-time is now more pressing than ever. Lead-time may be regarded as the time between when the customer makes an order and when the customer receives the finished product [12, 13]. The major components of non-value added lead-time are: wait time, move time and down time [11, 14]. Therefore focusing on MLT is important, because MLT is the sum of setup time, processing time, and non-operation time [9, 15]. Also reliable supply chain management requires a cost-optimal control and reducing MLT of all processes along the entire value-added chain. Internal supply chain optimization depended on three factors which

© Springer International Publishing AG 2017

G. Campana et al. (eds.), *Sustainable Design and Manufacturing 2017*, Smart Innovation, Systems and Technologies 68, DOI 10.1007/978-3-319-57078-5_38

are; purchasing, manufacturing, and distributing [16, 18]; every sub-process of the internal supply chain has its own goals. The purchasing department wants to keep inventory low, while the production team strives to use its resources efficiently [17].

2 Supply Chain Management and Lead Time Reduction

Supply-chain management is the integration of the activities that procure materials and services, transform them into intermediate goods and the final product, and deliver them to customers. It is also the management of the links between an organization and its suppliers and customers to achieve strategic advantage [1, 19, 24]. It is actually focused on managing of the physical flows of goods, the flows of information, collaboration, cooperation and communication between all activities. Improving internal supply chain collaboration by increasing information and processing capacity also reducing MLT is very important factor which has significant impact on internal supply chain management [21]. If the overall view of the internal supply chain process is missing, conflicts are inevitable and an optimal performance cannot be achieved [7, 20]. A smooth and integrated supply chain planning process can only be achieved when all factors are integrated into a single, comprehensive plan stretching across all divisions (Fig. 1).

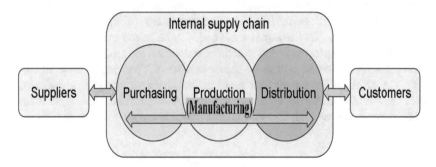

Fig. 1. The internal supply chain consists of three major processes [1]

The main issue discussed in Capgemini's "2016 Future Supply Chain" report is the impact of lead-time reduction on improving supply chain networks. Supply chain integration should be considered because in a supply chain context, integration is defined as a process of interaction and collaboration in which companies in a particular supply chain work together to arrive at mutually acceptable outcomes [23]. By closely integrating with suppliers and customers, firms can reduce costs, improve quality and shorten lead times in order to remain competitive therefore lead-time consists of two consecutive components through a supply chain, the order information pipeline and the material flow. A time-based company is only as good as their fellow players in the supply chain. The importance of reducing lead-time comes from the ability to answer the market demand on time. In this way, market responsiveness increases along with the shelf availability of the product. Several previous researchers have suggested improving demand chain performance, is better for managers in a supply chain to focus first on lead time reduction, or instead concentrate on improving the transfer of demand information upstream in the chain. Even though the theory of supply and demand chain management suggests that lead time reduction is an antecedent to the use of market mediation (i.e., adjusting production to fit actual customer demand as it materializes) [1, 22].

3 Methodology

This work aims to use a simple hypothetical model of a manufacturing system to illustrate the basic factors that determine manufacturing throughput time and explain why each factor occurs and how it influences the internal supply chain. The main purpose of this approach is how planning techniques provide the basic groundwork for lead-time reduction and identifying simple strategies for reducing lead time in the area of production/none-productive and operations. The strategies fall into two general categories: (1). The basic factors that determine MLT and manufacturing throughput time must be clearly understood by developing mathematical models for quantifying the effect of those factors on MLT (2). The factors that influence manufacturing throughput time, the actions that can be taken to alter each factor, and their interactions, by dual approach such as technical & theoretical of manufacturing management will provide potential for lead-time reduction that means that the supply chain optimization depended on MLT reduction (Fig. 2). The first stage of the procedure involves survey-based research. This research was needed to be traced over a period of time to reflect manufacturing processes and lead time, and to find the greatest number of factors which had a significant impact on reducing lead-time. The second stage of the procedure was to consider experiential case study. The case study was designed and located in the Zhala Plastic Pipe factory. A face-to-face interview and workshop procedure were carried out to examine the procedure that enables the production

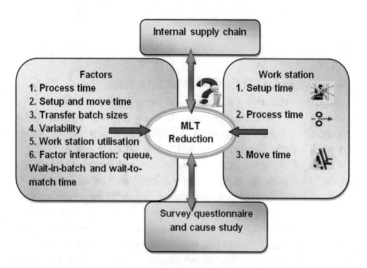

Fig. 2. Road map of the research area

planner to move the work between time periods to smooth the load or at least to bring the manufacturing system within capacity. In this research study, statistical techniques, various tools and techniques were used to determine how to reduce manufacturing lead-time and its effect on internal supply chain.

4 Survey Results, Case Study Results and Discussion

The objective of the questionnaire was to gain insight into the MLT being studied, to identify the defects which cause an increase in lead-time and to identify improvement opportunities for reducing lead-time. The responses to each question were assigned to the manufacturing assessment, which asked the respondents to answer the assessment questionnaire. The response rate was 100%. The summary of 160 responses indicates that all the participants have different primary functions. Among the 160 staff were 22 engineers, 54 technicians, 29 supervisors, 10 managers and 45 staff with different, unspecified jobs. The survey shows that 76% of them have experience of between one and three years and 24% of them have more than three years' experience. The level of experience and the participants' functions are the most important factors in their knowledge of the level of system performance. In addition, these factors inform these staff members' opinions and suggestions to improve internal supply chain and lead-time reduction towards zero defect manufacturing. The survey shows that the 90% of respondents informed their customers that the orders expected to be late from the eight factories; this is the indicator of long MLT and internal supply chain was inefficient.

Figure 3 shows that most of the eight factories have not provided enough professional training and feedback to employees and that most of them do not have enough Quality Assurance, Quality Control and Traceability in place in their company procedures. Only 13% of the participants mention fully supporting company procedures. Therefore, these statements provide the major direction for reducing defects in each step, measures which should be taken before the next step. 72% of the responses mentioned that the company maintained stock production. This affects decisions about batch size for products because 90% of respondents noted in the survey that the company informs their customers when orders are expected to be late. The respondents were also asked to rate their companies on job organization. 52% of respondents referred to an average situation and none of the 160 respondents mentioned an excellent situation. The assessment questionnaire shows that employees have a limited ability to work because they referred to serious shortages in the following areas: labour skill (69%); quality management and layout strategy for operation management (95%); planning for lot or batch sizes and the firm's policy (82%); and equipment, machinery and technology (88%). In addition, the respondents had limited current abilities, revealed by statements, such as a high production rate every day, stock supply and/or in time deliveries, lack of communication on the workshop floor, and sudden changes in production/transfer batch size decisions, which were mentioned by 88% of respondents as being serious issues while 82% mentioned shifts not being scheduled regularly each day. Management of the internal supply chain requires more than simply planning individual processes also optimal planning must be integrated where the optimization function is agreed by internal supply chain managers. Therefore, these areas of

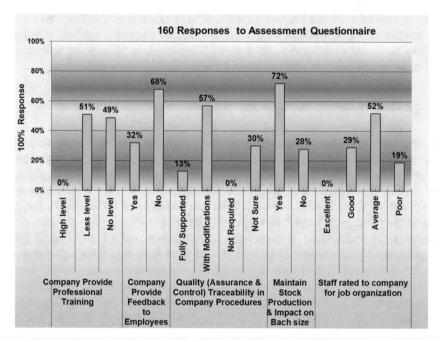

Fig. 3. Responses to management practices and human resources assessment

management practice, human resources and operation management have been identified as needing some attention and they should be considered and improved before starting to reduce lead-time or to improve internal supply chain process. Therefore, most of the statements indicated that the factories have not competes on quality, which means that defects and long lead-time could be expected during the process time as well as lack of communication, human resources and operations on the workshop floor are the main factors that affect internal supply chain management.

Figure 4 shows the respondents' answers regarding the above factors which have a significant impact on MLT reduction and that should be targeted by their companies. The average rating of 4.7 indicates that the general feeling among respondents is that process time has a major impact on lead-time reduction and their standard deviation is 0.30. This means that the largest average ranking indicates the top answer choice. Move time has an average rating of 4.4, indicating that the system performance needs a strategy for process and product layout procedure. Meanwhile, batch size, setup time, waiting time and time utilization received average ratings (3.8, 3.4, 3.4 and 3.3 respectively). The average rating of 4.7 indicates that the general feeling among respondents is that process time has a major impact on lead time reduction, also the Coefficient of Variation CV 10% is less value than others factors and the Z-Score, which is a six-sigma technique, indicated that the percentile rank of process time is 93% (Z-Score), which is more than other factors. Therefore, these factors have a major strategic role in reducing lead-time towards ZDM. They should be considered as guidance for future experimental case studies as well as guidance to industry

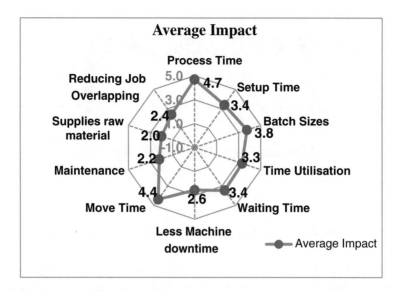

Fig. 4. Average ranking of factors that have a significant impact on MLT reduction

practitioners on how to reduce lead-time towards ZDM. Since no organization can excel in all these factors simultaneously, the decision to focus on one or more of these factors provides a unifying directional force for competitive advantage. If a company competes on quality without defects and lead-time, then it should be evaluated in terms of its ability to deliver high-quality products in a timely fashion, therefore focusing on (process time, move time, batch sizes, and waiting time) will provide for internal supply chain management more on control and efficiency as well as those factors that affect internal supply chain integration directly.

Figure 5 shows that the average rated between 2.13 and 2.34 meaning that respondents indicated that most of the factories have the causes of variability of workload. 35% of the responses suggested that the causes of variability of workload are seriously out of control and 64% indicated 'slightly' due to manufacturing variation not being controllable while 14% of the responses suggested that the causes of variability of workload are seriously under control and 85% are 'slightly' due to manufacturing variation being under control, therefore both controllable variation and random variation existed in the system. Those conditions of variation were uncontrollable variation because there are differences in the processing time of different parts due to design differences and inaccurate transfer batch size decisions will be taken at different times. Also, insufficient MRP will be applied and the amount of potential work in outstanding quotes is not known and is not used when forecasting shop loading due to insufficient operations management. There is always a longer lead time associated with manufacturing defects and variation; it is really a potential problem or defects for internal supply chain processes therefore a corrective action should be taken to avoid long MLT and late delivery time. All eight factories should use/or consider a formal job tracking system, therefore fewer changes to orders and production schedules in order to achieve higher manufacturing efficiency levels.

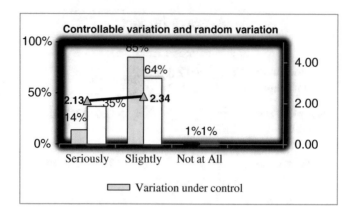

Fig. 5. The causes of variability of the workload in eight factories

Figure 6 shows the results of a case study which aims to provide a logical tactic for smoothing the load and minimizing the impact of changed lead-time. One such tactic includes order or lot splitting at Bamok—a factory that produces plastic pipes. Bamok's production planners have insufficient time to complete their production orders in the allotted seven days. Pipes should be ready for shipping so as to be delivered on time, but a problem occurs four times during the year. The pipes take 15 min each but only 480 min of production time is available in the factory each day, and employees only work five days a week. Figure 6 shows that the capacity was exceeded in the (A) period on day two and day five because the required capacities were 510 min for day two and 525 min for day five. As they have only 480 min available in the work center each day, they assigned two units from day two to day one's work, and two units from day five to day four's work and one unit to day three (or requested overtime). As shown for the period B, the average available capacity is adequate, at 480 min, and has become equal to the required capacity. Therefore by splitting the order, the production planner is able to utilize the factory's capacity more effectively and still meet order requirements. This tactic will lead to controlling and reducing lead-time also it is a tactic for smoothing the load and minimizing all units ordered in the requirement plan, meaning a trade-off between the capacity required (minutes) and capacity available (minutes). Therefore capacity available time with proper planning will lead to improve the internal supply chain. The observations show that manufacturing lead time reduced from 9 days to 7 days as MLT reduced by 22.22%. The next requirement task is to staff and balance the work because designing or developing a work balance chart is very important procedure for daily production which also puts internal supply chain under a proper planning. This procedure involves two steps: determining the takt time and the number of operators required because the factory expects 160 pipes delivered weekly.

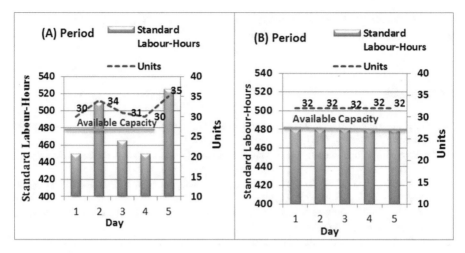

Fig. 6. (A) period resource requirements (B) period smoothed resource requirements for Bamok plastic pipes

5 Conclusion

The inductive approach of this research relies on the interpretive method. More specifically, the techniques applied in this work include a survey conducted in order to find the best opportunities for reducing lead-time. MLT is an indicator of PLT; as such, it should be considered a research area worth investigating. The Manufacturing Survey Questionnaire is one of the key tools for this purpose and is an effective assessment tool for helping to understand the problems of and opportunities for reducing MLT and also provides actionable insights that will allow manufacturers to make better decisions for improving internal supply chain. Finding the role of human resources in terms of level of experience, technicians and skills is very important and should be considered to improve the capacity of companies as well as to improve internal supply chain management. Motivation, commitment and training for the set of mechanisms should be established to develop and improve employees' knowledge and/or skills in job-related areas because they are important factors that affect internal supply chain integration. The survey also provided a list of areas in which improvement might be possible, based on their firm's responses to the survey-questionnaire. Finding that there is always a longer lead time associated with manufacturing defects and variation; is really an area where corrective action should be taken to avoid long MLT and late delivery times. Finally the survey indicated that the eight factories have direct influence on the management and may cause low flexible manufacturing, decreased resource efficiency, inaccurate demand forecast, disruptions and unreliable lead times. This will potentially affect critical delivery dates and internal supply chain management. Finding the splitting orders is the best tactic for smoothing the load and minimizing the impact of the changed lead time at the Bamok factory. Future experimental case studies should be considered for five factors, altering each factor related to Fig. 2 and will be designed in order to provide guidance to industry practitioners on how to reduce manufacturing lead-time and its effect on internal supply chain.

References

1. Heizer, J., Render, B.: Principle of Operations Management, 7th edn. Pearson Prentice Hall, Upper Saddle River (2008)
2. Gaither, N.: Production and Operations Management, 6th edn. The Dryden Press, Orlando, New York (1994)
3. Silver, E.A., Pyke, D.F., Peterson, R.: Inventory Management and Production Planning and Scheduling. Wiley, New York (1998)
4. Suri, R.: Quick Response Manufacturing: A Companywide Approach to Reducing Lead-Times. Productivity Press, New York (1998)
5. Jaff, T., Ivanov, A.: Manufacturing lead-time reduction towards zero-defect manufacturing. In: The SDM2014 International Conference on Sustainable Design and Manufacturing, Cardiff, UK, 28–30 April, pp. 628–640 (2014). http://sdm-14.kesinternational.org
6. Fahimnia, B., Luong, L.H.S., Motevallian, B., Marian, R.M.: Analysing and formulation of product lead-time. Int. J. Appl. Math. Comput. Sci. 3(4), 221–225 (2007)
7. Johnson, D.J.: A framework for reducing manufacturing throughput time. J. Manuf. Syst. 22(4), 283–298 (2003)
8. Hoppe, W.J., Spearman, M.L.: Factory Physics: Foundation of: Manufacturing, 2nd edn. Irwin/McGraw-Hill, Boston (2001)
9. Groover, M.P.: Automation, Production Systems, and Computer-Integrated Manufacturing, 3rd edn. Pearson Prentice Hal, Upper Saddle River (2008)
10. Gerchak, Y., Parlar, M.: Investing in reducing lead-time randomness in continuous-review inventory models. Eng. Costs Prod. Econ. 21(2), 191–197 (1991)
11. Wang, Y., Gerchak, Y.: Input control in a batch production system with random yields, lead times and due dates. Eur. J. Oper. Res. 126(2), 371–385 (2000)
12. Ray, S., Gerchak, Y., Jewkes, E.: The effectiveness of investment in lead time reduction for make-to-stock products. IIE Trans. 36(4), 333–344 (2004)
13. Karki, P.: The impact of customer order lead time-based decisions on the firm's ability to make money. Ind. Manag. 25, 18–46 (2012). Acta Wasaensia, 257
14. Ketokivi, M., Heikkila, J.: A strategic management system for manufacturing: linking action to performance. Prod. Plann. Control 6, 487–496 (2003)
15. Hoppe, W.J., Spearman, M.: Practical strategies for lead time reduction. Manuf. Rev. 3(2), 78–84 (1990)
16. You, F., Ignacio, G.: Design of responsive supply chains under demand uncertainty. Comput. Chem. Eng. 32(12), 3090–3111 (2008)
17. Suri, R.: QRM and POLCA. Technical report, Centre for Quick Response Manufacturing, USA (2003). www.apics-nwie.org/images/uploads/QRM.pdf
18. Kuhlang, P., Edtmayr, T., Sihn, W.: Methodical approach to increase productivity and reduce lead time in assembly and production-logistic processes. CIRP J. Manuf. Sci. Technol. 4, 24–32 (2011)
19. Karmarkar, U.S.: Manufacturing lead times. In: Graves, S.C., RinnoyKan, A.H.G., Zipkin, P.H. (eds.) Logistics of Production and Inventory. Handbooks in Operations Research and Management Science, vol. 4. Elsevier Science Publishers B.V., North-Holland (1993)

20. Halpin, J.F.: Zero Defects: A New Dimension in Quality Assurance. McGraw-Hill, New York (1996)
21. Crosby, P.B.: Quality Is Free: The Art of Making Quality Certain. McGraw Hill, New York (1979)
22. Calvin, W.: Quality control techniques for 'zero defects'. IEEE Trans. Compon. Hybrids Manuf. Technol. 6(3), 323–328 (1983)
23. Jayal, A.D., Badurdeen, F., Dillon Jr., O.W., Jawahir, I.S.: Sustainable manufacturing: modeling and optimization challenges at the product, process and system levels. CIRP J. Manuf. Sci. Technol. 2, 144–152 (2010)
24. Eyers, D., Dotchev, K.: Technology review for mass customization using rapid manufacturing. Assembly Autom. 30(1), 39–46 (2010)

Remanufacturing as Pathway for Achieving Circular Economy for Indonesian SMEs

Yun Arifatul Fatimah[1(✉)] and Wahidul Biswas[2]

[1] Engineering Faculty, Universitas Muhammadiyah Magelang,
Jl. Bambang Sugeng km5, Magelang, Central Java, Indonesia
yun.fatimah@ummgl.ac.id
[2] Sustainable Engineering Group, Curtin University, 6 Sarich Way,
Technology Park, Bentley, Perth, Australia

Abstract. Remanufacturing could potentially offer economic and environmental benefits for Indonesian SMEs. The objective of this research is to explore as to how remanufacturing strategy could attain greater resource efficiency through resources consumption reduction and waste minimization. An assessment on resources efficiency has been performed for Indonesian remanufacturing SMEs producing auto parts. The value, contributions and limitations of remanufacturing for achieving resources efficiency were explored and some key issues including organizational, consumer behaviors and government incentives issues have been identified. Accordingly, national resources policy recommendations have been made. Future potential business value that enhances Indonesian economic, social and environmental pillars of sustainability through remanufacturing has been discussed in this paper.

Keywords: Remanufacturing · Circular economy · Resource efficiency

1 Introduction

A global sustainable development challenge arises all over the world due to unsustainable consumption and production leading to the depletion of natural resources and the increase in harmful environmental impacts. Circular economy is an industrial framework which is giving new strength to design out of waste, to use few resources, to extract value from resources, and to recover and regenerate materials and products cycling products and materials, and making better use of existing component/parts [1]. Remanufacturing can be regarded as a potential pathway to achieve by transforming used products into as good as new through core collection, inspection, cleaning, disassembly, recondition, reassembly and final testing activities [2]. The application of remanufacturing strategy is potential to improve resources efficiency as it avoids energy and materials requirements, reduce waste generation and GHG emission in processing raw materials [3]. The objective of this research is to explore as to how remanufacturing can be applied in Indonesian SMEs to attain circular economy through greater resource efficiency, resources consumption reduction and waste minimization. A thorough observation has been conducted in a SME remanufacturing auto parts to investigate how the SMEs could decouple revenues from energy, material, and other

© Springer International Publishing AG 2017
G. Campana et al. (eds.), *Sustainable Design and Manufacturing 2017*, Smart Innovation,
Systems and Technologies 68, DOI 10.1007/978-3-319-57078-5_39

resources input in their circular business operations. Some key issues and policy recommendations were also presented accordingly to determine future potential business value of the remanufacturing strategies.

2   Literature Reviews

2.1   Circular Economy

Japan is an example for circular economy, because the country had no other alternative except for applying 6Rs to address its resource scarcity while maintaining the economic growth. Accordingly, financing mechanisms and policies have been developed by the Japanese government to apply 6Rs. For example, the metal recycling rate is up to 98% in Japan that avoided landfill area by 95%, and recovered up to 89% of the electronic and electrical products [4]. Circular economy has also been considered in the planning and development in China. China's Government has developed 50-year sustainable plan, strategies and law to overcome the growth of resource consumption and production to address its population and economic growths. Based on these approaches, China could reduce 62% of energy consumption/GDP in 2010, increase 45% of waste water treatments and reduce 45% of resources consumption [4].

However, in Indonesia, circular practices have been considered even at minimum level in the economic development framework [5]. Even though the concept has been applied in some large and medium enterprises due to economic reasons and competitiveness advantages, the concept of circular economy practices has not been yet implemented, especially in small industries. In Indonesia, remanufacturing strategies have already been applied to heavy equipment, auto parts, toner cartridge, electronics products, car battery, and tires. Auto parts including alternators and starters are one of the most dynamic markets in Indonesia. In 2015, there are about 1,500 automotive companies including auto parts remanufacturing SMEs that produce vehicle parts. About 40% of the auto parts are supplied to OEMs and 60% of auto part was sent to aftermarket [6]. Research conducted by Fatimah [5] found that the remanufacturing auto parts and electronic products offer very potential economic, social and environmental benefits for Indonesian manufacturing industry.

2.2   Resource Efficiency Through Remanufacturing

Globally, the concept of circular economy has improved industrial business and delivered significant resource efficiency [4]. Remanufacturing which is defined as a series of manufacturing steps acting on EoL parts or products to return it to like new or better performance with warranty to match is the ultimate strategy of the circular practice in which waste generation and new material consumption are significantly reduced [7]. The cores (used products) are restored to useful life, passing a number of remanufacturing operations (i.e. inspection, disassembly, part reprocessing, reassembly, and testing) to meet the desired product standards [2]. In the world metabolism, remanufacturing would maintain products in circulation and place them in long term store [8].

Remanufacturing unlocks at least fifth opportunities including energy and material use minimization, waste and emissions reductions, financial performance enhancements, new technology encouragement, and job creation developments [1]. Remanufactured products are manufactured with less energy consumption and small portion of virgin materials thus enabling the remanufactured products to be sold at affordable prices through cost reductions in remanufacturing plants [9]. Furthermore, remanufacturing industries contribute to reducing product life cycle cost, machine down time and supply chain network through product reuse and recycling [3].

Waste and emission are economic burdens for an industry. It can be excluded from landfill or emission taxes, if they are managed properly through remanufacturing. Remanufacturing protects environment from hazardous materials (e.g. plastic, steels) and conserves natural resources and energy significantly. A number of studies show that remanufacturing provides better or same quality and reliability to new products, supported by an appropriate warranty period with reduced energy consumption (i.e. 50% to 80%) and less cost (20% to 80%). It reduced material consumption by 26–90% and solid waste generation by 65–88% [10, 11]. Cost reductions associated with the maximum use of reused materials and the minimum use of virgin materials and energy consumption may lead to a significant reduction in price (35–40% of the new product price) while achieving a satisfactory profit margin (20%) [10, 12]. In addition, GHG emission during remanufacturing process (e.g. compressors) could potentially be reduced by 89.4–93.1% [13].

3   Methodology

A case study on SME remanufacturing auto parts (i.e. alternators) in Java Island, Indonesia was conducted to develop comprehensive assessments of resource efficiency achievements. The SME was selected as one of leading industry which has main business on remanufacturing, refurbishing and repairing auto parts (i.e. alternators, starters). Alternator was chosen as the subject as it is the highest products remanufactured and sold in the SMEs. A comprehensive research through intensive interview and direct observation was conducted to gather data and information in associated with alternator remanufacturing process. The data and information including number of products, materials (i.e. cores, raw and auxiliary materials), energy (i.e. electricity, fuels), water and pollution (i.e. waste) were collected based on 1 year time period to assess the resource consumption. The SME is categorized as medium enterprise supported by semi-automatic technology involved in the process. Labor is still as the main resource for disassembly and reassembly processes while cleaning, coating and testing were mainly conducted by machines.

To evaluate resource efficiency in the remanufacturing SME, three resource indicators (i.e. energy, water and material) were used. In addition, three pollution indicators (i.e. carbon, solid waste and wastewater) were also used to determine the pollution intensity. The indicators were selected on the basis that they are considered as the most important environmental aspects of SME remanufacturing process and keys of manufacturing competiveness [14]. Resource productivity is used to measure product output per unit of resource consumption. It is calculated by dividing the product outputs (i.e. remanufactured products) by the amount of respective resources used in remanufacturing process in year. Pollution intensity is used to measure the amount of pollution per unit of remanufactured product output. Pollution intensity is estimated by dividing the amount of pollution created from the remanufacturing process by number of remanufactured product produced. The formulas for resource productivity and pollution intensity calculations are adopted from [15].

A comparative analysis is then conducted to evaluate the resource efficiency and pollution intensity of the existing remanufacturing process in comparison with the threshold values of remanufactured products and new manufactured products. The data for the threshold value and new manufactured products are estimated based on literatures discussing auto part products. A comprehensive analysis was then performed to identify the issues to achieve the resources efficiency.

4   Results and Discussion

4.1   Resource Efficiency in Remanufacturing SMEs

The assessments on the SME remanufacturing auto parts highlighted that the SME has adopted circular principles throughout their business process from collecting of cores, transforming the cores into as similar as new and marketing them to local and national markets. Direct selling to customers at their shops has been identified as the main marketing strategy and there was no online shopping. The price of remanufactured auto parts is between 40% and 50% of the price of new part and the warranty provided was 1 year. The SME employs local people whom majority of them are working through their experience and skills gathered from their vocational school.

Tables 1 and 2 present the detail resource efficiency assessment of the remanufactured auto parts in comparison with new one. The value of new auto parts and threshold value were calculated using the resource and energy consumptions figure from literature reviews [16–19]. These literatures were selected as they represent the energy and resource used for new and manufactured alternators and auto parts.

Table 1. Resource productivity of remanufactured and new auto parts

Resource productivity	Remanufactured auto parts	New auto parts	Threshold value
Energy productivity [in unit of product/MJ energy]	4,500/344,250 = 0.013 unit/MJ	4,500/1,192,500 = 0.004 unit/MJ	4,500/262,350 = 0.017 unit/MJ
Materials productivity [in unit/kg]	4,500/4,257 = 1.06 unit/kg	4,500/9,900 = 0.46 unit/kg	4,500/3,465 = 1.30 unit/kg
Water productivity [in unit of product/liter water]	4,500/22,500 = 0.20 unit/L	4,500/90,800 = 0.05 unit/L	4,500/13,500 = 0.33 unit/L

The results from the assessment in Table 1 show that the resource productivity of the remanufactured auto part is higher than the new one. The increases of resource productivity offered significant value of the auto parts and minimize cost through better

Table 2. Pollution intensity of remanufactured and new auto parts

Pollution intensity	Remanufactured auto parts	New auto parts	Threshold value
Emission (Carbon) intensity [in kg CO_2-eq/unit of product]	18,450/4,500 = 4.1 $kgCO_2$-eq/unit	47,922/4,500 = 10.65 $kgCO_2$-eq/unit	5,750/4,500 = 1.28 $kgCO_2$-eq/unit
Solid waste intensity [in kg/unit of product]	495/4,500 = 0.11 kg/unit	9,000/4,500 = 2.2 kg/unit	1,980/4,500 = 0.44 kg/unit
Waste water intensity [in liter/unit of product]	22,500/4,500 = 5 L/unit	90,800/4,500 = 20.18 L/unit	13,500/4,500 = 3 L/unit

manufacturing stability of the SME. The increase of resource productivity shows that the SMEs could reduce the amount of energy, material and waters consumption significantly. The type of materials conserved are steels, aluminum, and copper, and energy resources including oil, coal and gas which were required to produce electricity could potentially avoided due to this remanufacturing operation, thus enhancing intergenerational social equity. The SME reduced the energy they use in their production up to 71% of new product, materials use up to 57% of new product while increasing recycling and reuse potentials and water consumption up to 75% of new product. Accordingly, there is great economic opportunity for the SMEs by cutting the manufacturing costs through reusing materials and components.

In addition, it can be seen in Table 2, that the increase of resource productivity has positive correlation with environmental performance of the SMEs. The SME could reduce the amount of pollution and emission created from their production process. The SME could reduce the amount of GHG emission by 61.5% of new product which is equivalent to taking 2,073 cars of the road, the solid waste by 95% of new product which is avoiding the amount of solid waste goes to landfill area and waste water also by 75.2% of new product.

4.2   Key Barriers to Achieve Resource Efficiency

The aforementioned results show high performance of the resource productivity and pollution intensity of the remanufactured product. However, the resource productivity and pollution intensity of the SME are still far lower than the threshold value, which means that threshold value of sustainable manufacturing was not achieved. Hence, an identification of the potential issues was conducted through direct observation and questionnaire to the SME. The observations from the industry visit are as follows:

They exists lack of knowledge and understanding about the application of remanufacturing among the workers as most of them do not even have basic educational qualification. Majority of the labors have high school/vocational school level qualification with no prior knowledge or skill in remanufacturing. Secondly, there are very limited number of training facilities and workshops to provide labors with up to date knowledge and equip them with technical skills to operate modern machinery and advanced machine tools.

Indonesian people are culturally used to with using these products until the EoL so the quality of used product for remanufacturing is not up to the mark. This not only affects reliability and durability of these remanufactured products, but also affects the competitiveness in the local market. While in the developed world, the products that were considered for remanufacturing are used for a certain amount of time so that they can be turned into high quality remanufactured products. There are some companies (e.g. Interface company, Leaseurope) [22, 23] recovering their product and achieving zero waste from their business. They lease their product rather than selling their product so that they can get back their products after a certain period of time for recycling or remanufacturing purposes.

The lack of investment is a critical limitation for remanufacturing SME to develop their industry, to adopt advanced technology, to train workers and to provide sufficient infrastructure. Even though, a number of governments funding and incentive (e.g. kredit usaha kecil – KUK) have been provided by Indonesian Government to the SME, however, limited financial documents and information has become the main barrier to access the funding, thus the SME financial needs are fully funded by family.

Cores (used alternators) are getting harder to be collected due to scarcity of quality and quantity cores. The remanufacturing SME collects used products from collection centers, users and scavengers in Java Island on the basis of availability of cores. These suppliers sent the used products to the SMEs regularly about 400 used products per month in average. However, in Indonesia, products are likely used until end of life product with very low probability to be remanufactured, which creates difficulties for remanufacturer to collect quality cores from local customers. Lack of consumer awareness for collecting used products to be remanufactured is also another challenge due to high value price offered by scavengers to directly recycle them.

The lack of infrastructure is critical issue experiencing by the remanufacturing SME and creating high remanufacturing cost. Provision of infrastructure (e.g. transportation facilities, ICT) to support remanufacturing activities have not been addressed to reduce costs of transportation and logistics during recovery processes. In addition, incentives return of recovery processes have not been found to be applied in Indonesia. Remanufactured products are not attractive for high end customers as the customers are prefer to purchase new products. Unfortunately, market often treats remanufactured products as secondhand products with low quality. Even though the price of remanufactured products is cheaper than the price of new products, low trust from customer has become a border for marketing the remanufactured products.

4.3   National Resource Strategy and Policy

A number of resource strategy and policy recommendation are purposing in this re-search as presented in the following session (Table 3).

Remanufacturing needs special skills that are specific to the requirement remanufacturing jobs. Strengthen labor skills through technical training and education on the subject of remanufacturing process and remanufactured products and establishing partnership program among the supply chain of remanufacturing area can increase the understanding and knowledge of labors. The supports from Government and private sectors play important roles to successfully enable the SME in developing their labors and adopting the technology in affordable manner through soft loans or flexible financial mechanism. In addition, adopting green and advanced technology (i.e. organic

Table 3. Strategy and policy

Issues	Strategy and policy
Lack of understanding and knowledge due to low educational qualification	Strengthen labor special skills on the subject of remanufacturing process and remanufactured products through formal technical training and education, government and education institution collaborations, establishing partnership program among the supply chain of remanufacturing area to update with modern machineries skills
Lack of investment and technology	Government and private sectors financial supports (soft loan, flexible financial mechanism, sister company program with OEM, green and advanced technology procurements, resource and energy efficiency standard
Lack of quality and quantity core and products	Global implementation on cores import regulation, leasing program for remanufactured products, developing partnership with different enterprises, smart financing mechanism on product recovery, cores and product standardizations, warranty and guarantee
Lack of regulation and infrastructure	Global implementation of resource and energy efficient regulations, infrastructure provision improvements (e.g. transportation facilities, Information Computer Technology), incentive and tax reduction
Lack of market competitiveness	Regulation implementation of 80% local contents, secondary market penetration market, public awareness developments through campaign, seminar, workshop, national green products competition

solvent, jet, and electrolytic cleaning technologies) is a potential strategy to ensure low energy consumption instead of to achieve efficient remanufacturing process.

To meet the needs of cores, Indonesian government allows the import of goods/used products for remanufacturing proposes through regulation of the Ministry of trade No. 27/M-DAG/PER/5/2012. The regulation can help remanufacturers to provide quality and quantity cores thus will encourage domestic automotive industry to produce finished goods with the support of local spare parts. Involving OEM as supplier of cores, leasing programs and partnership through the value chain and smart financing mechanism on product recovery could be great strategies to satisfy sufficient quantity and quality of cores. In addition, labeling program on resource efficiency and conservation, certification standards for Pre Used products and warranty as new product can maintain the quality of remanufactured products in the global market.

Government has developed policies and regulation for promoting SMEs in achieving resource efficiency such as president regulation no. 5/2006 on national energy policy directions, national energy vision 25/25, Energy and Mineral Resources Ministry (EMR) Regulation No. 13/2012 focusing on electricity saving, and No. 14/2012 focusing on energy management and No. 01/2013 focusing on fuel oil saving. These policies and regulations implementations can encourage SME remanufacturers in achieving resource and energy efficient. Furthermore, developing partnership with different enterprises for example, car workshop/auto parts retails or shops can educate the customers to use efficient resource.

To strengthen the global market, the implementation of "80% local content" followed by clear regulation or initiative to encourage the development of part manufacturer (e.g. engine, alternator), will create markets for local auto parts. Selling remanufactured products at low income secondary market can increase the marketability of the remanufactured products. Creating public awareness including campaign, seminar, workshop, national green products competition could be developed through comprehensive collaboration among the remanufacturing stakeholders. Furthermore, education program through Government portals (e.g. television, news, web site) can develop customer awareness. Furthermore, an identification of the consumer's habits and preferences to address the question of the potential resource efficiency during the life cycle of auto parts need to be conducted through further comprehensive research and education projects.

4.4 Future and Cross Cutting Strategies

The circular economy through remanufacturing SME is expected to grow more business and customer awareness in Indonesia. However, it must be concerned that there are phenomenon including the development of advanced manufacturing processes, the application of digital virtual and resources-efficient manufacturing, the adoption of dynamic technology and innovation (e.g. Internet of things - IoT, 3D printing, additive technology), which can impact and drive changes in the remanufacturing SMEs.

Product design has been quoted as the most essentials factor for enabling remanufacturing process [19]. The implementation of advanced product design approach such as 3D printing can help remanufacturing SME to reduce more substantial materials, transportation time, production cycle and delay risks [20]. The other benefit of the 3D printing includes slow moving part which could be printed on site thus no need suppliers and less transportation cost which can dramatically reduce energy and CO_2. Addictive manufacturing is emerging technology in manufacturing and can be potentially applied in the remanufacturing SME. The technique offers a possibility of developing "real material" component based on a Computer Aided Design model [21] by transforming the end of life component without returning to raw material.

In addition, the use of ICT through Internet of things (IoT) can improve efficient collaboration, knowledge sharing and quality logistic process among the SME remanufacturing stakeholders. IoT plays important role to interconnect all stake-holders in a remanufacturing business system, from supplier until the customers. The interconnection offers possibility of remanufacturing SME to efficiently track their cores and products where the data are processed, and to open new business of remanufacturing through digitalization (e.g. marketing, sharing network). For example, the use of radio frequency identification (RFID) can help remanufacturers to track used products (cores), whereabouts and condition of remanufactured products and components, thus reduce the remanufacturing process cost. Based on the aforementioned strategies, it is expected that in the next future, remanufacturing will be capable to adapt physical and intellectual infrastructures technology development, will be able to adapt faster and responsive to global markets and closer to customers, and will be consistent in quality creating higher business value.

5 Conclusion

The research presents the environmental and resource conservation benefits of the use of remanufacturing strategies to achieve circular economy by reducing the energy, material and water consumption in the SME auto parts. The resource productivity of the remanufacturing SME is higher than the manufacturing, and also the pollution intensity of the former is much lower than the new one. However, the research also confirms that the performances of resource productivity and pollution intensity of the remanufacturing SMEs have failed to meet the threshold values. The barriers to achieve these threshold values are lack of knowledge, technology and investment, regulation, quality of used product and market competition. Therefore, the strategies to overcome these barriers are an active communication to customers and society about the advantages of remanufactured products, a development leasing products than selling, government regulation on the adoption of circular strategies (e.g. energy policy direction) and rewards (e.g. tax reduction), which are accompanied by adopting adaptive and smart approaches such as addictive technology and 3D printing.

References

1. Preston, F.: A Global Redesign? Shaping the Circular Economy. Chatham House, London (2012)
2. Sundin, E.: Product and Process Design for Successful Remanufacturing. Department of Mechanical Engineering, Linköping University, Linköping (2004)
3. Biswas, W., Duong, V., Frey, P., Islam, M.N.: A comparison of repaired, remanufactured and new compressors used in Western Australian SMEs in terms of global warming. J. Remanufact. 3(4), 1–7 (2011)
4. Ellen MacArthur Foundation: Towards the Circular Economy - Accelerating the Scale-up Across Global Supply Chain. World Economic Forum, Geneva (2014)
5. Biswas, W., Fatimah, Y.: Remanufacturing as a means for achieving low-carbon SMEs in Indonesia. Clean Technol. Environ. Policy 18(8), 2363–2379 (2016)
6. Murrali, T.: Auto Industry to Fuel Indonesia's Economic Growth. AutoPartsAsia (2015)
7. Ijomah, W.L.: The application of remanufacturing in sustainable manufacture. Waste Resour. Manag. 163, 157–163 (2010)
8. Blok, K., Hoogzaad, J., Ramkumar, S., Srivastav, P., Tan, I., Terlouw, W., De Wit, M.: Implementing Circular Economy Globally Makes Paris Targets Achievable (2016)
9. Fatimah, Y., Biswas, W.: Sustainable assessment of remanufactured computers. Procedia CIRP 40, 150–155 (2016)
10. Smith, V.M., Keoleian, G.A.: The value of remanufactured engines: life-cycle environmental and economic perspective. J. Ind. Ecol. 8, 29 (2004)

11. Matsumoto, M., Umeda, Y.: An analysis of remanufacturing practices in Japan. J. Remanuf. **1**(2), 1–11 (2011)
12. Nasr, N., Hughson, C., Varel, E., Bauer, R.: State of the Art Assessment of Remanufacturing. National Center for Remanufacturing and Resource Recovery, Rochester Institute of Technology, Rochester (1998)
13. Biswas, W., Rosano, M.: A life cycle greenhouse gas assessment of remanufactured refrigeration and air conditioning compressors. Int. J. Sustain. Manufact. **2**(2–3), 222–236 (2011)
14. European Commision: Communication from the Commission to the European Parliament, the Council, the European Economic and Social Committee and the Committee of the Regions. European Commission, Brussels (2014)
15. UNEP and UNIDO: Enterprise Levels Indicators for Resource Productivity and Pollution Intensity, Vienna (2010)
16. Liu, S.C., Shi, P.J., Xu, B.S., Xing, Z., Xie, J.J.: Benefit analysis and contribution prediction of engine remanufacturing to cycle economy. J. Cent. South Univ. Technol. **12**(2), 25–29 (2005)
17. Mukherjee, K., Mondal, S.: Some studies on remanufacturing activities in India. In: Surendra, A.J.D., Gupta, M. (eds.) Environment Conscious Manufacturing, pp. 446–466 (2008)
18. Severengiz, S., Skertos, S.J., Selinger, G., Kim, H.: Economic and environmental assessment of remanufacturing. In: 15th CIRP International Conference on Life Cycle Engineering, Sydney, NSW (2008)
19. All Party Parliamentary Sustainable Resource Group: Remanufacturing Towards a Resources Efficient Economy, London (2014)
20. Kungl. Ingenjörsvetenskapsakademien, IVA: Resource Efficiency – Pathways to 2050. A Report from IVA Project Resource Efficient Business Models. The Royal Swedish Academy of Engineering Sciences (2015)
21. Van-Thao Le, P.H., Mandil, G.: Using additive and subtractive manufacturing technologies in a new remanufacturing strategy to produce new parts from End of Life parts. In: 22eme Congres Francais de Mecaniq. UE, Grenoble, France (2015)
22. Interface: Interface Europe reaches sustainability milestones, achieving 90 percent carbon reduction (2017). http://www.interfaceglobal.com/. Accessed Feb 2017
23. Leaseurope: The European Federation of Leasing Company Associations (2017). http://www.leaseurope.org. Accessed Feb 2017

Challenges and Opportunities of Clean Technology in Production Engineering

Cross-Functional Mapping to Link Lean Manufacturing and Life Cycle Assessment in Environmental Impact Reduction

Jun T. Leong and Wai M. Cheung[(✉)]

Department of Mechanical and Construction Engineering,
Faculty of Engineering and Environment,
Northumbria University, Newcastle upon Tyne, NE1 8ST, UK
leongjt77@gmail.com, Wai.m.cheung@northumbria.ac.uk

Abstract. In industry, carbon emissions are mainly produced from the amount of energy used in the manufacturing processes due to the burning of fossil fuels, material of products and transportation. The aim of this paper reports the synergy of integrating life cycle assessment (LCA) and Lean manufacturing to reduce the negative environmental impacts of a plastic injection moulded product. A cross-functional mapping method is used because a number of functional areas such as Lean manufacturing and LCA are involved. This work demonstrated that the adaptation of lean thinking and LCA could minimise negative environmental impacts of a product significantly.

Keywords: Carbon emissions · Plastic · Injection moulding · Lean manufacturing · Life cycle assessment

1 Introduction

Industrial waste generation and raw material consumption are critical concerns given their impact on the environment, especially as the global population grows and demand for products increases. Material efficiency relates to the amount of material used in manufacturing a product can be improved by manufacturing practices using less materials per product and/or generating less waste per product [1]. In response to the need for increased material efficiency in manufacturing, many strategies have been developed, for example, material flow cost accounting, eco-efficiency and cleaner production [2–4].

Lean manufacturing aims to eliminate waste and non-value-added activities and is a common underlying principle in many major businesses and production facilities [5, 6]. Lean continuously improves resource productivity, therefore decreasing products' intensity in both materials and energy [7]. The lean production paradigm can be accomplished by applying a wide variety of lean manufacturing tools such as Kanbans, First In-First Out (FIFO), Value Stream Mapping (VSM), Takt time, Just In Time (JIT), Single Minute Exchange of Die (SMED), and 5S principles [5, 8]. A well-known method for visualizing time waste in a manufacturing system is VSM. VSM seeks to highlight waste in a manufacturing system with the ultimate goal of reorganizing

© Springer International Publishing AG 2017
G. Campana et al. (eds.), *Sustainable Design and Manufacturing 2017*, Smart Innovation,
Systems and Technologies 68, DOI 10.1007/978-3-319-57078-5_40

production practices to align with "Lean thinking" and establish plans for future improvement [5]. This analysis represents the time taken to complete a process, with a particular emphasis on time that does not add value to the product [8]. Lean production on environmental impacts, in particular, VSM can be used to identify the environmental impacts of production processes. 5S can be useful for improving waste management. Cellular manufacturing can lead to a decrease in electricity consumption, whereas TPM can help to reduce several impacts of the machines, such as oil leakage and emissions of dusts and chemical fumes into the atmosphere [9].

Life cycle assessment (LCA) is a systematic method for evaluating the environmental burdens associated with a product, process or activity, by identifying and quantifying energy and materials consumed and wastes released to the environment [10]. It is concerned with the environmental impact of industrial operations or a system. A full LCA study involves four stages: goal and scope definition; life cycle inventory (data gathering); environmental impact assessment and, finally, interpretation (including recommendations) [11]. There are many LCA studies have been undertaken, however research on LCA has traditionally focused on methodological issues. Few papers investigate whether the use of LCA leads to the raising of awareness, learning and/or reconceptualization of product systems [12]. Cross-functional mapping approach is used because a number of functional areas (i.e. Lean manufacturing and LCA) are involved. Lean manufacturing techniques are used to identify area of wastes within a production process by shortening throughput. LCA is a technique to analysis the amount of carbon emissions produced as a result of changes in the manufacturing processes due to the level of usage in energy, raw materials and transportation. Therefore, the overall contribution of this work is to develop a cross-functional mapping approach of integrating lean production and LCA to reduce negative environmental impacts of a plastic injection product. The remainder of this paper is organised as follows: Sect. 2 describes the proposed methodology and implementation; Sect. 3 presents relevant case studies; Sect. 4 discusses the overall results and finally the conclusion and future work.

2 Method of Linking Lean Manufacturing and Life Cycle Assessment for Environmental Impact Reduction

The proposed solution of this research is shown diagrammatically in Fig. 1. It is considered that the application of lean manufacturing coupled with life cycle assessment techniques could minimise the negative environmental impacts of a product. Lean thinking helps companies to better manage their products throughout their lifecycle from material requirements, scheduling to waste generation etc. VSM is a functional method or visual flow chart by which the production process can be represented as a set of processes connected in time [9]. The method excels at showing the time dimension, particularly the non-value-added or waste time. It is therefore the lean method of choice for industries where costs are mostly determined by time or where a shorter production cycle confers competitive advantage. VSM can map an entire process, supply chain network, or the subtasks within a single process. It therefore readily scales

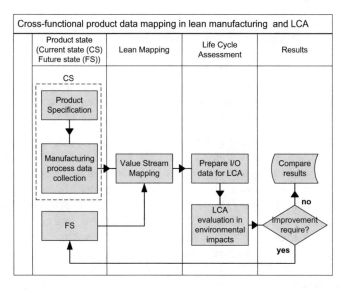

Fig. 1. The proposed method of incorporating lean thinking and Life Cycle Assessment in manufacturing environmental impact analysis

hierarchically. In addition it maps both the material flow and the information that controls production [9].

LCA system will perform the evaluation of the environmental impacts based on data input from manufacturing processes such as energy requirements, time intervals and materials of a product. Life Cycle Assessment is a set of procedures for compiling and examining the inputs and outputs of materials and energy, associated with environmental impacts directly attributed to the function of a product or service system throughout its life cycle. Life cycle Assessment will assess the potential impacts a product can have on the environment by:

- Collecting an inventory of relevant inputs and outputs.
- Evaluating the potential environmental impacts associated with those inputs and outputs.
- Interpreting results of inventory and impacts phases in relation to the objectives.

Figure 1 illustrates a cross-functional representation of product mapping in lean manufacturing and LCA. The method is functioned as follows:

- Current-State (CS) in this context is referred as a product's specification stage. Based on the product specification a set of relevant manufacturing data such as processes, materials requirement and time intervals can be modelled visually using lean's VSM technique.
- Based on the VSM layout, a set of input and output data such as material type, requirements, total weight of materials and process energy requirements can be identified. These data type will act as the input values for LCA evaluations.

- If improvement of CS is required, and this can be identified through the CS's VSM and hence a new VSM can be mapped and this is referred as 'future-state' (FS) of the product. Creating a second set of VSM allows key processes to be identified so that unnecessary waste can be identified.
- Based on the FS's VSM layout, a new set of input and output data will be formulated and hence a subsequent LCA evaluation can be performed.
- The final stage of the method is to compare LCA results of CS and FS.

The lean techniques are incorporated into a visual stream mapping process for the CS and FS stages. The future state analyses the improvement of the current process cycle to recommend changes which should be made to make the product more sustainable. Modelling of a plastic injection moulding product has been created for the study to focus upon improving the product and the life cycle analysis of this process based on the literature review. The software Simapro V8 was used to carry out the data analysis for the "current state" and "future state" of a product specification. The main focus categories are the climate change, human toxicity, photochemical oxidant formation, Terrestrial acidification and Terrestrial ecotoxicity. Both set of results are compared to show where improvements can be made in the manufacturing process to make it more environmentally friendly.

3   Case Study of Modelling a Plastic Injection Moulding Product

3.1   Product Specifications

The case study is a plastic casing and the monthly forecast arranged on the material requirement planning system is 4,000 pieces. The finished product will be placed into a returnable tray which carries 5 pieces per tray and 12 trays on a pallet. The products will be shipped to customers on every Tuesday and Friday by a lorry truck with a size no greater than 7 tons. The manufacturer's normal working day is 8 h, with 2 working shifts per day. The planned production for the day is to produce 360 pieces, therefore 180 pieces per shift. In order to produce 4000 pieces monthly the amount of time requires are 12 working days with 23 shifts. Each shift will require 198 kg of raw material; this is from the quantity' of products multiplied by the part weight, 1.10 kg * 180 = 198 kg. The total waste during the plastic injection moulding process is calculated as follow:

- The hopper dryer machine's (part of the granular plastic pre-heating process) material waste rate (failure rate) is 0.15%, hence the dryer hoper would contribute 198 * 0.15% = 0.3 kg per shift,
- The machine set-up/test runs waste rate is 0.25% and therefore it will contribute 198 * 0.25% = 0.5 kg
- The part moulding defects flashes waste rate is 0.0125%, hence the part moulding defects contributes 198 * 0.0125% = 0.025 kg, per part weight.
- The total amount of waste materials is 0.825 kg.

For the required 4000 pieces per month, the total material required per shift is calculated as: total number of shifts multiplied by the total required materials, i.e, 22.22 * 198.825 kg = 4417.9 kg.

3.2 Manufacturing Process Data Collection

The manufacturing process of top housings and their process requirements data are summarised in Table 1.

Table 1. Top housing process requirement summarized list

Process requirement										
Process#	Time					Material/transportation			Machine/equipment	
	Op.	Inv.	L/T day	C/T sec	C/O sec	Item#	Usg '+'	Usg '−'	Item#	Usg '+'
Material acquisition (receiving)	N/A	N/A	5	N/A	N/A	Polycarbonate (PC)	5 t			
						>7 ton Lorry truck	750 tkm			
Granular plastic pre-heating	2	0	N/A	2400	0	Raw material granular plastic (PC)	4417.9 kg	6.63 kg	200 kg plastic hopper dryer machine	144.43 kWh
Part injection moulding	2	1500	7.5	70	5	Pre-heated Polycarbonate (PC)	-	11.04 kg	650 tonnage Hydraulic injection moulding machine	7388.2 kWh
Flashes deburring	1	950	4.75	5	5	Mould part	-	0.53 kg		
Gate runner cutting	1	1300	6.5	10	5	Deburr part	-	80.00 kg	Bench drill machine	7.78 kWh
Packing & kitting	1	1100	5.5	10	0	Completed part	-			
Shipping customer	N/A	N/A	N/A	N/A	N/A	Packed finish good	-			
						> 7 ton Lorry truck	215.98 tkm			

Op. - Operator, Inv. - Inventory, L/T - Lead time, C/T - Cycle time, C/O - Changeover time, Usg "+" = value added, Usg"−" = non value added

3.3 LCA Analysis of Product Current State

3.3.1 I/O Setup for LCA Evaluation

Figure 2 illustrates the LCA's system boundary of the CS. The system boundary includes all necessary inputs which represent the overall amount of materials and energy requirements used in the production of 4000 pieces of the plastic house casings. This LCA analysis adopts the ISO 14044/44 standard and is focused on climate change (GWP), Human toxicity (HTP), photochemical oxidant formation (POFP), Terrestrial acidification (TAP) and Terrestrial eco-toxicity (TETP).

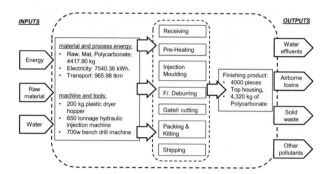

Fig. 2. System boundary for I/O LCA data for CS

3.3.2   Product Current-State Environmental Impact Results

The LCA result of CS's environmental impacts is shown in Fig. 3. It can be seen from the result that a product's material has contributed the largest impact among all the categories. Hence, the reduction of material usage will improve these negative environmental impacts. The second factor to be considered will be the manufacturing process and its associated energy consumptions reduction.

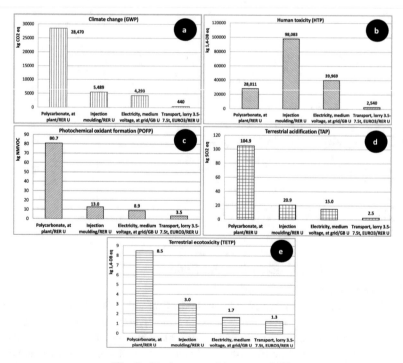

Fig. 3. Environmental impacts of CS

3.4 Product Improvement for Environmental Impact Reduction

3.4.1 The Plastic Injection Moulding Process Improvement

A feasibility study has been conducted using lean manufacturing's VSM technique to identify where improvements on injection moulding process can be carried out. These improvement include production lead times, machine failure, set-up time, non-value added material waste, and most significantly material usage in the product. Changing of a customer's order strategy has affected the production plan and material requirements. The production process will therefore to be rearranged and the data of this rearrangement are shown in Table 2.

Table 2. New improvement top housing manufacturing process requirement summarized list

Process requirement									
Process#	Time				Material/transportation			Machine/equipment	
	Op.	L/T day	C/T sec	C/O sec	Item#	Usg '+'	Usg '−'	Item#	Usg '+'
Material acquisition (receiving)	N/A	1.5	N/A	N/A	Polycarbonate (PC)	4 ton			
					>7 ton Lorry truck	600 tkm			
Granular plastic pre-heating	2	0	2,352	0	Raw material granular plastic (PC)	2645.60 kg	1.32 kg	200 kg plastic hopper dryer machine	101.92 kWh
Part injection moulding	2	1	68.6	2.75	Pre-heated Polycarbonate (PC)	-	1.32 kg	650 tonnage Hydraulic injection moulding machine	4344.64 kWh
– Flashes deburring – Gate runner cutting – Packing & kitting	1	0.5	23.75	0	Moulded part	-	48.264 kg	Bench drill machine	4.4 kWh
Shipping customer	N/A	N/A	N/A	N/A	Packed finish good	-			
					>7 ton Lorry truck	54 tkm			

Op. - Operator, L/T - Lead time, C/T - Cycle time, C/O - Changeover time, Usg "+" = value added, Usg "−" = non value added

- Raw material acquisition implemented the Kanban pull system technique in order to manage the new method. The previously 6 week forecasts and order of 5 tons of materials in monthly basis has changed to daily delivery in a single standard pack size of 200 kg. Order quantity has been reduced, the raw material supplier produced and deliver lead time has been completed in 1.5 days. Transportation remains the same with a truck size no greater than 7 tons and the distance taken is 150 tkm from supplier to manufacturer. The new process requires 5 packs to be delivered per week and therefore requires a total of 600 tkm.
- Granular plastic pre-heating utilising the 5S and TPM techniques and resulted an improvement of 2% on the cycle time which is equivalent to 39.2 min (2,352 s). In order to complete the new process, the total energy consumption used by the plastic hopper dryer is 101.92 kWh. The hopper dryer failure allowance is 0.1% which contributes to 1.32 kg of waste material.
- Part injection moulding process also used the 5S and TPM techniques and resulted an improvement of 2% on the cycle time which has reduced the cycle time to 69.6 s.

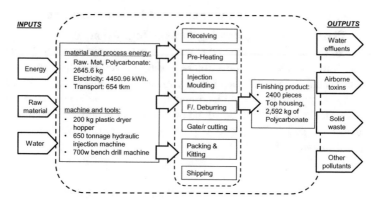

Fig. 4. System boundary for I/O LCA data for FS

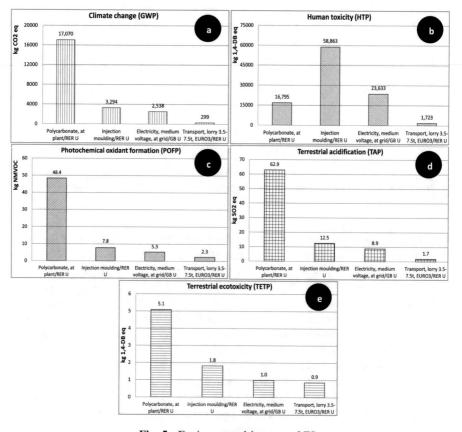

Fig. 5. Environmental impacts of FS

Total energy consumption for the injection moulding is 4344.4 kWh. The machine set-up has a failure allowance of 0.1% which contributes to 1.32 kg of materials as non-value added (waste).

- By using the cellular manufacturing method, the individual process of flashes deburring, gate runner cutting, packing and kitting have been grouped together at the same work station which requires only one operator. The overall cycle time has been improved by 5% and the total cycle time has been reduced to 23.75 s. The total non-value added to the combined process is determined as follows:
  - A material flash deburring has an allowance rate of 0.01% which is equal to 0.264 kg.
  - The gate runner removal allowance rate is 1.8% which is equal to 48 kg.
- Therefore the total non-value added to this process is 48.264 kg. Energy for the drilling machine for gate runner removal process total 4.40 kWh.
- Shipping and deliveries to customers have changed from every Tuesday and Friday to daily basis. The lorry size is less than 7 ton and the distance between manufacture to customer site is 50 km. The new shipping arrangement is equivalent to 54 tkm.

Table 3. The overall analyse results and the outcomes

Items#	Unit	Current State [CS]	Future State [FS]	Improved %
Production lead time	day	29.25	3	89.74%
Process time	s	2,495.00	2,444.35	2.03%
Operators	-	7	5	28.57%
Material usage (value added)	kg	4,319.7	2,592.00	
Material usage (non-value added)	kg	98.2	48.26	40.24%
Energy consumption	kWh	7,540.36	4,450.96	40.97%
Transportation (from supplier)	t/km	750	600	20.00%
Transportation (to customer)	t/km	215.98	54	32.30%
Climate change (GWP)	kg CO_2 eq	38,691.97	23,201.26	40.04%
Human toxicity (HTP)	kg 1,4-DB eq	16,8602.78	10,1013.38	40.09%
Photochemical oxidant formation (POFP)	kg NMVOC	106.10	63.82	39.85%
Terrestrial acidification (TAP)	kg SO_2 eq	143.35	86.04	39.98%
Terrestrial eco-toxicity (TETP)	kg 1,4-DB eq	14.46	8.76	39.44%

3.4.2   LCA of the Product Future State (FS)

The revised system boundary of the LCA I/O is shown in Fig. 4 and the results are discussed as follow.

Results shown that the usage of material contributes to the largest impact. Hence, the reduction of material usage has improved the negative environmental impacts among all the categories as shown in Fig. 5. Table 3 presents a comparison of the overall results of the current state (CS) and the improved future state (FS).

4   Conclusions and Further Work

The results and finding show that significant improvements of the environmental impacts have been achieved. By implementing the kanban's pull system to control customer and supplier order strategies, the total operations time have been shortened significantly. Due to the change of delivery requirements, the total material usage has been reduced and this has a direct influence in terms of reducing manufacturing emissions by 40% in climate changes (GWP). Furthermore, implementation of TPM, 5S and Cellular Manufacturing methods have contributed a significant saving in energy and electricity consumptions to the production process and thus, this represents a sizeable reduction of human toxicity. In conclusion, lean production tools are considered as an essentially part in implementing sustainable manufacturing. It is recommended that future work should carry out an in depth cost analysis in order to find a compromise between environmental reduction and cost effectiveness.

References

1. Shahbazi, S., et al.: Material efficiency in manufacturing: Swedish evidence on potential, barriers and strategies. J. Cleaner Prod. **127**, 438–450 (2016)
2. Despeisse, M., Mbaye, F., Ball, P.D., Levers, A.: Emergence of sustainable manufacturing practices. Prod. Plan. Control **23**(5), 354–376 (2012)
3. Cheung, W.M., et al.: Towards cleaner production: a roadmap for predicting product end-of-life costs at early design concept. J. Cleaner Prod. **87**, 431–441 (2015)
4. Cheung, W.M., Pachisia, V.: Facilitating waste paper recycling and repurposing via cost modelling of machine failure, labour availability and waste quantity. Resour. Conserv. Recycl. **101**, 34–41 (2015)
5. Abdulmalek, F.A., Rajgopal, J.: Analyzing the benefits of lean manufacturing and value stream mapping via simulation: a process sector case study. Int. J. Prod. Econ. **107**(1), 223–236 (2007)
6. Bortolini, M., et al.: A reference framework integrating lean and green principles within supply chain management. World Acad. Sci. Eng. Technol. Int. J. Social, Behav. Educ. Econ. Bus. Ind. Eng. **10**(3), 884–889 (2016)
7. Moreira, F., Alves, Anabela C., Sousa, Rui M.: Towards eco-efficient lean production systems. In: Ortiz, Á., Franco, R.D., Gasquet, P.G. (eds.) BASYS 2010. IAICT, vol. 322, pp. 100–108. Springer, Heidelberg (2010). doi:10.1007/978-3-642-14341-0_12

8. Chiarini, A.: Sustainable manufacturing-greening processes using specific lean production tools: an empirical observation from European motorcycle component manufacturers. J. Cleaner Prod. **85**, 226–233 (2014)
9. Roosen, T.J., Pons, D.J.: Environmentally lean production: the development and incorporation of an environmental impact index into value stream mapping. J. Ind. Eng. **2013** (2013). Article ID 298103. https://www.hindawi.com/journals/jie/2013/298103/
10. Tait, M.W., Cheung, W.M.: A comparative cradle-to-gate life cycle assessment of three concrete mix designs. Int. J. Life Cycle Assess. **21**(6), 847–860 (2016)
11. Curran, M.A. (ed.): Life Cycle Assessment Student Handbook. Wiley, Hoboken (2015)
12. Coelho, C.R., McLaren, S.J.: Rethinking a product and its function using LCA—experiences of New Zealand manufacturing companies. Int. J. Life Cycle Assess. **18**(4), 872–880 (2013)

Sustainable Materials: Renewable and Eco Materials, Bio-polymers, Composites with Natural Fibres

Developing Fiber and Mineral Based Composite Materials from Paper Manufacturing By-Products

Cynthia Adu and Mark Jolly[(✉)]

Manufacturing Department, Cranfield University, Cranfield MK43 0AL, UK
{c.e.adu,m.r.jolly}@cranfield.ac.uk

Abstract. Developing valuable materials from the by-products of paper industry can help to address some environmental and economic issues associated with traditional synthetic composites. Particularly, the management of paper mill sludge (PMS) waste remains an economic and environmental challenge for the pulp and paper industry. 11 million tons of PMS is generated annually in Europe from the wastewater treatment (WWT) process of paper mills. PMS is mostly used in low value applications. However, PMS contains fibers and minerals with physio-chemical properties that exhibit a high potential to substitute some conventional materials in other industries. The research presented in this paper aims to explore new directions for further investigation on PMS material applications by reviewing the literature on PMS materials and subsequently characterizing sludge from 6 different mills. The study shows the technical feasibility, opportunities and technological readiness of fiber and mineral based composites obtained from PMS, such as; cementitious products, polymer reinforcement and fiberboards.

Keywords: Paper mill sludge · Waste recovery · Sustainable materials · Cellulose

1 Introduction

It is expected that by 2020 over 60% of paper products will be recycled to aid the reduction of environmental impacts caused by the pulp and paper industry. However, the waste water treatment process of recycled paper mills produces a solid waste by-production large quantities known as paper mill sludge [1]. Paper mills are pressured by stringent environmental legislations, limited landfill space and taxation costs of £82.60/ton of waste [2]. Moreover, the environmental impact of 1 ton of PMS in landfill releases 2.69 tons of CO_2 and 0.24 tons of Methane [3]. Over the years, the industry has investigated valorization options to manage PMS. The most common adopted means is energy recovery and agricultural land spreading.

PMS is recovered for energy by incineration which still generates 50% PMS waste ash. The use of PMS on agricultural land helps in improving soil nutrition and condition, the PH of the soil is also ameliorated by the $CaCO_3$ in the paper sludge. Paper sludge is rich in carbon though lacking in nitrogen and phosphorus, thus fertilizers are still required to satisfy total plant requirement [4]. These applications are a critical stepping

G. Campana et al. (eds.), *Sustainable Design and Manufacturing 2017*, Smart Innovation, Systems and Technologies 68, DOI 10.1007/978-3-319-57078-5_41

stone away from landfill however there remains scope to optimize valorization. According to the circular economy principles; to fully valorize waste, its maximum value must be extracted before considering energy recover or soil restoration [5].

2   Characteristic of Paper Mill Sludge Waste

PMS comprises of cellulose fibers (A) and mineral fillers (B), shown in Fig. 1. The prime constituents of natural cellulose fibers are: cellulose; hemicellulose and lignin. However, during the processing of cellulose fibers into paper form, lignin is almost completely removed.

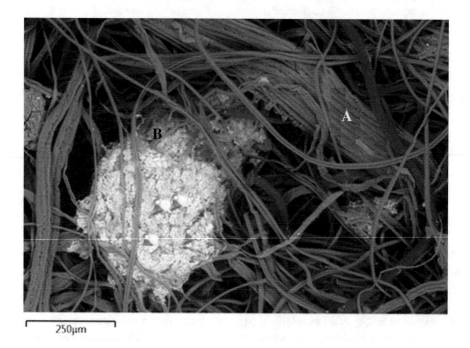

250μm

Fig. 1. SEM micrograph of PMS showing minerals enclosed with cellulose fibers at 100 x

Cellulose exhibits a crystalline structure with amorphous-like regions of hemicellulose, providing structure and strength to plant cell walls. These crystals are in nanoscale also known as cellulose nanocrystals (CN) or nanoparticles. The chains of hydrogen bonds allow cellulose to obtain an orderly molecular stacking, thus rendering cellulose suitable for reinforcing materials [6]. Cellulose nanocrystals possess mechanical properties similar to engineering materials. They have similar elastic modulus (150 GPa) as carbon fiber and a tensile strength of 7.5 GPa, which is higher than steel wire and Kevlar-49 fibers [7]. Lignin is an amorphous polymer also present in plant cell wall, the removal of lignin is required before producing Nano cellulose materials. This process is highly energy intensive, whereas in paper production lignin is already dissolved during the pulping process. This suggests PMS as an option for processing cellulose materials as it contains less lignin. Advances in cellulose materials research reveal its desirable properties such as high tensile strength and light weight [6]. This implies that PMS may contain some untapped resources for material applications. Therefore, a future perspective is necessary to explore innovative value adding solutions to improve the use of this byproduct.

A biological process breaks down the organic matter to generate secondary sludge, containing a higher content of minerals, proteins, microorganisms and cellulose fibers. Some paper mills combine both processes for ease of dewatering. Mills using recycled paper produce deinking sludge that primarily consists of high inorganics, ink and paper coating materials. The combination of these wastes consists of organics and inorganics as seen in Fig. 2. The organics content of the sludge is from the woody fibers contained in the pulp. The fibers have a crystalline structure with amorphous-like regions of hemicellulose. Lignin acts as a strengthening agent in the plant cell wall. Extractives are non-structural parts of wood such as fats and waxes. Lignin and extractives in PMS are present in very small traces (<2%) as they are mostly dissolved during pulping process. The inorganic minerals contained in PMS form compounds such as aluminum oxide (Al_2O_3), silicon dioxide (SiO_2), iron oxide (Fe_2O_3), calcium (Ca) and magnesium oxide (MgO) which are common chemicals present in Portland cement [8]. The general characteristics of PMS are identified based on content of; ash, cellulose, chemical composition, fiber length, extractives, lignin and particle size. The contents depend on the paper mill operations, WWT and final paper product. The amount of these constituents also determines the overall properties of its end product. Thus it is of much significance to quantify these PMS contents in order to understand the mechanical properties of PMS based materials.

2.1 Analysis of PMS from Different Paper Mills

Most studies have reported on PMS based materials obtained from one mill to develop specific materials applications [9–11], this lacks a thorough investigation of the PMS potentials. Table 1 shows the characterization of PMS obtained from 6 different mills for an in depth study of their possible material applications. PMS was oven dried at 105 °C for 24 h to reduce moisture content below 3%. Ash refers to the inorganic matter consisting of chemicals and minerals used in the paper. Ash content was determined based on TAPPI[1] 211 methods for ash in wood, pulp, paper and paperboard [12]. The sludge is ignited in a furnace at 525 °C, volatile organic compounds are burnt off to leave the inorganics. Based on the ash analysis, the data table presents two groups of sludge types; low ash sludge (LAS) < 30% of ash (mill 6, mill 5 and mill 1) and high ash sludge (HAS) > 30% ash (mill 3, mill 2 and mill 4). This gives an approximate indication of the ratio of cellulose fiber content to mineral content. Elemental oxides such as calcium oxide CaO and silicon di oxide SiO2 were identified with ESEM chemical analysis. The HASs show a higher percentage of CaO compared to LASs. The LASs contain more SiO, this indicates they have a higher organic content as a result of silicon dioxide being a natural compound commonly found in biomass.

---

[1] Technical Association of Pulp and Paper Industry.

Table 1. Characteristics of PMS from 6 paper mills

Mill Code	Mill 1	Mill 2	Mill 3	Mill 4	Mill 5	Mill 6
Feedstock	Virgin fibers, Coffee cups	Office paper	Office paper	Office paper	Virgin fibers	Virgin fibers, Fine PE[a]
Color	Black	Grey	Grey	Grey	White	Brown
Ash (%)	25	54	66	70	4	7
Cellulose (%)	47.40	30.70	17.49	11.89	81.61	77.20
Extractives (%)	17.11	15.23	16.51	18.11	14.39	15.80
Ca	35.5	55	52.8	52.4	7.8	16.1
SiO_2	20.7	5.3	4.1	5.2	46.5	34.9

[a]Polyethylene

Cellulose was determined with 17.5% NaOH solution [13]. Extractives appear soluble in acetone which can be determined in accordance with the TAPPI 204 solvent extraction of wood and pulp method [14]. The classification of fiber lengths was carried out based on TAPPI 203 standard using a Bauer McNett classifier. PMS was diluted in water and evenly dispersed using a pulp disintegrator at 3000 rpm. The analysis gives the weighted average fiber lengths for the following mesh screen sizes; 14(1.19 mm), 28(0.595 mm), 48(0.297 mm) and 65 (0.149 mm) as reported in Fig. 2. Fines are contents <0.149 which contain both minerals and tiny fibers. The mill 6 sludge shows a high fiber content of 81% of ≥1.19 mm fibers and 14% fines. Sludge from mill 2, mill 1, mill 5 show an even distribution of different fiber lengths and fines. Mill 5 sludge contains mainly cellulose fibers between 1.19–0.595 mm with 20% fines. Sludge's from mill 3 and mill 4 contain mainly fines (86.22% and 87.10%) respectively.

3   Material Applications from Paper Mill Sludge and Their Mechanical Properties

The literature on PMS materials covers two groups; fiber applications and mineral applications. The technological readiness for these applications range from analytical concepts (TRL 3) to commissioning (TRL 8) as illustrated in Fig. 2 below.

It is possible to choose suitable material applications of PMS depending on their characteristics by referring to Table 1 and Fig. 2. Sludge from mill 1, 2 and 5 which contain a mixture of both fiber and fillers are suitable for board materials such as hardboard and medium density fiberboard (MDF). The high fiber percentage of mill 5 sludge and it being free from impurities renders a possibility for processing nanocellulose. Mill 6 sludge which contains 81% of fibers with similar lengths, is suitable for board applications requiring loads. To form fiber matrix for MDF and pallets. Sludge from mill 3 and 4 which show high inorganic content and contain more calcium have the ability to derive pozzolanic properties when reacted with water, for applications such as concrete, cement, mortar and bricks.

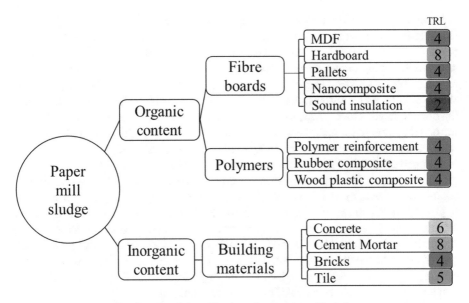

Fig. 2. Material applications from paper mill sludge

3.1 Mineral Based PMS Materials

Minerals from PMS are chemically compatible with traditional materials used to produce cement and concrete. Hollow concrete blocks were substituted with 50% PMS which met Thailand standards (TIS109-2517) for non-load bearing [15]. Cement in concrete can be replaced with up to 10% of PMS which met EU standards for indoor or outdoor plaster, though increase in PMS decreases compressive strength [16]. The lightweight property of PMS ash is attractive for the production of aggregates for cement mortars and plaster blends. PMS ash and Expanded Polystyrene (EPS) was mixed to make lightweight mortar with insulation properties, although the thermal conductivity and compression strength of mortar reduced by 50% and 20% respectively compared to control mortar. The PMS/EPS mortar met EU standards [17]. Hydrophobic coatings can be used in applications such as self-cleaning surfaces and applications for water repellence in concrete. PMS was dry milled for eight hours with stearic acid to create a superhydrophobic powder which maintained a water contact angle of approximately 153° [18]. 12.5% cellulose fibers were used as a substituted in brittle gypsum board which successfully improved the boards impact strength and modulus of rupture. PMS tiles were produce on a pilot scale. PMS was bonded with Methylene diphenyl di-isocyanate (MDI) and Polyurethane (PU) then hot pressed to produce ceramic tiles. The tile was bonded with MDI and PU was used as a surface finish.

3.2 Fiber Based PMS Materials

Fiber board materials. Particle board or fiber board is typically manufactured by binding wood fibers together to form MDF. PMS was used to develop MDF using urea formaldehyde (UF) and methylene-diphenyl di-isocyanate (MDI) as binding agents. The characteristic of the paper sludge used contained; (27.76%) Ash, (20.83%) cellulose, (17.41%) Lignin and (6.37%) extractives. UF bonded samples and MDI bonded samples with the following ratios of PMS to wood fibers (0:100, 15:85, 30:70 and 45:55) were produced by hot pressing formed fiber mats [6]. They were tested for bending strength, sheer strength, water absorption and thickness swelling (TS). The bending strength of the boards ranged from 7 Mpa to 18 MPa, EN standards 312-2 minimum requirement for bending strength in boards is 12 MPa. The 3 layer 15:85 UF bonded sample fully satisfied the requirements set for EN standards for general uses. However, the mechanical properties of the other boards decreased as PMS content increased.

The major problem affecting the other samples were as a result of inorganic material content (ash content) and water absorption. The addition of wax, had no effect on improving the thickness swelling of the boards. However, a study showed that ultrasonic wave processing combined with an alkali treatment of cellulose fibers, improved the water absorption of the fibers by 50%, this also had an effect on increasing the surface area of the cellulose fibers [19] PMS with the following characteristics; ash 19.50% and cellulose content 62.7% was mixed with 100% virgin spruce pine fibers (SPF) at different ratios bonded with UF resin [20]. The elastic modulus and modulus of rupture for the virgin SPF was 2250 MPa and 28 MPa respectively. The PMS/SPF board of ratio 3:7 showed elastic and rupture modulus of 2000 MPa and 24 MPa respectively, the fiber board was tested to meet ANSI A208.2-2002 MDF standard for interior applications. Urea-formaldehyde (UF) is a common binding agent used in the particle board industry which leads to formaldehyde emissions. A major environmental advantage of PMS in fiber boards is the reduction of formaldehyde (HCHO). Addition of secondary sludge (SS) for particle board production showed a reduction of HCHO by an average of 48% the effect occurs due to the proteins contained in SS which react with the formaldehyde [21].

Paper mill reject fibers were used for partial replacement of virgin wood fibers in hardboards. The optimal fiber content to maintain the mechanical properties was at 10% and has been used to manufacture commercial building boards in the Netherlands. However, the facility was later shut down. Natural fiber from kapok, cotton, milkweed and basalt fibers have been studied for sound insulation panels [22, 23] which improved sound reduction. PMS fibers could be used in a similar study to determine their acoustic properties.

Wood plastic composite. Wood Plastic Composites (WPC) are environmental friendly options used in non-structural building components such as roof tiles, fencing, decking [10]. PMS was used as a reinforcement with Maleic anhydride-grafted polypropylene (MAPE) for PMS-WPC [24]. The materials were extruded, milled and injection molded to form test specimens. The composite was proven to have a higher young's modulus although lower tensile strength and deformation at break. The mechanical properties showed that it is feasible to substitute conventional mineral fillers in PP composites with PMS for composites in low mechanical applications such as panels, furniture and fencing. PMS-PP composites were compared against polypropylene co-polymer material currently being used in a commercial fence. 10%wt of PMS showed no effect on the mechanical properties, though when Wood fiber was replaced with 50%wt of sludge the flexural strength and young modulus decreased and the impact strength increased. The inorganic contents in the PMS is said to have affected the flexural strength of the composite as the fibers were enclosed in $CaCO_3$. Increase in cellulose fibers has a positive effect on the flexural and tensile strength, although a negative impact on Thickness Swelling (TS) and water absorption. The WPC composites were not limited by the ash content as the ash may have played the role of mineral fillers typically used in plastic composites. The fibrillation of the fibers will cause a better interlocking with the polymer. The study concluded PMS could play a role in substituting wood flour used in WPC as a reinforcing fiber.

Pallets. Pallets used in the transportation of goods are made from materials such as wood, plastic and metal. Lightweight pallets have been found to be beneficial for logistics operations. PMS was mixed with wood particles using UF as a binder and hardened with NH_4CL and cold pressed to form pallets. The characteristics of the sludge used was 27.76% ash, 20.83% cellulose and 6.37% extractives. The PMS pallet met European standards of minimum requirements for green pallet manufacturing, when 10wt% PMS is used. The water absorption property of the pallet was not significantly affected by 10%wt paper sludge thus no thickness swelling was observed. Inorganic materials such as kaolin clay and calcium carbonate resulted in weaker adhesion between the PMS and wood particles [25]. Understanding variables such as particle size and inorganic content which affect the pallets mechanical properties will improve research on PMS for pallets. Pallets could also be manufactured from PMS-WPC and compared against standards for green pallets.

Polymer reinforcement and rubber composites. Cellulose fibers have been investigated in recent studies as a reinforcement to improve the mechanical properties of polymers. PMS fibers and Polyamide (PA) were extruded and injection molded into thin parts to develop a biodegradable thermoplastic composite. The average modulus of the samples was higher than the control samples and had good moisture resistance. This lead to further investigations of cellulose fibers reinforced in rubber. Typically, toxic materials such as Carbon black (CB) and silica are used as fillers in rubber composites. These fillers were partially replaced with short cellulose fibers from PMS to produce a hybrid rubber composite. Two samples, PMS carbon black and PMS silica were tested for fatigue, tensile properties and curing [26] The PMS/CB rubber composite samples had a better fatigue life and loading capacity than PMS/Silica rubber composite. Though the PMS/CB had lower curing and tensile properties than the control samples. Micro Crystalline Cellulose (MCC) was used as partial replacement of the silica fillers in rubber composite with up to 18% of silica replaced showing no effect on the mechanical properties of the rubber. In some cases, (5% MCC) the tensile strength of the rubber increased. The use of MCC also reduced the energy consumption of the process [27]. Thus the production of MCC from PMS could be of interest to be used as a filler in rubber composites

Nanocomposites. Nanocellulose are ideal materials for reinforcement in biopolymer composites materials as a result of their high aspect ratio, high stiffness and low density

[28, 29]. Nano cellulose was obtained from PMS using high pressure defibrillation by steam exploding the fibers at 138 kPa and chemical purification process with NaOH and acetic acid. The nanofibrils were successfully separated to uniform diameter below <20 nm. The nanofibrils also preserved the original crystalline cellulose structure. The fibers were used in preparation of polyurethane (PU) nanocellulose composites which showed increase in mechanical properties after an addition of 4wt% of cellulose. The PU nanocellulose composite exhibited 45.6 MPa tensile strength and a modulus of 152.63 MPa, ordinary PU specimen tested was at 17.5 MPa and 37.5 MPa respectively, this represents over 50% increase in mechanical properties [30]. Nanocellulose composite derived from PMS cellulose will be essential for lightweight materials. The nanofibers were bundles of cellulose fibers with 5 to 30 nm in width; this method of obtaining PMS nanofibers could be essential for improving applications of PMS.

4   Conclusions

Paper mill sludge shows a high potential for valorization into material applications. However, it is of no doubt that these applications require further research to meet technical standards for use in commercial products. The major issue with sludge in concrete is its effect on compressive strength and water absorption above 10%. Fiber board applications of PMS are generally affected by the increasing thickness swelling caused by water absorption, poor internal bonding between the fibers and inorganic content which results in low tensile and flexural strength. Further research into obtaining nanofibers from PMS for use in PMS board applications will improve their mechanical strength. Thermogravimetric analysis is required to study the effects of the hot press on the behavior of PMS fiber board composite due to the thermal decomposition of cellulose. Hydrophobic powder generated from PMS could be used to reduce the water absorption of the boards and concrete mixes. The type of resin used also has an effect on the properties of the PMS fiber composite which requires further investigation. PMS and recent biodegradable polymers such as Poly-lactic acid (PLA) or Polybutylene succinate could be investigated to form biodegradable fiber boards. Overall the research requires a holistic study on various sludge compositions used in multiple applications to assess how the material properties are affected by different variables. These considerations will assist the continuing development of recycling options for a range of sludge compositions.

References

1. Likon, M., Saarela, J.: The conversion of paper mill sludge into absorbent for oil spill sanitation - the life cycle assessment. Macromol. Symp. **320**, 50–56 (2012)
2. HM Revenue and Customs: A general guide to landfill tax (2015)
3. Likon, M., Cernec, F., Svegl, F., et al.: Papermill industrial waste as a sustainable source for high efficiency absorbent production. Waste Manage. **31**, 1350–1356 (2011)
4. Nemati, M., Caron, J., Gallichand, J.: Using paper de-inking sludge to maintain soil structural form field measurements. Soil Sci. **64**, 275–285 (2000)

5. Ellen MacArthur Foundation: Towards the circular economy, vol. 1 (2013). http://www.ellenmacarthurfoundation.org/publications/towards-the-circular-economy-vol-1-an-economic-and-business-rationalefor-an-accelerated-transition

6. Eichhorn, S.J., Dufresne, A., Aranguren, M., et al.: Review: current international research into cellulose nanofibres and nanocomposites. J. Mater. Sci. **45**, 1–33 (2010)

7. Sabapath, J.G.S.N.: Cellulose nanocrystals: synthesis, functional properties, and applications. Nanotechnol Sci. Appl. **8**, 45 (2015). doi:10.2147/NSA.S64386

8. Bajpai, P.: Management of Pulp and Paper Mill Waste. Springer, Switzerland (2014)

9. Taramian, A., Doosthoseini, K., Mirshokraii, S.A., Faezipour, M.: Particleboard manufacturing: an innovative way to recycle paper sludge. Waste Manage. **27**, 1739–1746 (2007)

10. Huang, H.B., Du, H.H., Wang, W.H., Shi, J.Y.: Characteristics of paper mill sludge-wood fiber-high-density polyethylene composites. Polym. Polym. Compos. **16**, 101–113 (2012). doi:10.1002/pc.22287

11. Davis, E., Shaler, S.M., Goodbell, B.: The incorporation of paper deinking sludge into fiberboard. For. Prod. Ind. **53**, 46–54 (2003)

12. TAPPI T: 211 om-02; Ash in wood, pulp, paper and paperboard: combustion at 525 °C. TAPPI test methods 2005, pp. 3–6 (2004)

13. Rabemanolontsoa, H., Ayada, S., Saka, S.: Quantitative method applicable for various biomass species to determine their chemical composition. Biomass Bioenerg. **35**, 4630–4635 (2011)

14. Tappi: TAPPI Slovent Extractives of wood and pulp (T204 cm-97), pp. 7–10. Tappi (2007)

15. Kaosol, T.: Reuse water treatment sludge for hollow concrete block manufacture. Energy Res. J. **1**, 131 (2010)

16. Nazar, A.M., Abas, N.F., Mydin, M.A.: Study on the Utilization of Paper Mill Sludge as Partial Cement Replacement in Concrete (2014)

17. Ferrándiz-Mas, V., Bond, T., García-Alcocel, E., Cheeseman, C.R.: Lightweight mortars containing expanded polystyrene and paper sludge ash. Constr. Build. Mater. 61, 285–292 (2014)

18. Wong, H.S., Barakat, R., Alhilali, A., et al.: Hydrophobic concrete using waste paper sludge ash. Cem. Concr. Res. **70**, 9–20 (2015)

19. Guo, X., Jiang, Z., Li, H., Li, W.: Production of recycled cellulose fibers from waste paper via ultrasonic wave processing. J. Appl. Polym. Sci. **132**, 6211–6218 (2015)

20. Geng, X., Zhang, S.Y., Deng, J.: Characteristics of paper mill sludge and its utilization for the manufacture of medium density fiberboard. Wood Fiber Sci. **39**, 345–351 (2007)

21. Xing, S., Riedl, B., Deng, J., et al.: Potential of pulp and paper secondary sludge as co-adhesive and formaldehyde scavenger for particleboard manufacturing. Eur. J. Wood Wood **71**, 705–716 (2013)

22. Ganesan, P., Karthik, T.: Development of acoustic nonwoven materials from kapok and milkweed fibres. J. Text. Inst. **5000**, 1–6 (2015)

23. Moretti, E., Belloni, E., Agosti, F.: Innovative mineral fiber insulation panels for buildings: thermal and acoustic characterization. Appl. Energy **169**, 421–432 (2016)

24. Soucy, J., Koubaa, A., Migneault, S., Riedl, B.: The potential of paper mill sludge for wood-plastic composites. Ind. Crops Prod. **54**, 248–256 (2014)

25. Kim, S., Kim, H.J., Park, J.C.: Application of recycled paper sludge and biomass materials in manufacture of green composite pallet. Resour. Conserv. Recycl. **53**, 674–679 (2009)

26. Ismail, H., Rusli, A., Azura, A.R., Ahmad, Z.: The effect of partial replacement of paper sludge by commercial fillers on natural rubber composites. J. Reinf. Plast. Compos. **27**, 1877–1891 (2008)

27. Bai, W., Li, K.: Partial replacement of silica with microcrystalline cellulose in rubber composites. Compos. Part A Appl. Sci. Manuf. **40**, 1597–1605 (2009)
28. Moon, R.J., Martini, A., Nairn, J., et al.: Cellulose nanomaterials review: structure, properties and nanocomposites. Chem. Soc. Rev. (2011). doi:10.1039/c0cs00108b
29. Potulski, D.C., De Muniz, G.B., Klock, U., De Andrade, A.S.: Green composites from sustainable cellulose nanofibril. Sci. For Sci. **40**, 345–351 (2014)
30. Leao, A.L., Cherian, B.M., Narine, S., Sain, M.: Applications for nanocellulose in polyolefins-based composites. In: Polymer Nanocomposites Based on Inorganic and Organic Nanomaterials, pp. 215–228 (2012)

Sustainable Carbododiimine and Triazine Reagents as Collagen Cross-Linking Agents in the Presence of PAMAM Dendrimers

V. Beghetto[1,2(✉)], L. Agostinis[1,2], V. Gatto[1,2], R. Sole[1], D. Zanette[1], and S. Conca[2]

[1] Dipartimento di Scienze Molecolari e Nanosistemi,
Università Ca' Foscari Venezia, Venice, Italy
beghetto@unive.it
[2] Crossing S.r.l., Piazza delle Istituzioni n. 27 ed. H, 31100 Treviso, Italy

Abstract. This work reports a general outline on sustainable technologies for the stabilization of collagen and comparative study of 1-ethyl-3-(3-dimethyla-minopropyl)carbodiimide (EDC) versus 4-(4,6-dimethoxy [1, 3, 5] triazin-2-yl)-4-methyl-morpholinium chloride (DMTMM) as cross-linking agents of collagen powder. The cross-linking efficiency of these agents on collagen matrixes in the presence of different polyamidoamine dendrimers (PAMAM) has been tested in order to determine the influence of steric hindrance and aminic groups abundance.

Keywords: Cross-linking · Innovative · Sustainable · Zero-length · Carbodiimides · Triazine

1 Introduction

Collagen, one of the most common biomaterials, has been employed for a wide range of different applications in the biomedical field, for drug delivery systems, wound dressing for tissue engineering, due to its excellent biocompatibility and biodegradability [1]. A very important challenge for biotechnologies today, is to reconstruct organic tissues such as corneal tissues, tendons, cartilage, etc. [2]. However, native collagen is subject to fast biodegradation and low thermal stability, therefore in many cases it does not match the demand of in vitro and in vivo applications [3]. Cross-linking is a well known technique adopted to slowdown the biodegradation rate of native collagen for the production of collagen scaffolds [4]. Although physical treatments may be employed, such as ultraviolet irradiations or dehydrothermal treatments [5], chemical modification is by far the most widely employed method for collagen stabilization [6].

Dendrimers, such as polyamidoamine (PAMAM), poly-propyleneimine (PPI), poly-lysine, polyester and glycol dendrimers [7], are employed for a wide range of applications [8]. In particular, PAMAM, are among the most widely used dendrimers as auxiliaries for collagen stabilization to produce biomaterials and for therapeutic applications. Dendrimers are hyper-branched macromolecules having different core and structure, possessing a high density of free peripheral functional groups [9]. For example, in

G. Campana et al. (eds.), *Sustainable Design and Manufacturing 2017*, Smart Innovation, Systems and Technologies 68, DOI 10.1007/978-3-319-57078-5_42

Fig. 1 are depicted the structures of 2G0 and 2G1 generation PAMAM dendrimers having 4 and 8 free primary amine groups on the surface.

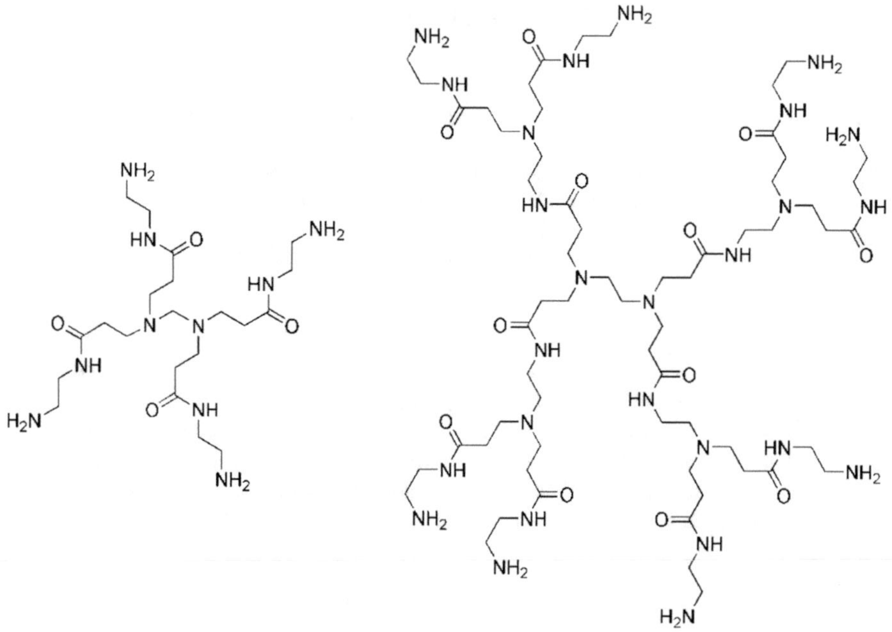

2G0 PAMAM   2G1 PAMAM

Fig. 1. Structure of 2G0 and 2G1 PAMAM dendrimers

Dendrimeric auxiliary contribute to collagen stabilization by amide bond formation, leading to cross-linking between the protein matrix and the dendrimer. This reaction needs to be carried out in the presence of an activating agent such as an aldehyde, acyl azide, etc. Alternately, a commonly employed cross-linking agent is 1-ethyl-3-(3-dimethylaminopropyl)carbodiimide (EDC), a water soluble carbodiimide used in combination with N-hydroxysuccinimide (NHS) (Fig. 2) [10].

EDC/NHS has been extensively studied as cross-linking agent for collagen in the presence of different dendrimers [7], while insofar, to the best of our knowledge, no similar example has been reported employing 4-(4,6-dimethoxy [1, 3, 5] triazin-2-yl)-4-methyl-morpholinium chloride (DMTMM, see Fig. 2), that can be considered a good candidate to be used in combination with dendrimer molecules.

In this work we wish to give a general outline on the cross-linking of collagen matrixes. Moreover, we will report our preliminary studies on the cross-linking of collagen in the presence of PAMAM dendrimers and two different activating agents such as EDC/NHS or DMTMM.

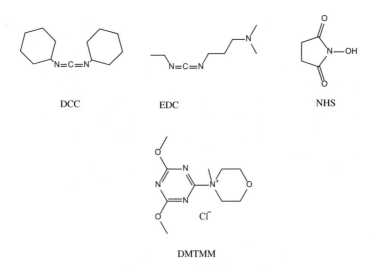

Fig. 2. Structure of cross-linking agents

2   Results and Discussion

The stabilization of collagen can be achieved by physical or chemical protocols. Although several physical techniques have been reported [5], chemical cross-linking remains the preferred method to stabilize collagen matrices. Many different chemical compounds may be used for the scope, such as aldehydes, acyl-azides, anhydrides, carbodiimides etc. [11], which react with collagen to increase the cross-linking degree of the protein. The nature and strength of these bonds formed is different depending on the reactant used and the functional groups of the collagen matrix involved.

In our work we were interested to study the cross-linking of collagen due to the formation of an amide bond between the free carboxylic and aminic functionalities present in the protein. In fact, for every 1000 amino acids present in collagen type I, an average of 120 free carboxylic groups and 80 amine groups are available for bonding.

Amide formation may be achieved at high temperatures (160–180 °C) [11]. This protocol is, however, incompatible with the stability of native collagen which has a shrinkage temperature (T_S) of ca. 55–60 °C. Thus, for cross-linking to occur under mild reaction conditions, chemical activators are needed.

Carbodiimides are widely used for peptide synthesis obtained by amino acid condensation. These reagents are basic organic compounds, due to the presence of two nitrogen atoms, which react with a carboxylic group to form the corresponding O-acylurea, a highly reactive intermediate species, which in the presence of an amine leads to the formation of the amide [11].

One of the most common carbodiimides used is dicyclohexylcarbodiimmide (DCC, see Fig. 2). When DCC is employed as condensing agent the reaction needs to be carried out in an organic solvent and leads to the formation of N,N'-dicyclohexylurea (DCU), a toxic co-product, which needs to be carefully removed at the end of the reaction.

Alternatively, EDC may be used as condensing and cross-linking agent, in organic and aqueous medium; in this case the 1-(3-(dimethylamino)propyl)-3-ethylurea formed as co-product is easily washed away at the end of the reaction. EDC requires to be employed in the presence of at least a stoichiometric amount of N-hydroxysuccinimide in order to be active, as reported by Luyn et al. which studied the reaction mechanism of amide condensation in the presence of EDC/NHS system [12].

In alternative, another condensing agent reported in literature is a derivative of 1,3,5-triazine and in particular DMTMM, which can activate amide formation reactions both in organic and water medium. DMTMM generates as reaction co-product 2-hydroxy-4,6-dimethoxy-1,3,5-triazine (DMT-OH) and a quaternary ammonium salt, both highly water soluble, which can be recovered and recycled at the end of the reaction (Fig. 3) [13].

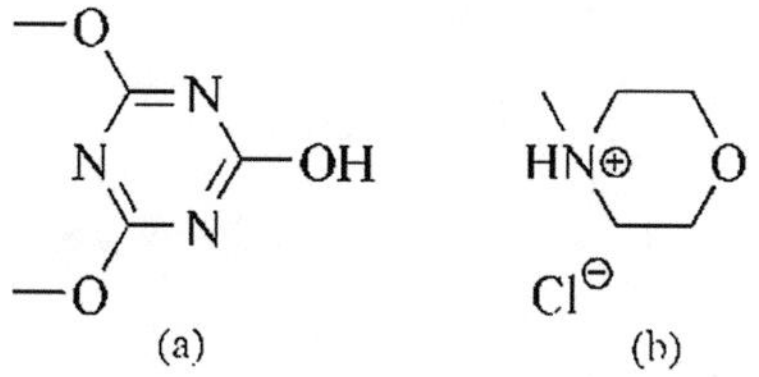

(a)   (b)

Fig. 3. Structure of DMT-OH (**a**) and N-methyl morpholinium chloride salt (**b**)

EDC/NHS and DMTMM activate condensation reactions which leave no trace at the end of the reaction i.e. they are zero length cross-linking agents [11]. This latter characteristic together with the possibility to use water as reaction medium, their ease of removal and recyclability at the end of the reaction, make these reagents the best candidates as sustainable reaction agents for the cross-linking of collagen matrices.

Both EDC/NHS and DMTMM provide the reaction between the free carboxylic groups of aspartic and glutamic acid and the amine free groups of lysine and arginine available for reaction in the collagen matrix [7]. The cross-linking activity of these reagents is thus limited by the amount of free NH_2 groups present in the collagen protein which are present in sub-stoichiometric amount compared to the COOH (80/120 for every 1000 amino acids).

According to literature it is expected that the presence of PAMAM dendrimers, increasing the number of amine linker groups available for the cross-linking of collagen, should increase the thermal stability (T_S) of the protein.

However, according to literature, an optimum NH_2/COOH ratio is noted and a balance needs to be achieved between the degree of cross-linking and the number of amine groups available in the polyamine, suggesting that the addition of free amine groups is not in itself enough to improve cross-linking density. In fact, short diamines such as, for example, ethylenediamine or tris-(2-aminoethyl)amine are known to give very poor cross-linking [7] in the presence of EDC/NHS, while bigger size dendrimers having a higher number of free peripheral amine groups (above 32), may result in no substantial increment of the T_S value [16]. Moreover, dendrimers used for cross-linking having an average size above 1.1 nm, generally present steric and toxicity issues [14].

Thus our preliminary studies were focused on PAMAM dendrimers of generation 0 and 1 (see Fig. 1).

In previous work by us we have demonstrated that collagen powder can be used as a standard substrate for routine tests [15]. In particular, a systematic study has been carried out to test T_S of collagen powder and dermal cow collagen in the presence of DMTMM and a PAMAM dendrimer showing that when a $COOH_{coll}$/DMTMM/ PAMAM molar ratio of 1/1/0.25 is employed, T_S of the stabilized collagen reaches values higher than 81 °C [16]. Differential scanning calorimetry and thermal gravimetric profile of the collagen stabilized by treatment with EDC/NHS are reported in Fig. 4 below. Measures were carried out according to standard procedure as reported in the literature [7].

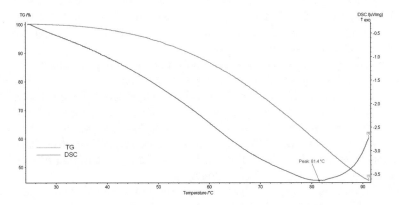

Fig. 4. Differential scanning calorimetry (blue) and thermal gravimetric profile of a sample of collagen powder after cross-linking with EDC/NHS

The data obtained with this protocol provide comparable values of T_S with an error of a few degrees between the powder and the skin collagen. Although thermal stabilization is not the only parameter which requires to be evaluated for bio-engineering and medical applications, nevertheless T_S measurements are crucial to verify the stability of the cross-linked collagen [2]. Thus, we deemed it possible to adopt this protocol as an easy, straightforward method for the preliminary systematic study for the use of new cross-linking agents, dendrimers, etc. in alternative to liquid collagen.

Our preliminary studies have shown that when EDC/NHS is used as cross-linking agent in the presence of dendrimers having four free peripheral NH_2, an increase in T_S is achieved. In fact, T_S in the presence of EDC/NHS and 2G 1.0 PAMAM dendrimers allows to achieve T_S of over 85 °C, while EDC/NHS alone gives T_S of 80 °C. Also in this case there is an optimum NH_2/COOH ratio so that if higher generation dendrimers are employed no further improvement in Ts is achieved and in some cases the Ts is comparable to the value measured in the absence of the dendrimer ($T_S \leq 80$ °C).

Further experiments were carried out for the cross-linking of collagen in the presence of DMTMM. Preliminary experiments have shown that although DMTMM is a very efficient active agent for the condensation of amine and carboxylic groups [15], no relevant difference on T_S are achieved in the presence of PAMAM dendrimers. Further investigations are being carried out to better understand the anomalous behavior of DMTMM/PAMAM system.

3   Experimental

All the reagents used were purchased from Sigma Aldrich, except when otherwise specified. Powder collagen is limed bovine pelt, dried under gentle conditions and milled to fibers of 2–4 mm.

The PAMAM dendrimers used for the present studies have been synthesized according to standard procedure [17]. A diamine core and methylacrilate are allowed to react to give an ester identified as half generation dendrimer (starting from G - 0.5), which is converted to G 0.0 by subsequent reaction with a diamine. Repetition of these two reactions (divergent method) gives higher generation PAMAM leading to an exponential increase of amine surface groups.

Cross-linking experiments were carried out on powder collagen suspended in a water solution to which a % by weight of the different PAMAM is added. The pH of the solution was adjust to 5.5 for EDC/NHS. For DMTMM no pH monitoring is needed, after 30 min the cross-linking agent is added and left to react for 4 h. The collagen powder is filtered, air dried and analyzed by Differential Scanning Calorimetry according to standard protocol [7]. The EDC and NHS ratio remained constant as 1:1, such as collagen and amidation agent ratio, both for EDC/NHS and DMTMM, for the latter different ratios were tested with no relevant difference founded.

All cross-linking experience were carried out with a standard protocol, and for example in the presence of EDC/NHS and a PAMAM dendrimer:

In a 50 mL beaker are added 115 mg (0.6 mmol) of EDC, 69 mg (0.6 mmol) of NHS and 25 ml of distilled water, 250 mg of collagen powder (corresponding to 0.3 mmol of free carboxylic groups) and 5 ml of a dendrimer solution corresponding to 0.3 mmol of amine groups (dendrimer G 0.0: 0,075mmoli). After 4 h the suspension is filtered and washed with 50 ml of distilled water. The treated collagen is then analyzed by Differential Scanning Calorimetry. $T_S = 81$ °C (see Fig. 4).

4   Conclusions

In the current study, an overview on the stabilization of collagen in the presence of PAMAM dendrimers and sustainable active cross-linking agents has been reported. In particular, our investigations focused on the use of EDC/NHS or DMTMM, which, for their specific characteristics are among the more sustainable reactants available today on the market for the stabilization of collagen matrixes. Our preliminary results indicate that cross-linking of collagen in the presence of amidation agents such as 1-ethyl-3-(3-dimethylaminopropyl) carbodiimide (EDC) or 4-(4,6-dimethoxy [1, 3, 5] triazin-2-yl)-4-methyl-morpholinium chloride (DMTMM) give highly stable matrices with T_S as high as 85 °C.

EDC/NHS has proved to be a very suitable amidation agent for cross-linking of collagen also in the presence of PAMAM dendrimers. Preliminary data available show that both the COOH/NH_2 ratio of free groups available and the size of the dendrimer employed seam to play an important role in the stabilization of collagen, as evidenced by tests with 2G0 and 2G1 dendrimers. Molecules with higher steric hindrance, do not affect overall stabilization temperature, probably due to their scant ability to penetrate inside the protein structure. The results of the current study suggest that dendrimers can be successfully applied for powder collagen cross-linking to produce stable matrices which could be used for the production of tissue-engineering scaffolds for biomedical applications. Further studies will be carried out to investigate the differences in activity between EDC/NHS and DMTMM.

Acknowledgements. The authors acknowledge funding support from Crossing S.r.l. and University Ca' Foscari Venice.

References

1. Parenteau-Bareil, R., Gauvin, R., Berthod, F.: Materials **3**, 1863–1887 (2010)
2. Duan, X., Sheardown, H.: Biomaterials **27**, 4608–4617 (2006)
3. Chena, F.-M., Liuc, X.: Prog. Polym. Sci. **53**, 1–148 (2015)
4. Ma, L., Gao, C., Mao, Z., Zhou, J., Shen, J.: Biomaterials **25**, 2997–3004 (2004)
5. Haugh, M.G., Jaasma, M.J., O'Brien, F.J.: J. Biomed. Mater. Res. Part A **89A**, 363–369 (2009)
6. Ma, L., Gao, C., Mao, Z., Shen, J., Hu, X., Han, C.: J. Biomater. Sci. Polym. Ed. **14**, 861–874 (2003)
7. Duan, X., Sheardown, H.: J. Biomed. Mater. Res. **75A**, 510–518 (2005)
8. Chan, J.C.Y., Burugapalli, K., Naik, H., Kelly, J.L., Pandit, A.: Biomacromol **9**, 528–536 (2008)
9. Kinberger, G.A., Cai, W., Goodman, M.: Tetrahedron **62**, 5280–5286 (2006)
10. Albericio, F., Chinchilla, R., Dodsworth, D.J., Najera, C.: Organ. Prep. Proced. Int. **33**, 205–303 (2001)
11. El-Faham, A., Albericio, F.: Chem. Rev. **111**, 6557–6602 (2011)
12. Zeeman, R., Dijkstra, P.J., Wachem, P.B., Luyn, M.J.A., Hendriks, M., Cahalan, P.T., Feijen, J.: Biomaterials **20**, 921–931 (1999)
13. Kunishima, M., Hioki, K., Wada, A., Kobayashi, H., Tani, S.: Tetrahedron Lett. **43**, 3323–3326 (2002)
14. Wang, S., Li, Y., Fan, J., Wang, Z., Zeng, X., Sun, Y., Song, P., Ju, D.: Biomaterials **35**, 7588–7597 (2014)
15. WO 2016/103185 A2 V. Beghetto
16. WO 2015/044971 A2 V. Beghetto, G. Pozza, A. Zancanaro
17. Tomalia, D.A., Baker, H., Dewald, J.R., Hall, M., Kallos, G., Martin, S.: Polym. J. **9**, 117–132 (1985)

Banana Fiber Processing for the Production of Technical Textiles to Reinforce Polymeric Matrices

Zaida Ortega[1(✉)], Mario Monzón[2], Rubén Paz[2], Luis Suárez[2],
Moisés Morón[3], and Mark McCourt[1,2,3]

[1] Departamento de Ingeniería de Procesos, Universidad de Las Palmas de Gran Canaria,
Las Palmas de Gran Canaria, Spain
zaida.ortega@ulpgc.es

[2] Departamento de Ingeniería Mecánica, Universidad de Las Palmas de Gran Canaria,
Las Palmas de Gran Canaria, Spain
{mario.monzon,ruben.paz,luis.suarez}@ulpgc.es

[3] Acondicionamiento Tarrasense Asociación – LEITAT, Terrassa, Spain
mmoron@leitat.org

Abstract. Banana fibers have been extracted by mechanical means from banana tree pseudostems, as a strategy to reevaluate banana crops residues. Extracted long fibers are cut to 45 mm length and then immersed into an enzymatic bath for their refining. Conditions of enzymatic treatment have been optimized to produce a textile grade of banana fibers, which have then been characterized. This fiber has then been transformed into yarns and woven to produce a technical textile with different textile structures. Woven material was then used to produce a composite by compression molding, using polypropylene (PP) as polymeric matrix.

Once the composite was prepared, mechanical testing was carried out (tensile, flexural and impact tests). Results were compared to those obtained with parts made only of PP and with results composite made with a commercial woven product made of flax.

Keywords: Banana fiber · Technical textile · Composite · Circular economy

1 Introduction

Natural fibers interest in composites sector is clearly demonstrated by the high number of published papers in this field, using wood fibers, flax, hemp, sisal, cotton, abaca, etc. [1–3]. Most of researches focus on the use of short fiber composites, with a ratio of use between 5 up to over 40%. The use of short natural fibers usually results in an increase of mechanical properties of the composite, although with some problems due to moisture absorption or UV-radiation.

A strategy to increase mechanical properties is the use of technical textile products, with a predefined orientation, which allows producing a composite with higher mechanical properties in a preferred direction [4]. Some studies related to the production of natural fiber composites by compression molding have been found [4], although no references about the use of banana fiber technical textile products in the composites industries are available.

© Springer International Publishing AG 2017
G. Campana et al. (eds.), *Sustainable Design and Manufacturing 2017*, Smart Innovation,
Systems and Technologies 68, DOI 10.1007/978-3-319-57078-5_43

Banana fibers are extracted from the pseudostem of banana plants, once the fruits have been harvested [4]. They are mainly made of cellulose (63–64%), hemicellulose (10%) and lignin (5%) [1]. Different chemical treatments can be applied in order to increase the fiber affinity with the polymeric matrix or its properties [1–3]. Most usual treatments to achieve a physical modification of the fiber surface and to increase the fiber-matrix compatibility, are corona-discharge or steam explosion treatments [1]. Chemical treatments usually led to an increase of fiber-matrix adhesion and a reduction in moisture absorption, providing higher stability of the composite; some common treatments are mercerization, silanization or acetylation [1], while a number of different treatments can be found in bibliography. Enzymatic treatments are currently presented as a more sustainable alternative to chemical ones; these treatments have been applied to flax or abaca fibers [2], but also to banana fibers [4]. More specifically, banana fibers have been treated with pectinase and hemicellulase, and with poligalacturonase, obtaining better refining results and higher thermal stability with the last one [4].

Banana treated fibers were introduced in a conventional textile process for yarn production, mixing it with wool to obtain better processability, as it was not possible to process it alone due to its high stiffness [4]. Paper here presented focuses on the production of technical textile structures and composites production and characterization, using polypropylene (PP) as matrix.

2 Materials and Methods

2.1 Materials

Banana yarn was produced under the BANTEX project (MAT2013-47393-C2-1-R), as specified in a previously published paper [4]. In summary, banana fibers were extracted from banana trees pseudostems, treated in an enzymatic bath of poligalacturonase for 6 h at 45 °C and pH = 4.5. Treated fibers were blended with wool fibers to produce the yarn using a conventional lab-scale spinning plant. The final yarn consisted in a 70% banana fiber and 30% wool fibers.

Polymeric matrix is made of polypropylene from Total Petrochemicals (PPH 9069), in pellets form.

For comparison purposes, Flaxdry BL (300 g/m^2) from Lineo was also used. This is a 2 × 2 twill structure.

2.2 Methods

Technical textile production. Three different textile structures were produced: plain, twill 3 × 1 and basket. Intended weight is 300 g/m^2 in conventional lab-scale weaving devices, as shown in Fig. 1:

Technical textile characterization. Textile structures were characterized by mechanical testing, following EN ISO 13934-1:2013 standard, obtaining maximum tensile strength and elongation, using an Instron constant elongation gradient dynamometer (VCA), class 0.5.

Fig. 1. Fabric production

Tests were conducted at 20 ± 2 °C with 65% ± 4% relative humidity, using Zwick manual jaws, at a test rate of 100 mm/min, with 100 mm of distance between nippers and a preload of 5 N. Specimens were cut to 50 mm width. 5 specimens of each type of textile structure were tested. Structures were tested in two directions: in the warp and in the weft direction, as they are no symmetrical.

Composite production. Composites were produced by using one layer of textile product and one layer of PP in a Collin Press P 200 PM hot-plate compression molding machine. Plates were heated up to 165 °C with the plates closed, but applying no pressure. Once this temperature was achieved, a pressure of 35 bar was applied while heating up to 190 °C; this conditions were hold for 3 min, followed by a degasification stage of 10 s, and 3 more minutes of holding 190 °C and 35 bar. After that, the cooling stage started, keeping the previous pressure, 35 bar, until reaching 50 °C, where the plates were opened to obtain the composite sheets.

PP sheets for the compression molding were produced following the same thermal cycle, using a 1 mm – thick frame, so that 1 mm PP sheets were used for the compression molding composites production.

Composite characterization. Specimens were machined from obtained sheets to produce normalized test bars, following ISO 527-2:2012. 5 specimens of each composite were tested to characterize their tensile, flexural and impact behavior.

3   Results and Discussion

3.1   Technical Textile Structures Characterization

Weight of the different produced structures (produced as shown in Fig. 2) was calculated by weighting 190×190 mm^2 squares, after drying for 24 h at 105 °C. Results for all structures used in this paper are shown in Table 1, showing that an approximate weight of 300 g/m^2 has been achieved.

Fig. 2. Banana fabric production

Table 1. Weight of different textile structures (in g/m^2)

Banana plain	Banana basket	Banana twill
279	291	284

Mechanical tests provide results summarized in the following Table:

Table 2 shows mechanical behavior of different technical textiles tested under two directions (Warp and weft). The variation coefficient represents the typical deviation divided by the average value. From this Table, it can be clearly concluded that banana textile samples have lower mechanical behavior than flax ones and that also show higher elongation. This is due to the 30% content of wool fiber in the yarn. It can also be seen that there are no important differences among the three different structures used in the weaving process.

Table 2. Mechanical behavior of different technical textile structures

	Test direction	Maximum force	Variation coefficient	Elongation max. force	Variation coefficient
Lineo Flaxdry	Warp	1969.3 N	1.3%	12.8%	2.7%
BL 300 g/m²	Weft	1594.1 N	2.0%	4.1%	1.8%
Banana plain	Warp	915.6 N	2.3%	30.3%	4.2%
	Weft	676.6 N	2.5%	22.1%	11.9%
Banana basket	Warp	878.5 N	2.7%	21.9%	0.4%
	Weft	633.2 N	0.3%	20.6%	2.3%
Banana twill	Warp	910.8 N	2.1%	23.5%	0.2%
3 × 1	Weft	651.0 N	0.9%	21.1%	9.0%

The effect of wool is clearly demonstrated by the comparison of mechanical properties of enzymatic treated fiber with the obtained yarn; from 36.8 cN/tex (as the lowest value obtained for banana treated fibers in previous research [4]) to 5.5 cN/tex in the yarn (containing 30% of wool).

3.2 Composites Characterization

The following picture (Fig. 3) shows an example of compression cycles performed to obtain the composite sheets (Fig. 4):

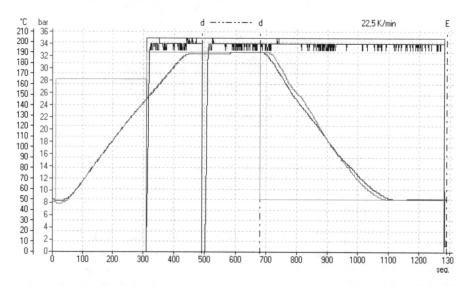

Fig. 3. Compression cycle for composites production (dark red line shows real pressure while red one represents nominal pressure; dark blue and green lines indicate real temperature for the upper and lower plates, while light blue and green ones refers to nominal temperatures)

Fig. 4. Fabrics, composite sheets and test bars

Mechanical behavior of composites made with one layer shows that commercial flax textile increases both flexural and tensile mechanical properties, while drastically reducing impact behavior. On the other hand, banana textile products can increase flexural modulus of net polypropylene, with a slight reduction of tensile properties. Impact behavior is similar for the four series of products containing textiles. In the Table 3, LN300 refers to Lineo 300, P-TX to banana twill textile, P-PL to banana plain fabric, P-BK to banana basket structure, and, finally, PP to net polypropylene.

Table 3. Mechanical behavior of composites

	% fiber	Tensile tests				Flexural tests		Impact testing	
		Modulus (MPa)	Variation coefficient	Max. stress (MPa)	Variation coefficient	Modulus (MPa)	Variation coefficient	kJ/m²	Variation coefficient
LN300	37,37	3225,49	6,13%	79,89	2,54%	2350,10	19,94%	3,96	29,49%
P-TW	39,58	1402,16	17,77%	23,40	5,44%	1631,29	4,71%	3,83	11,63%
P-PL	41,94	1322,64	16,86%	23,82	6,07%	1489,09	15,51%	3,66	13,37%
P-BK	41,51	1473,33	16,80%	25,82	5,49%	1680,69	11,2%	3,76	11,36%
PP	0,00	1600	–	32	–	1500	–	17	–

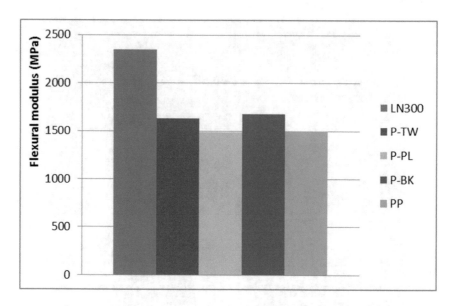

Fig. 5. Flexural modulus average values comparison between different series

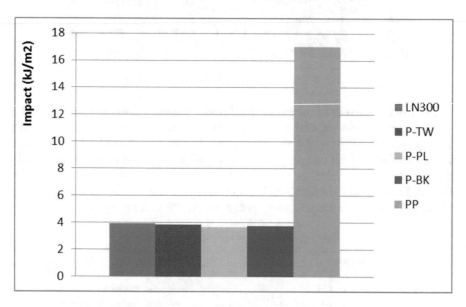

Fig. 6. Impact behavior of the different parts (average values)

This lower behavior of banana fabric is attributed again to the presence of wool in the banana yarn. It is observed that the same differences found in the textile testing are reproduced in the composite mechanical properties, that is, flax has between 2 and 3 times higher tensile strength than banana/wool fabrics. Beyond the negative effect of the wool in the composite, it is clear that optimal compatibilizing treatment fiber-matrix should be investigated for this particular composite.

Figure 5 shows the average values of flexural modulus obtained for the 5 series tested. It is clearly observed how LN300 provides the higher values, while banana fabrics are similar or slightly higher than net polypropylene. On the other hand, Fig. 6 shows the similar impact properties obtained for all parts containing natural fibers, which is reduced to approximate 75% of the value obtained for net PP.

4   Conclusions

- As a summary, it can be concluded that banana fibers have been used to produce a technical textile product using three different structures (basket, plain and twill).
- Results obtained in the fabrics characterization show that banana fabrics have lower mechanical properties than commercial flax one. This is due to the use of wool for the yarn production.
- Similar behavior to those observed for the mechanical properties of the fabric were obtained in the composites characterization: flax fabric give better mechanical properties than banana ones, while impact properties are similar for all parts containing natural fibers.
- Banana fabric is more sustainable than flax one, as it uses banana fibers, coming from an agricultural residue, and no resources have been employed for them to grow: banana plants are cultivated to produce banana fruits and not fibers, being these an added value; on the other hand, flax needs to be cultivated only for the fiber obtaining.
- Future research is focused on the characterization of multiple layer composites and, which results more interesting, in the production of a yarn made 100% of banana fiber. Also, chemical treatments to improve fiber-matrix compatibility should be investigated.
- Results presented in this paper are going to be used to validate simulation software for composite materials.

Acknowledgements. The authors acknowledge the Spanish Ministry of Economy and Competitiveness, as well as Fondo Europeo de Desarrollo Regional (FEDER) funds, as research was conducted under the BANTEX project (code: MAT2013-47393-C2-1-R).

References

1. Gurunathan, T., Mohanty, S., Nayak, S.K.: A review of the recent developments in biocomposites based on natural fibres and their application perspectives. Compos. Part A **77**, 1–25 (2015). doi:10.1016/j.compositesa.2015.06.007
2. Pickering, K.L., Aruan, M.G.: Efendy, T. M. Le. A review of recent developments in natural fibre composites and their mechanical performance. Compos. A **83**, 98–112 (2016). doi: 10.1016/j.compositesa.2015.08.038
3. Faruk, O., Bledzki, A.K., Fink, H.P., Sain, M.: Biocomposites reinforced with natural fibers: 2000–2010. Prog. Polym. Sci. **37**, 1552–1596 (2012). doi:10.1016/j.progpolymsci.2012.04.003
4. Ortega, Z., Morón, M., Monzón, M.D., Badalló, P., Paz, R.: Production of banana fiber yarns for technical textile reinforced composites. Materials **9**, 370 (2016). doi:10.3390/ma9050370

Experimental Investigation into the Use of Natural Rein-forcements for Sustainable Composite Materials

Michele Del Borrello, Matteo Secchi, Giampaolo Campana$^{(\boxtimes)}$, and Mattia Mele

Department of Industrial Engineering, University of Bologna,
Viale del Risorgimento 2, 40136 Bologna, Italy
{michele.delborrello,matteo.secchi2}@studio.unibo.it,
{giampaolo.campana,mattia.mele3}@unibo.it

Abstract. The present work investigates the manufacturability of different composites reinforced with natural fibres that are hemp and bamboo vinyl. Feasibility of different processes were studied and mechanical properties of obtained composite materials were examined by using non-destructive analysis and tensile tests to assess the manufacturability.

Hemp fibres were manufactured by using braiding technology to produce differently oriented textiles. Bamboo reinforcement were employed in form of vinyl sheets. Matrices were made of traditional or innovative polymeric materials. The composite materials were processed through Resin Transfer Moulding (RTM) or Resin Powder Moulding (RPM). Applicability onto different shapes were assessed.

The obtained material performances have been compared with both analytical and numerical models to evaluate their applicability.

Eventually, results are discussed pointing out foreseen opportunities for the industrial application of investigated materials.

Keywords: Hemp fibres · Bamboo vinyl · Composite materials · Manufacturability

1 Introduction

Advanced composites are an important family of materials because of the opportunity to achieve excellent stiffness-on-weight ratios. However, these materials present limits related to the high environmental impact. More in depth, both the production process of raw materials and disposal of final products at the end of life, generally represent critical phases in terms of environmental sustainability. In fact, non-recyclable raw materials are usually employed for both fibres and matrices and their separation for disposing is neither possible nor efficient.

The use of natural fibres as a reinforcement has been proven to be an efficient alternative to improve environmental sustainability in comparison with currently employed materials [1]. Some examples of natural reinforcements, which are investigated nowadays are: kenaf, sisal, flax, banana, hemp, bamboo and others [2–5]. The opportunity to

© Springer International Publishing AG 2017
G. Campana et al. (eds.), *Sustainable Design and Manufacturing 2017*, Smart Innovation, Systems and Technologies 68, DOI 10.1007/978-3-319-57078-5_44

substitute glass-reinforcements by the use of natural ones is nowadays a very interesting field of research where already satisfactory results have been obtained [6].

The opportunities for the application of natural fibre reinforced composites affect the most different domains of design [7]. As an example, automotive industry utilises these materials as an efficient way to improve environmental performances of products [8].

The present work investigates the use of hemp fibres and bamboo vinyl reinforcements for composite materials. Furthermore, a particular attention to the relation between manufacturing and the obtainable characteristics has been paid. The aforementioned reinforcements can be applied with a number of manufacturing processes in order to obtain composites with good mechanical performances and high-quality surfaces. Compression moulding, long fibre injection, vacuum infusion processing and Resin Transfer Moulding (RTM) are the most used and common processes.

The present paper focuses on the production of: (a) hemp fabrics by using a braiding technology and then hemp reinforced composites by employing RTM or Resin Powder Moulding (RPM); (b) Vinyl bamboo reinforced composites by RPM. Section 2 provides an overview onto materials and processes that have been investigated. Section 3 reports the results obtained for the textile and the final composite parts. Section 4 briefly describes an analytical and a numerical model with the aim to compare simulated results with the experimental ones. Section 5 briefly discusses obtained results.

2 Materials and Methods

2.1 Materials

Hemp is a variety of the cannabis species that, in comparison with the flax, presents: longer fibres, higher mechanical properties, lower growing cost, lower processing cost even if the fibres production requires a larger amount of labour [9].

Bamboo is one of the fastest growing plants on earth with documented growth rates up to 1.0 m per day, remarkable environmental benefits in terms of erosion control, water conservation, land rehabilitation and carbon sequestration [10].

To produce the matrix, two different solutions have been investigated in order to compare an innovative polymeric material with a traditional one. In the first configuration, the matrix was produced by an epoxy resin. An innovative polymeric powder, developed at the *Institut für Leichtbau und Kunststofftechnik* of the TUD (*Technische Universität Dresden*), was the second employed material.

2.2 Manufacturing of Reinforcements

Bamboo vinyl was produced through a milling process of bamboo trees. The result consisted in adjacent flat bundles (with a diameter of approximately 80 microns) arranged in the shape of a paper sheet with size $300 \times 6000 \times 0.25$ mm^3.

While bamboo vinyl sheets are ready-to-use, the hemp fibre requires a manufacturing process for the production of a textile reinforcement structure. Braiding technology is one of oldest textile manufacturing processes in the world. For the braiding process of hemp fibres, radial braiding machine equipped with small bobbins has been employed, as in Fig. 1. The textile was produced by braiding the fibres around a cylindrical core and with a certain orientation as respect the rotation axis (braiding angle). After the braiding, the structure was pull out and cut in smaller pieces.

Fig. 1. (Left) Manufacturing of the reinforcement on the braiding machine. (Right) Preparation of the mould for Resin Transfer Moulding

Several technological combinations of the process parameters of the braiding process have been employed in order to manufacture textiles items with different orientation:

- Braiding angle: 30°, 45° and 60°.
- Core diameter: 40.0, 60.0 and 80.0 mm.
- Tissue thickness: 0.6–1.4 mm.

2.3 Composite Production

The textiles obtained by hemp braiding and the bamboo vinyl have been used as reinforcements through RTM and RPM to produce the final composite material. RTM was based on the use of a mixture between 800 g of standard Epoxy L resin as matrix and 350 g EPH 294 as curing agent that were injected in a hot steel mould. The mould was warmed up to 40 °C. Once the infiltration of the resin was completed, the curing of the resin started: the mould was heated up to 60 °C, under 6 bar vacuum for 15 h (Fig. 1, right side).

A RPM process was used for the cure of the innovative polymeric material A.S.SET.-Powder 01 (15 g) into a hot compression mould. Reinforcements and the A.S.SET. matrix have been alternatively positioned into the press that was warmed up to 120 °C.

Fig. 2. Hemp fabric stacking sequence with A.S.SET: (left) first powder layer; (centre) second layer of fabric and third layer of powder; (right) fourth layer of fabric and fifth layer of powder.

The curing process of the powder has been realized by maintaining a fixed pressure for 8 min. Figures 2 and 3 show the stacking procedure for the hemp textile and bamboo vinyl reinforcement, respectively.

Fig. 3. Bamboo vinyl stacking sequence with A.S.SET: (left) first powder layer; (centre) second layer of Bamboo vinyl and third layer of powder; (right) fourth Bamboo vinyl of fabric and fifth layer of powder.

The obtained sheets were machined with the aim to obtain specimens in the form of normalized test bars, according to the specifications in [11]. Six specimens for each combination of matrix and reinforcement have been produced and tested to characterize the tensile strength of the resulting composite material.

2.4   Composite Production of Non-flat Shapes

In order to better investigate the possible fields of application for hemp reinforced composites, several non-flat shapes have been manufactured in the form of pipes by RTM and air bag cases by RPM. A cylindrical mould was used to manufacture a pipe made of multi-layer hemp reinforced composite (Fig. 4 left). An open mould has been used to produced airbag cases reinforced with bamboo vinyl (Fig. 4 right) or hemp braided fabrics (Fig. 4 centre). The mentioned samples showed a good infiltration ratio and an optimal surface finishing. Moreover, no failures of the reinforcement have been observed.

Fig. 4. Hemp reinforced pipes (left) and airbag cases (centre). Bamboo vinyl reinforced airbag case (right)

3   Characterisation of Composites

3.1   Weight of Reinforcements and Classification of the Composites Samples

The hemp braided structures that were obtained around a cylindrical core with a diameter equal to 40 mm and a braiding angle equal to 45° were considered. A $300 \times 125 \times 1.0$ mm^2 sheet of fabric has been weighted and density equal to 0.40 g/cm^3 was measured.

The weight of the bamboo vinyl was measured by weighting a $300 \times 300 \times 0.25$ mm^2 squares samples of bamboo vinyl. The weight was referred to the surface and its value was equal to 0.40 g/cm^3.

Bamboo vinyl reinforced samples have been marked as follows: "8L_0_90_10kN_RPM". "8L" indicates the number of 8 layers, "0_90" means the direction of even and odd layers, "10_kN" the manufacturing pressure and "RPM" specifies the manufacturing process.

Hemp fibres reinforced composites have been indicated with the following notations:

- "2L_30 kN_RPM". "2L" indicates the number of layers, "30 kN" the manufacturing pressure "RPM" specifies the manufacturing process.
- "4L_RTM". "4L" indicates 4 layers, "RTM" specifies the manufacturing process.

A number of configurations have been taken into consideration and shown in the following Table 1. 6 specimens have been machined and tested for each configuration in order to achieve a statistical value of the performance characteristics.

Tab. 1.   Specimens configuration

Reinforcements		Bamboo vinyl	Hemp fabric
Nr. of layers	(adim.)	8	2,4
Pressure	(kN)	10, 20, 30,50	10,20
Process	–	RPM	RPM, RTM
Sheet orientation	–	0_0, 0_90	/
Nr. of configurations	(adim.)	6	4

3.2   Mechanical Characterization by Tensile Test

2 mm thickness composites sheets were manufactured for the mechanical characterization by tensile test. For each manufacturing condition and composite.

Figure 5 shows some significant results obtained by testing bamboo vinyl reinforced composites produced with RPM.

By observing the result obtained for unidirectional oriented composites (8L_0_0_10_kN_RPM and 8L_0_0_20_kN_RPM samples), it is possible to notice that the highest resistance value was achieved and an increase in the applied pressure during RPM implies a relevant increase of the ultimate strength (about 58%); the strain remains almost constant (about 1%).

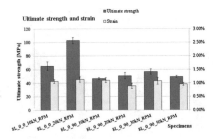

Fig. 5. Ultimate strength and strain of bamboo vinyl reinforced composites

In the case of bi-directional reinforced sheets, moving from 8L_0_90_10_kN_RPM specimen to 8L_0_90_30_kN_RPM (i.e. increasing the manufacturing pressure from 10 to 30 kN) a similar behaviour can be observed with the increase of the ultimate strength (about 22%). The microscopical analyses, shown in Fig. 6, confirm the previous results: samples manufactured with a 30 kN pressure allowed a better infiltration of the matrix and achieved higher mechanical properties than the samples manufactured with a 10 kN.

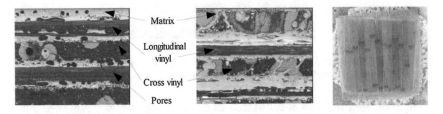

Fig. 6. Bamboo vinyl bi-directional reinforced composites manufactured by RPM with a 10 kN (left) and a 30 kN (centre) manufacturing pressure. Damaged bamboo vinyl reinforced plate by RPM, working pressure 50 kN (right).

Conversely, increasing the manufacturing pressure up to 50 kN led to a decrease around 15% of ultimate strength: this behaviour is due to damages that were induced in the internal reinforcement structure subject to an excessive pressure. Figure 6 shows the correspondent plate, in which the distortion of fibres is visually appreciable.

Figure 7 shows results obtained in terms of ultimate tensile strength and strain under different conditions in case of hemp fibres reinforced composites (i.e. manufacturing process and amount of reinforcement). It can be noticed that the manufacturing pressure has a slighter effect compared to the previously commented case of bamboo vinyl rein-forcements. It is worth mentioning that an increase the working pressure from specimens 4L_30kN_RPM to 4L_40kN_RPM determines an increase of in mechanical properties in terms of both strength (+12.1%) and strain (+9.2%). These results have been confirmed by the computer tomography analysis: the higher the manufacturing pressure the lower the porosities content (Fig. 8).

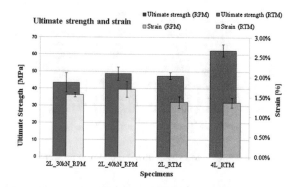

Fig. 7. Ultimate strength and strain of hemp reinforced composites manufactured with Resin Powder Moulding (RPM) and Resin Transfer Moulding (RTM)

Fig. 8. Hemp fibres reinforced composites manufactured by RPM with a 30 kN (left) and a 40 kN (right) manufacturing pressure observed through a computer tomography

The amount of reinforcements in case of RTM, revealed to be an extremely influencing factor on the ultimate strength of the material while not affecting the strain. This behaviour can be observed in Fig. 7 by comparing 2L_RTM and 4L_RTM specimens, which own 2 and 4 reinforcement layers, respectively.

4   Virtual Models of Hemp Fibres Reinforced Composites

Both analytical and numerical models for the simulation of hemp fibres composites are proposed and validated by comparison with the obtained experimental results. These models have been implemented by using the necessary data that were obtained through bibliographical sources [12, 13].

Two analytical models, corresponding to the Maximum Stress Theory (MST) and the Tsai-Hill Theory (THT), have been considered for the single laminate composite by implementing an on-purpose script. Then, a numerical simulation has been conducted through a FEM (Finite Element Method) commercial code for the "4L_RTM" hemp specimen under an applied stress equal to 63 MPa in the longitudinal direction that corresponds to the experimentally measured ultimate strength.

4.1 Analytical Models

According to the MST, failure occurs when any stress in the principal material direction is equal to or greater than the corresponding ultimate strength [14]. These requirements can be expressed by the following Eq. (1):

$$X_c < \sigma_{xx} < X_T \wedge Y_c < \sigma_{yy} < Y_T \wedge -S < \tau_{xy} < S \tag{1}$$

where X and Y represent stresses (subscript c stands for compression and t for tension) along longitudinal and transversal directions of the fibre, respectively. S is the allowable in-plane shear stress; σ and τ represent the applied stress and shear along directions indicated by subscripts, respectively. Only the upper tensile limits have been considered because of the load investigated condition.

Figure 9 (left) shows the results obtained applying this criterion where the load condition in a three-dimensional space is shown by a red marker.

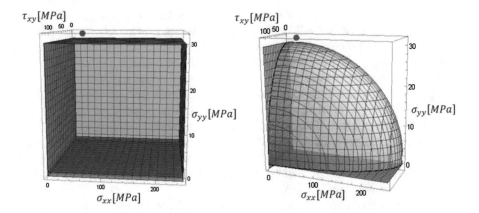

Fig. 9. Results of the MST analysis (left) and of the THT analysis (right) for the specimen

The THT failure condition, which is more conservative than the MST one, can be expressed (with the same notation as in the previous equation) by Eq. (2):

$$\frac{\sigma_{xx}^2}{X_T^2} - \frac{\sigma_{xx}\sigma_{yy}}{X_T^2} + \frac{\sigma_{yy}^2}{Y_T^2} + \frac{\tau_{xy}^2}{S^2} = 1 \tag{2}$$

In Fig. 9 (right), a three-dimensional representation of the analysed case compared to the boundary surfaces of the failure criterion is provided. Also in this case, the analysis only investigates the space of positive tensile stresses.

As it can be noticed, the load condition is slightly over the upper limits of the failure criteria (i.e. $\sigma_{yy} > Y_T$) for both the representations. This behaviour well simulates the one observed in the experimentation; in fact, the specimen failed under the considered stress

4.2 Numerical Model

The geometry of the specimen ($250 \times 25 \times 2$ mm^3) was also modelled using 6250 shell elements. Four layers of Hemp/Epoxy (with volume fractions of 35% and 65%, respectively) were disposed at [+45/−45]. The load condition imposed was the one exposed above. Simulation pointed out a maximum displacement of 2.77 mm; the measured one was 3.42 mm. A difference of the 19% exists between the two results.

A second simulation has been performed by moving maximum dimensions of mesh elements from 1 to 0.5 mm (25000 shell elements) but no differences have been appreciated in the calculation results.

5 Conclusions

Summarising: (a) Hemp fibres appear to be an extremely flexible material for the quick production of textiles using braiding technologies. (b) RTM and RPM curing processes revealed their suitability in the application for the production of the final composite material by using epoxy or AS.SET - Powder 01 polymers, respectively. This suitability has been proven in the case of flat sheets and extended to more complex shapes such as pipes and airbag cases for the automotive field. (c) Mechanical characterization of bamboo vinyl reinforced composites pointed out how the increase of curing pressure during the RPM process leads to an increase of the ultimate strength until the achievement of an optimal point, after which the effect of internal inducted damages prevails. As well, the RPM pressure did not manifest any significant influence on strain performances of the final composite material. The main influence factor onto the mechanical properties of the composite material (in terms of both ultimate tensile strength and strain) revealed to be the amount of reinforcement material. This conclusion is supported by similar works in literature (e.g. [6]).

As a general conclusion, the use of hemp fibres and bamboo vinyl seems to be a very promising alternative to the use of synthetic fibres as reinforcements in composite material. The lower cost and the higher environmental sustainability of these materials compared to traditional reinforcements promote their employment in application on components that are not heavily burdened. According to results presented in terms of obtainable shapes and effectiveness of manufacturing parameters onto the mechanical properties of finished product, further research in this direction seem to be definitely promising.

Furthermore, a good correlation between experimental and analytical results has been found. This promotes the employment of virtual model for the simulation of these composites materials with an expected reduced gap from the real behaviour of the composite material, further extending the opportunities for adoption in design solutions. To reduce the gap found in the numerical experimentation, further characterizations of the specific materials employed must be conducted.

Acknowledgments. We would like to thank the ILK (*Institut für Leichtbau und Kunststofftechnik*) of the TUD (*Technische Universität Dresden*), and particularly Prof. Werner Hufenbach, for stimulating and supporting the presented research activity.

References

1. Joshi, S.V., Drzal, L.T., Mohanty, A.K., Arora, S.: Are natural fiber composites environmentally superior to glass fiber reinforced composites? Compos. Part A **35**, 371–376 (2004). Elsevier
2. Summerscales, J., Dissanayake, N.P.J., Virk, A.S., Wayne Hall, W.: A review of bast fibres and their composites. Part 1 - Fibres Reinforcements Compos. Part A **41**, 1329–1335 (2010)
3. Summerscales, J., Dissanayake, N., Virk, A., Wayne Hall, W.: A review of bast fibres and their composites. Part 2 – Composites. Compos. A **41**, 1336–1344 (2010)
4. Summerscales, J., Virk, A., Wayne Hall, W.: A review of bast fibres and their composites: Part 3 – Modelling. Compos. A **44**, 132–139 (2013)
5. Xua, X., Jayaramana, K., Morinb, C., Pecqueux, N.: Life cycle assessment of wood-fibre-reinforced polypropylene composites. J. Mater. Process. Technol. **198**, 168–177 (2008)
6. Wambua, P., Ivens, J., Verpoest, I.: Natural fibres: can they replace glass in fibre reinforced plastics? Compos. Sci. Technol. **63**, 1259–1264 (2003). Elsevier
7. Pil, L., Bensadoun, F., Pariset, J., Verpoest, I.: Why are designers fascinated by flax and hemp fibre composites? Compos. Part A **83**, 193–205 (2016). Elsevier
8. Wotzel, K., Wirth, R., Flak, M.: Life cycle studies on hemp fibre reinforced components and ABS for automotive parts. Die Angewandte Makromolekulare Chemie 272, 121–127 (Nr. 4763), WILEY-VCH Verlag GmbH, Weinheim, Fed. Rep. of Germany (1999)
9. Sen, T., Jagannatha Reddy, H.N.: Various industrial application of hemp, kinaf, flax and natural fibers. Int. J. Innov. Manag. Technol. **2**(3), 192–198 (2011)
10. Zhou, B., et al.: Ecological functions of bamboo forest: research and application. J. Forestry Res. **16**(2), 143–147 (2005). Springer
11. ISO 527–4:1997: Plastics – Determination of tensile properties – Part 4: Test conditions for isotropic and orthotropic fibre-reinforced plastic composites (1997)
12. Mallick, P.K.: Fiber-Reinforced Composites: Materials, Manufacturing, and Design. CRC Press, USA (2007)
13. Shahzad, A.: A study in physical and mechanical properties of hemp fibres. Adv. Mater. Sci. Eng. **2013**(1–2), Article ID 325085 (2013)
14. Gay, D., Hoa, S.V., Tsai, S.W.: Composite Materials Design and Applications. CRC Press, USA (2003)

The Effects of the Industrial Processing on Commercial Polyhydroxyalkanoates

Laura Mazzocchetti[✉], Tiziana Benelli, Emanuele Maccaferri, and Loris Giorgini

Department of Industrial Chemistry "Toso Montanari", University of Bologna, Bologna, Italy
{laura.mazzocchetti,tiziana.benelli,emanuele.maccaferri3,
loris.giorgini}@unibo.it

Abstract. In the last years polyhydroxyalkanoates (PHAs) gained increasing attention as potential sustainable replacement of current plastics from fossil sources. PHAs are biodegradable polymers which can be produced from bacteria exploiting waste as feedstock. In order to bring these polymers on the market, it is necessary to know their properties and assess the effects of industrial processing stages on the final material. For this reason, in this work different industrial PHAs were investigated. These materials were structurally characterized and their thermal properties were assessed. Furthermore, they were subjected to different thermal and thermo—mechanical treatments (various time, temperature and frequency of rotation in the extruder) in order to study the effect of the applied conditions on the properties of the final material.

Keywords: Polyhydroxyalkanoates · Processing · Thermal stability · Thermal degradation · Thermomechanical degradation

1 Introduction

The replacement of fossil-based non-biodegradable plastics with alternative bio-based polymers that can be (bio)degraded, and possibly composted, after their disposal, has become an industrial, social and environmental priority [1]. The ability to obtain such polymers via green routes is also of paramount importance. A number of biodegradable, aliphatic polyesters and polycarbonates are reported to be produced via pseudo living enzymatic catalysis with good control over performances and architecture of the macro-molecules [2–4], and their properties can be modified and/or tuned, by blending or using their fiber forming ability to produce fully green biocomposites [5, 6]. Poly(hydroxyal-kanoate)s (PHAs), in particular, are a peculiar class of such linear polyesters produced by bacteria through aerobic fermentation of various carbon sources [7]. They have attracted much attention for over two decades [8] owing to their biodegradability into harmless organic products. Moreover their production can be based on agro-industrial (renewable) waste streams [9], such as cheese whey, as starting raw materials with the twofold advantage of depleting wastewaters from organic residues and producing added valued plastic materials. Polyhydroxybutyrate (PHB) has been often taken as a model polymer for crystallinity studies, owing to its perfectly regular (100% isotactic) structure and to its very low nucleation density that promotes growth of very large spherulites

© Springer International Publishing AG 2017
G. Campana et al. (eds.), *Sustainable Design and Manufacturing 2017*, Smart Innovation,
Systems and Technologies 68, DOI 10.1007/978-3-319-57078-5_45

upon crystallization from the melt: this feature however, makes PHB too fragile for common applications in the plastic industry, requiring either its blending with other possibly biodegradable polymers [10] or the use of its copolymers [11]. According to their co-monomers composition, which in turn depends on the selected bacterial strains and their nutrients, PHAs can display properties ranging from elastomers to thermoplastic. Despite the promising bright future of this family of polymers, their industrial development is still hampered, among other factors, by their high cost (roughly 5 times more expensive than standard petrochemical plastics, i.e. ~ 5–6 €/kg vs 1 €/kg) [12]. Additionally, it is worth to note that among the most commonly applied purification techniques, HCl or NaClO treatments are often applied to destroy bacterial cell walls and remove cell debris [13]. Such an approach, however, might detrimentally affect the overall polyester properties, thus leading to a commercial product with lowered performances [13, 14]. Another drawback still preventing PHAs widespread as application commodity materials rather than niche-market high-value products comes from the problems in their processability. Indeed, PHAs decomposition temperature, T_d, tends to be quite low: PHB for example has a T_d which is slightly above its melting ($T_m = 180 \,°C$) making it difficult to process it in a common extruder. Moreover, when heated, they become very sensitive to humidity and easily hydrolyze, so the drying step represent a crucial issue in the processing of such materials. Thus, while in principle the common transformation techniques used for thermoplastics can be satisfactorily applied to polyhydroxyalkanoates, the actual working conditions must be carefully tuned in order to avoid significant deficiencies in the final product's performance. While some work has been recently reported on the effect of the industrial processing on the PHA performances, they are usually carried out to validate some specific polyester batch in terms of potential scale up [15]. Since the possibility of exploiting wastes as secondary raw materials, brings on the market a number of new PHAs producers, it is hence important to realize that the average condition of waste-based PHAs production might result in inhomogeneous and poorly characterized materials. The impact of such situation on the overall reproducibility of polymers properties and on the processing conditions might thus be quite dramatic. Since the great interest raised around PHAs, they are the object of a great number of studies concerning mainly their thermal degradation, carried out using lab-scale samples. However, in order to be processed in industrial facilities on a large scale, PHAs should be able to further endure melt processing techniques such as extrusion and/or injection, where the macromolecules suffer not only for the high temperature input, but also because of the mechanical stresses. Indeed, the simple pelletization of such polyester might induce some degradation in the polymer, which is usually not even considered when working on the lab-scale powdered materials. Hence in order to raise some awareness in the field of PHAs industrial processing, in the present work, two different commercial PHAs, namely a Poly(hydroxybutyrate-*co*-hydroxyvalerate) (PHBV) and a Poly(3-hydroxybutyrate-*co*-4-hydroxybutyrate) (P3,4HB) were analyzed and characterized. In particular, samples were heated and treated in an industrial extruder, and the effects of these two treatments on the molecular weight distribution and on the thermal stability were assessed. These two different treatments (heat, and heat + mechanical stress in the extruder) were applied in order to better evaluate the mechanical stress component in the behavior of the PHAs.

2 Experimental

2.1 Materials

Commercial testing samples of Poly(hydroxybutyrate-*co*-valerate) with an indication of 3-hydroxyvalerate content of about 4–6% molar content and Poly(3-hydroxybutyrate-*co*-4-hydroxybutyrate) were obtained as powders. The same PHBV polymer was also provided in from of pellets obtained from the producer from the very same supplied powder. Chloroform (HPLC grade) and chloroform-d were also obtained from Aldrich Chemical Co and used without further purification.

2.2 Instrumentation

^{1}H NMR spectra were obtained at room temperature, on 5–10% w/v CDCl$_3$ or DMSO solutions, using a Varian MercuryPlus VX 400 (^{1}H, 399.9; ^{13}C, 100.6 MHz) spectrometer. Chemical shifts are given in ppm from tetramethylsilane (TMS) as the internal reference. Infrared spectroscopy was carried out in ATR (attenuated total reflectance) mode on a Bruker Alpha spectrophotometer equipped with ATR accessory with a diamond window [16]. The average number and weight molecular weights together with the polidispersity index (M_n, M_w and PDI = M_w/M_n respectively) of polymers were measured by gel permeation chromatography (GPC) using an HPLC Lab Flow 2000 apparatus working with a 1 mL min-1 flow, equipped with an injector Rheodyne 7725i, a Phenomenex Phenogel 5u 10E6A column and a RI detector Knauer RI K-2301. Calibration curves were obtained using several monodisperse polystyrene standards. Chloroform was used as the eluent at 1 ml/min flow rate and polymer molecular weights were determined based on a conventional calibration curve generated by narrow polydispersity polystyrene standards from Aldrich Chemical Co. TGA measurements were carried out using a TA Instruments STD Q600 (10 °C/min, room temperature to 600 °C, nitrogen flow) [17]. Thermal stability was defined according to two different parameters: $T_{d1\%}$, defined as the temperature at which the weight loss is 1% of the initial weight, and $T_{max\,deg}$, which represents the temperature at which the polymer reaches the maximum degradation rate (weight loss rate).

3 Results and Discussion

While it is well renown that PHB for example, the most studied of the PHAs family, is characterized by a very narrow processing window from a thermal point of view, the mechanical stimulus is also contributing to the degradation of the polymeric backbone: the overall result of the melt processing can thus be a further reduction of the Molecular Weight Distribution. Based on these findings, the present work aims to characterize the degradation caused by heating and/or shearing resulting from the melt processing of PHAs, namely a PHBV and a P3,4HB, the former both in powder and in pellets form. The aspect of the analyzed polymers is quite different passing from the powdered PHBV material to the relative pellet (Fig. 1), with a strong yellowing of the pellets clearly evident, while no significant evidence of degraded groups appears in the relative NMR spectrum. The latter phenomenon can be associated with a degraded polymer, probably owing to an unoptimized palletization procedure. Moreover, the P3,4HB powder appears slightly darker (in particular more yellow) than the powdery PHBV. Owing to the yellowish color of both the powdery materials, FT-IR spectra were recorded, in order to detect the presence of contaminants, for example of proteic nature. In such a case, a typical absorption of the amide carbonyl stretching would appear in an otherwise clear zone [18]. None of these signals however is detected in the analyzed samples, accounting for protein residues free polymers.

Fig. 1. The analyzed PHAs: PHBV in powder and pellets, and P3,4HB in powder form.

In order to define, or confirm, the chemistry and the composition of such polymers, ^{1}H-NMR spectra were recorded for the three samples of Fig. 1. The ^{1}H-NMR spectrum recorded for PHBV powder is reported in Fig. 2, where signals appear that are attributed to the comonomer 3-hydroxybutyric acid (HB) [19]: a multiplet at 1.25 ppm belonging to the methyl residue of the 3-hydroxybutyric residue, a couple of double doublets at 2.4–2.6 ppm attributed to the diastereotopic protons in the α position with respect to the carbonyl group and, at 5.3 ppm, a signal accounting for the oxygen bearing CH moiety. Chemical shifts accounting for the presence of the hypothesized 3-hydroxyvaleric comonomers, instead, are not observed, leading to the conclusion that such comonomer's content, if any, is well below 1%, based on the instrument sensitivity. Hence, the so-called PHBV is in fact a homopolymer poly-3-hydroxybutyrate. An identical spectrum is obtained also for the pellets: in this case, no additional signals are observed, such as those coming from the crotonic groups that can form upon high temperature degradation, either because they are not present ore because they are too diluted, representing just the chain ends. The ^{1}H NMR of P3,4HB shows instead all the expected signal of the two units: 3-hydroxybutyrate (3HB) chemical shifts, as described above, and 4-hydroxybutyrate (4HB), CH_2-O at 4,1 ppm, hydrogens in α position to the carbonyl group at about 2,3 ppm and the backbone CH_2 (position 5 in Fig. 2B) at 1,9 ppm. It was thus possible to define the copolymer composition, which is 17%mol in 4HB.

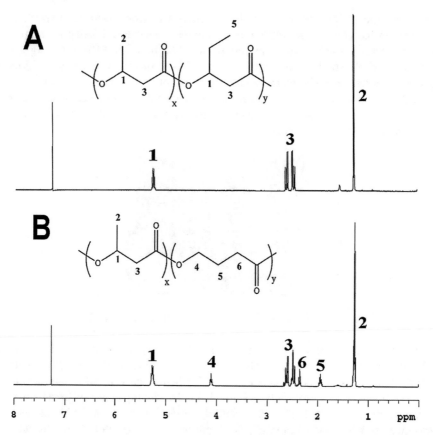

Fig. 2. ^{1}H NMR spectra of PHBV (A) and P3,4HB (B) powdery samples.

The GPC analysis of the starting polymers (Table 1) provides information about the starting Molecular Weight Distribution of the commercial pristine polymers. The significant PDI values account for the complex situation of waste-based bacterial PHAs production. It is interesting to observe the impact of pelletization process on the MWD of the PHBV sample: while the starting powder has characteristics that well compares with the average PHB [20], the first extrusion process more than halves the $\overline{Mw}$. This observation brings in relevant issues about the stability of mechanical properties in the final products obtained after a second melt processing step. P3,4HB displays instead already a lower starting $\overline{Mw}$.

Table 1. Molecular weight distribution characterization of the analyzed pristine samples

	$\overline{Mn}$ (g/mol)	$\overline{Mn}$ (g/mol)	PDI
PHBV powder	209000	1168000	5,6
PHBV pellet	123000	466000	3,8
P3,4HB powder	165000	694000	4,2

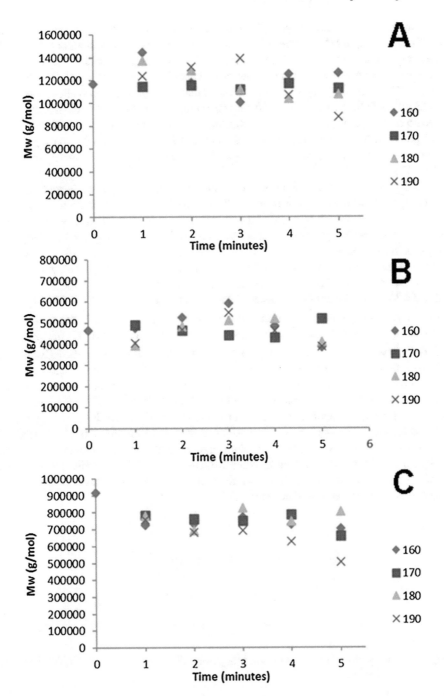

Fig. 3. $\overline{\text{Mw}}$ trend of PHBV in powder (A) and pellets (B) and P3,4HB (C) upon different heating treatments.

In order to verify thermal stability of these polymers TGA scans were carried out in oxidizing atmosphere (air). Thermal stability of P3,4HB is the lowest among the investigated candidates ($T_{d1\%} = 237°C$), while, surprisingly, the degradation temperature of the PHBV pellets is higher than the relative powder ($T_{deg\,1\%} = 254°C$ and $242°C$ respectively). This behavior might be due to the fact that upon processing, while the average molecular weight decreases, the PDI also decreases possibly owing to the narrowing of the molecular weight distribution that reduces also the low molecular relative fraction that possibly decompose during the treatment. In this frame, while the starting degradation is retarded, $T_{max\,deg}$ is, as expected, higher for the higher MWD powdery material, (285 °C vs. 276 °C).

Different samples of the three candidates were thus singularly heated in oven reaching temperatures ranging from 160 °C to 190 °C (every 10°) for a different time (1', 2', 3', 4' e 5') for a total of 20 different heating treatments applied for each polymer type. Such treated polymers were thus analyzed first gravimetrically, then via GPC and TGA. PHBV pellets do not show any substantial weight loss upon heating, while the two powders mass, display some decrease which, however, is always well within 1%.

GPC results (Fig. 3) show that all the PHBV samples are stable at 160 °C, 170 °C, and 180 °C for all the applied timespan. The harsher treatment instead (190 °C) induces some degradation of the $\overline{\text{Mw}}$ for longest heating segments (4' and 5'). Worth noting is that beside some decrease of molecular weight occurring at 190 °C, the powdery PHBV's $\overline{\text{Mw}}$ does not even get close to the one of pelletized polymer. P3,4HB instead, shows a slight decrease of the molecular weight recorded for the heat-treated polymers compared to the pristine one. The heated P3,4HB are not affected too much by the different temperature treatments up to 180 °C, while the 190 °C tests carried out for 4' and 5' again leads to some more significant degradation. The analysis of the single GPC chromatograms (Fig. 4) provides some more useful information. It can indeed be observed that, while usually the molecular weight distribution peak stays constant, what happens is that the low molecular weight fraction increases, probably owing to the random chain scission with a crotonate chain end formation that that is recognized as the degradation process affecting PHAs [21].

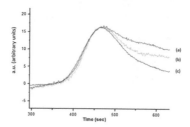

Fig. 4. Comparison of GPC traces recorded for powder PHBV treated 5' at 190 °C (a), at 160 °C (b) and as received (c).

TGA results, providing the trend for degradation temperatures, display a similar trend, confirming such results.

When PHBV pellets are treated in an industrial extruder (Fig. 5) at analogous temperature, i.e. summing up the shearing effects to the thermal input, molecular weight (Fig. 6) and thermal stability of the polymer are further affected. Depending on the applied extruding conditions, indeed, $\overline{Mw}$ is more than halved again, showing that such materials are more significantly affected by the concomitant action of heat and mechanical stresses. Hence, the simple thermal characterization of such materials is not enough to foresee some criticalities in the industrial processing.

Fig. 5. PHBV pellets extruded at 180 °C for 5'

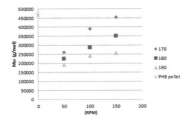

Fig. 6. $\overline{Mw}$ trend of PHBV pellets extruded in different conditions (temperature and RPM).

4   Conclusion

The present investigation highlights the difficulties associated to the industrial production of PHAs when they come from waste streams as raw materials. Such products come to the market with a poor a sometimes-misleading characterization, and this issue might lead to a number of problems during the processing. Moreover, it was demonstrated that the most significant influence on the stability of the analyzed PHAs processed at low/medium temperatures is due to the application of mechanical stresses: this observation accounts for some cautions to be undertaken when starting to design the processing cycle for this kind of Polyesters.

Acknowledgements. Authors wish to acknowledge the financial support from project no. F/ 0264/01-02/X26 funded by MISE (mise.AOO_IAI.REGISTRO INTERNO.R.0000727. 09-02-2016 on 09-02-2016).

References

1. Mooney, B.P.: The second green revolution? Production of plant-based biodegradable plastics. Biochem. J. **418**, 219–232 (2009)
2. Mazzocchetti, L., Scandola, M., Jiang, Z.: Random copolymerization with a large lactone enhances aliphatic polycarbonate crystallinity. Eur. Polym. J. **48**, 1883–1891 (2012)
3. Mazzocchetti, L., Scandola, M., Jiang, Z.: Copolymers of ethyl glycolate and ω-Pentadecalactone: enzymatic synthesis and solid-state characterization. Eur. Polym. J. **47**, 942–948 (2011)
4. Mazzocchetti, L., Tsoufis, T., Rudolf, P., Loos, K.: Enzymatic synthesis of amylose brushes revisited: details from X-ray photoelectron spectroscopy and spectroscopic ellipsometry. Macromol. Biosci. **14**, 186–194 (2014)
5. Ortega, Z., Morón, M., Monzón, D.M., Badalló, P., Paz, R.: Production of banana fiber yarns for technical textile reinforced composites. Materials **9**, 370 (2016)
6. Sisti, L., Belcari, J., Mazzocchetti, L., Totaro, G., Vannini, M., Giorgini, L., Zucchelli, A., Celli, A.: Multicomponent reinforcing system for poly(butylene succinate): composites containing poly(l-lactide) electrospun mats loaded with graphene. Polym. Testing **50**, 283–291 (2016)
7. Doi, Y.: Microbial synthesis and properties of polyhydroxy-alkanoates. MRS Bull. **17**, 39–42 (1992)
8. Mozejko-Ciesielska, J., Kiewisz, R.: Bacterial polyhydroxyalkanoates: still fabulous? Microbiol. Res. **192**, 271–282 (2016)
9. Bengtsson, S., Karlsson, A., Alexandersson, T., Quadri, L., Hjort, M., Johansson, P., Morgan-Sagastume, F., Anterrieu, S., Arcos-Hernandez, M., Karabegovic, L., Magnusson, P., Werker, A.: A process for polyhydroxyalkanoate (PHA) production from municipal wastewater treatment with biological carbon and nitrogen removal demonstrated at pilot-scale. New Biotechnol. **35**, 42–53 (2017)
10. Gazzano, M., Mazzocchetti, L., Pizzoli, M., Scandola, M.: Crystal orientation switching in spherulites grown from miscible blends of poly(3-hydroxybutyrate) with cellulose tributyrate. J. Polym. Sci. B Polym. Phys. **50**, 1463–1473 (2012)
11. Di Lorenzo, M.L., Raimo, M., Cascone, E., Martuscelli, E.: Poly(3-hydoxybutyrate)-based copolymers and blends: influence of a second component on crystallization and thermal behavior. J. Macromol. Sci. B. **40**, 639–667 (2001)
12. Chanprateep, S.: Current trends in biodegradable polyhydroxyalkanoates. J. Biosci. Bioeng. **110**, 621–632 (2010)
13. Samorì, C., Abbondanzi, F., Galletti, P., Giorgini, L., Mazzocchetti, L., Torri, C., Tagliavini, E.: Extraction of polyhydroxyalkanoates from mixed microbial cultures: impact on polymer quality and recovery. Bioresour. Technol. **189**, 195–202 (2015)
14. Samorì, C., Basaglia, M., Casella, S., Favaro, L., Galletti, P., Giorgini, L., Marchi, D., Mazzocchetti, L., Torri, C., Tagliavini, E.: Dimethyl carbonate and switchable anionic surfactants: two effective tools for the extraction of polyhydroxyalkanoates from microbial biomass. Green Chem. **17**, 1047–1056 (2015)
15. Pachekoski, W.M., Dalmolin, C., Agnelli, J.A.M.: The influence of the industrial processing on the degradation of poly(hydroxybutyrate) - PHB. Mater. Res. **16**, 327–332 (2013)

16. Angiolini, L., Benelli, T., Giorgini, L., Raymo, F.M.: Chiroptical switching based on photoinduced proton transfer between homopolymers bearing side-chain spiropyran and azopyridine moieties. Macromol. Chem. Phys. **209**, 2049–2060 (2008)

17. Bicciocchi, E., Chong, Y.K., Giorgini, L., Moad, G., Rizzardo, E., Thang, S.H.: Substituent effects on raft polymerization with benzyl aryl trithiocarbonates. Macromol. Chem. Phys. **211**, 529–538 (2010)

18. Angiolini, L., Caretti, D., Salatelli, E., Mazzocchetti, L., Willem, R., Biesemans, M.: Synthesis and characterization of new functional polystyrenes containing tributyltin carboxylate moieties linked to the aromatic ring by a trimethylene spacer. J. Inorg. Organomet. Polym Mater. **18**, 236–245 (2008)

19. Martínez-Sanz, M., Villano, M., Oliveira, C., Albuquerque, M.G.E., Majone, M., Reis, M., Lopez-Rubio, A., Lagaron, J.M.: Characterization of polyhydroxyalkanoates synthesized from microbial mixed cultures and of their nanobiocomposites with bacterial cellulose nanowhiskers. New Biotech. **31**, 364–376 (2014)

20. Samorì, C., Galletti, P., Giorgini, L., Mazzeo, R., Mazzocchetti, L., Prati, S., Sciutto, G., Volpi, F., Tagliavini, E.: The green attitude in art conservation: polyhydroxybutyrate-based gels for the cleaning of oil paintings. ChemistrySelect **1**, 4502–4508 (2016)

21. Kawalec, M., Sobota, M., Scandola, M., Kowalczuk, M., Kurcok, P.A.: Convenient route to PHB macromonomers via anionically controlled moderate-temperature degradation of PHB. J. Polym. Sci. A Polym. Chem. **48**, 5490–5497 (2010)

Pyrolysis of Low-Density Polyethylene

Giorgio Zattini[1(✉)], Chiara Leonardi[1], Laura Mazzocchetti[1,2], Massimo Cavazzoni[3], Ivan Montanari[3], Cristian Tosi[3], Tiziana Benelli[1,2], and Loris Giorgini[1,2]

[1] Department of Industrial Chemistry "Toso Montanari",
University of Bologna, Bologna, Italy
{giorgio.zattini3,chiara.leonardi3,laura.mazzocchetti,
tiziana.benelli,loris.giorgini}@unibo.it
[2] Interdepartmental Center for Industrial Research on Advanced Applications in Mechanical Engineering and Materials Technology, CIRI-MAM, University of Bologna, Bologna, Italy
[3] CURTI S.p.A., Divisione Energia, Modena, Italy

Abstract. Pyrolysis of low-density polyethylene in an innovative batch pilot plant, with a hydraulic guard ensuring a safe process, was performed. The influence of process temperature on yield, distribution and composition of products was investigated. The oil/waxes were analyzed by gas chromatography coupled mass spectrometry, while pyrolysis gas was monitored online during the process by micro-gas chromatography. Pyrolysis were carried out at 450, 500, 550 and 600 °C. Results obtained show that low temperatures yield a greater amount of oil/waxes, and a gas enriched in carbon oxides and C_{3+} hydrocarbons. At higher temperatures, the gas fraction, riche in methane and hydrogen, is predominant over liquid products. This process has proved to be a versatile way to recover polyethylene wastes into valuable oils (rich in aliphatic and simple aromatic hydrocarbons) or gas, to be used as petrochemical feedstock or fuel, thus providing a sustainable method for material and energy recovery of waste packaging.

Keywords: Pyrolysis · Polyethylene · LDPE · Material recovery · Waste management

1 Introduction

As the rate of consumption of plastic materials in the world is continuously expanding, more plastic wastes are generated. In 2014, 25.8 million tons of post-consumer plastic waste ended up in the waste upstream in Europe (EU28+2), mainly arising from packaging [1]. In this context, replacement of fossil-based non-biodegradable plastics with alternative bio-based polymers that can be (bio)degraded, and possibly composted, after their disposal, has become an industrial, social and environmental priority. In fact, a number of biodegradable, aliphatic polyesters [2, 3] and polycarbonates [4, 5] are reported to be produced, also via pseudo living enzymatic catalysis [6, 7], and their properties can be modified and/or tuned by blending, or using their fiber-forming ability to produce entirely green bio-composites [8, 9].

© Springer International Publishing AG 2017
G. Campana et al. (eds.), *Sustainable Design and Manufacturing 2017*, Smart Innovation, Systems and Technologies 68, DOI 10.1007/978-3-319-57078-5_46

Amongst technical end-of-pipe solution to waste management, European approach is oriented to promote material recycling and energy recovery from waste, discouraging landfill disposal of Plastic Solid Wastes (PSW), which can be recovered in other ways. In order to harmonize the different national measures concerning the management of PSW, and to prevent or reduce their impact, the Packaging and Packaging Waste Directive (94/62/EC) has been adopted in 1994. The directive has been subsequently amended in different steps, for the establishment of increasingly improved objectives. In 2004, for instance, the recovery targets of packaging wastes have been raised up to at least 60 wt% for incineration with energy recovery, and between 55 and 85 wt% for material recycling. Despite this, different transpositions have brought wide differences in plastic waste disposal in each European country. Amongst the different routes for recovering material from wastes, chemical, or tertiary recycling, involves processes able to convert plastic macromolecules into gas or liquid products, addressable as fuels or polymer feedstock. This direct conversion of PSW into fuel or raw materials has become very attractive to obtain valuable and useful chemicals. Pyrolysis, a thermal degradation in an inert atmosphere, is the one of the most attractive techniques capable to process back PSW to petrochemical feedstock. Amongst PSW, packaging materials, largely consisting of polystyrene and polyolefins (LDPE, HDPE and PP), are the main target for pyrolysis processes, since their cracking results in desirable products [10]. Pyrolysis of polyolefins is a thermal cracking reaction, thermo-sensitive in the 400–850 °C range [11, 12]. The use of catalysts, such as zeolites, allows lower process temperature and a consequent reduction of energy consumption, and permits control on the product distribution [10, 13]. Despite these advantages, many researchers have put their efforts into carrying out the reaction in absence of catalysts, modifying process variables, in particular temperature, in order to drive the reaction to the desired products [11, 12]. Additional key variables are feedstock size, residence time and reactor geometries. Regarding the latter, several types of reactors have been used to perform pyrolysis of PSW: fluidized bed [12, 14], fixed bed [15], batch [10, 11], conical spouted bed [13], screw kiln [16], horizontal tube [17] and microwave assisted [18] reactors.

In our previous work, we have developed a fixed bed batch pyrolysis system that employs an innovative reaction chamber, plunged in a water tank which behaves as a hydraulic guard, ensuring the sealing of the chamber and an extremely safe process. The plant, in a pilot scale, has been used with the purpose of investigating the viability of pyrolysis on composite materials such as tires [19], thick CFRPs [20, 21] and fiberglass [22], which are otherwise hardly recoverable wastes. The research work has shown that plant innovations made pyrolysis of tires an environmentally and economically sustainable process, and resulted in an industrial-scale designed patented plant. In this context, owing to the great potential demonstrated by the innovative designed configuration, the present study is aimed to evaluate the feasibility of recovering chemical feedstocks from LDPE scraps. The paper focuses on process yields and product characterization as a function of process temperature, with a particular focus on gas characterization. Scraps of LDPE, coming from the packaging industry, have been pyrolyzed in the 450–600 °C range.

2 Experimental

2.1 Materials

Scraps of commercial low density polyethylene (LDPE), provided by Progeny S.p.A, has been used as raw material. Scraps has been delivered in form of thin sheets (Fig. 1). All samples were pyrolyzed without any further treatment: no shredding nor crushing were applied.

Fig. 1. Typical LDPE scraps arranged on the sample plate

2.2 Fixed Bed Pyrolysis Reactor

A diagrammatic scheme of the experimental apparatus is shown in Fig. 2. The pyrolysis reactor is a batch pilot plant, consisting of two parts: the lower is a tank containing water, which acts as hydraulic guard, while the upper is a double walled mobile bell, hosting electric resistances with a total power of 21 kW. The reactor, provided by Curti S.p.A., has an internal volume of 5.5 m^3 and is able to treat up to 10 kg of polyethylene scraps per pyrolysis cycle. It is worth noting that the reactor allows the loading of non-shredded scraps up to a diameter size of 2 m.

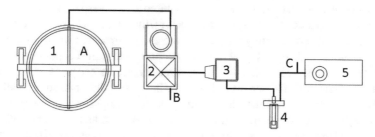

Fig. 2. Schematic layout of the experimental pyrolysis pilot plant: pyrolysis chamber (1); demister (2); acid gas scrubber (3); fan (4); burners (5); collection point of solid residue (A); collection point of pyrolysis oil and water (B); bleed point of pyrolysis gas (C)

In a typical experiment, the bell is lifted to allow the loading of 10 kg of LDPE scraps. The lid is then closed and the reaction chamber is flushed with nitrogen to remove air, ensuring the inertness of the reaction environment, then the system is heated with a rate of 8 °C/min up to the set point, and then kept at the desired temperature for a total residence time of 150 min, which has been considered the end-point of the test. Then, the resistances are switched off and the reactor left to cool down. When the temperature decreases below 100 °C, the reactor is opened to remove any eventually solid residue present inside the reactor.

All the gases generated in the process are extracted from the reactor and partially condensed in a water-cooled coil. The obtained liquid fraction, composed by oil and waxes, is collected via a demister and subsequently characterized. During the heating phase, water from the hydraulic guard facing the inner chamber evaporates, forming steam that adds up to the pyrolysis products flow, hence requiring a separation step of the vaporized water from the oily products in the demister. The permanent gas proceeds to the gas bleed point for online characterization via micro-gas chromatography (μ-GC) prior to reach the burners at the end of the line.

2.3 Methods

Thermogravimetric Analysis (TGA) was carried out by means of a TA Instruments SDT-Q600 apparatus [23, 24]. Preliminary pyrolysis experiments were carried out on 10–15 mg of material in inert atmosphere (N_2, 100 ml/min gas flow) from Room Temperature (RT) to 700 °C with a heating rate of 10 °C/min. For each pyrolysis temperature, at least 3 tests have been performed. The collected solid residue has been dried and weighted. Subsequently, these samples have been treated at 600 °C for at least 24 h in oxidizing atmosphere (i.e. air) to evaluate pyrolytic carbon content.

The peculiar plant design adopted involves the co-condensation of oil, wax and process water. For this reason, water has been separated from the hydrocarbon fractions with a centrifuge operating for 6 min at 4200 rpm.

The amount of gas released has been estimated by difference between the initial weight of LDPE and the amount of obtained liquids (oil + waxes) and solids. The gaseous phase was monitored in situ by means of an Agilent 490 μ-GC gas-chromatographer. The instrument has been placed on-line in the pilot plant so no active sampling is required. Two columns were used: a Molsieve 5 Å (MS, 20 m) at 80 °C with argon as carrier gas to separate H_2, O_2, N_2, CO and CH_4, and a PoraPLOT U (PPU, 10 m) at 90 °C with helium as carrier gas, to separate CO_2, C_2H_4, C_2H_6 and C_3/C_4. The detector is a Thermal Conductivity Detector (TCD). Quantification has been performed with a standard gas mixture. The Gross Calorific Value (GCV) of the gas was calculated on the actual obtained composition, according to UNI 7839.

3   Results and Discussions

3.1   Pyrolysis Batch Experiments

A preliminary thermogravimetric analysis (TGA) of LDPE scraps has been carried out to assess the conditions for the pilot plant pyrolysis experiments. The degradation process appears almost complete at 435 °C with a residue of approximately 0.3 wt%. Thus, 450 °C has been selected as the lowest batch process temperature. On the other hand, considering the dimensions of the plant and the quantity of sample treated per batch, 600 °C has been identified as the maximum temperature to be investigated. With the aim to investigate the effect of the temperature, experiments in the pilot plant have been carried out at 450, 500, 550 and 600 °C.

The main streams obtained by the pyrolysis of LDPE are gaseous and liquid products. Their relative distribution depends on the applied reaction conditions (such as temperature and residence time). A negligible amount of solid residue has been also registered. For each process temperature, at least 3 batch tests have been performed. Due to the formation of waxes during the pyrolysis process, as also stated in literature [11, 12, 14], the reactor and pipeline system have been entirely cleaned with steam at the end of each set of experiments conducted at the same temperature. The retrieved waxes have been reunited and added to the oil fraction.

3.2   Pyrolysis Yields

The average yields of the obtained solid residue, oil and gas (the latter calculated by difference) as a function of the process temperature are shown in Fig. 3.

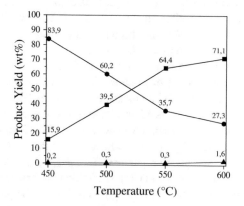

Fig. 3. Products distribution at different pyrolysis temperature: (▲) solid residue, (■) pyrolysis gas and (●) oil fractions

The solid residue obtained from pyrolysis batches performed at 450, 500 and 550 °C has been quantified being approximately 0.3 wt%, confirming TGA results. These results seem to prove that the lowest process temperature is sufficient to ensure the complete sample decomposition. The average quantity of solid residue obtained from the batches performed at 600 °C was higher, approximately 1.6 wt%.

More interesting is the variation of oil and gas distribution when increasing process temperature. The main product of the pyrolysis of LDPE at 450 °C is the liquid fraction (83.9 wt%), which is distributed as follow: 50.2 wt% is the oil spilled from the demister while 33.7 wt% consists of waxes. The gas yield has been calculated, by difference, and resulted to be the 15.9 wt%. The experiments performed at 500 °C show an average total liquid fraction amounting to 60.2 wt% (36.8 wt% and 23.4 wt% the oil and waxes, respectively), thus 39.5 wt% of gas has been produced. Further increasing the process temperature results in gas yield increase, reaching 64.4 wt% (550 °C) and 71.1 wt% (600 °C) of the obtained products. At these process temperatures, liquid products have been entirely collected from the demister since any wax has been removed from the reactor or the piping. The oil obtained at 550 and 600 °C was 35.6 and 27.3 wt%, respectively.

Several authors have reported similar trends regarding the gas fraction increasing with the process temperature [11, 12, 14]. Experimental data and the aforementioned literature suggest that pyrolysis of polyethylene is very sensitive to process temperature. In particular, the gas fraction tends to increase with the increasing process temperature. Our experiments have produced on average a greater amount of gaseous products in relation to data reported by other authors.

3.3 Solid Residue

The solid residues obtained from the tests performed at each temperature have been heated up to 600 °C in oxidizing atmosphere (i.e. air) for at least 24 h, in order to determine the amount of pyrolytic carbon. The results show that no significant difference between the samples can be identified. The weight loss during oxidation is randomly distributed in the 10.7–16.1 wt% range and the pyrolytic carbon content is limited under 0.17 wt% of the initial sample weight. These results demonstrate that a complete thermal degradation is obtained at all process temperatures investigated. It is worth to note that, since the samples subjected to pyrolysis are real LDPE scraps from the industry, the filler content is rather unknown and unpredictable. This could be a possible cause for the random trend registered.

3.4 Pyrolysis Gas

The permanent gas evolving during the pyrolysis tests has been continuously monitored on-line through a μ-GC in order to obtain a correlation between the gas composition and the applied process conditions. The gases detected are carbon oxides (CO and CO_2), hydrogen (H_2), methane (CH_4), ethane (C_2H_6), and ethylene (C_2H_4). Furthermore, C_3 is the sum of propane and propylene, while the sum of 1-butene, n-butane, 1,3-butadiene and iso-butylene is defined as C_4. Finally, with the terms "Others", indeterminate incondensable species, mainly composed of C_{4+}, are indicated.

The unexpected, although minimum, presence of carbon oxides in the produced gas is possibly due to the persistence of air pockets inside the reactor which are not completely removed during the nitrogen purging step prior to test. At the early stage of each test, when the process temperature reaches approximatively 400 °C, the oxygen reacts forming carbon oxides. The gas composition recorded during the entire test time, for each process temperature is shown in Fig. 4 and reported in Table 1. Nitrogen is not reported and is quantifiable as the complementary percentage to 100 mol%. Since this is a discontinuous process, during the test a great variation in nitrogen concentration occurs, with maximums at the beginning and end of each test. At the lowest process temperature (i.e. 450 °C) the amount of produced gas (15.9 wt%) is not sufficient to fill the plant piping. This causes a delay in the detection of the gas, which reaches the μ-GC sampling point only after the final nitrogen purging. Starting from 500 °C, the amount of produced gas allows the complete filling of the reaction chamber and pipeline. Hence, in the first 100 min only nitrogen and a small amount of residual oxygen are detected, then a marked increase in produced gas species is observed. Carbon oxides reach their maximum concentration in the first minutes of gas outflow after which they are considerably reduced to few percentage points. Due to the higher amount of gas produced with increasing temperature, both the gas concentration and its detection window are increased at the higher process temperatures.

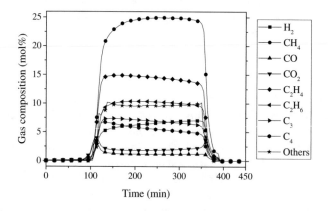

Fig. 4. Typical gas composition pattern during a test at 600 °C

The average composition of the pyrolysis gas produced at different pyrolysis temperatures, and the relative Gross Calorific Value (GCV) calculated for each composition are reported in Table 1. It is worth noting that the main evolved compounds at 450 °C, are carbon oxides. This fact is also explained by the evolution of low amounts of H_2 (4.2 vol%), CH_4 (8.5 vol%) and other hydrocarbons (mostly C_4 compounds), which causes a scarce dilution of undesired carbon oxides. At 500 °C carbon oxides concentration (2.2 vol% CO and 3.0 vol% CO_2) decreases considerably and, contextually, a significant increase in CH_4 (19.4 vol%), ethylene and ethane (respectively 19.7 and 12.4 vol%) is noticeable. A further increase in pyrolysis process temperature results in a gas enriched in lighter molecular weight compounds (H_2 and CH_4) at the expense of heavier species, particularly C_3/C_4 compounds. These data suggest that higher process temperatures are associated with the evolution of lighter molecular weight compounds, especially H_2 and CH_4, as reported by other authors [12].

Table 1. Chemical composition and GCV of the gas produced by pyrolysis of LDPE scraps at different process temperature

Pyrolysis gas component (vol%)	Process temperature (°C)			
	450	500	550	600
H_2	4.2	4.5	5.9	8.1
CH_4	8.5	19.4	24.1	29.7
CO	22.5	2.2	0.9	1.8
CO_2	20.0	3.0	1.4	3.1
C_2H_4	4.4	19.7	19.8	17.9
C_2H_6	5.8	12.4	13.0	12.3
C_3	9.3	11.2	10.5	8.5
C_4	14.0	14.6	11.8	7.0
Others	11.4	13.2	12.8	11.7
GCV (MJ/Nm^3)	46.5	66.6	65.3	58.8

Since it is known that the calorific value of a gas is a function of its chemical composition, the data obtained by μ-GC analysis were used to calculate the GCV of the gaseous fractions according to the standard method UNI 7839 (Table 1). The calculated GCV has been observed to increase from 46.5 to 58.8 MJ/Nm^3 from 450 to 600 °C, reaching its maximum at 500 °C (66.6 MJ/Nm^3). A possible explanation for the GCV value at 500 °C could be attributed to the relatively low concentration of carbon oxides (with practically null calorific values), hydrogen and methane (respectively 12.7 and 39.8 MJ/Nm^3) and the simultaneous highest content in C_3 and C_4 species (with assigned calorific values of 97.3 and 93.7 MJ/Nm^3) (Fig. 5).

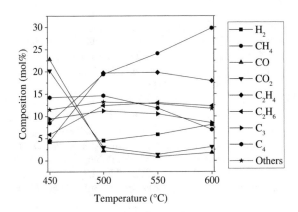

Fig. 5. LDPE pyrolysis gas composition trends as a function of the batch temperature

3.5   Oil/Wax Fraction

The liquid fraction, consisting of oil and highly condensable waxes, follows an opposite and complementary trend compared to the gas fraction. Greater liquid yields are obtained at low process temperatures with an inversion of the liquid: gas ratio occurring between 500 and 550 °C (Fig. 3). Liquid products yield range from 83.9 to 27.3 wt%, from 450 to 600 °C. The composition of these oils/waxes consists mainly of aliphatic hydrocarbons. However, an aromatic hydrocarbons fraction is present, with compounds like benzene, toluene, xylenes (BTX fraction) and naphthalene. The composition does not vary significantly amongst the different temperatures tested.

4   Conclusions

Pyrolysis of LDPE scraps has been conducted in a fixed bed batch pilot plant that employs an innovative hydraulic guard, to ensure an airtight reaction chamber and an extremely safe process. Temperatures in the 450–600 °C range have been investigated, with a particular attention on the correlation between temperature and products yields. For temperatures up to 500 °C, liquid products (oil + waxes) are predominant over the gas, while an opposite trend is seen at 550 °C and higher. Moreover, waxes are not formed at all at the higher temperatures. Solid residues are neglectable at all temperatures investigated. The gas composition, as a function of temperature, has also been evaluated. A higher process temperature produce a gas enriched in lighter compounds such as hydrogen and methane, due to more intense cracking reactions occurring at higher temperatures. The present work confirms the feasibility of LDPE pyrolysis, as well as the flexibility of the pilot plant and process previously developed and tested with other waste materials.

References

1. Plastics - The Facts 2016, Plastics Europe. http://www.plasticseurope.org/
2. Samorì, C., Basaglia, M., Casella, S., Favaro, L., Galletti, P., Giorgini, L., Marchi, D., Mazzocchetti, L., Torri, C., Tagliavini, E.: Dimethyl carbonate and switchable anionic surfactants: two effective tools for the extraction of polyhydroxyalkanoates from microbial biomass. Green Chem. **17**, 1047–1056 (2015)
3. Samorì, C., Abbondanzi, F., Galletti, P., Giorgini, L., Mazzocchetti, L., Torri, C., Tagliavini, E.: Extraction of polyhydroxyalkanoates from mixed microbial cultures: impact on polymer quality and recovery. Bioresour. Technol. **189**, 195–202 (2015)
4. Mazzocchetti, L., Scandola, M., Jiang, Z.: Random copolymerization with a large lactone enhances aliphatic polycarbonate crystallinity. Eur. Polym. J. **48**, 1883–1891 (2012)
5. Cortecchia, E., Mazzocchetti, L., Scandola, M.: Organic-inorganic interactions in poly(trimethylene carbonate)–titania hybrids. Macromol. Chem. Phys. **210**, 1834–1843 (2009)
6. Mazzocchetti, L., Scandola, M., Jiang, Z.: Copolymers of ethyl glycolate and ω-pentadecalactone: enzymatic synthesis and solid-state characterization. Eur. Polym. J. **47**, 942–948 (2011)

7. Mazzocchetti, L., Tsoufis, T., Rudolf, P., Loos, K.: Enzymatic synthesis of amylose brushes revisited: details from X-ray photoelectron spectroscopy and spectroscopic ellipsometry. Macromol. Biosci. **14**, 186–194 (2014)

8. Sisti, L., Belcari, J., Mazzocchetti, L., Totaro, G., Vannini, M., Giorgini, L., Zucchelli, A., Celli, A.: Multicomponent reinforcing system for poly(butylene succinate): composites containing poly(l-lactide) electrospun mats loaded with graphene. Polym. Test. **50**, 283–291 (2016)

9. Ortega, Z., Morón, M., Monzón, M., Badalló, P., Paz, R.: Production of banana fiber yarns for technical textile reinforced composites. Materials **9**, 370 (2016)

10. Miskolczi, N., Bartha, L., Deák, G.: Thermal degradation of polyethylene and polystyrene from the packaging industry over different catalysts into fuel-like feed stocks. Polym. Degrad. Stab. **91**, 517–526 (2006)

11. Onwudili, J.A., Insura, N., Williams, P.T.: Composition of products from the pyrolysis of polyethylene and polystyrene in a closed batch reactor: effects of temperature and residence time. J. Anal. Appl. Pyrolysis **86**, 293–303 (2009)

12. Mastral, F.J., Esperanza, E.: García, P., Juste, M.: Pyrolysis of high-density polyethylene in a fluidised bed reactor. Influence of the temperature and residence time. J. Anal. Appl. Pyrolysis **63**, 1–15 (2002)

13. Elordi, G., Olazar, M., Lopez, G., Amutio, M., Artetxe, M., Aguado, R., Bilbao, J.: Catalytic pyrolysis of HDPE in continuous mode over zeolite catalysts in a conical spouted bed reactor. J. Anal. Appl. Pyrolysis **85**, 345–351 (2009)

14. Williams, P.T., Williams, E.A.: Fluidised bed pyrolysis of low density polyethylene to produce petrochemical feedstock. J. Anal. Appl. Pyrolysis **51**, 107–126 (1999)

15. Achilias, D.S., Roupakias, C., Megalokonomos, P., Lappas, A.A., Antonakou, E.V.: Chemical recycling of plastic wastes made from polyethylene (LDPE and HDPE) and polypropylene (PP). J. Hazard. Mater. **149**, 536–542 (2007)

16. Serrano, D.P., Aguado, J., Escola, J.M., Garagorri, E.: Conversion of low density polyethylene into petrochemical feedstocks using a continuous screw kiln reactor. J. Anal. Appl. Pyrolysis **58**, 789–801 (2001)

17. Angyal, A., Miskolczi, N., Bartha, L., Valkai, I.: Catalytic cracking of polyethylene waste in horizontal tube reactor. Polym. Degrad. Stab. **94**, 1678–1683 (2009)

18. Zhang, X., Lei, H., Yadavalli, G., Zhu, L., Wei, Y., Liu, Y.: Gasoline-range hydrocarbons produced from microwave-induced pyrolysis of low-density polyethylene over ZSM-5. Fuel **144**, 33–42 (2015)

19. Giorgini, L., Benelli, T., Leonardi, C., Mazzocchetti, L., Zattini, G., Cavazzoni, M., Montanari, I., Tosi, C.: Efficient recovery of non-shredded tires via pyrolysis in an innovative pilot plant. Environ. Eng. Manag. J. **14**, 1611–1622 (2015)

20. Giorgini, L., Benelli, T., Mazzocchetti, L., Leonardi, C., Zattini, G., Minak, G., Dolcini, E., Cavazzoni, M., Montanari, I., Tosi, C.: Recovery of carbon fibers from cured and uncured carbon fiber reinforced composites wastes and their use as feedstock for a new composite production. Polym. Compos. **36**, 1084–1095 (2015)

21. Giorgini, L., Benelli, T., Mazzocchetti, L., Leonardi, C., Zattini, G., Minak, G., Dolcini, E., Tosi, C., Montanari, I.: Pyrolysis as a way to close a CFRC life cycle: carbon fibers recovery and their use as feedstock for a new composite production. Pyrolysis as a way to close a CFRC life cycle: carbon fibers recovery and their use as feedstock for a new composite production, 354–357 (2014)

22. Giorgini, L., Leonardi, C., Mazzocchetti, L., Zattini, G., Cavazzoni, M., Montanari, I., Tosi, C., Benelli, T.: Pyrolysis of fiberglass/polyester composites: recovery and characterization of obtained products. FME Trans. **44**, 405–414 (2016)

23. Angiolini, L., Caretti, D., Salatelli, E., Mazzocchetti, L., Willem, R., Biesemans, M.: Synthesis and characterization of new functional polystyrenes containing tributyltin carboxylate moieties linked to the aromatic ring by a trimethylene spacer. J. Inorg. Organomet. Polym Mater. **18**, 236–245 (2008)
24. Angiolini, L., Benelli, T., Giorgini, L., Raymo, F.M.: Chiroptical switching based on photoinduced proton transfer between homopolymers bearing side-chain spiropyran and azopyridine moieties. Macromol. Chem. Phys. **209**, 2049–2060 (2008)

Business Model Innovation for
Sustainable Design and Manufacturing

Sustainable Business Models of Small-Scale Renewable Energy Systems: Two Resource-Scarce Approaches for Design and Manufacturing

Tatu Lyytinen[✉]

Department of Management Studies, Aalto University, Helsinki, Finland
tatu.lyytinen@aalto.fi

Abstract. We need to pay attention to both the design and manufacturing and business model approaches when analysing the sustainability of firms. Though there is increasing literature on the sustainable business model, little attention has been paid to solution design's implication for the sustainability of the business model. In this article, I compare the solution design and business model approaches of two similar small-scale bioenergy solutions (using the high-income context in a developed country and the low-income context in a developing country). The sustainability perspective is integrated into the business model framework, and the implications of technological solutions to business models are analysed. I demonstrate in this study that while a high-tech solution in the high-income context has been able to integrate technological and organisational sustainability into its business model, a low-tech solution in the low-income context has mainly focused on social sustainability, has not paid attention to ecological sustainability and is struggling with financial sustainability.

Keywords: Sustainability · Business model · Design and manufacturing · Bioenergy · Developing country · Developed country

1 Introduction

Seventy years ago in Finland, resources were scarce. Cars and buses used to run on wooden biomass fuels. Then, when the country started to become more prosperous, these solutions were forgotten, and fossil fuels gained popularity. Since oil prices started to peak 10 years ago, these century old technologies have been taken to use again – this time in the form of small-scale, off-grid energy production. At the same time, these technologies have been used in another context: to solve the energy poverty challenge; four billion people lack sustainable access to energy. These two contexts have many implications for sustainable design and manufacturing when we observe them from the lenses of sustainable business models.

The deployment of renewable energy technologies is critical for the ability of countries to move towards sustainable energy systems in the future. In this transition, small-scale renewable energy technologies that utilise biomass are one of

© Springer International Publishing AG 2017
G. Campana et al. (eds.), *Sustainable Design and Manufacturing 2017*, Smart Innovation, Systems and Technologies 68, DOI 10.1007/978-3-319-57078-5_47

the key sectors that can deliver mini-grid and off-grid solutions to urban and rural high-income, middle-income and low-income markets. Panwar et al. [1] argue that renewable energy technologies provide an excellent opportunity for the mitigation of greenhouse gas emissions so as to reduce global warming and respond to sustainable economic and social development. Becker and Fischer [2] argue that fossil fuels continue to dominate the energy markets, as they are believed to be cheaper than renewable energy sources. However, at the same time, renewable energy technologies are claimed to be more sustainable. These technologies can simultaneously contribute to social, economic and ecological development and climate change mitigation, bypassing fossil fuel-based development models [3].

The sustainable design and manufacture of renewable energy solutions can be related to the scarcity of resources. Recently, literature on frugal innovations has emerged to describe resource-scarce solutions in the contexts of developed and developing countries [4–6]; these innovations are often claimed to be sustainable [7–10]. However, there is little empirical evidence how sustainable frugal innovations are from the global perspective. Thus, the purpose of this study is to explore what implications design and manufacturing have on the sustainability of high-tech and low-tech solutions in the contexts of developed and developing countries. I first study the sustainability of one renewable energy solution in the low-income context, which is often claimed to be frugal and sustainable. I then compare it to a similar solution in the developed country context. To do this, I review the literature on sustainable design and manufacturing and apply the sustainability perspective as an analytical framework to the business model concept in order to explore the implications of sustainability in two different contexts of small-scale bioenergy solutions.

1.1 Literature Review

1.1.1 Sustainable Design and Manufacturing

Sustainability is a multidimensional phenomenon that can be integrated into firm activities in various forms. Wells [11] argues that integrating sustainability into business activities requires interplay between product design, manufacturing process and business model design. Companies can become more competitive (or greener) either by developing new products and services based on new technology and/or developing new business models [12, 13]. Therefore, we need to pay attention to both technological solutions and business models when analysing the sustainability of firms.

Sustainable design and manufacturing are connected to the triple bottom line perspective of interplay between the economy, society and environment. Beltramello et al. [14] argue that greener products and services enable the reduction of environmental pollution, optimise the use of natural resources and increase energy efficiency by providing new sources of economic growth. Levänen et al. [8] also emphasise the social and economic aspects of sustainability. Though there have been many attempts to conceptualise the aspects of sustainability, it can be argued that there is no comprehensive classification available. Thus, sustainability needs to be understood as a context-specific phenomenon.

Gassmann et al. [15] argue that different development, production and sales activities are needed for different customer segments in the developed and developing worlds. While concepts such as green products, Cleantech and industrial symbiosis have emerged to describe ecological sustainability in the developed market context, the concept of frugal innovation describes similar innovations in the emerging middle-income and lower-income markets [6]. The concept of frugal innovation highlights features such as simple technologies, lower cost and ease of use, in addition to ecological sustainability features, to make solutions affordable for the four billion people living in different levels of poverty globally [5, 16]. To understand how sustainable design and manufacturing make their way into the market and have ecological and social impacts, we need to turn to the business model concept that provides the analytical framework for this study.

1.1.2   Business Model as an Analytical Framework

A business model can be defined as a bounded analytical framework that describes how a firm does business [17]. The business model also describes how value is created for the customers and then captured for the focal firm [18]. It consists of a component structure to enable the modelling of key business activities of the focal firm [23, 24]. While Osterwalder's business model canvas has become the most popular component structure among practitioners, scholars do not seem to agree on an ideal component structure or exactly what defines a business model [19]. For example, in their recent review, Wirtz et al. [20] divided a business model into strategic, customer and market and value creation components and their subcomponents. Thus, there is no general consensus on an ideal component structure. Each scholar seems to define their own context-specific component structures.

Baden-Fuller and Morgan [21] argue that these structures can be driven from theoretical and/or practical worlds. Zott and Amit [22] argue that a business model considers elements that describe the architecture of an activity system. Thus, a business model is not a complete description of every activity a firm does but something more general that goes beyond a particular context [23]. Business model analysis includes defining the component structure to create boundaries, identifying the key activities to provide content and abstracting the activities to more general business model configurations.

In this study, the sustainable business model literature is applied to create boundaries and content for the component structure. Sustainable business models are closely related to material efficiency [24] and green [25] business models. A sustainable business model, as a concept, refers to environmentally benign business models. However, the same term could, and perhaps should, also comprehend the social and inclusive aspects of a business model [26, 27]. Based on the theoretical world of business models [17, 20, 27] and the practical world of renewable bioenergy solutions [28, 29], four business model elements are selected for this study: offer, customer interface, infrastructure and financial model. To integrate the sustainability dimension into these selected business model elements, I apply the framework of Bocken et al. [30], where they divide sustainable business model archetypes into technological, social and organisational groupings. Technological and ecological sustainability relates to the offer element, social sustainability relates to the customer interface element and organisational sustainability relates to the infrastructure and financial model elements (Table 1).

Table 1. Business model component structure and sustainability perspective

BM components	Traditional questions	Sustainability questions
Offer	What products and services are offered to the customer?	How does offering promote material and energy efficiency, create value from waste and substitute non-renewables?
Customer interface	Who are the customers, and how are relationships with customers organised?	How does the customer relationship promote the well-being of the customer?
Infrastructure	How are relationships with suppliers and human resources organised?	How do supply chain management and operations and maintenance promote ecological and social sustainability?
Financial model	How are costs and benefits managed to make a business profitable?	How does the integration of sustainability aspects into other elements promote financial performance?

2 Methods and Data

The comparative case study method is used to explore sustainable business models of small-scale bioenergy solutions in high-income and low-income contexts. The empirical part of this study uses an abductive approach to create a theory-driven iterative process between empirical and theoretical worlds, systematically combining the empirical world, framework, theory and case [31]. In this study, small-scale bioenergy solutions define the context of the empirical world, the business model concept provides the framework, the sustainability concept constructs the theoretical foundations and two small companies from Finland and India act as empirical cases.

In this study, I focus on analysing combined heat and power (CHP) biomass gasification technologies and their potential in the mini-grid and off-grid markets of developed and developing countries. Dong et al. [32] argue in their literature review that CHP gasification technologies can replace traditional energy production systems and increase energy savings, reducing greenhouse gas emissions and improving energy security. However, Kirkels and Verbong [33] argue in their review (of 30 years of biomass gasification) that there seems to be an overly optimistic advocacy of the potential for the development of small-scale gasification technology. Thus, two cases with similar small-scale biomass gasification solution providers were selected in this study, one from a developed market and one from a developing market. Energy production in the gasification process has three basic phases:

1. Pyrolysis, where solid biomass is charred and gasified in produced gas;
2. Gas purification, with filters and cooling; and
3. Electricity and heat from gas, using a combustion engine.

Volter was founded in 1998 when the prime minister of Finland (a former business-man) was looking for a self-sufficient electricity production solution for his remote cottage to replace the diesel engine. Volter offers a fully automated biomass gasification solution that utilises wood chips with energy production capacities of 30 kW electrical and 80 kW thermal output. Their first solution was piloted in 2009 in Kempele Ecovillage in Northern Finland, and by 2016, around 50 solutions were sold globally.

Husk Power System (hereafter, HPS) has been one of the most cited frugal energy innovations in recent years [8]. HPS was founded in 2007 when two wealthy Indians working in the U.S. decided to seek an affordable energy solution for their home village in the state of Bihar in Northern India. Bihar is one of the poorest state in India, and around 90% of Bihar's 100 million population

are not connected to the reliable electricity grid. HPS offers a low-cost biomass gasi-fication solution that utilises rice husk waste with an energy production capacity of 35 kW electrical output. In 2016, HPS had 68 operational plants in India and Africa, serving around 200 000 low-income customers. While Volter targets their solution to the wealthy, HPS targets the poorest of the poor in rural India, making the comparison between these two cases interesting.

The data was collected in two phases. In 2012–2013, I participated in a research project at the VTT Technical Research Centre of Finland with the case company Volter and conducted interviews and field visits during that period. The results of the research project, which are published in one report and seminar presentations, are used for this study [34, 35]. Emails related to the sustainability of Volter's solution were exchanged in October and November of 2016 with the CEO of Volter and a bioenergy expert from VTT. A field trip to India was made in February 2016, where key personnel of Husk Power System were interviewed in Patna, India, and two days were spent in one HPS gasification plant in the Mahjoria village. Notes were taken from personal discussions with HPS employees at the site, and photos were taken of the gasification plant. In addition, I have conducted systematic documentary material analysis on HPS, which includes 86 documents. The data was coded and analysed by identifying the key activities of the business model elements of both cases.

3   Results

In this section, the results of analysis of two case companies are presented, and comparison between two different approaches for sustainable design and manufacturing is made. The business model configurations of two case companies are described in Table 2, and after, a more detailed analysis of the business model activities is conducted.

Table 2. General business model configurations of the case companies

	Finland case (volter)		India case (HPS)	
	High-tech model	High-tech sustainability	Low-tech model	Low-tech sustainability
Offer	High investment cost, uncompromised quality, functionality and efficiency	Formal recycling of side streams and high pollution control	Low investment cost, compromised quality, functionality and efficiency	Informal reuse of side streams and medium pollution control
Customer interface	Reliable and seamless service	Green energy, self-sufficiency and security	Service pricing and door-to-door service	Raising awareness of renewable energy
Infrastructure	Automated and maintenance-free	Effortless operations and maintenance	Manual operations and maintenance intensive	Local employment and capacity building
Financial model	Revenues from green customers and costs from context-specific wood chip supply	Sustainability oriented customers and green image building	Donor dependency, high cost of operations and uncertain revenues	Poverty reduction and inclusive energy trends

3.1   High-Tech Resource-Scarce Business Model

In this section, Volter's solution design and manufacturing, business model and sustainability approach are analysed. The gasification process has many technical challenges created by the ash and tar content of the wooden biomass, meaning that there are high maintenance requirements. However, Volter has been able to develop a solution that functions autonomously without high maintenance requirements. This makes the solution viable in the high-income market context, where the labour force is expensive. Volter's solution model is visualised in Fig. 1, and the sustainability implications of its business model are analysed below.

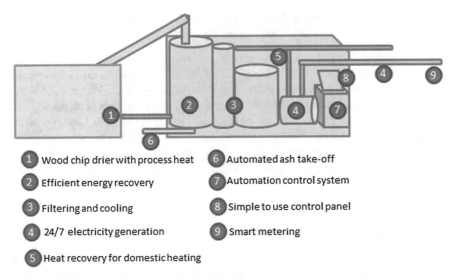

1 Wood chip drier with process heat   6 Automated ash take-off

2 Efficient energy recovery   7 Automation control system

3 Filtering and cooling   8 Simple to use control panel

4 24/7 electricity generation   9 Smart metering

5 Heat recovery for domestic heating

Fig. 1. Volter solution system model with key functionalities

Offer: 24/7, fully-automated green heat and electricity

Volter offers its customers a renewable combined heat and electricity solution that functions 24/7 regardless of weather conditions. This enables customers to become self-sufficient regarding local energy production and improves energy security when local raw materials are available. Volter has optimised their gasification, combustion and purification process to make gas clean, meaning that air pollution is minimised. Side stream ash is used as fertiliser or combusted in other processes. Volter has also been able to eliminate tar waste generation in the process; before, a couple of litres of condensate water were produced per day and delivered to a wastewater treatment plant.

Customer interface: Smart electricity and heating system

Volter's customers are high-income people living in remote areas, people willing to become self-sufficient regarding energy production and organisations that want a greener image. Volter manages its customer relationships by identifying environmentally aware customers and providing them energy with market prices and reliable customer service. Volter's solution improves customer well-being by reducing the air pollution and providing them with energy security.

Infrastructure: Local raw materials and maintenance-free

Volter utilises local wood chips, which have high moisture level requirements, to maintain an efficient gasification process. Volter has worked with research institutes and universities to improve the gasification and filtering processes and evaluate the ecological impacts. Volter has been able to automate its operations and maintenance, meaning that minimal involvement in terms of human labour is needed. Volter's plant requires basic 'housekeeping' weekly and monthly mechanical level maintenance. However, Volter employs highly qualified engineers for design and manufacturing and installation services. A locally designed and manufactured solution provides employment, and a locally procured biomass provides income for forest industry actors.

Financial model: Subsidised green energy

Volter's financial sustainability is dependent of available subsidies, cost of alternative energy sources and cost of wood chips. For example, in Finland, Volter's solution is only financially sustainable in domestic off-grid production, as there is no feed-in tariff for under 100 kW and no need to pay the electricity tax or transfer fee. Costs are related to the availability of suitable raw materials, making its solution only sustainable in certain contexts. Thus, promoting sustainability through offering, customer relationships and business infrastructure activities is the basis for Volter's financial performance.

3.2   Low-Tech Resource-Scarce Business Model

In this section, HPS's solution design and manufacturing, business model and sustainability approach are analysed. The gasification process has many technical challenges created by the high ash and tar content of the rice husk waste biomass, meaning that there are high maintenance requirements. HPS has, however, been able to develop a solution that functions daily when manual operations and maintenance is done. This makes the solution viable in the low-income markets, where labour force is cheap. HPS's solution model is visualised in Fig. 2, and the sustainability implication of its business model is analysed below.

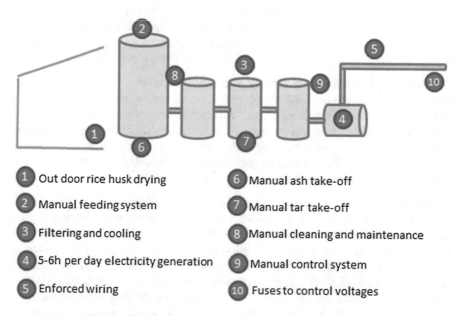

1 Out door rice husk drying   6 Manual ash take-off

2 Manual feeding system   7 Manual tar take-off

3 Filtering and cooling   8 Manual cleaning and maintenance

4 5-6h per day electricity generation   9 Manual control system

5 Enforced wiring   10 Fuses to control voltages

Fig. 2. HPS solution system model with key functionalities

Offer: Low investment cost and locally-produced night electricity

HPS offers locally-produced, affordable night energy to villages without access to electricity to replace their expensive kerosene-based lighting. In place of this inefficient lighting, HPS offers low-cost LED lights to its customers. Because of the high ash content of rice husk waste, HPS is only able to provide 5–6 h of electricity, as the system needs to be cleaned daily. It has also simplified the solution to reduce the initial investment cost. HPS is able to offer a solution with the same energy output as Volter but with almost one tenth of the up-front investment cost. HPS also reuses the ashes to produce incense sticks in its local factory, where they train and employ local women. For the most toxic side stream tar, HPS has not found a recycling or disposal solution yet. Also, HPS does not use the heat generated from the process, meaning that 70% of its total energy output is wasted.

Customer interface: Raising awareness with door-to-door service pricing

HPS's customers are living below the absolute poverty level (less than 2 USD per day), have low levels of education and lack awareness of the benefits of sustainable energy. Rural villages in India often lack access to electricity, and if there is main grid connection, it is normally only functional a couple of hours per day and never during the night when there is the highest demand for electricity. HPS has had to install fuses to control the misuse of electricity, enforce distribution lines to avoid the stealing of electricity and organise a door-to-door fee collection to improve revenue streams. Moreover, HPS had to build its own bamboo post distribution lines in order to gain informal electricity producer status. HPS intends to actively increase its rural population's awareness on the health harms of indoor air pollution (kerosene light) and to promote more sustainable solutions (LED light).

Infrastructure: Local raw materials and employment

HPS procures the local abundant rice husk waste for a fair price from local rice millers and utilises the waste as fuel. In the state of Bihar, there are no other uses for rice husk waste, as there are no process industries, such as cement industries, nearby. HPS employs four people in each plant from local villages to operate its plants and organises training for them in its own training centre. In addition, HPS employs and trains maintenance engineers in the training centre to serve various plants in the region. Finally, HPS employs local women in incense stick factories.

Financial model: Donor-based impact

HPS utilises a service-based pricing model. With less than 2 USD per month, villagers can use the lighting, charge their cell phones and use one fan for 6–8 h per day. By raising local awareness in the customer interface, employing locally and generating additional revenues using side streams in the incense stick factory, HPS is able to generate revenues so that some of its plants has become self-sustainable. Most importantly, by integrating the sustainability into offer, customer interface and infrastructure elements, HPS has been able to attract significant amounts of charity investment to donate to plants in villages and finance their operations. However, operations and maintenance costs have been high, and uncertain revenue streams and investors' requirements for profitability have increased after the series-A round in 2013. Since then, HPS has shut down 30 unprofitable power plants.

4 Discussion and Conclusions

In this study, two approaches for the sustainable design and manufacture of small-scale bioenergy solutions were explored through lenses of the business model framework in developed and developing countries. The two cases demonstrate the need for radically different design and manufacturing approaches in order to promote sustainability in different parts of the world. Comparison of similar solutions in different contexts enabled viewing both cases from multiple perspectives.

Even though the HPS solution can be argued to be more sustainable than existing kerosene and diesel solutions in low-income rural contexts in Bihar, India [8], global comparison to Volter's solution provides a slightly different picture. The HPS business model can arguably be claimed as socially sustainable, but some challenges remain for ecological sustainability compared to Volter's solution: (1) While Volter captures and utilises heat in its process, heat output is not utilised in the HPS process, making the process inefficient, as around 70% of energy content of gasification process is wasted; (2) Even though part of the ash waste in the HPS process is utilised through incense stick production, there is no disposal for the remaining ashes and especially for toxic tar in condensate water. Instead, Volter has been able to eliminate tar waste by improving the gasification process and is recycling the ash waste. At the same time, HPS is having challenges with financial sustainability, as upfront investment is dependent on donations, operation and maintenance cost remains high and there is an uncertain revenue stream from low-income customers. Volter also faces financial challenges in Finland caused by institutional constraints, but they have found other markets that are more favourable for small-scale bioenergy production.

This study provides three main contributions to the discussion of sustainable design and manufacturing. First, the frugal innovation perspective is integrated into sustainable design and manufacturing so that frugal principles [5] promote sustainability, especially in the context of low-income markets. Second, the framework of comparative studies for traditional and sustainable business models is developed. Finally, we provide insight on the implication of two design and manufacturing approaches regarding sustainability and highlight the challenges in the low-income context.

The business model concept was applied in this study to provide an analytical framework to explore the sustainability of design and manufacturing approaches in two different contexts. In this study I separated the traditional and sustainable business model questions and these should be explored further. There is no ideal framework for the business model concept, and flexible use of this concept can act as a powerful tool in an analytical work when applied correctly and connected to a theoretical discussion.

To conclude, at the end of the Pikokaasu (small-scale gasification in Finnish) project in 2013, many experts in Finland argued that small-scale gasification solutions do not have significant commercial potential, which is why public support for these solutions was discontinued. However, Volter has proven this wrong, and in 2016, they announced that they have closed a deal in Japan for 25 bioenergy plants. At the same time, the donor community has advocated the sustainability of HPS, with many praising evaluation reports [36, 37]. It seems that, in industrial community in Finland, there has been over-scepticism, while the donor community has been overly optimistic, focusing on social sustainability. Thus, Volter has been able to make the small-scale gasification process greener while HPS has only been able to romanticise it.

References

1. Panwar, N.L., Kaushik, S.C., Kothari, S.: Role of renewable energy sources in environmental protection: a review. Renew. Sustain. Energy Rev. **15**(3), 1513–1524 (2011)
2. Becker, B., Fischer, D.: Promoting renewable electricity generation in emerging economies. Energy Policy **56**, 446–455 (2013)
3. Akella, A.K., Saini, R.P., Sharma, M.P.: Social, economical and environmental impacts of renewable energy systems. Renew. Energy **34**(2), 390–396 (2009)
4. Zeschky, M., Widenmayer, B., Gassmann, O.: Frugal innovation in emerging markets. Res. Technol. Manage. **54**(4), 38–45 (2011)
5. Rao, B.C.: How disruptive is frugal? Technol. Soc. **35**(1), 65–73 (2013)
6. Radjou, N., Prabhu, J.: Frugal Innovation: How to do More with Less, 1st edn. Profile Books Ltd., London (2014)
7. Rosca, E., Arnold, M., Bendul, J.C.: Business models for sustainable innovation – an empirical analysis of frugal products and services. J. Cleaner Prod., 1–13 (2016) (In Press). http://www.sciencedirect.com/science/article/pii/S0959652616002122
8. Levänen, J., Hossain, M., Lyytinen, T., Hyvärinen, A., Numminen, S., Halme, M.: Implications of frugal innovations on sustainable development: evaluating water and energy innovations. Sustainability **8**(1), 1–17 (2015)
9. Basu, R.R., Banerjee, P.M., Sweeny, E.G.: Frugal innovation. J. Manage. Global Sustain. **1**(2), 63–82 (2013)
10. Brem, A., Ivens, B.: Do frugal and reverse innovation foster sustainability? introduction of a conceptual framework. J. Technol. Manage. Growing Economies **4**(2), 31–50 (2013)
11. Wells, P.: Sustainable business models and the automotive industry: a commentary. IIMB Manage. Rev. **25**(4), 228–239 (2013)
12. Teece, D.J.: Business models, business strategy and innovation. Long Range Plan. **43**(2–3), 172–194 (2010)
13. Bisgaard, T., Henriksen, K., Bjerre, M.: Green Business Model Innovation - Conceptualisation, Next Practice and Policy. Nordic Innovation, Oslo (2012)
14. Beltramello, A., Haie-Fayle, L., Pilat, D.: Why New Business Models Matter for Green Growth. OECD Publishing, Paris (2013)
15. Winterhalter, S., Zeschky, M.B., Gassmann, O.: Managing dual business models in emerging markets: An ambidexterity perspective. R&D Management (2015). http://onlinelibrary.wiley.com/doi/10.1111/radm.12151/full. Accessed

16. Prahalad, C.K.: Bottom of the pyramid as a source of breakthrough innovations: BOP as source of innovations. J. Prod. Innov. Manage **29**(1), 6–12 (2012)
17. Osterwalder, A., Pigneur, Y., Tucci, C.L.: Clarifying business models: origins, present, and future of the concept. Commun. Association Inform. Syst. **16**, 1–25 (2005)
18. Zott, C., Amit, R., Massa, L.: The business model: recent developments and future. Res. J. Manage. **37**(4), 1019–1042 (2011)
19. DaSilva, C.M., Trkman, P.: Business model: What it is and what it is not. Long Range Plan. **47**(6), 379–389 (2014)
20. Wirtz, B., Pistoia, A., Ullrich, S., Göttel, V.: Business models: origin, development and future. Res. Perspect. Long Range Plan. **49**(1), 36–54 (2016)
21. Baden-Fuller, C., Morgan, M.S.: Business models as models. Long Range Plan. **43**(2–3), 156–171 (2010)
22. Zott, C., Amit, R.: Business model design: an activity system perspective. Long Range Plan. **43**(2–3), 216–226 (2010)
23. Baden-Fuller, C., Mangematin, V.: Business models: a challenging agenda. Strateg. Organ. **11**(4), 418–427 (2013)
24. Halme, M., Anttonen, M., Kuisma, M., Kontoniemi, N., Heino, E.: Business models for material efficiency services: conceptualization and application. Ecol. Econ. **63**(1), 126–137 (2007)
25. Roos, G.: Business model innovation to create and capture resource value in future circular material chains. Resources **3**(1), 248–274 (2014)
26. Yunus, M., Moingeon, B., Lehmann-Ortega, L.: Building social business models: lessons from the Grameen experience. Long Range Plan. **43**(2–3), 308–325 (2010)
27. Boons, F., Lüdeke-Freund, F.: Business models for sustainable innovation: State-of-the-art and steps towards a research agenda. J. Clean. Prod. **45**, 9–19 (2013)
28. Richter, M.: Business model innovation for sustainable energy: German utilities and renewable energy. Energy Policy **62**, 1226–1237 (2013)
29. Gupta, R., Pandit, A., Nirjar, A., Gupta, P.: Husk Power Systems: Bringing light to rural India and tapping fortune at the bottom of the pyramid. Asian J. Manage. Cases **10**(2), 129–143 (2013)
30. Bocken, N., Short, S., Rana, P., Evans, S.: A literature and practice review to develop sustainable business model archetypes. J. Clean. Prod. **65**, 42–56 (2014)
31. Dubois, A., Gadde, L.-E.: Systematic combining: an abductive approach to case research. J. Bus. Res. **55**(7), 553–560 (2002)
32. Dong, L., Liu, H., Riffat, S.: Development of small-scale and micro-scale biomass-fuelled CHP systems – Literature review. Appl. Therm. Eng. **29**(11–12), 2119–2126 (2009)
33. Kirkels, A.F., Verbong, G.P.J.: Biomass gasification: still promising? a 30-year global overview. Renew. Sustain. Energy Rev. **15**(1), 471–481 (2011)
34. Lyytinen, T.: Perspectives on the international business strategies of small Finnish technology companies in developing countries: The case of small scale gasification. VTT Technol. **150** (2014). ISBN 978-951-38-8087-3
35. Pikokaasu 2013. Final presentations of VTT seminar: Electricity and heat from wood chips – Development and future of small-scale gasification technologies. http://www.vtt.fi/medialle/tapahtumat/vtt-seminaari-puuhakkeesta-s%C3%A4hk%C3%B6%C3%A4-ja-l%C3%A4mp%C3%B6%C3%A4-pienen-kokoluokan-kaasutustekniikan-kehitys-ja-tulevaisuus
36. Husk Power Systems: Lighting up the Indian rural lives. http://oikos-international.org/wp-content/uploads/2013/11/oikos_Cases_2013_Husk_Power.pdf. Accessed 14 July 2015
37. Inclusive business models: Guide to the inclusive business models in IFC portfolio: Client case studies. http://www.ifc.org/wps/wcm/connect/3af114004cc75b599498b59ec86113d5/Pub_002_IFC_2011_Case%2BStudies.pdf?MOD=AJPERES. Accessed 14 July 2015

Co-design for Resilience: Solutions, Services and Technologies for Urban Spaces

Valentina Gianfrate[✉], Jacopo Gaspari, and Danila Longo

Department of Architecture, University of Bologna, viale Risorgimento 2, Bologna, Italy
{valentina.gianfrate,jacopo.gaspari,danila.longo}@unibo.it

Abstract. Resilience design strategies anticipate significant detrimental climate change to create optimal conditions to face the continuous and deep changes of urban environment, acting on the causes (mitigation) and on the effects (adaptation). The paper illustrates a methodology that combines technological and social aspects for the transition to resilient districts and communities with the aim to co-deliver city-based solutions and services and to drive sustainable growth in vulnerable contexts.

Keywords: Resilience design strategies · Climate change · Urban transition · Technological and social infrastructures · Living Lab

1 General Framework

Several international research studies [1–5] outlined how the high concentration of population and the density of the urban environment contribute in cities' warming at local and global level creating remarkable differences with the surrounding rural areas [6]. The physical conformation of the urban lands, the anthropic emissions, the extreme events – such us flooding and heat waves – have relevant impacts on the urban environment and on the communities [7]. Following the International Panel for Climate Change (IPCC) previsions, phenomena connected to climate change will increase their intensity in the next decades improving the related risks at social and ecological level. Thus many efforts of the scientific world are spent on these issues with the aim to develop new and reliable predictive methods and tools to produce realistic scenarios [8]. As emerged during COP21 in Paris, the debate is moving from the identification of long-term objectives to the definition of short-term priorities to face climate change impacts.

Agenda 21 and ICLEI (Local Governments for Sustainability) network played an important role in increasing the awareness at national and international level on these topics and several Cities realized a first database concerning climate change effects with the purpose to reduce social, environmental and economic risks.

In this general framework, open and public spaces play the relevant role of social and urban infrastructure [9]. The associated functions are often at the core of regeneration policies to create optimal conditions to face the continuous and deep changes of urban environment. The open public space concept is currently linked to new issues: soil permeability, water management, air quality, psycho-physical health of people,

G. Campana et al. (eds.), *Sustainable Design and Manufacturing 2017*, Smart Innovation, Systems and Technologies 68, DOI 10.1007/978-3-319-57078-5_48

urban heat island [UHI] mitigation, new mobility solutions. At the same time, it assumes new values: social significance, inclusive design, co-maintenance of public spaces [10]. These new values are at the basis of new collaborative initiatives, in which communities of practice can contribute to achieve the transition of urban spaces in low-carbon and resilient realms.

At local level, governments are exploring the possibility to directly involve their citizen in transformation processes, allowing the exchange of value and knowledge among an ecosystem of actors [11, 12].

At European level, the EC perspective is to transform cities into Urban Living Labs, supporting the process of policy innovation at the municipal level through local community empowerment and through the promotion of partnership with enterprises [13]. Amsterdam and Barcelona, for instance, are seen as laboratories where companies can test their products in a living environment before commercialization. This process entails small scale testing – usually in a limited city area – and then, the implementation in the whole city and in other cities. In this case, the municipality enables companies that are developing innovative solutions in various fields (energy, mobility, lightening, urban planning, etc.) to test them in a specific district (for instance Passeig the Gracia or the 22@Barcelona or the IJburg, Zuidoost and Nieuw-West districts in Amsterdam) through pilot trials. Users can become part of the design team as 'expert of their experiences', but in order for them to take on this role, they must be given appropriate tools for expressing themselves.

2   Macro and Micro Design: Co-design for Climate Change

According to Brown [14], Tyler [15] and Davoudi [16] resilience is assumed for the purpose of this paper as the ability of a system to react to stresses and shocks with relation to cities and climate change. Therefore it is the capacity of urban systems to adequately react to unexpected events (extreme phenomena) and softer events (heat islands, energy peaks demand, etc.) as Pelling [17] reminds.

Resilience assumes that climate change is occurring, recognizing uncertainty, change and crisis as normal but instead of aiming to sustain the status quo, tries to improve the ability of individuals, communities, or systems to recognize and adapt to disturbances, to overcome them and eventually come out stronger and transformed, changing the stability landscape [18] and creating new system pathways when ecological, economic or social structures make the existing system untenable, as elaborated by Walker, Falke et al. [19, 20].

The genealogy of the concept of resilience has evolved, from its initial focus on the persistence of ecological system functions, through an emphasis on the adaptability of coupled social-ecological systems, to its most recent reorientation towards addressing the transformability of society in the face of global change [21, 22].

The Stockholm Resilience Institute identifies "principles" for building resilience in coupled socio-ecological systems, such as maintaining diversity and redundancy, encouraging learning, managing connectivity, promoting polycentric governance systems and broadening participation, calling municipalities to seek multiple modes of governing for climate change [23]. The physical dimensions of a resilient city are inherent to urban form, infrastructure, systems and services in the ways these impact coupled socio-ecological system [24] and require dedicated programming, knowledge and funding. With relation to the built environment, three predominant approaches can be out-lined: ecological resilience (a systems-based approach focusing on "the magnitude of disturbance that can be absorbed before the system changes its structure"), engineering resilience (that focuses on the stability and constancy within the system that ensures the protection of physical or human assets), and the emerging concept of evolutionary resilience (that views climate change as an element that is introducing a number of new stressors into the system that will transform how we live [16]). Resilience is increasingly being viewed as a combination of all of these [25].

Design for climate change follows precautionary approaches acting on both the causes - by reducing CO2 emissions deriving from human activities, slowing down their storage in atmosphere (mitigation) - and the effects - limiting the territorial and socio-economic vulnerability (adaptation). These are complementary strategies: more effort in mitigation actions corresponds to less requirements of adaptation and vice versa [26]. The activation of a virtuous path to resilience affects the three dimension of the city:

1. Environmental Dimension, producing a reduction of the risk level linked to flooding events, the control of soil reduction, the increase of water quality, the decrease of UHI effect.
2. Economic Dimension promoting the creation of green jobs, the transition to an attractive urban environment for new entrepreneurship, significant cost of maintenance's decrease involving directly the communities in the management of public open space, etc.
3. Social dimension, improving the urban environmental quality, the management of ecosystem services as common goods; creating link between communities and their awareness in climatic risks.

A systemic perspective is adopted to realize a multi-scale network of green/blue infrastructures and the built environment elements, taking into account the different target groups that live in that specific environment. The combination of urban quality objectives for the microclimate improvements and the mitigation of UHI, with the risk reduction and the urban comfort, will be supported by technological solutions (permeable surfaces, greening, reduction of thermal radiation) and management innovations. Musco [27] assumes cities as places addressed to test the adaptation capacity of the urban system to the effects of climate change: on the one side the urban systems have an active/ negative role in the production of negative externalities, on the other one they can be assumed as places to test innovate mitigation practices. Transformational change at smaller scales enables resilience at larger scales: critical events can be perceived as windows of opportunity innovation, and for recombining different sources of knowledge and experience, critical in periods of social-ecological transition [28]. The context of innovation is human settlements and the spatial-social conditions of their communities. This can be addressed on different scales, from neighborhoods (micro-scale) to cities (macro-scale). Examples include self-managed services for children and/or elderly care;

new forms of exchange and mutual help; community mobility systems; community gardens; networks linking consumers directly with food producers, etc. [29].

Design for resilience implies the inclusion of communities in the transition process, to act macro-scale strategies and micro-scale initiatives.

The first refers to big portions of the city involved in transition project, realizing new low-carbon settlements, transforming existing neighborhood in climate resilient districts, with a strong effort in terms of costs, stakeholders involved, community engagement, combining the physical intervention with a new urban economy.

The second is based on the adoption of micro-design initiatives in mitigation projects as effective solution in high-density urban areas (historic city centres, very dense downtowns, etc.), where the creation of green corridors or climate infrastructures could be difficult due to the need of preservation/conservation of the existing contest, or where budget constraints affect macro-scale interventions. Micro scale interventions are often carried out directly by the communities becoming the grassroots for a resilient transition.

Therefore, the city is considered as a fluid [30] and resilient system, capable of including flexible places, privileged contexts in which social relations are facilitated, scenarios that – despite having specific functions – become versatile and adaptable [31]. Multipurpose and temporary uses facilitate multiple forms of accessibility, enhancing the "educational function" of the city, inviting its citizens to the knowledge and exploration, encouraging renewals and co-designed transformations, to a more responsible behavior towards the environment, taking advantage from the educational potential of the landscape, culture and society [32].

This focus on cities, as distinct from conventional sustainable urban design and planning which focuses on urban form, urban growth, liveability, walkability, energy reduction and place-making separately and sustainable architecture which focuses on individual buildings, finds its ground in theoretical framings of cities as complex adaptive systems [33]. Framing cities as complex adaptive systems requires under-standing and taking into account the interrelationships between technologies, ecosystems, social and cultural practice and city governance in design decisions [34].

The micro-scale is based in the design of a multiplicity of interconnected and diverse experiments to generate changes in large and complex systems [35].

3   Technological and Social Infrastructures: Bologna Living Lab

The Department of Architecture of the University of Bologna is currently involved, with the Municipality of Bologna in an international project (H2020 project: ROCK project Regeneration and optimization of Creative and Knowledge cities) to find new drivers for the sustainable growth and the improvement of the resilience in the city. The project is specifically addressed to the historic downtown of Bologna with the aim to drive a transition of this urban area, following an approach that provides temporary and permanent transformations, in a perspective of sustainability and resilience as a "potential to create opportunities for new measures aimed to innovation and development" [23]. The great potential is to arrange the urban spaces as places to enable the creation of resilient communities capable of creatively responding to the changes [36], and to co-produce new civic services/products to enable the transition and an economic growth, for a new city-based manufacturing. Historically characterized by art craft production and ateliers, the area of intervention is currently afflicted by alterations and degradation phenomena due to social decay, lack of security, non-effective management of the spaces, difficulties in the application of mitigation and adaptation measures in the urban fabric, lack of social cohesion and environmental awareness, underused open spaces, low engagement of the communities. All these barriers increase the vulnerability of the whole city, caused by the intersection of human systems and the built environment. Through specific actions, the connection between the elements of sustainable district by soft (adoption of ICT, apps, participatory learning, co-design of regeneration solutions) and hard (physical modification of the space for the creation of new local services connected with the city, the creation of infrastructure for greening and mitigation, etc.) measures is pursued, to transform the area into a sustainable, creative and cultural district, supported by a knowledge based economy. Bologna will follow the examples of other European cities, which are currently involved in a transformation process that include a shift from an industrial vocation to a cultural/knowledge based one. In these cities physical changes are accompanied by outcomes on the urban economy and social aspect, such as in Bristol (http://www.rdmrsc.org.uk), in Liverpool (https://www.liverpoollep.org), in Turin (http://www.comune.torino.it/torinoplus/english/Portrait_nuovo/artecultura/index.shtml): specifically, these last two cities are involved in a mentoring program (part of ROCK project) to assist Bologna in this regeneration process.

The research is addressed to develop, at neighborhood level, social infrastructure and light and widespread technology, aimed at encouraging the active participation of citizens, associations, creative communities and private entities in the protection, management and regeneration of the local heritage, to increase the sense of belonging and responsibility that can become simultaneously ecological conversion engine and resilience factor for the entire urban system. The high concentration of people, industrial and cultural activities, also means new opportunities that can make the real city like innovation laboratories [37], where it is possible to catalyze insurgent strategies to reduce greenhouse gas emissions, test adaptation solutions, such as warning systems in case of disasters and mutual aid networks, where improve the socio-economic standards and reduce vulnerability to the impacts of climate change through mitigation and adaptation measures, for sustainable urban regeneration, and where to use knowledge as the prominent landmark and driver for socio-economic and technological dynamics. To bridge the gap between the existing urban environment and contemporary issues, such as sustainability, competitiveness, social cohesion and creativity, with a cross-disciplinary spatial approach, the methodological approach is based on horizontal integration, a mix of top-down planned elements and emergent, self-organized activities coalescing into a model of local development, creating a local added value able to promote transformative urban management and local business development including new businesses, service models and start-ups for urban promotion, accessibility, monitoring and living experiences based on a strong stakeholder engagement approach (see Fig. 1). The focus is on the entire urban morphology, with the aim to provide the opportunity to revitalize not only the physical dimension of a city, but also the economic and social ones. The priority steps of the social infrastructure design are: to identify the priority characteristics of resilience for a target community; to assess the communities' achievement of these characteristics also during crisis/disaster events; to identify the characteristics and strategies of resilient historic city; to identify the most highly rated interventions or services in building local resilience.

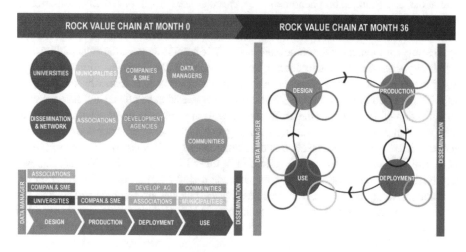

Fig. 1. ROCK project value chain.

The preparation phase is crucial to establish the target groups to be assessed; to identify the lead and supporting entities (Municipalities, Urban agencies, stakeholders) responsible for undertaking the assessment; to understand current perceptions of resilience (or non-resilience); and to assess the geographic coverage of the community. At the beginning of the project, an inventory of significant players and social factors have to be drafted, then characterization models for translating social and inventory data into subcategory (based on competences, role, communication, resources, etc.) results will be developed. The players will be involved with the community of practices in the Bologna Living Lab to experiment, explore and support the scaling-up of grassroots social innovations. This Living Lab realized in collaboration with students, communities, creative and cultural entrepreneurs, and minority groups, will be addressed to co-create urban transformation's solutions in a sustainable perspective, (facing urban regeneration of underused spaces, improvement of sustainable mobility systems and other issues), to develop new ideas and new start-up, and to increase the sense of belonging of citizens. The Living Labs, will be virtual and physical spaces for operational meetings that will allow to share with the local participants initiatives and decisions to regenerate the cultural heritage of the city. Interviews (anonymously managed and providing aggregated information) to final users (citizens, institutions, etc.) organized during focus groups or other validation instruments (conferences, role games, etc.) will be used during the assessment phase together with data coming from technological devices implemented in the LLs and all feedbacks and data will be filed in a local digital platform. Bologna Living Lab will work on: sustainable mobility, security, "green transition policies"; production and fruition of cultural events (new usages and scopes for public spaces, according to the principles of sustainability, re-use and re-cycle). The first action of Bologna Living Lab is the development of a "climate-social experiment" in the strategic nodes of the analyzed neighborhood, through the introduction of collaborative technological and social services, aimed at increasing the resilience of the area. The experiment involves:

- mapping and improving accessibility of public and semi-public open areas (as courts), considering physical accessibility and social inclusion;
- greening actions such as mitigation strategy related to climate change effects;
- the co-design and co-production of services/products to increase pedestrian mobility, promoting a "walkable neighborhood", way-finding system at the neighborhood scale, bike mobility, optimizing the security of the paths, equipping the area with protection systems (physical and ICT), capable of reducing phenomena of theft and damage; systems for resource efficiency in the area, by adopting solutions for water saving, energy-efficient lighting, intelligent management, to tackle climate risk.
- the test of the co-delivered solutions.

This experiment will be then implemented in a long-term plan to engage citizens of neighborhoods of Bologna in order to: build up consensus on the project and implement a process of citizens democratic participation in its realization; analyze the user acceptance of the proposed sustainable and resilient solutions; co-design with citizens and stakeholders ideas for new solutions to face specific neighborhood challenges.

The social infrastructure will exploit the main principles of co-creation and will rely on the implementation of localized infrastructures like networks of stakeholders that share a problem and actively co-produce its solution/s by exploiting different forms of innovation and economies: social and technological innovation, sharing and collaborative services, collaborative decision making.

4   Future Research Branches and Conclusion

The methodology is tested in the operational environment, identifying the most effective strategies and tools to improve sustainability and resilience in the pilot neighborhood. Permanent and temporary initiatives will follow an integrated management plan for sustainable urban context (i.e. carbon neutral cultural events, slow mobility, sustainable approaches for community-led regeneration). The combination of technological and social infrastructures and services supports the integration between environmental and sustainable growth to enhance the local development and the regeneration process, moving from a proactive engagement of the communities.

The design of social and technological infrastructure is based on a match between enabler-driven and user-driven approach [38], fostering a multi-party cooperation and a co-creation of value with the communities. This match will produce different outcomes:

- new cooperation opportunities between PA and Citizenship with specific agreements for the management of the urban spaces as a common good (following the Common Goods Regulation of Bologna Municipality);
- effective and shared policies able to accelerate the regeneration of vulnerable districts in Bologna;

- improvement of accessibility and social cohesion support (i.e. active and visible participation of women and ethnic minorities in temporary events and initiatives);
- increased awareness and participation in local decision-making and wider civic engagement in historic city (i.e. immigrants networks and women associations working with stakeholders in the regeneration of public spaces; collective initiatives under the Local Common Good agreements);
- increasing in the attractiveness and sustainability of the areas;
- improvements in employment opportunities;
- new financing opportunities (shared business collaboration platform; PPPs creation).

During ROCK project, Bologna Municipality and its local partners become agents of change, innovators capable of activating actions to trigger a positive spiral of interaction with external parties, stakeholders and end-users. Certainly this mechanism is not simple and in this paper we shall argue the points associated to the implementation or rather, the potential institutionalization of design-thinking for social innovation aiming at renovation and valorization of public spaces.

Further research branches linked to the activation of this combined infrastructure envisage the definition of an impact evaluation method to check how this approach interacts and influences social, environmental and economic sides, and to test the replication potential of the methodology adopted at regional and national level.

References

1. Dirmeyer, P.A., Niyogi, D., de Noblet-Ducoudré, N., Dickinson, R.E., Snyder, P.K.: Impacts of land use change on climate. Int. J. Climatol. **30**(13), 1905–1907 (2010)
2. Pataki, D.E., Emmi, P.C., Forster, C.B., Mills, J.I., Pardyjak, E.R., Peterson, T.R., Thompson, J.D., Dudley-Murphy, E.: An integrated approach to improving fossil fuel emissions scenarios with urban ecosystem studies. Ecol. Complex. **6**, 1–14 (2009)
3. C40 Cities, Unlocking Climate actions in megacities, C40 (2016). http://www.c40.org/researches/unlocking-climate-action-in-megacities
4. UNEP, Climate Change. The role of cities. Involvement, Influence, Implementation, UNEP and UN-Habitat (2009). http://www.unep.org/urban_environment/PDFs/RoleofCities_2009.pdf
5. McDonald, R.I.: Global urbanization: can ecologists identify a sustainable way forward? Front. Ecol. Environ. **6**, 99–104 (2008)
6. Oke, T.R.: Boundary Layer Climates, Routledge (2002)
7. Willems, P., Arnbjerg-Nielsen, K., Olsson, J., Nguyen, V.T.V.: Climate change impact assessment on urban rainfall extremes and urban drainage: methods and shortcomings. Atmos. Res. **103**, 106–118 (2012)
8. IPCC, Climate Change 2014: Impacts, Adaptation, and Vulnerability. Part A: Global and Sectoral Aspects. Contribution of Working Group II to the Fifth Assessment Report of the Intergovernmental Panel on Climate Change, Cambridge University Press, Cambridge, United Kingdom and New York, NY, USA, pp. 1–32 (2014)
9. http://ipcc-wg2.gov/AR5/images/uploads/WG2AR5_SPM_FINAL.pdf
10. Gill, S.E., Handley, J.F., Ennos, A.R., Pauleit, S.: Adapting cities to climate change: the role of the green infrastructure. Built Environ. **33**(1), 115–133 (2007)
11. van der Buitendag, W.J., Malebane, T., de Jager, L.: Addressing knowledge support services as part of a living lab environment. Informing Sci. Inform. Technol. **9**, 221–241 (2012)

12. Feurstein, K., Hesmer, A., Hribernik, K.A., Thoben, K.D., Schumacher, J.: Living Labs – A New Development Strategy in European Living Labs - A New Approach for Human Centric Regional Innovation, Chapt. 1, pp. 1–14. Wissenschaftlicher Verlag, Berlin (2008)

13. Nesti, G.: Urban living labs as a new form of co-production. Insights from the European experience. In: ICPP - International Conference on Public Policy II Milan, 1–4 July 2015

14. Brown, E.D., Byron, W.K.: Resilience and resource management. Environ. Manage. **56**, 1416–1427 (2015)

15. Shaw, K.: "Reframing" resilience: challenges for planning theory and practice. Plan. Theory Pract. **13**(2), 308–312 (2012)

16. Davoudi, S.: Resilience: a bridging concept or a dead end? Plan. Theor. Pract. **13**(2), 299–307 (2012)

17. Pelling M.: Adaptation to climate change: from resilience to transformation. Routledge (2010)

18. Gallopín, C.: Linkages between vulnerability, resilience, and adaptive capacity. Glob. Environ. Change **16**(2006), 293–303 (2006). Elsevier Ltd.

19. Walker, B., Holling, C.S., Carpenter, S.R., Kinzig, A.: Resilience, adaptability and transformability in social–ecological systems. Ecol. Soc. **9**(2), 5 (2004)

20. Folke, C., Carpenter, S.R., Walker, B., Scheffer, M., Chapin, T., Rockström, J.: Resilience thinking: integrating resilience, adaptability and transformability. Ecol. Soc. **15**(4), 20 (2010)

21. Keck, M., Sakdapolrak, P.: What is social resilience? lessons learned and ways forward. erdkunde – archive of scientific. Geography **67**(1), 5–19 (2013)

22. Frerks, G., Warner, J., Weijs, B.: The politics of vulnerability and resilience, Ambiente & Sociedade, Print version ISSN 1414-753X Ambient. soc., vol. 14 no. 2, São Paulo, July/December 2011

23. Boeri, A., Gianfrate, V., Longo, D., Lorenzo, V.: Resilient communities. Social infrastructures for sustainable growth of urban areas. a case study. Int. J. Sustain. Dev. Plan. **12**(2), 227–232 (2017)

24. Stockholm Resilience Centre, Applying resilience thinking. Seven principles for building resilience in social-ecological systems. www.stockholmresilience.org/download/18.10119fc11455d3c557d6928/1398150799790/SRC+Applying+Resilience+final.pdf

25. Wholey, F.: A framework for assessing and communicating the costs and benefits of resilient design strategies. Res. J. **7**(1), 7–18 (2015)

26. Parry, M.L., Canziani, O.F., Palutikof, J.P., van der Linden, P.J., Hanson, C.E.: IPCC, Contribution of Working Group II to the Fourth Assessment Report of the Intergovernmental Panel on Climate Change. Cambridge University Press, Cambridge (2007). http://www.ipcc.ch/pdf/assessment-report/ar4/wg2/ar4_wg2_full_report.pdf

27. Musco, F., Patassini, D.: Mitigazione e adattamento ai cambiamenti climatici: valutazioni di efficacia di piani e politiche in Usa, in Europa e in Italia, Maggioli (2012)

28. Folke, C., Carpenter, S.R., Walker, B., Scheffer, M., Chapin, T., Rockström, J.: Resilience thinking: integrating resilience, adaptability and transformability. Ecol. Soc. **15**(4) (2010). http://www.ecologyandsociety.org/vol15/iss4/art20/

29. Meroni, A.: Creative Communities. People Inventing Sustainable Ways of Living, Edizioni Polidesign, Milan, Italy (2007)

30. Carta, M.: Il paradigma della città fluida, in M. Carta (a cura di), L'Atlante dei Waterfront. Visioni, paradigmi, politiche e progetti integrati per i waterfront Siciliani e Maltesi, Palermo, DARCH (2013)

31. De Angelis, A., Izzo, M.V.: Lo spazio pubblico acceleratore e generatore del rinnovo della città resiliente, in Lo spazio pubblico acceleratore e generatore del rinnovo della città resiliente, a cura di Sbetti, F., Rossi, F., Talia, M., Trillo, C., Urbanistica Dossier 4, INU Edizioni (2013)

32. Armato, F.: Pocket Park. Spazi tra gli edifici, Tesi di Dottorato, Università degli Studi di Firenze, Firenze (2013)
33. Bettencourt, L., West, G.: A unified theory of urban living. Nature **467**, 912–913 (2010)
34. Ceschin, F., Gaziulusoy, I.: Evolution of design for sustainability: From product design to design for system innovations and transitions. Des. Stud. **47**, 118–163 (2016)
35. Manzini, E., Rizzo, F.: Small projects/large changes: Participatory design as an open participated process. CoDesign **7**(3–4), 199–215 (2011). Taylor & Francis
36. UN-HABITAT, Cities and Climate Change: Global Report on Human Settlements 2011, Series: Global Report on Human Settlements, Earthscan (2011)
37. Leminen, S., Westerlund, M., Nyström, A.G.: Living labs as open-innovation networks. Technol. Innovation Manage. Rev. (TIM) **2**(9), 6–11 (2012). http://timreview.ca/article/602
38. Hillgren, P.-A., Seravalli, A., Emilson, A.: Prototyping and infrastructuring in design for social innovation. CoDesign **7**(3–4), 169–183 (2011)

Digital Redistributed Manufacturing (RdM) Studio: A Data-Driven Approach to Business Model Development

Christopher Turner[1(✉)], Ashutosh Tiwari[1], Jose Luis Rivas Pizarroso[1], Mariale Moreno[1], Doroteya Vladimirova[2], Mohamed Zaki[2], and Martin Geißdörfer[2]

[1] Manufacturing Department, Cranfield University, Bedfordshire, MK43 0AL, UK
{c.j.turner,a.tiwari,J.Rivas-Pizarroso,m.moreno}@cranfield.ac.uk
[2] Institute for Manufacturing (IFM), University of Cambridge, Cambridge, CB3 0FS, UK
{dkv21,mehyz2,ml733}@cam.ac.uk

Abstract. The theme of Redistributed Manufacturing (RdM) has gained in interest over recent years. While much research has taken place into the effects of RdM on current manufacturing models very few people have proposed new business models for this concept. The RdM studio is a new approach to business model development that will allow future users to dynamically incorporate data and experiment with new redistributed manufacturing scenarios. An RdM System Dynamics (SD) model is illustrated (as a potential constituent model of the RdM studio) with a case study called ShoeLab that explores RdM scenario generation through parameter sets utilising the SD modelling method. This research provides a valuable platform on which future models and scenarios may be derived.

Keywords: Redistributed manufacturing · Business model development · System dynamics

1 Introduction

The theme of Re-Distribution in manufacturing has gained in interest over recent years. The ability to manufacture products through the participation of a number of geographically dispersed production sites is a central feature of this paradigm [1]. Xu [2] acknowledges the role of distributed manufacturing in his promotion of the cloud manufacturing concept which envisages the use of cloud technologies to change the operational geography of production. The UK Engineering and Physical Sciences Research Council (EPSRC) [3] have a working definition of Re-Distributed Manufacturing (RdM) as 'Technology, systems and strategies that change the economics and organisation of manufacturing, particularly with regard to location and scale'. Moreno and Charnley [4] corroborate this definition and state that RdM "enables a connected, localised and inclusive model of consumer goods production and consumption that is driven by the exponential growth and embedded value of big data". It is certain that developments in technology, that have progressed automation and virtualisation of business activities in the service industry, when applied in manufacturing will have a transformative effect and enable new forms of production. Both Industry 4.0 and Industrial Internet Consortium visions of future manufacturing recognise the role of distributed manufacturing in

© Springer International Publishing AG 2017
G. Campana et al. (eds.), *Sustainable Design and Manufacturing 2017*, Smart Innovation,
Systems and Technologies 68, DOI 10.1007/978-3-319-57078-5_49

delivering flexible, personalized and on demand production in 21st century [5, 6]. It is still an open question as to how industry will respond in the task of developing new business models and adapting existing ones to maximize the opportunities presented by this new paradigm. It is the aim of the RdM Studio project to provide a new model development environment for the manufacturing industry enabling the active data driven development of new business models utilising System Dynamics simulation (and potentially other modelling formats in the future). The provision of an SD business model for RdM is a unique in literature, building on contributions in the area of SD business model development. This paper details the background research in the area that has influenced this work and provides an RdM business model utilising a System Dynamics approach (for use in the RdM studio) and illustrates its use with a case study and As-Is and To-Be scenarios based on experimentation with parameter settings.

2   Background Literature

In addition to their aforementioned definition Moreno and Charnley [4] identify three different types of RdM: Distributed Production and Services; Connected Production and Services; Localised Production and Services. The concept of Servitisation has also influenced RdM creating a need for distributed production in order to efficiently provide combined product service offerings. According to Baines et al. [7] servitisation is defined as "the evolution of product identity based on material content to a position where the material component is inseparable from the service system". Product Service Systems (PSS) are a particular type of servitisation that include one or more product functionality with one or more associated service functionality. While a company can decide to offer PSS from the start, the usual path towards such an offering is that a company that already provides either products or services adds the missing component to its offerings [8]. System Dynamics (SD) as a simulation medium for business models is not a new field [9] however its application to the field of RdM is new. The work of Abdelkafi and Täuscher [10] extends SD business model development to take in environmental sustainability factors. The move towards more sustainable forms of production is another factor linked with the development and adoption of RdM. The circular economy is an expression of the need for a holistic treatment of a product's production, use and disposal, recycling or reuse to maintain economic growth and environmental protection. A circular economy aims at the realisation of closed-loop resource flows in the entire economic system [11, 12]. It can be argued that the manufacturing strategies put forward as Industry 4.0 and the Industrial Internet provide compelling visions of future production systems while integrating RdM and the Circular Economy practice as components and potential goals. With Industry 4.0 production will be transformed: isolated and individually optimised cells will become a fully integrated, automated and holistically optimised production system. This will result in more efficient production network relationships not only between agents but also in human-machine interaction [13, 14]. The convergence of the virtual and physical worlds has given rise to the Smart Factory. This integrates artificial intelligence, machine learning, automation of knowledge work and machine-to-machine communication with the manufacturing process. These advances in the way machines and other objects communicate, and the movement of some types of decision making from humans to technical systems mean that manufacturing processes become "smarter" [15]. Enabling RdM as a business model has to this point has attracted little attention in literature. This paper aims to address this gap and provide a business model for RdM taking into account developments highlighted in this review.

3   Framework and Methodology for the RdM Studio

As mentioned earlier, the RdM Studio project aims to provide a new tool for the manufacturing industry that is enabling the active data driven development of new business models by utilising simulation. As a contribution to this aim this paper details the development of a business model for RDM based on the ShoeLab case study utilising System Dynamics (SD) as the format for model. While currently in development Fig. 1 illustrates the methodology underpinning the studio as an environment for the development of RdM business models. Current business model design relies on the knowledge and intuition of industry experts and while such experience is not redundant in the RdM Studio the ability to incorporate quantitative measures within the model as parameters so expediting this process. The complete RdM Studio will allow users to interact with data and develop RdM business models suitable for their organization's needs. The research in this paper details the development of the initial RdM SD business model for the studio tool. A case study drawn from the ShoeLab project is used to develop the initial model.

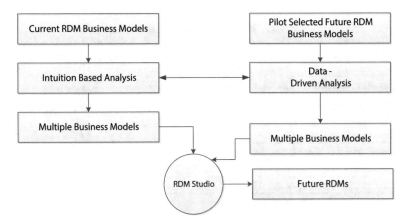

Fig. 1. RdM studio guiding methodology

Based on the development of smart and sustainable shoes, utilising digital intelligence, redistributed manufacturing, and a product-service system approach, ShoeLab is the product of a feasibility study funded by the EPSRC RECODE network [16]. The methodology for the development of the initial RdM SD model involves a qualitative and quantitative analysis followed by the development of the model its evaluation and validation by a focus group that is composed of researchers in industrial sustainability. The Qualitative Quantitative analysis involved both literature review and dialog with ShoeLab project experts to determine appropriate mathematical models and parameter sets for use in the model. An As-Is scenario has been developed with the RdM SD model to describe the current business model envisaged for ShoeLab. The To-Be scenario utilises a set of parameters with given settings (shown in Sect. 4 as Table 1) as an initial experimentation set. The model utilises the AnyLogic [17] implementation of the System Dynamics (SD) simulation method and utilises Agent Based modelling to describe customer segmentation and simulate the client's demand for services.

4   RdM Model Development

From an initial qualitative analysis it was possible to identify the initial actors within the model:

- **Producer:** Manufactures the Shoes according to the customer demand. This element integrates the supply chain, fabrication and transport.
- **Retailer**: Is in charge of the retail and delivery of the product. It delivers the product to the client and receives the subscription fee paid by them.
- **Service provider:** Ensures services such as repairing or refashioning of the shoe. Again, responding to the customer's demand for these services.
- **Data Manager:** Pays a fee for the data gathered from the clients' shoes.

Table 1.   RdM SD model input parameters

Name	Value
Advertisement effectiveness (relative measure, $1.0 = 100\%$)	0.009
Adoption fraction (of customers adopting)	0.15
Product batch size	200
Daily salary (£)	10
Equipment capacity (£)	12
Number of machines (Fabrication)	3
Elastomer cost per shoe (£) (base material for 3D printed shoes)	9
Revenue per person per day of data sharing	0.005
Monthly subscription fee (£)	15
Cost: new shoe (£)	7
Number of operators (workers)	$7 \rightarrow 1$
Population size (Potential customer base)	1000
Recycling factor (share of materials recycled)	$0 \rightarrow 0.7$
Cost: repair shoe (£)	4
Cost: refashion shoe (£)	4
Cost: replace shoe (£)	4
Technology cost per shoe (£)	8
Transport capacity (£)	25
Transport cost per item (£)	1
Transport necessary (for goods and materials)	True $\rightarrow$ False

Recycling Partner: Processes the material coming from 'thrown-away' shoes.

The RdM SD simulation of the ShoeLab case study has five main objectives: Obtain the temporal response (retard) of the system to the client demand; Analyse the cost implications for the PSS approach and its profitability; Make recommendations on product prices and capacity requirements for ShoeLab; Enable data-driven experimentation to allow for multiple scenarios. All the input parameters and variables take the value showed in Table 1.

4.1 The Model

The model is composed of five main subsystems (Marketing, Production, Customer Service, Material Supply and Accounting) modelled with a System Dynamics approach and one additional subsystem describing Customers modelled with an Agent Based approach. The marketing subsystem details how publicity and word of mouth affect the rate of adoption of the product service system; in effect the demand for hiring the service is modelled here. The marketing subsystem is then linked to the production subsystem which how manufacturing and transport capabilities affect the lead time and delivery time. The raw material resources of the system are modelled by the material supply subsystem which includes inventory management, the material supply and the recycling of wasted products with the potential to provide feedback loops into other subsystems. The subsystems are now described in more detail.

Marketing Subsystem
The aim of this subsystem is to model the demand for hiring the service. Publicity and word of mouth affect the rate of adoption of the product. Figure 2 shows the simplified system flow and causal loops for this subsystem. The "Order/Adoption Rate" depends on the adoption due to "Advertising" plus "Word of Mouth" (WOM); the "Fulfilment Rate" depends on the Production, described below. This subsystem enables the modelling of the system transient response to the demand (retard), driven by the "Fulfilment

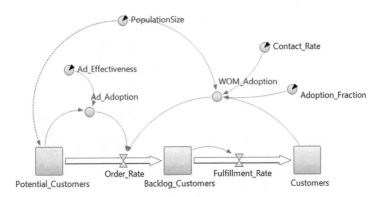

Fig. 2. Marketing subsystem

Rate". The customer demand subsystem focusses on satisfying customer demand for services rather than products and is linked into the production subsystem.

Production Subsystem

The purpose of this subsystem is to measure how manufacturing and transport capabilities affect the lead time and delivery time. The primary output of this model will be the "Fulfilment Rate". Figure 3 depicts a simplified version of the SD diagram for this part. In this case, the stocks represent products, going from "Backlog Orders", which represent the pending orders computed by the Marketing Subsystem; to the products fulfilled to the "Retailer".

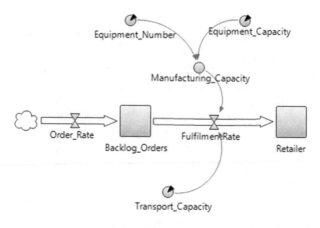

Fig. 3. Production subsystem

Customer Demand, Accounting and Material Supply and Customer subsystems

The customer demand subsystem acts as analogue to the production subsystem but focusses on satisfying customer demand for services rather than products. In this subsystem, the stocks contain services, where "Backlog Services" includes the pending services to be completed (shown in Fig. 4). The raw material resources of the system

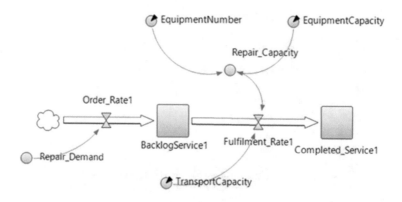

Fig. 4. Customer service subsystem

are modelled by the material supply subsystem which includes inventory management, the material supply and the recycling of wasted products. An accounting subsystem (not shown) is provided to calculate the costs within the model and output a set of statistics to quantify the financial impact of the scenario. It collects data from the primary cost driver and income sources. An Agent-Based Model (ABM) is used to describe the customer subsystem. The aim of this subsystem is to model a population of customers by means of statecharts.

The statecharts determine the state of every customer in terms of service requirements. The transition between states is triggered by timeouts or conditions, varying on each client depending on the distribution of the three ShoeLab customer types ("Fashionable", "Active" or "Body Builder").

Validation and Scenario Generation
Two different scenarios ("as-is" and "to-be") were created in the simulation. The as-is scenario was based on the Shoe Lab business model. The "to-be" scenario was obtained by applying data-driven experimentation to the "as-is" scenario. From this starting point, in order to predict a future RdM business model, it was assumed that 3 different sources of data would be subject to change in the future:

- **Recyclability**. Recyclability of raw materials will be a factor of increasing importance in the future of RdM in order to ensure circularity.
- **Transportation**. RdM aspires to minimise shipping costs by fabricating on site.
- **Level of automation.** Smart Factories will reduce the share of human workforce in production processes.

4.2   Results

The dynamic response measures how a system reacts to a given input such as the manufacturing/service system reaction to the fulfilment of customer demand. In this case, the input studied for the SD model will be the output from the Customer subsystem.

With the to-be model the "new product demand" starts with an abrupt increase with the release of the product to the market, this demand cannot be fulfilled instantaneously by the system until its stabilisation. A similar observation is true for the "service demand" (triggered by the Agent Based Customer subsystem). In the as-is model, with the planed £15 monthly one year is required to compensate the initial investment costs for manufacturing the shoes and reach break-even, from where the revenues begin to exceed the costs (shown in Fig. 5). The total cost is higher in the beginning, when the new customers start demanding their shoes. Once the "new product demand" has been fulfilled, the cost increase slows down, despite continuing to grow due to the services provided. With the to-be scenario labour costs are drastically reduced due to process automation, requiring just supervision and maintenance of the machines. The reduction of material costs accrued is explained by the recycling activities. Operational costs decrease due to minimisation of transport requirements and the automation of processes.

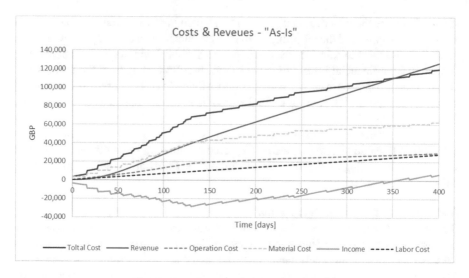

Fig. 5. Cost-Revenue histogram ("As-Is")

This improvement in costings allows for an earlier break-even point (210 days) and higher profit margins (shown in Fig. 6) and a reduction is costs of 40%. The output of the to-be scenario highlights potential benefits for future RdM, providing quantitative measures in terms of cost savings and income improvements.

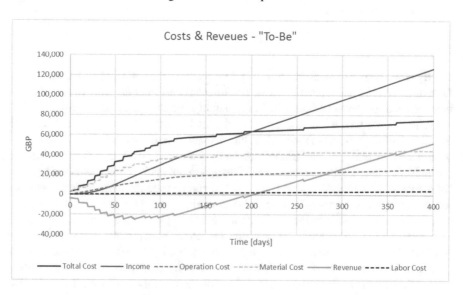

Fig. 6. Cost-Revenue histogram ("To-Be")

5   Discussion

It has been shown how ShoeLab can benefit from scenario generation through the use of SD models. The RdM SD model allows a quantitative analysis that can be used to make recommendations and help design distribution networks and production capabilities. There are at present a number of limitations in the model as a number of assumptions have had to be made when developing the RdM SD model about the market and related parameters (based on the ShoeLab case study). The operation of RdM in practice is based on current understanding in the research community, future interpretations of this concept influenced by rapidly changing technology may differ in their implementation. It has also been assumed that the technologies employed are at maturity and some parameters were estimated based on comparative observations with other case studies. Future research will involve the establishment of more accurate figures for the costs of transportation and 3-D printing for incorporation to the model. A Monte Carlo analysis would be a good addition to the model to evaluate the risk level encountered in different scenarios. Overall the developed RdM SD model provides a new level of flexibility for RdM scenario generation and experimentation. This research provides a template that can be extended and developed further to provide a set of SD (and other simulation type) models for the RdM Studio environment.

6   Conclusions

An initial SD simulation model has been created that enables data-driven experimentation for RdM. ShoeLab has been identified as a suitable case study for the simulation. The SD simulation model scenarios have succeeded in describing the ShoeLab environment, representing its value chain and connections. In order to build the model, it was necessary to identify the primary constraints and elements that were going to be most relevant when simulating an RdM manufacturing environment. This task has been completed successfully as the tool enables the identification and modification of input parameters that have the most impact. Data driven decisions have been used to simulate different RdM scenarios. To develop the "to-be" scenario, it has been necessary to select the main parameters whose change could predict the future of RdM. The model shows how RdM will help future manufacturing. The simulation and comparison between the current and the future scenario show promising results: moving towards an RdM environment will contribute to capacity improvements with the potential for cost savings. Acquiring finer grained input data would contribute to higher level of model accuracy when extracting information in terms of capacity planning, inventory management and finance analysis. This research provides a valuable platform on which future models and scenarios may be derived within an RdM Studio environment.

Acknowledgements.   The Engineering and Physical Sciences Research Council (EPSRC- EP/M017567/1) have funded this research as the feasibility study 'Digital Re-Distributed Manufacturing (RdM) Studio' under the Network on Re-distributed Manufacturing Consumer Goods and Big Data (RECODE) project.

References

1. Srai, J.S., Kumar, M., Graham, G., Phillips, W., Tooze, J., Tiwari, A., Ford, S., Beecher, P., Raj, B., Gregory, M., Tiwari, M.: Distributed manufacturing: scope, challenges and opportunities. Int. J. Prod. Res. **54**(23), 6917–6935 (2015)
2. Xu, X.: From cloud computing to cloud manufacturing. Robot. Comput.-Integr. Manuf. **28**(1), 75–86 (2012)
3. EPSRC 2016. https://www.epsrc.ac.uk/newsevents/pubs/re-distributed-manufacturing-workshop-report/. Accessed 11 Nov 2016
4. Moreno, M., Charnley, F.: Can re-distributed manufacturing and digital intelligence enable a regenerative economy? An integrative literature review. In: Setchi, R., Howlett, R., Liu, Y., Theobald, P. (eds.) Sustainable Design and Manufacturing 2016, vol. 52, pp. 563–575. Springer International Publishing, Cham (2016)
5. German Federal Government, the new high-tech strategy innovations for Germany (2016). https://www.bmbf.de/pub/HTS_Broschuere_eng.pdf. Accessed 17 Aug 2016
6. Posada, J., Toro, C., Barandiaran, I., Oyarzun, D., Stricker, D., De Amicis, R., Pinto, E.B., Eisert, P., Dollner, J., Vallarino, I.: Visual computing as a key enabling technology for industrie 4.0 and industrial internet. Comput. Graph. Appl. IEEE **35**(2), 26–40 (2015)
7. Baines, T.S., Lightfoot, H.W., Evans, S., Neely, A., Greenough, R., Peppard, J., Roy, R., Shehab, E., Braganza, A., Tiwari, A., Alcock, J.R.: State-of-the-art in product-service systems. Proc. Inst. Mech. Eng. Part B: J. Eng. Manuf. **221**(10), 1543–1552 (2007)
8. Kuijken, B., Gemser, G., Wijnberg, N.M.: Effective product-service systems: a value-based framework. Ind. Mark. Manag. **60**, 33–41 (2016)
9. Cosenz, F.: Supporting start-up business model design through system dynamics modelling. Manag. Decis. **55**(1), 57–80 (2017)
10. Abdelkafi, N., Täuscher, K.: Business models for sustainability from a system dynamics perspective. Organ. Environ. **29**(1), 74–96 (2016)
11. Lieder, M., Rashid, A.: Towards circular economy implementation: a comprehensive review in context of manufacturing industry. J. Clean. Prod. **115**, 36–51 (2016)
12. Macarthur, E.: Towards the Circular Economy: Opportunities for the Consumer Goods Sector. Ellen MacArthur Foundation (2013)
13. The Government Office for Science 2013 Future of manufacturing: a new era of opportunity and challenge for the UK. https://www.gov.uk/government/publications/future-of-manufacturing Accessed 11 Nov 2016
14. Rüßmann, M., Lorenz, M., Gerbert, P., Waldner, M., Justus, J., Engel, P., Harnisch, M.: Industry 4.0: The Future of Productivity and Growth in Manufacturing Industries. Boston Consulting Group (2015)
15. The CRO Forum: The Smart Factory – Risk Management Perspectives (2015). http://www.thecroforum.org/the-smart-factory-risk-management-perspectives/. Accessed 11 Nov 2016
16. EPSRC 2016 The RECODE Network http://www.recode-network.com/. Accessed 11 Nov 2016)
17. AnyLogic 2016. http://www.anylogic.com/. Accessed 11 Nov 2016

Exploring Disruptive Business Model Innovation for the Circular Economy

Anna Aminoff[1(✉)], Katri Valkokari[2], Maria Antikainen[2], and Outi Kettunen[1]

[1] VTT Technical Research Centre of Finland, Espoo, Finland
{anna.aminoff,outi.kettunen}@vtt.fi
[2] VTT Technical Research Centre of Finland, Tampere, Finland
{katri.valkokari,maria.antikainen}@vtt.fi

Abstract. Recently the concept of the Circular Economy (CE) has attracted growing interest as a novel economic model aiming to foster sustainable economic growth, boost global competitiveness, and generate new jobs. A system-wide disruptive innovation shaping new ecosystems and changing the whole process of value creation is needed to tackle the current challenges and transformation to the CE. This paper asks how disruptive business model innovations work as a change mechanism for the CE. The paper develops a conceptual framework for shaping the industrial systems towards CE ecosystems and proposes how value circles and co-creation of value with a variety of partners are crucial aspects in enabling CE. The paper highlights that the concept of value circles would be beneficial in clarifying the difference to linear value chain models and the co-existence of several overlapping value circles.

Keywords: Circular economy · Business model innovation · Value circles · Business ecosystems

1 Introduction

The dominant linear economic model is running out of road, with non-renewable natural resources dwindling and causing volatility in prices. Recently the concept of the Circular Economy (CE) has attracted growing interest as a novel economic model aiming to foster sustainable economic growth, boost global competitiveness, and generate new jobs [1]. In the concept of CE, recovery and valorization of waste allow for reusing materials back into the supply chain, thereby decoupling the economic growth from environmental losses [1]. A system-wide disruptive innovation that shapes markets and the processes of value creation [2] is needed to tackle the current challenges, as part of the transformation towards a CE [3]. However, this system-level change sets big challenges for established companies, and might even disrupt the usefulness of their existing capabilities, networks, and business models [3, 4]. Well-known examples of disruptive business models based on a sharing economy, such as Uber and AirBnB, are already changing business logics and ecosystems. As a system-level phenomenon, a CE business model requires interaction between all ecosystem actors, including both the core-business network and other stakeholders [5].

© Springer International Publishing AG 2017

G. Campana et al. (eds.), *Sustainable Design and Manufacturing 2017*, Smart Innovation, Systems and Technologies 68, DOI 10.1007/978-3-319-57078-5_50

Current CE research is still scattered, and mainly focusing on resource-efficiency or recycling [1, 6, 7]. In their systematic literature review, Lieder and Rashid [6] identified 156 articles related to a CE between 1991–2015 and observed that a major part of these attempts has been lacking a systematic approach to the CE. The critical viewpoints necessary to create an in-depth understanding of the phenomenon and to transform industrial systems towards a CE have been neglected. To answer this call, the aim of the paper is to explore: *How does a disruptive business model (co-) innovation work as a mechanism that changes industrial systems towards CE ecosystems? How to support business model co-innovation that utilizes opportunities created by the breakthroughs of a CE?* The paper creates a conceptual framework for shaping industrial systems towards CE ecosystems. Ever since the seminal work of Christensen [2], the discussion of disruption has mainly focused on disruptive technologies, whereas our aim is to explore phenomena related to re-configuring entire industrial systems, i.e. their change towards CE ecosystems, rather than being simply a technology push. This paper approaches the phenomena from the meso level - the viewpoint of management of a company - and emphasizes that managers need to understand the mechanisms to be able to lead the transformation. This paper is conceptual in nature.

2   Circular Economy

Closing the loop and circularity are not new concepts, and attempts to respond to challenges of resource scarcity, environmental impact or economic benefits have been made [1]. The concept of CE can be traced back to different schools of thought [6, 8], but the research on CE implementation has been and is still mainly rooted in industrial ecology [1]. ReRecently, the CE discourse has started to emphasize the economic aspect, seeing CE business models as enablers to create a competitive advantage [1, 9]. As of today, CE is commonly defined as an industrial system that is restorative or regenerative by intention and design [6, 9]. This definition considers both environmental and economic benefits simultaneously under the notion of "regenerative performance requiring high-quality circulation of technical nutrients while ensuring safe entry of bio nutrients into the biological sphere" [6 p. 37]. We avoid here repeating the literature review of a CE which can be found in earlier contributions [10], and focus on the aspects related to transforming industrial systems towards a CE.

The transformation towards a CE requires changes at all levels, i.e. micro, meso and macro levels [1, 11]. Research on such institutional change is based on understanding how institutions emerge, transform, and are extinguished and how institutional processes can influence a change toward sustainable business [12, 13]. Thus, transforming industrial systems towards the CE depends, on the one hand, on policymakers and their decisions. On the other hand, it depends on the actions of companies and company networks to introduce circularity into their business models, since developing circular business creates the need for new business models, including new revenue models, in a more complex system of actors [14]. However, currently, there is only limited understanding of how the proclaimed shift in the economy will translate into new business models and value chains [14]. Unlike in today's buy-and-consume economy, durable products are leased, rented or shared wherever possible [5, 9]. The shift towards product service systems (PSS) is suggested to be one of the key solutions in accelerating the transformation, as companies have incentives to create products that have a long service life, which are used intensively and which are also cost- and material-effective [15, 16], and creating a much-needed resource revolution. However, on the negative side of servitization, a "rebound" effect might appear, referring to a systemic responses to a measure taken to reduce environmental impacts that offsets the effect of the measure [17].

3   Research Design

To answer the research questions, a conceptual framework is advanced by deploying an in-depth literature review, which combines several streams of literature. The process was divided into three steps (see Fig. 1):

(1) Identification of the state of the art
(2) Identification of the main elements and categorization of the initial body of literature according to the identified main elements
(3) Synthesis and development of the framework

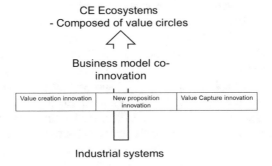

Fig. 1. Disruptive business model co-innovation as a mechanism that changes industrial systems towards CE ecosystems

(1) The first step identified the body of knowledge needed to obtain the answers for the research questions in the next steps. The literature search was conducted using keywords in the following academic databases: Google Scholar, Ebsco, Emerald, and Science-Direct Elsevier. These databases cover the majority of scientific journals of interest. Key words included variations on terms such as circular economy, value networks, disruptive innovation, co-innovation, business model, business ecosystem, value chains and combinations of those. Finally, a snowballing technique was applied, and specifically, previous literature reviews were used to identify additional articles falling outside the database searches. Also, a complementary manual search was conducted on the websites of contributors to CE to identify other relevant papers, reports and books. (2) In the second step, we identified the main elements of disruptive business model co-innovation that are needed when shaping industrial systems towards CE ecosystems. Steps one and

two were iterative, as we went back to the literature search again. Next, we categorized the literature according to the main elements. (3) The last step was synthesis and development of the framework.

4   Conceptual Framework

This section reports the results of the literature review as categorized into identified main elements (Sects. 4.1, 4.2 and 4.3). Then it synthesizes the literature into a framework (Sect. 4.4).

4.1   Disruptive Innovation and Co-innovation

Both disruptive and radical innovations are architectural and system-level in their nature, i.e. they are new to the firm, its customers and suppliers, industry and/or the world. Therefore, such innovations can be competence destroying or destructive, when reflecting their impact on markets, firms, and industries [18] and the whole processes of value creation [2]. It is commonly held that such technological innovation must be supported by a corresponding evolution of social arrangements and institutional support structures [19]. Reflecting theme of this paper, most of the recent studies on innovation management highlight the collaborative nature of innovation, i.e. co-innovation [20–22].

Since Christensen's [2, 23] seminal work on disruptive innovation, the discussion has mainly focused on disruptive technologies. Still, the previous literature related to sustainability has already used a variety of concepts in this emerging research area, from environment/environmental [24] to green [25] and eco- [26] innovations. These prior studies recognize the importance of the non-technical side of innovation and point out the need to innovate across the full life-cycle. For instance, Arundel and Kemp [27] have highlighted that eco-innovations may be technological or non-technological, i.e. either organizational, institutional or marketing-based. Conceptual framing of management within system-wide disruptive innovations within industrial systems, i.e. shaping new markets and changing the whole processes of value creation in industrial, is recognized as a rather neglected area of research in prior studies. This is noted for instance by Andersen [28], "environmental innovation research is still in its early phase, and there are worldwide very few actual innovation researchers working with environmental issues".

4.2   Inter-organizational Relationships and Value Co-creation in a CE

Without collaboration, achieving circularity is barely possible [4], and previous contributions have emphasized the role of in-depth collaboration between key partners [11, 29]. Based on their literature review, Lieder & Rashid [6] highlighted how collaborative business model innovation for CE involves a re-thinking of partnerships. Concurrently, a paradigm shift towards value co-creation highlights interactions, joint resource bases, engagement platforms and ecosystemic approaches as new means for business [30], which is highlighted in a CE. Current discussion about markets and value [31], and competitive strategy [32], as well as strategic sustainable business development [33] has emphasized how co-creation of value with partners, other stakeholders as well as consumers is a crucial and strategic element in maintaining competitive advantage.

However, the previous literature that discusses "inter-organisational relationships" in a CE has been limited mainly to researching remanufacturing, closed-looped and reverse supply chains based on the current material flows and manufacturing processes [1, 6]. The traditional concept of value chain is important in a CE since it requires the key actors involved in value flows to be defined [34]. However, it is not enough, and instead of seeing themselves as isolated entities and only taking responsibility for their part of the value chain, companies need to be aware of themselves as part of a bigger network and cooperate along the chain [35, 36]. Thus, a company in a networked business environment is part of various value chains, possibly with different and concurrently changing roles. In other words, the same actors can be each other's suppliers and customers in parallel value chains. Network-level governance structures are what bring actors into working together—the process, rules and norms by which the network enables individuals to influence the network's operations and decision-making [37].

4.3 Business Model Innovation

Business models define as structural templates how firms run and develop their business on holistic and system-levels [38], while companies commercialize new ideas and technologies through their business models [39]. There is a widespread agreement that the business model is (1) a new unit of analysis that is distinct from the product, firm, or network; and it is centered on a focal firm, but its boundaries are wider than those of the firm; (2) business models emphasize a system-level, holistic approach to explaining how firms "do business"; (3) the activities of a focal firm and its partners play an important role, and (4) business models seek to explain both value creation and value capture [38]. Three main dimensions that explain a firm's business model are value creation, value proposition, and value capture (interalia [38, 40]). A good overview of business model definitions, perspectives and components in the literature can be found for instance from Teece, Wirtz et al., Zott and Amit and Zott [38, 41, 42].

Business model innovation considers the business model instead of products or processes as the subject of innovation [43]. Clauss [40] proposes a more detailed concept of business model innovation. The value creation innovation would include new capabilities, new technology/equipment, new partnerships, new processes, while a new proposition innovation would consists of a new offering, new customers and markets, new channels and new customer relationships. Finally, a value capture innovation would include new revenue models and value cost structures.

As new business models are a powerful tool in the transformation towards a CE, novel knowledge on designing circular business models is needed to foster implementation of the CE [11]. A circular business model can be defined as the rationale of how an organization creates, delivers, and captures value with and within closed material loops [44]. Circular business models are by definition networked, and require different actors to work together towards common objectives [4]. As in a CE, there are often game-changing alterations in business models [45], where a consideration of value for the broad range of actors and stakeholders become vital. The challenge of co-creation, required in CE, is to find the 'win-win-win' setting. For this reason, the value for each of these stakeholder groups needs to be identified [46]. Several approaches have been developed to support circular BMI. For example Lewandowski, [11] and Antikainen and Valkokari [4] develop a framework for CBMI based on Osterwalder's [47] business model canvas. Boons and Lüdeke-Freund [48] identified four normative requirements for sustainable business model innovation: (1) A value proposition that reflects the balance of economic, ecological and social needs; (2) Supply chain that engages suppliers in sustainable supply-chain management (materials cycles); (3) A customer interface that motivates customers to take responsibility for their consumption; (4) A financial model that reflects an appropriate distribution of economic costs and benefits among the actors involved in the business model. Given that value propositions are considered one of a firm's most important organizing principles [49], firms need more understanding of how to integrate environmental elements into value propositions. A sustainable value proposition is defined as a "promise on the economic, environmental and social benefits that a firm's offering delivers to customers and society at large, considering both short-term profits and long-term sustainability" [50] (p. 146). However, the extant research usually considers value propositions in terms of the economic benefits, or monetary value, that the supplier's offering delivers to its customers [50].

4.4   The Framework for Shaping Industrial Systems Towards CE Ecosystems

Our conceptual framework approaches the phenomenon from the meso-level, that is, from the viewpoint of the managers of an industrial company. The framework emphasizes that to be able to lead the transformation, managers need to understand the mechanisms to shape industrial systems towards CE systems. In our framework, we suggest the disruptive business model (co)-innovation as a mechanism that shapes industrial systems towards CE ecosystems (Fig. 1).

From Industrial Systems to CE Ecosystems. As discussed above, not enough attention is paid to the opportunities provided for businesses by being part of larger systems over and above the traditional direct value chain of a company [35]. The change drives the need for different models and affects the terminology utilized in the practice as well as in the literature. Thus, introducing an ecosystem approach to the CE would bring novel understanding in the re-conceptualization of value chains and industrial systems. In CE ecosystems, actors integrate resources in interaction with each other, and therefore ecosystems reconfigure themselves, i.e. they are dynamic and potentially self-adjusting [31, 51]. Based on the original concept of Moore [52], a business ecosystem is typically viewed as a layered constellation with direct and indirect interactions between the involved actors. It consists of numerous actors that can be categorized based on their respective roles, such as suppliers, competitors, and customers, and together they form the business ecosystem [53]. Here, we define CE ecosystems as co-evolving, dynamic and potentially self-organising configurations [31], in which actors integrate resources and co-create circular value flows in interaction with each other.

From Value Chains to Value Circles: Enabling a CE can be considered as creating different flows of resources, knowledge and value [46]. Therefore, we see that the concept of value circles would be beneficial in highlighting the difference to typically linear value chain models and the co-existence of several overlapping value circles. Although the concept is mentioned in some articles/reports [3, 54], it is not an established concept. In this paper we define the concept of value circles as follows, (deriving from the sustainable value chain concept, interalia [32, 55]): The value circle includes the full range of activities, performed by different actors, which are required to bring a product or a service to a user and back to the system. A CE ecosystem composes of various value circles. The value is shared across different value circle actors through innovative governance and collaboration. In reflecting the regenerative and restorative principles of a CE, the value is considered mutually beneficial for value circle actors, the environment and society. CE as a system-level phenomenon requires sense-making of broader collaborative settings than the one within the single value circle.

Disruptive Business Model (co-) Innovation as a Mechanism that Changes Industrial Systems Towards CE Ecosystems. The framework is based on the assumption that business model innovation plays a critical role in creating new (disruptive) business ecosystems and opening up new markets for a CE [11, 56]. Furthermore, the co-innovation of business models is identified to be crucial for the system-level changes required for CE [4, 6, 11, 29]. Table 1 presents the main elements of disruptive business model co-innovation that are needed when shaping industrial systems towards CE

Table 1. The main elements of disruptive business model co-innovation that are needed when shaping industrial systems towards CE ecosystems

Value creation innovation	New proposition innovation	Value capture innovation
Searching for new value circle actors and re-thinking partnerships within a CE ecosystem [6, 11, 29, 34]	Engaging users in co-innovation [21]	New revenue & costs models in value circles [4]; a financial model reflecting an appropriate distribution of economic costs and benefits among value circle actors – a 'win-win-win' setting [46, 48]
Mapping value to different value circle actors (suppliers, partners, competitors, and customers) [4, 53, 57]	Implementing CE principles in the value proposition (sustainable value proposition) [50, 57], for example, related to PSS [11, 58]	New revenue models based on, for example, selling product-based services as performance-based services [11, 15]
Composing new value circles	New channels, new markets and customers [40]	
Creating novel flows of resources, knowledge and value [46]		
Engaging suppliers into CE principles [48]		

ecosystems. These are adopted from the previous literature as presented in previous sectors and modified, as necessary, towards CE principles. We apply a categorization as suggested by Clauss [40], and divide business model innovation into value creation innovation, new proposition innovation, and value capture innovation. Disruptive business model innovations are architectural and system-level in their nature, and have implications for the business ecosystem level, and thus, are often based on value creation innovation. In other words, they require companies to re-think their current value flows, as well as their partnerships with customers and suppliers. Therefore, such innovations can be competence destroying or destructive, when reflecting on their impact on markets, firms, and industries [18] and the whole processes of value creation [2].

5   Conclusions

The CE paradigm introduces a new perspective for looking at the industrial ecosystem, where economic growth is decoupled from resource consumption and pollutant emissions as end-of-life materials and products are conceived as resources rather than waste [59]. Most of the discussion towards CE development is from a resource scarcity and environmental impact perspective, leaving out the economic benefits of industrial actors [6]. The CE concept itself might create some barriers for implementations, as it might be perceived as being too radical or too complex to be fully adopted. The language of the concept is rather abstract, and the definition and interpretation of the term needs to be worked out by all the actors involved.

This paper contributes to these aspects of the CE discussion by developing a conceptual framework that proposes that value chains become value circles, while industrial systems are transformed into CE ecosystems. It emphasizes the importance of shaping current industrial systems into CE ecosystems, and further it defines CE ecosystems. Ecosystem thinking can be also described as a disruptive innovation, as the managers typically focus on their own strategies and do not consider the impacts at the ecosystem level. We suggest that the CE ecosystem is composed of value circles, and we highlight that the concept of value circles would be beneficial in clarifying the co-existence of several overlapping value circles and how they differentiate from linear value chain models. In defining the value circle we emphasize governance innovation, since network-level governance structures are what bring actors into a working relationship together [37]. The value-circle approach contrasts sharply with the mind-set embedded in most of today's industrial operations, where even the terminology—value chain, supply chain, end user—expresses a linear view. The value circle approach is in line with the basic principle of sustainability of the Earth, since the biosphere, which is a closed system, functions in cycles, which vary for instance in their length, duration, participants and their roles. In other words, adopting a CE approach requires alterations in ecosystems "to allow products to flow through value circles, allowing products and materials to remain in the economic cycle for longer and [where] the quality of the material input is conserved" [60]. The framework proposes a disruptive business model of (co)-innovation as a mechanism that shapes industrial systems towards CE ecosystems. As central to the framework, we identify the critical elements of disruptive business model co-innovation as categorized into the following categories: value creation innovation, new proposition innovation, and value capture innovation. The results reveal that networked business model innovation is highlighted in the transition towards a CE. The paradigm shift towards value co-creation highlights interactions, joint resource bases, engagement platforms, and ecosystemic approaches as new means for business [30].

We acknowledge that this conceptual framework is only one step in the theory building. We do not, therefore, see it as a final product, and thus it needs to be further validated and tested before being considered a concise theoretical framework [61]. Also, there is a need to drive the research forward to provide managers with a deeper understanding of the mechanisms of co-evolution and self-organizing in CE ecosystems. Such further development would also contribute to the previous literature on business ecosystem management, which has typically focused on the different strategies that companies may have in their ecosystem [62], while giving only scant consideration to the question of business ecosystem management, including the composing and orchestrating of such ecosystems. Moreover, this paper approaches the phenomena from the viewpoint of management of a company, but research is needed also in other levels (micro and macro). Although the promises of CE are appealing, CE research would need more critical analyses. Several research streams, including degrwoth [63] and ecologically dominant logic [64], would offer interesting avenues for this.

References

1. Ghisellini, P., Cialani, C., Ulgiati, S.: A review on circular economy: the expected transition to a balanced interplay of environmental and economic systems. J. Clean. Prod. **114**, 11–32 (2016)
2. Christensen, C.M.: The Innovator's Dilemma: When New Technologies Cause Great Firms to Fail. Harvard Business School Press, Boston (1997)
3. Stahel, W.R.: The business angle of a circular economy. Higher competitiveness, higher resource security and material efficiency. In: Foundation, E.M. (ed.) A New Dynamic. Effective Business in a Circular Economy (2014)
4. Antikainen, M., Valkokari, K.: A framework for sustainable circular business model innovation. Technol. Innov. Manag. Rev. **6**, 5–12 (2016)
5. Bocken, N.M.P., Pauw, I. de, Bakker, C., Grinten, B. van der: Product design and business model strategies for a circular economy. J. Ind. Prod. Eng. (2016), In press
6. Lieder, M., Rashid, A.: Towards circular economy implementation: a comprehensive review in context of manufacturing industry. J. Clean. Prod. **115**, 36–51 (2016)
7. Schenkel, M., Krikke, H., Caniëls, M.C.J., van der Laan, E.: Creating integral value for stakeholders in closed loop supply chains. J. Purch. Supply Manag. **21**, 155–166 (2015)
8. Sauvé, S., Bernard, S., Sloan, P.: Environmental sciences, sustainable development and circular economy: alternative concepts for transdisciplinary research. Environ. Dev. **17**, 48–56 (2015)
9. Ellen MacArthur Foundation: Towards the Circular Economy Vol. 1: An Economic and Business Rationale for an Accelerated Transition (2013)
10. Govindan, K., Soleimani, H., Kannan, D.: Reverse logistics and closed-loop supply chain: a comprehensive review to explore the future. Eur. J. Oper. Res. **240**, 603–626 (2014)

11. Lewandowski, M.: Designing the business models for circular economy—towards the conceptual framework. Sustainability **8**, 43 (2016)

12. Thompson, N.A., Herrmann, A.M., Hekkert, M.P.: How sustainable entrepreneurs engage in institutional change: insights from biomass torrefaction in the Netherlands. J. Clean. Prod. **106**, 608–618 (2015)

13. Dacin, M.T., Goodstein, J., Scott, W.R.: Institutional theory and institutional change: introduction to the special research forum. Acad. Manag. J. **45**, 43 (2002)

14. Planing, P.: Business model innovation in a circular economy reasons for non-acceptance of circular business models. Open J. Bus. Model Innov. **1**, 1–11 (2015)

15. Tukker, A.: Product services for a resource-efficient and circular economy – a review. J. Clean. Prod. **97**, 76–91 (2015)

16. Beuren, F.H., Gomes Ferreira, M.G., Cauchick Miguel, P.A.: Product-service systems: a literature review on integrated products and services. J. Clean. Prod. **47**, 222–231 (2013)

17. Hertwich, E.G.: Consumption and the rebound effect: an industrial ecology perspective. J. Ind. Ecol. **9**, 85–98 (2008)

18. Schilling, M.: Strategic Management of Technological Innovation. McGraw-Hill Publishing, New York (2013)

19. Freeman, C.: The greening of technology and models of innovation. Technol. Forecast. Soc. Change. **53**, 27–39 (1996)

20. Lee, S.M., Olson, D.L., Trimi, S.: Co-innovation: convergenomics, collaboration, and co-creation for organizational values. Manag. Decis. **50**, 817–831 (2012)

21. von Hippel, E.: Democratizing Innovation. MIT Press, Cambridge (2005)

22. Chesbrough, H.W.: The era of open innovation. MIT Sloan Manag. Rev. **44**, 35–41 (2003)

23. Bower, J.L., Christensen, C.M.: Disruptive technologies: catching the wave. Harward Bus. Rev. **73**, 43–53 (1995)

24. Porter, M.E.: America's Green Strategy. Sci. Am. **264**, 168 (1991)

25. Rennings, K., Zwick, T.: Employment impact of cleaner production on the firm level: empirical evidence from a survey in five european countries. Int. J. Innov. Manag. **6**, 319–342 (2002)

26. OECD: eco-innovation in industry: enabling green growth. OECD (2009)

27. Arundel, A., Kemp, R.: Measuring ecoinnovation. In: United Nations University – Maastricht Economic and Social Research and Training Centre on Innovation and Technology. Maastricht, UNU-MERIT #2009-017 (2009)

28. Andersen, M.: Eco-innovation: towards a taxonomy and a theory. In: Entrepreneurship and Innovation DRUID Conference, Copenhagen, 17–20 June 2008

29. Bocken, N.M.P., Short, S.W., Rana, P., Evans, S.: A literature and practice review to develop sustainable business model archetypes. J. Clean. Prod. **65**, 42–56 (2014)

30. Ramaswamy, V., Ozcan, K.: The Co-creation Paradigm. Standford Business Books, Standford (2014)

31. Vargo, S.L., Lusch, R.F.: It's all B2B…and beyond: toward a systems perspective of the market. Ind. Mark. Manag. **40**, 181–187 (2011)

32. Porter, M.E., Kramer, M.R.: Creating shared value. Harv. Bus. Rev. **89**, 62–77 (2011)

33. Robèrt, K.-H., Schmidt-Bleek, B., Aloisi de Larderel, J., Basile, G., Jansen, J.L., Kuehr, R., Price Thomas, P., Suzuki, M., Hawken, P., Wackernagel, M.: Strategic sustainable development — selection, design and synergies of applied tools. J. Clean. Prod. **10**, 197–214 (2002)

34. Peltola, T., Aarikka-Stenroos, L., Viana, E., Mäkinen, S.: Value capture in business ecosystems for municipal solid waste management: comparison between two local environments. J. Clean. Prod. **137**, 1270–1279 (2016)

35. Tsvetkova, A., Gustafsson, M.: Business models for industrial ecosystems: a modular approach. J. Clean. Prod. **29–30**, 246–254 (2012)
36. Aminoff, A., Kettunen, O.: Sustainable supply chain management in a circular economy - towards supply circles. In: Setchi, R., Howlett, R., Liu, Y., Theobald, P. (eds.) Sustainable Design and Manufacturing Smart Innovation, Systems and Technologies, vol. 52, pp. 61–72. Springer, Cham (2016)
37. Valkokari, K., Padmakshi, R.: Towards sustainability governance in value networks. In: Liyanage, J.P., Uusitalo, T. (eds.) Value Networks in Manufacturing: Sustainability and Performance Excellence, pp. 43–63. Springer, Switzerland (2016)
38. Zott, C., Amit, R., Massa, L.: The business model: recent developments and future research. J. Manage. **37**, 1019–1042 (2011)
39. Chesbrough, H.: Business model innovation: opportunities and barriers. Long Range Plann. **43**, 354–363 (2010)
40. Clauss, T.: Measuring business model innovation: conceptualization, scale development, and proof of performance. R&D Manag. (2016), In press
41. Teece, D.J.: Business models, business strategy and innovation. Long Range Plann. **43**, 172–194 (2010)
42. Wirtz, B.W., Pistoia, A., Ullrich, S., Göttel, V.: Business models: origin, development and future research perspectives. Long Range Plann. **49**, 36–54 (2016)
43. Baden-Fuller, C., Haefliger, S.: Business models and technological innovation. Long Range Plann. **46**, 419–426 (2013)
44. Mentink, B.: Circular business model innovation: a process framework and a tool for business model innovation in a circular economy (2014). http://repository.tudelft.nl/assets/uuid:c2554c91-8aaf-4fdd-91b7-4ca08e8ea621/THESIS_REPORT_FINAL_Bas_Mentink.pdf
45. Beattie, V., Smith, S.J.: Value creation and business models: refocusing the intellectual capital debate. Br. Account. Rev. **45**, 243–254 (2013)
46. Aminoff, A., Valkokari, K., Kettunen, O.: Mapping multidimensional value(s) for co-creation networks in a circular economy. In: Afsarmanesh, H., Camarinha-Matos, L., Lucas Soares, A. (eds.) Collaboration in a Hyperconnected World. IFIP Advances in Information and Communication Technology, vol. 480, pp. 629–638. Springer, Cham (2016)
47. Osterwalder, A.: Clarifying business models: origins, present and future of the concept. Commun. Assoc. Inf. Syst. **16**, 1–25 (2005)
48. Boons, F., Lüdeke-Freund, F.: Business models for sustainable innovation: state-of-the-art and steps towards a research agenda. J. Clean. Prod. **45**, 9–19 (2013)
49. Payne, A., Frow, P.: Developing superior value propositions: a strategic marketing imperative. J. Serv. Manag. **25**, 213–227 (2014)
50. Patala, S., Jalkala, A., Keränen, J., Väisänen, S., Tuominen, V., Soukka, R.: Sustainable value propositions: framework and implications for technology suppliers. Ind. Mark. Manag. **59**, 144–156 (2016)
51. Palomäki, K., Valkokari, K., Hakanen, T.: From channel management in sales and distribution to co-evolving service eco-systems. In: Proceedings of 26th Annual RESER Conference, Naples, Italy, 8–10 September 2016. European Association for Research on Services, RESER (2016)
52. Moore, J.F.: Predators and prey: a new ecology of competition. Harward Bus. Rev. **71**(3), 75–86 (1993)
53. Adner, R., Kapoor, R.: Value creation in innovation ecosystems: how the structure of technological interdependence affects firm performance in new technology generations. Strateg. Manag. J. **31**, 306–333 (2010)

54. Ellen MacArthur Foundation: Towards the circular economy vol. 3: accelerating the scale-up across global supply chains (2014)
55. Fearne, A., Garcia Martinez, M., Dent, B.: Dimensions of sustainable value chains: implications for value chain analysis. Supply Chain Manag. Int. J. **17**, 575–581 (2012)
56. Thompson, J.D., MacMillan, I.C.: Business models: creating new markets and societal wealth. Long Range Plann. **43**, 291–307 (2010)
57. Bocken, N., Short, S., Rana, P., Evans, S.: A value mapping tool for sustainable business modelling. Corp. Gov. Int. J. Bus. Soc. **13**, 482–497 (2013)
58. Tukker, A., Tischner, U.: Product-services as a research field: past, present and future. Reflections from a decade of research. J. Clean. Prod. **14**, 1552–1556 (2006)
59. Elia, V., Gnoni, M.G., Tornese, F.: Measuring circular economy strategies through index methods: a critical analysis. J. Clean. Prod. **142**, 2741–2751 (2016)
60. Bechtel, N., Roman, B., Völkel, R.: Be in the loop: circular economy & strategic sustainable development (2013)
61. Choi, T.Y., Wacker, J.G.: Theory building in the OM/SCM field: pointing to the future by looking at the pasT. J. Supply Chain Manag. **47**, 8–11 (2011)
62. Zahra, S.A., Nambisan, S.: Entrepreneurship in global innovation ecosystems. AMS Rev. **1**, 4–17 (2011)
63. Demaria, F., Schneider, F., Sekulova, F., Martinez-Alier, J.: What is degrowth? From an activist slogan to a social movement. Environ. Values. **22**, 191–215 (2013)
64. Montabon, F., Pagell, M., Wu, Z.: Making sustainability sustainable. J. Supply Chain Manag. **52**, 11–27 (2016)

Business Models for Sustainability: The Case of Repurposing a Second-Life for Electric Vehicle Batteries

Na Jiao[✉] and Steve Evans

Institute for Manufacturing, University of Cambridge,
17 Charles Baggage Road, Cambridge, CB3 0FS, UK
{nj268,se321}@cam.ac.uk

Abstract. The rapid development of electric vehicles (EVs) has caused a problem for the industry: what happens to the batteries at the end of their useful life in EVs? Repurposing those batteries for a less-demanding second-life application, e.g. stationary energy storage, could provide a potential solution to extract more value than just recycling or disposal. This study examines the battery second use (B2U) business models being developed by various actors that generate value through different second-use applications. Based on empirical interview data from stakeholders involved in B2U, this paper presents a typology of current B2U business models – standard, collaborative and integrative business models – and offers implications for designing business models that incorporate sustainability at the core.

1 Introduction

The electric vehicle (EV) industry holds great promises for future sustainable transportation systems that reduce greenhouse gas emissions [1]. Despite the rapid increase in EV market share in recent years, the high initial cost is still considered to be the major impediment to EV mass-market adoption. Research, technology development and manufacturing ramp-up are underway to reduce the battery cost which constitutes the single most expensive part of an EV [2]. Apart from that, efforts from research and industries are looking for ways to increase the value of the battery over its lifetime. When the battery fails to meet the criteria for automotive service upon its end-of-vehicle-life (EOVL), it could have around 70–80% of the initial capacity which has the potential to be utilized in less demanding stationary storage applications [3]. Repurposing EOVL batteries in post-vehicle applications for a second-life is termed battery second use (B2U) and has the potential to reduce battery cost for both the automotive and energy storage applications [2].

However, it is not easy to transform the old and less capable car batteries into stationary storage battery systems. Effective business models are needed that bring together cross-sector actors and generate value from delivering stakeholder benefits. The full potential value of second-life batteries, which not only includes the economic value, but also social and environmental value, should be examined for strategic decision making. This study draws on the business model literature and empirical case studies of

© Springer International Publishing AG 2017

G. Campana et al. (eds.), *Sustainable Design and Manufacturing 2017*, Smart Innovation,
Systems and Technologies 68, DOI 10.1007/978-3-319-57078-5_51

the current B2U business models in practice. A typology of existing B2U business models is presented and the implications for designing business models for sustainability are analysed.

2   Literature Review

2.1   Business Models and Sustainability

The concept of business model has been extensively discussed in both the academia and the industry, especially with the advent of the Internet boom in the 1990s [4, 5]. Much of the business model literature focuses on defining business models and their elements [6–8], but there is still no general agreement on the definition. Business models have been described as firm's business logic [9, 10], market devices [11], and focusing devices [12]. One of the commonly accepted description is that business model explains *how the enterprise creates, delivers and captures value* [7, 13]. Recent studies take a somewhat broader scope that extends the firm boundary and incorporate a system-level thinking, conceptualizing business model as *"a system of interdependent activities that transcends the focal firm boundaries"* [14].

With increasing global sustainable pressures, there is a growing interest in business models that could improve the sustainable performance of companies. Schaltegger et al. (2016) define a business model for sustainability as *(i) a company's sustainable value proposition to its customers, and all other stakeholders, (ii) how it creates and delivers this value, (iii) and how it captures economic value while maintaining or regenerating natural, social, and economic capital beyond its organizational boundaries* [15]. Some researchers have seen business model innovation as an essential way to improve the sustainable development of the company and society [16]. Although still under-researched, the literature on business model for sustainability show some common features including a wider range of stakeholders, sustainable value perspective and a life cycle thinking [17–19].

2.2   Research Context: Battery Second Use

The idea of B2U is not new to the industry, and the benefits and concerns are widely discussed in the academic literature and industrial reports [20–22]. Researchers have been investigating the technical and economic feasibility of B2U in various stationary storage applications [20, 22–29], quantifying the effect of B2U on EV cost reductions [30], developing frameworks/tools to inform optimum operating conditions to maximize battery value [31, 32], analysing the environmental feasibility of second-life batteries [33] and so on. Researches show that although most B2U strategies are viable in terms of battery performance and economics, there are uncertainties in battery degradation and new battery price reduction. Most of these studies are quantitative and require a detailed breakdown of technical, economic or environmental parameters of each process. Due to the complexity and uncertainty of parameters in the nascent stage of B2U, these studies tend to be limited in certain boundary conditions and estimate assumptions [34].

Neubauer et al. from the National Renewable Energy Laboratory (NREL) did comprehensive work on predicting battery degradation and availability for second use, illustrating repurposing procedures and calculating repurposing cost, and evaluating the potential markets for second-life batteries [2]. Critical barriers were identified and recommendations for major B2U stakeholders were proposed. It was suggested that although B2U has little ability to reduce EV upfront cost, the overall benefits to the society can be quite large, which should be recognized and valued. Apart from that, little work has been done so far to comprehensively study the secondary use of EV batteries as a new market. Especially, there is a lack of research on the business model perspective to understand the full potential value of second-life batteries. In this study, the author draws on the literature on business model for sustainability and empirical case studies to understand B2U business models that increase the sustainable value of second-life batteries.

3 Methodology

The research question that this paper tries to answer is 'how companies across sectors develop business models for increased value of second-life EV batteries?' To answer this question, a case study method is used to provide a contextual understanding of the topic. Semi-structured interviews are conducted with CEOs, managers/directors of companies across automotive and energy industries. The use of qualitative case study research methodology is justified by the uncertainty of the emerging industry studied, along with the uncertain and exploratory nature of the research topic [35].

In the nascent stage of B2U, only a handful of cases can provide substance to study business models at the commercial level. Most of the B2U projects are still in the planning, piloting or demonstration phases and are more focused on the technical or economic aspects. The case studies are selected from the existing B2U markets that have passed the phases of pilot or demonstration projects to reach the early commercialization stage.

4 Findings and Discussion

4.1 A Typology of B2U Business Models

The business models examined from the case studies could be categorised into three types: standard business model, collaborative business model and integrative business model. This categorization is based on the interaction between automotive OEMs and another key stakeholder in the energy sector – the B2U system providers who actually provide the products or services based on second-life batteries. The schematic typology is shown in Fig. 1. The boxes in each type of the business models represent the key stakeholders involved in B2U: the OEM, the B2U system provider and the end-customer. The red and green arrows represent the battery ownership flow and the service flow respectively. And the yellow dash line describes the trend of the growing degree of OEM

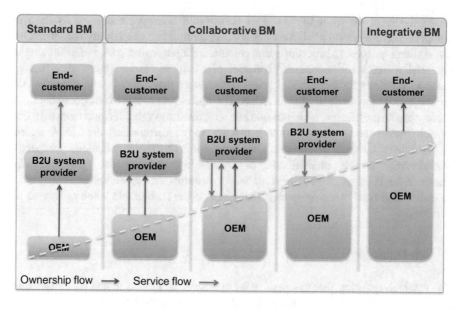

Fig. 1. A typology of B2U business models (BMs)

engagement in B2U from standard to collaborative and then to integrative business models.

Standard Business Model. The standard business model is where the OEMs just sell second-life batteries to their customers who then build their own final solutions with the batteries. The interaction between the OEM and the B2U system provider is just supplier-customer relationship, like in most business models. In this case, the purpose of OEMs involvement in B2U is to get additional revenue from selling the property in the free market. And there is no collaboration and service flow between the OEM and the B2U system provider. This type of business model requires little OEM engagement in the final solution development for the end-user and is very vulnerable to competitors.

Collaborative Business Model. Most of the companies doing B2U businesses fall in the second type – collaborative business model. In this category, OEMs collaborate with B2U system providers to develop the final solutions for the end-customers. In collaborative business models, OEMs collaborate with B2U system providers, e.g. energy companies in different ways and to different extent. In general, there are three types of collaborative business models in which the OEMs' degree of engagement gradually increases: (a) the OEMs sell the batteries as well as provide knowledge about the batteries to better fit them into the final solutions developed by the B2U system providers; (b) the OEMs sell the batteries but also cooperate with the B2U system providers to jointly develop the final solutions for the end-customers; (c) the OEMs retain the ownership of the batteries and get the B2U system providers onboard to develop the final solutions for the batteries. The degree of OEM engagement could be evaluated on the range of activities they are involved in: co-ideation, co-design, co-testing, co-production and co-marketing. OEMs in this situation are trying to maximize the residual value of second-life batteries for certain applications and capture more benefits. Instead of just selling the asset to the free market, OEMs collaborate with certain B2U system providers to optimize the value of the batteries according to the applications they want to fit the batteries into. There are different forms of collaboration as well as various service flows between OEMs and B2U system providers.

OEMs in most cases still sell the batteries, but they are also committed to putting knowledge and resources into the final products or services. Through various collaborative activities, OEMs add to the value of second-life batteries. OEMs benefit more from delivering batteries to B2U system providers that are better fitted into the final solutions. Thus, it can be regarded that the OEMs are providing batteries plus services to B2U system providers in order to add to the value of the batteries in the final systems built by the later. In cases that the OEMs co-develop and co-market the final solution with the B2U system providers, there are bi-directional service flows between the two actors because the final products or services are their mutual goals and outcomes. In one of the case studies, however, the OEM retains the ownership of the battery and gets its partners together to develop and market the final solutions for them. Costs, revenues and risks are shared in this partnership. Retaining the ownership of the batteries allows the OEM to continuously extract value throughout the second-life of the batteries. In this case, the B2U system provider is providing services to the OEM in terms of developing and marketing the final solution. In a nutshell, the collaborative business model requires OEMs to engage in the final solution development and integrate expertise and resources from various actors in the network.

Integrative Business Model. The integrative business model is an extreme case of collaborative business models, where the OEM vertically integrates the functions of all the other actors and develops the final solutions for the end-users. In this case, the purpose of the OEM is to use its own and partners' resources and capabilities to maximize the value that they can get from delivering that final product or service to the end-users. It can be regarded that the OEM integrates all the services provided by its partner networks. The integrative business model requires high OEM engagement and high-diversified resources and capabilities of the company, and is restricted to certain applications due to OEMs' limited access to certain markets (e.g. grid-scale applications).

This typology compares existing B2U business models in practice through examining the relationships and interactions between key stakeholders involved in B2U businesses that add to the value of second-life batteries. The factors that cause the differences in the business models include the second-life applications and OEM's purpose which determines OEM's degree of engagement and battery ownership model.

4.2 A Sustainability Perspective for Understanding B2U Full Potential Value

B2U has attracted extensive attention from both the academia and industry but the full potential value of second-life batteries is poorly understood. It is commonly accepted that OEMs are interested in B2U due to its potential to reduce initial cost of EVs through additional revenue streams. However, with new battery cost decreasing rapidly in the past few years, it will be increasingly difficult for second-life batteries to compete in the energy storage market only from the cost perspective. It is crucial to understand the value of second-life batteries from a sustainability perspective, namely the sustainable value analysis, life cycle thinking and multi-stakeholder perspective.

Sustainable Value Analysis. The benefits of developing second-life batteries are insufficient to sustain B2U if we only think from economic aspect. Currently OEMs are incentivised to develop second-life batteries because they have cost advantage over new batteries, but how about in the future when they are no longer cost competitive? The social and environmental benefits should also be integrated into the B2U value analysis. For example, apart from getting additional revenue streams B2U could improve companies' reputation as being more environmentally friendly. As commented by one of the interviewees, "customers would be happy to know that their used batteries could be applied in a next application. The promotion of our company image is a very big effect for us". For the society as a whole, feeding cost-effective batteries into the energy storage industry could generate enormous social value through facilitating renewable integration and grid balancing, thus contributing to a cleaner, smarter and securer grid system. A comprehensive understanding of the sustainable value of developing second-life batteries could provide a basis for better decision-making.

Life Cycle Thinking. The concept of increasing the value of second-life batteries is embedded in increasing the battery value over its lifetime. Second-life battery value could not be optimized only during second-life, without a life cycle thinking. The performance of second-life batteries, unlike new batteries, depends not only on the cell chemistry but also on how they were designed for second-use and how they were used during their first-life in vehicles. Optimizing the battery system design for second-life without sacrificing its vehicular performance and adding additional cost could substantially reduce the repurposing cost and increase the predictability of second-life battery performance. It is also viable to educate and incentivise EV customers to utilize the battery optimally to ensure a better and longer second-life for the battery. The end-of-life (EOL) liability issues and recycling responsibilities should be made clear among stakeholders to close the loop when the battery reaches the very end of its life and could not be used anymore.

Multi-stakeholder Perspective. Automotive OEMs play a pivotal role in the nascent stage of B2U because they hold the key to the initial battery design and EOVL battery strategies. However, transforming second-life batteries into the final solutions requires various value-adding activities by cross-sector stakeholders, for example, efficient logistics and battery grading during repurposing, safety and performance standards setting, system integration and the final solution development and marketing. It is also crucial that regulators understand the value of batteries in the energy market and create market mechanisms that enable the energy market to be more open and flexible.

4.3   Implications for the B2U Business Model Design

The selection of B2U business models depends on many factors, both internal and external. Since B2U is most expected to be a demand-driven market, the foremost is finding the right applications for the targeted market. OEM's purposes of B2U are also essential in determining the business models for second-life batteries: whether they just want to sell the asset and get some additional revenues or they want to maximise the value of second-life batteries. Those factors determine the battery ownership model and the OEM's engagement in B2U and thus, the according business models for B2U. For example, if the OEM wants to maximise the value of second-life batteries through deploying second-batteries for large-scale, grid-related applications to provide grid services, the OEM might want to retain the ownership of the battery and get continuous benefits throughout the second-life of the batteries.

5   Conclusions

This paper set out to explore business models of second-life batteries. Through empirical case studies with various stakeholders involved in B2U, a typology of B2U business models is proposed: standard business model, collaborative business model and integrative business model. The relationships and interactions between major stakeholders are examined and the implications for embedding sustainability in business models and the selection for B2U business models are analysed.

As always, there are certain limitations to this study, which mainly relate to the collection of the empirical data. Firstly, although case studies were conducted with companies that have already started commercializing second-life batteries, the number of cases is small and the typology may not include all the possible business model types. Second, the B2U market is at a very nascent stage and case studies were conducted in a relatively short period of time. The business models examined are subject to the changing market of both the EV and energy storage sectors. Follow-up studies on the B2U business models would provide richer insight into the increased value for second-life batteries.

References

1. IEA, Global EV Outlook: Understanding the Electric Vehicle Landscape to 2020 (2013)
2. Neubauer, J., Smith, K., Wood, E., Pesaran, A.: Identifying and overcoming critical barriers to widespread second use of PEV batteries (2015)
3. Cready, E., Lippert, J., Pihl, J., Weinstock, I., Symons, P.: Technical and economic feasibility of applying used EV batteries in stationary applications (2003)
4. Timmers, P.: Business models for electronic markets. Electron. Mark. **8**, 3–8 (1998)
5. Pohle, G., Chapman, M.: IBM's global CEO report 2006: business model innovation matters. Strateg. Leadersh. **34**, 34–40 (2006). doi:10.1108/10878570610701531
6. DaSilva, C.M., Trkman, P.: Business model: what it is and what it is not. Long Range Plann. **47**, 379–389 (2014)

7. Teece, D.J.: Business models, business strategy and innovation. Long Range Plann. **43**, 172–194 (2010)
8. Zott, C., Amit, R., Massa, L.: The business model: theoretical roots, recent developments, and future research. IESE Res. Pap. **3**, 45 (2010). doi:10.1177/0149206311406265
9. Osterwalder, A.: The business model ontology: a proposition in a design science approach (2004)
10. Magretta, J.: Why business models matter? Harv. Bus. Rev. **80**, 86–92 (2002)
11. Doganova, L., Eyquem-Renault, M.: What do business models do? Innovation devices in technology entrepreneurship. Res. Policy **38**, 1559–1570 (2009). doi:10.1016/j.respol. 2009.08.002
12. Chesbrough, H., Rosenbloom, R.S.: The role of the business model in capturing value from innovation: evidence from Xerox Corporation's technology spin-off companies. Ind. Corp. Chang. **11**, 529–555 (2002)
13. Richardson, J.: The business model: an integrative framework for strategy execution. Strateg. Chang. **17**, 133–144 (2008). doi:10.1002/jsc.821
14. Zott, C., Amit, R.: Business model design: an activity system perspective. Long Range Plann. **43**, 216–226 (2010)
15. Schaltegger, S., Hansen, E.G., Lüdeke-Freund, F.: Business models for sustainability: origins, present research, and future avenues. Organ. Environ. **29**, 3–10 (2016). doi: 10.1177/1086026615599806
16. Bocken, N.M.P., Short, S.W., Rana, P., Evans, S.: A literature and practice review to develop sustainable business model archetypes. J. Clean. Prod. **65**, 42–56 (2014). doi:10.1016/ j.jclepro.2013.11.039
17. Short, S.W., Rana, P., Bocken, N.M.P., Evans, S.: Embedding sustainability in business modelling through multi-stakeholder value innovation. In: Emmanouilidis, C., Taisch, M., Kiritsis, D. (eds.) APMS 2012. IAICT, vol. 397, pp. 175–183. Springer, Heidelberg (2013). doi:10.1007/978-3-642-40352-1_23
18. Rizzi, F., Bartolozzi, I., Borghini, A., Frey, M.: Environmental management of end-of-life products: nine factors of sustainability in collaborative networks. Bus. Strateg. Environ. **22**, 561–572 (2013). doi:10.1002/bse.1766
19. Matos, S., Silvestre, B.S.: Managing stakeholder relations when developing sustainable business models: the case of the Brazilian energy sector. J. Clean. Prod. **45**, 61–73 (2013). doi:10.1016/j.jclepro.2012.04.023
20. Neubauer, J.S., Pesaran, A., Williams, B., Ferry, M., Jim, E.: A techno-economic analysis of PEV battery second use: repurposed-battery selling price and commercial and industrial end-user value (2012)
21. Faria, R., Marques, P., Garcia, R., Moura, P., Freire, F., Delgado, J., et al.: Primary and secondary use of electric mobility batteries from a life cycle perspective. J. Power Sources **262**, 169–177 (2014). doi:10.1016/j.jpowsour.2014.03.092
22. Heymans, C., Walker, S.B., Young, S.B., Fowler, M.: Economic analysis of second use electric vehicle batteries for residential energy storage and load-levelling. Energy Policy **71**, 22–30 (2014). doi:10.1016/j.enpol.2014.04.016
23. Wolfs, P.: An economic assessment of "second use" lithium-ion batteries for grid support. In: Universities Power Engineering Conference (AUPEC), 20th Australasian. IEEE (2010)
24. Starke, M.R., Andrews, G.: Final report economic analysis of deploying used batteries in power systems (2011)
25. Beer, S., Gómez, T., Member, S., Dallinger, D., Momber, I., Marnay, C., et al.: An economic analysis of used electric vehicle batteries integrated into commercial building microgrids. IEEE Trans. Smart Grid **3**, 517–525 (2012)

26. Lih, W.-C., Yen, J.-H., Shieh, F.-H., Liao, Y.-M.: Second use of retired lithium-ion battery packs from electric vehicles: technological challenges, cost analysis and optimal business model. In: 2012 International Symposium on Computer Consumer and Control, pp. 381–384 (2012). doi:10.1109/IS3C.2012.103
27. Warner, N.A.: Secondary life of automotive lithium ion batteries: an aging and economic analysis (2013)
28. Elkind, E.N.: Reuse and repower: how to save money and clean the grid with second-life electric vehicle batteries (2014). https://www.law.berkeley.edu/files/ccelp/Reuse_and_Repower_--_Web_Copy.pdf
29. Foster, M., Isely, P., Standridge, C.R., Hasan, M.M.: Feasibility assessment of remanufacturing, repurposing, and recycling of end of vehicle application lithium-ion batteries. J. Ind. Eng. Manag. 7, 698–715 (2014). doi:10.3926/jiem.939
30. Neubauer, J., Pesaran, A.: The ability of battery second use strategies to impact plug-in electric vehicle prices and serve utility energy storage applications. J. Power Sources 196, 10351–10358 (2011). doi:10.1016/j.jpowsour.2011.06.053
31. Viswanathan, V.V., Kintner-Meyer, M.: Second use of transportation batteries: maximizing the value of batteries for transportation and grid services. IEEE Trans. Veh. Technol. 60, 2963–2970 (2011)
32. Keeli, A., Sharma, R.K.: Optimal use of second life battery for peak load management and improving the life of the battery. In: 2012 IEEE International Electric Vehicle Conference, pp. 1–6 (2012). doi:10.1109/IEVC.2012.6183276
33. Ahmadi, L., Yip, A., Fowler, M., Young, S.B., Fraser, R.A.: Environmental feasibility of re-use of electric vehicle batteries. Sustain. Energy Technol. Assess. 6, 64–74 (2014). doi: 10.1016/j.seta.2014.01.006
34. Bowler, M.: Battery second use: a framework for evaluating the combination of two value chains (2014)
35. Eisenhardt, M.: Building theories from case study research. Acad. Manag. Rev. 14, 532–550 (1989)

Circular Economy Business Model Innovation Process – Case Study

Maria Antikainen[1(✉)], Anna Aminoff[1], Outi Kettunen[1],
Henna Sundqvist-Andberg[1], and Harri Paloheimo[2]

[1] VTT Technical Research Centre of Finland, Espoo, Finland
{maria.antikainen,anna.aminoff,outi.kettunen,
henna.sundqvist-andberg}@vtt.fi
[2] CoReorient, Espoo, Finland
harri.paloheimo@coreorient.com

Abstract. The concept of the Circular Economy has recently caught the attention of academia as well and businesses and decision makers offering an attractive solution for an environmentally sustainable economic growth. Companies need to consider how to close material loops, reduce the resources needed and think more about how materials and products are kept in the loop as long as possible. In order to do that, companies need to find new collaboration partners and reconsider the value offered for stakeholders. To solve that, we need new or modified innovation tools and processes to guide businesses in their innovation journey resulting in novel business models in a circular economy. Thus, the aim of this study is to increase our understanding of the circular business model innovation process. Our main focus is to explore what kind of mixed methods create value in circular business model innovation and what kind of challenges there are related to each method and how is it possible to overcome those challenges. The paper highlights the importance of involving different perspectives, stakeholders and using mixed methods during the innovation process.

Keywords: Circular Economy · Business model innovation process · Consumers · Methodology · Rapid business model experimentation · Case study

1 Background

1.1 Introduction

In order to solve the challenges related to exceeding the biosphere's regenerative capacity, rising consumption and companies' decreasing competitiveness due to volatility of prices and tightening legislation, a new economic model must be found. Instead of the linear "make-buy-use-dispose" model, there is a viable solution called the Circular Economy (CE). The main idea of CE is to find a solution for how to maximize the value of products and materials in order to reduce the usage natural resources and create positive societal and environmental impacts [1]. In fact, the discussion about a novel economic model is not new: Lovins et al. [2] launched a term "National Capitalism" to focus on the issue that business strategies built around the radically more productive use

© Springer International Publishing AG 2017
G. Campana et al. (eds.), *Sustainable Design and Manufacturing 2017*, Smart Innovation, Systems and Technologies 68, DOI 10.1007/978-3-319-57078-5_52

of natural resources will solve many environmental problems at a profit. After them, McDonough and Braungart [3] continued this discussion in their book "Cradle to Cradle" by suggesting that industry should preserve and enrich ecosystems and nature's biological metabolism while also maintaining a safe, productive technical metabolism for the high-quality use and circulation of organic and technical nutrients. Recently, the Ellen MacArthur Foundation [4] made a major contribution in presenting the concept "Circular Economy" for businesses, politicians, and also consumers.

We know that succeeding in transformational, fundamental and system-wide innovation is much more challenging than making incremental innovations. A well founded assumption is that as environmental, social, political and technological changes continue to shift the foundations of our current business models, incremental innovation will become less effective in enabling companies, industries and whole economies to adapt and succeed. Hence, it is stated that there is an urgent need for radical system-wide innovation changing the processes of value creation [5].

CE has the potential to change dramatically the innovation environment of companies. Companies need to consider how to close material loops, reduce the resources needed and think more about how materials and products are kept in the loop as long as possible. In order to do that, companies need to find new collaboration partners and reconsider the value offered for stakeholders. Closing material loops often affects multiple, if not all aspects of the current business models of companies [6]. Thus, we also need new or modified innovation tools and processes to guide businesses in their innovation journey resulting in novel business models in a circular economy.

1.2   The Research Problem

The business model innovation process is quite often a complex process for companies involving several steps. Thus, the aim of this study is to increase our understanding of the circular business model innovation process. Our main focus is to explore what kind of methods create value in circular business model innovation.

We pose the following research questions:

1. What kind of value can be obtained by different methods in the innovation process?
2. What are the challenges related to each method and how is it possible to overcome those challenges?

In the first question we analyse the value of each of the used methods while in the second question our focus is to assess the challenges we have confronted related to our process and how to resolve them. As an outcome we offer a concept of circular business model innovation process, including detailed information about the steps, tools, perspectives and participants. We also discuss the identified challenges and advantages of using the gained knowledge.

2   Sustainable and Circular Business Model Innovation Process

2.1   Prior Research on Sustainable and Circular Business Model Innovation

A business model represents the rationale of how an organization creates, delivers, and captures value [7]. Derived from the existing extensive business model literature, new literature streams have emerged on sustainable business models (SBM) and sustainable business model innovation (SBMI). Instead of concentrating purely on creating economic value, SBM and SBMI literature include a consideration of other forms of value for a broader range of stakeholders and this way contributes to sustainable development of the company and society [8].

Currently, many companies are dealing with the same questions: How can their business models be transformed into a circular form or how can their existing business be transitioned into circular forms? Despite the active discussion on CE-related business models, the adoption of CE-based business models has been rather slow [9]. In fact, usually no business can be alone circular, but circular business models are part of the larger network that together closes the loop [10]. In this context the need for open innovation and co-creation is even more needed than before. It is also known that business model innovation and implication can take multiple steps or several innovation rounds. Instead of concentrating on tools, it is also important to look at the whole process including various steps, different perspectives, tools and methods and also multiple participants in order to gain the big picture.

2.2   Framework for Innovation Process

Antikainen and Valkokari [11] proposed a framework for sustainable circular business model innovation (CBMI). Their framework was built on the basis of the Österwalder and Pigneur's [7] business canvas adding also elements from other tools developed to add sustainability [12] and circularity [10] perspectives. In order to provide a more holistic approach to the tool Antikainen and Valkokari [11] added three perspectives: recognizing trends and drivers at the ecosystem level; understanding the value to partners and stakeholders within a business; and evaluating the impact of sustainability and circularity. The multiple stakeholder value perspective is also included in the value mapping tool developed by Bocken and colleagues [12], introducing three forms of value (value captured, destroyed, and missed) and value opportunities for major stakeholder groups (environment, society, customer, and network actors). One of the main benefits of the Bocken et al. [12] tool is to raise awareness of the potential for unintended impacts on external stakeholders, as well as to propose alternative solutions that might offer greater alignment between stakeholder interests.

Three reasons can be listed for why it is important to rethink what kind of participants should be involved in the innovation processes. First, the model of CE is based on the idea that the loop is closed within the ecosystem, and thus companies need to collaborate and understand each other's business. Second, there is a need for radical new innovations, and creativity usually increases when new insights are added into the process. Third, in CE there is a need to rethink how value is created, delivered and captured in networks. For example, moving from product orientation towards services changes the value creation model.

3   Methodology

3.1   Case Study Approach

In the AARRE-project, exploring user-driven circular economy business models, close collaboration occurred with multiple Finnish companies. Our data for this study is derived from three companies. One of them operates in the recycling and waste management field and is interested in focusing on consumer markets with their innovative products made from recycled materials. The second case company is a startup focusing on digitalization of recycling centres. The third one is also a startup having multiple ideas related to crowdsourcing and novel consumer services. During the project the researchers have been participating and facilitating different phases of their innovation process from ideation to implementation and adding also different tools from business model innovation to foresight. All the phases were not conducted with each company. During the processes different stakeholder groups were involved in the innovation process. All interview data and materials have been recorded and documented. The data was analysed with Atlas.ti software. The business model innovation process is presented in Fig. 1. The arrow depicts the objectives while the steps below describe the methods we used.

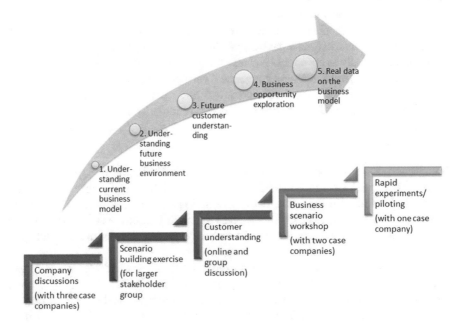

Fig. 1. Business model innovation process in AARRE-project

3.2   Outline of the Innovation Process Steps

The first step was about understanding the changing business environment in a circular economy and its impact to the business model innovation. The second step consisted of understanding the future business environment which was realized by a scenario building process. The third step aimed at understanding future customers through consumer discussions and online discussions. In the fourth step, company-specific business scenario workshops were facilitated to seek new business models. The fifth step consisted of the implementation of the business idea through rapid experimentation or a more complex pilot.

4   Results

4.1   Steps

The First Step: Understanding the Future Business Environment and Its Impact to the Business Model Innovation. This was started by setting up a meeting with companies' key persons in order to together go through the current business model of each company. In the meeting the early version of the tool presented by Antikainen and Valkokari [11] was used. The idea was to understand how company key persons see the current and forthcoming business environment affecting their own business model, what are the main elements of their business model, how do they consider the sustainability impact and how do they constantly evaluate circularity when modifying their business model or developing new products and services.

Because of the practical issues we had reserved only two to three hours for the meeting, so it was not a surprise that we had a lack of time. In all the meetings, the value creation for especially to customers and also for other stakeholders was one of the most time-taking tasks. Instead, our case companies' entrepreneurs seemed to have a sophisticated understanding of the current ecosystem and trends as well as the impact of sustainability and how to evaluate sustainability and circularity. However, entrepreneurs stated that there is a lack of tools for holistic and systematic evaluation of sustainability and circularity.

The Second Step: Scenario Building Exercise to Understand Alternative CE Futures. Forward looking stories and narratives can facilitate anticipatory thinking and actions that could change the future. Scenario building is an iterative process to generate a common understanding of the future. This scenario building exercise focused on creating complementary future images of a circular economy in order to understand the future business environment.

The scenario building process included a literature search on trends, drivers and weak signals, a stakeholder scenario workshop, trend and driver analysis, and scenario writing. The workshop focused on identification of drivers of change, enablers and barriers in the operating environment affecting how target vision, the sustainable circular economy is reached. The workshop was organized with 26 participants from key stakeholder organisations; research organisations, private companies and public organisations, such as ministries and funding agencies. The workshop participants were divided into three groups and each group was facilitated by an AARRE project researcher. The PESTEL framework was used as a tool, and each group focused on different parts of the framework. The workshop assignment was kept flexible, which put significant emphasis on the role of facilitator to keep the discussions within the given topics. The approach was found rather challenging but productive. Bringing together actors from different backgrounds, views, values and objectives resulted in lively, informal discussions and produced diverse future-oriented information and insight. The workshop discussions were recorded and notes transcribed. The results were analysed and used both in trend and driver analysis and scenario drafting.

As a result of the process three complementary scenarios were drafted: (1) Factory of the future, (2) Experiential service economy, and (3) New tribes. The generic main drivers relevant for all scenarios were resource efficiency, raw material fluctuation and scarcity of resources. Digitalization was the key enabling technology driver. Scenarios had different emphasis on the role of key actors, orientation towards value creation, and had different impact on economic growth. Factory of the future scenario focused on the industrial and manufacturing side of the circular economy with emphasis on narrowing and closing material loops, whereas the experiential service economy scenario highlighted the service business side of the story by bringing access and performance over ownership to the focal point in economic actions. The New tribes scenario emphasized the role of people; citizens, consumers, prosumers and communities, and fair sharing economy within the circular economy.

The results of the scenario building exercise were utilised during the business model workshops (the fourth step). Yet, to fully benefit from this kind of forward-looking exercise the scenario building should be even more integrated with the whole business model innovation process from the beginning. For example, a business model ideation could be done in a workshop based on the complementary CE scenarios.

The Third Step: Understanding the Future Customers Through Consumer Discussions and Online Discussions. A consumer discussion panel was organized with 42 consumers. The consumers were divided into five groups, and each group was facilitated by a researcher. The discussions lasted for three hours and they were recorded. The consumers were selected so that there would be a wide representation regarding age, education, profession, type of living and attitudes towards environmental issues. The discussion issues covered five topics: relationship to ownership, recycling, secondary raw material products, lending and sharing of products, and CE services. As a result, data was obtained concerning factors that influence on the willingness to use services instead of buying products as well as boundary conditions for preferring recycled or secondary raw material products. The discussion within the panel was vivid and it remained structured, which helps the utilization of the data. The background of each facilitator might have influenced the course of the discussion to some extent, but on the other hand it was possible to meet this quite well by defining beforehand the time limit for each topic.

Consumer discussions were also held online within an online space called Open Web Lab (Owela). The space facilitates companies to co-create products and services together with their users. The online discussions were focused on paving stones which contain waste wood as a raw material, and are thus products made out of secondary raw materials. The discussion was open for two weeks and there were 35 participants. When necessary, it was facilitated by a researcher. The discussion included four topics: garden construction in general, ecology in garden construction, wooden paving stones and their use, and comparison of materials. The aim was to gain an understanding of consumers' attitudes towards garden construction products made out of secondary raw materials. How do the consumers feel about using them? Are there some doubts related to them and what kind of doubts? What do they appreciate about them? What should the price be compared to products made out of virgin materials? The discussion active, and both research and the case company obtained valuable data for understanding consumers better in a CE. Owela works well as long as success was achieved in attracting a sufficient amount of participants. This can be done, e.g. by sharing information about upcoming discussions actively in social media.

The Fourth Step: Business Scenario Workshops to Seek New Business Models.
The scenarios created during the first step were used as an inspirational background material during business scenario workshops for companies. The aim was to endorse proactive thinking, and move the workshop participants beyond the current situation and to facilitate new business opportunity identification and business model innovation. As a pre-workshop assignment participants were asked to brainstorm and identify preliminary business ideas inspired by step 1 future scenarios.

Three workshops were organized for two companies: two for a company operating in recycling and waste management, and one for a company focusing on digitalization of recycling centres. The workshops lasted for three hours with 1–4 company representatives and 2–3 researchers as facilitators.

The goal of the workshop was to identify business opportunities in a circular economy. The main emphasis was on the anticipatory network value creation element of the business model. The approach was based on a framework for sustainable circular business model innovation [11] and a value mapping tool [12]. The first workshop sessions covered four topics: preliminary vision building; anticipated value to be captured, destroyed or missed; identification of enablers and barriers of value capture; and value opportunities for relevant stakeholders (company, customer, network actors, society and environment).

Half-a-day workshops were considered to be good in content but too short in time to receive optimal results. Therefore, a second workshop session was organized in order to refine and develop further the preliminary business ideas. Due to the practical issues this was done at this stage only with one company.

Even though the general feedback was good and new business ideas managed to be created, the forward-looking approach was rather limited. In addition, the new business ideas had relatively short-term focus and had a low level of innovativeness in the sense of business model innovation. Workshops did not manage to move participants significantly beyond the current situation and optimally facilitate new business opportunity identification and business model innovation. As SME and especially start-up managers are often overwhelmed with solving day-to-day management and business issues, it is understandable that short and sudden forward-looking exercises can be rather challenging. Therefore, more time and effort should be allocated to this topic during the workshop sessions. Since value in circular economy is co-created in value networks, involving key stakeholders and identifying the value for each of them would be essential addition to this step.

The Fifth Step: The Implementation of the Business Idea Through Rapid Experiment or More Complex Pilot. As a fifth step, one of the business ideas emerged during the innovation process was implemented through piloting. This was possible because one of our case companies, the Finnish startup CoReorient (www.coreorient.com) was active in designing and running the pilot with the collaboration of Finnish hardware store and AARRE project. The pilot called Liiteri (www.liiteri.net, www.townhall24.fi), was designed and implemented during September–December 2016 including the design and set-up phase as well as crowdfunding experiment at the end. At Liiteri consumers can rent tools and house cleaning equipment. The gear can be picked-up at the 24/7 Liiteri self-service point (intelligent container) at Teurastamo, Helsinki, which is accessible by public transport. Consumers can also order home delivery with the crowdsourced PiggyBaggy (www.piggybaggy.com) service. Liiteri pilot also offers other services such as virtual shoe repairing service and bike repairing service.

The researchers co-designed with the start-up entrepreneur the research part of the pilot. This was started by using the cards aimed for designing and measuring rapid experiments developed by Strategyzer (www.strategyzer.com). However, soon it was found out that there were challenges to use this ready-made tool in our explorative and qualitative approach. Hence, we modified the cards by adding an explorative approach. The multi-methodological approach was built by combining survey, interviews, Owela discussions and diary kept by the Liiteri personnel. The data was collected from Liiteri users and also consumers who had not used Liiteri. Survey and interviews aimed at gathering data on consumers' opinions, benefits and sacrifices on buying services instead of owning in general. Owela discussions were aimed at concentrating more on ideating and developing further the Liiteri concept. The diary was used to gather the instant comments related to Liiteri gathered at the customer interface.

As a result we got rich data related to consumer's views on the service usage that is also partly applicable to a wider context of circular consumer services. Piloting as a method seems to offer a very valuable data for entrepreneurs but also researchers. Yet, designing and implementation of pilots needs a lot of resources that sets limitations for their usage in companies. In this light implementing a series of limited rapid business model experimentations that do not test or explore the whole business model but some predefined part of it, is easier to design and implement. However, certain minimum level of user experience has to be met in case consumers are involved in the testing. Otherwise poor user experience may influence the business model testing and thus distort the results.

5   Conclusions and Contribution

Our process of circular business model innovation consisted of five steps (Fig. 1). Each step was designed to provide some new, valuable insight into the innovation process. Yet, the following challenges were also confronted. First, there were challenges related to the lack of time and resources as limiting factors. This can be of course prevented by reserving more time for workshops, which might mean dividing them in two sessions.

Second, our aim was to emphasize the forward-looking nature of the business model innovation process. Scenario development method was used as a tool in challenging conventional thinking by reframing perceptions, changing mind-sets and helping to identify future opportunities. The scenario development revealed multi-faceted images of the circular economy. However, it was difficult to link the scenario process with the short term business model development which appeared to be more relevant for the case companies, and on the other hand to move business workshop participants significantly beyond the present stage and mind-sets. Thus, this is one perspective that should be emphasized more in the further studies.

Third, although stakeholders were involved, there is an urgent need to integrate them more to the whole process. After initiating the novel business model or models, there is need for joint discussion concerning the value creation for each stakeholder.

Fourth, in designing and implementing the pilot we confronted challenges related to our explorative approach that hindered us from using simple test cards (for example www.strategyzer.com) and lead us to create our own approach. There is also a risk that finding joint hypotheses with the entrepreneur and researchers might be challenging if the objectives are different. Yet, in our case this was not considered as a risk because of joint understanding based on our close collaboration.

Finally, quite often the lack of active participation is one challenge related to consumer studies, yet we did not confront that problem. One explanation for the active participation might be the fact that consumers found the discussion topic interesting and important to them personally.

The paper aims at creating understanding on circular business model innovation process based on the real case. Our main aim is not to concentrate on detailed steps, but rather take a more holistic viewpoint on the process. However, our knowledge and insights are added with the short description of the used tools and methods in each step. Contribution is also made by introducing our ideas on conducting rapid experiments and pilots; a rather neglected area in business innovation literature.

The process is based on the idea of using co-creation and open innovation actively in multiple stakeholder groups. As such, the study contributes to open innovation, co-creation and business model innovation literature. The study also contributes to the growing literature related to circular business and sustainable business model literature.

Although our ideas and theories are based on academic knowledge, this study has a very strong practical orientation as it describes and analyses the innovation processes of three different case companies and the gained benefits and challenges of the used innovation process model. This is an explorative study, and our next aim is to continue by providing more insightful data focused on each step presented in this study.

Acknowledgements. This research has been conducted as a part of the AARRE (Capitalising on Invisible Value – User-Driven Business Models in the Emerging Circular Economy) project. The authors would like to express their gratitude to the Green Growth Programme of the Finnish Funding Agency for Innovation (Tekes), Technical Research Centre of Finland (VTT), case companies and other parties involved in the AARRE project.

References

1. Kraaijenhagen, C., van Oppen, C., Bocken, N.M.B.: Circular business. Collaborate & Circulate. Circular Collaboration, Amersfoort, The Netherlands (2016). www.circular collaboration.com
2. Lovins, A., Lovins, L., Hawken, P.: A road map for natural capitalism. Harv. Bus. Rev. **77**(3), 145–158 (1999)
3. McDonough, W., Braungart, M.: Cradle to Cradle: Remaking the Way We Make Things, p. 195. North Point Press, New York (2002)
4. Ellen MacArthur Foundation: Towards the Circular Economy. Economic and Business Rationale for an Accelerated Transition, vol. 1 (2012)
5. Boons, F., Montalvo, C., Quist, J., Wagner, M.: Sustainable innovation, business models and economic performance: an overview. J. Clean. Prod. **45**, 1–8 (2013)
6. Stahel, W.R.: The business angle of a circular economy. Higher competitiveness, higher resource security and material efficiency. In: Ellen MacArthur Foundation (ed.) A New Dynamic. Effective Business in a Circular Economy (2014)
7. Osterwalder, A., Pigneur, Y., Clark, T.: Business Model Generation. A Handbook for Visionaries, Game Changers, and Challengers. Wiley, Hoboken (2010)
8. Boons, F., Lüdeke-Freund, F.: Business models for sustainable innovation: state-of-the-art and steps towards a research agenda. J. Clean. Prod. **45**, 9–19 (2013)
9. Sommer, A.: Managing green business model transformations. Dissertation. Springer Verlag (2012)
10. Mentink, B.: Circular business model innovation, a process framework and a tool for business model innovation, in a circular economy. For the degree of Master of Science in Industrial Ecology at Delft University of Technology & Leiden University (2014)
11. Antikainen, M., Valkokari, K.: A framework for sustainable circular business model innovation. Technol. Innov. Manag. Rev. **6**(7), 5–12 (2016). Carleton University
12. Bocken, N.M.P., Rana, P., Short, S.W.: Value mapping for sustainable business thinking. J. Indus. Prod. Eng. (2015). doi:10.1080/21681015.2014.1000399

Resource and Energy Efficiency for Sustainability Advances in Process Industries

Combining Process Based Monitoring with Multi-layer Stream Mapping

Daniela Fisseler[1(✉)], Alexander Schneider[1], Emanuel J. Lourenço[2],
and A.J. Baptista[2]

[1] Fraunhofer Institute for Applied Information Technology, Sankt Augustin, Germany
{daniela.fisseler,alexander.schneider}@fit.fraunhofer.de

[2] INEGI – Instituto de Ciência e Inovação em Engenharia Mecânica
e Engenharia Industrial, Campus da FEUP, 4200 Porto, Portugal
{elourenco,abaptista}@inegi.up.pt

Abstract. For a company it is important to improve resource and eco-efficiency in order to save money, the environment and to improve the company's image. We present a new approach combining Multi-layer Stream Mapping (MSM) and a Business Process Based Monitoring and Control Framework to monitor relevant process variables and use the values as an input for MSM to reduce waste and costs. This combination supports the decision making process and allows to identify major inefficiencies and provides means for more sustainability.

Keywords: Multi-layer Stream Mapping · BPMN · Process based monitoring · Resource efficiency · Waste reduction · Domain modelling

1 Motivation

For a company increasing resource efficiency (such as reducing energy consumption for the same output) and eco-efficiency (such as improving economic value and/or environmental impact) does not only mean to save money, but it can help protect the environment and at the same time positively influence the image of this company. A survey, which was published by the European commission[1], indicates that environmentally-friendly production becomes more and more important for consumers and for many it is even more important than the price. But in order to optimize both resources and the eco efficiency, one first has to monitor and analyze values relevant for the resource consumption. A first step is to identify the process variables which are relevant and define Performance Indicators (PIs). But it is not enough to identify and assess these values as an off-line analysis basis, they also have to be measured or monitored in order to get an in-line overview of the current status and enable continuous analysis capabilities.

Also, in order to be able to improve the use of resources and consequently to reduce waste in a process, the process itself needs to be known in detail and be somehow documented. One way is to model business processes with a specific language, such as the Business Process Modelling and Notation (BPMN). BPMN is a graphical language

---

[1] http://ec.europa.eu/public_opinion/flash/fl_367_en.pdf, 11.01.2016.

© Springer International Publishing AG 2017
G. Campana et al. (eds.), *Sustainable Design and Manufacturing 2017*, Smart Innovation,
Systems and Technologies 68, DOI 10.1007/978-3-319-57078-5_53

developed to help as a means of communication between technical and non-technical personnel [1]. BPMN models can also be executed in process engines and it is one of the de-facto standards used today in industry. In order to be able to analyze data in a much deeper way, such data coming from sensors and production systems needs to be mapped to process steps and annotated automatically. This way the process can be monitored in real-time and the data can be used for a deeper analysis.

In order to assess the overall resource and eco efficiency of a given production system, the so called Multi-layer Stream Mapping (MSM) is applied in this work. MSM can be traced in its base fundamentals in the traditional Value Stream Mapping as basis, but it allows an overall efficiency assessment of production systems. In this paper we present a new approach which combines this MSM with a monitoring and control framework for business processes. The latter will provide the data for performing the MSM methodology calculations.

The next sections are structured as follows: Sect. 2 gives a brief overview of the state of the art, Sect. 3 explains in detail the MSM approach.

Section 4 explains how business process based monitoring can be used as an input for MSM. Section 5 depicts an example from real-life and how the before described approaches would be implemented within this scenario. Finally, Sect. 6 gives a conclusion and an outlook on possible future works.

2 State of the Art

In the past lean production principles and tools helped improve productivity notably by optimizing human resources and used materials. One of these tools is Value Stream Mapping (VSM) which originates in the Toyota just-in-time production system [2]. The core idea is to visualize the value stream and analyze where wastes, i.e. non-value added actions, can be reduced. The value stream analysis should include the whole process, that means a collection of all actions (actions that add value and actions that do not) which are required to bring a product or a group of products through the main flows, starting with raw material and ending with the customer which means the inclusion of raw materials to delivery [3] including the material and information flows. During the analysis, for each relevant process step it will be determined how much of the time needed is value added and how much is non-value added. In a welding station, for example, the welding itself would be value adding while the transport from and to the welding station and possible waiting states would not be non-value adding, or in other words: time in value adding actions is spent in a productive manner while in non-value adding actions time is spent in a non-productive manner as it does not add value to the examined product or service.

The proportion of value adding activities compared to the total efforts consumed gives a good indication of potential optimization opportunities.

There are extensions of this concept such as Quality Value Stream Mapping (QVSM), which addresses "the issue of a suitable integration of testing processes within the process chain" [4], yet, this method only emphasizes quality aspects which can be seen as a shortcoming, since it has a reduced spectrum to meet the current industrial challenges, particularly in terms of resource efficiency. Moreover, several tools address the extension of VSM to incorporate additional criteria, for instance, energy-related and sustainable aspects focused on environmental performance. For instance, the Sus-VSM, is used to evaluate economic, environmental and social sustainability performance in manufacturing [5]; while the Sustainable Manufacturing Mapping approach focuses on the application of a VSM based assessment as an integrated visualization and monitoring method for environmental impacts and production control [6].

However, these approaches lack the means to easily gain an overall efficiency assessment of production systems. They also lack the assessment of efficiency perform- ance of individual process steps and multi process parameters, which is important since efficiency assessments are based on the principle that the efficient material and energy use can reduce natural resource inputs and waste or pollutant outputs.

Moreover, there is a clear need to adapted and integrated modern technologies for process monitoring and optimization for a wider and facilitated adoption of state-of-the- art tools, namely MSM, for resource efficiency, enabling the support of the decision making process on the spot and avoid errors related to human mistakes managing the data.

Another way to get an overview of the current status of the system is to define and monitor Key Performance Indicators (KPIs), such as cycle times or reject ratios. There are a lot of different systems available on the market to fulfill this task. One possibility is to use Business Activity Monitoring (BAM). Its goal is to monitor processes and control certain aspects in real-time [7]. The data is gathered from multiple systems but due to the different abstraction levels of the data and the heterogeneity of the systems involved, it is often quite challenging to get access to the relevant data. It must be designed and implemented individually for every specific use case. Although an approach has been described on how to combine BAM with BPMN [8] this is still lacking improvements of our approach described in this paper e.g. support through standard tools for creating the models (UML and BPMN diagrams) and the integration of the different models in order to configure the involved software components for the auto- matic data monitoring and annotation.

3   Multi-layer Stream Mapping

Multi-layer Stream Mapping follows the VSM logics, and gives an overview of value adding and non-value-adding elements. But MSM overcomes some of the shortcomings of VSM, in particular it was designed to enable an overall efficiency assessment of production systems and the single process steps, which is not possible with VSM. Addi- tionally, the MSM provides the means to evaluate the costs related to misapplications and inefficiencies in a disaggregated form (valuable costs and wasteful costs).

There are 4 phases to this method, which need to be executed after each other. In the first phase a value stream map needs to be created. It assesses the value adding actions versus non- value adding actions. For this, all the actions or process steps which are needed to produce a given product have to be known and listed. In Fig. 1, these actions would be P_1, P_2, and P_3 on top of the graphic. Actions are either marked as value adding

(VA) or non-value adding (NVA). When VA actions executed, there might still be periods of time which are NVA, for example if there is a waiting state in between. In Fig. 1 this is indicated by VA and NVA: For the process variable "Time" PT_1 is value adding, while everything spent in WT_1 is waste or non-value adding.

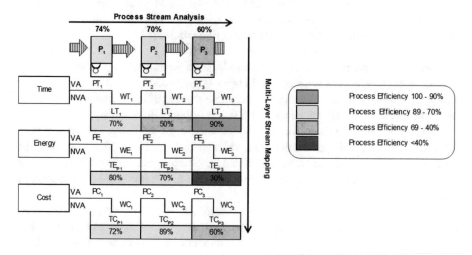

Fig. 1. MSM expanded diagram with visual management attributes [9]

In the second phase variables and KPIs are to be systematically evaluated through efficiency ratios, which is the value adding part proportionally to the overall value. The values of the variables should always be between 0 and 100% and is calculated as follows:

$$\Phi = \frac{\text{Value added fraction}}{\text{Value added fraction} + \text{Non-value added fraction}} \tag{1}$$

For this all variables which effect the value chain need to be identified, including KPIs. The latter should be designed in a way that they are to be maximized. Variables which might be relevant could be for example raw material, electrical energy consumption or CO_2 emission reduction. In Fig. 1 efficiency ratios of the process variable "Time" are 70% for the first action P_1, 50% for P_2, and 90% for P_3.

In the third phase the results will be color-coded in order to distinguish efficiency ratios more quickly. Green is used for high efficiency (90–100%), yellow for 70–89% efficiency, orange for 40–69% efficiency, and red for high inefficiency (below 40%) as can be seen in Fig. 1, which depicts how the process, its process variables, and efficiency ratios will be displayed within MSM expanded diagram with visual management.

During the fourth phase the efficiency of the selected variables are aggregated, this will give place to the unit process efficiency. Figure 2 shows how such as scorecard will look like and how it can be included with the work done in the first three phases. The global efficiency is the average of the single process efficiency values.

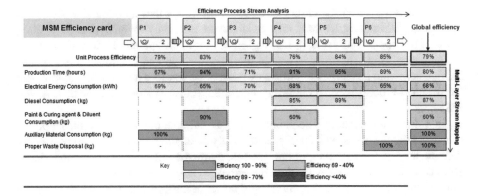

Fig. 2. MSM efficiency scorecard

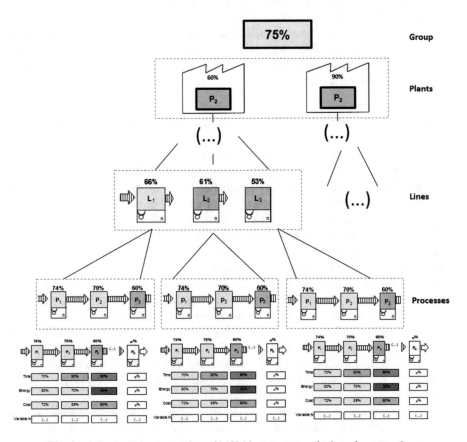

Fig. 3. Schematic representation of MSM bottom-up analysis and aggregation

This outcome is useful to assess the efficiency as well as the misuses and waste within the process. Ultimately, MSM is intended to be used not only for analytical evaluation, but also to support the decision making process, namely greenfield design or online systems monitoring, in order to enable the identification and quantification of major inefficiencies and keep track of efficiency progress. Therefore it is sensible to combine the Business Process based monitoring and control framework described in Sect. 4 with MSM.

In order to implement this combination as an online monitoring system, set points should be carefully defined via a specific analysis of the processes in study, i.e., by defining the VA figure that represents the best figure theoretically attainable in order to eliminate/reduce the NVA activities. Nevertheless, target setting strategy should keep in mind that the set-points (targets) should be set in a manner that does not affect the collaborators' motivation. After the definition of the internal set-point, i.e. VA value for each process variable in each process step, the framework can provide the total value of the process variable and consequently the efficiency ratio can be automatically calculated, providing near to real time efficiency results and scorecards.

Moreover, if data is collected at the most elementary level, for instance data form the machine on the shop-floor, it is possible to consecutively aggregate the efficiency along production system, processes, lines, sectors, (…), or even plants, adopting a bottom-up analysis as depicted in Fig. 3.

4   Business Process Based Monitoring and Control Framework

To apply the MSM methodology it is necessary to gather the data for the selected variables first. Unfortunately, this can be a time-consuming and difficult task. The data could be received from a Business Activity Monitoring Software, but then it is still unclear where the data comes from and maybe also how to get access to it. Another possibility to solve this challenge is to use the framework for processed-based monitoring presented in [10, 11]. This framework helps defining the data sources, to gather and monitor the data and associating it in real-time with process information and additionally enables automated reactions in case unwanted events occur.

First, a domain model is created which includes all the relevant machines, sensors, actuators, and other data sources. Then a process model is created with all the relevant process steps which are included when using the MSM methodology using Business Process Modelling and Notation (BPMN). For MSM one important aspect is to determine which actions are value adding and which actions are non-value adding. In the BPMN model the process step can be marked like that (by defining a corresponding field for the activity), which ensures that only the values which are required will be processed.

The data sources from the domain model are then mapped to the individual process steps. The general architecture is shown in Fig. 4 and consists of several layers.

The data sources depicted in the lowest level provide the information needed for the variables and KPIs identified during the implementation of the MSM methodology. The data is then distributed, for example in the form of MQTT events. The application layer annotates the event with the metadata of the process step according to the mapping done during design-time. Additional calculations will be performed by the application logic layer as well, if needed. The functionality of the analytics layer and the interface layer will be provided by the MSM methodology as presented in the previous section. The analytics layer will assess the variables in regards to their efficiency and aggregate values for the overall efficiency calculations. The interface layer then provides means to display the results in an easy to understand manner, such as the efficiency scorecard from Fig. 2.

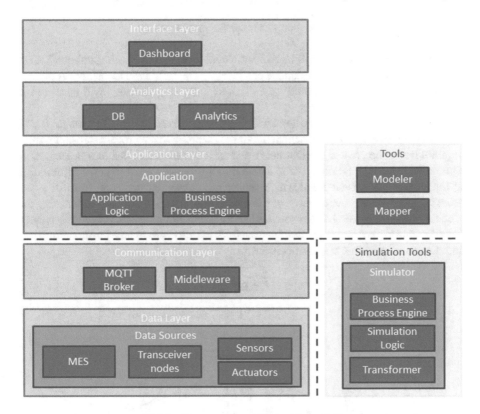

Fig. 4. Framework architecture from [11]

The tools depicted in Fig. 4 are only needed during design time, not during run time. They help to actually create the models and to map them together.

The simulation tools can be used to produce simulated data, in case the MSM is to be used to evaluate new set-ups or changes in the production process.

5   Use Case: Injection Moulding Process

In our chosen use case, we look at a company which produces coffee capsules for coffee machines. The plastic used, needs to be brought in the right form before it can be further processed. First, the plastic will be heated in order to be injected into the casting mould.

Then the mould is closed and the liquid plastic will be injected. After this, the plastic needs some time to cool down again, before the mould can be opened and the plastic capsule can be extracted. When modelling the process with BPMN, it could look like in Fig. 5.

Fig. 5. Injection moulding process model

For each process step, it will be defined whether the action is value adding (VA) or non-value adding (NVA). VA actions are "Heat", "Inject", and "Cool". NVA actions are "Close mould", "Open mould" and "Extract". These parameters will be stored as parameters in the BPMN model.

The next step is to model the domain. In Fig. 6 a Unified Modeling Language (UML) class diagram of the domain is depicted. For the mapping to the process step an instance diagram has to be created, so that every sensor instance is mapped to a process step. This enables the annotation of the received monitoring data point with the information of the current process step and which sensor the data point was read from. It is not necessary to connect every sensor of the machine so that one can start monitoring the process with just a few key sensors and if sufficient optimization potential has been identified further sensors could be added easily. This allows to reduce the modeling time and therefore make this step more efficient.

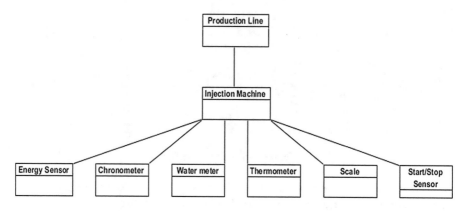

Fig. 6. Domain Model for injection moulding

With this approach it is possible to measure various KPIs of the processes within the Injection machine and the related systems such as an Enterprise-Resource-Planning (ERP) system. It is possible to measure energy consumption during specific process steps, cycle time for each process step, water consumption and water infeed temperature, and the mass of the material. In addition, the machine can also detect planned and unplanned downtime, the overall time in use as well as the effectiveness of meeting the deadlines of customer orders.

Next, the KPIs are selected in order to create the MSM scorecard. In Fig. 7 the first conceptual creation of the MSM matrix can be seen. Finally, the data will be collected, the selected KPIs will be calculated during runtime and displayed in the digital representation of the MSM scorecard.

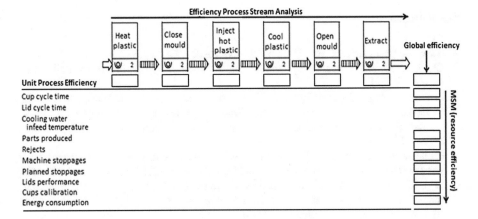

Fig. 7. Structure MSM scorecard for injection moulding process

6 Conclusion and Outlook

This paper presented the conceptual steps needed in order to combine Multi-layer Stream Mapping (MSM) and business process based monitoring and control framework. While MSM is a good tool to assess overall performance and efficiency in order to reduce waste and costs, the process oriented monitoring and control framework provides real real-time monitoring as well as the automatic annotation of the data required for the calculation of the MSM results. This combination supports the decision making process and enables "on the spot" decisions. It also helps to identify major inefficiencies and consequently underpins sustainability. Moreover, automated data collection for the energy and resource consumption, as well as other process related activities in near-real-time along with the definition of an internal threshold for each process variable in each process step, will enable MSM to "trigger alarms", therefore managers and operators can take action immediately.

The next goal will be to install the process oriented monitoring and control framework and apply the MSM methodology for the described use case in a real production environment. This way, the process can be monitored in real time and the necessary data for applying MSM is collected. We also envision to refine the process diagram to increase the detail level and enable a much more refined analysis. This will be done by decomposing the process steps into more fine-granular ones so the ratio of value added vs. non-value added work can be calculated with increased precision. If changes in the process are to be simulated, the framework can also provide the simulated data so that new process layouts can be tested.

Acknowledgements. Work presented in the paper has been supported by European Union's Horizon 2020 research and innovation program under grant agreement No. 680570 "MAESTRI - Energy and resource management systems for improved efficiency in the process industries".

References

1. Business Process Model and Notation (BPMN) Version 2.0 (2011)
2. Ohno, T.: Toyota Production System: Beyond Large-Scale Production. Productivity Press, Portland (1988)
3. Rother, M., Shook, J.: Learning to See: Value Stream Mapping to Add Value and Eliminate Muda, 2nd edn., Lean Enterprise Institute, Brookline, Massachusetts, USA (1999)
4. Haefner, B., Kraemer, A., Stauss, T., Lanza, G.: Quality value stream mapping. Procedia CIRP **17**, 254–259 (2014)
5. Faulkner, W., Badurdeen, F.: Sustainable value stream mapping (Sus-VSM): methodology to visualize and assess manufacturing sustainability performance. J. Clean. Prod. **85**, 8–18 (2014)
6. Paju, M., Heilala, J., Hentula, M., Heikkilä, A., Johansson, B., Leong, S., et al.: Framework and indicators for a sustainable manufacturing mapping methodology. In: Proceedings of the 2010 Winter Simulation Conference (2010)
7. http://www.gartner.com/it-glossary/bam-business-activity-monitoring. Accessed 02 Dec 2015
8. Friedenstab, J., Janiesch, C., Matzner, M., Müller, O.: Extending BPMN for business activity monitoring. In: 2012 45th Hawaii International Conference on System Science (HICSS), pp. 4158–4167 (2012)
9. Lourenço, E.J., Baptista, A.J., Pereira, J.P., Dias-Ferreira, C.: Multi-layer stream mapping as a combined approach for industrial processes eco-efficiency assessment. In: Nee, A., Song, B., Ong, S.K. (eds.) Re-engineering Manufacturing for Sustainability, pp. 427–433. Springer, Singapore (2013)
10. Fisseler, D., Reiners, R.: Prozessorientierte Überwachung in der Produktion. In: Cunningham, D., Hofstedt, P., Meer, K., Schmitt, I. (Hrsg.) INFORMATIK 2015 Lecture Notes in Informatics (LNI), pp. 917–927. Gesellschaft für Informatik (2015)
11. Fisseler, D., Reiners, R., Kemény, Z.: Monitoring and control framework for business processes in ubiquitous environments. In: Proceedings of 2016 International IEEE Conferences on Ubiquitous Intelligence & Computing, Advanced and Trusted Computing, Scalable Computing and Communications, Cloud and Big Data Computing, Internet of People, and Smart World Congress (2016, in press)

Virtual Sector Profiles for Innovation Sharing in Process Industry – Sector 01: Chemicals

Hélène Cervo[1(✉)], Stéphane Bungener[2], Elfie Méchaussie[2], Ivan Kantor[2],
Brecht Zwaenepoel[3], François Maréchal[2], and Greet Van Eetvelde[1,3]

[1] INEOS Europe, Lavéra, France
eposproject@ineos.com

[2] EPFL Valais Wallis, Industrial Process and Energy Systems Engineering, Sion, Switzerland

[3] Energy and Cluster Management, Faculty of Engineering and Architecture,
Ghent University, Ghent, Belgium

Abstract. Production data in process industry are proprietary to a company since they are key to the process design and technology expertise. However, data confidentiality restrains industry from sharing results and advancing developments in and across process sectors. Using virtual profiles that simulate the typical operating modes of a given process industry offers an elegant solution for a company to share information with the outside world. This paper proposes a generic methodology to create sector blueprints and applies it to the chemicals industry. It details the profile of a typical chemical site based on essential units and realistic data gathered from existing refineries and chemical plants.

Keywords: Virtual profile · Sector blueprint · Industrial data · Chemicals sector · Refineries · Data confidentiality · Industrial symbiosis · Energy efficiency

1 Introduction

Given the confidential nature of industry, handling process data typically implies non-disclosure agreements and – often exclusive – intellectual property rights granting protection to a company's background assets. More particularly, industrial data require protection because they are central to a company's plant or process design, crucial to their technology expertise and investments, essential in trade secrets, and most of all vital to feed innovation. Therefore, a good management of intellectual property is key to a company's competitiveness.

Securing and protecting data also has drawbacks, however, as industries are restrained from sharing best practices, advancing developments in and across sectors, transferring technology, building industry databases and benchmarks, feeding sector positions, creating synergies, etc. Additionally, it complicates the outreach of findings or results from research and innovation (R&I) projects, even when funded through public grants. Horizon 2020 is a key example of such R&I programme implementing the European Innovation Union. Through public-private partnerships such as SPIRE [1], companies are encouraged to participate in R&I projects but the issue of data confidentiality

G. Campana et al. (eds.), *Sustainable Design and Manufacturing 2017*, Smart Innovation,
Systems and Technologies 68, DOI 10.1007/978-3-319-57078-5_54

remains the biggest hurdle for process industry to enter a project as full partner with a site-specific role.

Publications resulting from industrial research in collaboration with universities are intrinsically bound to a degree of public availability. Today, emerging concepts such as open source and open innovation, or industry-intrusive policies such as closing the loop of resources using circular economy principles, invigorate the need for answers on how to disclose information and facilitate communication across industry, academia and society. However, there is still a lack for publically available and reliable databases gathering information about industrial sectors operations and providing case studies to test and verify scientific developments. Furthermore, investigations with respect to data anonymisation techniques such as k-anonymisation, generalisation and Bucketisation, have been mainly carried in the field of individuals' data protection [2, 3] and have not been developed for industrial datasets.

The need to disclose industrial information and share project results has thus triggered the idea of creating a typical chemical site and a virtual sector profile. Anonymising industrial data as to enable communication between project partners and across process sectors demands for standardised methodologies. Therefore, typical industry sites and profiles per sector should be based on reality, generated through a recognised method and present representative data that are acceptable to all parties, industry and sector associations such as CEFIC for the chemical industry [4].

Building from the experience above and reinforced by the challenge of industrial symbiosis in the SPIRE project EPOS [5], this article presents a methodical and industry-proof approach to communicate on the chemicals sector by creating a virtual chemicals sector profile. This first section of the paper is followed by an analysis of the data needs and the structure of a virtual site in Sect. 2. Section 3 shows the results of the methodology applied in the chemicals industry, followed by applications of such virtual profile in Sect. 4. A final section summarises the conclusions of this work.

2 Data Sourcing

As described in the introduction, typical industrial sites and profiles per sector should be based on reality and accepted by all parties, including sector associations. They should have identifiable units, use verifiable data and provide the means to present solutions matching the operating conditions of a real plant. A compromise should thus be found between the resolution and the relevance of data so as to select a representative set of data.

This section develops the methodology for building the profile of a typical chemical site based on essential units and realistic data gathered from existing refineries and chemicals plants.

2.1 Generic Site Layout

The generation of a virtual chemical site profile implies the identification and characterisation of the major mass and energy flows across and within the site's boundaries, as well as their connections with the different building blocks composing the plant. Following a top-down approach, three levels of details to represent a plant layout can be defined: *black box*, *site map* and *process* unit representations.

Black box. The lowest level of detail is called the *black box* representation and is represented on Fig. 1 by the external black rectangle. Any cluster, site or plant can be characterised by its input and output flows of material and energy. This representation of a plant is the most general form and only gives insight into the main requirements or outputs. Each input and output flow can be used as an interface to connect with other sites or can be viewed as opportunities for internal usage as well. Nonetheless, this view on a site cannot provide details on opportunities for internal recycling or information about the real needs of the site in terms of material and energy unless it is assumed that the site is internally optimised and no further improvements are possible.

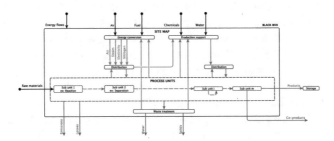

Fig. 1. Plant layout – representation levels [6]

Site map. The next level of detail decomposes the site into its major sections, related to the production process, utility system or waste treatment and production support areas. This representation allows external observers to become familiar with the general layout of the plant and provides additional detail on the site operations. It enables understanding the main transformation steps onsite, from feedstock to product, including the major flows of materials and energy. Likewise, the utility demand and supply is represented per plant section. The boundaries of the sections are chosen according to the views of site engineers and the common understanding of the site.

This representation of an industrial site gives a good overview of the process units onsite, the supporting infrastructure and their interconnections. Numerous indications can be derived, such as the main energy consumers, the energy and mass balances per unit, the cold and hot utility distribution profiles or also the units' specific energy consumption. However, the process units' energy requirements (e.g. temperature levels and heat loads) are unknown, therefore it is not yet possible to generate a sufficiently detailed site profile using this second level of detail.

Process units. For drawing a virtual profile, a third level of detail is required which enters the production process per unit. Via block flow diagrams the transformation steps per process unit are visualised. Each process unit groups a number of steps to transform material and energy inputs into useful outputs or final products.

The links between the process units are systematised and the logical flow of materials between the process units is captured. Again the units are defined from an engineering point of view by identifying the transformative operations and defining the boundaries for each process unit. While this is the most detailed site representation, the units remain common to chemical processes across the sector. As an example, steam-methane reforming (SMR) processes consist of similar transformation steps at most manufacturing sites.

With the deepest level of detail defined as the process unit, the next step is to select and allocate the necessary data to build the virtual profile of a chemical site.

2.2 Simulated Data and Streams

Energy flows. The energy vectors for the different representations identified in Sect. 2.1 will vary according to the level of detail; for example, the black box level may show an input of natural gas while the subsequent levels show the conversion of gas into steam using a utility boiler. In order to build the virtual energy profile, data should first be collected with respect to the utility production and distribution (e.g. steam boiler and steam network). In a second step information about the utility consumption should be gathered. Mass and energy balances on the utility network should be closed at all points, considering losses inherent to large chemical sites.

While the utilities mass and energy flows are characterised by direct thermodynamic properties (temperature, pressure, low heating value, etc.), the utility consumption in process units is defined according to the real energy requirements at the interface between the process and the utility usage. This *dual representation* facilitates the definition of the utility requirements at the process unit level, reducing the complexity of data collection and enabling the identification of indirect/direct internal and external schemes for heat recovery (improve existing heat exchangers network, install external hot water circuit, etc.), heat pumping and cogeneration. Dedicated sources provide for more detailed explanations on data required to characterise process and utility streams [7].

Mass flows. Mass flows differ from energy flows due to the variable requirements of different processes and the plethora of possibilities for chemical conversion. Options for conversion of mass streams are almost unlimited.

For a plant represented by a black box, the interface with the rest of the world exists as the feedstock requirement and the production rate of the outputs which is often enough to understand the main purpose of the site and explore generic new business opportunities. However, mass requirements should be represented in a more detailed way since it could reveal material needs which do not exist at the higher levels, an example being a requirement for hydrogen in a process unit while the site imports only natural gas and generates hydrogen in a utility SMR unit.

A further separation of mass flows is made to distinguish process needs from the transformation of inputs to outputs. *Materials flowing in a network* within a site can be defined at each point by their pressure, temperature, quality and flowrate such that the utility system is defined to provide these specifications (e.g. water network [8], hydrogen network [9]), they can be cascaded between process units or accessed at various points throughout the connected nodes. *Material feed* and *product flows* are defined similarly, though they are typically used in fewer points throughout the plant and are immediately transformed. These feedstock and product streams are the basis for the size of the process/plant in the virtual profile and are the basis by which the utility and production support streams are scaled.

Simulation method. Virtual profiles must be based on realistic and relevant data. However, companies are reluctant to disseminate specific data on their sites as they wish to retain the technical expertise and process specificities giving an advantage over their competitors. Thus, it is necessary to start from real plant data and apply a method that does not discern the specificities of the plant operations and process units. In the generation of a virtual chemical site profile, the anonymisation of data is coming from several aspects.

First of all, only the major mass and energy flows are included, even at the deepest level of detail. Sensitive specificities of processes linked to competitive advantages, mostly related to equipment design and operation, catalyst characteristics or process-process integration (direct heat exchange between two or more process streams), are left out of the profile construction process and can therefore not be determined from the resulting representation.

The characterisation of the utility consumption within process units is made at the interface between process and utility streams, limiting greatly the disclosure of internal heat recovery schemes. For simplicity and generalisation purposes, the focus is being put on the most important heat requirements. A Pareto approach [10] can be followed to characterise 80% of the energy consumption, or a targeted approach can be preferred, focusing on the most significant energy use as defined by the ISO 50001 [11].

Finally, the construction of the profile implies the consideration of several process units, therefore all of them are merged together making it hard to precisely identify their specificities. No relevant information is lost in that process since the optimisation is made at the site level.

3   Typical Chemical Sites

This section introduces a virtual profile for the chemicals sector. The profile is obtained by applying the methodology described in Sect. 2 and is further elaborated to produce a typical heat and power profile of a chemical site. The methodology can be extended to material streams or can focus on specific resources, waste streams, carbon emissions, etc. It is also applicable to various units giving the virtual profile some requisite flexibility (PUs can be added or removed if needed).

The case study presented in this paper concerns an industrial site made up of a refinery (site R) and a chemical site (site C); it is referred to as a Typical Industrial Site (TIS) [6].

3.1   Typical Process Units

There are thousands of different chemical products, hence the profile that is built requires a helicopter view. It should be complete enough to give a general idea of the different conversion pathways that exist for a given chemical, while remaining sufficiently detailed to share information or discuss solutions across industries or with external actors.

The chemicals industry can be divided into two major subsectors, both either petroleum or bio-based [12]: refineries converting raw materials such as crude oil or biomass into fuels, chemicals and power and chemical sites which convert olefins, aromatics and natural gas into higher value products [13].

The refining and chemicals sectors are briefly described below as well as their main process units (PUs) and process flows.

Refineries.   Figure 2 shows the main PUs constituting a refinery as well as their main input and output flows. The conversion efficiency of a refinery can be as high as 93%, meaning that 7% of the initial mass of crude oil is either used to provide energy to the PUs or evacuated as a waste product [14]. Waste streams are typically small as low value petrochemical fractions can be transformed into higher value products in PUs.

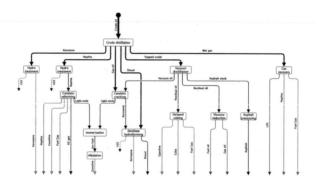

Fig. 2.   Refinery main process units and mass flows [6]

Chemical sites.   Figure 3 illustrates the conversion pathways of oil fractions and methane into some of their end products. Each conversion of a feedstock into intermediary product implies significant energy input or output as well as many different types of units/processes: reactors, distillation columns, strippers, etc. The PUs selected for the creation of the profile are typical of a petrochemical site having a cracker unit. They are highlighted in red in Fig. 3.

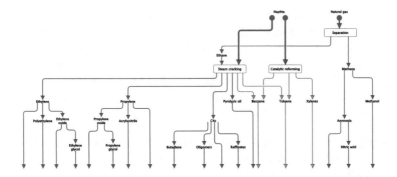

Fig. 3. Schematic of petrochemistry conversion pathways [6]

3.2 Typical Energy Profiles

Refineries and chemical sites are significant energy consumers. The utility system as well as production support units deliver streams such as demineralised water, steam and electricity which are essential for the PUs operations.

Heat profile. A Total Site Analysis (TSA) is performed in order to obtain the typical heat profile of the TIS. TSA [15] is a methodology where the utility system is integrated in the analysis and heat can indirectly be exchanged between processes using intermediate utility networks (ex. steam, hot water). In order to apply TSA, specific data are required to obtain the enthalpy profiles of both the PUs and utility system.

The systematic approach developed in [7] is followed and the required data set about the steam and cooling demand are integrated into the virtual profile. The *steam demand* corresponds to the amount of steam that must be produced by the utility network to supply PUs and utility demand with steam. It may include: boiler preheating, demineralised water degassing, tank tracing and steam turbine activation amongst others. Thermal losses and physical losses (leaks and condensation) also contribute towards the steam demand of an industrial site and should be included. The *cooling demand* is usually required after separation or to remove heat from exothermic reactions. The principal utilities used for cooling are air and water cooling.

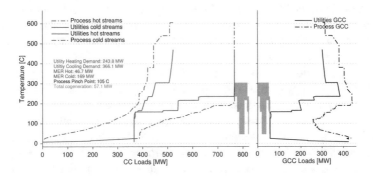

Fig. 4. CCs and GCCs of typical industrial site for a given time period [6]

The Composite Curves (CCs) and Grand Composite Curves (GCCs) of the TIS are shown in Fig. 4 for one specific point in time. However, the data are made available for multiple periods in order to account for the different site operating modes and the entire system variations. A specific methodology for clustering typical operation periods was developed in [16]. Since refineries and chemical sites are not necessarily coupled, CCs and GCCs can also be generated separately for site R and C, in order to analyse each of the site heat profile independently.

Electricity profile. Electricity is used as a resource in many production steps and it is critical to the safety of the system. A significant amount of the total power used by a site is often generated on-site (50-100%). Two types of electricity production are typically found: *internal combustion* and *turbines*. The electrical consumption on a chemical site can be broken down into two main categories: pumping (50%) and compressing (43%). It is important to note that many processes in the chemicals industry have longer time constants than the electrical grid, resulting in a potential for demand response programmes [17].

The backbone of the TIS electricity system is the Medium Voltage (MV) rail system. This system is composed of redundant 3-phase rails with a voltage between 3 kV and 36 kV. These rails are fed by High Voltage (HV) transformers, coupling the site to the external electricity grid with a working voltage between 70 kV and 400 kV. The switchyard is the central node in the sites electricity system, which connects all units onsite. From the switchyard, electricity is fed into the site. Transformers feed distribution boards from this backbone. Large units might have a direct MV connection, e.g. large steam turbines. MV is used to distribute and transport the bulk of the electrical energy around the site. The MV system is also duplicated to provide redundancy and maintainability of the electrical system. Different paths can be selected from the switchyard to the distribution boards. In distribution substations onsite, the MV is finally transformed to low voltage (between 1 kV and 400 V). This is the working voltage of most small (<100 kW) machines and equipment.

4 Virtual Profile Applications

Once the concept of virtual sector profiles is adopted by industry and endorsed by the sector associations, the blueprints can be used to share best practices, screen new business potential across process industries or to challenge new policies. The virtual sector profiles are made to gain generic insights, produce a holistic site view, point to potential solutions or suggest possible ways forward for a plant or a sector.

Such an application can be illustrated via the use of the virtual energy profile [6]. First it shows the heating and cooling requirements of typical processes that can be found in the chemical industry while illustrating the way utilities deliver or remove the process energy. It thus enables to estimate the generic and internal optimisation potential of the system. On the other hand, this representation also highlights how much excess heat is available and transferable to potential external partners (district heating). Finally, it can also be used to display the opportunities for creating a common utility network with neighbouring sites in order to optimise the cogeneration and the energy integration.

However, it should be noted that for each specific company or site case, an in-depth study must be completed to reach the required profundity for a grounded decision.

EPOS as an example. Projects such as EPOS [5], also prove the broad utilisation of virtual profiles. Prior to finding interactions and exchanges across process sectors, the sector blueprints are first used to create knowledge and a basic understanding of each process industry involved in the project (chemicals, steel, cement and minerals). Typical industrial sites are secondly used to map opportunities as well as barriers for cross-sectorial industrial symbiosis. In essence, the typical profiles of each sector are matched to explore industrial symbiosis (IS) potential without the need to disclose confidential data. Once an IS opportunity has been detected, the process is internalised and the involved industrial partners assess and validate the option using their own resources and site-specific data sets. When an opportunity results in a factual symbiosis, the economic, environmental, social and overall sustainability gains are calculated and extrapolated to the sector level. The process also works in reverse, starting from societal targets and pointing to new business opportunities across sectors by using the simulated sector profiles. Finally using virtual profiles also ensures a broad dissemination of the solutions detected through the entire EPOS project.

5 Conclusion

A generic methodology for building sector blueprints has been proposed and is applied to the chemicals sector. The virtual chemical site profile, based on essential units and realistic data gathered from existing refineries and chemical plants, provides a welcome solution for a company to share best practices and innovative solutions with the outside world without disclosing confidential information.

Data required for building a sector profile must be gathered at process unit level and that a trade-off should be made between the resolution and the relevance of information. The method is proven for building a typical heat and power profile of the chemicals sector and can be extended to mass flows covering resources and specific utilities. The profile is flexible since typical PUs can be added or removed while following the same methodology. At all times the virtual profile should be representative of the various operating modes of a real site in order to generate realistic solutions.

The use of virtual profiles opens up a range of applications. They enable sector benchmarks, substantiate sector positions, fuel prospection and impacts studies, etc. Similarly they grow the understanding of the chemicals sector, accelerate discussions between chemical companies and facilitate interactions across process sectors. Sectors blueprints also provide a welcome means to publish and present project results. They enable *researchers* to challenge scientific methodologies and tools and validate integrated solutions. Finally, they provide a common ground to *policymakers* for trying new strategies or, most of all, increasing the outreach of innovation projects funded through public grants.

Acknowledgment. The EPOS project has received funding from the European Union's Horizon 2020 research and innovation programme under grant agreement No. 679386. This work was supported by the Swiss State Secretariat for Education, Research and Innovation (SERI) under contract number 15.0217. The opinions expressed and arguments employed herein do not necessarily reflect the official views of the Swiss Government.

References

1. SPIRE. https://www.spire2030.eu/
2. Mogre, N.V., Agarwal, G., Patil, P.: A review on data anonymization technique for data publishing. Int. J. Eng. Res. Technol. (IJERT) **1**(10) (2012)
3. Cormode, G., Srivastava, D.: Anonymized data: generation, models, usage. In: Proceedings of the ACM SIGMOD International Conference on Management of Data (SIGMOD 2009), Providence (2009)
4. CEFIC – The European Chemical Industry Council. http://www.cefic.org/
5. EPOS – Symbiosis in Industry. https://www.spire2030.eu/epos
6. Bungener, L.G.S.: Energy efficiency and integration in the refining and petrochemical industries. Ph.D. thesis, Ecole Polytechnique Fédérale de Lausanne (2016)
7. Méchaussie, E., Bungener, S., Maréchal, F., Van Eetvelde, G.: Methodology for streams definition and graphical representation in total site analysis. In: 29th International Conference on Efficiency, Cost, Optimization, Simulation and Environmental Impact of Energy Systems (ECOS 2016), Portorož (2016)
8. Natural Resources Canada, Pinch Analysis: For the Efficient Use of Energy, Water & Hydrogen (2003)
9. Hallale, N., Liu, F.: Refinery hydrogen management for clean fuels production. Adv. Environ. Res. **6**(1), 81–98 (2001)
10. Juran, J.: Pareto, lorenz, cournot, bernoulli, juran and others. In: Industrial Quality Control, p. 25 (1960)
11. ISO 50001:2011, Energy management systems – Requirements with guidance for use
12. De Jong, E., Higson, A., Walsh, P., Wellisch, M.: Bio-based Chemicals: value added products from biorefineries. IEA Bioenergy Task 42 Biorefinery (2010)
13. Petrochemistry – Association of Petrochemicals producers in Europe. http://www.petrochemistry.eu/flowchart.html
14. Wang, M., Lee, H., Molburg, J.: Allocation of energy use in petroleum refineries to petroleum products implications for life-cycle energy use and emission inventory of petroleum transportation fuels. Int. J. Life Cycle Assess. **9** (2003)
15. Dhole, V.R., Linnhoff, B.: Total site targets for fuel, co-generation, emissions and cooling. Comput. Chemicals Eng. **17**, 101–109 (1993)
16. Bungener, L.G.S., Van Eetvelde, G., Maréchal, F.: A methodology for creating sequential multi-period base-case scenarios for large data sets. Chem. Eng. Trans. **35**, 1231–1236 (2013)
17. Noor, I.M., Thornhill, N.F., Fretheim, H., Thorud, E.: Quantifying the demand-side response capability of industrial plants to participate in power system frequency control schemes. In: 2015 IEEE Eindhoven PowerTech, Eindhoven (2015)

A Heuristic Approach to Cultivate Symbiosis in Industrial Clusters Led by Process Industry

Amtul Samie Maqbool[(✉)], Giustino Emilio Piccolo, Brecht Zwaenepoel, and Greet Van Eetvelde

Energy and Cluster Management, Faculty of Engineering and Architecture, Ghent University, Ghent, Belgium
ecm@ugent.be

Abstract. This paper introduces a heuristic approach for industrial symbiosis (IS) facilitators to investigate and instigate better energy and resource management via synergies across process industries. The proposed method studies the industrial system at three levels; regional, cluster and company. At the company level, in-depth information is collected using a pentagonal LESTS (Legal, Economic, Spatial, Technical, Social) survey, which is formulated after weighing the regional effects on the whole system. At the cluster level, an inventory of technological and organisational opportunities is produced, offering leverage for IS activities. A gap analysis between the IS potential of the cluster and the IS appreciation on the industrial sites is visualised via LESTS pentagons. The coupled investigation at company and cluster level results in a list of realisable IS activities, which is then translated into business strategies for each participating company using a SWOT analysis.

1 Introduction

Process industry transforms material resources into intermediate or end-products and holds an important place at the core of every value-chain [1]. In Europe alone, the process industry represents about 20% of the manufacturing industry (employment and turnover) [1]. This provides the opportunity to significantly improve resource and energy efficiency and the global environmental footprint of industrial activities. Current systems of resource efficiency and recycling, however, are not enough to replace extraction of virgin materials to keep at par with the increasing needs of the present society. Such improvement can only result in 50% to 75% improvement in environmental performance [2], but in order to bring a substantial reduction in environmental burdens radical innovations are needed.

Global material consumption has doubled between 1950 and 2010 and has even accelerated during the last decade [3]. According to an estimate in 2005, globally recycled material only contributed to 6% of the total processed material, with 13% in Europe [4]. Owing to the rate at which human society is pushing the planetary boundaries [5], businesses require improved management and utilisation of energy and resources, reduction of waste and finally a circular economy. Especially in Europe, current business

© Springer International Publishing AG 2017
G. Campana et al. (eds.), *Sustainable Design and Manufacturing 2017*, Smart Innovation, Systems and Technologies 68, DOI 10.1007/978-3-319-57078-5_55

structures, cultures and practices are evolving, opening windows for new ideas and innovations [6] within and across process sectors.

In the past, collaborations between companies of a sector have grown organically, starting from a wish, a need or a duty to collaborate [7, 8]. **Industrial Symbiosis** (IS) examines cooperative management of resource flows [9] between businesses. IS engages traditionally separate entities in a collective approach to competitive advantage [10] involving, on one hand, physical exchange of materials, energy, water, and by-products, and on the other hand social tactics at the firm and multi-organisational level [11]. Based on the principle that businesses working together can strive for a collective economic and ecological benefit that is greater than the sum of individual benefits each company can achieve [10, 12], the European Commission adopted an ambitious package on circular economy in mid-2016. The circular economy package includes roadmaps and action plans towards sustainable economic development including concrete measures to promote reuse of materials and to stimulate IS [14], hence also introducing legislative changes related to waste registration and handling of resources.

Despite the incentives, practical implications to achieve symbiosis activities are complex due to the intra-cooperative level at which the resources are treated and the way regulations are built around this process [15]. Legal inconveniences, economic barriers such as the lack of financial profit, as well as spatial, technological and social boundaries, can all lead to failing cooperation [7]. Hence, there is a need to bridge the gap between bottom-up appreciation of IS by businesses and the top-down policies at national or international level.

In support of tackling these challenges, this paper presents a heuristic approach to cultivate cross-sectorial industrial symbiosis. Building on consolidated studies in the field of business park management [7], a systemic methodology to identify IS opportunities across process industries is developed. The proposed methodology is applied to a selection of three self-organising cross-sectorial industrial clusters in Europe, under the H2020 project 'EPOS' [16]. These clusters involve two or more process industries from five different sectors: steel, chemicals, minerals, cement and engineering. Taking into account the characteristics of the process industries analysed, the industrial clusters are placed in a system's perspective with the aim to investigate, instigate and initiate IS collaborations, and to turn the industrial clusters into eco-clusters.

The paper first introduces the methodology and then defines three systemic levels, each with a unique set of impacts and precursors for IS. An exemplary result is discussed for one EPOS cluster, followed by a conclusion with recommendations for applying the methodology.

2   LESTS Methodology

In order to propose an approach based on existing industrial clusters and to deal with the complexity of the industrial clusters, the frame is set in systems perspective and LESTS (Legal, Economic, Spatial, Technical and Social) approach is applied to capture the multidisciplinary aspects that influence industrial symbiosis.

2.1   LESTS Approach

The pentagonal approach of LESTS was developed to assess the appreciation of existing resources at a business park and provide a set of guidelines for better park management. The LESTS book series [7] has served as a practical basis for building the LESTS methodology for IS facilitators in industry clusters.

In the adapted LESTS approach, the IS facilitator assumes the role of park or site manager and follows a roadmap to identify IS opportunities for the process industries in a (cross-)sectorial cluster. The proposed methodology incorporates the LESTS considerations into IS identification and initiation process and hence brings a holistic view of the system, by adding both technological and non-technological information under one umbrella as shown in Fig. 1. Within each LESTS domain the framework for managing cluster activities is investigated and the level of collaboration is scored from a top-down as well as bottom-up perspective. When the data is collected at the company level, the bottom-up approach is used to investigate the potential to start symbiosis activities (as used in this study). When the information is collected from a park manager or a cluster manager, it follows a top-down approach of implementing symbiosis in the industrial cluster. The arrows suggest the progressive stages of an industrial network (scored 0–5) when symbiotic relationships are formed and mature over time. The intermediate stages of industrial clusters are detailed in Table 1.

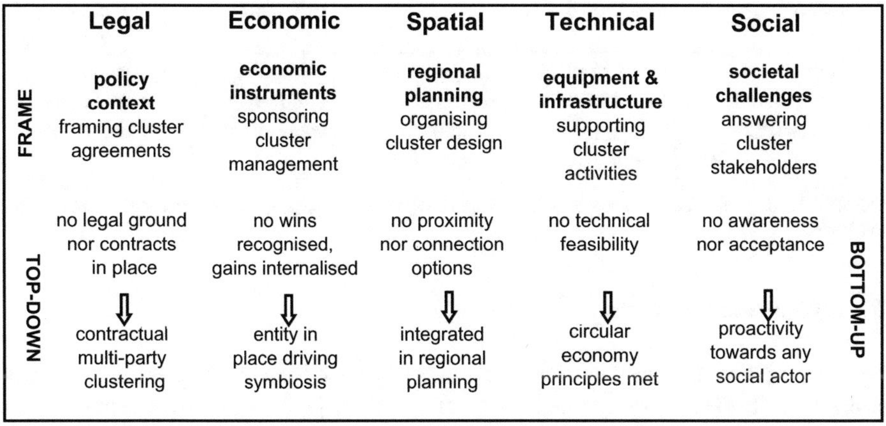

	Legal	Economic	Spatial	Technical	Social
FRAME	policy context framing cluster agreements	economic instruments sponsoring cluster management	regional planning organising cluster design	equipment & infrastructure supporting cluster activities	societal challenges answering cluster stakeholders
TOP-DOWN	no legal ground nor contracts in place ⇩ contractual multi-party clustering	no wins recognised, gains internalised ⇩ entity in place driving symbiosis	no proximity nor connection options ⇩ integrated in regional planning	no technical feasibility ⇩ circular economy principles met	no awareness nor acceptance ⇩ proactivity towards any social actor **BOTTOM-UP**

Fig. 1.   LESTS framework adapted from [7]

Aiming for sustainability progress in industrial systems is a challenging task. Interactions across companies or process sectors make up the industrial cluster, which is embedded in a regional system, which in turn is surrounded by the global system. The IS facilitator should understand these hierarchies and envision system trends that promote a more sustainable future.

Table 1. Ranking system for visualisation of LESTS pentagon

	Legal	Economic	Spatial	Technical	Social
0	no legal ground nor contracts in place	no wins recognised, gains internalised	no proximity nor connection options	not technically feasible	no awareness nor acceptance
1	unilateral contracts in place for diverse activities	incentives for collective wins but not in place	local potential but not in place	similar needs* but organised separately	no joint action towards common stakeholders
2	unilateral contracts in place for similar activities	simple win-win clustering in place	local/individual connections* in place	local needs* jointly answered	joint actions* focused on own employees
3	bilateral contracts in place, between parties/ providers	generic win-win collaborations managed	site design supportive of clustering*	basic opportunities* shared via clustering	cluster-integrated actions* & platforms
4	multi-lateral agreements in place	entity in place that offers basic & optional wins	district approach to clustering	opportunities* jointly optimised	active stakeholder networks (cluster, district)
5	contractual multi-party clustering	entity in place that drives IS (3rd party beneficiary)	integrated in regional planning	principles of circular economy met	proactivity towards any social actor

*All options in the ranking assume interactions & exchanges between parties in clusters, focused on materials, infrastructure and services.

3   System Hierarchies

For investigating industrial clusters, a systems approach is adopted. A system is viewed as a composite of subsystems that form a unitary whole [17]. The core conceptualisation in systems thinking is that by understanding the structure of the system (elements and relations) its behaviour can be explained [18], understood and managed. An industrial cluster is a complex system. It includes a high number of companies and actors, which interact among each other in many different and unpredictable ways. In addition to being complex, an industrial cluster is also dynamic and adaptive. Being dynamic means that industrial cluster evolves through a non-linear path, highly influenced by the contextual framework (global effects) and hard to predict based only on the initial condition of the system [19].

A typical relation in an industrial cluster is not a linear cause and effect. The variables, the different system's components and the elements of each component (e.g. people, businesses, governments), are constantly interacting and changing, in response to each other and to external pressures, creating non-linear feedback loops (Sanders, 2008). This results in an emergent behaviour [13] and defines the unique properties that distinguish each cluster from the others. In this study, three different EPOS industrial production sites, operated and managed by multi-national companies are considered as the basic elements of the industrial system. These industrial sites are defined by a system boundary, across which they communicate with their environment, as shown in Fig. 2. This can be exchanging materials, energy, utilities, information, culture, people, or sharing infrastructure, services, resources with other companies, with society or with the ecosphere [20].

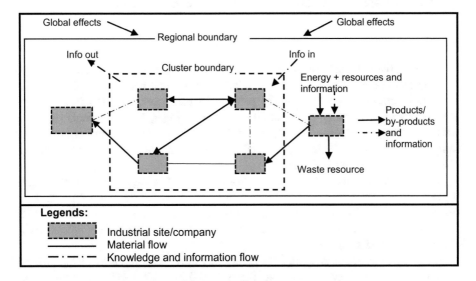

Fig. 2. System hierarchies and the interactions within

4   System Hierarchies Applied to Case Study

To capture the intrinsic complexity and dynamism of each industrial cluster, the LESTS methodology is applied in the clusters to investigate IS possibilities around the selected process industries. The system effects are studied at three levels; regional, cluster and company in the following manner.

4.1   Regional

At the regional level, two LESTS aspects are most prominent; the policy context and regional planning rules. These two aspects define the envisioned development direction for the region and outline the system drivers or hurdles towards resource and energy efficiency. Since the industrial system exists within a global system, it is effected by contextual elements that are factored through weighing. Examples are bio-based or circular economies, pull & push market forces, availability of utility networks, planning instruments for renewable energies, the geo-political state of affairs affecting the supply chain, etc. This helps to identify the possible constraints and drivers of IS and streamline the focus of the surveys developed for the individual site analysis.

For the EPOS clusters, the evolving European regulation defines the threshold for the impact of all proposed IS projects. The literature review of technological advancements of individual process industry established the drivers and bottlenecks for IS projects that were proposed to individual industry. Theoretically this stage of system study is followed by a cluster level analysis; in this study the company level analysis preceded the cluster level analysis. For practical reasons, the data collected at company level were also used to draw inferences for the characteristics of the clusters.

4.2 Cluster

At this level, the IS facilitator carries out an inventory of the cluster elements. Used and unused; material and energy resources, business and social networking platforms, IS facilitation opportunities, potential IS partners in the vicinity, trust building events between potential partners, information exchange linkages, access to local public or private (financial and non-financial) incentives, common linkages between individual companies, stakeholder knowledge, social networks, and other characteristics intrinsic to the cluster, are taken into account.

Based on this inventory and feedback from the LESTS surveys and in-depth studies at the company level, a gap analysis is carried out to exhibit the unused potential of available resources. For the EPOS industrial clusters, the surveys and interviews carried out at the company level helped to draw inference about the level of interconnectivity of businesses in each cluster.

A holistic perspective is applied for identifying the system trends that support IS by using the LESTS framework. The ranking provided in Table 1 helps to visualise the results of the gap analysis by plotting the existing potential of IS against the realised IS activities. For each of the five aspects, a different set of leverages is presented, which could be exploited by the actors within the industrial cluster. Each aspect was translated into a survey question for the companies. The survey answers and interview responses were coupled with the literature review to infer the connectivity level of industries within their respective clusters. These inferences were plotted on a pentagon, showing the level that can be achieved by the cluster and the level that was actually in place at the time of data gathering.

4.3 Industrial Site/Company

Company or an individual industrial site, is the key level for IS activities. The actors at the company level have much more knowledge of the technological and organisational drivers and barriers of IS and hence an in-depth analysis is possible. Quantitative methods for material consumption, such as Material Flow Analysis of the industrial site, Life-cycle Analysis can be applied to gather information about the hotspots for IS. In this study, qualitative data collection via an LESTS survey and semi-structured interviews was carried out, regarding all activities on an industrial site that may support or hinder IS with other partners.

The framework of LESTS survey was built around three major activities on an industrial site, which includes; (1) production of by-products, waste and emissions; (2) utilisation of energy, water, equipment and infrastructure; and (3) requirement of packaging, logistics and storage. Related to each factor, information was collected on existing collaborations with neighbouring companies; nature and length (time) of these collaborations and finally; a possibility to extend these collaborations to other industry partners. The LESTS survey was supplemented with semi-structured interviews with key company representatives (often, site manager, energy/environmental manager) to collect information about the interaction and consultation between stakeholders at the cluster level. The interview focused on gaining insight into the decision making power of the site, stakeholder engagement at the cluster level, communication with neighbouring communities. These interviews helped to gain an overview of stakeholder perception about possibilities of IS activities.

In essence, LESTS surveys lead to a longlist of potential exchanges and interactions between the companies - and even municipal communities - in the region. Via consecutive SWOT analyses, indicating strengths and weaknesses (internal factors) that induce opportunities or point to threats (external factors) viable symbiosis opportunities are put forth, consequently, building a framework to identify and formulate business strategies. The potential IS opportunities are further developed into IS business cases and calculated for their sustainability improvements, considering the three pillars of sustainability [7]. To identify the opportunities of IS that are sustainable in nature, leverage is created when the principles of cleaner production are in place, a commitment to CSR has been incorporated into company policy, company is inclined to minimising its footprint or when good management practices are already in place.

5 Results

The described LESTS methodology resulted in identifying a number of IS activities within each cluster. To provide an exemplary result on the applicability of this methodology, IS opportunities from one cluster are presented in Fig. 3. The identified opportunities have been communicated to the respective industries and are currently under study for techno-economic feasibility.

The application of the heuristic method presented in this paper provided two sets of results. One for the cluster level and one for the companies within the cluster, willing to involve in an industrial symbiosis. The results for the cluster level visualise the gap between possibilities to engage in IS and the realisation of these possibilities by actors in the cluster as shown in Fig. 4. The outer pentagon indicates the leverage provided by the unique properties of the cluster, which arise due to the presence of diverse actors and the linkages formed between them. The inner pentagon exhibits the level to which these opportunities are availed by the actors in the cluster to form symbiosis activities.

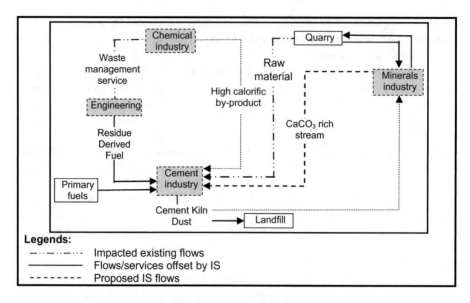

Fig. 3. Identified IS opportunities in the UK cluster of EPOS

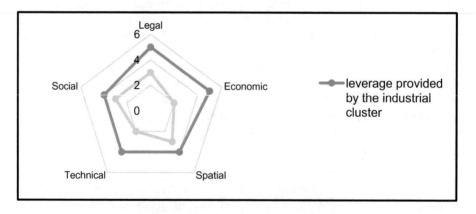

Fig. 4. The gap analysis visualised in an LESTS pentagon

The second set of results for the individual companies is provided in a list of IS possibilities and SWOT strategies to realise the presented IS opportunities. The SWOT analysis of individual companies are communicated individually to each company and are handled under confidentiality agreement between the IS facilitator and the company.

6   Conclusion

The proposed methodology for facilitating industrial symbiosis answers to the needs of process industries active in (cross-)sectorial clusters and willing to improve their energy and resource profile through collaboration. The approach takes a system perspective: it provides flexibility, desired by industrial actors and leads to stability, required by the global society. The LESTS surveys, coupled with semi-structured interviews, proved adequate to identify IS options in cross-sectorial clusters. Still cluster level analysis is perceived as a daunting task and requires many resources (time and costs). Likewise, the identification of key site actors for interviews is a crucial element for the success of the survey and the consecutive implementation of symbiosis activities. With an IS facilitator appointed and key contacts per industrial site identified, a good understanding of the industrial cluster, its actors and potential can be reached through regular exchange of information. Even so, socio-economic benefits are essential to attract and hold a company's interest in symbiosis activities.

Due to the EPOS project boundaries and consortium agreements, only the project partners in each industrial cluster were approached for interviews and surveys. To fully assess an industrial cluster however, it is recommended to include a maximum number of cluster actors in the analysis. The methodology provides enough flexibility to be coupled with different analyses at the company level (product life cycle), as well as, at the cluster level (material flow, environmentally extended Input Output, Social Network).

By applying the methodology at each industrial site in the EPOS project, a longlist of IS activities was identified. In a next step, techno-economic feasibility studies will be performed and the expected environmental improvement will be calculated per industrial site as well as for each cluster as a whole. Current research explores the possibility to model the complex adaptive systems of industrial clusters and predict/ascertain the behaviour of the individual actors with agent based modelling.

Acknowledgment. The research leading to these results has received funding from the European Union's Horizon 2020 research and innovation programme under grant agreement no. 679386, EPOS project (Enhanced energy and resource Efficiency and Performance in process industry Operations via onsite and cross-sectorial Symbiosis). The sole responsibility of this publication lies with the authors. The European Union is not responsible for any use that may be made of the information contained therein.

References

1. Tello, P., Weerdmeester, R.: SPIRE Roadmap, SPIRE (2013)
2. Tukker, A.: Leapfrogging into the future: developing for sustainability. Int. J. Innov. Sustain. Dev. **1**(1), 65–84 (2005)
3. Schaffartzik, A., Mayer, A., Gingrich, S., Eisenmenger, N., Loy, C., Krausmann, F.: The global metabolic transition: regional patterns and trends of global material flows, 1950–2010. Glob. Environ. Change **26**, 87–97 (2014)
4. Haas, W., Krausmann, F., Wiedenhofer, D., Heinz, M.: How circular is the global economy?: an assessment of material flows, waste production, and recycling in the European union and the world in 2005. J. Ind. Ecol. **19**(5), 765–777 (2015)
5. Rockström, J., et al.: Planetary boundaries: exploring the safe operating space for humanity. Ecol. Soc. **14**, 472–475 (2009)
6. Grin, J., Rotmans, J., Schot, J.: Transitions to Sustainable Development: New Directions in the Study of Long Term Transformative Change. Routledge, New York (2010)

7. Van Eetvelde, G., et al.: Groeiboeken Duurzame BedrijvenTerreinen juridisch, economisch, ruimtelijk, tech-nisch bekeken. Vanden Broele Grafische Groep, Brugge (2005)

8. Van Eetvelde, G., Deridder, K., Segers, S., Maes, T., Crivits, M.: Sustainability scanning of eco-industrial parks. In: Presented at the In 11th European Roundtable on Sustainable Consumption and Production (ERSCP) (2007)

9. Lowe, E.A., Evans, L.K.: Industrial ecology and industrial ecosystems. J. Clean. Prod. **3**(1), 47–53 (1995)

10. Chertow, M.R.: Industrial symbiosis: literature and taxonomy. Ann. Rev. Energy Environ. **25**(1), 313–337 (2000)

11. Puente, M.C.R., Arozamena, E.R., Evans, S.: Industrial symbiosis opportunities for small and medium sized enterprises: preliminary study in the Besaya region. J. Clean. Prod. **87**, 357–374 (2015). (Cantabria, Northern Spain)

12. Jacobsen, N.B.: Industrial symbiosis in Kalundborg, Denmark: a quantitative assessment of economic and environmental aspects. J. Ind. Ecol. **10**(1–2), 239–255 (2006)

13. Chertow, M., Ehrenfeld, J.: Organizing self-organizing systems. J. Ind. Ecol. **16**(1), 13–27 (2012)

14. European Commission, "Circular Economy Strategy," European Commission - Environment, Brussels, Belgium, EU Action plan COM/2015/0614 final (2016)

15. Golev, A., Corder, G.D., Giurco, D.P.: Barriers to industrial symbiosis: insights from the use of a maturity grid. J. Ind. Ecol. **19**(1), 141–153 (2015)

16. "EPOS", Enhanced energy and resource Efficiency and Performance in process industry Operations via onsite and cross-sectorial Symbiosis|SPIRE. https://www.spire2030.eu/EPOS. Accessed 26 Jan 2017

17. Von Bertalanffy, L.: General System Theory: Foundations, Development, Applications. Braziller, New York (1968)

18. Frantzeskaki, N.: Dynamics of societal transitions; driving forces and feedback loops, TU Delft, Delft University of Technology (2011)

19. Albino, V., Fraccascia, L., Giannoccaro, I.: Exploring the role of contracts to support the emergence of self-organized industrial symbiosis networks: an agent-based simulation study. J. Clean. Prod. **112**(5), 4353–4366 (2016)

20. Wallner, H.P.: Towards sustainable development of industry: networking, complexity and eco-clusters. J. Clean. Prod. **7**(1), 49–58 (1999)

IMPROOF: Integrated Model Guided Process Optimization of Steam Cracking Furnaces

Marko R. Djokic[1](✉), Kevin M. Van Geem[1], Geraldine J. Heynderickx[1],
Stijn Dekeukeleire[1], Stijn Vangaever[1], Frederique Battin-Leclerc[2], Georgios Bellos[3],
Wim Buysschaert[4], Benedicte Cuenot[5], Tiziano Faravelli[6], Michael Henneke[7],
Dietlinde Jakobi[8], Philippe Lenain[9], Andres Munoz[10], John Olver[11],
Marco Van Goethem[12], and Peter Oud[12]

[1] Laboratory for Chemical Technology, Ghent University, Ghent, Belgium
{marko.djokic,kevin.vangeem,geraldine.heynderickx,
stijn.dekeukeleire,stijn.vangaever}@ugent.be
[2] Centre National de La Recherche Scientifique, Nancy, France
frederique.battin-leclerc@univ-lorraine.fr
[3] Dow Benelux B.V., Terneuzen, The Netherlands
bellos@dow.com
[4] Cress B.V., Breskens, The Netherlands
Wim.Buysschaert@cressbv.nl
[5] European Centre for Research and Advanced Training in Scientific Computation,
Toulouse, France
benedicte.cuenot@cerfacs.fr
[6] Politecnico di Milano, Milan, Italy
tiziano.faravelli@polimi.it
[7] John Zink International Luxembourg SARL, Dudelange, Luxembourg
Michael.Henneke@johnzink.com
[8] Schmidt + Clemens GmbH +CO. KG, Lindlar, Germany
d.jakobi@schmidt-clemens.de
[9] Ayming Belgium, Brussels, Belgium
plenain@ayming.com
[10] AVGI, Ghent, Belgium
andres.munoz@avgi.be
[11] Emisshield Inc., Blacksburg, VA, USA
john.olver@emisshield.com
[12] Technip Benelux B.V., Zoetermeer, The Netherlands
{mvangoethem,poud}@technip.com

Abstract. IMPROOF will develop and demonstrate the steam cracking furnace of the 21st century by drastically improving the energy efficiency of the current state-of-the-art, in a cost effective way, while simultaneously reducing emissions of greenhouse gases and NO_X per ton of ethylene produced by at least 25%. Therefore, the latest technological innovations in the field of energy efficiency and fouling minimization are implemented and combined, proving that these technologies work properly at TRL 5 and 6 levels. The first steps to reach the ultimate objective, i.e. to deploy the furnace at the demonstrator at commercial

© Springer International Publishing AG 2017
G. Campana et al. (eds.), *Sustainable Design and Manufacturing 2017*, Smart Innovation,
Systems and Technologies 68, DOI 10.1007/978-3-319-57078-5_56

scale with the most effective technologies, will be discussed based on novel pilot scale data and modeling results.

Keywords: Industrial steam cracking furnace design · Increased energy efficiency · Reduced coke formation · Reduced emissions of greenhouse gasses · Increased time on stream

1   Introduction

Ethylene is a basic building block of the chemical industry produced by steam cracking process, and is the link between chemical companies and petroleum refiners. Steam cracking is the most energy-consuming process in the chemical industry and globally uses approximately 8% of the sector's total primary energy (Ren et al. 2008). Additionally it is responsible for massive amounts of CO_2 emissions (Ren et al. 2006). Improving the energy efficiency has an immediate pay-out because energy cost counts for a substantial part of the production costs in typical ethane or naphtha based olefin plants (Ren et al. 2006). In Europe primarily naphtha is used as a feedstock, which is substantially more expensive than cheap shale-derived ethane. This makes that the European petrochemical industry is under threat, and increase the risk that this pillar of the industry will gradually become uncompetitive in a globalized World economy. If no action is taken, it is not unthinkable that its fall will affect the complete European chemical industry, which produces 14.7% of the world's chemicals, employs 1.2 million workers and contributes €519 billion to the European Union economy (CEFIC 2016).

Although steam cracking is considered to be a mature technology, the complexity of the process and the harsh operating conditions allow the implementation of technology developments towards substantial heat transfer enhancement. One of the most important ways to reduce the energy input in steam cracking furnaces per ton ethylene produced is to reduce coke formation on the reactor wall of the long tubular reactors that are mounted in the furnaces (Muñoz Gandarillas et al. 2014). That is primarily why steam is added to the feedstock, also improving the selectivity towards the desired light olefins, i.e. ethylene and propylene. Typically, the outlet temperature of the tubular reactor is very high, around 820–890 °C (Albright et al. 1983). In modern cracking furnaces, the residence time is reduced to a few hundred milliseconds in order to improve the yield of desired light olefins. After the cracking temperature has been reached, the gas is quickly quenched to stop the cracking reaction in a transfer line heat exchanger to recover as much energy as possible (Dhuyvetter et al. 2001). Typically, ethylene furnaces have to be decoked after 30–60 days to remove the coke that collected in the coil. When the furnaces are decoked, production of the desired products is stopped for approximately 48 h (Zhang and Albright 2010). During the course of one run, deposited coke can reduce the heat-transfer efficiency of the firebox by 1–2%, resulting in a 5% increase in fuel consumption (Zimmermann and Walzl 2000). The use of either advanced coil materials, combined with 3D reactor designs, improved process control, and more uniform heat transfer could increase run lengths (Zhang et al. 2015), reducing simultaneously CO_2 emissions and the lifetime of the furnaces. It has been proven that improved metallurgy of radiant coils can reduce catalytic coking (Muñoz Gandarillas et al. 2014), and that advanced 3D coil design such as the swirl flow reactor design

(Schietekat et al. 2012) and the SCOPE design (by Schmidt+Clemens GMBH+CO. KG) can mitigate coke deposition leading to heat transfer improvement. It is believed that the current design of these 3D reactors is still far from optimal, and that Computational Fluid Dynamics can lead to the design of radiant coils with even better performance (Reyniers et al. 2015). Improved geometries can result in smaller radial temperature gradients inside the coil, and thus lower wall temperatures where the coke formation occurs. Advanced 3D reactor simulations (see Fig. 1) are extremely useful for that purpose (Van Cauwenberge et al. 2016). The latter will be translated to significant improvement of energy consumption per mass of produced ethylene. In addition, lower tube wall temperature – which results from the improved geometry - will extend the run length resulting in lower energy consumption for decoking the radiant coils on annual basis.

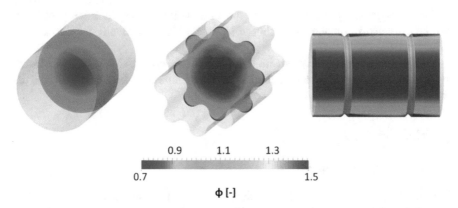

Fig. 1. Three-dimensional representation of the studied reactor designs with contour plots of the corresponding local fluid age correction factors, from left to right: a bare tubular reactor, a longitudinally finned reactor and a transversally ribbed reactor (Van Cauwenberge et al. 2016).

An important challenge for the petrochemical industry is the upcoming stronger environmental regulations, in particular related to NO_X and CO_2. The high temperatures needed in the furnace to "crack" the gaseous and liquid hydrocarbon feeds such as naphtha, Liquefied Petroleum Gas (LPG), condensates and recycled ethane, also results in substantial NO_X emissions. The total NO_X emissions of all the furnaces currently operating in the EU are approximately 16.3×10^3 tpa. To reduce NO_X emissions the use of advanced oxy-fuel combustion is demonstrated on pilot plant scale. The advantage is clear; because no nitrogen is added (apart from leakages), almost no NO_X is produced (Olajire 2010). An additional advantage is that the produced flue gas is a concentrated CO_2 stream, that can be more easily captured, stored or used for other applications (Carbon Capture and Storage), for example in chemical looping (Adanez et al. 2012).

Bio-gas and bio-oil are renewable fuels, and hence, decrease net CO_2 production (Oasmaa et al. 2015). It is expected that these fuels will become available in substantial amounts in near future in Europe and therefore can be used as fuels for steam cracking furnaces. Nevertheless, a logical concern that rises for bio-oils is the possible presence of fuel-bound nitrogen, of which 30%–50% typically converts to NO_X depending on the

O_2 levels and temperature in flames. Therefore, hydro-treated bio-oils need to be considered as a fuel for the firebox (Venderbosch and Prins 2010).

A final point for improvement is related to the radiant section of a steam cracking furnace, where the major part of heat transfer occurs by radiation. The radiation is emitted by the refractory walls towards the process radiant coils. Application of high emissivity coatings on the external surface of the radiant coils could improve the energy consumption in several ways (Stefanidis et al. 2008). In addition, improved emissivity coatings can be applied to the furnace walls (Holcombe and Chapman 1999). Less firing is required to reach the same process temperatures in the radiant coils. This will reduce fuel gas consumption and CO_2 emissions by an anticipated 10 to 15%, at the same time reducing the high pressure steam generation in the convection section. In addition to higher heat absorption, coating the external surface of radiant coils can improve the surface homogeneity and eliminate hot spots on the tube walls. An additional benefit will be extended run lengths resulting in less energy spent annually for decoking the radiant coils.

IMPROOF project will demonstrate the advantage of combining all these technological innovations, with an anticipated reduction of emissions, and increase of the time on stream and energy efficiency.

2 Methodology

2.1 Consortium Structure

The strongly industrially-oriented consortium is composed of five industrial partners (IND), two Small and medium-sized enterprises (SME), two Research and Technology Organizations (RTO), and two universities (UNI) (see Fig. 2). IND include DOW Benelux B.V. (The Netherlands), TECHNIP Benelux B.V. (The Netherlands), John Zink International Luxembourg SARL (JZHC, Luxembourg), Schmidt+Clemens GmbH+CO. KG (S+C, Germany), and Ayming Belgium (Belgium). SME are CressBV (The Netherlands) and AVGI (Belgium). RTO include Centre National de la Recherche Scientifique (CNRS, France) and European Centre for Research and Advanced Training in Scientific Computation (CERFACS, France). UNI are Ghent University (UGENT, Belgium) and Politecnico di Milano (POLIMI, Italy). The partnership shows a clear and strong path to the industrial and economical world with the involvement of industrial end-users.

CNRS and POLIMI will propose kinetics models to optimize the use of the novel alternative fuels in industrial conditions. **CRESS, S+C and JZHC** will contribute to the project by developing and implementing different technologies to the whole furnace. **CRESS** is a refractory company and supplier of high emissivity coatings. **S+C** is the world market leader in the supply of spun cast tubes and fittings for steam crackers. **JZHC** is one of the World's leaders in providing advanced combustion technologies to the ethylene production market.

CERFACS aims to improve and adapt the numerical methods to steam cracking furnace applications targeted by the project. The research at **UGENT** will be focused on the kinetics of industrial reactions and coupling that with advanced CFD. **UGENT** will also develop, commission and host the Technology Readiness Level (TRL) 5 pilot unit. **AVGI** is a

Belgian SME aimed at the development of software for modelling petrochemical processes, and particularly steam cracking of hydrocarbons. In this context, AVGI will provide simulations to the other project partners, in which the effect of their developments (materials, burner improvements, etc.), can be evaluated in terms of yields of desired products, and minimization of energy consumption.

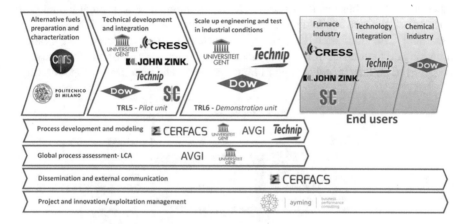

Fig. 2. IMPROOF consortium along the value chain of the ethylene steam cracking process industry.

The commissioning and engineering of the complete solution to the industrial world will be led by **TECHNIP** and **DOW**. **TECHNIP** is a world leader in project management, engineering and construction for the energy industry. They will work hand in hand with DOW to upscale the solution from the laboratory pilot scale to the integration of the novel furnace in DOW's chemical production line. **DOW** is one of the largest ethylene producers in Europe. DOW will beneficiate from the project with a novel and enhanced steam cracking process to produce ethylene.

In order to ensure a close collaboration between the different consortium members and a proper project management, **a project manager from AYMING** will work in close collaboration with the project coordinator, namely **UGENT**.

2.2   Concept and Approach

To answer to the market need and implement novel technology to the ethylene steam cracking furnaces, the project will tackle four technology blocks (see Fig. 3):

1. Alternative fuels – Oxy-fuel
2. Advanced high temperature alloys, and novel emissive and refractory materials
3. Novel furnace design
4. 3D modelling

Alternative fuels - Oxy-fuel. The primary objectives for the alternative fuel are:

(a) Development of detailed kinetic mechanisms; and

(b) Kinetic post-processing of industrial equipment for evaluating the emissions of nitrogen oxides.

This will require the development, validation, and tuning of one or more detailed kinetic mechanisms able to describe the pyrolysis, oxidation and combustion of the selected fuels. This is essential for burner development, furnace optimization emission reduction and last but not least energy efficiency improvement through model based process design. Once available, the developed kinetic mechanisms will be adapted for performing CFD simulations of the selected industrial burners and furnaces. This allows to investigate the impact of operating conditions, geometry, etc. on the formation and emissions of pollutants and to support both the design and operation of the furnaces.

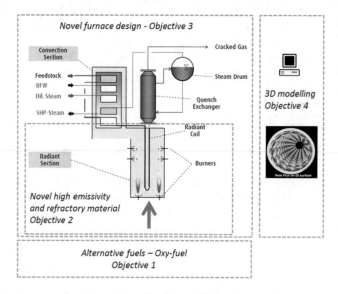

Fig. 3. General concept overview linked to the project objectives.

Novel High Emissivity and Refractory Material. One of the most important objectives of this project is to increase the energy efficiency of the radiation section of a steam cracking furnace, minimizing exergy losses. Not only will the radiation section be studied, but the furnace as a whole. A key element to realize this is the use of high emissivity coatings manufactured and studied under real furnace conditions in the project (Adams and Olver 2015). Coating different sections of the furnace will be considered, such as radiant walls/floor of furnace, radiant roof panels, burners & burner blocks, process coils/tubes, as well as convective section tubes and walls. In addition to higher heat absorption, coating the external surface of radiant coils can improve the surface homogeneity and eliminate hot spots on the tube walls. An additional benefit will be extended run lengths resulting in less energy spent annually for decoking the radiant coils. Measurements of emissivities and temperature uniformity are planned to prove that indeed a uniform heating of the tubes, with reduced hot spots and non-uniform heat flux, is attained. Moreover, the pilot tests at TRL5 should demonstrate that amount of coke that is formed should be significantly less, more granular and thus easy to remove.

Novel Furnace Design. IMPROOF will develop and demonstrate the furnace of the future. Therefore, not only the radiation section will be studied, but the furnace as a whole. The UGENT pilot plant, equipped with a convection section, a radiation section and a transfer line heat exchanger will provide a unique dataset to assess all the practical implications of the proposed technological innovations before they are implemented at the demonstrator. One of the most expensive parts of the cracking furnaces ($\approx 1/3$ of the furnace cost) are the reactors, the so-called coils. Radiant coils will be selected in light of the following considerations: operating temperature, tube service life, cost, carburization resistance, creep-rupture strength, ductility, and weldability. Recent innovations in the development of new coil materials, especially 45Ni35Cr steels, or materials with an increased content of aluminum ($\sim 4\%$) that can be operated up to 1200°C will be considered. IMPROOF will provide data that show if there is a benefit or not of using these more advanced but more costly materials. The best materials will be tested on TRL5 level in combination with the high emissivity coatings before going to TRL6. Moreover, the best material will be combined with another recent innovation, the 3D Swirl flow reactor design or the SCOPE technology. The use of advanced coil materials, combined with 3D reactor designs, improved process control, and more uniform heat transfer could increase run lengths, reducing simultaneously CO_2 emissions and the lifetime of the furnaces. It is believed that the current design of these 3D reactors is far from optimal, and that CFD can lead to the design of radiant coils with even better performance. Also coating the external reactor wall with a high emissivity material seems to have potential for a more even heat distribution.

3D Modelling and Optimization. New manufacturing techniques such as three-dimensional printing have the potential to drastically transform the chemical industry. Novel, complex reactor geometries can be created that were previously either impossible or required dedicated facilities to make. Driven by process intensification and in combination with the enormous power of high performance computing (HPC), this project will take full advantage of these new opportunities to enhance heat and mass transfer in advanced chemical reactors by using multiscale modelling and experimentation.

One main objective of IMPROOF is to propose a methodology to evaluate existing and propose new reactor designs. To reach this objective, several tools will be employed:

– In terms of cracking chemistry, a dynamical reduction method will be applied to drastically decrease the number of species required and therefore the CPU cost of the simulations. Several existing techniques - quasi-steady-state (QSS), partial equilibrium (PEQ) or on-the-fly flux-based – will be tested to determine which one enables to derive the most accurate chemistry for a reasonable CPU cost.
– In terms of numerical simulations, first high-fidelity aerothermal wall-resolved Large Eddy Simulations (LES) including reduced chemistry will be performed using AVBP to produce reference solutions in some selected reactor geometries. LES will be validated by comparison with industrial data or possibly experimental validation on lab scale or pilot plant scale, in terms of pressure drop, heat transfer enhancement and

product selectivity. Then, classical optimization techniques will be applied to coarse LES or Reynolds-Averaged Navier-Stokes (RANS) simulations to optimize and propose new reactor designs. The starting point of this geometry optimization process will be ribbed reactors such as MERT reactor coils. These tubes are said to offer a 40% improvement on the heat transfer characteristics of a bare tube, which is reported to lead to a halving of the coking rates inside these reactors. The strong turbulence induced by the spiral fin and the breaking of the laminar boundary layers, however, also greatly increases wall shear and pressure drops.

2.3   Strategic and Technical Objectives

In the frame of the IMPROOF project, the **strategic objectives** are:

- Energy consumption:
    - Emissive coating emitting in the non-absorbent flue gas spectrum;
    - Enhanced heat transfer between flue gas and the process;
    - Novel radiant coil alumina forming alloy tubes that lower the coking rate leading to a lower outside tube temperature.
- Operating costs:
    - Increase production rates per ton olefins produced by increasing furnace combustion efficiency and by advanced process simulation;
    - Increase production rates per ton olefins produced by selectivity increase using 3D reactor technology;
    - Increase furnace availability and production rates per ton olefins produced because of reduced coke formation using advanced materials and 3D coil design.
- Reduce CAPEX and OPEX costs of the furnaces:
    - Double refractory life, longer process tube life and shorter furnace downtime;
    - OPEX: 88% is related to feed cost, so the optimization is there next to availability.
- Reduction of NO_x and CO_2 emissions:
    - Decrease NO_x and CO_2 emissions by considering new burner designs, and oxy-fuel combustion;
    - Decrease net CO_2 emission by using bio-gas and bio-oil as fuel.

In the frame of the IMPROOF project, the **technical objectives** are:

1. Demonstrate the individual impact of novel emissive, reactor and refractory materials on pilot scale (TRL5).
2. Demonstrate the power of advanced process simulation (high performance computing and CFD) for furnace design and optimization:
    - Methodology demonstration and patent at least one new 3D reactor design for reduced coking and improved ethylene and propylene yields;
    - To design and develop systematic, logic extended, new sophisticated innovative enhanced, ultrafast, powerful molecular inspired, parallel, model-based algorithms for simulation and optimization.
3. Demonstrate the technical economic and environmental sustainability of the IMPROOF furnace at TRL6:

- Increase energy efficiency of the radiative section by 20% by combining advanced coatings, 3D reactor designs, and novel refractory and reactor materials;
- Provide coatings that provide ideal heat flux profile to the coil in combination with advanced modelling, leading to decreased fouling and increased time on stream, decreasing decoking time by a factor 3 on year basis;
- Optimized coating formulation depending on the chosen burners and fuel (low NO_X versus oxy-fuel).

4. Novel combustion technology using alternative fuels and oxy-fuel combustion:
   - Proof reduced net CO_2 emissions by using bio-gas and (hydrotreated) bio-oil by 15% compared to the state of the art at TRL 5 and TRL 6;
   - Demonstrate the power of Large Eddy Simulation by model guided fuel optimization and burner design.

5. Coke formation reduction and real time optimization:
   - Development of a validated 3D model that is able to simulate fouling, cracking reaction kinetics, heat transfer and flow in steam cracking reactors. Incorporate the influence of the coil metallurgy on the cokes formation, increasing run length by a factor 3;
   - Demonstrate the RTO's ready technology to improve process control for furnace availability and overall production yields, reduce CAPEX and OPEX per ton ethylene produced by at least 5%.

3   Implementation

The work-package (WP) distribution among partners is described in Fig. 4. The objectives of the WP1 are to investigate the kinetic of combustion of different fuels also in oxygen-rich environment, with a particular attention to the pollutant formation (CO, NO_X and eventually SO_X). Different fuels will be investigated. Both the fossil fuels (natural gas) and renewable fuels, like bio-gas and bio-oil. To this goal both experimental and modelling activity will be performed.

WP2 will demonstrate all the technologies individually at TRL5 level to assess their individual performance. This includes emission measurement for oxy-fuel combustion with classical fuels but also bio-gas and bio-oil on pilot scale, benchmarked to BAT. Also testing of the five different high emissivity coatings, the high performance alloys in the UGENT pilot units, and the 3D swirl flow and S+C SCOPE® 3D reactor technology is foreseen. As these technological improvements are not mutual exclusive, combinations of the best performing technologies will be selected to illustrate the multiplier effect.

In WP3 the kinetic models developed in WP1 for combustion will be implemented in the CFD tools of respectively partners. They will be validated using the pilot data obtained in WP2. On the other hand, advanced modelling on the reactor will allow further optimizing existing 3D reactor designs and developing novel 3D geometries. The latter will require accurate gas phase kinetic models and coking models to determine olefin selectivity as function of the time on stream, i.e. the run length.

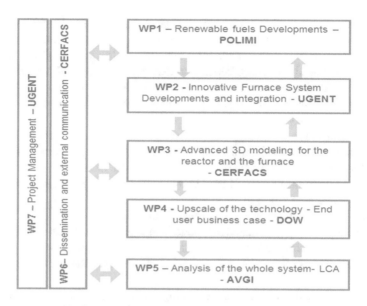

Fig. 4. Graphical representation of the WP structure with related WP leaders.

The objective of the WP4 is to deploy the demonstrator at integrated commercial scale (TRL6) with the most effective technologies improving heat transfer of ethylene furnaces.

The goal of WP5 is to evaluate the impact of the different technological improvements, and combinations thereof based on relevant data of the complete integrated furnace platform. Ultimately, this WP will determine the furnace improvements as a whole as function of the considered developments, to evaluate more accurately the environmental, technical and economic potential of the whole IMPROOF platform concept.

The objectives of WP6 are to continuously monitor and provide means for the IMPROOF partners to share their knowledge within the consortium and to integrate the research activities as well as to communicate and disseminate the results to the scientific community and to the wider audience. WP7 will carry out an efficient project management adapted to the specificities of IMPROOF. The aims are to reach the planned ambitious program results and goals but also position the project durably in its environment and maximize its breakthrough potential/innovations and impacts.

4   Conclusions

The objective of the project is to drastically improve the energy efficiency of steam cracking furnaces, and this in a cost effective way, while simultaneously reducing emissions of greenhouse gases and NO_x per ton ethylene produced with ideally 25% or more. Advanced oxy-fuel combustion will be applied in order to reduce NO_x emissions. The crucial element to increase the energy efficiency of the radiant section of a steam cracking furnace is the use of high emissivity coatings on the inner wall of the refractory. Moreover, applying the high emissivity coatings on the external surface of the reactor tubes can additionally improve the surface homogeneity and eliminate hot spots. Combining the advanced coil materials, novel 3D reactor designs, improved process control, and more uniform heat transfer could increase run lengths as well as lifetime of the furnace, while simultaneously reducing CO_2 emissions. In addition, it is believed that the current design of the 3D reactors is far from optimal, and that CFD can lead to the design of radiant coils with intensified performances. IMPROOF will aim at selecting the correct, systematic, specific, strategic objectives for sustainable implementation in complex plant-wide and industrial data-intensive process systems, including all the parameters involved in real-plant conditions.

Acknowledgment. The work leading to this invention has received funding from the European Union Horizon H2020 Programme (H2020-SPIRE-04-2016) under grant agreement n°723706.

References

Adams, B., Olver, J.: Impact of high-emissivity coatings on process furnace heat transfer. In: AIChE Spring Meeting 2015 (2015)

Adanez, J., Abad, A., Garcia-Labiano, F., Gayan, P., de Diego, L.F.: Progress in chemical-looping combustion and reforming technologies. Prog. Energy Combust. Sci. **38**, 215–282 (2012)

Albright, L.F., Crynes, B.L., Corcoran, W.H.: Pyrolysis, Theory and Industrial Practice. Academic Press, New York (1983)

CEFIC: The 2016 Cefic European Facts & Figures (2016). http://www.cefic.org/Facts-and-Figures/. Accessed 27 Oct 2016

Dhuyvetter, I., Reyniers, M.-F., Froment, G.F., Marin, G.B., Viennet, D.: The influence of dimethyl disulfide on naphtha steam cracking. Ind. Eng. Chem. Res. **40**, 4353–4362 (2001)

Holcombe, C.E., Chapman, L.R.: High emissivity coating composition and method of use. Google Patents (1999)

Muñoz Gandarillas, A.E., Van Geem, K.M., Reyniers, M.-F., Marin, G.B.: Coking resistance of specialized coil materials during steam cracking of sulfur-free naphtha. Ind. Eng. Chem. Res. **53**, 13644–13655 (2014)

Oasmaa, A., Van de Beld, B., Saari, P., Elliott, D.C., Solantausta, Y.: Norms, standards, and legislation for fast pyrolysis bio-oils from lignocellulosic biomass. Energy Fuels **29**, 2471–2484 (2015)

Olajire, A.A.: CO_2 capture and separation technologies for end-of-pipe applications – A review. Energy **35**, 2610–2628 (2010)

Ren, T., Patel, M.K., Blok, K.: Olefins from conventional and heavy feedstocks: Energy use in steam cracking and alternative processes. Energy **31**, 425–451 (2006)

Ren, T., Patel, M.K., Blok, K.: Steam cracking and methane to olefins: Energy use, CO_2 emissions and production costs. Energy **33**, 817–833 (2008)

Reyniers, P.A., Schietekat, C.M., van Cauwenberge, D.J., Vandewalle, L.A., van Geem, K.M., Marin, G.B.: Necessity and feasibility of 3D simulations of steam cracking reactors. Ind. Eng. Chem. Res. **54**, 12270–12282 (2015)

Schietekat, C.U., Van Goethem, M.M., Van Geem, K.T.W., Marin, G.: 3D swirl flow reactor technology for pyrolysis processes: Hydrodynamic and computational fluid dynamic study. In: XX International Conference on Chemical Reactors (2012)

Stefanidis, G.D., Van Geem, K.M., Heynderickx, G.J., Marin, G.B.: Evaluation of high-emissivity coatings in steam cracking furnaces using a non-grey gas radiation model. Chem. Eng. J. **137**, 411–421 (2008)

Van Cauwenberge, D.J., Van Dewalle, L.A., Reyniers, P.A., Van Geem, K.M., Marin, G.B., Floré, J.: Periodic reactive flow simulation: Proof of concept for steam cracking coils. AIChE J. (2016)

Venderbosch, R.H., Prins, W.: Fast pyrolysis technology development. Biofuels Bioprod. Biorefin. **4**, 178–208 (2010)

Zhang, Y., Qian, F., Schietekat, C.M., van Geem, K.M., Marin, G.B.: Impact of flue gas radiative properties and burner geometry in furnace simulations. AIChE J. **61**, 936–954 (2015)

Zhang, Z.B., Albright, L.F.: Pretreatments of coils to minimize coke formation in ethylene furnaces. Ind. Eng. Chem. Res. **49**, 1991–1994 (2010)

Zimmermann, H., Walzl, R.: Ethylene. In: Ullmann's Encyclopedia of Industrial Chemistry. Wiley-VCH Verlag GmbH & Co. KGaA (2000)

Conceptual Analysis of Eco-Efficiency and Industrial Symbiosis: Insights from Process Industry

Yan Li[✉], Maria Holgado, Miriam Benedetti, and Steve Evans

Institute for Manufacturing, University of Cambridge, 17 Charles Babbage Road,
Cambridge CB3 0FS, UK
{yl483,mh769,mb2132,se321}@cam.ac.uk

Abstract. The interior relationship between Industrial Ecology, Eco-Efficiency and Industrial Symbiosis has been scarcely investigated in literature. We identify three main aspects linking the concepts, which are 'Actions', 'Stakeholders' and 'Value', and use them to drive the conceptual analysis. Considering the application and implementation, authors conduct a conceptual comparison between Eco-Efficiency and Industrial Symbiosis by using Industrial Ecology as the leading concept. A conceptual framework is developed to uncover the relationship of Industrial Ecology, Eco-Efficiency and Industrial Symbiosis, from a firm level perspective.

Keywords: Industrial Ecology · Industrial Symbiosis · Eco-Efficiency · Sustainable value · Process industry · Energy efficiency · Resource efficiency · Firm analysis

1 Introduction

Innovations oriented to sustainability pursue real and substantial improvements by developing superior production processes, products and services and by exercising large market influence as well as social or political influence [1]. Industrial Ecology (IE) is a leading concept for sustainability-oriented innovations. It is strongly connected to the idea of closing material loops, thus, emphasizing on materials and energy flows and life cycle perspectives at firm, inter-firm and regional/global levels [2]. IE related innovations drive increasing attention to the development of ecologically benign, clean resources, technologies and new products [3]. Still a relatively new field, IE is "*a cluster of concepts, tools, metaphors and exemplary applications and objectives*" [2]. Among the myriad of concepts within the IE frameworks, Eco-Efficiency (EE) and Industrial Symbiosis (IS) are highlighted as firm level and inter-firm level guiding concepts [4, 5]. Whereas they are considered as key parts of IE, there is no current research effectively explaining the interior relationship of these concepts (EE, IS, IE) in one context. Some authors have connected IS and EE through the use of EE indicators to assess the impact of the application of IS [6, 7]. This research aims at investigating the connections between EE and IS under the frame of IE as guiding concept.

© Springer International Publishing AG 2017
G. Campana et al. (eds.), *Sustainable Design and Manufacturing 2017*, Smart Innovation, Systems and Technologies 68, DOI 10.1007/978-3-319-57078-5_57

The research focuses on the process industry in order to provide a context for our exploratory reasoning at conceptual level and as a key industry with strong impact on sustainability at European and global level. Nine industry sectors are considered as part of the process industry [8]: chemicals, food, glass, paper and pulp, pharmaceuticals, metal, rubber and plastics, textile and building materials. The process industry is often characterized as energy-intensive and resource-intensive, with a significant contribution to GHG emissions and a high dependence on resources availability [9]. This makes it especially a good target for implementation of new strategies and methods for increasing resource and energy efficiency. The potential gains of implementing IE oriented innovations would have a huge impact on environmental and societal aspects.

Authors define our research question as follows: *How firms in process industry can implement better EE and IS concepts?* There are three steps defined to answer this research question. Initially, an analysis at conceptual level of EE and IS has been performed and this provided an initial conceptual framework for our research. A second step looks at in-company analysis to understand the necessary capabilities for EE and IS implementation and possible synergies between them. Finally, our research aims at developing tools and methods to support companies in their EE and IS implementation activities. This paper presents our results regarding the first research step. An extensive literature review has been performed to develop the initial conceptual framework; besides, case studies from secondary data sources are analyzed to illustrate the potential effectiveness of the conceptual framework.

2 Background

2.1 Eco-Efficiency

In the 1970s, EE was first suggested as a concept of environmental efficiency. In the 1990s, environmental factors compelled a new interest in manufacturing sectors. At this stage, the role of industry has changed from being the cause of environmental degradation to a driver for sustainability. Therefore, as a business links to sustainable development, the concept of environmental efficiency has been extended to EE [10]. The World Business Council for Sustainable Development (WBCSD) initially defines the concept of EE as "a management philosophy that encourages business to search for environmental improvements which yield parallel economic benefits" [11]. Besides, WBCSD also details seven key principles of EE, which are: reduction in the material intensity of goods or services, reduction in the energy intensity of goods or services, reduction in the dispersion of toxic materials, improved recyclability of materials, maximum use of renewable resources, greater durability of products, production materials and equipment, increased service intensity of goods and services [11, 12].

Yu et al. discussed that EE is the main strategy for promoting sustainability through living within global resource carrying capacity [13]. EE is also recognized as a significant tool to evaluate environmental and economic challenges at the same time [14]. Specifically, it indicates that a firm operates in a good financial performance with less environmental impact or a high quality product with added value [15].

Mickwitz et al. argued that eco-efficiency could be viewed from many perspectives, such as the macro-economic (national economy), the meso-economic (region) and the micro-economic (company) levels [16]. In addition to the micro-economic level, EE has also been applied in corporate level, process level, and product level [17, 18]. EE is an essential component of corporate social performance. It also acts as forward-looking measures of firm financial performance for both researchers and practitioners [19]. Mickwitz et al. illustrated that EE is capable of reducing the environmental impact and natural resources, as well as maintaining or increasing the value of the output [16]. Thus, improving EE requires producing more desirable outputs (GDP), while reducing the consumption of resources and adverse ecological impacts [20].

Currently, EE is becoming an increasingly organizational performance measurement [21]. It is widely accepted as a means for both increasing economic value and reducing environmental affects [22]. It is also defined as the ratio of resource inputs ad waste outputs to final product [23]. It is considered an instrument for sustainability analysis, showing the empirical economic relationship between environmental cost or value and environmental input [16].

It is worth to mention that an exact definition of EE does not exist [24]. In this research, authors selected Sorvari et al. concept to define EE which is "to create more value with fewer resources and less negative impact" [25].

2.2   Industrial Symbiosis

IS has been positioned within the IE field as a concept engaging "traditionally separate entities in a collective approach to competitive advantage involving physical exchange of materials, energy, water" [26]. The concept of IS was inspired by the observation of the Kalundborg network, in Denmark, in which exchanges of waste, by-products, and energy occur among closely situated companies over a period of more than 20 years [27, 28].

IS is seen as a means to progress towards a more eco-efficient industrial system [29]. Considering the system as a whole, the overall environmental performance would be higher than the performance at each individual the factory level [27]. At company level, IS brings additional opportunities to increase revenues through by-products sales or cost savings [30]. Moreover, IS based business strategies are currently enabling new business models that create value from waste and additionally allows to repurpose for society and environment at multi-organizational level [31]. Thus, providing a broader set of benefits, other than economic value, and for a broader range of stakeholders.

The key activities for IS applications are the recovery, reuse and recycle of waste (materials, water, or energy) from one facility as alternative input in a neighbouring facility [32]. Therefore, the final quantity of waste being disposed can be significantly reduced or even eliminated. Whereas there is a reduction of waste disposed, there are some additional benefits for the companies receiving the waste. Waste and byproducts can replace raw materials and fossil fuels in industrial processes [9], reducing supply costs for receiving companies.

In the present work, the definition given by Chertow in 2000 [26] and reported at the very beginning of this section has been taken as a reference.

3   Conceptual Framework

Based on literature and definitions introduced in previous sections, authors identified three main aspects linking the concepts of EE and IS, which are 'Actions', 'Stakeholders' and 'Value'. These three aspects are present in most of the definitions identified and have therefore been used to drive the analysis of the interior relationship among EE and IS. IE has been considered as the leading concept for both EE and IS, and a comparative analysis of these drivers for IE and EE and IE and IS has been carried on.

"Actions" are all of the activities carried on by practitioners to either improving the performance of existing technologies, or creating new technologies [33]. At the firm level, IE is then analogous to EE [2]. In other words, the target of IE and EE is the same, i.e. to increase the value of the product while reducing the environmental impact through recycling, reusing and reducing. Specifically, examining definitions in the Oxford Dictionary, the action of recycling is "to convert (waste) into reusable material", reusing is defined as "the action of using something again", while reducing is "to make something become smaller or less in size, amount, or degree" [34]. The Cambridge dictionary adds a definition to recycling, which is "to use something again for a different purpose" [35]. In EE literature, the action of reusing is to repeat the usage of production wastes; recycling is to reuse raw materials and correctly dispose the items that cannot be reused; reduce is to decrease materials and energy intensity as well the dispersion of toxic substances [11, 12, 36]. In IS literature, reuse and recycle are the most important actions to reduce negative impacts such as the usage of oil, emissions of carbon dioxide and the quantities of waste disposal [37]. For instance, a recycling action for an industry might be to use waste products as an alternative energy source. This approach helps to relieve the community need to process this waste and also helps to limit CO2 emissions [38].

"Stakeholder" indicates all the individuals or groups that affect or are affected by the corporate actions [39]. For instance, companies, industries, regulators, interest groups, consumers, households and local communities, regions or countries [40, 41]. In IE, EE and IS, the configuration of the industrial system is created by many different actors or agents under a variety of coordination mechanisms such as organizations, markets, policy, and regulation. It's a context-based field of research and the solutions are strongly determined by contextual factors, where the detailed advantages to each party are not necessarily well understood [41–43]. In practical, there is no general differentiation among stakeholders involved in EE and IS, as it highly depends on the specific case and context.

"Value" is intended as an extensive set of benefits for different stakeholders. It entails different meanings from different stakeholders' point of view, which should be as aligned as possible to enable the realization of EE and IS implementation possibilities. In the sustainability perspective, value includes monetary profit, social and environmental aspects [44, 45]. In this research, value has been considered in following perspectives: in IE, practitioners increase resources' value by closing resource loops; in EE, practitioners increase the value of the product from the customers' perspective; in

IS, practitioners increase the value of waste and byproducts, as well as resources' value by creating mutually beneficial transactions.

Considering what above concluded, it is possible to say that actions usually undertaken in EE and IS are very similar, with the only exception of "reduce". Whereas, IE and EE characterized by all three types of actions, which are "reduce", "reuse" and "recycle". Therefore, "Action" is not suitable to be selected as the main driver in a conceptual framework. 'Stakeholders' is a highly context-related driver. It depends on the specific situation and the definition relies on multi-party, which is why it will also not be considered as main driver in the conceptual framework. 'Value' is the only aspect that links the concepts of IE, EE and IS by the resources, customers and waste perspective. Therefore, authors decided to use 'Value' as the main driver and to use 'Actions' as subsidiary feature to build the conceptual framework. Figure 1 represents a synthesis of the conceptual framework proposed by the authors.

In this framework, as previously discussed, IE is selected as the leading concept, as it includes all actions, stakeholders and value associated with both EE and IS. IE is also usually referred to a higher level compared to EE and IS, i.e. regional or global level rather than intra-firm (which is the level used for EE) or both intra-firm and inter-firm (as exchanges are realized between processes, IS can occur between two different processes in a single firm or between different firms [41]). Based on the leading concept, the framework then makes a comparison between EE and IS in terms of actions and value generation. The framework clarifies the different actions undertaken in EE and IS, and also value management alternatives, that mainly differ in waste management strategies. Considering EE in production processes, the main aim of waste management is to produce less waste. The quantity of waste is decreased and the value is almost the same. In contrast, through IS implementation; the 'waste' would become by-product. This means that the value of the waste is increased and the quantity of the waste does not necessarily have to decrease in order to reduce negative impact and increase value.

The next section describes four cases of EE and IS basing on main concepts presented in the framework. It is worth to mention that for sake of brevity, authors will only describe four cases; deeper analysis will be conducted in further researches.

4   Industrial Cases

4.1   Eco-Efficiency Cases

4.1.1   Eco-Efficiency Case 1

Ozturk et al. carried out a study aimed at reducing the environmental impact of a cotton/polyester fabric finishing-dyeing process in a textile mill located in Denizli, Turkey [46]. The mill is a mid-sized outsourcer dye house with two production lines, bleaching 2412 tons/year and dyeing 6682 tons/year. Main resources consumed by the plant are water, thermal energy and electricity. Water demand is provided from groundwater sources and mainly used in facility cleanings and finishing/dyeing processes; wastewater recovery and reuse techniques had never been implemented in the mill [46]. Thermal energy consumption, provided by coal and gas, is mainly consumed

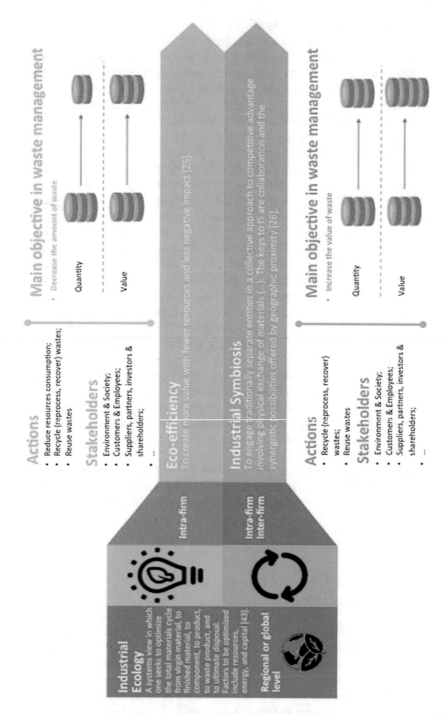

Fig. 1. The conceptual framework

by steam and hot-oil boiler systems, while electricity is supplied from electricity grid and used for electric engines, lightings and other devices. Chemical usage is intensive at almost every stage of production processes. After an initial assessment, 92 different improvement actions were identified on the basis of Best Available Techniques defined by the European Commission (2003), then reduced to 22 after a feasibility study and prioritization process. In particular, systems to reuse/recovery washing wastewater and dye bath were introduced, as well as practices regarding the insulation of hot surfaces (tank, pipe, etc.), the optimization of boiler units, the heat recovery from separated hot wastewater and from flue gas and stenters. In addition, chemical consumption was reduced by removing iron from fabric surfaces before scouring process and by recovering caustic from mercerization process wastewater by membrane techniques and chemical substitution. As a result of the implementation of these improvement actions, a consistent reduction in the consumption of resources as well as in the production costs was achieved. The environmental impact was sensitively lowered by the chemical substitution in particular.

4.1.2 Eco-Efficiency Case 2

A modern beef and lamb processing plant based in Northern Ireland can slaughter and process 1800 cattle and 4000 lambs per week, supplying meat products to major supermarket chains such Tesco, M&S, Dunnes, Centra and Musgrave Supervalu. The company is highly committed to reduce its environmental impact and the amount of waste sent to landfill. The eco-efficiency actions implemented along these lines are the reduction of resources required for the process, reuse and recycle of wastes. In particular, their focus over the last years has been to improve the waste management and handling practices. In fact, the company was already recycling its cardboard and wood packaging waste, but had never created a system to segregate and recycle plastic packaging waste and cans. With the help of a consulting company and of a recycling company, the processing plant was able to start recycling 27 tons of plastic packaging and cans, saving up to £5,700 and reducing its carbon dioxide emission of about 280 tons [47]. This, consequently, lowered significantly manufacturing costs in the plant.

4.2 Industrial Symbiosis Cases

4.2.1 Industrial Symbiosis Case 1

A fruit juice concentrate producer in Iskenderen Bay, Turkey, generated 12,000 tons of fruit pulp waste each year as a by-product of its process and wanted to find a way to reuse the material rather than sending it to costly landfill. A team of researchers from the Faculty of Agriculture at Cukurova University, financed by the Turkish government, tested out potential ways to treat the pulp, making it suitable for reuse. In addition, a mining company, which wanted to find an outlet for its waste heat from lime production, was contacted as a potential partner. Researchers at the University carried out tests to dry the fruit pulp waste using the waste heat from the lime production process, successfully transforming the material into animal feed. The nutrient composition and energy value analysis proved that the quality of the animal feed end product was high, which was critical to the full commercialisation of the scheme.

Implementing this three-way synergy redirected 115 tons of petroleum coke waste heat, reduced carbon dioxide emissions by 3,500 tons, reused 12,000 tons of waste pulp each year and produced 1400 tons of animal feed for reselling [48].

4.2.2   Industrial Symbiosis Case 2

A Scottish global manufacturer of alcoholic drinks receives the aromatics for its gin in hessian sacks, which were going to landfill after being emptied. The company was

Table 1. Cases analysis

Case	Actions	Stakeholders	Changes in waste quantity and value	Results
Case 1 EE	Reduce, Reuse, Recover	Firm, landfills, chemical providers, energy providers customers, technology providers, consultants, customers, environment, operators, population	The quantity of waste has been drastically reduced by reducing resource consumption	Reduced costs, reduced resources consumption, reduced environmental impact, value created for the company
Case 2 EE	Recycle	Firm, landfill, consultants, customers (B2C), customers (B2B), environment, recycling company, population	The quantity of waste has been drastically reduced by paying a recycler	Reduced costs, reduced environmental impact, value created for the company
Case 1 IS	Recycle (implement exchanges)	Government, university, 3 firms participating in the exchange, population, operators, technology providers, customers, consultants, landfill	The value of waste has been increased	Increased revenues, reduced resources consumption, reduced environmental impact, value created for three companies
Case 2 IS	Recycle (implement exchanges)	2 firms participating in the exchange, consultants, government, customers, sacks providers, landfill, population	The value of waste has been increased	Increased revenues, reduced costs, reduced resourced consumption, reduced environmental impact, value created for two companies

committed to finding a more sustainable alternative to landfill. A new business start-up based in Fife uses the staves from old whisky barrels to make firewood. The start-up was buying hessian sacks to package the firewood it was selling. The quality of hessian sacks landfilled by the drinks manufacturer was assessed and they were found clean and durable enough to be ideal to be used as firewood packaging. The sacks were declared as waste on the Pollution prevention and control license, so the drinks manufacturer was not immediately able to divert the sacks, but had to seek the help of a consultancy company, who started a dialogue with Scottish Environment Protection Agency. Eventually, it was possible to re-categorize the sacks from a waste to a by-product. This synergy has saved the start-up over £20,000 in packaging costs, which has helped them considerably as a newly funded company. The drinks manufacturer is also seeing cost benefits, as they no longer have to pay to dispose the sacks to landfill [47].

4.3   Discussion About the Cases

The following Table 1 synthesizes key concepts of previously illustrated industrial cases. Actions vary slightly among the four cases, while stakeholders involved are different in all of them. In EE cases the quantity of waste is reduced and value is created for the company (and subsequently for the customer, as it will potentially lead to lower selling price), while in IS cases the waste is revaluated by giving it further use and, therefore, value is created for all the companies participating in the exchange.

5   Concluding Remarks

Based on the leading concept of IE, this research conducts a higher-level comparison between EE and IS. In theoretical, authors identify a conceptual framework and generalized the relationship of EE and IS. Specifically, EE is focus on the intra-firm while the processes of IS works on both intra and inter level of a firm. In the value perspective, EE is to reduce the amount of waste, which would reduce the cost of waste disposal. In contrast, IS is to increase the value of the waste, which increase the profit to the firm.

Case analysis reflects a recurring barrier to IS: pioneer companies that find a higher value solution for their waste (normally non-hazardous but categorized as waste by legislation) frequently struggle with current regulations and standards. Regulation can be in fact a key success factor if well managed but is often rather an obstacle for companies starting a symbiotic exchange [49]. In addition, case analysis suggested that IS implementation often requires the involvement of a third-party or consultant in order to be effective. This is not usually true as well for EE cases. Finally, it is often verified that IS exchanges do not happen because companies keep considering waste only as waste and not as valuable product, while in IS waste ceases to be waste, as "a waste product might no longer be waste when it is marketable as a useful and environmentally safe product" [31].

It is worth to mention that there is not one way to reduce environmental impact and increase economic value applicable universally. Tools and methods are needed to support the analysis of the most adequate strategy in each case; thus, prior to implementation plans, practitioners should have a comparison with other optimal mechanisms to work out the most effective solution in their case [41].

Further research will firstly address the generalizability of the framework. This will be conducted by analyzing a more extensive set of case studies on EE and IS applications. Additionally, this research will address an analysis of the capabilities for EE and IS to identify the possible synergies between them. This stage will actively involve manufacturing companies into the research activities; thus, a participatory research approach will be taken. Finally, this research will focus on the development of tools and/or methods to support companies willing to apply more effectively both EE and IS into their operations. These research results will provide them with guidelines to understand how IE can be implemented at firm level and to have a better performance on the implementation of EE and IS.

Acknowledgements. This work was supported by the European Union's Horizon 2020 research and innovation program (grant no. 680570) and the EPSRC Centre for Innovative Manufacturing in Industrial Sustainability (grant no. EP/I033351/1).

References

1. Schaltegger, S., Wagner, M.: Sustainable entrepreneurship and sustainability innovation: categories and interactions. Bus. Strategy Env. **20**(4), 222–237 (2011)
2. Lifset, R., Graedel, T.E.: Industrial ecology: goals and definitions. In: Ayres, R.U., Ayres, L. (eds.) Handbook for Industrial Ecology (2001)
3. Huber, J.: Towards industrial ecology: sustainable development as a concept of ecological modernization. J. Env. Policy Plann. **2**(4), 269–285 (2000)
4. Van Berkel, R., The role of eco-efficiency in industrial ecology. In: Proceedings of the III International Conference on Industrial Ecology for a Sustainable Future, Stockholm, pp. 12–15, June 2005
5. Boons, F., Spekkink, W., Mouzakitis, Y.: The dynamics of industrial symbiosis: a proposal for a conceptual framework based upon a comprehensive literature review. J. Clean. Prod. **19**, 905–911 (2011)
6. Park, H.-S., Behera, S.K.: Methodological aspects of applying eco-efficiency indicators to industrial symbiosis networks. J. Clean. Prod. **64**, 478–485 (2014)
7. Salmi, O.: Eco-efficiency and industrial symbiosis – a counterfactual analysis of a mining community. J. Clean. Prod. **15**, 1696–1705 (2007)
8. Stindt, D., Sahamie, R.: Review of research on closed loop supply chain management in the process industry. Flex. Serv. Manuf. J. **26**(1–2), 268–293 (2014)
9. Garetti, M., Taisch, M.: Sustainable manufacturing: trends and research challenges. Prod. Plann. Control **23**(2–3), 83–104 (2012)
10. Schaltegger, S., Sturm, A.: Ökologische Rationalität (Ecologic Rationality). Unternehm 4, 273–290 (1990). (available only in German)
11. World Business Council for Sustainable Development (WBCSD). Eco-efficiency: Creating More Value with Less Impact, Geneva (2000)
12. WBCSD, Cross cutting themes: Eco-efficiency, World Business Council for Sustainable Development (1992)

13. Yu, Y., Chen, D., Zhu, B., Hu, S.: Eco-efficiency trends in China, 1978–2010: decoupling environmental pressure from economic growth. Ecol. Ind. **24**, 177–184 (2013)
14. Carvalho, H., Govindan, K., Azevedo, S.G., Cruz-Machado, V.: Modelling green and lean supply chains: an eco-efficiency perspective. J. Res. Conser. Recycl. **120**, 75–87 (2017)
15. Nikolaou, I.E., Matrakoukas, S.I.: A framework to measure eco-efficiency performance of firms through EMAS reports. J. Sustain. Prod. Consumption **8**, 32–44 (2016)
16. Mickwitz, P., Melanen, M., Rosenström, U., Seppälä, J.: Regional eco-efficiency indicators: a participatory approach. J. Clean. Prod. **14**, 1603–1611 (2006)
17. Hahn, T., Figge, F., Liesen, A., Barkemeyer, R.: Opportunity cost based analysis of corporate eco-efficiency: a methodology and its application to the CO2-efficiency of German companies. J. Environ. Manag. **91**(10), 1997–2007 (2010)
18. Kerr, W., Ryan, C.: Eco-efficiency gains from remanufacturing: a case study of photocopier remanufacturing at Fuji Xerox Australia. J. Clean. Prod. **9**, 75–81 (2001)
19. Guenster, N., Bauer, R., Derwall, J., Koedijk, K.: The economic value of corporate eco-efficiency. Eur. Financ. Manag. **17**(4), 679–704 (2011)
20. Huang, J., Yang, X., Cheng, G., Wang, S.: A comprehensive eco-efficiency model and dynamics of regional eco-efficiency in China. J. Clean. Prod. **67**, 228–238 (2014)
21. Davéa, A., Salonitisa, K., Ball, P., Adams, M., Morgan, D.: Factory eco-efficiency modelling: framework application and analysis. CIRP **40**, 214–219 (2016). 13th Global Conference on Sustainable Manufacturing - Decoupling Growth from Resource Use
22. Sun, Y.Y., Pratt, S.: The economic, carbon emission, and water impacts of chinese visitors to taiwan: eco-efficiency and impact evaluation. J. Travel Res. **53**(6), 733–746 (2014)
23. Schmidheiny, S.: Changing Course: a Global Business Perspective on Development and the Environment. MIT Press, Cambridge (1992). Business Council for Sustainable Development
24. Koskela, M.: Measuring eco-efficiency in the Finnish forest industry using public data. J. Clean. Prod. **98**, 316–327 (2014)
25. Sorvari, J., Antikainen, R., Kosola, M.-L., Hokkanen, P., Haavisto, T.: Eco-efficiency in contaminated land management in Finland: barriers and development needs. J. Environ. Manag. **90**, 1715–1727 (2009)
26. Chertow, M.R.: Industrial symbiosis: literature and taxonomy. Annu. Rev. Energy Env. **25**(1), 313–337 (2000)
27. Ehrenfeld, J.R., Gertler, N.: Industrial ecology in practice: the evolution of interdependence at Kalundborg. J. Ind. Ecol. **1**(1), 67–79 (1997)
28. Jacobsen, N.B.: Industrial symbiosis in Kalundborg, Denmark: a quantitative assessment of economic and environmental aspects. J. Ind. Ecol. **10**(1–2), 239–255 (2006)
29. Teresa, D., Michael, D.: The role of embeddedness in industrial symbiosis networks: phases in the evolution of industrial symbiosis networks. Bus. Strategy Env. **20**(5), 281–296 (2011)
30. Paquin, R.L., Busch, T., Tilleman, S.G.: Creating economic and environmental value through industrial symbiosis. Long Range Plan. **48**, 95–107 (2015)
31. Ruiz Puente, M.C., Romero Arozamena, E., Evans, S.: Industrial symbiosis opportunities for small and medium sized enterprises: preliminary study in the Besaya region (Cantabria, Northern Spain). J. Clean. Prod. **87**, 356–374 (2015)
32. Van Berkel, R.: Comparability of industrial symbioses. J. Ind. Ecol. **13**(4), 483–486 (2009)
33. Eco-efficiency action project, Methods, Models, & Analytics for Sustainable Operations Management (2014). http://eco-efficiency-action-project.com
34. Oxford dictionary. https://en.oxforddictionaries.com. Accessed Nov 2016
35. Cambridge dictionary. http://dictionary.cambridge.org/dictionary/english/recycling. Accessed Nov 2016

36. Mohanty, C.R.C.: Reduce, reuse and recycle (the 3Rs) and resource efficiency as the basis for sustainable waste management. In: Synergizing Resource Efficiency with Informal Sector towards Sustainable Waste Management (2011)

37. Hopewell, J., et al.: Plastics recycling: challenges and opportunities. Philos. Trans. R. Soc. B (2009). doi:10.1098/rstb.2008.0311

38. Industrial ecology and recycling, Sustainability report (2011). http://www.lafarge.com/05182012-publication_sustainable_development-Sustainable_report_2011-industrial-ecology-uk.pdf. Accessed Nov 2016

39. Edward Freeman, R.: Stakeholder theory of the modern cooperation, General Issues in Business Ethics (2012)

40. Posch, A.: Industrial recycling networks as starting points for broader sustainability-oriented cooperation? J. Ind. Ecol. 14(2), 242–257 (2010)

41. Holgado, M., Morgan, D., Evans, S.: Exploring the Scope of Industrial Symbiosis: Implications for Practitioners. In: Setchi, R., Howlett, Robert J., Liu, Y., Theobald, P. (eds.) Sustainable Design and Manufacturing 2016. SIST, vol. 52, pp. 169–178. Springer, Cham (2016). doi:10.1007/978-3-319-32098-4_15

42. Dijkema, G.P.J., Basson, L.: Complexity and industrial ecology foundations for a transformation from analysis to action. J. Ind. Ecol. 13, 157–164 (2009). http://onlinelibrary.wiley.com/doi/10.1111/j.1530-9290.2009.00124.x/full

43. Chertow, M.R.: Industrial symbiosis: literature and taxonomy. Ann. Rev. Energy Env. 25, 313–337 (2000)

44. Rana, P., Short, S., Evans, S.: D2.5 – Lessons Learned Report, Documenting the Impact from Use of the Tools & Methods and Areas for Improvement (2013)

45. State-of-practice in business modelling and value-networks, emphasising potential future models that could deliver sustainable value. http://www.sustainvalue.eu/publications/D2_1_Final_Rev1_0_web.pdf. Accessed Nov 2016

46. Ozturk, E., Koseoglu, H., Karaboyaci, M., Yigit, N.O., Yetis, U., Kitis, M.: Sustainable textile production: cleaner production assessment/eco-efficiency analysis study in a textile mill. J. Clean. Prod. 138, 248–263 (2016)

47. Invest Northern Ireland: Industrial Symbiosis - Improving productivity through efficient resource management - Guide for Businesses in Northern Ireland. https://secure.investni.com/static/library/invest-ni/documents/industrial-symbiosis-guide-for-businesses-in-northern-ireland.pdf. Accessed Nov 2016

48. Iskenderun Bay Industrial Symbiosis. http://www.international-synergies.com/projects/iskenderun-bay-industrial-symbiosis/. Accessed Nov 2016

49. Costa, I., Ferrao, P.: A case study of industrial symbiosis development using a middle-out approach. J. Clean. Prod. 18(2010), 984–992 (2010)

Integration of Eco-Efficiency and Efficiency Assessment Methodologies: The Efficiency Framework

A.J. Baptista[1], E.J. Lourenço[1(⊠)], E.J. Silva[2], M.A. Estrela[2], and P. Peças[3]

[1] INEGI – Instituto de Ciência e Inovação em Engenharia Mecânica e Engenharia Industrial, Campus da FEUP, 4200 Porto, Portugal
{abaptista,elourenco}@inegi.up.pt

[2] ISQ - Instituto de Soldadura e Qualidade, 2740 Oeiras, Portugal
{EJSilva,MAEstrela}@isq.pt

[3] IDMEC, Instituto Superior Técnico, Universidade de Lisboa, 1049 Lisbon, Portugal
ppecas@tecnico.ulisboa.pt

Abstract. The overall aim of the Efficiency Framework is to encourage a culture of continuous improvement and sustainability within manufacturing and process industries. The framework presented supports informed decision-making processes and helps to define strategies for continuous performance improvement. The proposed innovative Efficiency Framework, materialized through the integration of concepts and results provided by eco-efficiency methodology, namely Eco-Efficiency Integrated Methodology for Production Systems (ecoPROSYS) and the lean based resource efficiency assessment method, Multi-layer Stream Mapping (MSM). Thus, the framework assesses simultaneously the environmental, economic and efficiency performance of complex production systems, which helps to identify major inefficiencies and circumstances of low eco-efficiency performance, consequently leading to the definition of improvement priorities. Ultimately, this framework aims to facilitate the overall efficiency performance assessment, by an integrated multi-dimensional analysis, presented as the Total Efficiency Index. The logic behind this index is to combine, for each unit process and for the overall production process, two fundamental efficiency aspects, namely eco-efficiency and operations efficiency.

Keywords: Operations efficiency · Eco-efficiency · Resource efficiency · Total efficiency framework · Environmental and economic performance · ecoPROSYS · MSM

1 Introduction

European process industries are largely dependent on resources imports from international markets that are hampering the industry's access to globally traded raw materials, due to the increased political instability in many regions of the globe, which

© Springer International Publishing AG 2017
G. Campana et al. (eds.), *Sustainable Design and Manufacturing 2017*, Smart Innovation, Systems and Technologies 68, DOI 10.1007/978-3-319-57078-5_58

is perfectly visible due to a sharp increase in raw material prices during recent years. Moreover, European industry has also accounted for more than a quarter of total energy consumption in 2010 in Europe with a significant portion of that used within the process industry [1].

This represents both an opportunity and responsibility of this sector contribution to the sustainability challenges of European societies, being imperative to drastically reduce the environmental footprint, and increase competitiveness and production systems efficiency by "doing more with less", namely by enhancing resource and energy efficiency in a sustainable manner [1]. However, to successfully implement sustainability in manufacturing and process industries, a holistic, multidimensional and systematic approach is required. With this in mind, the aim is to propose a flexible and scalable Efficiency Framework based on a holistic approach, which combines different assessment methods and tools, the overall purpose of the framework is to support improvement on a continuous basis and increase eco-competitiveness by fostering sustainability in routine operations. Nevertheless, despite the environmental, economic and social improvement potentials, it is essential to understand and assess resource and energy efficiency in order to optimize production systems. Moreover, the increased availability of ultra-modern technologies for process monitoring and optimization should be carefully adapted and integrated for a wider and facilitated adoption of state-of-the-art tools and methodologies for operations efficiency and eco-efficiency.

1.1 Environmental Concerns and Tools

Concerns for the environmental impact and depletion of resources as a result of unlimited economic growth, have stimulated engineers, for many years, to reduce the impact of product lifecycles.

Regarding eco-efficiency assessments, several evaluations have taken place for various industries and with various approaches [2, 3]. For instance, BASF performed an eco-efficiency assessment to quantify the sustainability of products and processes, and to support decision making [4]. Côté assessed the eco-efficiency performance of several SME using a self-developed eco-efficiency checklist [5]. Kharel and Charmondusit [6] evaluated the eco-efficiency of an iron road industry, using an empirical assessment that considers energy, material and water consumption and emissions, as the environmental aspects.

Despite eco-efficiency and environmental performance assessment of industrial processes being powerful tools for achieving sustainability, especially when considering the decompiling of the economic performance from environmental burdens, the assessments presented in literature, have no clear link between the eco-efficiency principles and the environmental aspects. Additionally, no evaluation of the significance of the environmental aspects is done. Furthermore, the lack of transparency on how efficiently resources are being used, in the eco-efficiency assessments mentioned-above, is a clear gap to underpin sustainability.

The efficiency assessments are based on the principle, that the efficient and effective material and energy use can reduce natural resource inputs and waste or pollutant outputs, thus avoiding environmental degradation [7]. Yet, the efficient use of materials, energy or resources principle does not take into direct account the related environmental impacts, as it focuses on the efficiency of resource consumption, and consequently never incite companies to look for alternatives and improvements that can enhance economic and environmental performance.

With this in mind, the proposed innovative Efficiency Framework was developed to assess, simultaneously, the environmental, economic and efficiency performance of complex production systems, in order to support in the identification of major inefficiencies and circumstances of low eco-efficiency performance. Moreover, it will support the decision making process and enabling managers to take actions that will improve both, efficiency and eco-efficiency performance.

2 Eco-Efficiency and Efficiency Assessment Methods and Tools

The eco-efficiency and efficiency assessment methods and tools are the core part of the Efficiency Framework. The proposed framework consists in the integration of two innovative methodologies, namely Eco-Efficiency Integrated Methodology for Production Systems (ecoPROSYS), which is an integrated methodology for evaluation and assessment of eco-efficiency performance, and Multi-layer Stream Mapping (MSM), a lean based method, developed to assess overall efficiency of a system. The next sections aim to briefly describe these methodologies.

2.1 Eco-Efficiency Assessment Method - ecoPROSYS

The ecoPROSYS approach relies on the use of a systematized and organized set of indicators easy to understand/analyse, aiming to promote continuous improvement and a more efficient use of resources and energy. The goal is to assess eco-efficiency performance in order to support decision making and enable the maximization of product/processes value creation while minimizing environmental burdens.

Eco-efficiency, the base concept of ecoPROSYS, measures the relationship between environmental and economic development of activities as sustainability aspects that evidence more value from lower inputs of material and energy and with reduced emissions. Eco-efficiency is commonly expressed by the ratio between value and environmental influence.

$$\text{Eco-Efficiency} = \frac{\text{Production or Service Value}}{\text{Environmental Influence}} \tag{1}$$

According to the WBCSD the two most common goals of eco-efficiency assessments are: (i) measuring progress and (ii) internal and external communication of economic and environmental performance. In order to improve overall performance, the WBCSD established seven principles: Reduce material intensity; Reduce energy intensity; Reduce dispersion of toxic substances; Enhance recyclability; Maximize use of renewable resources; Extend product durability; and Increase service intensity [8].

ecoPROSYS, is aligned with the eco-efficiency goals and principles defined by the WBCSD. From a conceptual point of view, in this methodology the indicators are generated by a combination of three components: (1) Environmental Performance Evaluation (EPE) (2) Life Cycle Assessment (LCA), and (3) Cost and Value Assessment. The interaction between the different modules leads to the decision support indicators and to the environmental, value and eco-efficiency profiles. In addition, by connecting environmental influence with the inventory data and the goals defined by the organization for eco-efficiency principles, the ecoPROSYS methodology enables also the simulation of alternative scenarios and the evaluation of these goals and objectives [9].

2.2 Efficiency Assessment Method - MSM

The MSM is a lean based resource efficiency assessment methodology, which takes into account the base design elements from the VSM. Namely the value streams, in order to identify and quantify, at each stage of the process system, all "value added" (VA) and "non-value added" (NVA) actions, as well as, all types of waste and inefficiencies along the production system [10]. Therefore the basic principle of the MSM relates to Lean Principles, i.e. clear definition between value and waste. The MSM approach intends to encourage achieving maximum efficiency, (i.e. 100%) and continuous improvement mind-set. Moreover, unlike the VSM that focuses mainly on the VA and NVA of the time dimension, the aim of the innovative approach, MSM, is to assess the overall performance, by taking into account the efficiency of each process parameter (e.g. time, energy, water, raw material) associated to one or more process steps. The goal is to: (i) provide an efficiency integration analysis; (ii) identify inefficiencies in a very direct and visual manner; (iii) support decision making; and (iv) help prioritize the implementation of improvement actions [11].

Other tools that address the extension of Value Stream Mapping (VSM) to incorporate additional criteria, namely, energy-related and sustainable aspects focused on environmental performance are utilized to evaluate economic, environmental and social sustainability performance in manufacturing [12], but without the flexibility and multi-domain integration provided by MSM along an original and modular dashboard.

The MSM methodology resembles a matrix (m × n), where "n" is the number of process parameters evaluated and "m" the number of process steps. In order to apply the MSM, the following steps should be carried out:

- Identification of the system boundaries;
- Identification of the process steps;
- Identification of all relevant process parameters;
- Definition of the associated KPI to each parameter, always to be maximized and with values ranging between [0–100%];
- Analysis of the results and identification of the process parameters and process steps with lower efficiency results;
- Study and prioritization the improvement actions;
- Implementation of improvement actions and assessment of the efficiency gains evolution and cost reductions.

One of the cornerstones of the methodology involves the systematic non-dimensionalization of the process parameters that characterize the production system, with the base ratio between the portion of the "fraction that adds value" to the product transformation and the "total of the amount that enters the unit process" (see Eq. 2). Such particularity of the method, enables consecutive aggregation of the efficiency ratios along production system, sectors, or even plants, adopting a bottom-up analysis.

$$\text{Efficiency} = \frac{\text{Value added Fraction}}{\text{Value added fraction } + \text{Non-value added fraction}} \tag{2}$$

3   Efficiency Framework

The outline of the integration of ecoPROSYS and MSM concerns the exchange of information between efficiency and eco-efficiency assessments, which corresponds to the central objective of the Efficiency Framework. The approach followed to integrate ecoPROSYS and MSM is primarily through the combination of the eco-efficiency and efficiency results as opposed to a fusion of results. Such approach enables to obtain, besides the efficiency and eco-efficiency stand-alone, results to support decisions, and new integrated results, namely the:

- Total Efficiency Index (TEI) - New metric obtained by integrating results from ecoPROSYS and MSM, i.e. combining eco-efficiency with efficiency metrics;
- Environmental and Value Performance - Indicators based on production system performance regarding environmental and value target figures;
- Normalized Eco-efficiency - Representing the current performance and the improvement margin;
- Environmental Influence and Costs of the VA and NVA activities: These are obtained by integrating results from ecoPROSYS and MSM.

3.1   How Results Are Combined for Integration

As mentioned before, the outcomes of the Efficiency Framework will enable a simultaneous eco-efficiency and efficiency performance assessment by using the direct results from ecoPROSYS and MSM, respectively. These results besides providing support for a more complete and informed decision making process regarding sustainable development, also play an important role when it comes to reporting results through scorecards/dashboards within the Efficiency Framework.

The combination of the efficiency and eco-efficiency results will enable to assess the effectiveness of the eco-efficiency performance improvement. This will be done by monitoring deviations between the current production system eco-efficiency (real economic value over real environmental influence) and targeted eco-efficiency (target economic value over target environmental influence). Moreover, the goal is to evaluate if eco-efficiency performance variation is due to higher or lower environmental influence, or due to higher or lower economic value.

The definition of targets for environmental influence and value is required in order to obtain the novel indicators. Eco-efficiency targets are defined by the VA figure from the efficiency assessment. One major consideration for presenting the ratios and the targets is that these should always be self-contained (per process step). This means that the results for a certain process step are only dependent on the variances that occur within that process step. Therefore, major efficiency, environmental or value variations occurring in a certain process step, will not affect positively nor negatively the results of other process steps, thus ensuring a robust decision making process.

The Efficiency Framework results and dashboards should show different information according to the profile (e.g. manager, director, operator, team leader, etc.), and should enable the unfolding/breakdown of the dashboards through the organisational or functional hierarchy of the company. This means that logical links should be implemented from raw data (shop floor data, e.g. energy consumption from a specific machine), to process steps, sectors, all the way to Plant, Company or Group Level presenting the global aggregated data (overall business aspects).

3.2   Efficiency Analysis

In the analysis the outcomes from MSM are combined and integrated as direct results (process step efficiency results in %), where the VA helps to set the maximum reduction (i.e. NVA is zero) and the NVA helps to identify improvement priorities (reduce NVA).

The target setting is uses the result from MSM's VA actions. This figure is considered since the goal is to eliminate/reduce the NVA activities, being the VA the best value theoretically attainable. Yet, the targets can also be defined via an internal set point, or even by a sectorial set-point, as benchmarking amount for a given industrial sector. When using targets based on sector benchmark, this assessment should be performed preferable as "off-line" analysis and could be used for companies to position themselves, always keeping in mind that benchmark just provides a sectorial reference, and is not company specific. Nevertheless, depending on the company's operations and goal setting strategy, the direct benchmark reference targets can be applied if it does not affect the collaborators' motivation.

Regarding the technological aspect, the targets should be set for the existing technology in operation at the company. In case the technology is updated, the targets must also be updated. In terms of the management strategy, an "off-line" analysis with benchmark targets can be relevant to measure the company's competitiveness in respect to the state-of-art technology available.

Figure 1 depicts the results from the MSM methodology, to be considered for integration. From the integration point of view, the MSM results are also used in order to distinguish all VA and NVA activities, enabling ecoPROSYS modules, namely the LCA and Process-Based Cost Models (PBCM), to assess the VA and NVA for environmental influence and costs (and value). For instance, in order to define the targets for eco-efficiency, the MSM approach is used considering the VA (e.g. (a) 5 kg of steel

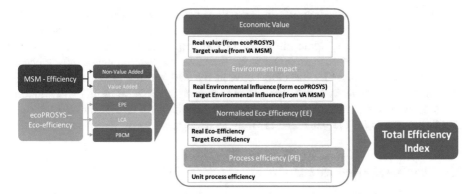

Fig. 1. Total Efficiency Index diagram calculus structure

adds value, therefore the target for the impact and monetary value should consider 5 kg of steel; (b) 1 h of machining adds value, therefore the target for the impact and monetary value should consider 1 h of machining). Besides these results being used for setting targets and consequently to quantify target eco-efficiency ratios for the normalization of the eco-efficiency performance, they are used to quantify the avoidable environmental impacts and value.

One important remark is that the MSM approach is founded on the principle that "more is better" i.e. closer to 100%, which means higher performance. Contrarily, when analysing the integrated results, one must keep in mind that, for eco-efficiency, higher environmental influence (Pt) and costs (€) are worse, but higher added value (e.g. Gross Value Added (€)) and lower impact points are better. Thus, the KPI formula must be well understood by the user. Nevertheless, the great objective is to improve the figure for eco-efficiency ratio where "more is better", which is aligned with the MSM principle.

3.3   Normalised Eco-Efficiency

Concerning the eco-efficiency results and their normalisation, it is necessary to calculate within the Efficiency Framework, the ratio between the real and the target eco-efficiency for each process step. This means that the real and target figures regarding the value dimension (€) and the environmental influence (Pt) should be quantified for each process step. Figure 1 depicts the input results from the eco-PROSYS methodology. The targets for value and environmental influence are a clear result from the integration of ecoPROSYS and MSM. As mentioned above, these results are the outcome of the quantification of environmental influence and costs using the VA and NVA dichotomy. The normalised eco-efficiency result, for each process step, is determined by the ratio between the real and the target eco-efficiency ratio. The normalised eco-efficiency result represents the percentage (%) of how much the process step fulfils the eco-efficiency performance for a predefined combination of targets.

During the normalisation of the eco-efficiency results, within the Efficiency Framework, the economic or monetary value and environmental influence efficiencies are also calculated. The value effectiveness (see Eq. 7) is determined by the ratio between the real and the target value, while the environmental influence effectiveness (see Eq. 8) is calculated by the ratio between the real and the target environmental influence. Both efficiency results are given as percentages. These results will support the decision making process, by helping on the quantification of the gap between real and target results for the value and environmental aspects.

3.4   The Total Efficiency Index

Regarding the ultimate novel outcome from the Efficiency Framework, the MAESTRI metric called Total Efficiency Index, it is calculated for each process step of the production system under analysis, but can be successively aggregated for the complete production process of a given product (through the organisational or functional hierarchy of the company). In quantitative terms, the TEI is obtained by multiplying the normalized eco-efficiency and the efficiency assessment results from MSM (see Eq. 3). The normalized eco-efficiency ratio represents, in percentage, the relation between real and targeted eco-efficiency performance (see Eq. 4) and is calculated dividing the real eco-efficiency ratio (see Eq. 5) by the targeted eco-efficiency ratio (see Eq. 6).

$$\text{Total Efficiency Index } (\%) = \text{Normalized eco-efficiency} \times \text{Process efficiency} \tag{3}$$

$$\text{Normalized eco - efficiency } (\%) = \frac{\text{Real eco - efficiency ratio}}{\text{Target eco-efficiency ratio}} \tag{4}$$

$$\text{Real eco-efficiency ratio } (\%) = \frac{\text{Real value } (\text{€})}{\text{Real environmental influence } (\text{Pt})} \tag{5}$$

$$\text{Target eco-efficiency ratio } (\%) = \frac{\text{Target value } (\text{€})}{\text{Target environmental influence } (\text{Pt})} \tag{6}$$

$$\text{Value effectiveness } (\%) = \frac{\text{Real value } (\text{€})}{\text{Target Value } (\text{Pt})} \tag{7}$$

$$\text{Environmental influence effectiveness } (\%) = \frac{\text{Real environmental influence} (\text{Pt})}{\text{Target environmental influence} (\text{Pt})} \tag{8}$$

The logic behind this index is to combine two fundamental efficiency aspects, namely eco-efficiency, which considers the ecology and economy, with ecoPROSYS, and resource and operational efficiency, which considers the NVA and VA activities aligned with the Lean Principles from MSM.

Consequently, TEI main outcome is providing the ability of evaluating if eco-efficiency performance variation is due to higher or lower environmental influence, or due to higher or lower economic value. In practice, this results from the distribution variance of TEI results that occur on two major axes: the efficiency and eco-efficiency. This distribution is presents in a graphical way in Fig. 2.

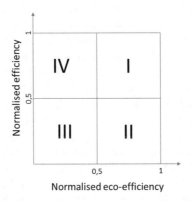

Fig. 2. Theoretical distribution of TEI values

The main characteristics and insights related to the distribution of the TEI results distribution (Fig. 2), are:

- Quadrant I - The production system generates good economic value considering the environmental impacts caused by its activities and high operations efficiency. Then, considering the current technological settings, it tends to have a low improvement potential.
- Quadrant II - The acceptable eco-efficiency performance is likely to be led by low environmental impacts related to the production system activities. Improvement actions to increase the economic value generation are advisable, namely by improving process efficiency.
- Quadrant III - The production system presents low performance, both low operations efficiency and eco-efficiency. The implementation of improvement actions is extremely recommended.
- Quadrant IV - Despite presenting good efficiency, the production system is not generating the expected value or high environmental impacts. Technological and procedural changes are recommended to decrease environmental impacts.

4 Conclusion and Outlook

The proposed Efficiency Framework concept aims at the integration of different aspects' results: eco-efficiency, operational efficiency, value and waste, etc. For this reason, the new approach was developed, which enables the Efficiency Framework to be used in an integrated way to assess eco-efficiency and efficiency performance, and overall production system efficiency. As its main objective, the integration of the

Efficiency Framework should empower a flexible and straightforward analysis, facilitating a company's implementation and use of the integrated framework. Its originality is founded not only on the integration of results of two original methodologies (ecoPROSYS and MSM), but also on the creation of a relationship and explicit perspective for the concurrent assessment of eco-efficiency and operational efficiency, with focus on the creation of practical and user-friendly analysis tools. From the integration of results from both methodologies, a new metric has been defined - "Total Efficiency Index" (TEI). The Efficiency Framework supports the decision making process and keeps track, simultaneously, of eco-efficiency and efficiency performance, enabling managers to take actions that will improve both, efficiency and eco-efficiency performance. It enables quantifying how effectively resources are being used, while evaluating eco-efficiency performance, therefore overcoming one of the major shortcoming of eco-efficiency assessment methods found in the literature. For this reason, it is imperative to assure their correct implementation to provide an accurate determination of the above-mentioned index (TEI) considering the existing settings of different sectors of process industries. Efficiency Framework usefulness and reliability highly depends on the data quality and on the assumptions that can be made in order to define the value and the non-value adding portions, and consequently the targets. If these are not well established and justified the results interpretation/evaluation may be misleading, as the the input data will influence the results. Future research work will take place in order to mitigate the limitations of the framework, namely by integrating modern technologies for process monitoring.

Acknowledgment. This work was supported by the European Union's Horizon 2020 research and in-novation program through the MAESTRI project (grant no. 680570).

References

1. Sustainable Process Industry: Multi-annual Roadmap for the Contractual PPP Under Horizon 2020. Publications Office of the European Union, Luxembourg (2013). ISBN:978-92-79-31250-2, doi:10.2777/30452
2. Herrmann, C., Blume, S., Kurle, D., Schmidt, C., Thiede, S.: The positive impact factory–transition from eco-efficiency to eco–effectiveness strategies. Manufact. Procedia CIRP **29**, 19–27 (2015). http://dx.doi.org/10.1016/j.procir.2015.02.066
3. Bi, Z.: Revisiting system paradigms from the viewpoint of manufacturing sustainability. Sustainability **3**(9), 1323–1340 (2011)
4. Saling, P., Kicherer, A., Dittrich-Krämer, B., Wittlinger, R., Zombik, W., Schmidt, I., Schrott, W., Schmidt, S.: Eco-efficiency analysis by BASF: the method. Int. J. Life Cycle Assess. **37**(23), 5340–5348 (2002)
5. Côté, R., Booth, A., Louis, B.: Eco-efficiency and SMEs in Nova Scotia, Canada. J. Cleaner Prod. **14**(6–7), 542–550 (2006)
6. Kharel, G.P., Charmondusit, K.: Eco-efficiency evaluation of iron rod industry in Nepal. J. Cleaner Prod. **16**(13), 1379–1387 (2008)
7. Despeisse, M., Ball, P.D., Evans, S., Levers, A.: Industrial ecology at factory level – a conceptual model. J. Cleaner Prod. **31**, 30–39 (2012)

8. Lehni, M., Schmidheiny, S., Stigson, B.: Eco-efficiency: creating more value with less impact. World Business Council for Sustainable Development, Geneva (2000)
9. Baptista, A.J., Lourenço, E.J., Pereira, J.P., Cunha, F., Silva, E.J., Peças, P.: ecoPROSYS: an eco-efficiency framework applied to a medium density fiberboard finishing line. Procedia CIRP **48**, 170–175 (2016)
10. Shook, J., Rother, M.: Learning to See: Value Stream Mapping to Add Value and Eliminate MUDA (1999)
11. Lourenço, E.J., Baptista, A.J., Pereira, J.P., Dias-Ferreira, C.: Multi-layer stream mapping as a combined approach for industrial processes eco-efficiency assessment. In: Nee, A.Y.C., Song, B., Ong, S.-K. (eds.) Re-engineering Manufacturing for Sustainability, pp. 427–433. (2013)
12. Faulkner, W., Badurdeen, F.: Sustainable value stream mapping (Sus-VSM): methodology to visualize and assess manufacturing sustainability performance. J. Cleaner Prod. **85**, 8–18 (2014)

Toward Industry 4.0: Efficient and Sustainable Manufacturing Leveraging MAESTRI Total Efficiency Framework

Enrico Ferrera[1]([✉]), Rosaria Rossini[1], A.J. Baptista[2], Steve Evans[3],
Gunnar Große Hovest[4], Maria Holgado[3], Emil Lezak[5],
E.J. Lourenço[2], Zofia Masluszczak[6], Alexander Schneider[7],
Eduardo J. Silva[8], Otilia Werner-Kytölä[7], and Marco A. Estrela[8]

[1] Istituto Superiore Mario Boella, Via Pier Carlo Boggio 61, Turin, Italy
{ferrera, rossini}@ismb.it
[2] Instituto de Ciência e Inovação em Engenharia Mecânica e Engenharia
Industrial, Rua Dr. Roberto Frias, Campus da FEUP, 400,
4200-465 Porto, Portugal
{abaptista, elourenco}@inegi.up.pt
[3] Institute for Manufacturing, University of Cambridge,
17 Charles Babbage Road, Cambridge CB3 0FS, UK
{se321, mh769}@cam.ac.uk
[4] ATB Institute for Applied Systems Technologies, Wiener Straße 1,
28359 Bremen, Germany
gr-hovest@atb-bremen.de
[5] IZNAB Sp z o. o., ul. Klobucka 23, 02-699 Warsaw, Poland
emil.lezak@iznab.pl
[6] Lean Enterprise Institute Polska, Muchoborska 18, 54-424 Wrocław, Poland
zofia.masluszczak@lean.org.pl
[7] Fraunhofer Institute for Applied Information Technology FIT,
Schloss Birlinghoven, 53757 Sankt Augustin, Germany
{alexander.schneider,
otilia.werner-kytoelae}@fit.fraunhofer.de
[8] Instituto de Soldadura e Qualidade,
Avenida do Professor Doutor Cavaco Silva 33, Oeiras, Portugal
{EJSilva, MAEstrela}@isq.pt

Abstract. This paper presents an overview of the work under development within MAESTRI EU-funded collaborative project. The MAESTRI Total Efficiency Framework (MTEF) aims to advance the sustainability of manufacturing and process industries by providing a management system in the form of a flexible and scalable platform and methodology. The MTEF is based on four pillars: (a) an effective management system targeted at process continuous improvement; (b) Efficiency assessment tools to support improvements, optimisation strategies and decision support; (c) Industrial Symbiosis paradigm to gain value from waste and energy exchange; (d) an Internet-of-Things infrastructure to support easy integration and data exchange among shop-floor, business systems and tools.

© Springer International Publishing AG 2017
G. Campana et al. (eds.), *Sustainable Design and Manufacturing 2017*, Smart Innovation, Systems and Technologies 68, DOI 10.1007/978-3-319-57078-5_59

Keywords: Efficiency assessment · Eco-efficiency · Industrial symbiosis · Lean Management · Internet of Things · Sustainable manufacturing · Process industries · Industry 4.0 · MAESTRI · H2020 SPIRE

1 Introduction

Europe was the cradle of the manufacturing industry and it has traditionally led important industrial changes. Process industries are around 20% of the total European manufacturing industry, which include more than 450,000 individual enterprises (EU27), employment of around 6.8 million citizens and generation of more than 1,600 billion € turnover. These industries are largely dependent on resources imports from international markets that are hampering the industry's access to globally traded raw materials, due to the increased political instability in many regions of the globe, which is perfectly visible from a sharp increase in raw material prices during recent years. Moreover, European industry has also accounted for more than a quarter of total energy consumption in 2010 in Europe, with a significant portion of that used within the process industry. This represents both an opportunity and responsibility for this sector's contribution to the sustainability challenges of European societies, as it is imperative to drastically reduce the environmental footprint and increase competitiveness and production systems efficiency by "*doing more with less*". However, to successfully implement sustainability in manufacturing and process industries, a holistic, multidimensional and systematic approach is required. Having this in mind, the MAESTRI Total Efficiency Framework (MTEF), which is currently being developed, aims to advance the sustainability of manufacturing and process industries by providing a management system in the form of a flexible and scalable platform as well as an accompanying methodology. Based on a holistic approach, which combines different assessment methods and tools, the overall purpose of the MTEF is to support improvement on a continuous basis and increase eco-competitiveness by fostering sustainability in routine operations. Its conceptual approach will be based on a life cycle perspective, centered on models for dynamic simulation and optimization of both individual and complex systems, to better understand processes and the opportunities to add value. This life cycle approach is important to avoid problems (waste, environmental impacts, etc.) shifting from one life cycle stage to another.

It should also be noted that in order to enhance process' resource and energy efficiency processes, utilize waste streams and improve recycling in a sustainable manner, modelling and assessing all the interacting value chains is essential. However, despite the environmental, economic and social improvement potentials by sharing resources (e.g. energy, water, residues and recycled materials), it is essential to understand and assess resource and energy efficiency in order to optimize production systems. Moreover, the increased availability of modern technologies for process monitoring and optimization, pursuing the Industry 4.0 concept, should be carefully adapted and integrated for a wider and facilitated adoption of state-of-the-art tools and methodologies for efficiency and eco-efficiency. Such methodologies and tools should support waste and cost reductions in to both large and small companies.

This paper is organized as follows. Section 2 describes the author's vision about the four pillars for obtaining Total Efficiency with an overview about the state-of-the art. Section 3 addresses the tools and methods to be integrated and/or developed as the MAESTRI Total Efficiency Framework. Finally, conclusions and further steps for research are discussed in Sect. 4.

2   The Four Pillars of Total Efficiency: Background and Related Works

The authors believe that total efficiency concept is based on four pillars, aiming to fill the current gaps regarding the effective implementation of energy and resource management (see Table 1).

Table 1. Main gaps regarding the effective implementation of energy and resource management

Technical/Technological gaps	Lack of flexible, scalable and holistic tools to support decision making processes regarding resource and energy efficiency
	Lack of simple and integrated tools to assess and optimize resource and energy efficiency, crossing the different environmental and economic operational aspects
	Deficient knowledge to identify the potential use of wastes as resources (energy, resources, man-power, etc.)
Management gaps	Non-incorporation of sustainability aspects in company strategy and objectives;
	Non-implementation of structured management systems targeting resource consumption and energy efficiency
	Communication of process efficiency relevant data and information across different departments of the company
	Difficulty on the definition of clear and consistent KPIs, and their usage
Organizational gaps	Poor means for sharing resources (e.g. plants, energy, water, residues and recycled materials) through the integration of multiple production units of a single company or multiple companies on a single industrial production site
	Difficulty to collect and share information about all process flows (resource and energy inputs as well as waste and pollutant outputs)

The four pillars enable the total efficiency concept, encompassing the following aspects: *(a)* an effective management system; *(b)* efficiency assessment tools; *(c)* Industrial Symbiosis paradigm; *(d)* an Internet-of-Things infrastructure. The following paragraphs describe them in more details.

Effective Management System – Lean Management recommends the companies to maximize customer value while minimizing waste. The core of the concept was formulated in the form of five lean thinking principles [1] that should help the management: *(i)* precisely specify value for specific product/service; *(ii)* identify the value stream for each product/service; *(iii)* make value flow without interruptions; *(iv)* let the customer pull value; *(v)* pursue perfection. Lean Management has been proven to be a successful way to organize production operation, however in order to sustain changes the Lean Management System has to be implemented [2].

Efficiency Assessment – To have an exhaustive assessment of the production system is important to combine approaches for evaluating both operational efficiency and eco-efficiency perspective, encompassing the basic principles of eco-efficiency [3–5]. As a result of the combination of these two fundamental efficiency concepts, the Efficiency Assessment intends to encourage businesses not only to search for environmental improvements that yield parallel economic benefits, but to put emphasis on value creation and capturing. In this sense, by increasing value of goods, business tend to maximize resource productivity, gain bottom-line benefits, and reward shareholders, rather than simply improve efficiency or eco-efficiency performance, or minimize waste or pollution.

Industrial Symbiosis – Reusing waste from an industrial process as a resource for another industrial process is known as industrial symbiosis (IS). The symbiotic relationship is established between processes (in terms of resource exchanges) and companies (in terms of value exchanges) [6]. IS enables similar cyclic circulation of waste by promoting cooperation between companies, which show that the collective benefits of engaging in IS are greater than the benefits a single entity could achieve by itself [7]. Therefore, it represents a direct positive effect on the efficiency of the whole system and enables the decoupling between the value created through material processing or products' manufacturing, and the environmental impact associated to natural resource intake and waste/pollutants assimilation.

Internet-of-Things – Internet-of-Things technologies provide an ICT infrastructure with middleware functionality, which facilitates the interoperation of heterogeneous hardware and software components, which may be already operating inside the company or not. The aim is to extend pre-existing ICT infrastructure into a broader and more scalable network of generic "things" (i.e. shop-floor machinery, sensors, actuators, controllers, smart objects, mobile devices, servers, ERPs, MESs, third-party systems, cloud services, etc.). In the industrial domain, data sources will be hardware and software components in the shop-floor as well as business systems of the industrial companies. The ubiquitous use of sensors, the expansion of wireless communication and networks, the deployment of decision support systems has the potential to transform the way goods are manufactured. Many observers believe that IoT is the key technology for taking Europe to a new industrial revolution, considered to be the fourth and hence labelled "Industry 4.0". Industry 4.0 concept [8] is a tentative to answer questions, such as: *"How can we exploit the opportunities that digital technologies present for industry, administration, society and political participation, and how we master the relevant challenges?"* Industry 4.0 paradigm is considered a step forward towards industrial future of production, i.e. with highly flexible production environments, early-stage integration of business partners within value-creation processes, also providing support to decisions.

The consensus between researchers suggests that the industrial revisions require a long-time period of development and cover the following aspects, considered as the future manufacturing visions:

- *Factory*. Future factory is going to involve all manufacturing resources (sensors, actuators, machines, robots, conveyors, etc.) are connected and exchange information automatically in purpose of predict and maintain the machines, control the production process, and to manage the factory system – the paradigm known as a Smart Factory [9].
- *Business*. Future business is going to involve a complete communication network between companies, factories, suppliers, logistics, resources, customers, influencing one to other section in purpose of achieving a self-organizing status and provide the real-time response [10, 11].

The main features [12] of Industry 4.0 include the following: *Horizontal Integration* through value networks to facilitate inter-corporation collaboration, describing the cross-company and company-internal intelligent cross-linking and digitalization of value creation modules all over the value chain of the product life cycle and between value chains of adjacent product life cycles, *Vertical Integration* of hierarchical subsystems inside a factory, describes the intelligent cross-linking and digitalization within the different aggregation and hierarchical levels of a value creation module from manufacturing stations via manufacturing cells, lines and factories, also integrating the associated value chain activities such as marketing and sales or technology development to create flexible and reconfigurable manufacturing system.

3   MAESTRI Total Efficiency Framework – MTEF

The MTEF represents a flexible and scalable platform, which provides an effective management system that aims to advance the sustainability of manufacturing and process industries. It combines the four pillars in one holistic platform which enables an overall efficiency performance assessment from environmental (including resource and energy efficiency), value and cost perspectives. MTEF encompasses Environmental Performance Evaluation with Environmental Influence and Cost/Value assessment models through a life cycle perspective. The aim is to support the decision making process, by clearly assessing resource and energy usage (valuable/wasteful) of all process elementary flows, and the eco-efficiency performance. Decision support via value-adding optimization is also foreseen among the integration of the modules.

This basic concept of MTEF is depicted in Fig. 1. The central element is an IoT platform, which facilitates the data transfer from machines, systems, and sensors to end user software tools and applications at the industrial sites.

The IoT Platform is based on the LinkSmart middleware platform, which was originally developed in the previous EU project Hydra [13] and further extended in the course of a variety of research projects. LinkSmart provides interoperable interconnection of appliances, devices, terminals, subsystems, and services. To cover requirements that are not yet supported by LinkSmart, new software modules will be developed and work loosely coupled with existing modules in a service-oriented

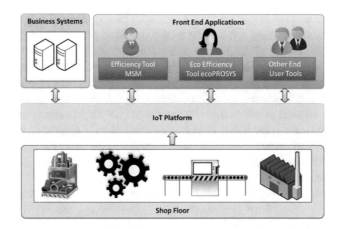

Fig. 1. MAESTRI total efficiency framework architecture

architecture (SoA). Each of the functional submodules of the architecture is explained in the following.

The *Shop Floor* will usually be the place where the major part of the relevant data is being produced. Device Connectors (DC) provide the means for devices to communicate with the rest of the framework regardless of the communication protocol it uses. DCs need to be developed specifically for each new device or protocol. *Business Systems* are the second type of data source for MTEF. ERP and MES systems can be connected to the IoT Platform in order to complement the data about shop floor activities. Besides being data sources for MAESTRI, the business systems could also receive data from the shop floor via the IoT Platform, therefore allowing a bi-directional connection would be possible between the IoT Platform and the Business Systems depending on the specific scenario. *Frontend Applications* represent all the end user software tools and applications, which are the main data consumers from the point of view of the IoT Platform. These include mainly ecoPROSYS and MSM, but could also be other mobile, web applications or simple KPI visualization dashboards, i.e. a GUI module that allows a simple visualization of streaming or historical KPI data.

The Eco-Efficiency Integrated Methodology for Production Systems (eco-PROSYS©) relies on the use of a systematized and organized set of indicators easy to understand/analyze, aiming to promote continuous improvement and a more efficient use of resources and energy. The goal is to assess eco-efficiency performance in order to support decision-making and enable the maximization of product/process value creation and minimization of environmental burdens. Considering the methodological description provided by [14], an algorithm was created, combining the results from three different modules: *(i)* Environmental Performance Evaluation, *(ii)* Environmental Impact Assessment and *(iii)* Cost and Value Assessment. As a consequence of the integration of these three components, the resulting decision support indicators intend to help companies on managing links between environmental and value performance. Their ultimate goal is to provide a clear vision of the production system baseline performance, and to assist on the implementation of improvement strategies by connecting the various levels of the system with clearly defined targets and benchmarks. For this reason, they can also be used to measure progress by evaluating trends or comparing the results along defined periods of time. As a component of the MTEF, ecoPROSYS' overall aim is to provide quantitative evidences to support decision making process pursuing both economic efficiency, which also has positive environmental benefits, and environmental efficiency, which also has positive economic benefits. In addition, it provides flexibility to the MTEF's application and adaptation to different sectors and companies, either from the modular approach applied for the complexity level of calculus, both for the economic value and cost assessment (from more simple cost assessment, to full LCC analysis), and for the environmental impact assessment (from simplified LCA to full LCA analysis).

The Multi-layer Stream Mapping (MSM) represents a method/tool able to achieve an overall efficiency/performance assessment of production systems [15] (see Fig. 2). It takes into account the base design elements from the well spread Lean tool Value Stream Mapping (VSM), in order to identify and quantify all "value adding" and "non-value adding" actions, as well as all types of waste and inefficiencies along the production process. Therefore, the great similarity to the VSM tool consists in the base lean mindset for the identification and quantification, at each stage of the process system, of "what adds value" and "what does not add value" to a product or service, for any given variable of the production system (multiple layers and multiple domains of variables, such as Resource Efficiency, Operational Efficiency, Flow Efficiency, etc.).

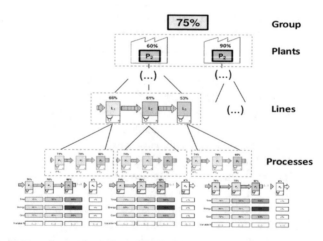

Fig. 2. Schematic representation of MSM© bottom-up analysis and aggregation.

Tools and methods that can support companies in identifying and evaluating the potential opportunities of waste streams that could be part of Industrial Symbiosis (IS) exchanges and to extend procurement approach to include secondary/alternative inputs sources are necessary to advance towards more sustainable operations in manufacturing industry [1]. Complementarily, examples and practical information can support companies on how to make the required mind-set and behavior shift to see waste as a potential resource and identify synergies within and outside their operations. This shift can only occur through the provision of a better understanding of IS. Based on the understanding of theoretical characteristics [16] and practical characteristics of IS, the following categories of possible challenges have been identified [17]: (i) finding embedded norms of exchange, culture and structure of enterprises; (ii) a lack of information on waste exchange; (iii) a lack of technical and financial support; (iv) increased cost incurred during the process of waste exchange (e.g. transaction costs, opportunity cost, labor costs, etc.); (v) a lack of guidance in institutional arrangements; (vi) a lack of inter-firm trust and unstable cooperation between participants in the IS network. The MTEF aims to support a wider application of IS in the process industry through a methodology that builds on a library of case studies, informing companies about possible uses of their waste streams and alternative input sources A stepwise process is to guide companies on their identification of exchanges, on valuing their resources and on building feasible IS-based synergies with other participants. The stepwise process is formed by four guiding questions that can lead companies' path towards creating higher value from IS applications: (a) *How to see waste* in company's processes and manufacturing operations, where the efficiency assessment tools will be applied (as MSM); (b) *How to characterize waste* in a way that allows to understanding the remaining value once a waste stream has been identified (also with the support of ecoPROSYS); (c) *How to value waste*, according to potentially interested stakeholders and waste exchanges opportunities; (d) *How to exploit waste* by selecting the opportunities bringing higher value and devising an action plan.

Eco Orbit View (EOV) methodology has been chosen by the MTEF in order to indicate areas in the production process where the company may focus an improvement activity in order to get simultaneous improvement of business and environmental performance. The Eco Orbit View analysis is performed in 4 steps: (i) identification of production process steps (for a selected product family); (ii) identification of Key Performance Indicators (KPIs) relevant for each process step; (iii) identification of Key Environmental Performance Indicators (KEPIs) or Environmental Aspects relevant for each process step; (iv) identification links and synergies between KPIs and KEPIs. In summary, the Eco Orbit View shows KPIs (reflecting company needs) and KEPIs (reflecting environmental needs) side by side for chosen process steps. The analysis results in the indication of potential improvement areas, reflecting the needs of the company to improve both the economic and environmental performance. Thus, the areas can be identified where the eco-efficiency of the company may be improved.

4 Conclusions and Research Agenda

The overall aim of the proposed MAESTRI Total Efficiency Framework, which is currently under development, is to support the sustainable improvement of industrial companies' environmental and economic performance, in particular the process industry ones.

The MAESTRI approach to support IS applications in the process industry builds on widening the knowledge on IS industrial cases as main source of information and on a stepwise methodology to support the path from waste identification and valuing to exchange definition and exploitation phases. This will support companies willing to engage with the implementation of IS concept in their business. MTEF aims to allow an innovative and practical integration of eco-efficiency analysis, including LCA and LCC, with process efficiency assessment (including both resource efficiency and operational efficiency) analysis from the perspective of Lean Manufacturing, thus implicitly evaluating "value added" and non-value added" (waste) actions and resource usage.

MTEF leverages a middleware platform, which facilitates the automatization of the data transfer from machines, systems, and sensors at the industrial sites to end user software tools and applications. It allows to seamlessly interconnect heterogeneous devices, systems and subsystems in order to achieve higher degree of interactions between the shop floor, the legacy management systems and the end users. The authors consider it an innovative and important instrument that acts as generic enabler of the hyper-connected factory, to support decision-making for the development and simulation of improvement strategies and to support the continuous analysis and monitoring of a companies' efficiency and eco-efficiency.

The further development and validation of the MTEF will be supported by its initial application in four real industrial settings across a variety of different sectors. Those four industrial pilot cases (two of the chemical sector, one of injection moulding, and one from the metalworking sector) represent one of the main sources for identification of the requirements upon the methodological approach and the platform's functionalities as well as for their continuous refinement based on the testing results.

Acknowledgements. This work was supported by the European Union's Horizon 2020 research and innovation program through the MAESTRI project (grant no. 680570).

References

1. Womack, J.P., Jones, D.T.: Lean Thinking: Banish Waste and Create Wealth in Your Corporation. Simon and Schuster, New York (1996). 2nd version
2. Mann, D.: Creating a Lean Culture: Tools to Sustain Lean Conversions, 2nd edn. CRC Press, Boca Raton (2010)
3. Herrmann, C., Blumea, S., Kurle, D., Schmidt, C., Thiede, S.: The positive impact factory – transition from eco-efficiency to eco-effectiveness strategies in manufacturing. In: Procedia CIRP – The 22nd CIRP Conference on Life Cycle Engineering, vol. 29, pp. 19–27 (2015)
4. Hauschild, M.Z.: Better – but is it good enough? On the need to consider both eco-efficiency and eco-effectiveness to gauge industrial sustainability. In: Procedia CIRP – The 22nd CIRP Conference on Life Cycle Engineering, vol. 29, pp. 1–7 (2015)
5. Ellen MacArthur Foundation: Growth Within: a circular economy vision for a competitive Europe. Cowes, Isle of Wight: Ellen MacArthur Foundation (2015)
6. Holgado, M., Morgan, D., Evans, S.: Exploring the scope of indus-trial symbiosis: implications for practitioners. In: Setchi, R., Howlett, R., Liu, Y., Theobald, P. (eds.) Sustainable Design and Manufacturing 2016. Smart Innovation, Systems and Technologies, vol. 52, pp. 169–178. Springer, Cham (2016)

7. Chertow, M.: Industrial symbiosis. Encyclopaedia Energy **3**, 407–415 (2004)
8. Lee, J.: Industry 4.0 in big data environment. In: German Harting Magazine, pp. 8–10 (2013)
9. Lucke, D., Constantinescu, C., Westkämper, E.: Smart factory - a step towards the next generation of manufacturing. In: Mitsuishi, M., Ueda, K., Kimura, F. (eds.) Manufacturing Systems and Technologies for the New Frontier, pp. 115–118. Springer, London (2008)
10. Kagermann, H., Helbig, J., Hellinger, A., Wahlster, W.: Recommendations for implementing the strategic initiative INDUSTRIE 4.0: securing the future of German manufacturing industry; Final report of the Industrie 4.0 Working Group, Forschungsunion, (2013)
11. Brizzi, P., Lotito, A., Ferrera, E., Conzon, D., Tomasi, R., Spirito, M.: Enhancing traceability and industrial process automation through the VIRTUS middleware. In: Proceedings of the Middleware 2011 Industry Track Workshop, p. 2. ACM, December 2011
12. Acatech.   http://www.acatech.de/fileadmin/user_upload/Baumstruktur_nach_Website/ Acatech/root/de/Material_fuer_Sonderseiten/Industrie_4.0/Final_report__Industrie_4.0_ accessible.pdf
13. Eisenhauer, M., Rosengren, P., Andantolin, P.: A development platform for integrating wireless devices and sensors into ambient intelligence systems. In: 6th Annual IEEE Communications Society Conference on Sensor, Mesh and AdHoc Communications and Networks Workshops, SECON Workshops 2009, pp. 1–3. IEEE (2009)
14. Baptista, A., et al.: Eco-efficiency framework as a decision support tool to enhance economic and environmental performance of production systems. In: eniPROD, ed. 3rd Workbook of the Cross-sectional Group 'Energy-related Technologic and Economic Evaluation' of the Cluster of Excellence eniPROD, pp. 11–20. Wissenschaftliche Scripten, Kemnitz (2014)
15. Lourenço, E.J., Baptista, A.J., Pereira, J.P., Dias-Ferreira, C.: Multi-layer stream mapping as a combined approach for industrial processes eco-efficiency assessment. In: Nee, A., Song, B., Ong, S.K. (eds.) Re-engineering Manufacturing for Sustainability, pp. 427–433. Springer, Singapore (2013)
16. Chertow, M., Park, J.: Scholarship and practice in industrial symbiosis: 1989–2014. In: Clift, R., Druckman, A. (eds.) Taking Stock of Industrial Ecology, pp. 87–116. Springer, Cham (2016)
17. Zhang, Y., Zheng, H., Shi, H., Yu, X., Liu, G., Su, M., Li, Y., Chai, Y.: Network analysis of eight industrial symbiosis systems. Front. Earth Sci. **10**(2), 352–365 (2016)

Manufacturing Technologies for Material Sustainability Throughout the Product Life-Cycle

Cryogenic Delamination and Sustainability: Analysis of an Innovative Recycling Process for Photovoltaic Crystalline Modules

M. Dassisti[1(✉)], G. Florio[1,2], and F. Maddalena[1,2]

[1] Dipartimento di Meccanica, Matematica e Management, Politecnico di Bari,
Via E. Orabona 4, 70125 Bari, Italy
michele.dassisti@poliba.it
[2] INFN, Sezione di Bari, 70126 Bari, Italy

Abstract. The increasing rate of production and diffusion of photovoltaic (PV) technologies for industrial and domestic applications urges improvement of the sustainability of their demanufacturing processes in order to reduce the amount of electronic wastes. Sustainability of demanufacturing processes concerns the reduction of energy consumption, the reduction of polluting substances as well as the reduction of the effort spent in recovery of the components. There is not an optimal process so far, provided a number of different approaches have been devised (see e.g. [1–3]). A promising choice relies on the use of thermo-mechanical treatments for inducing a *delamination* process where interfacial bonding between layers are weakened and, finally, broken inducing separation of the layers [4]. In this paper we present a preliminary industrialization study, based on Finite Element (FE) Analysis, to prove the validity of the new sustainable demanufacturing process endeavouring the delamination process. The analysis is performed searching the optimal thermally induced cycles at cryogenic temperatures. We finally show that it is possible to induce the delamination according to specific operating temperatures.

Keywords: Sustainability · Photovoltaic modules · Thermoelasticity · Demanufacturing · WEEE

1 Introduction

Product demanufacturing (namely disassembly, remanufacturing, recycling and/or recovery) is a promising economical activity because of its critical impact

---

The authors are deeply indebted with Antonio De Nobili for the support in providing pictures as well as the useful discussions and comments. G. Florio is supported by the Gruppo Nazionale per la Fisica Matematica (GNFM) of the Istituto Nazionale di Alta Matematica (INdAM) through "PROGETTO GIOVANI" and by INFN through the project "QUANTUM". F. Maddalena is supported by GNAMPA of the Istituto Nazionale di Alta Matematica (INdAM). This paper has been developed under the moral patronage of the "SOSTENERE" Group of the Italian Association of Mechanical Technologists (AITEM).

© Springer International Publishing AG 2017
G. Campana et al. (eds.), *Sustainable Design and Manufacturing 2017*, Smart Innovation, Systems and Technologies 68, DOI 10.1007/978-3-319-57078-5_60

on sustainability of manufacturing. Among the processes that fall within this class, a core problem to be solved is the detachment of layers in complex multi-layered, multi-material systems. Photovoltaic panels (PVP) are a typical example of such a product, which pose serious problems for their recycling or recovery [5].

Scope of the present paper is devising a sustainable process that allows to efficiently separate layers in multilayer products by inducing a physical phenomenon under controlled conditions: the delamination. At the core of some kind of PV panel (old generation mono- and poly-crystalline PVP) there are silicon cells encapsulated in a polymeric film made of Ethylene-vinyl acetate (EVA). This material exhibits a glassy transition occurring at a temperature lower than $-15\,°C$ [6–9]. In the glassy phase EVA is fragile and more prone to cracks [10]. Therefore, here we suggest the possibility to obtain an induced delamination process by detachment of EVA layers from the adjacent substrates by endeavouring the material state transition. The paper examine the underlying processes to understand in which conditions a proper thermal cycle can cause fractures and material failures in the EVA layer. We investigate this topic by using numerical simulations based on finite element (FE) analysis [11] to set a possible delamination theory of rigid layers. This method, giving approximate solutions to boundary value problems for partial differential equations, has been already applied for the study of PV panels at room (and higher) temperature (see [12] for instance).

The results of this analysis will be used for setting an industrial demanufacturing process and thus understanding the practical feasibility of a controlled delamination process through a thermo-mechanical treatment at low temperatures.

The paper is organized as follows. In Sect. 2 we introduce the layer structure of the PV sample we are going to model. In Sect. 3 we detail the model used in the simulations and the physical quantities considered in order to obtain information about the thermo-mechanical behavior of the sample. In Sect. 4 we will present two examples of thermal treatment that could induce material failure in the EVA layers. In Sect. 5 we draw some conclusions and indicate perspectives of this research topic.

2   Multi-layer Structure of a PV Panel

A typical PV panel is made of several layers and cells of different materials, each of them with specific mechanical and thermal properties. In the traditional PV panels metal layers (silicon and electrical connections) are intermitted with polymeric layers (ethylene-vinyl acetate (EVA) copolymer) to guarantee mechanical protection and electrical isolation. Figure 1(left) shows a sample extracted from a dismantled PV panel. In the following, we will refer to this specimen to study and simulate the thermo-mechanical behavior of a sample from a PV module.

In Fig. 1(right) a schematic section of the multi-layer structure of the sample is shown. In particular, from top to bottom, we have the following layers (for each material we have included a list of references for the values of mechanical and thermal parameters used in the numerical simulations):

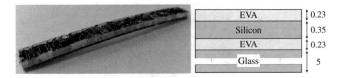

EVA	0.23
Silicon	0.35
EVA	0.23
Glass	5

Fig. 1. (Left) Sample module obtained from a real PV panel using water jet cutting. (Right) Representation of the transverse section of the PV sample used in the FE analysis with the indication of the thickness of the layers (in *mm*). The glass layer is not in scale in order to improve readability of the figure.

- Ethylene-vinyl acetate (EVA) [6–9]: the copolymer layer used as isolation and shock-absorber. Good properties of toughness, flexibility, adhesion and stress-cracking resistance are evident. This material exhibits a glassy transition [7,8] at critical temperatures (T_g) ranging from $-35\,°C$ to $-15\,°C$. In this condition the material is more vulnerable due to the increase of the storage and loss moduli [10]. We will use this result for our purposes. At room temperature the critical yield stress is $\sigma_b \simeq 6.5\,MPa$ [9]. We will consider this value as a threshold value for our considerations.
- Electronic-grade silicon [13–18]: the thermo-mechanical properties of silicon cells depend on their structure. Typical choices, also affecting the thickness of the layers, are mono or poly-crystalline silicon. The relevant physical parameters and their dependence on temperature are reported in literature.
- EVA: ibid.
- Glass [19–21]: float glass is a standard important element in a PV module in order to ensure mechanical protection against external agents and maximum efficiency of the optical transmission.

Tedlar (polyvinyl fluoride film), not shown in the figure, is used to improve electrical isolation and weather resistance. In our modeling this layer has been neglected because not useful for the demanufacturing purposes: this layer belongs to the not recyclable plastic and does need further treatments.

3 Numerical Simulation: Modeling and Measurements

3.1 Sample Used in the Simulation

As shown before, the PV sample modeled in this paper in Fig. 1, is a rectangular-shaped block of $100\,mm \times 10\,mm$ with $5.81\,mm$ thickness. As already stated, the sample is made of different layers, from top to bottom: EVA, silicon, EVA and glass. The FE modeling hypothesis are as follows:

- the mesh used, based on elements from the SOLID class (20 nodes), does not take into account the micro-structure of the silicon cells and metallic corrections. This choice is justified because, due to the delamination process we are investigating, we are interested in failures of the polymeric layer and not in the thermo-mechanical behavior of silicon; at the interfaces between different layers we have considered bonded contact surfaces;

Fig. 2. Magnification of the model of the sample; the mesh is more dense in the silicon layer, in the EVA layers and near the interface glass-EVA.

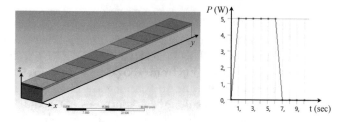

Fig. 3. (Left) Separation of the upper surface of the PV sample in ten elements. The heat source acts subsequently on each section with the power cycle shown in the picture on the right. (Right) Representation of the time dependence (in seconds) of the power P (in Watt) of the heat source applied to each element of the upper surface.

- the materials in the layers are considered homogeneus and isotropic;
- the mechanical and thermal parameters of the materials used in the simulation are temperature-dependent: their values have been derived from literature listed in the references. We consider a finer mesh (about $80\,\mu$m in thickness) in the layers with EVA, silicon and at the interface glass-silicon as in Fig. 2 where we show a magnification of the model of the block.
- As concerns the thermal cycles applied, the heat source is applied on the top EVA layer. In particular, we divided the upper surfaces in ten sections along the y direction as in Fig. 3(left). Each section is heated for a fixed time interval to provide a given amount of energy. As an example, we consider the function in Fig. 3(right) where the source in turned on from $t = 0\,$s to $t = 1\,$s using a linear ramp function. From $t = 1\,$s to $t = 5\,$s the source provides a power $P = 5$ Watts on the section. Finally, from $t = 6\,$s to $t = 7\,$s the source is turned off with a linear ramp function. During the turning off of the source in the i-th section we turn on another source in the $(i + 1)$-th section. A moving source has been simulated sweeping the upper surface in the y direction.

In the following section we will consider two different implementations of this scheme.

3.2   Physical Quantities Obtained from the Numerical Simulations

During the FE simulation we have monitored the behavior of several physical parameters: displacement, von Mises equivalent stress, temperatures. Each of them is of interest for the comprehension of the thermo-mechanical response of the PV sample in view of the realization of the delamination process. In the following we will list them and explain the rationale behind their use.

Displacement. Displacement detection, denoted as the vector $\mathbf{u} = (u_x, u_y, u_z)$, is fundamental in view of the experimental validation of the theory. As a consequence of the thermal load applied on the sample surface, layers will exhibit different deformation behaviors. The values of displacements in different directions can be evaluated theoretically either by analytical solution of the equations of thermo-elasticity or, this is our case, using FE analysis. These values can be, in principle, compared with experimental measurements performed with several techniques such as digital image correlation [23] or interferometric setups [24]. As we will see, the typical predicted displacement values for the process described in this paper are of the order of a few microns, thus posing a challenge for their experimental verification.

Von Mises Equivalent Stress. The von Mises equivalent stress [22] can be used to define a yield criterion to predict the insurgence of failure in ductile materials. In particular, this quantity connects values of uniaxial stress obtained from tensile testing to the multiaxial stress state of real structures, separated in hydrostatic stress and deviatoric stress. In general, one can consider a coordinate system parallel to the principal stress directions and the stress components σ_i with $i = 1, 2, 3$. It is thus possible to obtain a scalar invariant (von Mises equivalent stress) of the form

$$\sigma_e = \sqrt{\frac{1}{2} \left[(\sigma_1 - \sigma_2)^2 + (\sigma_2 - \sigma_3)^2 + (\sigma_3 - \sigma_1)^2 \right]}. \tag{1}$$

A criterion for detecting the failure of a material is the reference to σ_e: when this quantity exceeds the uniaxial yield stress of a material, failure is supposed to occur. We will use this to estimate the insurgence of detachments in the EVA layer and, therefore, as a forerunner of a delamination process.

Temperature. As mentioned above, the basic idea for implementation of our de-manufaturing process relies on the glassy transition of the EVA copolymer which makes this layer more fragile. Thus, the measurement of temperature is important in order to verify that the EVA layers are in their glassy phase.

4   Numerical Simulation Outcomes

4.1   Initial Conditions of the Simulations

In order to simulate the response of the system in cryogenic conditions, we have considered the PV sample to be in thermal equilibrium without internal or external stress, with the initial temperature set at $-150\,°C$. In these conditions the EVA polymer would be in its glassy phase and, thus, more fragile.

We have simulated the application to the upper EVA layer of heat sources with different energy flows and heating protocols. Having in mind Fig. 3(left), we have focussed on two cases:

1. (fast) heat source scanning the sample along the y direction from left to right with variable energy flow;
2. (slow) heat source scanning the sample along the y direction from left to right and back with fixed energy flow.

4.2   Case (1): Single Heating Scan of the PV Sample

For this case we considered a heat source applied to each of the ten sections of the upper surface in sequence, using a thermal cycle shown in Fig. 3(right) varying time and power. In particular, the first and second section experience a thermal field induced by a 10 Watts power source for six seconds (with a time overlap of 2 s); at the end of the sequence, the last section experiences a thermal field from a 8 Watts power source for three seconds. The duration of the entire thermal cycle is 50 s. Figure 4 reports the maximum and minimum displacements in the x, y and z directions as a function of time. We observe that transverse displacement (u_x) is rather symmetric because of the shape of the heat source. These values are localized in the EVA layers and correspond to the zones where the heat source is applied. The longitudinal (u_y) and height (u_z) displacements are not symmetric. This is due to the asymmetry of the source with respect to the sample geometry. The absolute values of this quantity range from a few microns to a few tenth of microns.

Figure 5 shows the equivalent stress at time $t = 5\,s$. The maximum value is on the upper EVA layer where $\sigma_e > 7.5\,MPa$ and at the interfaces EVA-silicon. This value exceeds the threshold stress $\sigma_b \simeq 6.5\,MPa$ previously mentioned. This result is confirmed in the rest of the process: we always obtain an equivalent stress close to 8 MPa.

The evaluation of the temperature (Fig. 6) in the sample shows that the application of a large power heat source induces an increase of the temperature, which can go beyond the critical value T_g for the glassy transition of the EVA polymer. On the other hand, this condition applies only at the end of the thermal cycle applied. For a time $t < 35\,s$ the sample remain under the critical temperature $-15\,°C$ with and equivalent stress larger than σ_b.

4.3   Case (2): Double Scan of the PV Sample

In the previous section we have seen that it is possible to thermally induce a stress in the EVA layer exceeding the critical value σ_b. On the other hand, we have obtained this result at the cost of a large power heat source and a large increase of the system temperatures. In the following simulations we decrease the energy flow but after scanning the PV sample from left to right, the heat

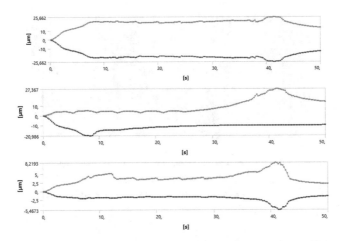

Fig. 4. Time dependence (in seconds) of maximum and minimum displacement (in μm) calculated for the PV sample in the x, y, z directions (top, center, bottom panel, respectively) for the single scan process.

Fig. 5. von-Mises equivalent stress in a section of the PV sample at time $t = 5$ s for the single scan process. The values ranges from $\sigma_e \simeq 2.3 \times 10^{-3}$ MPa in the glass layer to $\sigma_e \simeq 9$ MPa in the top EVA layer [25].

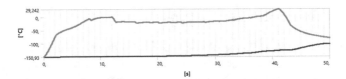

Fig. 6. Time dependence of maximum and minimum temperature calculated for the PV sample for the single scan strategy.

source is reversed and moves backwards: thus, each section of the upper surface experiences a power of 5 Watts for 5 s two times. The total duration of the thermal cycle is 100 s.

In this case, the displacements obtained (not shown) are smaller with respect to the previous case: their values range in the order of a few microns. The von

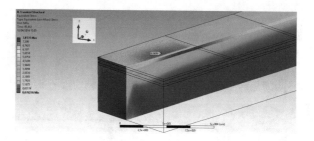

Fig. 7. von-Mises equivalent stress in a section of the PV sample at time $t = 95$ s for the double scan strategy. The values ranges from $\sigma_e \simeq 7.6 \times 10^{-2}$ MPa in the glass layer to $\sigma_e \simeq 7.85$ MPa in the silicon layer, $\sigma_e \simeq 6.87$ MPa on the top EVA layer [25].

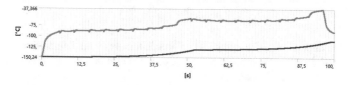

Fig. 8. Time dependence of maximum and minimum temperature calculated for the PV sample with the double scan process.

Mises equivalent stress can reach values about 6−7.8 MPa at the end of the procedure as shown in Fig. 7 and the maximum value is localized in the upper EVA layer and in the silicon (also close to the interface EVA-silicon). Typical values during the procedure are slightly below 6 MPa.

On the other hand (Fig. 8) the maximum temperature is far below the critical temperature T_g of the EVA glassy transition with a maximum value close to −60 °C. Accordingly, a value of the equivalent stress smaller is compensated by a sensibly lower temperature.

5   Discussion and Conclusions

In this paper we have analyzed the effects of a thermal cycle at cryogenic temperatures as a preliminary industrialization analysis of a new and sustainable de-manufacturing process [4] in which delamination, i.e. the separation of the layers in a laminated material, can be obtained inducing a sufficiently large stress and, thus a material failure in the layers. We have considered the possibility to use the structure of PV modules in order to recover the materials of the different layers. The application of proper thermal loads during the treatment at low temperature would be a change in the paradigma and largely improve the sustainability of the decommission process of PV panels. The simulations, based on the Finite Element Method, have considered two different thermal cycle conditions and can be seen as a proof of the industrial feasibility of the new de-manufacturing process. We have derived the dependence on time of temperatures, displacements and von Mises equivalent stresses. In particular, we have found that, due the induced stresses, the polymeric layer (EVA) can experience material failure and, possibly, delamination. This result has been obtained due to the fact that the sample in kept at a temperature lower that the critical one. Since the focus of the present paper relies on the delamination process, we have considered a simpler version of the PV sample with isotropic and homogeneous materials. We have also considered different sizes of the applied mesh for the simulation: the results are not affected from these choices. The simplified model of the PV sample used in this paper can be useful to have an initial estimation of the process behavior described by a mathematical model based on the equations of thermo-elasticity [22]. On the other hand, numerical analysis based on a more complete model of the sample will be investigated as a necessary steps towards industrialization. Moreover, a complete study of the sample behavior from the analytical point of view, using the extension to the tridimensional case of the modeling of adhesion-debonding phenomena in elastic system already studied in one and two dimensions [26–30] is presently under analysis. Finally, our results are at the basis of the experimental validation of the delamination process here proposed that will be implemented in the near future.

References

1. McDonald, N.C., Pearce, J.M.: Producer responsibility and recycling solar photovoltaic modules. Energy Policy **38**, 7041–7047 (2010)
2. Latunussa, C.E.L., Ardente, F.: Life Cycle Assessment of an innovative recycling process for crystalline silicon photovoltaic panels. Sol. Energy Mater. Sol. Cells **156**, 101–111 (2016)
3. Berger, W.S., et al.: A novel approach for the recycling of thin film photovoltaic modules. Resour. Conserv. Recycl. **54**, 711–718 (2010)
4. Dassisti, M.: (Patent WO2014141311) Thermo-mechanical controlled cryogenic delamination process for the full recovery of rigid mono-, polycrystalline or amorphous materials coated with plastic materials (2013)
5. Dong, A., Zhang, L.: Beneficial and technological analysis for the recycling of solar grade silicon wastes. JOM **63**, 23–27 (2011)
6. Elvax EVA resins for Adhesives, Sealants and Wax Blends; Thermal Properties of Elvax Measured by Differential Scanning Calorimeter (DSC). www.dupont.com
7. Agroui, K., Collins, G.: Characterisation of EVA encapsulant material by thermally stimulated current technique. Sol. Energy Mater. Sol. Cells **80**, 33–45 (2003)
8. Agroui, K., Collins, G., Farenc, J.: Measurement of glass transition temperature of crosslinked EVA encapsulant by thermal analysis for photovoltaic application. Renew. Energy **43**, 218–223 (2012)
9. Mishra, S.B., Luyt, A.S.: Effect of organic peroxides on the morphological, thermal and tensile properties of EVA-organoclay nanocomposites. Express Polym. Lett. **2**, 256–264 (2008)
10. Kempe, M.D., et al.: Ethylene-vinyl acetate potential problems for photovoltaic packaging. In: IEEE 4th World Conference on Photovoltaic Energy, vol. 2, pp. 2160–2163 (2006)
11. Kim, N.-H.: Introduction to Nonlinear Finite Element Analysis. Springer, New York (2015)

12. Chen, C.H., et al.: Residual stress and bow analysis for silicon solar cell induced by soldering (2009). http://140.116.36.16/paper/c38.pdf
13. Middelmann, T., et al.: Thermal expansion coefficient of single-crystal silicon from 7 K to 293 K. Phys. Rev. B **92**, 174113 (2015)
14. Hull, R. (ed.) Properties of Crystalline Silicon, EMIS Note No. 20, IET (1999)
15. White, G.K., Minges, M.L.: Thermophysical properties of some key solids: an update. Int. J. Thermophys. **18**, 1269 (1997)
16. Sparks, P.W., Swenson, C.A.: Thermal expansions from 2 to 40 K of Ge, Si, and four III-V compounds. Phys. Rev. **163**, 779 (1967)
17. Levy, M., Furr, L.: Handbook of Elastic Properties of Solids, Liquids and Gases-Vol. II. Academic Press, New York (2001)
18. Handbook of Chemistry and Physics 83rd edn. CRC Press Inc. (2002)
19. Proctor, B.A., et al.: The strength of fused silica. Proc. R. Soc. Lond. Math. Phys. Eng. Sci. **297**, 534–557 (1967)
20. Pedone, A., et al.: Molecular dynamics studies of stress-strain behavior of silica glass under a tensile load. Chem. Mater. **20**, 4356–4366 (2008)
21. Burge, J.H., et al.: Thermal expansion of borosilicate glass, Zerodur, Zerodur M, and unceramized Zerodur at low temperatures. Opt. Photonics News **38**, 10 (1999)
22. Gurtin, M.E., et al.: The Mechanics and Thermodynamics of Continua. Cambridge University Press, Cambridge (2010)
23. Sutton, M.A., et al.: Image Correlation for Shape, Motion and Deformation Measurements. Springer, New York (2009)
24. Bobroff, N.: Recent advances in displacement measuring interferometry. Meas. Sci. Technol. **4**, 907 (1993)
25. De Nobili, A.: Master thesis in Management Engineering, Advisor: Dassisti, M., Co-Advisor: Florio, G.: Politecnico di Bari (2016)
26. Maddalena, F., et al.: Mechanics of reversible unzipping. Continuum Mech. Thermodyn. **2**, 251–268 (2009)
27. Maddalena, et al.: Adhesive flexible material structures. Discr. Continuous Dynamic. Syst. B **17**, 553–574 (2012)
28. Maddalena, F., Percivale, D., Tomarelli, F.: Local and non-local energies in adhesion interaction. IMA J. Appl. Math. **81**(6), 1051–1075 (2016)
29. Coclite, G.M., Florio, G., Ligabó, M., Maddalena, F.: Nonlinear waves in adhesive strings. SIAM J. Appl. Math. **77**(2), 347–360 (2017)
30. Dassisti, M., Devillanova, G., Florio, G., Ligabò, M., Maddalena, F.: A model for thermally induced delamination (2017, In preparation)

Tuning Decision Support Tools for Environmentally Friendly Manufacturing Approach Selection

Giuseppe Ingarao[1(✉)], Paolo C. Priarone[2], Yelin Deng[3],
and Rosa Di Lorenzo[1]

[1] Department of Industrial and Digital Innovation, University of Palermo,
Viale delle Scienze, 90128 Palermo, Italy
{giuseppe.ingarao, rosa.dilorenzo}@unipa.it
[2] Department of Management and Production Engineering,
Politecnico di Torino, Corso Duca degli Abruzzi 24, 10129 Torino, Italy
paoloclaudio.priarone@polito.it
[3] Department of Mechanical Engineering, University of Wisconsin-Milwaukee,
Milwaukee, WI, USA
deng5@uwm.edu

Abstract. Awareness about the environmental performance of manufacturing approaches has arisen. Comparative analyses of different manufacturing approaches as well as decision support methods should be developed in the field of metal shaping processes. The present paper aims at tuning a decision support tool for identifying when mass conserving approaches (forming based) are actually preferable over machining processes for manufacturing aluminum based components. A full LCA is developed for comparing the environmental performance of forming and machining approaches as the batch size and geometry complexity hang. The impact of the used metric on the comparative results is analyzed. Results reveal that primary energy can be used as reliable metric for identifying environmentally friendly manufacturing processes.

Keywords: Sustainable manufacturing · Manufacturing approach comparison · LCA · Decision support tool

1 Introduction

A significant share of the greenhouse gas emissions is caused by material production. According to Worrel et al., producing materials caused about 25% of all anthropogenic CO_2 emissions [1]. Metals play a dominant role, as steel and aluminum account for 24% and 3% of worldwide industrial emissions, respectively (Allwood et al., 2011 [2]). Manufacturing processes selection, besides affecting the environmental load at the unit process level, has significant repercussions on the amount of the involved materials (Ingarao, 2016 [3]). Selecting the proper manufacturing approach, analyzing the whole material life cycle, is crucial for making actual environmentally friendly decisions. As concerns metals, over the last years, a few comparative manufacturing approaches have

© Springer International Publishing AG 2017
G. Campana et al. (eds.), *Sustainable Design and Manufacturing 2017*, Smart Innovation, Systems and Technologies 68, DOI 10.1007/978-3-319-57078-5_61

been developed. Most of the papers focus on the comparison between additive manufacturing and subtractive processes.

Morrow et al. (2007) compared laser-based Direct Metal Deposition (DMD) and CNC milling [4]. Serres et al. (2011) related the direct additive laser manufacturing approach (CLAD) with conventional machining [5]. Huang et al. (2016) applied the comparative analysis to five case studies, more specifically to five aircraft components [6]. Their paper focused on metallic materials (titanium as well as aluminum alloys), while Electron Beam Melting (EBM), Selective Laser Melting (SLM) and Direct Metal Laser Sintering (DMLS) were considered as additive manufacturing processes. Paris et al. (2016) matched the cumulative energy of machining and EBM processes to manufacture a titanium-based turbine [7]. Priarone et al. (2016) compared an EBM process and turning with varying part geometry complexity [8]. Ingarao et al. (2015) developed an environmental comparison between a hot extrusion process (bulk forming process) and a machining process [9].

These studies confirm that several factors contribute in worsening the environmental performance of a manufacturing approach; among them, material usage significantly affects the environmental performance of a given manufacturing approach and, subsequently, the result of the comparative analysis itself. This is the reason why manufacturing approaches comparison need to be developed by following a life cycle based analysis approach. In fact, some of the aforementioned studies apply full LCA analyses, others use simplified metrics such as primary energy instead. It is worth pointing out that despite that literature lacks decision support tools able to select the most environmentally friendly manufacturing strategy as the production scenario changes (kind of involved material, part geometry, and batch size). The authors (Ingarao et al., 2016 [10]) have recently proposed a decision support tool named 'Process Sustainability Diagram'. This graph enables to identify for which production scenario (i.e., batch size and amount of removed material) subtractive manufacturing approaches (as turning) are preferable over the mass conserving ones (as forming). The present paper aims at tuning the developed tool. Specifically, a full LCA analysis is carried out in order to analyze the impact of the selected metric on the decision support tool. In other words, the present research aims at identifying the most efficient metric to be selected for manufacturing processes practitioners. The present paper wishes to reply to the following research questions: is a full LCA analysis mandatory for identifying the most environmentally friendly solution among manufacturing approach alternatives? Are the simplified analysis/eco-audits (e.g., focusing on the primary energy consumption only) suitable for a reliable assessment?

2   The Analyzed Case Studies

In order to compare the manufacturing approaches with varying the component geometry, three different axi-symmetric product geometries which can be manufactured by both machining and forming processes have been considered (Fig. 1). The components were made of a high-strength AA-7075 T6 aluminum alloy. The basic idea was to analyze the environmental performance while changing the amount of process scraps for subtractive approaches. As a matter of fact, the change in shape from ID1 to ID3

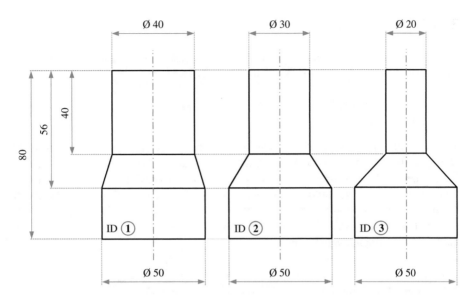

Fig. 1. Case study geometries (measures in mm). Adapted from [10].

results in a rise of machined-off material, which significantly affects the environmental performance of the machining approach. Table 1 reports the amount of resources involved in both the manufacturing approaches, as resulted from the life cycle inventory analysis previously developed by the authors (Ingarao et al., 2016 [10]). A good metric for geometry characterization is the solid-to-cavity ratio, which has been defined, according to Morrow et al. (2007), as the mass of the final part divided by the mass that would be contained within the bounding volumetric envelope of the part itself [4]. In other words, as the solid to cavity ratio decreases the product complexity increases along with the amount of material to be machined off. Such values are reported in Table 1 along with the amounts of material involved in tool manufacturing. It is worth pointing out that machining processes where developed on a CNC lathe and both roughing and finishing steps are envisaged for the three geometries. As concerns forming, one single forming step was designed to manufacture ID1 and ID2; two forming steps are necessary for ID3 instead. This is due to the high diameter reduction which causes an increase in die stresses.

Table 1. Masses involved in the material flows [10].

Geometry ID	ID1	ID2	ID3
Mass of the workpiece (kg of AA-7075)	0.35	0.27	0.21
Amount of material involved in die and punch manufacturing (kg of AISI H13)	7.23	6.2	5.88
Amount of machined-off material (kg of AA-7075)	0.1	0.17	0.23
Solid-to-cavity ratio	0.80	0.61	0.48

3 LCA Framework

In this section, the methodology and the results are reported following the LCA framework suggested by the ISO 14040 standards.

3.1 Goal and Scope

The aim of the research is to characterize the environmental performance of aluminum-based component manufacturing. Specifically, LCA analyses serve as a tool for comparing the environmental impacts of forming and machining approaches under different production scenarios. The present study aims at analyzing the influence of batch size and part geometry on comparative analysis results. Finally, the LCA results are used to setup and tune a decision support tools for environmentally friendly manufacturing approach selection.

3.2 Functional Unit

The functional unit for the LCA is the single component manufacturing via two different processes: forming and machining. One component production (within a defined batch size) was used as a basis for manufacturing approach comparison. It is worth pointing out that the environmental performance of the forming approach is affected by the size of the production batch. In fact, the contributions related to tooling (i.e., the punches and dies) have to be amortized over the number of parts to be manufactured.

3.3 System Boundaries

A cradle-to-grave system boundary is adopted with recycling selected to be the scenario at end-of-life (EoL). The impact of material production, product manufacturing, and recycling were evaluated. The common parts (use and transportation phases) were neglected as the research was carried out assuming that the components manufactured by both processes comply with the same product specifications (product manufactured either by forming or by machining serves the same customer and use). Concerning the materials involved in the analysis, the credit from recycling (end-of-life stage) was accounted for by implementing the substitution method (Hammond and Jones, 2010 [11]). As regards the manufacturing step, all the energy flows are considered: energy for processing (actual pressing or machining), tooling (materials and manufacturing energy) as well as heating (for the forming approach only).

3.4 Inventory and Impact Assessment Method

The inventory analysis of the considered processes has been developed by Ingarao et al. [10]. For providing a clear picture of the environmental performance of the analyzed manufacturing approaches, the primary energy along with fourteen impact categories

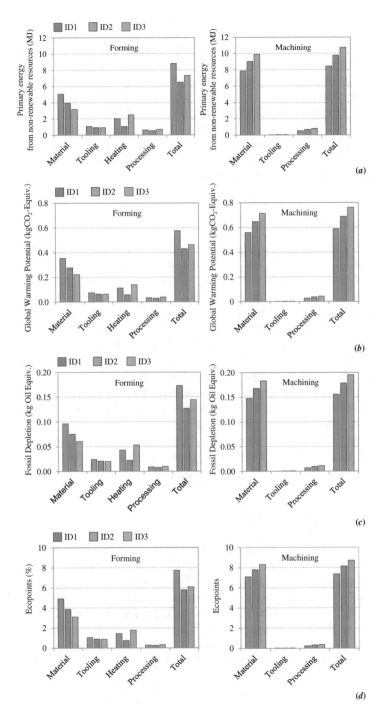

Fig. 2. Life cycle assessment results for different metrics: (a) Primary Energy, (b) Global Warming Potential, (c) Fossil Depletion Potential, (d) Ecopoint (%).

(Midpoints) and the Ecopoint (%) (single measure including all the midpoints) have been quantified. The midpoints are considered to be linked to the environmental mechanism of an impact category before the endpoints. Typical midpoint impacts include GHG emissions, Ozone depletion, photochemical oxidant formation, etc. The endpoints represent the damages to the areas of the targeted protection including human health, ecosystem and resource from all the midpoint environmental impacts. The impact categories were calculated by applying the ReCiPe method H/A. The ReCiPe method is selected because it is an update of the Eco-indicator 99 and CML 2002 method, both of them are widely applied impact assessment methods. In addition, the ReCiPe method integrates both the midpoint and endpoint impacts into a single framework, which best fits the goal of this study.

3.5   Interpretation of the Results

In this section the obtained results are discussed. Figure 2 depicts the output for both forming and machining. For the sake of clarity only the scenarios characterized by a production batch size equal to 100 and for only four metrics are reported. Specifically, results are presented for Primary Energy (Fig. 2a), Global Warming Potential (Fig. 2b), Fossil Depletion Potential (Fig. 2c), and Ecopoint (%) (Fig. 2d). Overall, it is possible to notice the significant impact of the material usage and the different role played by the tooling. Regardless the selected metric, it is possible to evaluate the performance of the two manufacturing approaches for each ID. Also, it is possible to notice that for ID1 the machining approach is the best option for all the considered metrics. On the contrary, for ID2 and ID3 the forming approach is always to be preferred. Such results are due to the worsening of the machining environmental performance as the amount of machined-off material increases. It is worth pointing out that, as the batch size increases, the suitability of forming approach increases (and vice-versa). In other words, for a small batch, the machining approach is the most environmentally friendly solution.

4   Decision Support Tool

With respect to the comments discussed in the previous section, for a given geometry, a batch size value for which the considered manufacturing processes are characterized by the same environmental performance could be identified (see also Ingarao et al., 2016 [10]). In this paper, the machining approach has been assumed to be independent from the batch size, as a fixed amount has been considered for the tooling allocation to each produced part. For a given geometry (IDi) the considered manufacturing approaches have the same value of a given midpoint (MDi) when Eq. 1 is verified. Therefore, the breakeven point can be calculated by applying Eq. 2.

$$MDi_M = MDi_{FF} + \frac{MDi_{FV}}{BPi} \tag{1}$$

$$BPi = \frac{MDi_{FV}}{MDi_M - MDi_{FF}} \qquad (2)$$

where:

- MDi_{FV} = Midpoint value for the batch size-dependent contribution of forming (i.e., tooling);
- MDi_{FF} = Midpoint for the fixed contribution of forming (i.e., processing, heating, and material usage);
- MDi_M = Midpoint value for machining.

The breakeven point changes as the geometry shape (solid-to-cavity ratio) and the considered metric change. Table 2 lists all the breakeven points per each considered condition.

Table 2. Breakeven points for different metrics and geometries.

Geometry ID	ID1	ID2	ID3
Primary energy	158	22	21
Climate change	85	20	18
Fossil depletion	353	29	29
Freshwater ecotoxicity	9	4	3
Freshwater	6	3	2
Human toxicity	5	2	2
Marine ecotoxicity	9	4	3
Marine eutrophication	100	45	33
Metal depletion	40	18	13
Ozone depletion	54	26	18
Particulate matter formation	73	34	24
Photochemical oxidant formation	203	86	64
Terrestrial	68	30	22
Terrestrial ecotoxicity	25	12	9
Ecopoint (%)	154	28	25

It is possible to notice a significant variability in breakeven point values across the considered metrics. Moreover, for a given metric (Midpoints, Ecopoint) the breakeven point values remarkably change. Specifically, as the solid-to-cavity ratio decreases the breakeven point value decreases too. The latter statement is due to the machining approach environmental performance worsening as the amount of the material to be machined-off increases.

By reporting the obtained results in a graph, and connecting them through a straight line, it is possible to highlight the trend of the breakeven point values as the solid-to-cavity ratio increases. Such graphs are plotted in Fig. 3 for some of the considered metrics. Such curves also represent a decision support tools named by the authors as '*Process Sustainability Diagram* (PSD)'. The area underneath the curve

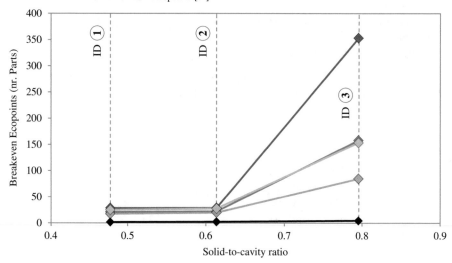

Fig. 3. Process Sustainability Diagram (PSD) with varying the considered metric.

contains all the factors of influence configurations for which the machining approach is preferable (for a given metric). Even though a strong difference among the different curves is visible, overall it is possible to highlight a common trend. The curves are characterized by a constant trend up to ID2 geometry, followed by a rise when shifting to ID3 geometry.

5 Discussion of the Results and Conclusions

The present research was aimed at analyzing the impact of the selected metrics on manufacturing approach comparison results. Specifically, an analysis with varying batch size and part geometry was developed. The change in terms of batch size breakeven values is highlighted when comparing machining and forming manufacturing approaches. The changes in the Process Sustainability Diagram with varying the considered metrics are also analyzed. Such diagram represents an eco-design tool for the specific case study. The designer, knowing the batch size and the amount of material that has to be removed, can easily identify the best manufacturing solution for a given metric. It is worth pointing out that Process Sustainability Diagram is the first decision support tool to the author's knowledge. Looking at Table 2 and Fig. 3, besides the strong value changes, the most relevant outcome concerns the (almost) matching results characterising primary energy and ecopoint metrics. In fact, a maximum difference of 6 can be observed for ID2. This result can be very useful for manufacture engineers. Actually, primary energy calculations can provide useful and reliable results for environmentally friendly manufacturing approach selection. It is worth pointing out that primary energy can be computed with available databases (e.g., Ashby, 2013 [12]) and complex full LCA analyses can be skipped making an acceptable breakeven point estimation error. Unless specific analysis on specific Midpoints are required, relying on primary energy demand analyses could be an effective procedure for identifying environmentally friendly metal shaping processes. Of course, the latter statement as well as the decision support tool has to be further developed with varying materials, shapes, and processes to prove the general validity of the present research.

References

1. Worrell, E., Allwood, J., Gutowski, T.: The role of material efficicency in environmental stewardship. Ann. Rev. Env. Resour. **41**, 575–598 (2016)
2. Allwood, J.M., Ashby, M.F., Gutowski, T.G., Worrell, E.: Material efficiency: a white paper. Resour. Conserv. Recy. **55**, 362–381 (2011)
3. Ingarao, G.: Manufacturing strategies for efficiency in energy and resources use: the role of metal shaping processes. J.Clean. Prod. **142**, 2872–2886 (2017)
4. Morrow, W.R., Qi, H., Kim, I., Mazumder, J., Skerlos, S.J.: Environmental aspects of laser-based and conventional tool and die manufacturing. J. Clean. Prod. **15**, 932–943 (2007)
5. Serres, N., Tidu, D., Sankare, S., Hlawka, F.: Environmental comparison of MESO-CLAD process and conventional machining implementing life cycle assessment. J. Clean. Prod. **19**, 1117–1124 (2011)
6. Huang, R., Riddle, M., Graziano, D., Warren, J., Das, S., Nimbalkar, S., Cresko, J., Masanet, E.: Energy and emissions saving potential of additive manufacturing: the case of lightweight aircraft components. J. Clean. Prod. **135**, 1559–1570 (2016)
7. Paris, H., Mokhtarian, H., Coatanéa, E., Museau, M., Ituarte, I.F.: Comparative environmental impacts of additive and subtractive manufacturing technologies. CIRP Ann. Manuf. Technol. **65**(1), 29–32 (2016)
8. Priarone, P.C., Ingarao, G., Di Lorenzo, R., Settineri, L.: Influence of material-related aspects of additive and subtractive Ti-6Al-4 V manufacturing on energy demand and carbon dioxide emissions. J. Ind. Ecol., doi:10.1111/jiec.12523 (2016, In Press)
9. Ingarao, G., Priarone, P.C., Gagliardi, F., Di Lorenzo, R., Settineri, L.: Subtractive versus mass conserving metal shaping technologies: an environmental impact comparison. J. Clean. Prod. **87**, 862–873 (2015)
10. Ingarao, G., Priarone, P.C., Di Lorenzo, R., Settineri, L.: A methodology for evaluating the influence of batch size and part geometry on the environmental performance of machining and forming processes. J. Clean. Prod. **135**, 1611–1622 (2016)
11. Hammond, G., Jones, C.: Inventory of Carbon and Energy (ICE), Annex B: How to Account for Recycling. The University of Bath, UK (2010)
12. Ashby, M.F.: Materials and the Environment: Eco-Informed Material Choice, 2nd edn. Butterworth Heinemann/Elsevier, Boston (2013)

Sustainability in Industrial Plant Design and Management: Applications and Experiences from Practice

Eco Orbit View – A Way to Improve Environmental Performance with the Application of Lean Management

Katarzyna Skornowicz[1(✉)], Malgorzata Fialkowska-Filipek[1,2], and Remigiusz Horbal[1]

[1] Lean Enterprise Institute Poland, Wrocław, Poland
{katarzyna.skornowicz,malgorzata.fialkowska-filipek,
remigiusz.horbal}@lean.org.pl
[2] Wroclaw University of Technology, Wrocław, Poland

Abstract. The purpose of this paper is to present two case studies from the application of Eco Orbit View, a method that extend the mind-set of Lean Management towards environmental performance, and to discuss the potential of broader implementation in medium and small sized manufacturing companies. Eco Orbit View method was developed during Eco Lean Compass project pursued in an international consortium including institution from Poland, Germany and Turkey. It applies low-cost improvements of production processes based on Lean Management in order to bring positive impact both on business and environmental performance. We verified the method on two case studies performed in medium-sized manufacturing companies from Poland and Turkey. In both cases the method brought positive impact on selected business and environmental metrics. Regarding business metrics, implemented improvements were inexpensive or even costless, but at the same time resulted in significant financial gains for the companies. When it comes to the environment, the main achievement was the significant reduction of physical wastes, electric energy and oil consumption. In next steps of Eco Lean Compass project we want to focus on delivering the method to a broad community of companies and develop the Eco Lean Transformation Program to assure the sustainability of the improvements.

Keywords: Lean management · Low cost improvement method · Environment · Lean & green · Manufacturing · Sustainability

1 Introduction

Companies around the world strive for more efficient practices to improve their competitive position in the market. The pressure to meet customer's needs and produce at the lowest costs is exerted as well as the pressure of society and governments to protect the environment. Every year the number of ecological regulation increase. Sustainability became an integral part of decision making process in industries since the 1970s, essentially influenced by the report "Limits to Growth" [1]. Nowadays, to become successful in business, environmental aspects should not be treated in separation from strictly business related measures, targets and objectives.

© Springer International Publishing AG 2017
G. Campana et al. (eds.), *Sustainable Design and Manufacturing 2017*, Smart Innovation, Systems and Technologies 68, DOI 10.1007/978-3-319-57078-5_62

Nevertheless many companies are convinced that the more environment-friendly they become, the more the effort will erode their competitiveness [2]. Managers, in most cases, are not strongly motivated to improve environmental performance, which have much lower priority then improvement of other indicators critical for business success (e.g. manufacturing cost, productivity, quality, delivery time). Moreover they believe that reduction of negative influence of manufacturing on the environment requires high investments, while Return On Investment (ROI) is uncertain. This is because commonly available methods and approaches for environmental improvement in fact require substantial amount of time and data, and in many cases result in innovations requiring significant investment (product modifications for Design for Environment and Life Cycle Assessment; technological innovations for ISO 14000 series and ISO 50001) [3]. There is still lack of simple solutions which are more and more desirable.

In order to achieve their business goals many companies implement Lean Management. Lean is an effective management approach, derived from Toyota Motor Corporation, used to organize the production and supporting processes in manufacturing industry. In Lean companies problems are perceived as opportunities for learning, development and improvements. Managers act as coaches, helping others in constructive problem solving and practicing ongoing continuous improvement in everything employees do. The core element of Lean is waste reduction. Based on common purpose, standardized process and respect for people, Lean is about creating more value for every customer while minimizing time, effort and resources (waste) [4]. These days, it is difficult to find a manufacturing industry which haven't implemented at least a few key Lean practices [5].

Thanks to combining Lean Management with "eco awareness" we developed Eco Orbit View method which tackles directly two of the main dimensions of sustainability: environment and economy. The method enhances the sustainment of prosperity and high standards of living and indirectly influences also the social dimension of sustainability. The purpose of this paper is to present two case studies from the application of Eco Orbit View, and to discuss the potential of broader implementation in small and medium-sized manufacturing companies.

2   Methodology

Eco Orbit View method was developed as a result of Eco Lean Compass project pursued in an international consortium including institution and industrial partners from Poland, Germany and Turkey. The method applies low-cost improvements of production processes based on Lean Management in order to bring positive impact both on eco-efficiency i.e. business and environmental performance [6].

The aim was to creating method where analysis will be fast, environmental aspects will be established and connected with business aspects and costs of improvements will be low (that is why lean philosophy is here main base).

Eco Orbit View follows five steps:

1. Select a family of products that will be reference point for the analysis. In this case a family of products are products which pass exactly through the same processes and machines and are relevant from the production volume point of view.
2. Look on the production process from afar and create simple process map. One process consists of certain operations, which result in raising the value of the product. This step is more about distinguish several main areas in production process than visualizing connections between them.
3. Distinguish the need of improvement for established areas using Selection Matrix tool. The Selection Matrix tool combines business aspects with environmental ones and determines the strength of connection between them[1]. The main goal of Selection Matrix is visualization of results (Fig. 1) and determining problematic business areas where improvements will be beneficial also from environmental point of view.

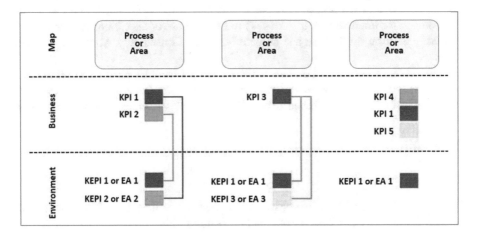

Fig. 1. Example of Eco Orbit View visualization

4. For selected operations use well-known Lean tool which is Gemba Walk and search for possibilities for improvements in desirable range.
5. If possibilities for improvement cannot be found by Gemba Walk make deeper analysis with use of other Lean-based tools like Resource Consumption Study or Eco A3 report.

We verified the method on two case studies performed in small and medium-sized manufacturing companies from Poland (Company A – medium-sized) and Turkey (Company B – small-sized). The key area of expertise of the Company A is manufacturing of fasteners such as wood, metal and plastic screws, set, self-taping and self-drilling screws, quick assembly collated screws, rivets. The company provides products for a number of international corporations such as household appliances, electronics, construction, automotive and furniture industries. The quality of production processes

---

[1] Detailed description of the Selection Matrix tool can be found here [7].

are ensured by certificates from the group of ISO 9001 and 14001, as well as ISO TS 16949:2009 for Automotive Quality Management System. The Company B produces dish drainers, kitchen racks, bath shelves, suction systems, sticker systems with any kind of coating like nickel, chrome, paint, stainless steel etc. The company provides products for organizations all around the world. The quality of production processes are ensured by ISO 9001 for Quality Management System.

The verification of the Eco Orbit View method was divided by three phases performed at intervals of 2–3 months. Each phase was made separately for the Company A and the Company B. First phase which includes three steps from the Eco Orbit View method was made during 2-days workshop on company premises. During the workshop selected group of managers, specialists and researchers worked side by side to choose the potential areas for improvement. The workshop was also the introduction to the vision of the project and a chance to get familiar with the method. Second phase which includes searching for possibilities for improvement of chosen operations was made during next 2-days workshop. Last phase was connected with deeper analysis when possibilities for improvement are not easy to find. In case of the Company A there was a need to perform third phase and prepare Resource Consumption Study for selected areas.

The scope of Eco Lean Compass project does not includes analysis of all defined connections. The target was to verify the ideas of focusing fast and low-cost improvements in most relevant areas based on both types of aspects: business and environmental.

3   Findings

3.1   Company A

During cooperation with the Company A two areas were chosen via Selection Matrix tool for further analysis: Heat Treatment and Heading (Fig. 2).

Heading

		Output vs. Plan	Downtimes	Defects	Life of tools
		2	3	0	3
Electric energy	3	15	0	0	0
Gas	0	0	0	0	0
Water	3	15	0	0	0
Oil and Emulsion	1	0	0	3	0
Scrap vs steel consumption	1	0	0	3	0

Tread Rolling

		Output vs. Plan	Downtimes	Defects	Life of tools
		2	0	0	3
Electric energy	3	15	0	0	0
Gas	0	0	0	0	0
Water	0	0	0	0	0
Oil and Emulsion	1	0	0	3	0
Scrap vs steel consumption	1	0	0	3	0

Heat Treatment

		Output vs. Plan	Downtimes	Defects	Life of tools
		0	3	0	0
Electric energy	3	0	18	0	0
Gas	3	0	18	0	0
Water	0	0	0	0	0
Oil and Emulsion	1	0	0	3	0
Scrap vs steel consumption	1	0	0	3	0

Fig. 2. Selection matrix for company A

Heat Treatment. Further actions in Heat Treatment, including Gemba Walk and preparation of Resource Consumption Study, show that the reduction of breakdowns on the furnace will allow to reduce energy consumption. Breakdowns require to turn the furnace off and reheat it again after repair or keep the temperature but stop the production (Fig. 3). This way of operating during breakdowns consumes 10% of all energy using by the furnace and is recognized as a main waste. As a low cost improvement solution in this case the implementation of Total Productive Maintenance (TPM) for the furnace was suggested because reduction of breakdowns, in this case, can resolve not only the problem with production plan realization but also decrease the electric energy consumption per one ton of product. TPM is a Lean technique which ensure that every machine in a production process is always able to perform its required tasks. It seeks total productivity of equipment by focusing on all of the six major losses that plague equipment: downtime, changeover time, minor stops, speed losses, and rework [8].

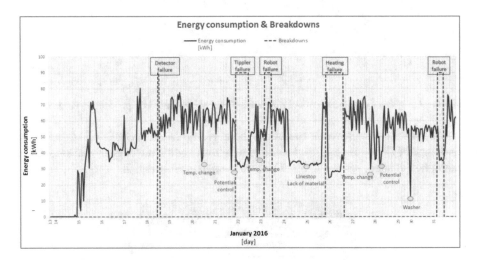

Fig. 3. Breakdowns on SAFET furnace in comparison to electric energy consumption

What is more, management of the Company A, after reviewing the analysis results, was so interested in elimination the main cause of the breakdowns that decided about implementation of high cost improvements such as change of atmosphere generation source from methane with nitrogen to endo atmosphere and changing external fan on larger and more efficient.

Heading. The results of analysis during second phase identified also this potential area of improvement. The increase in productivity in key machines would reduce overtime, which are necessary to fulfill orders on time. Moreover, reducing overtime will lead to the reduction of the consumption of electricity, oil and other utilities.

Two critical machines based on production plan have been identified - T14, T29. The hourly production control board were introduced to gathering data. Analysis of this data shows that the biggest waste for both machines are changeovers (Fig. 4).

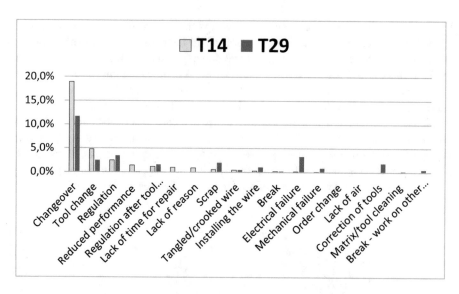

Fig. 4. Analysis of data from critical machines T14 and T29 based on production control board

As a low cost improvement solutions the implementation of SMED (Single Minute Exchange of Die) was suggested. During additional common workshop the analysis of exemplary changeover was made. The results are shown at Fig. 5.

Stage 1

		Changeover before improvement	
		Other machine	Changeover time
External	Internal		
4%	44%	52%	
0:07:00	1:13:19	1:26:36	2:46:55

Stage 2

		Internal to External change	
		Other machine	Changeover time
External	Internal		
17%	31%	52%	
0:29:11	0:51:08	1:26:36	2:46:55

Stage 3

		Improvements	
		Other machine	Changeover time
External	Internal		
17%	28%	55%	
0:24:12	0:39:28	1:18:56	2:22:36

Stage 4

		Work reorganization	
		Other machine	Changeover time
External	Internal		
17%	22%	3%	
0:24:47	0:31:58	0:03:50	1:00:35

Fig. 5. Results of SMED for Heading

The conclusion from Heading area was that small improvements during changeovers can reduce the operation time by 15% but in the same time reorganization of technician's work can reduce the operation time by additional 49%. The "eco" savings were calculating also for both stages of improvements implementation and the potential cost savings were presented to Management Board as an additional, significant benefits in favor of work reorganization. Currently Company A is changing the work system and implementing suggested small improvements connected with another Lean technique 5S, a 5-step-method of creating an organized and therefore effective workplace.

3.2 Company B

During cooperation with Company B also two operations were chosen via Selection Matrix for further analysis: Cutting+Bending and Assembly+Welding (Fig. 6).

Cutting+bending

	Business	Delivery time	Scrap reducing	Defect ratio
Eco		3	2	0
Electric energy	1	0	2	0
Physical waste	2	0	8	0
Noise	2	0	0	0

Welding

	Business	Delivery time	Efficiency of machines
Eco		3	1
Electric energy	1	0	2
Reactive electric energy	3	0	0
Noise	2	0	0
Air contamination	2	0	0

Cropping

	Business	Delivery time	Efficiency of machines	Tool lifetime
Eco		3	2	0
Electric energy	1	0	2	0
Physical waste	2	0	0	4
noise	2	0	0	0

Assembly Welding

	Business	Scrap reducing	Efficiency of machines	Defect ratio	Delivery time
Eco		0	0	0	0
Electric energy	1	3	2	0	0
Physical waste	2	12	0	8	0
noise	2	0	0	0	0

Packaging&Inspection

	Business	Scrap reducing	Efficiency of machines	Defect ratio	Delivery time
Eco		0	0	0	0
Electric energy	1	0	2	0	0
Physical waste	1	6	0	4	0

Fig. 6. Selection matrix for company B

In both Cutting+Bending and Assembly+Welding areas the main focus concerns on elimination of physical wastes.

Cutting+Bending. The main source of physical waste in this process was test production during adjustment after material change. This adjustment is made manually each time and is very dependent of experience of the technicians. As a solution to this problem the special scale was designed. This idea was suggested by company employees and it could be done with just company resources. In this particular case the tested solution was not implemented because the adjustment scale was not working as it was expected. Nevertheless the director of company was so interested in reduction of adjustment speed and metal scrap during this particular operation that is willing to make bigger investment in replacing adjustment mechanism for more advanced one.

Assembly+Welding. In this process the employees of the company suggested the solution for decreasing the amount of physical wastes – metal scrap. Small and simple change in design of the products was an answer on the problem. The trial version of new design was checked on 4 models and after one month no problems with durability were found. The quantity of metal scrap was reduced by 26%–28% (depends on the model) and in the same time company gain financial profits because they are able to produce more products from the same amount of raw material.

4   Discussion

The specifics of the companies are different. Not only from cultural but also business perspective. The Company A has various management systems implemented and maintained when the Company B, despite of ISO 9001 implementation, characterizes poor productivity culture. Because of that, the improvements in the Company B were quite obvious to find during Eco Orbit View, where many problems were unhidden. In case of the Company A, the additional deeper analysis was needed. Nevertheless in both cases the objectives of the project were achieved. It was noticed that Eco Orbit View is a powerful method which motivates towards eco-efficiency improvements. During workshops, the companies' employees have realized that it is possible to improve what is important for the company from business perspective but with the addition of environmental aspects. Moreover the suggested improvement solutions do not need large financial outlays or working hours to be implemented. Main advantages of this method such as:

- the short time between start of analysis and potential improvements identification,
- visualization of connections between business and environmental aspects,
- low-costs because of lean philosophy usage, were checked and confirmed during this case studies.

In both cases improvements have positive impact on environmental aspects such as physical wastes, electric energy or other utilities. It may be argued that the most of the environmental improvements arise from Lean Management's inherent focus on waste reduction [9]. Like other Lean wastes, environmental wastes do not add value from the customer's point of view. They also frequently represent costs to the company and society in general. Lean Management can lead to significant environmental benefits, since environmental wastes are related to Ohno's seven waste categories [10]: overproduction (e.g. more raw materials and energy consumed in making the unnecessary products), inventory (e.g. more packaging to store work-in-progress), transportation (e.g. emission from transport), defects (e.g. defective components require recycling or disposal), over processing (e.g. more parts and raw materials consumed per unit of production), waiting (e.g. wasted energy from heating, cooling, and lighting during production downtime). However, despite the relationship between Lean and eco-efficiency approaches, the companies often overlook opportunities to prevent or reduce environmental wastes and perceive the approaches as additive or complementary to each other. The awareness of the linkages between business challenges and environment may help to ensure that the reduction of lean wastes entails the reduction of environmental wastes. The Eco Orbit View method extends the mind-set of Lean towards environmental performance in conscious and systematic way. Decision concerning selection of the potential areas for improvement is based on business data as well as (or even more) on environmental aspects because only areas with the strongest connections are taking under consideration. Most of all, the influence on environmental aspects is exposed and became inherent element of sustainable development.

In modern word, it is a challenge to convince small and medium-sized companies to become environmental friendly, especially when still the highest priority is given to the cost reduction. Developed method proved that the both goals can be achieved in the same time and Lean Management is a perfect way to combine business and environmental targets when it is woven into the Eco Orbit View method.

5 Recommendation for Future Research

Presented case studies were a one-time application and limited to only two companies. In the next steps of Eco Lean Compass project we want to focus on delivering the method to a broad community of companies and develop the Eco Lean Transformation Program to assure the sustainability of the improvements.

The Eco Lean Transformation Program will be addressed to companies interested in development through the low-cost improvements connected with Lean Management. The proposed approach can be particularly useful for the companies with implemented ISO 14001, were it is an every year requirement to prepare improvement plan for environmental aspects. Another goal of the Transformation Program is to increase managers' ecological awareness through adding environmental aspects to the process analysis.

In the nearest future we plan to conduct pilot of the Transformation Program, which will involve participation of 8 Polish and 4 German small/medium-sized companies. During the 2 days workshops, the companies will be familiarized with the Eco Orbit View method. We will use special prepared material developed based on the case studies. Moreover, the Transformation Program will include assumptions of "double improvement loop" consisting of big and small improvement loops. Big loop is to perform the whole five-step Eco Orbit View method for checking the general status of the processes.

It is recommended to perform this loop one per year and based on the results establish processes, business and environmental aspects for next improvements. Small loop should be performed more often for areas defined as most significant during big loop and assumes continues observations, analysis and improvements. After workshop the companies will be monitored if taken improvement actions were satisfying and if the knowledge shared during workshop was sufficient.

6   Conclusions

In both cases the Eco Orbit View method brought positive impact on selected business and environmental metrics. Regarding business metrics, implemented improvements were inexpensive or even costless, but at the same time resulted in significant financial gains for the companies. When it comes to the environment, the main achievement was the significant reduction of physical wastes, electric energy and oil consumption. In next steps, the Eco Orbit View method will be delivered to a broad community of companies through the Eco Lean Transformation Program.

7   Funding

The described approach and methodology are result of research undertaken by the project ECO LEAN COMPASS co-financed by the National Centre for Research and Development as a part of program ERA-NET ECO-INNOVERA under agreement no ERA-NET-ECO-INNOVERA 2/2/2014. Lean Enterprise Institute Polska is the coordinator of the project and together with the Fraunhofer Institute for Production Engineering and Automation (IPA) of Germany constitutes as the main research partner. In addition, there are three production companies from Germany, Poland and Turkey involved. ECO LEAN COMPASS is an initiative aiming at development of methods and tools to reduce the negative impact of manufacturing companies on the environment thanks to the low budget improvement initiatives. More information on the project can be found at: https://www.eco-innovera.eu/2nd-call-projects-ecolean.

References

1. Meadows, D.H., Meadows, D.L., Randers, J., Behrens, W.W.: The Limits to Growth. Universe Books, New York (1972)
2. Nidumolu, R., Prahalad, C.K., Rangaswami, M.R.: Why sustainability is now the key driver of innovation. IEEE Eng. Manag. Rev. **43**(2), 85–91 (2015)
3. Hart, S.L., Ahuja, G.: Does it pay to be green? An empirical examination of the relationship between emission reduction and firm performance. Bus. Strateg. Environ. **5**(1), 30–37 (1996)
4. Womack, J.P., Jones, D.T.: Lean Thinking: Banish Waste and Create Wealth in Your Corporation. Simon & Schuster, New York (1996)
5. Stone, K.B.: Four decades of lean: a systematic literature review. Int. J. Lean Six Sigma **3**(2), 112–132 (2012)

6. Wbcsd, A.: Eco-efficiency: Creating More Value with Less Impact. World Business Council for Sustainable Development, North Yorkshire (2000)
7. Pawlik, E., Gutowska, D., Horbal, R., Maśluszczak, Z., Miehe, R., Bogdanov, I., Schneider, R.: Lean & green, how to encourage industries to establish pro-environmental behavior. In: 14th International Conference on Manufacturing Research, 6–8 September 2016, Loughborough University, UK (2016)
8. LEI: Lean Lexicon: A Graphical Glossary for Lean Thinkers, 5th edn. Lean Enterprise Institute, Cambridge (2014)
9. Zokaei, K., Lovins, H., Wood, A., Hines, P.: Creating a Lean and Green Business System: Techniques for Improving Profits and Sustainability. CRC Press, Boca Raton (2013)
10. EPA: The lean and environment toolkit (2007). http://www.epa.gov/lean

3D Printing Services: A Supply Chain Configurations Framework

Helen Rogers[1(✉)], Norbert Baricz[1], and Kulwant S. Pawar[2,3]

[1] Business Faculty, Technische Hochschule Nürnberg Georg Simon Ohm, Nuremberg, Germany
{helen.rogers,norbert.baricz}@th-nuernberg.de
[2] Centre for Concurrent Enterprise, Nottingham University Business School, Nottingham, UK
kul.pawar@nottingham.ac.uk
[3] Centre for Concurrent Enterprise, Nottingham University Business School, Ningbo, China
Kulwant.Pawar@nottingham.edu.cn

Abstract. This paper presents an extended framework for the classification and categorization of 3D printing services, using the findings of two previous studies (carried out by the authors) as a foundation. The work to date revealed that 3D printing services can be separated into three distinct categories based on the configuration of their design-related and manufacturing-related processes: generative, facilitative and selective services. This study examines in more detail 105 of the 558 originally identified 3D printing service providers in Germany, Austria, Switzerland and the Benelux countries, with the goal of further clarifying the main drivers, linkages, buildings blocks and modules that shape these three individual branches of services. These configurations will potentially have wide-ranging impact on the supply chain strategies, structures and operations of the future.

Keywords: 3D printing services · Supply chain configurations · Additive manufacturing · Distributed manufacturing

1 Introduction

3D printing (also known as additive, digital and rapid manufacturing) refers to not one, but multiple technologies and manufacturing processes that enable users to create a tangible object from a digital three-dimensional model [1–3]. The technology has undergone several phases of development in the past few decades, shifting from its original limited uses in prototyping to increasingly advanced applications in fields ranging from aerospace and healthcare to even food and customized consumer goods [4]. 3D printing is expected to significantly impact the business models, supply chain configurations and service provisions of the future, granting firms the ability to manufacture in a leaner, more sustainable and more energy-efficient manner [1–3, 5–7]. Nonetheless, despite relatively significant technological advances in 3D printing, widespread adoption of the technology remains hindered by a number of decisive factors, including high printer acquisition costs, lack of experience with the technology and the technical limitations of 3D printers [1, 3, 8, 9]. As identified by Rogers et al. [10, 11] these issues could, however, be addressed by the growing market for 3D printing services. Offering unique combinations of design-related and manufacturing-related processes, these services could not only close the gap until the

G. Campana et al. (eds.), *Sustainable Design and Manufacturing 2017*, Smart Innovation, Systems and Technologies 68, DOI 10.1007/978-3-319-57078-5_63

technology becomes mature enough for widespread adoption, but may even become a fixed key component in the supply chain configurations of the future. Over time, 3D printing services have the potential to become one of the foundational pillars of a 3DP-driven circular economy, creating unprecedented opportunities to repair, recycle and remanufacture products [7]. Several important questions pertaining to suitable 3D printing service provider selection, raw material sourcing strategies, sustainability impacts, optimal supply chain configurations, intellectual property rights, quality control, as well as ownership and liabilities nevertheless remain unanswered in the 3D printing service provision arena [3, 5, 7, 10]. These issues, coupled with the potentially significant impact of 3D printing services on companies' supply chain configurations pose interesting challenges for the future development of 3D printing services sector.

This study is the third in a series [10, 11] examining the size and composition of the 3D printing services industry in the DACH (Germany, Austria, and Switzerland) and Benelux (Belgium, the Netherlands, and Luxembourg) markets. This paper seeks to extend the original framework for 3D printing service supply chain configurations presented by Rogers et al. [10] by examining in more detail 105 of the 558 originally identified 3D printing service providers, with the goal of further clarifying the building blocks and connections that shape service providers' supply chains.

2   Research Methodology

The study at hand examines 105 consumer 3D printing service providers in the DACH (51 DE, 9 AT, 11 CH) and Benelux (21 NL, 11 BE, 2 LU) markets. This subset of data was collected between October and December 2015 in the context of a larger research project examining 558 service providers in these markets. Service provider data was collected by examining company websites, brochures, catalogues, marketing material and press releases, profiles from academic literature, conference exhibitor portraits, interviews, as well as from telephone conversations with the companies when data inconsistencies arose. Collected data points included all key aspects of the companies' product and service offering, infrastructure, channels, pricing model and business model.

3   The 3D Printing Service Supply Chain

The 3D printing service supply chain can be viewed as consisting of two distinct universal components: design-related processes and manufacturing-related processes [6]. Services can be separated into three distinct categories based on their configuration of these two supply chain segments: generative services, facilitative services and selective services [10]. These individual branches of services and their respective unique characteristics are examined in more detail in the following sections.

3.1   Generative Services

A total of 75 (71%) of the 105 identified 3D printing service providers offered generative services as of the end of the year 2015. Generative services include all on demand 3D printing services whose primary activity consists of creating a 3D model for a customer before subsequently 3D printing it. As illustrated in Fig. 1, these services can additionally be split into two distinct branches with unique characteristics: scanning services and construction services.

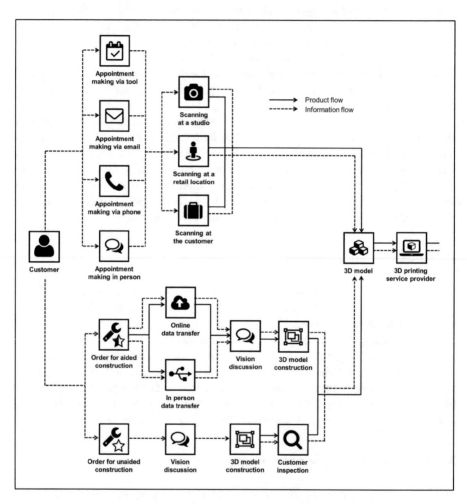

Fig. 1. Generative services supply chain

Generative scanning services employ 3D scanning equipment in order to create a 3D model for the customer. Their primary service consists of creating a digital copy of an existing object, with the goal of subsequently manufacturing its physical counterpart. The scanning process takes place either in a location belonging to the company that is not open to visitors without an appointment (a studio), a store with fixed opening hours (retail location), or at a sheltered location chosen by the customer (e.g. their home, their office or in the case of business customers even a company's event). These locations do not necessarily need to belong to the service provider, however. The German figurine manufacturer *3DMii* (3dmii.com) for example did not own a store with 3D scanning equipment at the time of writing, but instead entered a partnership with the electronics chain Saturn that allowed 3DMii to provide their services inside Saturn's branch in the German city Herford.

The scanning process is then carried out in a special sealed cabin that captures a 360° digital image of the desired object (most suitable for objects that can't stand perfectly still, like humans or animals) or by placing the object in a location with a neutral backdrop and scanning it using a handheld 3D scanner (more suitable for stationary objects). The time needed to scan an object can range anywhere between a few seconds (for cabins) and a few minutes (for handheld devices). The resulting 3D model is then edited and adjusted by an internal designer who corrects any visual errors that might have occurred in the scanning process. The scanning service is concluded by sending the optimized model to the manufacturing facilities of the service provider; however, most services also give the customer the option to only purchase the 3D model data.

The by far most popular use of scanning services is the creation of a 3D printed scale model of the customer (usually called a figurine). Out of the 71 companies that offered generative services at the end of the year 2015, a total of 55 (73%) specifically mentioned and marketed their capability to manufacture miniature versions of a customer or their pet. Unique extensions of this branch of services can be seen in firms like *Staramba* (staramba.com) and *Doob* (doob-3d.com) that also offer their customers the opportunity to have a figurine of their chosen celebrity be printed with the miniature version of themselves.

The only alternative available to customers who require a 3D model, but do not have a tangible precursor that can be scanned, is to employ construction services. Unlike scanning services, constructive services are theoretically not restricted in their ability to generate a 3D model tailored to the customer's vision, except for the skill level and creativity of the internal designer who will be constructing the model.

Constructive services can be classified as either aided, or unaided, depending on the degree to which the customer provides support to the design team responsible for the construction. Aided construction services usually request a series of two-dimensional sketches, drawings, pictures or paintings based upon which the 3D model can be constructed. Examples of services built on the principle of aided construction are *Uw huis 3D geprint* (uwhuis3dgeprint.nl) and *Ego3D* (ego-3d.de). Uw huis 3D geprint is a service provider in the Netherlands that produces miniature versions of a customer's house by turning two-dimensional pictures of the house into a three-dimensional model and subsequently printing it. Ego3D is a German company that uses customer photographs to construct a 3D model of their head that is then used to print a bust of the customer. More advanced applications of construction services based on pictures shift the attention away from 3D printing itself, focusing instead on the creation of a product that is tailored to the customer. A notable example of such an application can be seen in the Dutch company *Boulton Eyewear* (boultoneyewear.com) that creates customized 3D printed glasses by creating a 3D model of the customer's head from a series of pictures and tailoring the final shape to the model of the head. These more advanced applications could potentially become increasingly common in consumer-facing companies outside the 3D printing industry, offering a new level of customization to branches ranging from footwear design to prosthetics.

In contrast, unaided construction services generally do not request any pre-existing design material, but instead focus on transforming a customer's idea or concept initially into a 3D model and later on into a tangible object. Unaided construction can thus be considered the most demanding service model, requiring not only the largest dedication of resources to design, but also demanding the highest degree of customer involvement. In order to shift the resource burden outside of the company, a few of the largest 3D printing platforms worldwide, including for example *Shapeways* (shapeways.com) and *Materialise* (i.materialise.com) have chosen to crowdsource all construction-related activities, creating a market space in which designers can offer their services on demand. Smaller, more local organizations like the German company *trinckle* (trinckle.com) have however also chosen to adopt a similar approach, creating a forum that allows customers to post a description of their project and budget, to which designers can reply.

3.2 Facilitative Services

Facilitative services, offered by 71 (68%) of the identified firms, include all types of services that do not generate or provide a 3D model for the customer, but instead offer manufacturing-centered services to customers who already own or have access to a model they would like to have printed. As illustrated in Fig. 2, facilitative services, like

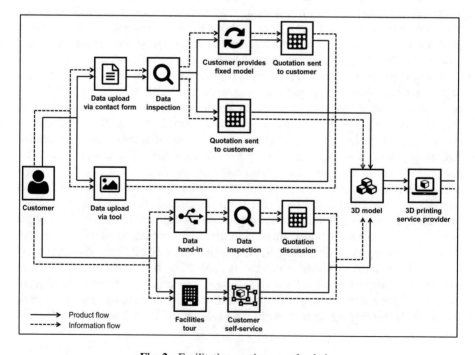

Fig. 2. Facilitative services supply chain

their generative and selective counterparts, begin with the customer. Customers may choose to either use an online service for their project, or to walk into a company's retail store to begin the process.

Online orders can be placed by using a service provider's contact form or upload tool. In contact-form-based ordering systems, customers fill out an online form with an explanation of how they desire to have their 3D model printed, attaching any relevant files to it, before subsequently submitting it. Service providers using such channels expect customers to handle all design-related activities surrounding the 3D model themselves before placing their orders. Models are generally required to have a stable weight distribution (i.e. that the resulting object does not topple over during 3D printing), fit within the maximum bounding box of the company's 3D printers, be watertight, and be saved in an acceptable file format (STL, OBJ, X3D, VRML, or others, depending on what printers and processes the company uses). If the model does not fulfil the service provider's requirements at this stage and does not pass the service provider's design inspection, the process is temporarily halted and the customer is notified that they need to take action before the chain can continue. If the customer provides a new model that passes the tests or if the original model was good to begin with, the company responds to the customer with a quotation for the entire process via email. The process is again halted until the customer confirms or cancels the order. If the offer is accepted, the service provider proceeds to 3D print the model, after which it is delivered to the customer in appropriate packaging. While less efficient time-wise than tools embedded into the ordering system, contact forms can be seen as the main channel for customer orders in most companies, serving as the only ordering channel in 67% of the identified online facilitative service providers, with 6% taking orders exclusively via email.

Standard uploading systems can significantly shorten this process by dynamically checking the 3D model's properties, as well as offering customers the opportunity to configure the dimensions and material of the resulting object and directly calculating the price of the resulting object for the customers. Uploading systems of this nature grant service providers the ability to automate the entire process to a large extent. By having such tools in place and subcontracting manufacturing activities, a 3D printing service can become operationally exceptionally lean, shifting the focus of its activities to improving the platform and creating a network of suppliers that can handle orders. Once such a network has been established and the uploading system has been configured to adequately handle all possible customer wishes, service providers become able to shift their focus completely to manufacturing, with service provider staff only being required at this stage for maintenance and platform improvements. One of the simplest and most efficient versions of such a platform can be seen on the website of the German company *MeltWerk* (meltwerk.com) that effectively reduces the entire process between the customer and the 3D printing process to just one step. More advanced systems, like the virtual workbench of *ubimake* (ubimake.com) even allow the customer to take full control of the data inspection process, enabling them to dynamically analyze, repair or even add structures to their model in real-time.

If the customer instead decides to handle their request in person, the entire process that would usually take place through email conversations (in systems without uploading platforms) is instead done in person at a service provider's store. Unlike the online model, the in-store system demands that every single store of the service provider be equipped with CAD workstations and be staffed with at least one individual who can carry out the data control process, as opposed to the online approach, where one individual could be dedicated to handling all customer orders (depending on the volume). The in-store model is therefore too costly to be run as a standalone service, but is nonetheless a suitable addition to other models that depend on a store, such as the scanning services described in the previous chapter. The data gathered on the service providers indicates that companies agree with the premise, as all identified in-store providers of facilitate services also rendered generative services in their stores at the end of 2015.

The in-store services also mimic the second chain of the online platforms in the form of self-service stations. Self-service models offer customers the chance to handle the entire process from data control to manufacturing itself by allowing them to reserve a section of the facilities for their projects. The standard procedure for self-service involves an introductory tour of the facilities, after which the customer is allowed full flexibility in their choice of workstation (generally a computer with the appropriate software and a set of basic tools), choice of 3D printer and choice of materials. Customers are generally not introduced to the process of 3D printing itself, as it is assumed that they are knowledgeable about the process and simply require the facilities to complete their project. The primary benefit of the self-service approach to the customer is that they not only get access to 3D printing facilities with which they can experiment, but also the fact that they receive their finished product immediately after the printing process is concluded, unlike online platforms and other in-store models that generally require one to two weeks to finish and deliver an order.

3.3   Selective Services

Selective services comprise the third and final category of 3D printing services. These services reduce the customer's involvement in the design and manufacturing process to a minimum. In selective service models, customers are restricted to choosing a 3D model from the service provider's database. Depending on the capabilities of the service provider, customers can then choose the color, material, resolution and size of the resulting 3D printed object. In more advanced service models, customers are even allowed to customize the model itself to a certain extent, i.e. by adding custom three-dimensional text to it or slicing and dicing it to their liking.

The module of selective services can be considered the shortest of the three branches discussed in the context of this study, as evidenced by its supply chain in Fig. 3. It is, however, not the least complex. Selective service models demand the highest degree of involvement from service providers in design. Unlike in generative service models, in which customers co-create 3D models with the service provider, in selective models, service providers must singlehandedly create an extensive database of 3D models in order to attract customers. Selective services, therefore compete primarily not through their manufacturing capabilities, but through the uniqueness, usefulness and overall attractiveness of their 3D model database. This database can be comprised either of models uploaded by the service provider themselves, or contain a collection of models designed by third parties, like other users of the platform or companies contracted by

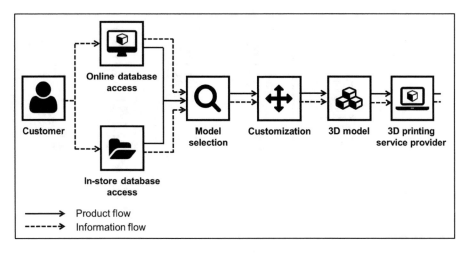

Fig. 3. Selective services supply chain

the service provider. Notable examples of such platforms include *ubimake* (ubimake.com), that lets customers browse and customize 3D models created by other users (building a community of designers), or *scope for design* (scopefordesign.de) that allows customers to modify a selection of 3D models created by the company (e.g. phone covers and jewelry). With only 13 (12%) of the identified firms offering such services, selective services can be considered the least common category.

The process works in a similar manner in chains that use retail stores, with the exception that the customer does not pick the model and customize it through a web browser on their device of choice but instead browses the company's catalogue in person at the store, discussing changes (like color, material or scale) with the staff of the store. Due to the similarity of selective services to facilitative services in terms of staffing and required facilities, it is unlikely that a service provider would chose to only offer selective services at a retail location as a standalone service.

3.4 Manufacturing Activities

The structure of the final stage of every 3D printing service (the manufacturing activities) depends to a great extent on the customer's choices during the design phase. Depending on internal capabilities, a service provider may choose to either use their own facilities or outsource the process to a third party, as illustrated in Fig. 4.

If a service provider chooses to employ their own 3D printing facilities, the actual manufacturing process can either be carried out by the service provider's staff, or in the case of a company that offers customer self-service facilities, by the customer themselves (under the supervision of staff). Alternatively, a service provider can also choose to outsource their manufacturing altogether, either through fixed agreements with a B2B 3D printing enterprise or by crowdsourcing the entire process. In nearly all identified cases, the 3D printing service providers exhibited a high degree of transparency about their internal production capabilities, specifically mentioning for example restrictions

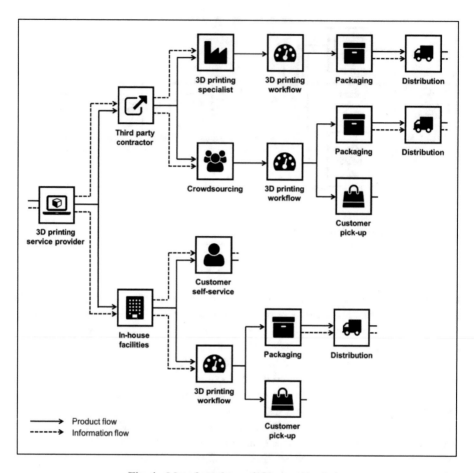

Fig. 4. Manufacturing activities supply chain

in terms of object size, material choice or 3D printing process choice, but only a handful of service providers chose to reveal whether they produce internally or through a third party. In addition to the aforementioned specifications, multiple service providers also chose to provide an online overview of the 3D printers that they own. A limited number of service providers chose to not reveal their limitations, instead informing customers that they will find a suitable contractor who can fulfil the customer's vision.

The final stage of the supply chain involves the packaging and shipment of the final product. Packaging for 3D printing goods tends to be tailored to their size and material of the end-product, with certain materials like ceramic porcelain requiring a higher degree of care. Distribution is then generally carried out by third party shipping services providers, such as DHL, TNT, Hermes or regional postal services. All of the identified companies use such services exclusively, with the exception of platforms that allow the customer to pick up their order at a retail location of the provider, service providers that employ elements of customer self-service and platforms like *3D Hubs* (3dhubs.com) that let 3D printing contractors handle shipping or pickup agreements.

While not covered by the supply chain model, it is important to note that a large part of service providers chose to not only offer 3D printing services, but instead also provide additional 3D printing-related products or services for their customers. Identified auxiliary services included, but were not restricted to, consulting services (25% of service providers), workshops and training seminars (18%), 3D printing equipment repair or rental (9%), as well as the sale of 3D printers (23%), accessories and components (17%) and filament or other manufacturing materials (21%).

4 A Framework for 3D Printing Service Supply Chain Configurations

The findings presented in the previous sections reveal that the 3D printing services are not as simple and homogeneous as the term would suggest, but instead mix and match a large number of modules with the scope of enabling consumers to turn their ideas and projects into tangible unique products. All of the 105 identified service providers opted for a bespoke combination of the previously discussed categories and subgroups. As illustrated in Fig. 5, the modular nature of the 3D printing service provider supply chain allows companies to set up their operations as lean or as complex as desired, and as specialized or diversified as required. Companies operating in this field can thus choose to focus on a very specific application, as seen in the example of the upload-and-print platform *MeltWerk* (meltwerk.com) that only offered SLS printing in white polyamide at the time of writing, or can opt to become 'all-rounders' for 3D printing services and products, as in the case of companies like *MrMake* (mrmake.de) that can offer nearly all subgroups of services. While individual service branches can be added and removed depending on demand and profitability, most companies nonetheless choose to focus on specific types of services or applications for their core business.

Advances in 3D printing technology will almost certainly lead to the development of additional service models and supply chain configurations. The recyclability of certain materials, coupled with the ability to produce at the point of consumption and the opportunity to maximize printer build chamber utilization across entire networks of manufacturers could, for example, support the development of sustainable circular service models. In such models, customers could choose to exchange old 3D printed products for new printed goods, paying only for the manufacturing process, and additional material (if needed). In this way waste and manufacturing costs could be significantly reduced in the process. Implemented on a regional, national or even global scale, such models could even serve as the 'tipping point' that enables societies to move towards a true circular economy [7].

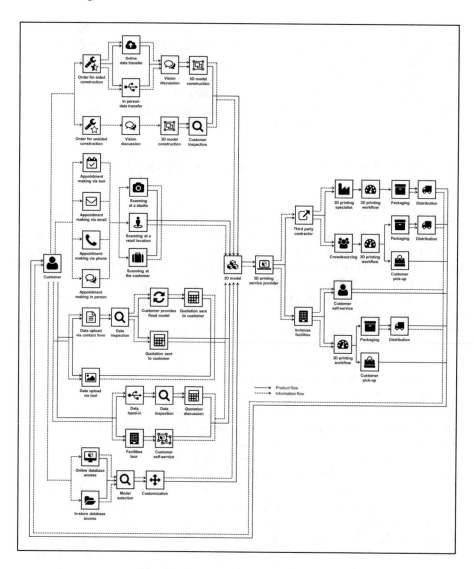

Fig. 5. 3D printing service supply chain

5   Conclusion

This paper has made a contribution to research on business and industry applications of
3D printing by being one of the first projects to investigate and categorize 3D printing
services and their unique supply chain configurations. It has built on previous studies
by the authors by mapping the service supply chains for three different types of services.
This resulted in the development of a framework for 3D printing service supply chains.
Further research is now warranted to test this framework, especially as the market

matures, user acceptability increases and the corresponding business models become clearer and more stable.

References

1. Hopkinson, N., Hague, R., Dickens, P. (eds.): Rapid Manufacturing: An Industrial Revolution for the Digital Age. Wiley, New York (2006)
2. Lipson, H., Kurman, M.: Fabricated: The New World of 3D Printing. Wiley, Indianapolis (2013)
3. Gibson, I., Rosen, D.W., Stucker, B.: Additive Manufacturing Technologies: 3D Printing, Rapid Prototyping, and Direct Digital Manufacturing. Springer, New York (2015)
4. Rayna, T., Striukova, L.: From rapid prototyping to home fabrication: how 3D printing is changing business model innovation. Technol. Forecast. Soc. Change **102**, 214–224 (2016)
5. Gebler, M., Uiterkamp, A.J.M.S., Visser, C.: A global sustainability perspective on 3D printing technologies. Energy Policy **74**, 158–167 (2014)
6. Rayna, T., Striukova, L., Darlington, J.: Co-creation and user innovation: The role of online 3D printing platforms. J. Eng. Tech. Manage. **37**, 90–102 (2015)
7. Despeisse, M., Baumers, M., Brown, P., Charnley, F., Ford, S.J., Garmulewicz, A., Knowles, S., Minshall, T.H.W., Mortara, L., Reed-Tsochas, F.P., Rowley, J.: Unlocking value for a circular economy through 3D printing: a research agenda. Technol. Forecast. Soc. Change **115**, 75–84 (2017)
8. Berman, B.: 3-D printing: The new industrial revolution. Bus. Horiz. **55**, 155–162 (2012)
9. PwC: 3D Printing comes of age in US industrial manufacturing (2016)
10. Rogers, H., Baricz, N., Pawar, K.S.: 3D printing services: classification, supply chain implications and research agenda. Int. J. Phys. Distrib. Logistics Manage. **46**, 886–907 (2016)
11. Rogers, H., Baricz, N., Pawar, K.S.: 3D printing services and their impact on supply chain configurations. In: The Proceedings of 21st International Symposium on Logistics, Kaohsiung, Taiwan, pp. 37–42 (2016)

On Reconciling Sustainable Plants and Networks Design for By-Products Management in the Meat Industry

R. Accorsi[1(✉)], R. Manzini[1], G. Baruffaldi[1,2], and M. Bortolini[1]

[1] Department of Industrial Engineering,
Alma Mater Studiorum – University of Bologna,
Viale Risorgimento 2, 40136 Bologna, Italy
riccardo.accorsi2@unibo.it
[2] Department of Management and Engineering, University of Padua,
Stradella S. Nicola, 3, 36100 Vicenza, Italy

Abstract. Population growth and rising per capita consumption of meat is growing and is expected to further accelerate in future. The production of beef is undoubtedly an high environmental stressor due to land-use change, water and energy consumption and by-products production. This paper focuses on the distribution and transportation processes of the beef slaughtering's by-products throughout their proper valorization chains. A methodology, inspired to the LCA, and encompassing data collection, simulation, and multi-scenario analysis is proposed and illustrated. This is applied to a real-world case study from the meat industry to showcase the importance of reconciling plant and network design to address both economic and environmental sustainability.

Keywords: By-products · Waste · Reverse logistics · Closed-loop · Meat industry

1 Introduction

Population growth and rising per capita consumption of animal products will double global food demand by 2050 [1]. While in western countries, meat demand has been roughly constant in the last 50 years (e.g., per capita consumption of 75 kg per year in US, 60 kg in Germany, and 45 kg in the other European countries), it is doubled in China (up to 40 kg per capita a year) following a trend expected to further accelerate in future [2]. The current global meat demand is about 106 million tons of poultry, 115 million tons of pork, and about 70 million tons of beef [3].

Among these, the production of beef is undoubtedly the highest environmental stressor. Different concerns contribute to the environmental impact associated to this industry. The land use change from forest to fields for pasturing reduces the CO_2 sink potential of the planet. The water and energy consumed for the primary feeding of cows as well as for the crops devoted to the production of feeds and nourishments (e.g., soy) significantly affect the environment and the natural resources, notwithstanding with the ethical concerns that this implies [4]. Indeed, the meat production already involves up

© Springer International Publishing AG 2017
G. Campana et al. (eds.), *Sustainable Design and Manufacturing 2017*, Smart Innovation, Systems and Technologies 68, DOI 10.1007/978-3-319-57078-5_64

to 30% of lands, and indirectly the 70% of croplands. The extant literature esteems that a kilogram of beef requires about 15 thousands liters of water over the livestock's lifecycle, whereas a kilogram of wheat needs just 1 thousand liters [5]. Lastly, the beef processing results in large flows of waste and by-products [6].

The yield of the slaughtering process is indeed about 65% for female and around 68% for male. This is the ratio between the weight of the cattle *after* and *before* slaughtering. The so-called *weight after* does not include the head, the extremities, the intestines, the heart sack, the skin and offal, the liver and the spleen. During the cutting processes, the remains are even deprived of bones, superficial fats, tendons, so that the fraction of meat to be sold is around half of the initial 450 kg per cow (See Fig. 1). Since a medium-size plant slaughters thousands of animals per day, the resulting flows of by-products are huge. Whether these (e.g., blood, skin, tendons, faeces) are not properly handled through energy valorization or transformation, might significantly pollute water, air and soil.

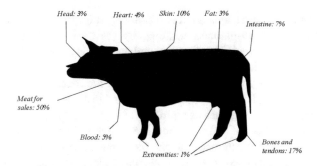

Fig. 1. Fractions of cattle after slaughtering

These concerns are far to be addressed given the intensification of beef production and consumption.

This paper analyzes the management of by-products in the beef industry, with the purpose of leading the practitioners toward the design of sustainable production and distribution systems, technologies and processes. This paper focuses on the distribution and transportation processes not for the primary products [7], but for collecting and delivering the flows of by-products throughout the proper valorization chains [8]. The meat producers are indeed responsible for the management of the by-products resulting by slaughtering. Nevertheless, most of the European slaughterhouses still do not implement procedures for the management of by-products, which are often disposed as organic fraction in the urban waste treatment system or directly landfilled. In this paper we will focus on the procedures necessary to comply with the Italian Rule *D.lgs.vo 152/2006* (i.e., Rule on waste management for environment care) and the European Rule *CE n. 1069/2009* (i.e., Rule on livestock by-products not devoted to human food), by highlighting the role of production and distribution best-practices in addressing the economic and environmental sustainability of by-products management.

A renowned and leading Italian company of the beef industry represents the case study for the application of the proposed methodology. This company sends the flows of the slaughtering by-products to anaerobic digesters for biogas generation or to plants devoted to production of compost. The proper planning of the collection, transport and delivery operations need to couple with the design of plants for by-products storage and management. The proposed methodology reconciles by-product generation, storage and distribution processes, assesses their economic and environmental performances, and suggests plants and network re-design toward sustainability.

The remainder of the paper is organized as follows. Section 2 presents the methodology of analysis and introduces the entities, flows and the operations of the observed case study. Section 3 illustrates and discusses the obtained results, while Sect. 4 draws the conclusions and presents suggestions for future researches.

2 Methodology

This paper studies the operations and the technologies for the processing, the storage and the distribution of by-products resulting by the beef slaughtering. In order to provide suggestions and guidelines for plant and operations re-design toward economic and environmental sustainability we built upon the well-known Life Cycle Assessment (LCA) methodology by considering specifically the operations throughout the by-products life cycle. The LCA is a methodology that assesses the environmental impacts associated with a product or a process over its life cycle (i.e., *from cradle to grave*) [9]. The adoption of LCA methodology aims to enhance the awareness of industry managers and practitioners about the impacts associated to the observed systems and is widely applied to study food supply chains and food systems [10]. In this paper we include both environmental and economic KPIs in the assessment of the operations for the by-products management.

The first step (i) of the proposed study is the data-collection, which cover the re-owned phases of the Life Cycle Inventory (LCI). The life cycle inventory (LCI) consists on the set of activities including the search, the collection, and inter-pretation of the amount of data necessary for the assessment the observed system (see an example in [11]). This analysis compels a deep and comprehensive inventory associated to the type and properties of the generated by-products, the characteristics (e.g., throughput, capacity) of the plants for beef processing at the slaughterhouse, the sets of primary (i.e., for beef production), secondary (i.e., for by-product collection) and tertiary (i.e., for by-product valorization) nodes involved in the observed network, the type of vehicle used to collect an delivery the by-products. This step results in building a database to aid data-driven analysis, process simulation and optimization.

The second step (ii) deals with the analysis of the flows and operations experienced by each category of the produced by-products throughout their lifecycle from the cattle slaughtering to the disposal or valorization. This results in drawing a set of flow-charts (or block-charts) that describe the connections and the interdependencies between inputs and outputs of the plants, the nodes, and the material flows.

Then, the process's block-charts defined at the (ii) are written as methods of a software applications (iii), which embeds the connection with the database of step i, a geographic information systems (GIS) required for the analysis of the transport and distribution operations, and a simulation tool that quantifies the economic and environmental KPI resulting by a scenario allows what-if multi-scenario analysis. The adopted language for the development of the software application is C#.NET.

The fourth step (iv), consists on applying the developed simulation tool to study the by-products' production and distribution systems under analysis. Two simulation analyses have been carried out. The first, is to analyze and benchmark the *as-is* production and distribution processes according to the collected life cycle inventory. The second simulates the impact of a plant re-design (i.e., *to-be*) on the costs and the environmental impacts associated to the transport phases, by providing suggestions and guidelines for plant improvements toward sustainability targets.

This methodology is applied to a real-world case study provided by a leading Italian beef industry. In particular, the case study deals with the network of slaughterhouses and the partners involved for the collection, management and valorization of the cattle processing's by-products.

3 Case Study

This case study is provided by an Italian leading company of the beef industry and is aimed to study the operations necessary for the by-products management and identify guidelines to improve their sustainability. As result of the slaughtering process (see Fig. 2), the plant generates around 56,000 tons of zoo-technical effluents per year, mostly composed by rumen, mud and blood. These by-products are currently sent to energy valorization (i.e., Heat 7.2 e6 kWh/year; Power 7.5 e6 kWh/year). The thermal power is used to heat the plant's sanitary water and for the drying station of the digestate. The remain flows that exceeds the digestor's capacity, as well as the dried fraction, are send to other plants for the proper storage, management and valorization of these special industrial wastes.

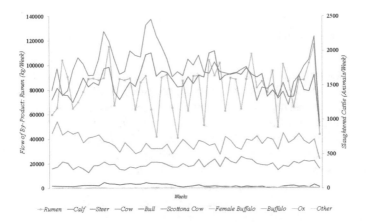

Fig. 2. Weekly throughputs of the slaughterhouse.

The management of the by-products flows and the associated collection chain throughout the partners' network is not the core business of the company. Nevertheless, given the large volumes of waste to be handled, these operations represent a relevant source of cost that should be properly addressed and managed.

The company is not aware about the impacts associated to the phases of by-products distribution and transport. Therefore, the illustrated methodology has been applied to map these phases, quantify their costs and environmental impacts, and aid the design of improved technologies and operations. The results from the LCI of *step i* regarding the network nodes, the used transport modes, and the by-products categories are summarized as follows in Tables 1 and 2.

Table 1. By-products description and adopted transport modes

By-product	Description
Rumen	The rumen is the group of organs of the animal's pre-stomach. Its content is mostly made by the digested semi-solid fraction
Hay and faeces	This flow is represented by the hay ans straw used to transport the cattle from the farms to the slaughterehouse. It is used to absorb faeses and urine and is completed re-placed at each trip
Mud	This is the semi-solid fraction obtained by the plant's sewage disposal. The solid part is about 8% of the total flow, and is thereby transported via tanker
Dried fraction	These flows result respectively by the drying and dehydrating processes carried out on the digestate fraction coming from the anaerobic digestor
Dehydrated fraction	These flows are devoted to the compost production
Compost	This results from the decomposition an humidification of organic waste and is obtained through a biologic anaerobic process. It is used as fertizeler for forms and crops

Vehicle name	Load capacity (kg)	Gross tonnage (kg)	Volume capacity (m^3)
Tanker	32000	47000	
Truck	32000	47000	35
Roll-off truck	19000	44000	22.5
Articulated lorry	33000	64000	44.5

Table 1 summarizes the main characteristics of the by-products involved in the analysis. These include rumen, hay and faeces, mud, dried and dehydrated fractions and compost. The load capacity of the adopted transport modes are reported in Table 1. In Table 2 the nodes of the network are listed and classified per type and role. The nodes are distinguished in slaughterhouses, plants devoted to the collection, consolidation and treatment of waste (i.e., waste management plants), anaerobic digestors, plants for compost production, and users of by-products and secondary products as agricultural holdings, farms, and fertilizer vendors.

The *step ii* of the methodology has been simplified in Fig. 3, where the flows of by-products across the network are depicted.

Table 2. Nodes of the network

Node Code	Type	NodeName	City
1	Slaughterhouse	INALCA S.p.a.	Castelvetro (MO)
2	Slaughterhouse	INALCA S.p.a.	Ospedaletto (LO)
3	Waste Plant	SARA S.r.l.	Nonantola (MO)
4	Agricultural Holding	Az. Agricola Buzzini	San Rocco al Porto (LO)
5	Slaughterhouse	Realbeef S.r.l.	Flumeri (AV)
6	Slaughterhouse	INALCA S.p.a.	Rieti (RI)
7	Fertilizer Plant	Letamaia Realbeef	Sant Angelo dei lombardi (AV)
8	Compost Plant	Elialpi	San Giorgio di lomellina
9	Compost Plant	Var	Belgioioso
10	Anaerobic Digestor	Caviro (Enomondo)	Faenza (RA)
11	Compost Plant	Agriflor S.r.l.	Perugia (PG)
12	By-product Client	Aguzzi Maria	Modena (MO)
13	Agricultural Holding	Az. Agr. Il Barchessino	Guastalla (RE)
14	By-product Client	CF S.r.l.	Sassuolo (MO)
15	By-product Client	Guasina Franco	Sant'Agata (BO)
16	Fertilizer Vendor	Fomet SpA	San Pietro di Morubio (VR)
17	Agricultural Holding	Soc. Agr. Corticella Srl	Spilamberto (MO)
18	Agricultural Holding	Az. Agr. Mioli	Sant'Agata (BO)
19	Agricultural Holding	Soc. Agr. Palazzone	Quartiere (FE)
20	By-product Client	Prandi e Ferraboschi	Novellara (RE)
21	By-product Client	Trombi	Guastalla (RE)
22	By-product Client	Campana Sergio	Campogalliano (MO)
23	By-product Client	Zoboli Omer	Nonantola (MO)
24	Fertilizer Vendor	Organazoto Fertilizzanti S.p.a.	Ponte a Egola (PI)
25	Fertilizer Vendor	ILSA S.p.a.	Arzignano (VI)
26	Agricultural Holding	Soc. Agr. Cavassi e Massari S.S.	Castel Bolognese (RA)
27	By-product Client	Borghi Norberto	Ravarino (MO)
28	Fertilizer Vendor	Antea Culture S.r.l.	Sassuolo (MO)

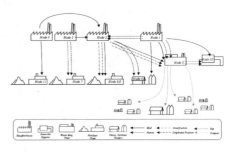

Fig. 3. Flows of the by-products throughout the network

The historical processing profiles at the slaughterhouse represented in Fig. 2 couple with the fractions generated by the each animal (see Fig. 1) to quantify the transport requirements for the by-products treatment (i.e., number of trucks and set of routes per day). These inputs fuel the software applications developed at the *step iii* and enable the simulation, the assessment and the comparison of different production-distribution scenarios, whose results are illustrated in Sect. 4.

4   Analysis and Results

The simulated scenarios reproduce the collection and delivery flows of by-products across the network illustrated in Fig. 3 over 22 months, within the period 1/1/2014 and 10/13/2015. This profile generates 2866 shipments which account 320,660 km and 63,075 tons of by-product. The developed software enables to show these flows as well as the network's connectivity on a map as in Fig. 4.

Figures 5 and 6 result by the simulation of the *as-is* scenario and quantify respectively the number of shipments carried out for the management of each type of by-product, and the average volume utilization (i.e., more relevant and constraining then the weight utilization) of the vehicles used for the transport phases. These figures showcase how the treatment of mud causes the highest number of shipments. Given the average saturation of the shipments, there is still room for further optimization of the transport operations. Particularly, the semi-liquid state of mud requires the use of tankers for its distribution. While the volume utilization of the tanker is 100%, the weight saturation is far from this target due to the low specific weight of mud and its high water content.

The second analysis simulates the impact of re-design the plant for the treatment of the muds at the slaughterhouse, on the resulting by-product distribution flows. Specifically, this scenario simulates the presence of a new belt press necessary for the

Fig. 4. By-product network connectivity.

pre-processing and the preparation of the mud before the shipment. The new plant compels a process re-design in two stages:

1. *Storage and Stabilization.* Two types of mud are treated. The primary results from the floating process, and the secondary made by the sewage collected within an oxidation tank. Both muds are stored in a basin for around 6 days and are continuously blended. Inside this basin the mud experiences an anaerobic digestion, whereas the blending boosts the stabilization process.
2. *Mechanical Dehydratation.* Once the mud is stable (i.e., no risk of rot), it is processed through a belt press in order to enhance the dry fraction from 5–6% up to 20–22%. The obtained dry fraction is palabile and can be shipped by roll-off trucks instead of tankers.

The developed simulation focuses on the slaughterhouse represented by Node 1 (see Table 2 and Fig. 3) and quantifies the resulting environmental and economic performances associated to the distribution of muds by roll-of trucks. These results refers to an horizon of 8 months and are summarized in Table 3.

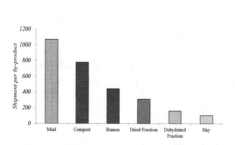

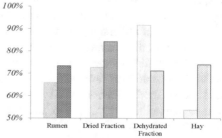

Fig. 5. Number of shipments per by-product

Fig. 6. Average utilization per transport mode: ▪ roll-off truck ▪ articulated Lorry

Table 3. Simulation scenarios: before vs. after plant re-design.

Scenarios	As-Is	To-Be
	Without belt press	With belt press
Trips	362	246
Travelling (km)	48926	32528
Carried load (tons)	4904	2928
GHGs emission (kg)		
CO2eq	40526	26169
CO	3.29	2.15
HC	0.67	0.44
NOx	147.43	95.55
N_2O	1.11	1.71
CH_4	0.03	0.04
PM	0.6	0.93
NH_3	0.24	0.16
SO_2	0.27	0.17
CO_2	40014	25836
Transport costs (€)	68,492.00 €	56,286.00 €

The obtained economic and environmental savings from the to-be scenario showcase the importance of reconciling the plant and processes design with the distribution operations in order to achieve sustainability targets.

5 Conclusion

This paper analyzes the management of by-products in the beef industry and leads the practitioners toward the design of sustainable production and distribution systems, technologies and processes. This paper focuses on the distribution and transportation processes necessary to collect and deliver the by-products throughout their proper valorization chains. A methodology, including data collection and the development a software application that aids multi-scenario simulation analysis is illustrated. This methodology is applied to a real-world case study provided by a leading Italian beef industry. The obtained results and, particularly the treatment of the muds resulting by the slaughtering process, demonstrates how the adequate design of the processing plant significantly affects the economic and environmental sustainability of the distribution and transport operations.

Acknowledgements. This research is part of the S.O.FI.A. Project that has received funding from the M.I.U.R under the Italian CLUSTER Funding Programme and Grant agreement code CL.A.N Agrifood CTN01_00230.

References

1. Koning, N., Van Ittersum, M.: Will the world have enough to eat? Curr. Opin. Environ. Sustain. **1**, 77–82 (2009)
2. Allievi, F., Vinnari, M., Luukkanen, J.: Meat consumption and production – analysis of efficiency, sufficiency and consistency of global trends. J. Clean. Prod. **92**, 142–151 (2015)
3. Food And Agriculture Organization of the United Nations (FAO): FAOSTAT, food balance sheets (2015). http://faostat3.fao.org/home/index.html
4. Elferink, E.V., Nonhebel, S., Moll, H.C.: Feeding livestock food residue and the consequences for the environmental impact of meat. J. Clean. Prod. **16**(12), 1227–1233 (2008)
5. Virtanen, Y., Kurppa, S., Saarinen, M., Katajajuuri, J.-M., Usva, K., Mäenpää, I., Mäkelä, J., Grönroos, J., Nissinen, A.: Carbon footprint of food - approaches from national input-output statistics and a LCA of a food portion. J. Clean. Prod. **19**(16), 1849–1856 (2011)
6. Bustillo-Lecompte, C.F., Mehrvar, M.: Treatment of actual slaughterhouse wastewater by combined anaerobic–aerobic processes for biogas generation and removal of organics and nutrients: an optimization study towards a cleaner production in the meat processing industry. J. Clean. Prod. **141**(10), 278–289 (2017)
7. Soysal, M., Bloemhof, J.M., Van der Vorst, J.G.A.J.: Modeling food logistics networks with emission considerations: the case of an international beef supply chain. Int. J. Prod. Econ. **152**, 57–70 (2014)
8. Okoro, O.V., Sun, Z., Birch, J.: Meat processing waste as a potential feedstock for biochemicals and biofuels – a review of possible conversion technologies. J. Clean. Prod. **142**(4), 1583–1608 (2017)
9. Accorsi, R., Manzini, R., Mora, C., Cascini, A., Penazzi, S., Pini, C., Pilati, F.: Life cycle modelling for sustainable food supply chain. In: Proceedings of 22nd International Conference on Production Research (ICPR 2013), pp. 1–5 (2013). (Scopus Code: 2-s2.0-84929359685)
10. Accorsi, R., Cascini, A., Ferrari, E., Manzini, R., Pareschi, A., Versari, L.: Life cycle assessment of an extra-virgin olive oil supply chain. In: Proceedings of 18th Summer School Francesco Turco, pp. 172–178, 11–13 September 2013. (Scopus Code: 2-s2.0-84982946700)
11. Accorsi, R., Versari, L., Manzini, R.: Glass vs. plastic: life cycle assessment of extra-virgin olive oil bottles across global supply chains. Sustainability **7**(3), 2818–2840 (2015)

Design of an Innovative Plant
for the Wastewater Recovery and Purification
in the Food & Beverage Industry

Marco Bortolini[✉], Mauro Gamberi, Francesco Pilati,
Alberto Regattieri, and Riccardo Accorsi

Department of Industrial Engineering,
Alma Mater Studiorum - University of Bologna, Bologna, Italy
{marco.bortolini3, mauro.gamberi, francesco.pilati3,
alberto.regattieri, riccardo.accorsi2}@unibo.it

Abstract. The food & beverage (F&B) industry is among the most water intensive sectors with thousands of litres per hour of raw water requirement. Starting from the statement of this issue, an overview of the evidences from the field and a quick survey of the existing technologies for the raw water saving through its local collection and treatment before discharge, this paper investigates the design of an innovative industrial plant for the water closed-loop recovery, purification and local reuse. Actually, a prototype of such a plant is working within a mid-size F&B company operating in the Emilia-Romagna region, Italy. The plant nominal capacity is of about 45,000 l/h of discharged wastewater. It integrates water ultra-filtration and reverse osmosis technologies. Details of the functional module design and of the logic of control are in the present paper. Finally, few preliminary evidences from the plant field-test are provided.

Keywords: Water saving · Food & beverage industry · Water purification · Local closed-loop water chain · Sustainable plant design

1 Introduction

Water is the key of life and its availability is crucial for the equal growth of communities [1]. Human activities require millions of litres of pure water per year. At a global scale, the most of the water use occurs in the agricultural sector even if high water volumes are necessary, also, in industry [2]. Focusing on the European Union (EU) area, the highest amount of water consumption is from industrial production plants. In the last decades, many investments are on water technologies. As example, every year around $150 billion are spent, worldwide, on wastewater treatment (industrial, residential and agricultural fields) and this figure is set to exceed $240 billion by 2020. In particular, in the wastewater treatment, $12 billion are spent every year for equipment for wastewater treatment, with an expected annual growth of 6% by 2020 and $4.2 billion are spent on membrane systems (+12% by 2020) [3]. Moreover, toward water saving are moving all the new regulations and laws at any level, from EU, to the national, regional and local Authorities.

© Springer International Publishing AG 2017
G. Campana et al. (eds.), *Sustainable Design and Manufacturing 2017*, Smart Innovation,
Systems and Technologies 68, DOI 10.1007/978-3-319-57078-5_65

Among all the industrial activities, Food & Beverage (F&B) is a very water intensive sector. Both foodstuff and beverage production and related auxiliary activities, e.g. plant washing, sanitation activities, steam generation, etc., require large amounts of pure water and, at the same time, generate thousands of litres of wastewater per hour. Consequently, improvements and innovations in such a sector, leading to the reduction of the required primary water, save thousands of litres of such a crucial resource contributing to a relevant reduction of the environmental impact of the F&B processes. Possible interventions for the conservation of the water resources belong to two major groups. The former deals with innovations to reduce the amount of the required primary water, the latter focuses on actions to recover wastewater and to treat it so that it becomes reusable within the creation of virtuous closed-loops.

The F&B sector is critic, also, because of water is subject to very high quality standards even in the case such a resource is used, simply, for the plant and machinery washing. The actual standards specify that the process water must always equals the quality of the drinkable water. Finally, the costs connected to water management in the F&B sector are out of being negligible. The main drivers include outcomes for water collection and/or purchase, the investments and costs for water treatment to guarantee the expected quality, the investments and costs for wastewater discharge and/or post-treatment and regeneration, etc.

Starting from this background, this paper focuses on the wastewater recovery and local treatment presenting the design of an innovative plant for the water collection, local purification and reuse leading to a significant saving of the extracted pure raw water. This system integrates a series of functional modules targeted to separate the typical pollutants from water, e.g. suspended and dissolved solids, bacterial charge, etc. A prototypal system, with a nominal capacity of about 45,000 l/h of input water, is described and actually works within a mid-size F&B company operating in the Emilia-Romagna region, Italy.

According to the paper goals, the reminder sections are the followings: Sect. 2 presents the state of the art in the field of water recovery and local reuse, Sect. 3 details the plant structure, size and target productivity, while Sect. 4 describes each plant functional module. Preliminary evidences from the plant run are in Sect. 5 before drawing the paper conclusions in the last Sect. 6.

2 Background

The overall purpose to locally recover, purify and reuse process wastewater in the F&B industry is forced by the following boundaries defining the area of influence and action of the adopted technologies. Particularly, the following two elements strongly influence the effectiveness and applicability of each water treatment plant:

– Existing regulations and restrictions on the water and wastewater collection, treatment and reuse within the F&B sector;
– Available technologies for water purification, their fit with the F&B process features and parameters, e.g. water flowrate, pollutants, level of automation and maintenance, etc.

The following sub-sections provide some details about such elements focusing, mostly, on the EU and Italian contexts because of the installation area of the prototypal plant under investigation.

2.1 Reference Regulations on Water and Wastewater Use

At the EU level, high attention is paid by the EU Commission and the central communitarian institutions on the quality and quantity of water to use. This is particularly true for the most water intensive sectors, as F&B.

The Regulation (Ec) No. 178/2002 of the European Parliament and of the Council provides the general principles and requirements of food law. It further establishes the European Food Safety Authority as the institution demanded to promote, apply and control the complex set of procedures in matters of food safety. The aim is to assure a high level of protection of the human health and to strength the consumer interest in relation to food, taking into account the diversity in the supply of food including traditional products.

Despite this regulation provides an ordered framework of the F&B sector at the EU level, previous EU directives set restrictions and references to the quality of drinkable water. As example, the Council Directive 98/83/EC on the quality of water intended for human consumption, known as the *'Drinking Water Directive'*, clearly states the general obligation of protecting human health from the adverse effects of any contamination of water intended for human consumption by ensuring that it is wholesome and clean. It further sets quality standards to achieve and forces the member states to follow the principles of planning, regulating, monitoring, informing and reporting toward the F&B stakeholders about the quality level of water for the human consumption they use.

Finally, starting from the late 2015 Eurostat data stating that up to 40% of the EU water extractions goes to industry and that a same percentage of the industry wastewater is not treated at all before discharge, the EU is forcing the member states to decrease their water footprint. This is particularly true within the F&B sector that presents an overall water footprint among the highest in industry together with textiles, papers, oil and basic metals.

At the Italian level, the following laws and decrees define the milestones and roadmap toward water quality and quantity control and saving:

- Legge 36/94 'Disposizioni in materia di risorse idriche' (Legge Galli) setting the concepts of water saving, recovery and reuse;
- D. Lgs. 152/99 'Disposizioni sulla tutela delle acque dall'inquinamento' (Testo unico sulle acque) transfering to regions the responsibility of setting rules for water saving, control and reuse;
- GAB/DEC/93/06 'Norme tecniche per il riutilizzo delle acque reflue' prohibiting the use of recovered and purified wastewater within F&B and pharmaceutical industries except in the case of a local recover, i.e. water collection and treatment inside the industry perimeter through local closed-loop chain;

– D. Lgs. 31/2001 'Attuazione della Dir. 98/83 del Consiglio Europeo relativa alla qualità delle acque destinate al consumo umano' actuating the EU Council Directive 98/83/EC and regulating the water quality control and responsibilities of industry and the control Authority.

The proposed wastewater recovery plant matches the regulations in force locally recovering and reusing the process water through the adoption of the purification technologies shortly revised in the following sub-section.

2.2 Water Purification Technologies

The water purification technologies aim at separating specific pollutant categories from the water flow so that the final water quality matches the standard forced by regulations and the process technical features.

Within the F&B industry, the key pollutant categories belong to suspended and dissolved solids (e.g. silica, salt ions, fungi, silt, rust, floc), microbiological contamination (e.g. bacterial colonies, pyrogenic contamination), minerals, heavy metals (e.g. lead, arsenic, cadmium, selenium, chromium) and dissolved gases. Among them, the first and second categories are strongly relevant in F&B wastewater, while the others are less critic because of initial raw water purification and very low contamination during the process.

A further element to consider when screening the water purification technologies to apply to the F&B sector is the continuity of the process and the standard flowrate to manage. This sector belongs to the process industry, i.e. continuous flow of water to feed the process, with an impressive water consumption, e.g. 30,000 to 80,000 l/h of raw water need. Consequently, the wastewater purification technologies to look for have to work coherently to the main process. As example, no batch processes are feasible despite of having uneconomic large tanks to store the water. Similarly, the purification speed, in terms of litres of purified water per hour of work, has to fit with the range of the wastewater production.

Given such boundary conditions and with reference to the target plant described in the following, the investigated purification technologies belong to the membrane group and to ultraviolet (UV) steriliser.

According to Wilf (2008) and Macedonio et al. (2012) [4, 5] membranes purify water according to the so called 'mechanical sieve theory' even if chemical reactions and biological pollutant degradation become possible depending on the membrane materials. Membrane technologies include filtration and reverse osmosis (RO). Filtration is distinguished among micro-, ultra- and nano-filtration depending on the dimensions of the membrane pores. The target pollutants removed by such technologies are the suspended and dissolved solids until the dimensions of 1 nm. The next Table 1 proposes the evidences of a short review of recent contributions about the application of the membrane technologies within the F&B industry.

Frequently, ultra-filtration and RO are coupled to remove the most of the suspended and dissolved solids from the water flow.

Table 1. Membrane technologies, F&B application review.

Year	Authors	Membrane technology				Ref.
		Micro-filtration [0.1–0.5] μm	Ultra-filtration [0.005–0.05] μm	Nano-filtration [0.0005–0.005] μm	Reverse osmosis	
2013	Kujawski et al.				✓	[6]
2013	Sudhakaran et al.		✓	✓	✓	[7]
2014	Sorlini et al.	✓	✓	✓	✓	[8]
2015	Alkaya and Niyazi		✓		✓	[9]
2015	Mahmoud et al.		✓		✓	[10]
2015	Mohammad et al.	✓	✓		✓	[11]
2016	Ghimpusan et al.				✓	[12]
2016	Meneses and Flores		✓		✓	[13]

Finally, because of the UV radiation inactivates the most of microorganisms, standard modular UV units are widely diffuse at highly competitive cost to tackle the water microbiological contamination. The degree of inactivation depends on the UV dose and time exposure (in $\mu W \cdot s/cm^2$).

3 Plant Overview and Structure

The design of the plant to recover and purify wastewater starts from the analysis of the existing local water sources, the available flowrates and the level of the pollutants to remove. After that, the overall concept of the plant is figured out, sized and its integration within the industrial operative environment is done.

In this context, the case of a mid-size F&B company producing soft drinks, non-carbonated beverages, juices and vegetable sauces is considered. The industrial plant is located in the Emilia-Romagna region, Italy and, actually, it supplies its water need by pumping underground pure water from five wells. The average annual consumption is of about 2.4 million m^3 of water managed in an open loop, i.e. discharged after the use. The key water streams are toward the coolers of the steam heating plants, the channels feeding the production lines with fruit and vegetables and, mainly, the RO unit producing the process pure water. Therefore, the stream of wastewater, in order of relevance, are from fillers (three lines per ~30,000 l/h, continuous), RO retentate (~15,000 l/h, continuous), cleaning in processes (~4,000 l/h, discontinuous), cooling towers (~2,000 l/h, continuous) and syrup room (~1,000 l/h, discontinuous). Among them, the first two streams are the most appealing in terms of quantity and continuity of the water so that the recovery and purification plant size is based on such two sources with a global flowrate of about 45,000 l/h. Pipes to connect the other streams to the input tank are provided to increase the overall recovery rate. Concerning the input water quality, lab analyses on water samples highlight the presence of suspended and dissolved solids, e.g. conductivity 1,625 µS/cm, total suspended solids 1,280 mg/l and turbidity 134 NTU.

3.1   Plant Functional Structure

The following Fig. 1 proposes the overall block diagram of the plant functional structure where *Niagara* is the nickname of the installed recovery and treatment plant. The RO unit (1) provides 15,000 l/h of retentate while the fillers (2–4) generate 6,000, 9,000 and 15,000 l/h of wastewater, respectively. Filler wastewater flows within the active carbon filter (5) removing the peracetic acid (hazardous for the membranes) before reaching the input tank (6). Such a tank collects eventual other wastewater flows (7). Units (8) and (9) are the pre-filtration and ultra-filtration modules, removing the most of the suspended solids. Their target efficiency is of about 4,000 l/h of waste per 45,000 l/h of entry water ($\sim 91\%$). After that, a tank (10) collects the filtered water, i.e. buffer, to decouple such a module from the RO unit (11) that removes the solutes (target efficiency of about 61%). Finally, the UV unit (12) reduces the microbiological charge before the water exits the system and it is collected in the output storage tank. Finally, the cleaning in place module (13) guarantees the periodic backwash of the pre-filtration, ultra-filtration and RO membranes according to the planned maintenance schedule. During cleanings, pure water flows backward through the membranes removing the collected impurities.

The overall target plant efficiency is of about 25,000 l/h of recovered water starting from 45,000 l/h of input wastewater ($\sim 55.6\%$).

3.2   Supervising System

The supervising system allows the autonomous run of the plant managing both the up time and the periodic cleaning in place cycles. The local custom PLC acquires the signals from the field, manages the plant and collects the run data to store and communicate via Ethernet. The sensors installed on the plant control the key parameters among flowrate, temperature, pH, redox, pressure drop and the turbidity. Finally, the quantity of the treated water and the overall efficiency are real-time updated.

4   Plant Functional Units

The following list is in accordance with the numbering of Fig. 1 and provides more details about the design features of the functional modules together with some pictures.

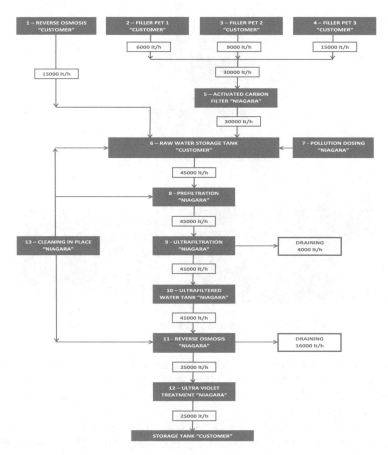

Fig. 1. Installed plant functional structure, block diagram.

1. Recovery tank from RO retentate. Flowrate 15,000 l/h. Equipped with a collection tank, pump and level control;
2. Recovery tank from filler No. 1. Flowrate 6,000 l/h. Equipped with a collection tank, pump and level control;
3. Recovery tank from filler No. 2. Flowrate 9,000 l/h. Equipped with a collection tank, pump and level control;
4. Recovery tank from filler No. 3. Flowrate 15,000 l/h. Equipped with a collection tank, pump and level control;
5. Activated carbon filter tank to remove peracetic acid. Capacity 6.65 m^3, active carbon NCL 1240, flowrate 30,000 l/h. Flow control in input and temperature, pH, and redox control in output;
6. Raw water storage tank. Capacity 30 m^3. Equipped with a pump, flowrate 45,000 l/h, and level control;
7. Other pollutant input flows. Max flowrate 2,000 l/h. Equipped with a level control;

Fig. 2. Pre-filtration modules.

Fig. 3. Ultra-filtration modules and membrane section.

8. Pre-filtration unit to pre-process the entry wastewater protecting the ultra-filtration membranes. Two units of 25,000 l/h each equipped with 150 μm self-washing membranes. Flow control in input and pressure drop control to manage self-washings (see Fig. 2);
9. Ultra-filtration unit. Parallel of eight modules. Flowrate 45,000 l/h. 75 nm PVDF membranes working both cross-flow and dead-end. Flow and turbidity control in input and output and pressure drop control to manage cleaning in place (see Fig. 3);
10. Storage tank. Capacity 8 m3. Equipped with a pump, flowrate 45,000 l/h, and level control and used for ultra-filtration backwash;
11. RO unit. Parallel of four modules. Flowrate 41,000 l/h to remove up to 15 mg/l of solutes. Flow and conductivity control in input and output, temperature control in input to protect the membranes and pressure drop control to manage cleaning in place;
12. UV unit. Flowrate 25,000 l/h. Reduction of the microbiological charge, wavelength 253 mm, 40 mJ/cm. Temperature and irradiation control.
13. Cleaning in place. Flowrate 25,000 l/h. Capacity 1.5 m^3, dosing of acid, caustic, antiscalant, sodium bisulphite, clorine. Flow, temperature and conductivity control.

5 Preliminary Field-Test

Despite a wide field-campaign to test the plant performances under different input wastewater quantity and quality is among the future activities, the evidences of a preliminary field-test after the plant installation and tuning are available to benchmark the adopted technologies against a typical wastewater produced within the considered F&B industry. The field-test goal is to purify a 0.1% sugar concentration wastewater coming from the syrup room. The water flowrate is limited to 35 m³/h due to primary plant operative contingencies so that the treatment plant works at a reduced power. The test is ∼ 2.5 h long and the total input wastewater is of about 90 m³. The ultra-filtration unit works in cross-flow mode.

Samples of the input and output water and of the process water after the ultra-filtration unit are collected and analysed by an independent lab. At this stage, field-tests validate the filtration technology so that the microbiological charge is not measured. The overall amount of recovered and purified water is of about 67 m³ with an overall efficiency of 74.4% (this value is in absence of any output pure water use for the plant cleaning in place). The following Table 2 proposes an extract of the physical and chemical properties of the entry and purified water together with a reference benchmark for EU drinkable water. The water analyses confirm the effect of the filtration units to remove a very relevant percentage of the suspended solids and solutes making the output purified water of interest for use within the F&B industrial plant, e.g. steam boilers, process uses.

Table 2. Field-test results.

	Input wastewater	Output purified water	Drinkable water limits
Total Hardness (CaCo3) [mg/l]	767	<2	150 ÷ 500
Oxidability (Kubel) [mg/l O2]	3530	2.1	5
Conductivity [µS/cm]	1625	20	2500
pH	4.1	5.9	6.5 ÷ 9.5
Redox potential [mV]	201	404	–
Total Suspended solids [mg/l]	1280	20	250
Turbidity [NTU]	134	0.6	<1
Total Alkalinity (CaCo3) [mg/l]	281	18.7	<85
BOD5 (O2) [mg/l]	1741	<5	5

6 Conclusions

This paper tackles the challenging issue of reducing the water footprint of the food & beverage (F&B) industry, known as a very water intensive sector, presenting the design and preliminary field-test of an innovative plant to recover and locally purify the process wastewater, decreasing the raw water use. The plant integrates an ultra-filtration, reverse osmosis (RO) and ultraviolet (UV) unit with a target overall efficiency of about 55.6%. The nominal capacity is of 45,000 l/h coherently to the wastewater flow of mid-size F&B industries, as the Italian company considered within the present industrial research. The preliminary field-test highlights results of interest for process reuse of the purified water, e.g. steam boilers, while the output water analyses highlights conditions within or close to the EU drinkable water limits. The next steps have to fully validate the proposed solution through a multi-scenario analysis of the plant performances and water quality varying the input conditions. Furthermore, the plant life cycle cost and life cycle assessment are of interest to quantify the global benefit of introducing such a system instead of open loop water practice.

Acknowledgments. This work is developed within the CIP Eco-innovation Project No. Eco/13/630314/LESS-WATER BEV.TECH, co-funded by the Eco-innovation Initiative of the European Union, in partnership with A DUE S.p.A. and CVAR Ltd.

References

1. United Nations (UN): The Millennium Development Goals Report 2011. Lois Jensen Edition, New York (2011).
2. United Nation (UN): The World Water Assessment Programme (2009). http://webworld. unesco.org/water/wwap/wwdr/wwdr3/pdf/WWDR3_Water_in_a_Changing_World.pdf
3. RobecoSAM: Sustainable Asset Management (2012). https://issuu.com/sam-group.com/docs
4. Wilf, M.: Membrane types and factors affecting membrane performance. National Water Research Institute, Fountain Valley, CA (2008)
5. Macedonio, F., Drioli, E., Gusev, A.A., Bardow, A., Semiat, R., Kurihara, M.: Efficient technologies for worldwide clean water supply. Chem. Eng. Process. **51**, 2–17 (2012)
6. Kujawski, W., Sobolewska, A., Jarzynka, K., Güell, C., Ferrando, M., Warczok, J.: Application of osmotic membrane distillation process in red grape juice concentration. J. Food Eng. **116**(4), 801–808 (2013)
7. Sudhakaran, S., Lattemann, S., Amy, G.L.: Appropriate drinking water treatment processes for organic micropollutants removal based on experimental and model studies—a multi-criteria analysis study. Sci. Total Environ. **442**, 478–488 (2013)
8. Sorlini, S., Gialdini, F., Collivignarelli, M.C.: Survey on full-scale drinking water treatment plants for arsenic removal in Italy. Water Pract. Technol. **9**(1), 42–51 (2014)
9. Alkaya, E., Niyazi, G.: Resources, conservation and recycling water recycling and reuse in soft drink/beverage industry: a case study for sustainable industrial water management in Turkey. Resour. Conserv. Recycl. **104**, 172–180 (2015)
10. Mahmoud, K.A., Mansoor, B., Mansour, A., Khraisheh, M.: Functional graphene nanosheets: the next generation membranes for water desalination. Desalination **356**, 208–225 (2015)
11. Mohammad, A.W., Teow, Y.H., Ang, W.L., Chung, Y.T., Oatley-Radcliffe, D.L., Hilal, N.: Nanofiltration membranes review: recent advances and future prospects. Desalination **356**, 226–254 (2015)
12. Ghimpusan, M., Nechifor, G., Nechifor, A.C., Dima, S.O., Passeri, P.: Case studies on the physical-chemical parameters' variation during three different purification approaches destined to treat wastewaters from food industry. J. Environ. Manage. (2016, in press)
13. Meneses, Y.E., Flores, R.A.: Feasibility, safety and economic implications of whey-recovered water in cleaning-in-place systems: a case study on water conservation for the dairy industry. J. Dairy Sci. **99**(5), 3396–3407 (2016)

A Methodology for the Identification of Confined Spaces in Industry

Lucia Botti$^{(\boxtimes)}$, Cristina Mora, and Emilio Ferrari

Department of Industrial Engineering, University of Bologna, Bologna, Italy
{lucia.botti5, cristina.mora, emilio.ferrari}@unibo.it

Abstract. Work in confined space is a high-risk activity posing a serious life-threatening hazard to workers who perform it. Accidents in confined spaces frequently lead to multiple fatalities. The cause of accidents and fatalities due to confined space work is related to the lack of awareness about the presence and the risks of such hazardous workplaces. This paper introduces a methodology for the identification of confined spaces in industry. The aim is to provide a useful tool for helping researchers and practitioners to recognize of confined spaces in industry. Four different characteristics of confinement are investigated: geometric features, access, internal configuration, and atmosphere and environment. The proposed methodology includes the definition of the Confined Space Risk Index (CSRI) for the analysis of the risk related to the investigated confined space. Finally, two case studies show the application of the proposed methodology to two suspected confined spaces in industry.

Keywords: Confined space · Confined space work · Risk assessment · Risk index

1 Introduction

Confined space work is a high-risk activity, posing a serious dangerous hazard to the workers. Hazards in confined spaces are difficult to evaluate and manage, due to the complex characteristics of such particular work environments [1]. Both the features of the confined area and the characteristics of the performed task have direct impact on the overall risk level of a specific confined space activity. Despite international efforts in defining consistent procedures and recommendations for safe confined space work, past and recent statistics show that fatal incidents still occur [2].

The 29 CFR 1910.146 standard of the American OSHA is widely known as the Permit-Required Confined Spaces (PRCS) Standard for confined space work in general industry [3]. Such standard provides a general definition of "confined space", together with requirements for practices and procedures to protect employees in general industry from the hazards of entry into permit-required confined spaces. The PRCS defines "Confined space" as a space that is large enough and configured that an employee can enter and perform work, has limited openings of entry or exit and is not designed for continuous occupancy [4, 5] Examples of confined spaces include silos, vessels, boilers, storage tanks, sewers and pipelines. Less common types of confined spaces are industry.

© Springer International Publishing AG 2017
G. Campana et al. (eds.), *Sustainable Design and Manufacturing 2017*, Smart Innovation, Systems and Technologies 68, DOI 10.1007/978-3-319-57078-5_66

Examples of such spaces are the interior areas of machines where operators access to perform maintenance tasks.

The PRCS Standard protects employees who enter confined spaces while engaged in general industry work. This standard has not been extended to cover employees entering confined spaces while engaged in specific industries, as construction work or confined space workers in agriculture because of unique characteristics of such worksites. Despite the numerous directions of the OSHA's standards, employers in general industry have difficulty determining if spaces are permit-required confined spaces. Several accidents and injuries related to confined space work showed that workers access to confined areas without proper training and personal protective equipment, exposing themselves to high levels of hazards [11, 12]. The lack of situation awareness is an underlying cause of human errors, especially when workers access to areas not designed for continuous occupancy as confined spaces. Rescue attempts in confined spaces are also hazardous situations, since emergency response is a low-frequency, high-risk operation. Many would-be rescuers perish while trying to rescue a victim after a confined space accident. Would-be rescuers deaths include trained fire-fighters and competent personnel who had years of experience, despite the requirements for training, planning and expertise with confined space rescue procedures. Data and statistics reveal that the 60% of confined space fatalities in U.S. occur among would-be rescuers [13]. The chain of would-be rescuer deaths is an on-going phenomenon globally challenging. The Canadian Centre for Occupational Health and Safety and the European Agency for Safety and Health at Work (EU-OSHA) state the same 60% statistic [14]. These data reveal a hidden phenomenon, i.e. both employers and workers fail to identify confined space work hazards [2, 15].

This paper introduces the structure of a tool for the identification of confined spaces in industry. The aim was to realize an effective tool to prevent workers entry into high-risk confined spaces. The tool addresses workers during the complex identification of high-risk confined spaces. Finally, the tool supports the mandatory risk assessment for confined spaces computing the risk index for the analyzed confined space and task.

2   Identifying Confined Spaces: A Challenging Task

The U.S. OSHA outlines the confined space features to help employers and employees in recognizing such hazardous workplaces. The PRCS outlines the boundary line between non-permit and permit-required confined spaces, i.e. a permit-required confined space is distinguished by the hazards present and the ability of the employer to eliminate them [16]. Particularly, a permit-required confined space contains or has potential to contain a hazardous atmosphere, contains a material that could potentially engulf a worker, has an internal configuration that could trap or asphyxiate a worker, or presents any other serious, recognized hazard. This definition has been repeated for years, mentioning the OSHA regulations and analyzing every detail. Design features of confined spaces increase risk to entrants. Such features include the physical configuration of entry and exit portals, structural weaknesses in walls and the absence of anchor points necessary for effecting emergency rescue [17, 18].

The tool for the confined space identification includes a simplified application of the PRCS definition. The characteristics of the confined space are gathered in four different categories: geometric features; access; internal configuration; atmosphere and environment.

Limited dimensions characterize confined spaces. Following the definition of the OSHA's standard 29 CFR 1910.146, a confined space is large enough and so configured that an employee can bodily enter and perform assigned work. The concept of limitation of a dimension is referred to the dimension of the human body, fully equipped to face the worst possible scenario. The minimum working area of a worker may be computed as the circumference drawn by his arm. Given the length of the arm of the 99th percentile man as equal to 800 mm [19] and an increase due to the PPE (e.g., protective gloves) of 5 mm, the minimum working area of a worker is equal to a circumference with a ray of 805 mm. Consequently, the space can be defined as "geometrically confined" if the circumference with a diameter of 1,800 mm (ray 900 mm) and center at the intersection between the transverse plane and the longitudinal axis is not completely clear and free from obstructions. Aggravating conditions of the geometric features of a confined space include the presence of hollowed areas and extensions far from the entry.

The international standards on anthropometric measures provide further useful dimensions of the human body [20–23]. Such standards allow determining the dimensions of the human body ellipse, which are 600 mm for the major axis (shoulder breadth) and 450 mm for the minor axis (body width). Consequently, the access of a space is confined if the diameter or the shortest dimension of the entry is smaller than 600 mm. The presence of a singular vertical access or the lack of protection and signal are aggravating conditions of the confined access.

Internal configuration refers to the internal characteristics of the space. The necessary condition of the confined internal configuration is that the space is not designed for continuous occupancy. The presence of material that has the potential for engulfing the entrant or residuals from previous operations increase the exposure of the worker to the risks of confined space work, aggravating the conditions of the confined space.

Finally, a space is atmospherically confined if it contains or has potential to contain a hazardous atmosphere. Specifically, the absence of a natural or artificial efficient ventilation system that ensures proper ventilation in every accessible point is the necessary condition to define the confined atmosphere of the space. Aggravating conditions of the atmospherically confined space include the characteristics of the expected operations, (e.g., hot and cold works, stock of heavy and bulky materials, and tests), high concentrations of explosive and toxic substances, and the presence of noise.

Each confinement category identifies a dimension of the confined space. Specifically, a space can be limited in four different dimensions, i.e. geometric features, access, internal configuration, and atmosphere and environment. Based on the described confined space characteristics, the following Sect. 3 shows the algorithm for the identification of confined spaces.

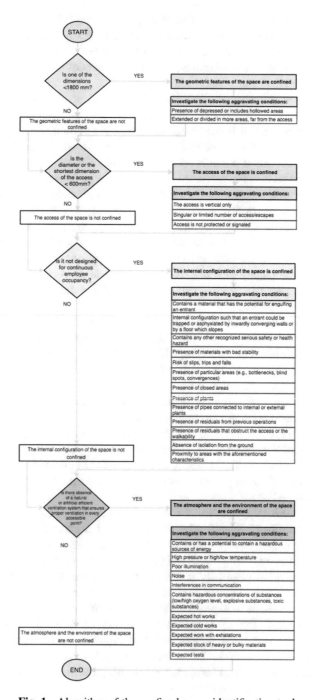

Fig. 1. Algorithm of the confined space identification tool

3   Algorithm for the Identification of a Confined Space

The aforementioned categories outline a structured representation of the characteristics of confined spaces. The presence of a confined space is confirmed when a necessary condition is verified. Therefore, the space can be confined in its geometric features, access, internal configuration, and atmosphere and environment. These categories define the structure of the algorithm for the identification of confined spaces in Fig. 1. The algorithm in Fig. 1 guides workers and practitioners through the process for the identification of confined spaces. Specifically, the answer "YES" to one of the four necessary conditions defines the presence of a confined space. The types of confinement refer to the specific categories for which the user answers "YES". In case of affirmative answer, the procedure suggests the investigation of the aggravating conditions. Based on the algorithm in Fig. 1, Fig. 2 shows a checklist including both necessary and aggravating conditions, for each category. The checklist is part of the risk assessment required by the current law. Workers and practitioners complete the checklist prior to perform the operations in the suspected confined space. A tick is assigned to each condition concerning the situation in the suspected confined space. The ticks in the four questions A1, B1, C1 and D1 on necessary conditions define the presence of the confined space. As an example, a tick in questions A1, B1 and D1 define a space with limited geometric features, limited access and atmospherically confined. The aggravating conditions contribute to the quantification of the risk index for the suspected confined space. When a necessary condition is not identified, the user skips the aggravating conditions for the corresponding category and moves to the next (e.g., if A1 does not concern the suspected confined space, skip A2 and A3, and go to condition B1). The definition of the risk index and the parameters for its calculation are in the following Sect. 4.

4   Confined Space Risk Index (CSRI)

The answers to the questions of the checklist in Fig. 2 contribute to the definition of the Confined Space Risk Index (CSRI). Specifically, a score of one is attributed to each answer with the tick. Questions without the tick have no score. The space is confined if at least one of the necessary conditions A1, B1, C1 or D1 is concerned. The aggravating conditions in each category contribute to increase the CSRI. The following Eq. (1) defines the CSRI.

$$CSRI = A1 \cdot \left[1 + \left(\frac{\sum_{i=A2}^{A3} Ai}{2}\right)\right] + B1 \cdot \left[1 + \left(\frac{\sum_{i=B2}^{B4} Bi}{3}\right)\right] +$$
$$C1 \cdot \left[1 + \left(\frac{\sum_{i=C2}^{C14} Ci}{13}\right)\right] + D1 \cdot \left[1 + \left(\frac{\sum_{i=D2}^{D12} Di}{11}\right)\right] \tag{1}$$

CRSI is a number between 0 and 8. The value of the CRSI is 0 if no necessary condition for confinement concerns the space investigated. The maximum value for the CRSI is 8 and it represents a confined space where all the necessary and aggravating conditions are verified. The value of the resulting index is then compared with the

CATEGORY	CONDITION	CONCERN
	IDENTIFICATION OF A CONFINED SPACE	
	Review each condition for the suspected confined space of interest, place a tick in the column on the right as appropriate.	
GEOMETRIC FEATURES	**A1. Geometrically limited if at least one of its dimensions <1800 mm**	
	A2. Depressed or includes hollowed areas	
	A3. Extended or divided in more areas, far from the access	
ACCESS	**B1. Diameter or the shortest dimension of the access < 600mm**	
	B2. The access is vertical only	
	B3. Singular or limited number of access/escapes	
	B4. Access is not protected or signaled	
INTERNAL CONFIGURATION	**C1. Not designed for continuous employee occupancy**	
	C2. Contains a material that has the potential for engulfing an entrant	
	C3. Internal configuration such that an entrant could be trapped or asphyxiated by inwardly converging walls or by a floor which slopes	
	C4. Contains any other recognized serious safety or health hazard	
	C5. Presence of materials with bad stability	
	C6. Risk of slips, trips and falls	
	C7. Presence of particular areas (e.g., bottlenecks, blind spots, convergences)	
	C8. Presence of closed areas	
	C9. Presence of plants	
	C10. Presence of pipes connected to internal or external plants	
	C11. Presence of residuals from previous operations	
	C12. Presence of residuals that obstruct the access or the walkability	
	C13. Absence of isolation from the ground	
	C14. Proximity to areas with the aforementioned characteristics	
ATMOSPHERE AND ENVIRONMENT	**D1. Absence of a natural or artificial efficient ventilation system that ensures proper ventilation in every accessible point**	
	D2. Contains or has a potential to contain a hazardous sources of energy	
	D3. High pressure or high/low temperature	
	D4. Poor illumination	
	D5. Noise	
	D6. Interferences in communication	
	D7. Contains hazardous concentrations of substances (low/high oxygen level, explosive substances, toxic substances)	
	D8. Expected hot works	
	D9. Expected cold works	
	D10. Expected work with exhalations	
	D11. Expected stock of heavy or bulky materials	
	D12. Expected tests	

Fig. 2. Checklist for the identification of a confined space

ranges in Table 1, which defines the corresponding risk level. Working in confined space poses or is likely to pose a risk to the safety and health of workers. In case of low risk, risk control measures as engineering controls, administrative controls and PPE should be taken. When the risk is high, workers should no enter the confined space. Tasks should be redesigned to avoid man entry and including the adoption of non-man entry technologies for work in confined spaces [24]. A possible alternative is the redesign of the workplace to eliminate the necessary conditions concerning the

Table 1. Ranges that define risk levels.

CSRI value	Risk level	Consequences
0	No risk	No confined space, no consequences
$1 \leq CSRI < 3$	Low risk	Acceptable: no significant consequences
$3 \leq CSRI < 5$	Medium risk	Improve structural risk factors or adopt risk control measures
$5 \leq CSRI \leq 8$	Significant risk	Redesign tasks and workplaces according to priorities. Avoid entry if possible.

identified confined space. The following Sect. 5 shows two practical applications of the checklist and the CSRI for two confined spaces in industry.

5 Quick Applications of the Checklist and CSRI Calculation

5.1 Case Study 1: Grain Silo

The first case study concerns the application of the proposed checklist for the risk assessment of silos for grain storage in an Italian mill. Silos have a rectangular section of 15 × 21 m. The height is 40 m. The internal surfaces have no openings, except for two manholes, which are on the top and on the lower part of the silo. The dimensions of the top manhole are 500 × 600 mm, while the lower manhole is 500 × 500 mm. Workers occasionally enter the silos for maintenance operations (e.g., unclog materials on the walls and inspect the grain). Such activities do not require specific equipment. Workers usually enter the space from the top manhole, with a shovel and a flashlight. Following the checklist in Fig. 2, the necessary conditions for confinement concerning the space are B1, C1 and D1. As a consequence, the investigated silo is a confined space as its entries, internal configuration and atmosphere are confined. Further conditions concerning the space are B3, B4, C2, C3, C5, C6, C7, C14, D4 and D11. Following Eq. (1) for the calculation of the risk index, the resulting CSRI is equal to 4.3. A medium risk level concerns the investigated confined space (Table 1) and the risk factors should be improved to reduce the risk level.

5.2 Case Study 2: Metal Tank in Filtration Plants

The manufacturing process of swimming pool filters requires workers to enter a cylindrical tank to perform welding of the metal components of the tank (e.g., top, lateral metal sheet, bottom and other small components). The tank diameter is 3 m. During welding operations, a positioning device sustains the tank with the diameter perpendicular to the floor. The width of the internal space where worker welds components is about 1.3 m, while the height is 3 m (tank diameter).

The worker enters the tank through a manhole of DN 500. All the four necessary conditions of the checklist concern the space, i.e. the tank is a confined space. Further aggravating conditions are B3, B4, C14, D2, D3, D4, D8 and D10. The resulting CSRI is 5.2, which involves significant risk for the investigated confined space. Workers should not enter the space and the redesign of the task is suggested. For example, an autonomous welding robot could enter the tank, while manual worker supervises the welding operations from the outside.

6 Conclusions

This paper has introduced a method to identify confined spaces in industry. The aim was to define an algorithm for the recognition of high risk confined spaces and prevent workers access. The analysis of fatalities due to confined space work showed that the lack of awareness about the presence and the risk of confined spaces is the main cause of accident. The proposed algorithm defines a structured framework for the recognition of workplace confinement characteristics in industry. Four categories of confinement have been defined to identify confined spaces: geometric features, access, internal configuration, and atmosphere and environment. A necessary condition and a set of aggravating conditions characterize each confinement category. The workplace concerning at least one of the proposed necessary conditions is a confined space. Based on the structure of the proposed algorithm, a checklist was developed to address workers and practitioners through the identification of confined spaces in industry. Finally, the Confined Space Risk Index (CSRI) analyses the risk of the confined space, defining the risk level. This study will be the basis for the development of an interactive tool for the identification of confined spaces in industry. The tool will have a user-friendly interface and it will accessible online from personal computers and other electronic devices (e.g., smart phones and tablets). Lastly, the CSRI will be improved including accident frequencies and the related risk factors.

Acknowledgements. The authors wish to thank Eng. Gastaldello Davide, Prof. Eng. Bragadin Marco Alvise and Prof. Eng. Berry Paolo of the School of Engineering and Architecture of the University of Bologna, Italy, for sharing their valuable documents and materials which were the basis of this research. Thanks to Eng. Bondioli Fabiano, Dr. Capozzi Maria and the "Confined Spaces Technical Group" of the *Solutions Database Project* (http://safetyengineering.din.unibo. it/en/banca-delle-soluzioni) for the technical support. The research was supported by Azienda Unità Sanitaria Locale (AUSL) of Bologna and Istituto Nazionale Assicurazione Infortuni sul Lavoro (INAIL). The authors are grateful for this support.

References

1. Nano, G., Derudi, M.: A critical analysis of techniques for the reconstruction of workers accidents. Chem. Eng. Trans. **31**, 415–420 (2014)
2. Burlet-Vienney, D., Chinniah, Y., Bahloul, A.: The need for a comprehensive approach to managing confined space entry: Summary of the literature and recommendations for next steps. J. Occup. Environ. Hyg. **11**, 485–498 (2014)
3. OSHA: Occupational safety and health standards. general environmental controls. Permit-required confined spaces. Publication No. 29 CFR 1910.146 (1993)
4. U.S. Department of Labor, Occupational Safety and Health Administration: Permit-required confined spaces. OSHA 3138-01R 2004 (2004)

5. U.S. Department of Labor: Confined spaces (2017). https://www.osha.gov/SLTC/confinedspaces/. Accessed Jan 2017
6. Taylor, B.: Confined spaces. common misconceptions & errors in complying with OSHA's standard, pp. 42–46 (2011)
7. OSHA: Safety and health regulations for construction. General safety and health provisions. Safety training and education. Publication No. 1926.21
8. OSHA: Safety and health regulations for construction. Confined spaces in construction. Authority for 1926 subpart AA. Publication No. 1926 Subpart AA (2015)
9. OSHA: Occupational safety and health standards. Special industries. Grain handling facilities. Publication No. 29 CFR 1910.272
10. OSHA: Occupational safety and health standards for shipyard employment. Confined and enclosed spaces and other dangerous atmospheres in shipyard employment. Publication No. 29 CFR 1915 Subpart B
11. Nano, G., Derudi, M.: Evaluation of workers accidents through risk analysis. Chem. Eng. Trans. **26**, 495–500 (2012)
12. Botti, L., Duraccio, V., Gnoni, M.G., Mora, C.: A framework for preventing and managing risks in confined spaces through IOT technologies. In: Proceedings of the European Safety and Reliability Conference on Safety and Reliability of Complex Engineered Systems, pp. 3209–3217 (2015)
13. NIOSH: NIOSH alert: Request for assistance in preventing occupational fatalities in confined spaces. Publication No. 86–110 (1986)
14. Muncy, C.: The sixty percent statistic. how to break the chain of would-be rescuer deaths in confined spaces. The Synergist, February–March 2013
15. Burlet-Vienney, D., Chinniah, Y., Bahloul, A., Roberge, B.: Occupational safety during interventions in confined spaces. Saf. Sci. **79**, 19–28 (2015)
16. Ye, H.: Atmosphere identifying and testing in confined space. In: 2011 First International Conference on Instrumentation, Measurement, Computer, Communication and Control, Beijing, pp. 767–771 (2011). doi:10.1109/IMCCC.2011.195
17. Wilson, M.P., Madison, H.N.: Protecting workers in industrial confined spaces. report to the los angeles district attorney and the california occupational safety and health standards board. School of Public Health, Center for Occupational and Environmental Health, University of California, Berkeley (2008)
18. Wilson, M.P., Madison, H.N., Healy, S.B.: Confined space emergency response: assessing employer and fire department practices. J. Occup. Environ. Hyg. **9**(2), 120–128 (2012)
19. Tilley, A.R., Henry Dreyfuss Associates: The Measure of Man and Woman: Human Factors in Design. Wiley, New York (2002)
20. UNI Ente Italiano di Normazione: UNI EN 547-1 Sicurezza del macchinario - misure del corpo umano - parte 1: Principi per la determinazione delle dimensioni richieste per le aperture per l'accesso di tutto il corpo nel macchinario (2009)
21. UNI Ente Italiano di Normazione: UNI EN 547-2 Sicurezza del macchinario - misure del corpo umano - parte 2: Principi per la determinazione delle dimensioni richieste per le aperture di accesso (2009)
22. UNI Ente Italiano di Normazione: UNI EN 547-3 Sicurezza del macchinario - misure del corpo umano - parte 3: Dati antropometrici (2009)
23. UNI Ente Italiano di Normazione: UNI EN ISO 7250 Misurazioni di base del corpo umano per la progettazione tecnologica (2000)
24. Botti, L., Ferrari, E., Mora, C.: Automated entry technologies for confined space work activities: a survey. J. Occup. Environ. Hyg. (2016, in press). doi:10.1080/15459624.2016.1250003

Sustainability of 3D Printing and Additive Manufacturing

Sustainable Small Batch Reproduction via Additive Manufacturing and Vacuum Casting: The Case Study of a Rhinoceros Toy Figure

Milan Sljivic[1(✉)], Ana Pavlovic[2], Jovica Ilić[1], and Mico Stanojevic[1]

[1] Faculty of Mechanical Engineering, University of Banja Luka, Bulevar Vojvode Stepe Stepanovića 71, 78 000 Banja Luka, Bosnia-Herzegovina
milan.sljivic@unibl.rs
[2] Department of Industrial Engineering, Alma Mater Studiorum University of Bologna, viale Risorgimento 2, 40136 Bologna, Italy

Abstract. This paper aims at clarifying the relationship between two unconventional manufacturing processes, the additive manufacturing and the vacuum casting, as a sustainable way in developing prototypes and small batches, even in the presence of complex geometries. The rapid reproduction of a rhinoceros toy figure was used as case study. Starting from the 3D CAD model, acquired by reverse engineering techniques, additive manufacturing and vacuum casting processes permitted to realize its replicas. Complex functional parts in small series were manufactured with high precision, accuracy and enhanced surface finish. Furthermore, significant reductions in time and costs, both for development or production comparing to other technologies were highlighted.

Keywords: Prototype · Replica · Sustainability · Additive manufacturing

1 Introduction

The more time is investing in realization of the product, the more occasions for profit are lost. This philosophy drives many industries during the development of new products. Recent advanced improvements in the area of Additive Manufacturing (AM) have been gradually permitting the use of this technology directly towards the manufacturing of products.

Additive manufacturing has a potential to provide a number of sustainability advantages. These advantages include: less waste during manufacturing; optimisation of geometries and creation of light weight components that reduce material consumption and energy consumption in use; reduction in transportation; waste reduction due to the ability to create spare parts on-demand. This technology is going to become one of the most sustainable technology in the future.

Rapid Prototyping (RP) of complex parts was the first field where the AM demonstrated all its potentialities [1]. Additive manufacturing involves a series of procedures that enable rapid fabrication of prototypes based on a 3D CAD model. Mechanical characteristics of material are studied in [2]. The great advantage of this process is the

G. Campana et al. (eds.), *Sustainable Design and Manufacturing 2017*, Smart Innovation, Systems and Technologies 68, DOI 10.1007/978-3-319-57078-5_67

timely detection of defects or quick correction of the errors during the processing of parts. Additive technologies (until recently referred to as RP technology) include forming parts layer by layer. In particular, AM represents an ideal technique to address the needs of industrial advanced realities such as aerospace, medical, orthodontics and orthotics. [3–5].

Even though some relevant investigations [6, 7] already demonstrated the advantages of using of this new technology for producing sustainable industrial products, its impact on sustainability is not yet fully explored. The adoption of additive manufacturing (AM) appears to herald a future in which value chains are shorter, smaller, more localised, more collaborative, and offer significant sustainability benefits. In brief, AM offers sustainability to products and production by:

– Improving resource efficiency: production and use phases as manufacturing processes and products can be redesigned for additive manufacturing;
– Extending product life: in repair, remanufacture and refurbishment, sustainable socio-economic patterns such as stronger person-product affinities and closer relationships between producers and consumers [8, 9];
– Reconfiguring value chains: shorter and simpler supply chains, more localized production, innovative distribution models, and new collaborations. In particular, the advantages of additive manufacturing and other similar technics, such as vacuum casting, lie in the ability to produce highly complex parts without tooling [10] and, thus, with a reduction in time and costs.

This paper aims at providing a tangible example of benefits offered by AM when coupled with other unconventional technologies of rapid tooling (e.g. vacuum casting) optimizing time and costs towards a sustainable production of small batches.

Nowadays, the quick development of new products and small lot productions represent a necessity in a large number of market segments. The more time spent on the product development, the more opportunities for profit are lost. It is this philosophy that drives modern industries during the development of new products [11]. Recognizing the need to create more prototypes of the same part, efforts have been made to develop a cost-effective system of rapid prototyping of replicas of this part. This system would allow users to check design, quality and other tests before starting the mass production and market launch.

Undoubtedly, RP technology has wide limitations [12], mainly related to the high costs of processing. These costs are generally related to the purchase, amortization and maintenance of delicate RP systems, purchase of materials and involvement of qualified personnel. In addition, one of the main drawbacks of RP technique is a limited number of materials that can be used for parts' production, considering that a specific material is used in each technique.

On the other hand, in 3D printing a wide range of materials is available, such as PA, TU and PTE filaments and the most commonly used ABS and PLA. In the present case of study, an extrusion-based 3D printer (Fused Deposition Molding - FDM) uses ABS plastic only. This material is often not the same or even similar to the one that a finished product needs to be made of. Also, mechanical and physical properties of the material generally do not coincide with the characteristics inherent of a final product [13–18].

For functional and final testing, it is of utmost importance that the physical and mechanical properties of the material used for making prototypes are the same or at least close to the characteristics of the materials used for the final product, which is a requirement that the materials used by the RP systems often do not meet.

At the stage of development and design of the product, it is often required to use a certain number of identical prototypes for testing. At this stage, therefore, it is necessary to produce a small series of identical prototypes, which by means of a 3D printer can be very slow, costly and inefficient. This is another shortcoming of RP techniques, and it also reveals the need to develop a system that allows the production of small series of identical prototypes.

The alternative is an integrated system using the 3D printer to produce a prototype of the product, which would be used in the next step as a master model for tools that would produce a series of prototypes, i.e. products made of different materials, with characteristics close to the final product.

2 Methods

The technology of vacuum casting and 3D printers based on the principle of FDM [15–17] as an integrated system of rapid prototyping and tooling shall fulfil the requirements of engineers, which perform the assessment of the functionality and characteristics of the prototype satisfying the following conditions (Fig. 1):

- The prototype should look like the final product in terms of dimensional accuracy, surface finish and colour.
- The prototype should be made of materials with the same or similar characteristics of the final product to be tested in real working conditions. Also, there should be no voids in the internal structure.
- The prototypes should be made in sufficient quantities to facilitate and accelerate the process of design optimization.

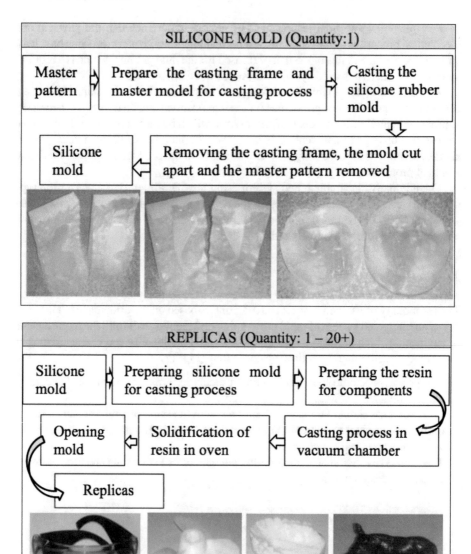

Fig. 1. An integrated system: additive manufacturing and vacuum casting

3   Results

The RP machine, which was specifically used for making master models of rhinoceros toy figures, is a 3D printer based on FDM principle (Stratasys - Dimension Elite) that plays a key role in the development of prototypes [19–23]. The 3D model of the rhinoceros figure is made using scanning step by Reverse Engineering technique. The advanced accuracy achieved by this 3D printer demands special attention to the positioning and orientation of the master model on the platform. The procedure by which this model is obtained can be described in the following steps:

- Product design in some of the CAD software packages;
- Conversion of CAD models in STL format that is recognized by a 3D printer;
- Transfer of STL files to the computer that controls the printer;
- Processing of STL files within the CatalystEX program in which all the parameters are set and adjusted according to the required model;
- Creating a three-dimensional model using additive technology;
- Further processing of created prototypes.

The layout of the rhinoceros STL model was designed in SolidWorks software package. Supplementary parameters were adjusted within the CalystEX software of the Stratasys Dimension Elite 3D printer, as shown in Fig. 2:

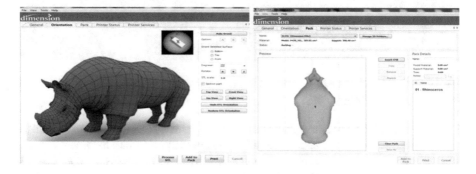

Fig. 2. Refining the 3D model provided by Reverse Engineering (by CatalystEX)

The technology of vacuum casting has become a widely accepted rapid tooling technique. Vacuum casting is a newer version of investment casting, differing in the process of making molds [15, 24, 25]. The most important feature of vacuum casting is a significant reduction in the time needed to produce a small series of parts as well as reducing the cost compared to traditional methods [26]. Generally, AM costs can be divided into the group of fixed costs and variable ones. This paper deals with the production costs of fabricated elements in case of small-scale production, and processing and post-processing, costs of enforcement, and material costs. The example of making silicon molds and casting replicas of a rhinoceros toy figure according to the master model developed in a 3D printer on the equipment for vacuum casting (MK-Technology, Germany) [15] is used to present the aforementioned manufacturing process, which consists of the following steps:

1. Preparation of negatives of the rhinoceros shape in 3D printer for the casting process;
2. Setting the parting tape on the negative of the toy figure for facilitating the separation of the silicon mold;

3. Bonding plastic gates on the part's negative, which has a role of forming an inlet channel in the silicon mold and provide the positioning and fixation of the negative in the silicon casting frame (Fig. 3a);

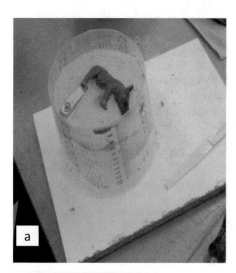
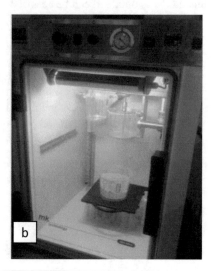

Fig. 3. (a) Casting silicon into the frame with the negative, (b) Removing the residual air from the silicon in a vacuum chamber.

4. Calculation of the necessary quantity of silicon mixture to form the mold to be used for the rhinoceros shape elements. The silicon mixture is poured into the frame with the fixed negatives that are subsequently submerged in silicon, placed in a vacuum chamber in order to remove residual air bubbles (Fig. 3b).

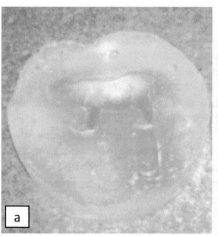
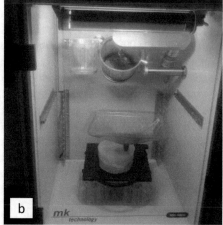

Fig. 4. (a) Silicon mold for casting, (b) the casting process in a vacuum chamber MK-Technology

5. Positioning of the silicon mold was made in a vacuum chamber. After solidification of silicon, the mold was cut to parting line, during which the negative was relieved and the mold for casting was created Fig. 4a;
6. There is a broad spectrum of multifunctional composite materials being used for finalization of replicas, depending on functional demands as well as mechanical and chemical characteristics of final parts. In the case of the rhinoceros toy figure, components made by Axson Technologies were used. Their physical and mechanical characteristics are displayed in Tables 1 and 2;

Table 1. Physical characteristics of components used for casting rhinoceros toy figure

Physical properties at 23 °C PX 223 HT		Part A	Part B	Mixing
Composition		**Isocyanate**	**Polyol**	
Mixing Ratio by Weight		100	80	
Aspect		Liquid	Liquid	Liquid
Colour		Colourless	Black	Black
Viscosity (mPa.s)		1,100	300	850
Density before mixing	ISO 1675	1.17	1.12	–
Density of cured mixing	ISO 1675	–	–	1.14
Pot life on 90 g (min.)				6–7

Table 2. Mechanical characteristics of components used for casting rhinoceros toy figure

Mechanical properties at 23 °C PX 223 HT				
Flexural modulus of elasticity	ISO 178:2001	PSI/(MPa)	334,000/(2,300)	
Flexural strength		PSI/(MPa)	11,600/(80)	
Tensile strength	ISO 527-2:1993	PSI/(MPa)	8,700/(60)	
Elongation		%	11	
Charpy impact resistance	ISO179/2D:1994	ft-lb/in^2/(kJ/m^2)	>29/(>60)[1]	
Izod impact – Notched	ASTM D256-05	ft-lb/in^2/(kJ/m^2)	3/(6)	
Izod impact – Unnotched	ASTM D256-05	ft-lb/in^2/(kJ/m^2)	>8/(>16)[1]	
Hardness	- at 73°F (23 °C)	ISO 868:2003	Shore D1	80
	- at 248°F (120 °C)			>65

7. After a certain quantity of the material needed for molding and the proportion of its individual components in a total amount is determined, then vacuum casting process follows. The casting process takes place in a vacuum chamber under conditions that are recommended for corresponding elements and components of the material, according to Fig. 4b.

After the solidification of the material in a vacuum chamber, mold halves are separated and, if necessary, post-processing of the molded item follows.

The rhinoceros toy figure, made in the integrated process of additive manufacturing technology and vacuum casting as a result of this study is shown in Fig. 5a, and silicon mold for replicas in Fig. 5b.

 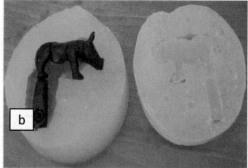

Fig. 5. (a) Rhinoceros toy figure made in the integrated process, and (b) silicon mold for replicas

The cost represents the value of all inputs used in the production process for the whole production time. Determination of the cost is of great importance for every economic decision-making. Total cost includes all the cash flows (fix and variable costs) necessary to produce a part. Fix cost of the equipment and software for modelling are the most significant while variable cost, that affect the most the total cost of the final part can be different [18]. Considering material costs and processing and post-processing time and costs, all calculations were performed in order to obtain the final average price of 0.5€ per piece.

4   Conclusion

The reproduction of a rhinoceros toy figure by integration of additive manufacturing and vacuum casting shows a remarkable example on how modern technologies permit a rapid development of prototypes and their replicas. Replicas can be realized using the original material, enlarging the spectrum of physical, technical and mechanical properties. It also means that the process under investigation permits the realisation of parts using a broad spectrum of materials including wood, plastics and composites.

Furthermore, inspections on these products demonstrated that characteristics, such as surface finish quality, transparency, rigidity, elasticity, strength, hardness, temperature resistance, colour and so on were comparable with market standards. Also in terms of production time and cost, every preliminary estimation proved that these process parameters were significantly reduced in respect to the use of traditional techniques (e.g. cutting tools) for production of prototypes or small batches. As demonstrated in several studies [22, 24], the additive manufacturing can represent the right option for a large gamma of industrial products. Moreover, additional savings could be realized using an integrated approach for the full cost optimization of the whole process, instead of moving toward single steps of reduction for additive manufacturing and, separately, vacuum casting. The benefits provided by these new techniques toward a modern manufacturing has to be also considered in term of eco-sustainability, evident when compared with conventional technological methods. It is important that society understand its potential positive and negative impacts so that positive impacts can be embedded and ensure that AM does not become a missed opportunity for improving sustainability.

In synthesis, this integrated technology can represent a sustainable response to the desire of fast innovation of products that characterizes many industrial fields such as automotive, aerospace, pharmaceutical, medical, food industry, wood industry and many others. Finally, the paper also provided empirical insights and practical implications on CAD modelling for 3D printing, Additive Manufacturing by Fused Deposition Molding and on vacuum casting.

References

1. Lavery, G., Pennell, N., Brown, S., Evans, S.: Next Manufacturing Revolution. http://www.nextmanufacturingrevolution.org/nmr-report-download/
2. Fragassa, C., Minak, G., Poodts, E.: Mechanical characterization of photopolymer resins for rapid prototyping. In: 27th Danubia-Adria Symposium on Advanced in Experimental Mechanics, DAS 2010, Wroclaw University of Technology, Wroclaw, Poland, 22 September 2010 through 25 September 2010, Code 125161
3. Guo, N., Leu, M.C.: Additive manufacturing: technology, applications and research needs. Front. Mech. Eng. **8**, 215–243 (2013)
4. Sandström, C.: Adopting 3D printing for manufacturing - evidence from the hearing aid industry. Technol. Forecast. Soc. Change **102**, 160–168 (2015)
5. Sljivic, M., Stanojevic, M., Djurdjevic, D., Grujovic, N., Pavlovic, A.: Implementation of FEM and rapid prototyping in maxillofacial surgery. FME Trans. **44**, 422–429 (2016)
6. Ford, S., Despeisse, M.: Additive manufacturing and sustainability: an exploratory study of the advantages and challenges. J. Clean. Prod. **137**, 1573–1587 (2016)
7. Gebler, M., Uiterkamp, A.J.M.S., Visser, C.: A global sustainability perspective on 3D printing technologies. Energy Policy **74**, 158–167 (2014)
8. Kohtala, C.: Addressing sustainability in research on distributed production: an integrated literature review. J. Clean. Prod. **106**, 654–668 (2015)
9. Kohtala, C., Hyysalo, S.: Anticipated environmental sustainability of personal fabrication. J. Clean. Prod. **99**, 333–344 (2015)
10. Lucisano, G., Stefanovic, M., Fragassa, C.: Advanced design solutions for high-precision wood-working machines. Int. J. Qual. Res. **10**, 143–158 (2016)
11. Despeisse, M., Ford, S.: The role of additive manufacturing in improving resource efficiency and sustainability. In: Umeda, S., Nakano, M., Mizuyama, H., Hibino, H., Kiritsis, D., Cieminski, G. (eds.) APMS 2015. IAICT, vol. 460, pp. 129–136. Springer, Cham (2015). doi: 10.1007/978-3-319-22759-7_15
12. Himmer, T., Stiles, E., Techel, A., Beyer, E.: PC$_{Pro}$ a novel technology for rapid prototyping and rapid manufacturing. Fraunhofer IWS. http://www.ccl.fraunhofer.org
13. Wohlers Report 2014: 3D Printing and Additive Manufacturing State of the Industry Annual Worldwide Progress Report. Wohlers Associates, Inc., Technical report (2014)

14. Grujovic, N., Borota, J., Sljivic, M., Divac, D., Rankovic, V.: Art and design optimized 3D printing. In: 34th International Conference on Production Engineering, Nis (2011)
15. MK Technology GmbH: Operating Instructions for Vacuum Casting System MK-mini. www.mk-technology.com
16. Gibson, I., Rosen, D.W., Stucker, B. (eds.): Additive Manufacturing Technologies, pp. 36–58. Springer LLC, New York (2010)
17. Forno, I.: Direct casting of rapid prototyping resins for luxury production: influence of burn-out and processing parameters on the final quality. Int. J. Eng. Sci. Innovative Technol. **3** (2014)
18. Grujovic, N., Pavlovic, A., Sljivic, M., Zivic, F.: Cost optimization of additive manufacturing in wood industry. FME Trans. **44** (2016)
19. Bassoli, E., Gatto, A., Iuliano, L., Violante, M.G.: 3D printing technique applied to rapid casting. Rapid Prototyping J. **13**, 148–155 (2007)
20. Ramos, A.M., Relvas, C., Simoces, J.A.: Vacuum casting with room temperature vulcanising rubber and aluminium moulds for rapid manufacturing of quality parts: a comparative study. Rapid Prototyping J. **9**, 111–115 (2003)
21. Salonitis, K., Pandremenos, J., Paralikas, J., Chryssolouris, G.: Multifunctional materials: engineering applications and processing challenges. Int. J. Adv. Manufact. Technol. **49**, 803–826 (2010)
22. Chhabra, M., Singh, R.: Rapid casting solutions: a review. Rapid Prototyping J. **17**, 328 (2011)
23. Dippenaar, D.J., Schreve, K.: 3D printed tooling for vacuum-assisted resin transfer moulding. Int. J. Adv. Manufact. Technol. (2012)
24. Singh, R.: Mathematical modelling of dimensional accuracy in vacuum assisted casting. J. Virtual Phys. Prototyping **7**, 129 (2012)
25. Jijotiya, D., Verma, P.L., Sanjay, J., Bajpai, L., Manoria, A.: Efficient rapid prototyping mechanism using vacuum casting process. Int. J. Emerg. Technol. Adv. Eng. **3** (2013)
26. Sljivic, M., Pavlovic, A., Stanojevic, M., Fragassa, C.: Combining additive manufacturing and vacuum casting for an efficient manufacturing of safety glasses. FME Trans. **44**, 393–397 (2016)

Assessment of Cost and Energy Requirements of Electron Beam Melting (EBM) and Machining Processes

Paolo C. Priarone[1](✉), Matteo Robiglio[1], Giuseppe Ingarao[2], and Luca Settineri[1]

[1] Department of Management and Production Engineering, Politecnico di Torino, Corso Duca degli Abruzzi 24, 10129 Turin, Italy
{paoloclaudio.priarone,matteo_robiglio, luca.settineri}@polito.it
[2] Department of Industrial and Digital Innovation, University of Palermo, Viale delle Scienze, 90128 Palermo, Italy
giuseppe.ingarao@unipa.it

Abstract. Additive Manufacturing is under the spotlight as potential disruptive technology, particularly for the production of complex-shaped structural metallic components. However, the actual AM process capabilities present some limitations in achieving the strict part quality requirements imposed by the aerospace and automotive sectors. Therefore, the integration of AM and conventional manufacturing represents an emerging scenario to be investigated. In this paper, a pure machining process and a hybrid production route (based on EBM and finish machining) are compared. The influence of material usage-related factors on costs and energy demand is discussed. The results prove that, despite precise process judgments are case-specific, the proposed methodologies are suitable to provide guidelines for identifying the optimal manufacturing route under multiple design objectives.

Keywords: Sustainability · Additive manufacturing · Machining · Cost · Energy

1 Introduction

Additive Manufacturing (AM), which builds parts layer-by-layer, allows creating complex geometries hardly achievable with cutting tools. Reductions of equipment/resource usage (as tools, lubricants, or molds) and savings in process wastes are expected. In addition, the so-called 'think additive' re-design of the components leads to significant weight reductions, with positive environmental impacts during the use phase [1]. Despite these advantages, the technological limitations imposed by AM in terms of surface quality have to be accounted for, therefore conventional material removal processes are unlikely to be overstepped, particularly for metal components. In this context, only a few researches have been addressed (*i*) to compare the economic and environmental footprint between AM and conventional machining, or (*ii*) to investigate their possible integrations. For instance, Paris and co-authors [2] proposed,

G. Campana et al. (eds.), *Sustainable Design and Manufacturing 2017*, Smart Innovation, Systems and Technologies 68, DOI 10.1007/978-3-319-57078-5_68

in 2016, a Life Cycle Assessment (LCA) method for analyzing the environmental impacts of additive and subtractive processes when producing an aeronautic turbine, while a cost comparison for aero engine parts was earlier presented by Allen [3]. Schröder et al. [4] applied the time-driven activity-based costing for the development of a business model suitable to be implemented as a Production Service System. Nevertheless, the roadmap enabling the greenest manufacturing approach selection (under multiple design objectives) is still quite incomplete.

The authors have recently introduced a life cycle-based approach to compare the energy and carbon footprint of conventional machining versus additive manufacturing plus finish machining processes. The production of differently shaped components made of a Ti-6Al-4V alloy was the main subject of the study. The results proved that, as far as primary energy demand and CO_2 emissions are concerned, the impacts related to material usage are usually dominant, even if the credit arising from metal recycling is accounted for [5]. The purpose of the present research is to extend the proposed methodology by including cost models for both the production routes, in order to identify whether additive manufacturing (i.e., EBM) processes are preferable (from both the economic and environmental perspective) over conventional machining.

2   Methodology and System Boundaries

A cradle-to-gate system boundary has been assumed, as shown in Fig. 1. The material, energy, and resource flows have to be assessed for the material production and the part manufacturing stages of the life cycle. The impact of transportation from the raw material production plant to the manufacturing plant has been included.

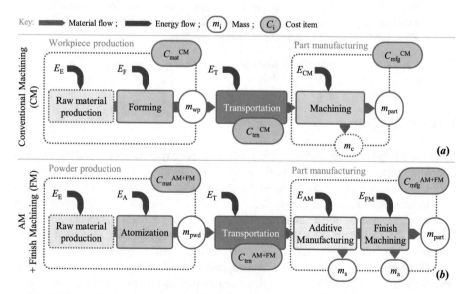

Fig. 1. Cradle-to-gate system boundaries for the (*a*) CM and (*b*) AM + FM approaches.

Concerning *conventional machining* (CM), the finished part (weighting m_{part}) is produced by removing from a workpiece (weighting m_{wp}) the excess material in the form of chips (weighting m_c). Such operation consumes primary energy (E_{CM}) due to the resource demands of cutting and auxiliary processes [6, 7], and results in a cost for manufacturing (C_{mfg}^{CM}). The workpiece is made out of an in-stock, available material (obtained from ores and feedstock and characterized by a certain value of embodied energy E_E, according to Ashby [8]), by means of a forming process (requiring an energy E_F) needed to achieve the geometry of the semi-finished product (i.e., a bar, a billet, a block, etc.). The workpiece has a cost (C_{mat}^{CM}) usually defined by the market rules. As concerns *additive manufacturing* (AM), the mass of the required metal powder (m_{pwd}) should comprise the masses of the finished part (m_{part}), the process scraps (m_s), and the allowance (m_a) for the post-processing *finish machining* (FM) operation. The energy for the metal powder atomization process (E_A) [9] has to be added to that of the raw material production (E_E) [1, 5]. In this case, the part manufacturing stage is carried out via two subsequent steps: the AM (i.e., EBM) process creates the near-net-shape component with a superimposed machining allowance, which has to be then removed via FM. Therefore, both the E_{AM} and E_{FM} energy requirements have to be analysed. The costs for the powder purchase (C_{mat}^{AM+FM}) and for the part manufacturing (C_{mfg}^{AM+FM}) represent the main cost drivers. In addition, the cost and energy footprint of transportation have to be accounted for both the manufacturing routes.

It is worth remarking that the CM and AM+FM routes described in Fig. 1 could be compared only if all the manufactured parts comply with the same product specifications, guaranteeing the same in-use performance. This hypothesis is supported by the literature, since EBM processes have proved to produce fully-dense parts, whose mechanical properties could be matched with those of traditionally manufactured components [5 and references therein]. The results presented in this paper focus on the assessment of cost and energy demand. Empirical models are presented in the following sub-sections. The CO_2 emissions could be considered as a proxy of the energy consumption, therefore comparable outcomes are expected [5, 10]. The system boundaries for the cost assessment have to include the main indirect costs (as listed in Table 1), that typically dominate the overall picture. Vice versa, the environmental impacts per produced part of machine tools, equipment and buildings are relatively small and usually neglected [7].

Table 1. Typical data required for the indirect cost assessment.

Production overhead	Rent, building area costs
Administration overhead	Amortization of hardware/software and consumables purchase
Production labor	Technician annual salary plus employer contributions
Machine costs	Amortization of machine and equipment purchase; Maintenance

2.1   Energy Demand Assessment for Conventional Machining (CM)

Several papers regarding the sustainability analyses of material removal processes (at different system levels) are available in literature, as reviewed in [11]. For instance, the effects of process parameters [12], machine tool architecture [13], and tool path strategy [14] on power/energy requirements have been recently studied. In general, for the CM approach in Fig. 1a, the total energy demand E_{tot}^{CM} (in MJ/part) could be computed according to Eq. 1 (adapted from [5]).

$$E_{tot}^{CM} = \overbrace{m_{wp} \cdot (E_E + E_F)}^{\text{Workpiece production}} + \overbrace{E_T \cdot m_{wp} \cdot d^{CM}}^{\text{Transportation}} + \overbrace{m_c \cdot S_E^{CM}}^{\text{Manufacturing}} \qquad (1)$$

All the main variables in the equation(s) are listed in Table 2. The energy for manufacturing can be obtained by multiplying the specific energy demand for CM (S_E^{CM}, expressed in MJ per kg of removed material) and the mass of the chips (m_c, in kg). The S_E^{CM} value, which refers to the primary energy demand, could be obtained using either black-box (as the SEC model proposed by Kara and Li [15]) or bottom-up approaches (e.g., Priarone et al. [6]). When accounting for the material production, the

Table 2. Nomenclature and main abbreviations.

C_{dir}^{CM}	(€/kg)	Total direct costs per unit mass (for CM)
$\dot{C}_{ind}^{CM}$	(€/h)	Total indirect cost rate (for CM)
C_{EE}	(€/kWh)	Cost for electric energy
C_{pwd}	(€/kg)	Cost of the powder for AM
C_T	(€/kg·km)	Cost for transportation
C_{wp}	(€/kg)	Cost of the workpiece material for CM
d^i	(km)	Distance travelled during transportation (for i = CM or AM + FM)
E_A	(MJ/kg)	Energy for powder atomization
E_E	(MJ/kg)	Embodied energy of the raw material
E_F	(MJ/kg)	Energy for forming the workpiece
E_T	(MJ/kg·km)	Energy for transportation
m_a	(kg)	Mass of the machining allowance (for AM + FM)
m_c	(kg)	Mass of the machined chips (for CM)
m_{part}	(kg)	Mass of the produced part (for both CM or AM + FM approaches)
m_{pwd}	(kg)	Mass of the metal powder (for AM: $m_{pwd} = m_{part} + m_s + m_a$)
m_s	(kg)	Mass of the support structures (for AM)
m_{wp}	(kg)	Mass of the workpiece (for CM: $m_{wp} = m_{part} + m_c$)
MRR	(kg/h)	Material removal rate
$S_C^{CM\ (or\ FM)}$	(€/kg)	Specific cost per kg of removed material for CM (or FM)
S_C^{AM}	(€/kg)	Specific cost per kg of deposited material for AM
$S_E^{CM\ (or\ FM)}$	(MJ/kg)	Specific energy demand per kg of removed material for CM (or FM)
S_E^{AM}	(MJ/kg)	Specific energy demand per kg of deposited material for AM
SEC^i	(kWh/kg)	Specific electric energy consumption (for i = CM or FM)

recycling benefit awarding should be included by applying the recycled content approach or the substitution method, as explained by Hammond and Jones [16].

2.2 Energy Demand Assessment for AM Plus Finish Machining (FM)

Only few researches concerning the sustainability analysis of AM processes for the production of metal parts have been published. It has to be emphasized that, unlike the conventional machining, AM powder-bed processes allow creating multiple parts simultaneously within the same build. Therefore, when assessing the energy and resource efficiency performance, different variables have to be accounted for, as the machine capacity utilization [17], the product demand [18], and the shape complexity [19]. For the AM + FM approach proposed in Fig. 1b, the total energy demand (in MJ/part) could be assessed as shown in Eq. 2 (adapted from [5]).

$$E_{tot}^{AM+FM} = \overbrace{m_{pwd} \cdot (E_E + E_A)}^{Powder\ production} + \overbrace{E_T \cdot m_{pwd} \cdot d^{AM+FM}}^{Transportation} + \overbrace{m_{pwd} \cdot S_E^{AM} + m_a \cdot S_E^{FM}}^{Manufacturing} \quad (2)$$

In addition to the energy needed to produce the metal powder and for its transportation to the processing plant, the energy contributions for (i) additively manufacture the near-net-shape metal part, and for (ii) removing the machining allowance m_a have to be both considered. Coherently with Eq. 1, these contributions can be computed by multiplying the specific energy demand values by the involved masses. The model in Eq. 2 implies the full recyclability of Ti-6Al-4V powder during the EBM process, according to [20, 21]. Therefore, m_s accounts only for the material losses attributed to the support structures. Further details can be found in [5].

2.3 Cost Estimation for Conventional Machining (CM)

The economics of machining processes, which are essential for achieving high dimensional accuracy and surface finish, have been widely analyzed in literature. The costs are mainly related to (i) machine tools, devices, fixtures and cutting tools; (ii) labor and overhead; (iii) material handling and movement; (iv) gaging for dimensional accuracy and surface finish; (v) cutting and non-cutting times. In machining, the total cost per produced part is the sum of fixed costs and variable costs depending on process parameters. The fixed contributes are unproductive costs due to setting up operations as well as loading/unloading and machine-handling. Variable contributes concern the machining cost (which decreases at high production rates) and the tooling cost (which becomes relevant at high production rates) [22].

Vila et al. [23] recently proposed a study aimed to identify the most convenient solution between face milling and surface grinding. Their model included the cost of machine depreciation, labor and consumables (i.e., the cutting tools). The cost for each tool and manufacturing process (excluding raw material cost and overhead) consisted of two contributes: (i) the machine costs associated with productive and non-productive times, and (ii) the cutting tool/grinding wheel wear costs. The unit cost (per part) has been obtained by dividing each cost by the product between the workpiece material volume and the number of machined parts [23]. The criteria for minimum energy consumption and minimum manufacturing cost usually do not coincide, as highlighted by Yoon et al. [24] while studying the micro-drilling process. Moreover, the cost of a machined part also depends on the surface finish requirements, since the manufacturing cost increases rapidly with finer surface finish. Schultheiss et al. [25] investigated the influence of the average surface roughness on the manufacturing costs when turning A48-40B, AISI 4140, AISI 316L, and Ti6Al4V. They proposed a cost model in which production rate loss, scrape rate, and downtime rate were introduced as factors of influence. Pusavec et al. [26] proposed a model for computing the part production costs when turning Inconel 718 under three different cooling conditions. Five contributes were accounted for: (*i*) machining cost, (*ii*) cutting tool cost, (*iii*) cooling/lubrication fluid cost, (*iv*) energy cost, and (*v*) cleaning cost. The cost for machining (i.e., due to machine-tool usage and labor cost rates) and cutting tool dominates the total cost since the contributes related to energy, lubricoolant, and cleaning were proved to be small. Therefore, on the basis of the existing literature, a simplified cost model is proposed in this paper (in Eq. 3).

$$C_{tot}^{CM} = \overbrace{m_{wp} \cdot C_{wp}}^{Workpiece\ purchase} + \overbrace{C_T \cdot m_{wp} \cdot d^{CM}}^{Transportation} + \overbrace{m_c \cdot S_C^{CM}}^{Manufacturing} \tag{3}$$

The specific cost for machining S_C^{CM} (in €/kg) could be defined as a function of the direct cost C_{dir}^{CM} (i.e., the cost for the consumed electric energy) and the total indirect cost rate $\dot{C}_{ind}^{CM}$, according to Eq. 4.

$$S_C^{CM} = \frac{\dot{C}_{ind}^{CM}}{MRR} + C_{dir}^{CM} = \frac{\dot{C}_{ind}^{CM}}{MRR} + C_{EE} \cdot SEC^{CM} \tag{4}$$

2.4   Cost Assessment for Additive Manufacturing (AM)

Models for assessing the well-structured costs of Additive Manufacturing (AM) processes have been developed in recent years. The comparison among them reveals potentially contrasting outcomes, as a result of the different assumptions and boundary conditions considered (e.g., in excess machine capacity estimation). Hopkinson and Dickens [27] proposed a cost analysis aimed to match three layer manufacturing processes (namely: Stereolitography, Fused Deposition Modelling, and Laser Sintering) with injection moulding. The unit cost for each part was computed (including material, machine, and labor costs) with varying production volumes. Results revealed that (*i*) the cost curve for injection moulding decreased when increasing the number of produced parts (since the cost for the moulds was amortized across the production volume), while (*ii*) the cost per part due to the layer manufacturing processes was supposed to be a constant unrelated to the production volume. Hopkinson and Dickens assumed that each AM machine would achieve 90% uptime, and produces only one part consistently for one year. This indirectly implies the maximization of the number of (equal) parts to be manufactured within a single build [27]. The idea of using the build volume of AM machine to produce multiple copies of the same part was applied also by Atzeni and Salmi [28], who evaluated the costs of a Direct Metal Laser Sintering (DMLS) process versus a High-Pressure Die-Cast (HPCD) for the production of metal components. Among the factors affecting the costs of a given technology, the authors focused on material and processing costs (i.e., machine amortization, part design and testing, labor, pre- and post-processing). The tooling costs were included for the conventional technology only. As a result, the AM cost was constant over production volume.

Ruffo et al. [29] detailed a cost model for AM in which all the different activity costs of the process were divided into direct and indirect costs. Only the material-related costs were considered to be a direct cost, while the machine costs, production labor, production overhead, and administrative overhead were counted among the indirect costs. The direct and indirect costs are allocated to each build by the involved mass and the build time, respectively. Both the main variables (material and time) could be related to the part shape and complexity. Again, the cost per each part is obtained by dividing the total build cost by the number of components (of equal geometry) that are included in the build. Unlike other models e.g., [27, 28], Ruffo et al. [29] achieved a saw-tooth shaped cost curve (*i*) showing a substantial deflection for low production volumes, (*ii*) that tends to stabilize at high production volumes (when the indirect costs are split on higher number of parts).

It might be argued that the cost models which consider builds made of identical parts do not fully meet the flexibility and the industrial potential of the process. In fact, AM can be used to produce in parallel (i.e., in the same build) multiple components with extremely different geometries. Also, the estimation of the part cost by dividing the total build cost by the number of parts has some limitations due to certain elements of cost that are independent on the quantity of parts contained in the build. In addition, in order to satisfy the technical efficiency requirements, the available capacity of the machine should be fully utilized, since the cost per part is expected to decrease when different components are mixed together in the same build [30]. With respect to that statement, Baumers et al. [18] have considered several builds made of industrial parts fairly dissimilar in shape and size (i.e., a bearing block, a turbine wheel, a belt link, a cap, and a small venturi tube). The authors proved that both the specific energy consumption as well as the specific production cost were significantly affected by the demand scenario and the realized part quantities. Overall, the results are influenced by the build volume floor area occupation. A total cost estimate for a build has been proposed by Baumers et al. [31] by adapting and extending the model by Ruffo et al. [29]. Comparable approaches have been recently applied (in 2016) by Hällgren et al. [32] and Manogharan et al. [33].

Overall, the total cost for the AM + FM approach could be defined as in Eq. 5 (according to the models proposed in Eqs. 2 and 3), where the cost for the post-AM finishing operation has been included. S_C^{AM} must involve both direct costs and indirect cost rate, as mentioned in Baumers et al. [31].

$$C_{tot}^{AM+FM} = \overbrace{m_{pwd} \cdot C_{pwd}}^{\text{Powder purchase}} + \overbrace{C_T \cdot m_{pwd} \cdot d^{AM+FM}}^{\text{Transportation}} + \overbrace{m_{pwd} \cdot S_C^{AM} + m_a \cdot S_C^{FM}}^{\text{Manufacturing}} \quad (5)$$

3   Case Study

The case study proposed by the authors in [5] has been considered also in the present paper. Different solid-to-cavity ratios (from ID 1 to ID 3, in Fig. 2) have been assumed in order to vary the amount of process scraps for subtractive approaches. The material flows in Fig. 1 are detailed in Table 3. The machining allowance has been imposed to be constant and 1-mm thick [5]. For AM, the mass of the support structures has been assumed equal to 15% of $(m_{part} + m_a)$ [32].

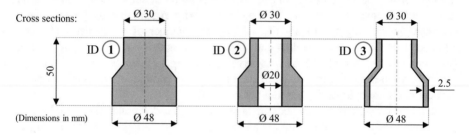

Fig. 2. Case study: geometries [5].

Table 3. Case study: masses involved in the material flows.

CM			AM+FM		
ID	m_{part} (kg)	m_c (kg)	m_{wp} (kg)	m_a (kg)	m_s (kg)	m_{pwd} (kg)
1	0.275	0.157	0.432	0.042	0.048	0.364
2	0.206	0.226	0.432	0.053	0.039	0.298
3	0.067	0.365	0.432	0.058	0.019	0.144

3.1   Life Cycle Inventory

The values applied in Eqs. 1–5 for the cost estimation and the primary energy demand assessment are listed in Table 4. The embodied energy of the raw material (E_E) is assumed to be either 685.0 MJ/kg (considering the primary material production from virgin sources [8]) or 206.6 MJ/kg (including the benefits deriving from material recycling at the end-of-life [5]). The energy for forming the material (E_F) is, on average, 14.5 MJ/kg [8], while the energy for atomization (E_A) is computed from the value of 6.6 kWh/kg of electric energy consumption proposed by Paris et al. [2], and considering a primary-to-electric energy conversion efficiency of 0.34 [6]. The costs of

Table 4. Life Cycle Inventory (LCI) for cost and energy estimation.

CM		AM+FM
Material	$E_E = 206.6$ or 685.0 MJ/kg	
	$E_F = 14.5$ MJ/kg	$E_A = 70.0$ MJ/kg
	$C_{wp} = 28.0$ €/kg	$C_{pwd} = 175.0$ €/kg
Transportation	$E_T = 0.94 \cdot 10^{-3}$ MJ/kg·km	
	$C_T = 0.025 \cdot 10^{-3}$ €/kg·km	
	$d^{CM} = d^{AM+FM} = 300$ km	
Process	$S_E^{CM} = 34.5$ MJ/kg	$S_E^{AM} = 332.0$ MJ/kg; $S_E^{FM} = S_E^{CM}$
	$S_C^{CM} = 31.3$ €/kg	$S_C^{AM} = 426.2$ €/kg; $S_C^{FM} = S_C^{CM}$
	$C_{EE} = 0.072$ (€/kWh)	

Ti-6Al-4V workpiece material and metal powder are equal to $C_{wp} = 28.0$ €/kg (from Hällgren et al. [32]) and $C_{pwd} = 175.0$ €/kg (commuted from the value of 156.97 £/kg mentioned in [31]). The energy penalty (E_T) and the cost (C_T) for the material transportation by means of an heavy-sized truck are extracted from Ashby [8] and commuted from Qu et al. [34] (who proposed a variable cost of 0.036 £/ton·mile). A travelled distance $(d^{CM} = d^{AM+FM})$ of 300 km is considered for both the manufacturing approaches [5]. For the EBM process, the specific values of S_E^{AM} and S_C^{AM} are obtained by handling the data from Baumers et al. [31], who realized (by means of an Arcam S12 system) a build made of different typical industrial components in Ti-6Al-4V, in order to achieve the full machine capacity utilization (93%). The AM process was characterized in terms of total direct/indirect cost and energy demand. Therefore, specific data per each unit of processed material volume (or mass) could be identified. For instance, a specific cost (including material and manufacturing) of 2.39 £/cm^3 was proposed for EBM [31]. Such values could be reasonably considered as representative of a process in which the maximum machine efficiency is exploited. For the conventional machining (CM) approach, S_E^{CM} is obtained from Priarone et al. [6] for a wet turning process when assuming $MRR = 2.5$ kg/h and $SEC = 9.9$ kWh/kg. A total indirect cost rate for CM (C_{ind}^{CM}) of 76.6 €/h (i.e., 85 USD/h, according to [25]) is hypothesized. The values concerning the finish machining (FM) are equalized to those of CM. The electric energy cost (C_{EE}) in Table 4 is commuted from 0.018 £/MJ [31].

3.2   Results and Discussion

The values collected during the LCI stage (in Sect. 3.1) have been applied to the energy and cost models described in Sect. 2. The main outcomes are summarized in Fig. 3. With respect to the costs (Fig. 3a), a noteworthy difference between the results for AM + FM and CM is noticed, and conventional machining appears to be the most advantageous solution. Analogous results have already been shown in literature (e.g., [33]). In fact, although the masses of material involved in the AM + FM approach are always lower than those of CM approach (despite the considered case study, as highlighted in Table 3), the workpiece and powder costs are significantly different $(C_{wp} = 28.0$ €/kg versus $C_{pwd} = 175.0$ €/kg). Moreover, while the material purchase is the main cost item for CM, for AM + FM the total cost per part is dominated by the EBM process, and the post-process finish machining operation (FM) cost appears to be small. The specific cost per unit mass of AM (S_C^{AM} = 426.2 €/kg) is one order of magnitude higher than that of CM (S_C^{CM} = 31.3 €/kg). This is mainly due to the longer duration of the AM process, since (*i*) the directs costs related to the electric energy consumption are basically negligible for both the approaches, and (*ii*) the assumed total indirect cost rate favors AM ($\dot{C}_{ind}^{AM}$ = 29.4 €/h, adapted from [31]) instead of CM ($\dot{C}_{ind}^{CM}$ = 76.6 €/h, adapted from [25]).

As for the energy requirements (Figs. 3*b* and *c*), the results presented in [5] are confirmed. Overall, both the energy demands and the costs are heavily affected by the material usage. When enabling significant material savings, the AM + FM approach appears to be the best strategy from the primary energy saving viewpoint. This outcome

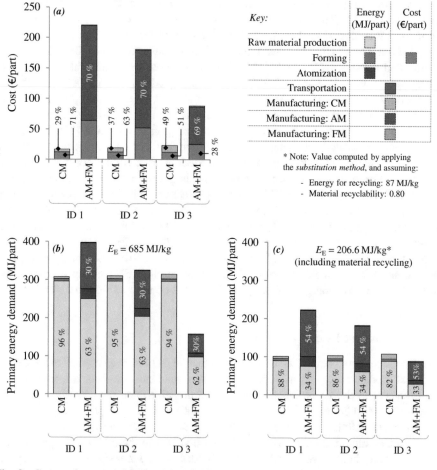

Fig. 3. Case study: cost estimation (*a*) and primary energy demand assessment (*b*, *c*), excluding (*b*) or including (*c*) the material recycling.

is particularly remarkable when the benefits of material recycling are not accounted for (Fig. 3*b*), since the impact of material usage dominates the energy breakdown. On the other hand, when a small amount of material has to be machined-off (i.e., for ID 1), the high energy intensity of the EBM process has a negative effect on the AM energy-intensity performance. The contribution of transportation appears to be negligible.

4 Conclusions and Outlooks

The study presented in this paper, which represents an extension of a previous research [5], was aimed to compare a pure machining process and a hybrid/integrated production route, based on an EBM process followed by a finish machining operation. Models for cost estimation and primary energy demand assessment have been presented. A simple case study has been introduced in order to verify the influence of material usage-related factors on the process performance. The results proved that, as far as energy-efficiency is concerned, the AM-based approach could be the best strategy when it enables a larger amount of material savings than that of machining. By contrast, as the costs are regarded, the AM + FM manufacturing route appears to be more costly. Obviously, despite the wide applicability of the proposed models, such conclusions are dependent on the numerical values applied in the models. The needed data have been extracted from the most recent literature, but it is difficult to generalize the achieved outcomes, which are extremely case-specific. However, this kind of studies is needed to promote the identification of suitable guidelines for selecting the optimum production scenario in view of the sustainable development.

References

1. Huang, R., Riddle, M., Graziano, D., Warren, J., Das, S., Nimbalkar, S., Cresko, J., Masanet, E.: Energy and emissions saving potential of additive manufacturing: the case of lightweight aircraft components. J. Clean. Prod. **135**, 1559–1570 (2016)
2. Paris, H., Mokhtarian, H., Coatanéa, E., Museau, M., Ituarte, I.F.: Comparative environmental impacts of additive and subtractive manufacturing technologies. CIRP Ann. - Manuf. Technol. **65**(1), 29–32 (2016)
3. Allen, J.: An investigation into the comparative costs of additive manufacture vs. machine from solid for aero engine parts. Manufacturing Technology Rolls-Royce plc, UK. Unclassified report RTO-MP-AVT-139 (2006)
4. Schröder, M., Falk, B., Schmitt, R.: Evaluation of cost structures of additive manufacturing processes using a new business model. Proc. CIRP **30**, 311–316 (2015)
5. Priarone, P.C., Ingarao, G., Di Lorenzo, R., Settineri, L.: Influence of material-related aspects of additive and subtractive Ti-6Al-4V manufacturing on energy demand and carbon dioxide emissions. J. Ind. Ecol. DOI:10.1111/jiec.12523 (In Press)
6. Priarone, P.C., Robiglio, M., Settineri, L., Tebaldo, V.: Modelling of specific energy requirements in machining as a function of tool and lubricoolant usage. CIRP Ann. - Manuf. Technol. **65**(1), 25–28 (2016)

7. Dahmus, J., Gutowski, T.: An environmental analysis of machining. In: Proceedings of the 2004 ASME International Mechanical Engineering Congress and RD&D Exposition (IMECE2004-62600), Anaheim, California, USA (2004)
8. Ashby, M.F.: Materials and the Environment: Eco-informed Material Choice, 2nd edn. Butterworth Heinemann/Elsevier, Oxford (2013)
9. Dawes, J., Bowerman, R., Trepleton, R.: Introduction to the additive manufacturing powder metallurgy supply chain. Johns. Matthey Technol. Rev. **59**(3), 243–256 (2015)
10. Ingarao, G., Priarone, P.C., Gagliardi, F., Di Lorenzo, R., Settineri, L.: Subtractive versus mass conserving metal shaping technologies: an environmental impact comparison. J. Clean. Prod. **87**, 862–873 (2015)
11. Duflou, J.R., Sutherland, J., Dornfeld, D., Herrmann, C., Jeswiet, J., Kara, S., Hauschild, M., Kellens, K.: Towards energy and resource efficient manufacturing: a processes and systems approach. CIRP Ann. - Manuf. Technol. **61**(2), 587–609 (2012)
12. Mativenga, P.T., Rajemi, M.F.: Calculation of optimum cutting parameters based on minimum energy footprint. CIRP Ann. - Manuf. Technol. **60**(1), 149–152 (2011)
13. Behrendt, T., Zein, A., Min, S.: Development of an energy consumption monitoring procedure for machine tools. CIRP Ann. - Manuf. Technol. **61**(1), 43–46 (2012)
14. Campatelli, G., Scippa, A., Lorenzini, L., Sato, R.: Optimal workpiece orientation to reduce the energy consumption of a milling process. Int. J. Precis. Eng. Man - Green Technol. **2**, 5–13 (2015)
15. Kara, S., Li, W.: Unit process energy consumption models for material removal processes. CIRP Ann. - Manuf. Technol. **60**(1), 37–40 (2011)
16. Hammond, G., Jones, C.: Inventory of Carbon and Energy (ICE), Annex B: How to Account for recycling. The University of Bath, Bath, UK (2010)
17. Baumers, M., Tuck C., Wildman R., Ashcroft I., Hague, R.: Energy inputs to additive manufacturing: does capacity utilization matter? In: Solid Freeform Fabrication Proceedings, Solid freeform Fabrication; an Additive Manufacturing Conference, pp. 30–40. University of Texas, Austin (2011)
18. Baumers, M., Tuck, C., Wildman, R., Ashcroft, I., Rosamond, E., Hague, R.: Transparency built-in: Energy consumption and cost estimation for additive manufacturing. J. Ind. Ecol. **17**, 418–431 (2013)
19. Baumers, M., Tuck, C., Wildman, R., Ashcroft, I., Rosamond, E., Hague, R.: Shape complexity and process energy consumption in electron beam melting. A case of something for nothing in additive manufacturing? J. Ind. Ecol. DOI:10.1111/jiec.12397 (in Press)
20. Petrovic, V., Niñerola, R.: Powder recyclability in electron beam melting for aeronautical use. Aircr. Eng. Aerosp. Technol. **87**(2), 147–155 (2015)
21. Nandwana, P., Peter, W.H., Dehoff, R.R., Lowe, L.E., Kirka, M.M., Medina, F., Babu, S.S.: Recyclability study on Inconel 718 and Ti-6Al-4V powders for use in electron beam melting. Metall. Mater. Trans. B **47**(1), 754–762 (2016)
22. Kalpakjian, S., Schmid, S.R.: Manufacturing Engineering and Technology, 6th edn. Pearson, Upper Saddle River (2009)
23. Vila, C., Siller, H.R., Rodriguez, C.A., Bruscas, G.M., Serrano, J.: Economical and technological study of surface grinding versus face milling in hardened AISI D3 steel machining operations. Int. J. Prod. Econ. **138**, 273–283 (2012)
24. Yoon, H.-S., Moon, J.-S., Pham, M.-Q., Lee, G.-B., Ahn, S.-H.: Control of machining parameters for energy and cost savings in micro-scale drilling of PCBs. J. Clean. Prod. **54**, 41–48 (2013)
25. Schultheiss, F., Hägglund, S., Ståhl, J.-E.: Modeling the cost of varying surface finish demands during longitudinal turning operations. Int. J. Adv. Manuf. Technol. **84**, 1103–1114 (2016)

26. Pusavec, F., Kramar, D., Krajnik, P., Kopac, J.: Transitioning to sustainable production–part II: evaluation of sustainable machining technologies. J. Clean. Prod. **18**, 1211–1221 (2010)
27. Hopkinson, N., Dickens, P.M.: Analysis of rapid manufacturing - using layer manufacturing processes for production. J. Mech. Eng. Sci. **217**(C1), 31–39 (2003)
28. Atzeni, E., Salmi, A.: Economics of additive manufacturing for end-usable metal parts. Int. J. Adv. Manuf. Technol. **62**, 1147–1155 (2012)
29. Ruffo, M., Tuck, C., Hague, R.: Cost estimation for rapid manufacturing - laser sintering production for low to medium volumes. Proc. IMechE Part B: J. Eng. Manuf. **220**, 1147–1427 (2006)
30. Baumers, M.: Economic aspects of additive manufacturing: benefits, costs and energy consumption. Doctoral thesis, Loughborough University (2012)
31. Baumers, M., Dickens, P., Tuck, C., Hague, R.: The cost of additive manufacturing: machine productivity, economies of scale and technology-push. Technol. Forecast. Soc. **102**, 193–201 (2016)
32. Hällgren, S., Pejryd, L., Ekengren, J.: Additive manufacturing and high speed machining - cost comparison of short lead time manufacturing methods. Proc. CIRP **50**, 384–389 (2016)
33. Manogharan, G., Wysk, R.A., Harrysson, O.L.A.: Additive manufacturing - integrated hybrid manufacturing and subtractive processes: economic model and analysis. Int. J. Comput. Integ Manuf. **29**, 473–488 (2016)
34. Qu, Y., Bektas, T., Bennell, J.: Sustainability SI: multimode multicommodity network design model for intermodal freight transportation with transfer and emission costs. Netw. Spat. Econ. **16**, 303–329 (2016)

Engineering a More Sustainable Manufacturing Process for Metal Additive Layer Manufacturing Using a Productive Process Pyramid

Paul O'Regan[1([☒])], Paul Prickett[1], Rossi Setchi[1], Gareth Hankins[2], and Nick Jones[2]

[1] Cardiff School of Engineering, Newport Road, Cardiff CF24 3AA, UK
{oreganp,Prickett}@cardiff.ac.uk
[2] Renishaw Plc, New Mills, Wotton-Under-Edge, Gloucestershire GL12 8JR, UK

Abstract. Sustainability within manufacturing is an increasingly important topic globally. One course of action being explored is to produce more parts 'right first time' so supporting an increasingly sustainable manufacturing process. This paper explores the Renishaw "productive process pyramid" and considers how it can be integrated into the ALM process. The pyramid is currently used to identify how layers of control can systematically remove variation from conventional machining processes. This application is focussed to consider how the variables that occur within the ALM manufacturing process can impact on the quality of the parts mechanically and geometrically. This approach can then inform the process foundation and process setting stages and enhance levels of in-process control.

Keywords: Additive layer manufacturing (ALM) · Selective laser melting (SLM) · Laser powder bed fusion (LPBF) · Sustainability · Process setting · Post-process monitoring · Metrology

1 Introduction

Today, more than ever, the environmental impact of manufacturing products has taken centre stage. In the UK for example the government introduced PAS2050 to assess the impact of manufacturing products on climate change. It provides an approach that can be used to assess a manufacturer's carbon footprint and the impact of their products. Such considerations can be applied to processes such as Traditional Computer Numerical Controlled (CNC) metal removal machining processes. These use billeted materials from which, through reductive manufacturing, a part is produced. Ratios as high as 20:1 have been reported for this process i.e. for every 1 kg of part produced, 19 kg of removed material enters the waste stream [1]. A reduction in the dimensions of the billet can reduce the waste and hence improve manufacturing sustainability, but this is limited to the dimensions of the part being produced.

Metal additive layer manufacturing (ALM) is a relatively new manufacturing process which has the potential to become a truly sustainable manufacturing technology. Laser powder bed fusion (LPBF) utilises only the amount of material it needs for the part and

© Springer International Publishing AG 2017
G. Campana et al. (eds.), *Sustainable Design and Manufacturing 2017*, Smart Innovation, Systems and Technologies 68, DOI 10.1007/978-3-319-57078-5_69

has the potential to reduce the life cycle material mass and energy consumed in relation to conventional subtractive techniques by eliminating scrap and also eliminating the use of sometimes harmful process enablers. Certain LPBF techniques have the capability to completely eliminate a supply chain. An LPBF machine can be set up in a company to produce parts using just in time (JIT) for the machinery they have on that site. This provides a consumer with the capability to create spares or enable repairs without having to order a part to be manufactured off site [2].

LPBF is an additive manufacturing process, the process begins with raw material being deposited on a build plate in the build chamber of an additive layer machine. A high power laser beam scans a geometry on to the powder layer. The energy from the laser is absorbed through radiation by the powder and the heat transfer produces a phase transformation. The powder changes from a solid to a liquid, forming a melt pool. Once the laser moves, this melt pool solidifies to produce a consolidated layer. When the scan finishes the geometry for the layer, the build bed is lowered and a fresh layer of powder is deposited. The process will then be repeated until the end of the build program and the part is finished. When the build is complete, the solid metal part will be embedded in powder. Once removed from the powder, the un-sintered powder is then sleeved and put back into the machine to produce a new part through the same operation [3]. This process has a minimal waste stream, making it a much more sustainable manufacturing process compared with traditional reductive methods.

Though this process is sustainable in terms of the manufacturing product to waste ratios, the process is not always "right first time" (RFT). Defects are commonly caused by three things; poor design, inadequate manufacturing process control and less than optimum materials being used in the process. Currently, the LPBF process has a number of variables that are not accounted for within the production cycle. Parts can thus be produced that are out of tolerance or damaged. To date there is no defect prevention built into the system; once the build starts, unless there is a serious failure in a part, the process does not stop. The product is produced within the machine without knowing if it is within specification. A user will only know it the part is "good" when post-process measurements are taken. If it is possible to control all of the processes during a build a part can be produced RFT every time making this process more sustainable. If defect detection can occur in-process it would assure a more sustainable and cost effective production method in comparison to traditional techniques. This paper will examine the variables that occur within the manufacturing process and how they can impact on the quality of the parts mechanically and geometrically. It will look at post-process monitoring and how this may be used to monitor in-process control and how it informs the settings for the process foundation and process setting.

2   The Productive Process Pyramid

The "productive process pyramid" comprises of four layers, shown in Fig. 1. The pyramid is used to identify how layers of control are used to systematically remove variation from machining processes. These layers build upon each other to deliver regularly conforming parts within a manufacturing process. The pyramid has been used to evaluate, adjust and bring reductive manufacturing processes under control by governing the variation within the processes. Users of CNC machines have used this process to become more consistent, sustainable and economical.

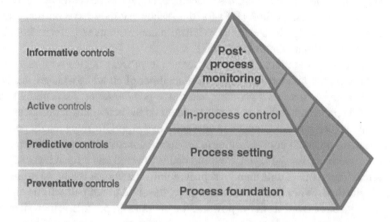

Fig. 1. Renishaw "Productive Process Pyramid" reproduces from [4].

2.1   Process Foundation

The process foundation layer acts on the machining environment, assessing its condition and adjusting it so that a stable manufacturing environment is provided. These are preventative controls that are introduced to reduce the number of sources of variance before manufacturing starts. It has been estimated that powder bed fusion (PBF) has more than 130 variables [5]. It has been stated that there are over 50 different selective laser melting (SLM) process parameters that impact on the quality of a part [6]. To help produce a part that is RFT every time, the part needs to be designed for manufacture. This is an approach that produces a product through the understanding of the current capabilities of the manufacturing operation and though best practice rather than trying to 're-invent the wheel'. There are a number of ALM predictive software programs, both being developed and available on the market, produced by; Netfabb [7], Virfac [8], Amphon [9], exaSIM [10] and Simufact Additive [11]. These programs can help operators and designers design for manufacture and maximise the advantages of using ALM to produce parts or prototypes. Using these programs can calculate the deformation of the final part and reduce/avoid distortion and minimise residual stress. The can optimise the build-up orientation and the support structure. They can also be used to predict the microstructure and indicate criteria-based part failure. They enable the conditioning of the part after heat treatment, base plate and support structure removal. Overall they can be used to reduce material and energy consumption cost and increase machine and manpower productivity while reducing unnecessary costs by replacing tests with simulations.

Control of process inputs involve the use of Failure Mode Effect Analysis (FMEA) and similar techniques to understand and control all the upstream factors that can affect machining process outcomes. For LPBF, this would include knowing the quality of the powder being used in the process, ensuring that the laser optics are clean, focused and aligned, that the heating element works, that it is able to uniformly heat the build bed area, that the gas flow is pure and that it has enough capacity to complete a build. If the conditions are consistent at the start of the process, they are more likely to be consistent at the end and predictable during the build cycle.

Environmental stability addresses those external sources of non-conformance that cannot be eliminated in advance, but which are inherent to the operating environment. These include changes in powder temperature within the elevator while parts are being produced, laser life management and gas flow velocity. Unexpected events should also be taken into consideration such as wiper blade crashes, power outages and software errors. There are solutions for some of these variation problems that have already been integrated in LPBF machines. For example some are fitted with sensors that monitor the power usage on the motor that drives the blade back and forth over the build bed. If the power required to move the blade increases then a problem may be flagged; the part may have curled, cracked or snapped away from the support and is protruding through the powder. The extra power being drawn would indicate greater resistance and therefore an obstruction in the build bed area.

Machine optimisation and regular condition monitoring of the machine is essential for the process foundation. An inaccurate machine cannot make consistently accurate parts and therefore cannot complete parts RFT all the time. When applying these measures to the LPBF machine a company can assure increased machine availability, increase process capability, improve quality and reduce overheads by focusing on proactive tasks.

2.2   Process Settings

The process setting layer deals with predictable sources of variation. For CNC processes, process setting is carried out in four ways: machine setting, probe setting, part setting and tool setting. For LPBF these predictable sources of variation include the percentage of oxygen within the build chamber and the elevator height. Though component construction should be produced with no oxygen present LPBF machines have and upper and lower oxygen content limit which could produce some variance in a component if the oxygen level fluctuates during a build. The part being produced will have major mechanical changes due to chemical reactions in the material. If the elevator moves down at varying rates the layer thickness will be different each layer. This could create areas on the bed which contain no powder. This is because the powder volume may not fill the new volume created by the elevator movement being too great therefore short feeding the build area. The opposite could be true if the elevator retracts and produces a shallow layer. This will cause overexposure to the layers below changing the characteristics within that layer and the preceding ones depending on the spot size and exposure time. Overdosing or short feeding in this instance would also contribute to the reduction in mechanical performance of the part you are producing because of the change in layer thickness. Building on the stability introduced by the process foundation layer, process setting controls help to eliminate human error by automating manual processes.

Due to the way LPBF works, standard process setting techniques cannot be utilised. For instance, for a standard milling machine a mechanical probe could be used to set up the process reducing the set up time. This is possible because the billet and cutter are objects that come into physical contact. The location of the billet and the wear on the cutter effects the final product. If the location or cutter diameter differs from the information the operator puts into the program then parts will be produced out of tolerance. A mechanical probe can be programmed so that any operator can complete the set up and not just a highly skilled operator. The LPBF process creates parts from powder so traditional setting up as previously mentioned cannot be completed.

To make sure that the LPBF is operating within optimum output the elevator position, laser spot size, focus and laser position need to be assessed. The build chamber elevator height is checked by attaching a build plate to the elevator. The elevator is then driven down a pre-set distance. The top surface of the build plate is then measured against the build chamber floor around its circumference. The measurement is completed with digital callipers. Builds that are completed after the initial setup then rely on the operator measuring the thickness of the build plate and updating this information into the LPBF machines elevator offset, so that when the machine is in operation the elevator movement moves the correct distance. The LPBF machine could be set up with an automated checking process that assesses the elevator accuracy and then assesses build plates that are fitted into the machines removing the human factor.

The laser power can be assessed using a NOVA II power meter. The laser is fired at a sensor that can be used to cross check the power being produced against the power requested on the LPBF machine. Different software programs can be utilised to communicate with the LPBF's laser PLC over a local network to re-map the power draw to deliver the correct power if the reading is incorrect. The laser is tested at every 20 watts to check that the Bitmap for the full range of the machine's power is correct. Once the power has been mapped correctly the laser spot size is configured. The beam focus is checked and altered if needed. This operation is usually carried out when an AM250 machine is moved or during a service. The laser is analysed using 12 different focus measurement increments. Once all 12 have been mapped the optimum focus is picked and programmed into the machine. The target spot size in the x and y is under 7 microns. The scan field accuracy is then checked using an anodised plate that is fitted on to the build plate. The anodised plate has etched markings printed on to its surface. These makings have been positioned in pre-set locations as a datum for the test. The laser in the ALM machine is then fired on to the plate. It etches new marks to pre-set x and y positions. These positions are compared to the marks already on the sheet and if the marks are within ±0.1 mm the machine is within tolerance. If the laser marks created by the laser are outside of this dimensional tolerance the difference is measured using a digital calliper by the operator and an offset inserted into the operation program to bring the laser into tolerance.

2.3   In-Process Control

The third pyramid layer is in-process control. This looks at the sources of variation that are inherent to the manufacturing process. Temperature variation, laser degradation and providing intelligent feedback to the process are examples here. This could be vitally important to LPBF because closed loops systems can compensate for in-process changes. If the laser starts to lose power during the process and this is picked up using sensors within the laser, the computer can extend the exposure time to compensate for the reduction in power. In-process control can only work if the process foundation layer and process setting layer have been implemented correctly. If these are not under control, then in-process control will be striving against variations that it is not possible to tackle. Attempting to automate a process in a chaotic environment with undefined manual processes will lead to greater variation. This means that in-process control is challenging to integrate for many manufacturing techniques. The idea of in-process control is to tackle the inherent sources of variation that occur during the operation of the machine. In traditional reductive manufacturing machines, the in-process control would cover tool wear during the machining process, part deflection, temperature and heat flows. In LPBF these would translate as laser operation, part heating, build-chamber temperature and gas flow. Some people believe that the only way you can truly monitor the 'build' is through the integration of an optical device to monitor the build bed or melt pool [12, 13]. The camera can either evaluate the material deposition on each new layer or the camera can be used to monitor the melt pool checking the size in x, y and z.

2.4 Post-process Monitoring

The top layer of the pyramid is currently used by many manufacturing companies to finally assess the part they produced and the process they have used. Post-process monitoring checks the process and the part against their respective specifications. The parts produced can be measured relative to the CAD models that are produced prior to manufacture. Post-process monitoring for LPBF could include the measurement of the parts' dimensional attributes prior to heat treatment and removal from the build plate. This can be carried out by the operator who may typically use a set of digital callipers to assess if the part produced is in specification dimensionally after grit blasting. Completing this by hand on a number of builds will not be consistent and may produce a large variation in the measurements taken. This variation could increase further if there are a number of different operators measuring parts being produced. The post process measurement system for all LPBF parts could be improved by automation. The build plate would need to be provided with a reference datum that can be transferred to a CMM. This could evaluate any part being produced using the CAD model. Producing a number of different parts means the same number of programs will be needed, making the process very labour, time and cost intensive.

If a test piece can be developed to fit on a build plate with the part you are constructing, an evaluation of the test part could be used to firstly dial in the machine and secondly as a mechanical and dimensional check. Figure 2 shows a developed test piece still attached to the build plate prior to heat treatment. This test piece has been developed to contain important geometric information that can relate to parts on the build within the chamber. If the test piece is measured and is within tolerance the information can be used to indicate that the other builds on that plate are all within tolerance, thus reducing the time needed to check every part on a build plate if the plate is full of complex parts. Checking the parts prior to them being processed further provides confidence that the part conforms to the specification.

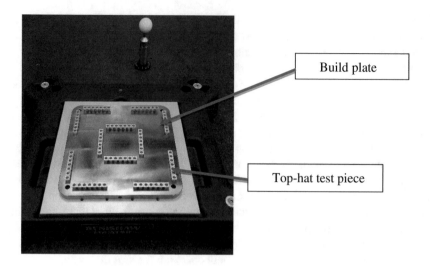

Fig. 2. Top-hat assemblies test piece samples

3 Process Variables and Interactions

The variables that occur within the LPBF process may be separated into categories: feedstock [14–17], build environment [18–21], laser [22] and melt pool [23]. The process factors can be divided into seven sections, as shown in Fig. 3: material, build bed, recoat blade, laser, environments, powder bed conditions and program software. The laser has been highlighted as it is one of the most significant contributions to part variation.

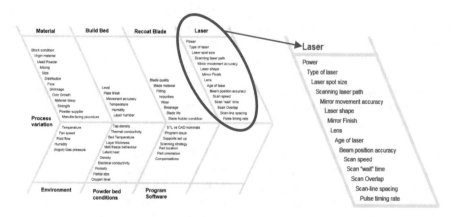

Fig. 3. Fishbone diagram show process variation contributors within LPBF

The laser power, exposure time, spot size and shape can influence the density of a part. If the laser does not put enough power into the material i.e. not fully melting it, unmelted particles will appear in the track and will produce less dense material which in turn has lower mechanical properties. Overexposure, which would definitely melt all the material, will penetrate more than three layers and will also increase the melt pool size creating geometric anomalies in the part you are producing. Understanding these relationships can be used to link post-process information to in-process signals to check if the manufacturing process is working correctly. Currently access is available to the build log that contains information on each layer being completed. Each layer records: layer time, laser position, laser power, time laser is on and off, oxygen level, pressure, temperature and elevator position etc. The full list of monitored parameters can be found in [3]. Currently this information is open loop; the loop can be closed by linking the geometric changes in a produced part with the information from the sensors installed within the machine. This can be done without adding new sensors.

4   Framework

The next step in producing parts RFT is producing a versatile test piece. Many manufacturing companies and processes use a known test component to check if their process is within specification. Such a test piece can be produced in the machine and checked for signs of geometric conformity on a comparator. This process can inform users of the operational efficiency of an LPBF machine. Integrating such a part into every build means that information can be extrapolated from it providing the user with real time data regarding the 'health' of the build. It could be used as a reference point to predict how the rest of the build will come out of the machine and if the parameters are within tolerance. The integration of a part that you already know can be used to advise the machine if alterations in the process need to occur in real time by comparing the part of the PLC log that refers to the part and not the full build.

The first step in completing a fully integrated system that can identify when a process is drifting out of tolerance is to identify a suitable test piece that will be able to monitor the key features of a build that can be affected greatly by small micro changes within the manufacturing eco-system. This test piece would need to be evaluated once the manufacturing process has finished, once again after it has been heat treated and then finally after it has been removed from the build plate to see if there is a correlation geometrically between each of these processes. As well as these three steps, an analysis of the variation across the build bed would need to take place. This will identify the resolution of the build across the build and how it changes depending on the orientation of the part. Understanding this will mean that the test piece can be positioned anywhere in a build to provide information that can inform the process and produce a part RFT.

Finally, if a link can be produced between heat-treated parts and parts that have been removed from the plate then the information gathered can inform the process foundation and process setting parts of the PPP. This information will provide key correlations between process parameters and machine variables which in turn will provide more consistent parts and make the process more sustainable. Using a test piece can reduce the amount of information that needs to be processed. Currently one layer of a medium dense build may contain roughly 40 Mb of information. The information in this file just states the path of the laser in the x, y axis. Adding to this by assuming that data from a thermocouple, oxygen sensor and laser power also produces a file that is 40 Mb each for a single layer size of data, the data available for analysis just for one layer becomes close to roughly 200 Mb. Considering that the build may be on average 1,000 layers thick there is a need to process 200 Gb of information. A build that uses the complete depth of the chamber could generate terabytes of information. Within the aeronautical industry there is a need to store all process information so that a component made using LPBF meet certification demands. It provides critical information that can be used to validate a production run or checked to see if the production process caused a component failure. If there is a parameter shift the information produced can provide indications of what these parameter shifts have affected. For instance if the laser power increases through a cycle a part could produce voids meaning that the mechanical strength of the component is compromised. By reducing the information needed to be processed to a known fixed amount means that in-process monitoring can be carried out virtually at the same time as the build takes place with only a small delay for processing. The added benefit being a large reduction in date storage needed per build.

Acknowledgments. This research is jointly funded by the Engineering and Physical Sciences Research Council and Renishaw under an iCASE award reference number 15220143.

References

1. O'Leary, R., Setchi, R., Prickett, P., Jones, N., Hankins, G.: An investigation into the recycling of Ti-6Al-4V powder used within SLM to improve sustainability. In: SDM 2015: 2nd International Conference on Sustainable Design and Manufacturing (2015)
2. Sreenivasan, R., Goel, A., Bourell, D.L.: Sustainability issues in laser-based additive manufacturing. Phys. Procedia **5**, 81–90 (2010)
3. O'Regan, P., Prickett, P., Setchi, R., Hankins, G., Jones, N.: Metal based additive layer manufacturing: variations, correlations and process control. Procedia - Procedia Comput. Sci. **96**, 216–224 (2016)
4. Renishaw Plc: Survival of the fittest - the process control imperative (2011)
5. Yadroitsev, I.: Selective Laser Melting: Direct Manufacturing of 3D-Objects by Selective Laser Melting of Metal Powders. LAP LAMBERT Academic Publishing (2009)
6. Spears, T.G., Gold, S.A.: In-process sensing in selective laser melting (SLM) additive manufacturing. Integr. Mater. Manuf. Innov. **5**, 2 (2016)
7. Autodesk: NetFabb (2016). https://www.netfabb.com/. Accessed 17 Oct 2016
8. Geonx and LPT: "Geonx," GeonX and LPT take modelling of additive manufacturing processes to a new level (2016). http://www.geonx.com/. Accessed 17 Oct 2016
9. Additiveworks: Simulation and Process Software for Additive Manufacturing (2016). https://additive.works/. Accessed 17 Oct 2016
10. 3DSIM: exaSIM (2016). http://3dsim.com/product/exasim/. Accessed 17 Oct 2016
11. Simufact: Simulation of manufacturing processes (2016). http://www.simufact.com/additive-manufacturing.html. Accessed 17 Oct 2016

12. Bechmann, J.: Printing 3D geometries versus molding them — new perspectives for design and function quality requirements. Mold Die j. **3**, 38–40 (2014)
13. Craeghs, T., Clijsters, S., Yasa, E., Bechmann, F., Berumen, S., Kruth, J.-P.: Determination of geometrical factors in Layerwise Laser Melting using optical process monitoring. Opt. Lasers Eng. **49**(12), 1440–1446 (2011)
14. Bremen, S., Meiners, W.: Selective laser melting a manufacturing technology for the future? Laser Tech. J. **9**, 33–38 (2012)
15. Carroll, P.A., Pinkerton, A.J., Allen, J., Syed, W.U.H., Sezer, H.K., Brown, P., Ng, G., Scudamore, R., Li, L.: The effect of powder recycling in direct metal laser deposition on powder and manufactured part characteristics. In: Proceedings of AVT-139 Specialists Meeting on Cost Effective Manufacture via Net Shape Processing. NATO Research and Technology Organisation (2006)
16. Spierings, A.B., Herres, N., Levy, G.: Influence of the particle size distribution on surface quality and mechanical properties in AM steel parts. Rapid Prototyp. J. **17**(3), 195–202 (2011)
17. Yasa, E., Deckers, J.: Investigation on occurrence of elevated edges in selective laser melting. In: … Symp. Austin, TX …, pp. 180–192 (2009)
18. Shiomil, M., Yamashital, T., Osakada, K., Shiomi, M., Yamashita, T., Abe, F., Nakamura, K.: Residual stress within metallic model made by selective laser melting process. Ann. CIRP **53**(1), 195–198 (2004)
19. Leuders, S., Thöne, M., Riemer, A., Niendorf, T., Tröster, T., Richard, H.A., Maier, H.J.: On the mechanical behaviour of titanium alloy TiAl6V4 manufactured by selective laser melting: fatigue resistance and crack growth performance. Int. J. Fatigue **48**, 300–307 (2013)
20. Sercombe, T., Jones, N., Day, R., Kop, A.: Heat treatment of Ti-6Al-7Nb components produced by selective laser melting. Rapid Prototyp. J. **14**(5), 300–304 (2008)
21. Dunsky, C.: Process monitoring in laser additive manufacturing. Industrial laser solutions (2014). http://www.industrial-lasers.com/articles/print/volume-29/issue-5.html. Accessed 28 Mar 2016
22. Gibson, I., Rosen, D.W., Stucker, B.: Additive Manufacturing Technologies. Springer, Heidelberg (2010)
23. Hua, T., Jing, C., Xin, L., Fengying, Z., Weidong, H.: Research on molten pool temperature in the process of laser rapid forming. J. Mater. Process. Technol. **198**(1–3), 454–462 (2008)

Sustainable Scenarios for Engaged Manufacturing: A Literature Review and Research Directions

Michael J. Ryan[✉] and Daniel R. Eyers

Cardiff Business School, Cardiff University,
Aberconway Building, Colum Drive, Cardiff CF10 3EU, UK
{ryanm6,eyersdr}@cf.ac.uk

Abstract. Additive Manufacturing (AM) is gaining increasing interest as a sustainable manufacturing technology. One important aspect of AM is the opportunities it presents for customization by increasing involvement of customer in the design and manufacturing process. This paper presents a review of literature examining proposed future scenarios for AM implementation, with particular interest in the way customers are engaged in the process, and the effect that this might have on the future of AM supply chains. 27 distinct scenarios for future AM implementations were identified, with Engineer-to-Order and Make-to-Order proving most popular approaches, alongside a trend towards localized production. Although structured methods are available, a "genius" approach was found to have been employed in developing the majority of the scenarios evaluated.

Keywords: Additive manufacturing · 3D printing · Engagement · Sustainable manufacture · Customization · Scenarios · Supply chains

1 Introduction

As Additive Manufacturing (AM) develops, an increasing amount of speculation around its future applications can be seen in both research and popular press articles. Sometimes known as Rapid Prototyping, Rapid Manufacturing, or 3D Printing, the technology was initially developed in the 1980s, and established a firm base in the production of one-off prototypes and low volume parts. In recent years however improvements in the machines and an increasing range of available materials have led to a growing interest in using AM for the production of end-use components [1].

AM represents a unique group of manufacturing technologies, and is defined as "the process of joining materials to make objects from 3D model data, usually layer upon layer, as opposed to subtractive manufacturing methodologies." [2]. Whereas traditional manufacturing methods are either subtractive or formative (i.e. material is ether removed or shaped to produce the final shape), AM operates by adding together material, such as fine powders or liquids, to produce the final part. This technique removes many of the restrictions often associated with conventional manufacturing approaches. As the process is additive, there is no longer a requirement for specialist tooling to be produced [3], drastically reducing the minimum economic batch quantities (as low as one in most cases) [4], and the need for batches of parts to be identical [5]. The additive process also

© Springer International Publishing AG 2017
G. Campana et al. (eds.), *Sustainable Design and Manufacturing 2017*, Smart Innovation, Systems and Technologies 68, DOI 10.1007/978-3-319-57078-5_70

removes many design restrictions, such as tooling or mold removal paths, allowing the designer much greater design freedom, and even allowing multiple parts to be combined into a single, working component.

With interest in AM increasing, driven somewhat by media coverage of the technology, a diverse range of possible applications are being proposed. These have examined various aspects of the technology, the products that it might produce and the effects that its implementation might have, on products, users and even on a global scale. With numerous possibilities and potential applications, there is clearly a great opportunity for the use of AM technologies in manufacturing. A large volume of literature now exists in this field, however, as much of it is unsubstantiated media hype it can be difficult to segregate those proposals which are merely speculative from those which provide realistic possibilities for the future.

In response to this currently unclear future for AM, this paper aims to collate information from literature to identify future scenarios for the future of AM supply chains, with a particular focus on the customer's involvement in the process. The research question tackled in this paper is therefore:

> *"What are the current commonly discussed future scenarios for customer engagement in sustainable AM systems?"*

1.1 Customer Engagement in AM

One of the defining characteristics of AM is its ability to involve the customer in the production of their own products by reducing many of the complexities involved in product design and realization. In principle customers can design new products (or customize existing ones) using 3D CAD software, without needing many of the skills typically associated with 'Design for Manufacture' [6]. As a result, the potential for customers to be more intimately involved in the manufacture of their product is increased, which we term 'engagement' in this paper.

The extent of customer integration within the production process affects the degree to which products can be tailored or customized [7], with earlier involvement supporting much more customer-bespoke production. From a supply chain perspective, this point at which the customer becomes involved in the order fulfilment process is known as the 'decoupling point' or 'order penetration point' [7–9]. Typically, the decoupling point concerns the part of the supply chain where speculative production (made to forecasts) meets actual customer orders [9], and those points in the supply chain are typically where strategic stocks are held [10]. Production is 'pushed' to the decoupling point by manufacturers, and pulled in response to customer orders. From Fig. 1, for AM, the ability for customers to design their own products can imply an "Engineer-to-Order" approach, but in theory any of the different points of engagement are feasible, and so in this paper we use the decoupling point to identify the extent of customer engagement.

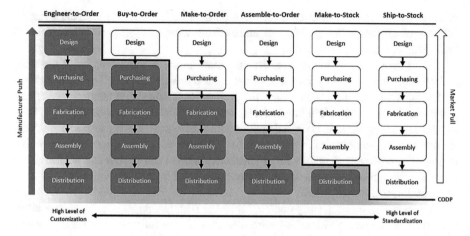

Fig. 1. Customer Order Decoupling Points (CODP, adapted from Gosling et al. [11])

1.2 Sustainable Additive Manufacturing

Issues of sustainability are complex, but are receiving mounting attention for AM; a review of the literature highlights that environmental factors of AM production are discussed with increasing frequency. This typically focuses around the design freedoms granted through the AM process, and the opportunities for the production of lightweight parts that stem from this. Nickels [12], for example, discusses weight savings of greater than 30% in aircraft brackets, reducing the weight and therefore the fuel consumption of the aircraft.

Aside from product design, operational and processing factors are also frequently discussed. Often this is viewed from a financial perspective, such as the "buy-to-fly" material ratios considered in aircraft production (discussed in detail in Qian and Froes [13]), or the cost of the energy used in the manufacturing process. For example, Senyana and Cormier [14] have examined the environmental effect of AM on the supply chain, producing models to compare centralized with distributed manufacture and showing that, particularly when considering low production volumes, there is a lower environmental impact with a distributed AM system. By extension, Gebler et al. [15] have taken a more holistic view, examining the economic, environmental, and societal impacts of AM technologies, predicting improvements in energy use and emissions of up to 5% by 2025 if AM were to remain the niche technology it is now. These studies suggest that AM presents some promising opportunities for sustainable manufacture. Although in many cases it could be said that the technology is "not there yet", reductions in material use and changes towards distributed manufacturing using AM show that AM technologies have the potential to form part a sustainable manufacturing system.

Sustainability need not only consider the manufacturing processes. From a customer's perspective, there is some speculation as to the effect of product customization for the sustainability of goods. AM creates a number of opportunities for customization of products that, by their nature, will require some input from the customer. There is the possibility that engagement in the design and manufacture process can create greater attachment to a product, potentially increasing its longevity [16], however easy customization could lead to frivolous disposal and replacement of goods driven by higher rates of obsolescence [17]. In this paper we investigate the nature of "Engagement", and the affect this might have on the manufacturing process, with a focus on employing sustainable AM systems to achieve this.

2 Research Method

2.1 Scenarios for AM

This study aims to identify the range of future scenarios for Engaged Additive Manufacturing, and therefore focusses on existing works which used scenario planning techniques. Scenario planning has been widely used as a method for aiding in future planning for companies, and was popularized by Schwarz [18] in "The Art of the Long View". The method differs from other techniques in that rather than attempting to determine a definitive future outcome, it produces a range of outcomes, allowing the company to position itself against a variety of possibilities. Scenario planning encourages in-depth thinking and creativeness in developing possible futures, and helps to avoid restrictions in suggesting possible outcomes which can sometimes come from experience working in a particular field.

Schoemaker and Mavaddat [19] describe the use of scenario planning and development in the introduction of disruptive technologies, emphasizing the technique's ability to examine multiple possible futures, and allow businesses to plan and best position themselves to adapt if any of these occur. The method is therefore highly suited to examining the way in which AM could be adopted in the future, and forms a strong base for ongoing research in the field. A detailed appraisal of the various scenario planning techniques is provided by Bishop and Hines [20], and this shows much variety in the methods employed. The various methods show a great deal of difference in the inputs required to successfully produce scenarios, ranging from "genius" type methods, which rely largely on the authors knowledge and judgment, to more structured, data-rich methods, such as "Probability Trees" and "Sensitivity Analysis". For this study to be successful in the scenarios which are produced, it is important that the scenario methods employed are noted when reviewing the literature so that appropriate data can be collected, and the most successful methods built upon.

2.2 Structured Literature Review

To build a firm knowledge base, and to collate information on common themes and characteristics of future propositions for AM, a literature search was necessary. A pilot literature review identified that a variety of futures for AM were proposed in literature, and so it was necessary to develop a comprehensive search of the literature, in order to identify the full range of potential future scenarios.

A structured literature review, using the structure defined in Fig. 2, was carried out.

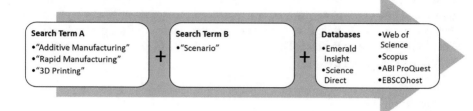

Fig. 2. Literature review structure

Each of the relevant papers identified in the literature search was coded for subsequent analysis. The coding criteria was developed as a result of the pilot study, using the features of the proposed scenarios that are common, yet clearly distinguished in each proposal: the supply chain strategy, the type of manufacturing operation, and the distribution of that manufacturing operation.

Customer engagement in the process was analyzed using the supply chain strategy, using Order Penetration Points (OPP), and the criteria based upon the work of Lampel and Mintzberg [7] (later expanded by, e.g. Olhager [21], Gosling et al. [11]) to examine the relationship between the consumer and the producer.

With regards the type of manufacturing operation proposed, this is segmented into three categories: craft operations, which use basic machinery which is largely operated by the end-user of the product; job shop, which remains low volume, but uses high-specification equipment and specialist operators; and factory, which produces high volumes, using specialist equipment and trained operators.

The distribution of manufacturing was identified geographically, distinguishing between personal (i.e. manufacturing at home), local, regional, and national centers of manufacturing. The criteria are shown in Fig. 3.

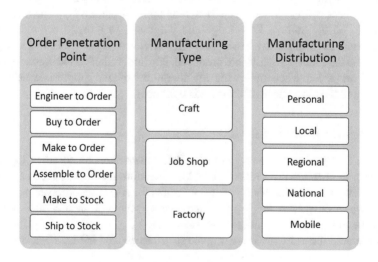

Fig. 3. Coding criteria for literature review

3   Results: Literature Analysis

This review produced 1,922 publications, of which 1,451 were found to be unique. Using the coding criteria identified earlier these results were analyzed, finding 203 unique scenarios for the future use of AM technologies.

In examining the results of the literature search, there is a clear, sharp increase in interest in the topic in recent years. Figure 4 shows that although there have been publications in the area as far back as 1985, the number of annual publications has increased drastically since 2009, coinciding with the expiry of a key FDM patent [22] which is often credited with heralding the advent of desktop 3D printers (e.g. [23]). The number of scenarios discovered remained a reasonably consistent percentage of the total papers reviewed each year, and so follows a similar pattern (also shown in Fig. 4).

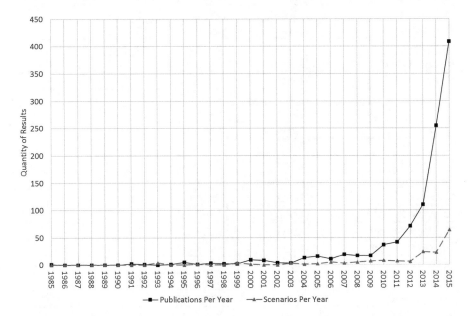

Fig. 4. Number of publications and scenarios by year (source: authors)

The coding criteria creates 90 possible combinations of the primary coding factors, however only 27 of the possible combinations are represented in the literature reviewed, following the coding exercise. These are shown in Table 1, along with the number of occurrences of each combination.

Table 1. Combinations of primary coding factors found in literature review with quantities (source: authors)

Order penetration point (OPP)	Distribution of manufacturing	Manufacturing operation type	Number of occurrences in literature
Make-to-Order	Local	Job shop	25
Engineer-to-Order	Local	Job shop	24
Engineer-to-Order	Home	Domestic	16
Engineer-to-Order	National	Job shop	13
Engineer-to-Order	Regional	Job shop	10
Engineer-to-Order	Local	Domestic	9
Make-to-Order	National	Job shop	9
Make-to-Order	Home	Domestic	8
Make-to-Order	Mobile	Job shop	8
Buy-to-Order	Local	Job shop	5
Engineer-to-Order	National	Factory	4
Make-to-Order	National	Factory	4
Buy-to-Order	Home	Domestic	3
Make-to-Order	Local	Domestic	3
Assemble-to-Order	National	Factory	3
Make-to-Stock	National	Factory	3
Engineer-to-Order	Mobile	Job shop	2
Buy-to-Order	Local	Domestic	2
Buy-to-Order	National	Factory	2
Make-to-Order	Local	Factory	2
Assemble-to-Order	Local	Job shop	2
Engineer-to-Order	Local	Factory	1
Engineer-to-Order	Regional	Factory	1
Buy-to-Order	Regional	Job shop	1
Buy-to-Order	National	Job shop	1
Make-to-Order	Regional	Job shop	1
Make-to-Stock	Local	Job shop	1
Incomplete scenarios			38

These combinations of results clearly show a strong tendency towards the customizable end of the OPP continuum (as proposed by Gosling and Naim [8]) when discussing possible futures for AM application. This is to be somewhat expected, since one of the principal advantages of AM that is often cited in literature is the ability to produce individual or customized products (e.g. [3]). Additionally, there is also a noticeable trend towards small, more localized production facilities, and away from the larger centralized factories that are normal in manufacturing at present. This is an interesting finding, and aligns well to the policies of various governments to support reshoring of manufacturing as a means of national competitiveness (e.g. [24]).

4 Conclusion

4.1 Findings of This Study

This detailed exploration of the literature has identified 27 distinct scenarios for future AM implementations, and highlighted the relative prevalence of each. It is apparent that Engineer-to-Order and Make-to-Order are most popular approaches (e.g. [17, 25]), which suggests that customers are likely to have much engagement in the creation of their products.

One of the most notable features of AM is its capability to enable customization in products. The technology has very few set-up costs as AM is a digital manufacturing technology, and any changes to design are made directly in CAD software on the computer [3]. This effectively removes the concept of economies of scale, as discussed by Schubert et al. [26], making low volume production cost effective, a concept also discussed by Petrick and Simpson [4] and in The Economist [5, 27]. The digital nature of the technology also reduces set-up and change-over times. This suggests that components can be produced when demanded, rather than kept in stock [5, 25, 26]. The on-demand capabilities of AM also allow the production of components to be delayed until late in the production process [27]. These factors build a strong case for the possibilities of using AM in customized production. It is also apparent that the scale of future AM implementations is unlikely to be particularly large, with a clear emphasis on job-shop or domestic implementations (e.g. [28, 29]). Production volumes are typically reported within the literature as low (often one-off), suggesting single machine instances may be commonplace. The size of the parts that will be produced are also likely to be small; they are constrained both by the physical capabilities of the AM machine, and the relatively high costs of materials. Furthermore, compared to conventional manufacturing, AM technologies are relatively small and require little space to operate in. Using OPP as an indicator of the level of customer engagement in the design and manufacture process, it is clear that current literature predicts a future where goods are produced on demand, and customized goods through AM, with high levels of input from the customer, and realized through localized production centers, will be commonplace.

In the majority of cases a "genius" approach was taken to scenario development, using the authors knowledge and judgment to build the scenarios, however some studies used more structured techniques, such as literature-based studies (e.g. [30]), case studies (e.g. [31]) and 2-axis methods (e.g. [29]).

4.2 Further Research

Although a good initial indication of customer engagement, it must be considered that OPP is a very simplified method for determining the level of customer engagement in the process. It provides us with an initial indicator as to whether there is an opportunity for any customer involvement or not, but gives little indication as to what that involvement might be. Production processes which fall into the Make-to-Stock or Ship-to-Stock categories have no customer engagement, as there is no interaction between the customer and the manufacture until after the product is finalized, however, there is nothing to determine that a customer has a high level of input to a product which might be considered Engineered-to-Order. It is already common practice, for example, to use AM to produce bespoke hearing aid shells [32] which are uniquely tailored to each customer, however beyond providing themselves as a model for the shape of the product the customer has very little input into the design process. It could equally be argued that someone with limited experience in product design could excerpt a large amount of effort in making the most minor of adjustments to designs obtained online, becoming more engaged in the process although there is a less obvious output.

One approach which may allow this to be analyzed is undertaken by Gosling et al. [33], in examining the OPP for Engineer-to-Order products in more depth, allowing greater detail of the input of the customer to be examined. Alternatively, Wikner and Rudberg [9] developed the Customer Order Decoupling Point (CODP) concept, beginning to segregate and consider Engineering CODPs and Production CODPs, adding granularity to the general CODP discussion. Further research activities could investigate the potential for developing an Engagement CODP as a means of explaining how the customer becomes involved, and the nature of their involvement in the process from start to finish. It is also planned that customer expectation in customized products will be investigated, not only to determine the potential demand, but to examine the likely effects on the longevity of the products, and the sustainability of the process.

References

1. Bak, D.: Rapid prototyping or rapid production? 3D printing processes move industry towards the latter. Assembly Autom. **23**(4), 340–345 (2003)
2. ASTM, F2792: Standard Terminology for Additive Manufacturing Technologies. ASTM F2792-10e1 (2012)
3. Berman, B.: 3-D printing: the new industrial revolution. Bus. Horiz. **55**(2), 155–162 (2012)
4. Petrick, I.J., Simpson, T.W.: 3D printing disrupts manufacturing how economies of one create new rules of competition. Res.-Technol. Manag. **56**(6), 12–16 (2013)
5. The Economist: Technology: Print Me a Stradivarius. The Economist (2011)
6. Gibson, I., Rosen, D.W., Stucker, B.: Design for additive manufacturing. In: Gibson, I., Rosen, D.W., Stucker, B. (eds.) Additive Manufacturing Technologies, pp. 299–332. Springer, Boston (2010). doi:10.1007/978-1-4419-1120-9_11
7. Lampel, J., Mintzberg, H.: Customizing customization. Sloan Manag. Rev. **38**(1), 21 (1996)
8. Gosling, J., Naim, M.M.: Engineer-to-order supply chain management: a literature review and research agenda. Int. J. Prod. Econ. **122**(2), 741–754 (2009)
9. Wikner, J., Rudberg, M.: Integrating production and engineering perspectives on the customer order decoupling point. Int. J. Oper. Prod. Manag. **25**(7–8), 623–641 (2005)
10. Naylor, J.B., Naim, M.M., Berry, D.: Leagility: integrating the lean and agile manufacturing paradigms in the total supply chain. Int. J. Prod. Econ. **62**(1–2), 107–118 (1999)
11. Gosling, J., et al.: Manufacturers' preparedness for agile construction. In: The IET International Conference on Agile Manufacturing, ICAM 2017. IET, Durham (2007)
12. Nickels, L.: Am and aerospace: an ideal combination. Met. Powder Rep. **70**(6), 300–303 (2015)
13. Qian, M., Froes, F.H.: Titanium Powder Metallurgy: Science. Technology and Applications. Butterworth-Heinemann, Oxford (2015)

14. Senyana, L., Cormier, D.: An environmental impact comparison of distributed and centralized manufacturing scenarios. In: Li, D., Zheng, D., Shi, J. (eds.) Materials Research and Applications, PTS 1–3, pp. 1449–1453 (2014)
15. Gebler, M., Uiterkamp, A.J.M.S., Visser, C.: A global sustainability perspective on 3D printing technologies. Energy Policy **74**, 158–167 (2014)
16. Diegel, O., et al.: Tools for sustainable product design: additive manufacturing. J. Sustain. Dev. **3**(3), 68 (2010)
17. Ford, S., Despeisse, M.: Additive manufacturing and sustainability: an exploratory study of the advantages and challenges. J. Clean. Prod. **137**, 1573–1587 (2016)
18. Schwarz, P.: The Art of the Long View: Planning for the Future in an Uncertain World. Currency Doubleday, New York (1991)
19. Schoemaker, P.J.H., Mavaddat, V.M.: Scenario planning for disruptive technologies. In: Wharton on Managing Emerging Technologies. Wiley, New York (2000)
20. Bishop, P., Hines, A., Collins, T.: The current state of scenario development: an overview of techniques. Foresight **9**(1), 5–25 (2007)
21. Olhager, J.: Strategic positioning of the order penetration point. Int. J. Prod. Econ. **85**(3), 319–329 (2003)
22. Crump, S.S.: Apparatus and Method for Creating Three-Dimensional Objects. Google Patents (1992)
23. Mims, C.: 3D Printing Will Explode in 2014, Thanks to the Expiration of Key Patents (2013). http://qz.com/106483/3d-printing-will-explode-in-2014-thanks-to-the-expiration-of-key-patents/. Accessed 27 Nov 2016
24. Moser, H.: Reshoring initiative spearheads effort to bring back us manufacturing. Mod. Appl. News **5**(12), 8 (2011)
25. Easton, T.A.: The design economy: a brave new world for businesses and consumers. Futurist **43**(1), 42–47 (2009)
26. Schubert, C., van Langeveld, M.C., Donoso, L.A.: Innovations in 3D printing: a 3D overview from optics to organs. Br. J. Ophthalmol. **98**(2), 159–161 (2014)
27. The Economist: Additive Manufacturing: Solid Print. The Economist (2012)
28. Bedinger, M., et al.: 21st century trucking: a trajectory for ergonomics and road freight. Appl. Ergon. **53 Pt B**, 343–356 (2016)
29. Birtchnell, T., Urry, J.: 3D, SF and the future. Futures **50**, 25–34 (2013)
30. Eyers, D.R., Potter, A.T.: E-commerce channels for additive manufacturing: an exploratory study. J. Manuf. Technol. Manag. **26**(3), 390–411 (2015)
31. Holmstrom, J., Partanen, J.: Digital manufacturing-driven transformations of service supply chains for complex products. Supply Chain Manag. Int. J. **19**(4), 421–430 (2014)
32. Eyers, D.R., Dotchev, K.: Technology review for mass customisation using rapid manufacturing. Assembly Autom. **30**(1), 39–46 (2010)
33. Gosling, J., Hewlett, B., Naim, M.M.: Extending customer order penetration concepts to engineering designs. Int. J. Oper. Prod. Manage. **37**(4), 402–422 (2017)

Design for Additive Manufacturing Using LSWM: A CAD Tool for the Modelling of Lightweight and Lattice Structures

Alessandro Ceruti[1(✉)], Riccardo Ferrari[2], and Alfredo Liverani[1]

[1] DIN Department of Industrial Engineering, University of Bologna, Bologna, Italy
alessandro.ceruti@unibo.it
[2] School of Engineering and Architecture, University of Bologna, Bologna, Italy

Abstract. This paper presents the development of a CAD conceived to support the modelling of lightweight and lattice structures just from the initial stages of the design process. A new environment, called LWSM (acronym of LightWeight Structures Modelling), has been implemented in Python programming language in an open-source CAD software to allow the fast modelling of several sandwich structures or the filling of solid parts with cubic and tetrahedral lattice structures which can be produced by Additive Manufacturing (AM) techniques. Several tests have been carried out to validate the tool, one of which is included in the paper. The design of a bracket component inside LWSM using a traditional dense geometry and a lattice structure is described. The use of Design for Additive Manufacturing (DfAM) functions helps the user in the design of innovative structures which can produced only with AM technologies. A significant change in the shape of the part respect to traditional solutions is noticed after the use of DfAM functions by experimenters: FEM analysis confirms a strong weight reduction.

Keywords: CAD · Lattice structure · Additive manufacturing · Lightweight structures · FEM analysis

1   Introduction

One of the challenges of modern engineering is to obtain lightweight structures and to compress as much as possible the "time to market" of products. Making optimized, light and easy to manufacture parts is needed for environmental and commercial purposes in all industrial fields, but in particular within manufacturers of transportation (air, sea, rail) systems, and biomedical application where it is important to replicate structures belonging to nature. For instance, there is a positive lever action in aircraft design when structural weights are reduced [1]: to spare the structural mass of 1 kg can lead to obtain a reduction of 3 or even more Kg of the aircraft total mass. In this way, a smaller propulsion system is required to fly and the fuel burned per passenger is reduced, with a beneficial effect on the environment pollutant emission and economy. Also other kinds of industry are interested in reducing structural weights: the cabin floor of the Italian Vivalto train, produced by Ansaldo Breda, is made of aluminum [2] to reduce the coaches weight and consequently the electric energy required for locomotion. Nature

© Springer International Publishing AG 2017
G. Campana et al. (eds.), *Sustainable Design and Manufacturing 2017*, Smart Innovation, Systems and Technologies 68, DOI 10.1007/978-3-319-57078-5_71

presents several examples in which a lightweight structure is obtained by using sparse truss based structures [3]: bones are typical examples of such a kind of structures. The engineering practice can benefit from the imitation of natural structures whose shape have been optimized by the evolution through a thousand of years long design. On the contrary, the traditional manufacturing processes are mainly based upon the removal of material, operation leading to weighty parts. Raw shapes like that obtained through the lamination process or the foundry are usually worked with chip removal operations (lathe, drill, milling machine) to obtain the shape of a part. Nowadays a new revolution is spreading in manufacturing: Additive Manufacturing processes for powders, like the Electron Beam Melting (EBM) or Selective Laser Sintering (SLS) allow the generation of complex and sparse structures obtained by melting material layer by layer. Structures based on the repetition of thin closed beam based elements (called lattice) [4], squared hole shapes, undercuts, sparse elements are now possible with very few limitations to the designer. Unlike traditional machines, there is no need to keep into account the working trajectory of a tool in the space, thus enabling the manufacture of more complex shapes. From a design point of view, AM is effective [5] when the concept "the design drives the shape" is adopted, since there is no limit to shapes which the designer can select. Complex geometries are usually obtained after an optimization process which is usually carried out to guess the lightest shape respecting the requirements on loads and constraints of the problem. On the other hand, traditional design requires to follow the rule "the shape drives the design" since it is not possible to obtain some shapes: the designer can model only parts which can be further properly machined and he must remember this while shaping a part. The challenge of the next years will be to develop a "new" class of tools and to train designers able to design parts exploiting the capability of AM to produce complex but structurally efficient shapes. CAD software packages which have been conceived to support traditional manufacturing processes must be improved to support the AM designer with new tools to allow the fast design and sketch of the structures typical of the AM process. New geometries and shapes can be produced with AM, so that among "Design for X" techniques, the concept of Design for Additive Manufacturing (DfAM) has been introduced to stress the attention on how AM has changed (and will further change in next years) the current product development cycle. From a research point of view, a bulk of literature deals with the design of lattice structures [6, 7] and topological optimization [8, 9]. However, several general purpose CAD systems do not still handle lightweight or lattice structures in an effective and efficient way. The aim of this paper is to describe the implementation of a CAD environment, called LightWeight Structures Modelling (LWSM), developed to support the designer while sketching lightweight and lattice structures. It is a matter of fact that current parts design workflow is based upon the CAD modelling of components conceived as to be obtained by chip removal machines; in the further, this parts are imported in external optimization environments [10] in which a topological optimization or the change from dense to lattice or graded structures is carried out. The LWSM tool we present has been written in Python programming language and implemented in the open source FreeCAD software in order to contribute to the development of tools able to support the everyday use of functions AM oriented, and to allow designers to think in an "Additive way" [11].

2   New Structures and CAD Systems

AM can be used to produce smart structures [12] based on three strategies: density graded materials, dense shapes obtained through topological optimization, lattice structures based parts. First two kinds of structures do not require particular attentions from a CAD perspective since they can be modelled by solid materials, eventually with a change in properties (like mass or elastic modulus) depending on the location in the body reference axis: the material can be considered locally isotropic, even if properties are position-dependent. Lattice structures require much more attention: when the size of the cells which are repeated to obtain the part is large, they can be sketched one by one, like usually done for nerves or thin appendices in solid bodies. Also in case of small cells there are no problems because the material can be considered both in CAD and in FEM as isotropic with a dense visualization. When the size of lattice structures cells are intermediate new functions are required. On the one hand the designer can't waste time to model a large number of small complex structures. On the other, a dense uniform visualization doesn't give an idea of the structure and can't be used to represent the component shape in a CAD system. There are problems also in FEM packages since the meshing of 3D structure based on small cellular elements requires a lot of nodes and solid elements to be discretized. Studies [13, 14] are focused on how to implement a sort of equivalent material to be applied to dense bodies to speed up analyses while describing in a proper way cellular structures. However, CAD systems must present the lattice structure in a realistic way and the modelling of these structures should be carried out in a very short time to reduce the user workload. A similar concept applies when sandwich structures based parts are modelled because also in this case tools and modelling functions to help the designer are required. From our point of view, lattice/lightweight modelling functions are required in CAD tools, so that parts can be shaped using the typical AM structures just from the beginning of the prototyping workflow. To optimize or to convert dense parts to lattice cells after the modelling in external tools can lead to losses in efficiency and can drive the final shaping towards sub-optimal solutions. A new class of structures, impossible to produce just a few years ago are now available to the designer: these geometries are usually based upon the repetition of small cells or elements and present a quite high geometrical complexity, so that the generation of a part conceived implementing DfAM concepts should be supported by specific functions developed to reduce the CAD user workload.

3   FreeCAD LWSM Environment

An environment to support the design of lightweight structures has been developed in the FreeCAD software. FreeCAD is an open source CAD in which it is simple to implement new additional environments respect to what already available in the downloadable release (part design, sketching, assembly, FEM analysis just to cite some of them). The programming of new environments is simple due to the capability of recording actions carried out by the user during a working sessions: commands are translated into a code called "macro" which can replicate in an automatic way what the human user did manually. In this way, it is possible to automate complex procedures or actions.

A palette function has been designed for the environment, called LWSM. In this first implementation of LWSM there are 4 functions to implement the lattice based discretization of a solid body, and 4 functions to sketch parametric sandwich panels. The Fig. 1 presents an image of the LWSM menu integrated in FreeCAD commands panel. There are two separate palettes in which 4 commands are grouped for each one. The first palette includes commands useful to modify the internal structure of already modelled 3D bodies, filling them with lattice/regular structures. They are: "3D Part filling with extruded hexagons", "3D Part filling with extruded cylinders", "3D Part filling with cubical lattice structures", "3D part filling with tetrahedral lattice structures". The following Figs. 2 and 3 present some examples based upon the filling of a solid shape given by the Boolean union of a cube with a 100 mm long edge and a cylinder with an height of 50 mm to show how these functions can work with whatever solid shapes.

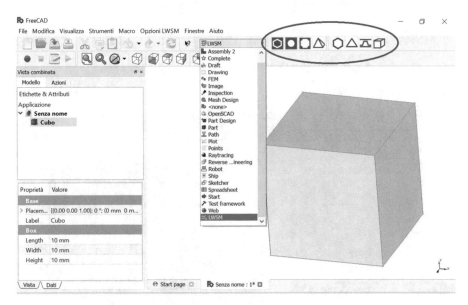

Fig. 1. LWSM menu palette with implemented functions

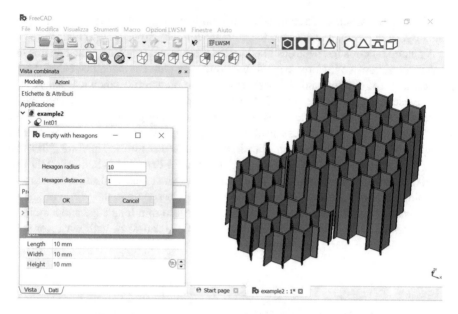

Fig. 2. "3D part filling with extruded hexagons" function

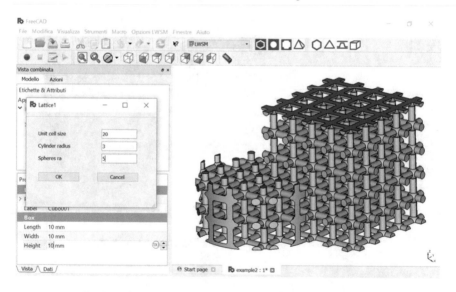

Fig. 3. "3D part filling with cubical lattice structures" function

This set of functions in which lattice/regular structures are applied to already modelled 3D dense parts works in the following way: the user selects the body whose solid structure must be changed; a box inscribing the object is computed by using functions available in FreeCAD; a "negative" lattice structure is sketched in all the box;

finally a Boolean intersection between the body and the "negative" lattice structure is carried out to obtain the final shape.

The command "3D Part filling with cubical lattice structures" for instance is based upon the discretization of the body with elements based on a cube in which spheres lies at the 8 corners, connected along the edges by cylinders. Both the element size and radii of spheres/cylinders can be set by the user. The second palette contains another set of functions for the automatic design of structures where the user selects through a menu dimensions and properties of the structure he/she wants to model. In this case "Create a sandwich with hexagonal cells", "Create a sandwich with triangular cells", "Create a Navtruss like sandwich", "Create a lattice box" are names of implemented functions whose symbols can be seen in Fig. 1 top-right corner.

Just to provide some examples, Figs. 4 and 5 show the control menu to set the data required for modelling and an example of the final structure which can be obtained (in transparency to show the internal structure features).

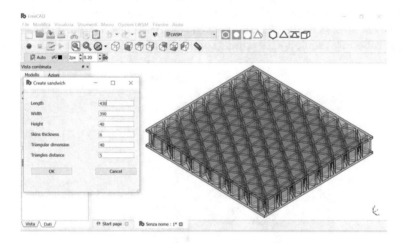

Fig. 4. "Create a sandwich with triangular cells" function

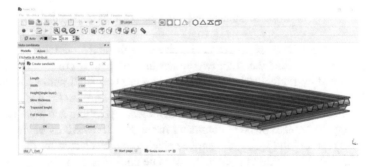

Fig. 5. "Create a Navtruss like sandwich" function

To give an idea of how these 4 functions have been modelled, they are based upon the implementation of the parametric modelling of structures imitating what carried out by the user during the sketch of sandwich/lattice structures. It is worth noting that all structures obtained through the LWSM tool are solid and can be used for FEM analysis, eventually using the FEA tool of FreeCAD or exporting the solid model in "step" file exchange format and importing it in external FEM codes in the further. Moreover, structures can be saved in STL extension, in order to be manufactured using AM machines.

4   Case Study

The tools we developed have been tested in several case studies to evaluate the effect of the implementation, just from the CAD modelling, of functions useful for the "DfAM". In this paper, we present an example related to the design of a bracket with three bosses, loaded with a lateral force of 1000 N. The position, diameter, and thickness of bosses is shown in Fig. 6, as well as the constraint and load positions. The material selected for the application are stainless steel powders which are commonly used in EBM machines.

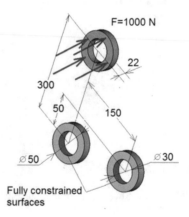

Fig. 6. Load, constraints, and dimensions for the case study

Aim of the design is to obtain a structure able to withstand a lateral load of 1000 N in the upper boss, provided the other two bosses are completely constrained. The part has been obtained by sketching bosses, adding a connecting shell with a thickness set equal to 20 mm and finally converting it into the lattice structure. At the end, bosses and the lattice shell have been glued together with a Boolean operation, and holes have been obtained using the "create pocket" command in the PART DESIGN environment of FreeCAD. As reported in Fig. 7a, the lattice cubic structure selected for the modelling presents cells with edge of 20 mm, equal to the plate thickness, and beams with a 4 mm diameter. A following FEM analysis, carried out in the FreeCAD Calculix solver, and in other FEM to be confident in the solution, showed a maximum stress of around 150 MPa (Fig. 7b). The FEM analysis has been carried out meshing the lattice structure (which is made of thin elements and requires a lot of nodes) and bosses with 3D tetra-hedral elements: a quite large number of nodes (around 350,000) has been obtained, but widely within commercial FEM computational capabilities.

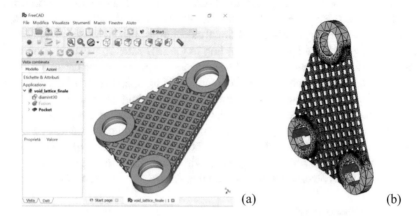

(a)   (b)

Fig. 7. Lattice based bracket: modelling and FEM analysis

If the model had been modelled as a dense metallic part it would have resulted as included in Fig. 8a where bosses have been welded to a plate with a constant thickness.

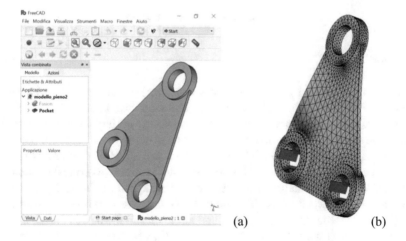

(a)   (b)

Fig. 8. Dense shell based bracket: modelling and FEM analysis

As an alternative, the milling of the component from a raw rectangular plate would be also possible. In this "traditional" scenario, the thickness of a plate assuring the same maximum stress of 150 MPa would have been defined after a set of FEM analysis, as did in this research. The Fig. 8b presents an image of one of the FEM analyses made to properly set the thickness of the plate, which resulted equal to around 7.4 mm to obtain a stress level similar to that obtained in the lattice based shape. The dense "traditional" model mass is around 5.2 kg (7.4 mm thickness), compared to the 4.4 kg of the lattice solution (20 mm lattice elements thickness). From a technological point of view, it is worth noting that the AM choice presents advantages over the traditional manufacturing. When the bracket is obtained by milling a thick block of steel a long time for machining should be assumed (together with high costs for machining). When it is obtained by welding bosses to a laser cut plate, additional calculations should be carried out to verify requirements on welding strength according to the ISO rules for welded components. In addition to this, the welding should be controlled with Non Destructive Inspections in case of high value components and a precise welding operation would require a tool to refer the components one another. Moreover, the welding could deform the component due to the high temperatures reached during the working. So, even if the AM can't be considered a cheap technology in case of small lots of components, it can be competitive due to the sparing in weights and the short "time to market" of the component. LWSD can be used also for the design of hybrid components in which a part in AM is added to dense parts obtained with traditional machines, as Fig. 9 shows.

Fig. 9. Hybrid AM with LWSM

5 Conclusions

The modern engineering is more and more concerned in reducing the environmental footprint of human habits and needs: both buildings and industrial products must contribute to reduce pollutant emissions and save energy during the whole lifecycle. At industrial level, the Additive Manufacturing can be an answer to the need for lightweight structures in engineering. Potentialities of AM are magnified when a proper and "AM oriented" design is carried out. Instead of using dense structures typical of subtractive technologies, graded or lattice structures should be adopted by the designer, just from the phase of component modelling within CAD software packages. A large part of current CAD systems can't effectively support the user in the design and sketching of this kind of new structures; parts, modelled as dense bodies are usually imported in external tools where the optimization or the conversion into lattice structure is carried out. An environment to support the design of lightweight structures has been implemented in the FreeCAD software and tested. Several functions have been implemented. After an evaluation phase, it is straightforward that this tool can be useful for students and professional designers to implement new structures which can benefit of AM capabilities. The user can "think additive" just from the modelling phase to completely exploit potentialities which AM can offer with proper shapes. A case study of a bolted plate that was modelled in traditional and lattice structure demonstrated the usefulness of the LWSM tool since it is possible to model complex structures in an effective way. As a matter of fact, using a traditional structure (mass = 5.2 kg) instead of a lattice based one (mass = 4.4 kg) an increase of the 18% in mass is noticed. Future works will add new functions to the tool and include the experimentation of the LWSM by structural designers when the weight reduction is of paramount importance as for transportation or biomedical applications.

References

1. Raymer, D.P.: Aircraft Design: A Conceptual Approach (Aiaa Education Series) 5th edn., ISBN-13: 978-1600869112 (2012)
2. Metra Consulting, Aluminum extrusions for the railway sector. http://www.metra.it/aluminium/EN/case-history-industry-unimaginable-sectors-vivalto-a-high-capacity-double-decker-train-railway-sector-259.aspx. Accessed Nov 2016
3. Boruciński, M., Królikowski, M.: Design of non-uniform truss structures for improved part properties. Adv. Manufact. Sci. Technol. **37**(4), 77–85 (2013)
4. Hadi, A., Vignat, F., Villeneuve, F.: Design configurations and creation of lattice structures for metallic additive manufacturing. In: 14eme Colloque National AIP PRIMECA, Mar 2015, La Plagne, France (2015)
5. Staiano G., Gloria A., Ausanio G., Lanzotti A., Pensa C., Martorelli M.: Experimental study on hydrodynamic performances of naval propellers to adopt new additive manufacturing processes. Int. J. Interact. Des. Manuf. 1–14 (2016). ISSN: 1955-2513, doi:10.1007/s12008-016-0344-1
6. Aremu, A.O., Brennan-Craddock, J.P.J., Panesar, A., Ashcroft, I.A., Hague, R.J.M., Wildman, R.D., Tuck, C.: A voxel-based method of constructing and skinning conformal and functionally graded lattice structures suitable for additive manufacturing. Add. Manuf. **13**, 1–13 (2017)
7. Rosen, D.: Computer-aided design for additive manufacturing of cellular structures. Comput. Aided. Des. Appl. **4**(5), 585–594 (2007)
8. Bendsoe, M.P., Sigmund, O.: Topological Optimization, Theory. Methods and Application. Springer-Verlag, Berlin (2004)
9. Taggart, D.G., Dewhurst, P.: Development and validation of a numerical topology optimization scheme for two and three dimensional structures. Adv. Eng. Softw. **41**, 910–915 (2010)
10. Altair Optistruct software. http://www.altairhyperworks.com/product/OptiStruct. Accessed Nov 2016
11. Ceruti, A. Liverani, A., and Bombardi, T.: Augmented vision and interactive monitoring in 3D printing process. Int. J. Interact. Des. Manuf. (IJIDeM) 1–11 (2016). doi:10.1007/s12008-016-0347-y
12. Martorelli, M., Gerbino, S., Lanzotti, A., Patalano, S., Vitolo, F.: Flatness, circularity and cylindricity errors in 3D printed models associated to size and position on the working plane. In: Eynard, B., Nigrelli, V., Oliveri, S.M., Peris-Fajarnes, G., Rizzuti, S. (eds.) JCM 2016. LNME, pp. 201–212. Springer, Cham (2007). doi:10.1007/978-3-319-45781-9_21
13. Ruegg, A.W.: Implementation of generalized finite element methods for homogenization problems. J. Sci. Comput. **17**, 671–681 (2002)
14. Gonella, S., Ruzzene, M.: Homogenization of vibrating periodic lattice structures. Appl. Math. Modell. **32**(4), 459–482 (2008)

Additive Manufacturing as a Driver for the Sustainability of Short-Lifecycle Customized Products: the Case Study of Mobile Case Covers

Paolo Minetola[1(✉)] and Daniel R. Eyers[2]

[1] Department of Management and Production Engineering, Politecnico di Torino, Turin, Italy
paolo.minetola@polito.it
[2] Cardiff Business School, Cardiff University, Cardiff, UK
EyersDR@cardiff.ac.uk

Abstract. Unlike subtractive manufacturing processes, by their own nature additive technologies offer the potential to reduce both raw material consumption and production waste. The positive impact of Additive Manufacturing on sustainability is more evident for those mass consumption products that are characterized by a short lifecycle. Nevertheless, in order to take full advantage of these technologies and let them achieve their full potential, a great change is needed in the behaviour and attitude of consumers. Compromises may be necessary to accommodate technological limitations, and customers may need to prioritize between functional and aesthetic product characteristics. The positive impact of additive manufacturing on sustainability is more evident for those mass consumption products that are characterized by a short lifecycle. The material flow and sustainability of 3D printing are described in this paper for the manufacturing of mobile case covers, and compared to those of mass production injection moulding.

Keywords: Additive manufacturing · 3d printing · Customization · Mobile cover · Recycling · Sustainability

1 Introduction

Recently Additive Manufacturing (AM) has been identified as the driver of the third industrial revolution [1, 2]. Additive technologies enable the fabrication of a product layer by layer directly from its virtual CAD model on a single machine without the need of any additional tools or equipment. Thus, AM implies deep changes in the way designers and manufacturers have approached production during the last 140 years since Taylor's second industrial revolution in the late 19th Century.

Researchers have been investigating the sustainability [3–10] of these relatively new manufacturing technologies, with attention extended to implications for both supply chains [11–14] and society [5, 15]. Various approaches have been taken to the assessment of sustainability and cost analysis, and in some cases this has been extended to the whole lifecycle of the product, with a comparison to conventional manufacturing processes. Nevertheless, while assessing the economic convenience of Additive

© Springer International Publishing AG 2017
G. Campana et al. (eds.), *Sustainable Design and Manufacturing 2017*, Smart Innovation, Systems and Technologies 68, DOI 10.1007/978-3-319-57078-5_72

Manufactured product, it is notable that the cost of externalities is not typically considered. In some respects, it must be acknowledged that externalities are difficult to evaluate, but they cannot be disregarded, particularly when related to waste management.

Two different kinds of studies are available in the literature concerning the sustainability of AM technologies. On the one hand, there are conceptual qualitative studies that thoroughly analyse a framework for implementation of AM by detailing advantages, shortcomings and still open issues [16–18]. On the other hand, quantitative analyses are related to material savings, energy saving, reduction of time to market, improvement of product's performances, but the studies disregard the end-of-life management of the components [19, 20].

The aim of this paper is the analysis of material flow to later develop a system dynamics model for quantitative assessment of the sustainability of 3D printing and pull production through comparison with that of injection moulding, i.e. push production.

The study is extended to the whole lifecycle of the material, including the recycling and disposal of cell phones case covers that are selected as case study.

Nowadays rampant consumerism of electronic products having a short life cycle, such as mobile phones, justifies low costs for the mass production of low-end accessories by conventional processes. These accessories, for instance case covers, become outdated soon after the obsolescence of the electronic device. The push production of the case covers and the retailed distribution worldwide of short-life cycle accessories is not sustainable and AM might replace the manufacturing model to improve sustainability and circular economy. Of course a change in consumers' habits is needed for an efficient shift from push production of mass consumption goods to on-demand production by means of AM. The shorter the market demand of the good, the greater the impact that make-to-order manufacturing would have on the sustainability. The waste issues of electronic products and their accessories is aggravated by fast innovation, which reduces the average product lifetime and accelerates product obsolescence.

2   The Case Study

Mobile phones represent an interesting challenge for sustainable manufacturing. Because users desire to substitute obsolete devices with newer and more innovative models, the average lifecycle of a mobile phone is estimated to be between 15 and 30 months [21, 22]. Mobile phones are often accessorised by their owners using either case covers for functional reasons (e.g. protection from damage), or aesthetic customization reasons. Case covers fit specific mobile phones, and as a result share similar obsolescence timescales.

In small shops or big consumer electronics retailers, there are racks full of plastic or silicone case covers for different models of mobile phones available on the market worldwide. Nowadays the production of mobile phone case covers is a push system. Case covers are typically made to stock, requiring entities in the supply chain to make forecasts on likely demand requirements, without actual demand data. Where demand is underestimated there is inadequate stock to satisfy demand. Where demand is overestimated, excess stock lingers in the supply chain. The production volume of mobile covers largely exceeds the real demand from the customers. This difference results in lots of unsold covers, that are out of the market earlier because their design does not meet users taste or later when the market is saturated or the model of the phone has become obsolete.

The result of this production model is an enormous waste of resources for accessories that are out the market in two to three years. The product life-cycle for mobile phone cases is heavily affected by fashion and individual customer preferences. Both of these change with time, and for certain demographics (e.g. teenagers), it is feasible that they may wish to change their phone case cover multiple times during their ownership of an individual phone [23]. The offer of case covers is huge and varied in order to meet customers taste, but also to keep with fashion trends.

Usually mobile case covers are sold in electronic shops, customer service points of mobile operators, dedicated shops (Fig. 1a). Moreover, it is possible to find and purchase mobile covers also in fashion accessories stores and sometimes in clothing stores. Market stalls (Fig. 1b) and fixed price discounter can be considered second-level retailers for the distribution of lower quality covers, or covers for mobile phones that have been succeeded by newer models and are no longer on the market. These retailers therefore provide a useful role of achieving revenue for obsolete stock and freeing-up capacity in distribution warehouses, albeit at far reduced profitability (or even loss).

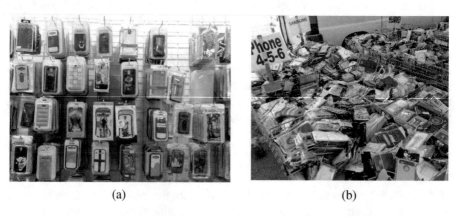

(a)   (b)

Fig. 1. Case covers for mobile phones on racks in shops (a) and in market stalls (b)

In the current make-to-stock production, since demand is not known in advance, case covers are fabricated with a large range of colors and designs the user can choose from. Typically, the case shell is produced by injection moulding, requiring an expensive tool to be developed for the exact phone shape, and hence necessitating large production runs to recover this fixed cost. The external surface of a cover may contain artwork (Fig. 2a) or printing applied by various printing processes (Fig. 2b). Other types of decorations include in-mould lamination and in-mould decoration (Fig. 2c). In this approach, individual mobile phone case designs are produced in high volumes, and the customer has no involvement in the design or configuration of their product, which are requisite for 'customized 'production' [24]. As a result, customer demand is satisfied by provision of variety, rather than by enabling customization.

(a)   (b)   (c)

Fig. 2. Example of injection moulded mobile case covers

An emerging opportunity to overcome this limitation, and to provide phone cases that meet individual customer requirements is offered by AM. Instead of producing millions of cases with different design to satisfy the larger number of consumers, AM can be used for a customization in mass production according to a make-to-order model. In AM, the absence of specific tools, dies or fixtures and the lack of geometric constraints enables the production of similar but not identical parts, that might be individually customized to suit customer's needs and desires [25, 26].

Examples of customized 3D printed covers for cell phones are shown in Fig. 3. Their customization is different from that of injection moulded covers, since no photo, image or 2D artwork can be easily printed on it. In order to fully exploit the advantages of AM technologies in terms of design freedom, AM customization should consider the fabrication of complex intricate 3D shapes or the enhancement of the cover functionality. For example, a triangular ruler is included in the design of the cover in Fig. 3a. The ruler can be detached for the cover and used separately to be later repositioned in place and carried along with the mobile phone. Alternatively, complex 3D organic patterns can be reproduced in the cover design (Fig. 3b) or small live assemblies, such as movable gears (Fig. 3c), can be manufactured as a unique part by combining one or more materials and colors.

Another benefit of tool-less manufacturing is that costs are cut down because no tool has to be fabricated and there is no raw material consumption for toolmaking. Thus AM technologies allows for the fabrication of unique pieces, small batches and large volume production with similar cost per part, since the tool cost is not divided among the number of fabricated products [27, 28].

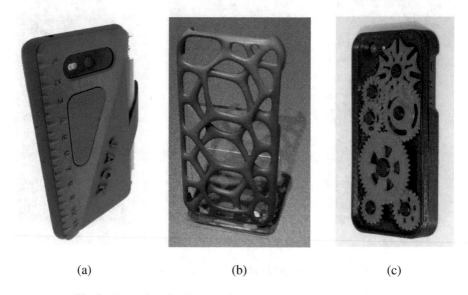

(a)   (b)   (c)

Fig. 3. Examples of AM customized case covers for mobile phones

The cover can be directly fabricated from its 3D CAD model immediately after the approval of the design. No time is wasted for the fabrication and delivery of the manufacturing tooling and fixtures that usually take some weeks. Currently a 3D printed cover can be ordered on-line or at fab-labs. The consumer can self-produce with his/her own 3D printer as well. In the latter case, the term prosumer was coned to indicate a consumer that produces the goods by him/herself [5]. In the case of fab-labs or prosumers, the production is localized close to the final user, so the manufacturing is distributed and decentralized. The potential re-localization of manufacturing through the adoption of AM has been described in the literature as the last event of a periodic historical cycle of localization-globalization-relocalization that perfectly comply with Vico's theory of recurrent historical patterns [29]. Distributed manufacturing results in simplification of the supply chains and improved efficiency and responsiveness in the fulfillment of the demand [8, 14, 19].

3   Model for Make-to-Stock Manufacturing

The material flow for injection moulded mobile covers is shown in Fig. 4. A steel mould is fabricated and used for the production of large volumes of identical covers starting from pellets of a polymeric material that is molten and injected into the mould by means of an injection moulding machine. In most of the cases the covers will be finished to improve the aesthetics and variety to meet the wider estimated demand from the customer. Covers are usually manufactured in China or emerging Asian countries and then distributed worldwide to electronic shops or dedicated shops that can be considered first level retailers. We assume that unsold items from first level retailers are then distributed to market stalls or fixed price discounter to have the opportunity to ideally saturate the market by second level retail aiming at achieving additional revenue for the obsolete stock. The end-of-life management of the covers depends on the user for sold items and on the second level retailers for unsold ones. If covers are correctly recycled, they will be converted into small plastic grains again, otherwise they will be disposed in incinerators or landfills.

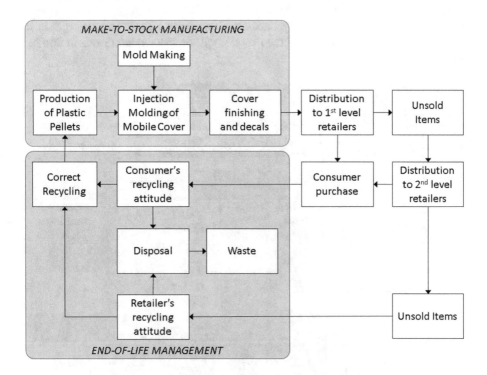

Fig. 4. Life-cycle of injection moulded covers for mobile phones

4   Model for Make-to-Order Manufacturing

Among AM technologies, the process of Fused Deposition Modelling, that is popularly renown as 3D Printing is taken into consideration as an alternative to traditional injection moulding for make-to-demand production of mobile covers. FDM belongs to the material extrusion processes according to the ISO/ASTM classification [30, 31] as it implies the extrusion of a molten polymeric filament and the subsequent deposition of the filament on the building platform [32]. 3D printing is the most diffused AM technology, thanks to the low cost of the FDM machines, whose price starts from about 250 euros for a kit that the user should self-assemble.

The material flow in the case of a 3D printed cell phone cover is summarized in Fig. 5. The raw polymer grains are first extruded to produce the filament that is then used to feed the 3D printer. The filament is usually stored in 1 kg spools that can be purchased on-line. Once the cover is 3D printed, it can be delivered to the customer or the customer can pick it up at the decentralized fab-lab or 3D printing shop. In the case of the prosumer the supply chain is even shorter because the final distribution is not necessary. Once the user does not need the cover anymore, this will be disposed. The prosumer might have an extruder to recycle 3D printed wastes and make new filaments directly. Discounts or recycling policies might also be proposed by fab-lab or 3D printing shop if the user return 3D printed wastes. Otherwise the cover might be recycled together with other plastic products for the generation of new pellets. In the worst case, the user disposes the cover incorrectly and the material will be incinerated or end its life in a landfill.

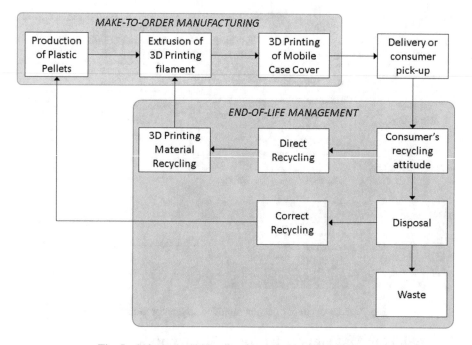

Fig. 5. Life-cycle of 3D printed covers for mobile phones

5 Conclusions

At the moment, sustainability implications of additive manufacturing are not correctly perceived by consumers. Although the diffusion of 3D printing technology is constantly increasing thanks to the affordability of low-cost 3D printers, many people are not aware of the potential impact of AM technologies on waste reduction. The shorter the product life cycle, the higher the impact of 3D printing, provided that the user correctly disposes the goods at their end of life.

However, in the literature waste management is often not considered in quantitative studies about the sustainability of 3D printing. Thus there is the need to assess the difference in sustainability between make-to-stock manufacturing and distributed make to-order production along the whole product life cycle, including the end-of-life management. For this reason, the material flow of injection moulded cell phone case covers and 3D printed covers is described was described in this paper as a preliminary analysis for a more complex system dynamics model to be developed in future research activities.

The approach of consumers to 3D printing solutions for short-life goods, also requires a shift to a lower level of aesthetical demand, that is balanced by the possibility to improve product functionality and customization. In fact, apart from limited material choices, one of the main shortcomings of additive technologies is the coarse achievable dimensional accuracy and scarce superficial finishing. The continuous surfaces of the 3D CAD model of the part undergo a first discretization process during the generation of the tessellated model of the STL (Solid To Layer) format that is used for the computation of the machine path in additive systems. The path is created through a 3D slicing operation that cuts the STL model into cross sections having the same thickness of the building layer, thus introducing a second approximation of the nominal geometry of the part. As a consequence of the use of a finite thickness for each layer, the as-build surfaces are characterized by a staircase effect that is more evident on vertical walls, inclines and curved faces. Finishing operations can be performed to improve the aesthetical appearance of additive manufactured products with an increase in costs.

In the craft sector, however, aesthetic imperfections are positively valued by the customer, since they are considered a distinguishing element for unique pieces manufactured by a craftsman. In such a context, together with the appreciation of the craftsman's skill, product's design and shape overcome the need of aesthetic perfection. The diffusion of additive technologies might benefit from a similar recognition by customers with a positive impact on sustainability. As long as AM parts will be compared to mass-produced products, consumers will maintain high expectations and requirements about the aesthetics. In terms of pleasing appearance, the added value of additive technologies should be recognised in the opportunity of creating an attractive design through shapes that could not be fabricated otherwise. The possibility to customize the product with personal distinguishing features is another key element to attract customer's affection. Sophisticated design and customization are two factors that can mitigate the lower aesthetic appearance of as-built additive manufactured parts without the need for additional superficial finishing. Moreover, in a framework of sustainability, the removal of any additional finishing operation has positive impact on resources and material savings.

References

1. Berman, B.: 3-D printing: the new industrial revolution. Bus. Horiz. **55**(2), 155–162 (2012)
2. The Economist: Manufacturing: The third Industrial Revolution. The Economist Special Report "The Third Industrial Revolution" (2012)

3. Baumers, M., Tuck, C., Bourell, D.L., Sreenivasan, R., Hague, R.: Sustainability of additive manufacturing: measuring the energy consumption of the laser sintering process. Proc. Inst. Mech. Eng. Part B J. Eng. Manuf. **225**(B12), 2228–2239 (2011)

4. Baumers, M., Tuck, C., Wildman, R., Ashcroft, I., Rosamond, E., Hague, R.: Transparency built-in energy consumption and cost estimation for additive manufacturing. J. Ind. Ecol. **17**(3), 418–431 (2013)

5. Chen, D.F., Heyer, S., Ibbotson, S., Salonitis, K., Steingrimsson, J.G., Thiede, S.: Direct digital manufacturing: definition, evolution, and sustainability implications. J. Clean. Prod. **107**, 615–625 (2015)

6. Faludi, J., Bayley, C., Bhogal, S., Iribarne, M.: Comparing environmental impacts of additive manufacturing vs traditional machining via life-cycle assessment. Rapid Prototyping J. **21**(1), 14–33 (2015)

7. Gebler, M., Uiterkamp, A.J.M.S., Visser, C.: A global sustainability perspective on 3D printing technologies. Energ. Policy. **74**, 158–167 (2014)

8. Kohtala, C.: Addressing sustainability in research on distributed production: an integrated literature review. J. Clean. Prod. **106**, 654–668 (2015)

9. Le Bourhis, F., Kerbrat, O., Hascoet, J.Y., Mognol, P.: Sustainable manufacturing: evaluation and modeling of environmental impacts in additive manufacturing. Int. J. Adv. Manuf. Tech. **69**(9–12), 1927–1939 (2013)

10. Mani, M., Lyons, K.W., Gupta, S.K.: Sustainability characterization for additive manufacturing. J. Res. Natl. Inst. Stan. **119**, 419–428 (2014)

11. Dawes, J., Bowerman, R., Trepleton, R.: Introduction to the Additive Manufacturing Powder Metallurgy Supply Chain Exploring the production and supply of metal powders for AM processes. Johnson Matthey Tech. **59**(3), 243–256 (2015)

12. Eyers, D.R., Potter, A.T., Gosling, J., Naim, M.M.: Supply chain flexibility for additive manufacturing. In: 20th EurOMA Conference 2013, Dublin, Ireland (2013)

13. Khajavi, S.H., Partanen, J., Holmstrom, J.: Additive manufacturing in the spare parts supply chain. Comput. Ind. **65**(1), 50–63 (2014)

14. Liu, P., Huang, S.H., Mokasdar, A., Zhou, H., Hou, L.: The impact of additive manufacturing in the aircraft spare parts supply chain: supply chain operation reference (scor) model based analysis. Prod. Plan. Control. **25**(13–14), 1169–1181 (2014)

15. Huang, S.H., Liu, P., Mokasdar, A., Hou, L.: Additive manufacturing and its societal impact: a literature review. Int. J. Adv. Manuf. Tech. **67**(5–8), 1191–1203 (2013)

16. Ford, S., Despeisse, M.: Additive manufacturing and sustainability: an exploratory study of the advantages and challenges. J. Clean. Prod. **137**, 1573–1587 (2016)

17. Mellor, S., Hao, L., Zhang, D.: Additive manufacturing: a framework for implementation. Int. J. Prod. Econ. **149**, 194–201 (2014)

18. Weller, C., Kleer, R., Piller, F.T.: Economic implications of 3D printing: market structure models in light of additive manufacturing revisited. Int. J. Prod. Econ. **164**, 43–56 (2015)

19. Wittbrodt, B.T., Glover, A.G., Laureto, J., Anzalone, G.C., Oppliger, D., Irwin, J.L., Pearce, J.M.: Life-cycle economic analysis of distributed manufacturing with open-source 3-D printers. Mechatronics **23**(6), 713–726 (2013)

20. Achillas, C., Aidonis, D., Iakovou, E., Thymianidis, M., Tzetzis, D.: A methodological framework for the inclusion of modern additive manufacturing into the production portfolio of a focused factory. J. Manuf. Syst. **37**, 328–339 (2015)

21. Paiano, A., Lagioia, G., Cataldo, A.: A critical analysis of the sustainability of mobile phone use. Resour. Conserv. Recycl. **73**, 162–171 (2013)

22. Jang, Y.C., Kim, M.: Management of used & end-of-life mobile phones in Korea: a review. Resour. Conserv. Recycl. **55**(1), 11–19 (2010)

23. Reed, M: The influence of mobile phones on teenagers. In: Australian Science. www.australianscience.com.au/technology/the-influence-of-mobile-phones-on-teenagers/
24. Duray, R., Ward, P.T., Milligan, G.W., Berry, W.L.: Approaches to mass customization: configurations and empirical validation. J. Oper. Manag. **18**(6), 605–625 (2000)
25. Wong, H.W., Eyers, D.R.: Enhancing responsiveness for mass customization strategies through the use of rapid manufacturing technologies. In: Cheng, T., Choi, T. (eds.) Springer Handbook on Innovative Quick Response Programs in Logistics and Supply Chain Management, pp. 205–226. Springer, Heidelberg (2010)
26. Eyers, D., Dotchev, K.: Technology review for mass customisation using rapid manufacturing. Assembly Autom. **30**(1), 39–46 (2010)
27. Atzeni, E., Iuliano, L., Minetola, P., Salmi, A.: Redesign and cost estimation of rapid manufactured plastic parts. Rapid Prototyping J. **16**(5), 308–317 (2010)
28. Atzeni, E., Iuliano, L., Marchiandi, G., Minetola, P., Salmi, A., Bassoli, E., Denti, L., Gatto, A.: Additive manufacturing as a cost-effective way to produce metal parts. In: 6th International Conference on Advanced Research in Virtual and Physical Prototyping, VR@P 2013, Leira, pp. 3–8 (2014)
29. Zeleny, M.: High technology and barriers to innovation: from globalization to relocalization. Int. J. Inf. Tech. Decis. **11**(2), 441–456 (2012)
30. ASTM Designation: F2792-12a - Standard Terminology for Additive Manufacturing Technologies, pp. 1–3
31. ISO/ASTM Standard: 52900:2015 – Additive manufacturing – General principles: Terminology
32. Calignano, F., Manfredi, D., Ambrosio, E.P., Biamino, S., Lombardi, M., Atzeni, E., Salmi, A., Minetola, P., Iuliano, L., Fino P.: Overview on additive manufacturing technologies. Proc. IEEE **105**(4), 593–612 (2017)

About the Use of Recycled or Biodegradable Filaments for Sustainability of 3D Printing

State of the Art and Research Opportunities

Jukka Pakkanen[1,2(✉)], Diego Manfredi[2], Paolo Minetola[1], and Luca Iuliano[1]

[1] Department of Management and Production Engineering, Politecnico di Torino, Turin, Italy
jukka.pakkanen@polito.it
[2] Center for Sustainable Future Technologies CSFT@PoliTo, Istituto Italiano di Tecnologia, Turin, Italy

Abstract. Additive Manufacturing (AM) and 3D printing are drivers for material savings in manufacturing. Owing to the continuous diffusion of 3D printing driven by low-cost entry-level material extrusion printers, sustainability of a so popular AM technology is of paramount importance. Therefore, recycling 3D printed wastes and 3D parts again at the end of their life is an important issue to be addressed. Research efforts are directed towards the improvement of the biodegradability of 3D printing filaments and the replacement of oil based feedstock with bio-based compostable plastics. The aim of this work is to describe the state of the art about development and use of recycled or biodegradable filaments in 3D printing. Beyond a critical review of the literature, open issues and research opportunities are presented.

Keywords: Additive manufacturing · 3D printing · Fused Deposition Modelling (FDM) · Biodegradability · Recycling · Bio-based filaments · Sustainability

1 Introduction

Recycling is an important topic brought up by the European Union this year with the circular economy initiative, wherein "the proposed actions will contribute to close the loop of product lifecycles through greater recycling and re-use, and bring benefits for both the environment and the economy" [1]. Unlike in subtractive manufacturing processes, parts are fabricated layer by layer in Additive Manufacturing (AM) with a minimum allowance for finishing operations. Since AM allows for greater material savings than traditional processes, 3D printing can be considered a distributed manufacturing technology for improving sustainability and circular economy worldwide [2].

Most of polymeric materials are produced from exhaustible resources. Around 4% of worldwide production of oil and gas is used as feedstock for plastics and a further 3–4% is used to supply energy for the transformation of polymeric materials [3]. On the other hand, unlike oil-based polymers, bio-plastics are derived from renewable sources i.e. sugars and natural fibres.

Among AM technologies, Fused Deposition Modelling (FDM) is a low-cost technique that uses a thermoplastic filament to build parts layer after layer [4]. The popularity of FDM started with expiration of the patent of Sir Scott Crump by Stratasys and the

G. Campana et al. (eds.), *Sustainable Design and Manufacturing 2017*, Smart Innovation, Systems and Technologies 68, DOI 10.1007/978-3-319-57078-5_73

subsequent open-source Reprap project [5], that ever since has become the preferred choice of makers. The FDM process is more popularly known as 3D printing and belongs to the material extrusion category according to the ISO/ASTM terminology [6]. FDM systems or 3D printers have a simple design consisting of a Cartesian structure with three controlled axes, up to three hot extrusion heads and a building platform. The price of consumer-end 3D printers starts from 100 USD and makes these machines affordable and appealing to many people. It can be forecasted that more and more people will adopt this technology in the near future. People will 3D print amazing pieces, but also generate a lot of waste if material is not correctly disposed or recycled at the end of its life.

In terms of AM sustainability, environmental impact of 3D printing should consider the usage of resources, energy, emitted emissions and waste. Among these energetic demand is especially critical with industrial FDM machines that use a hot sealed working volume [7].

Gebler et al. used a two-step model for calculating the possible environmental impact of all 3D printing technologies globally by 2025 [8]. They estimated that, in the best case in which improvements are applied to 3D printing and production efficiency increases, the use of AM for creating new parts can lead to savings of about 5% in energy and CO_2 emissions in the manufacturing industry worldwide.

Apart from the energetic footprint, the material management of FDM waste needs to be addressed. The waste and excess material from 3D printed parts (e.g. support structures, filament ends and scraps) is the second important environmental aspect of FDM. Waste management can be directly dealt by makers or final users by means of material recycling and production of recycled filaments.

As concerns to material sustainability in 3D printing, the aim of this paper is to review the state of the art about the use of recycled or biodegradable filaments, discussing open issues and related research opportunities. The organization of this work is the following: the different types of materials available as 3D printing filaments are presented in Sect. 2, the main properties of recycled and biocompatible materials are summarized in Sect. 3, Sect. 4 deals with the material sustainability and finally research opportunities from open questions are discussed in the conclusions.

2   Filaments for 3D Printing

Many polymeric materials are available for 3D printing, even though the choice is very limited when compared to that of injection moulding polymers. There are few similarities in the properties of all FDM materials: the material needs to have a low melting point and a reduced viscosity in order to flow out of the nozzle for deposition and adhesion to the previous layer under the low pressure applied by the extrusion mechanism. The key point for base plastics is to set the proper extrusion temperature to achieve the right viscosity for a good flowability, whereas rheological behavior of composite filaments is more complex. The main issues of composites are related to the type of filler and the amount of water content [9].

Acrylonitrile butadiene styrene (ABS) is one of the most common materials in 3D printing. Good mechanical properties and extrudability make it the preferred choice of makers. It is usually not biodegradable, but the company Enviro ABS claims to have produced a biodegradable ABS [10]. The downside of ABS is the lack of UV resistance. To overcome this limitation, acrylonitrile styrene acrylate (ASA) has been developed. The second most popular material is polylactic acid (PLA). PLA is made from polymerization of sugars and starches, so it is biodegradable and recyclable. Other commonly used polymers are polycarbonate (PC), polyamide (PA) family plastics, high density polyethylene (HDPE), polyethylene terephthalate (PET) and flexible thermoplastic polyurethane (TPU).

There are studies about the emission of fumes to close proximity while 3D printing ABS and PLA [11, 12]. The fumes are ultra-fine aerosol (UFA) or volatile organic compounds (VOC) particles that might be harmful for humans. In the order to reduce the unpleasant smells and related health risks, a closed printer design and good ventilation of the room is generally recommended. Bio-plastic filaments like PLA produce less fumes and smells than oil-based ones, e.g. ABS.

2.1 Bio-degradable FDM Filaments

Apart from PLA, other biodegradable plastics used for FDM filaments are polyhydroxy-alkanoates (PHA), polyvinyl alcohol (PVA), Polyethylene terephthalate (PET) and High impact polystyrene (HIPS). The properties of these materials are summarized in Table 1.

Table 1. Bio-degradable materials for 3D printing

Material	Produced from	Properties	Extrusion temperature	Pros	Cons
PLA	Plants starch	Tough, strong	160 ÷ 222 °C	Bio-plastic, non-toxic, odorless, low-warp	Low heat resistance, brittle
PVA	Petroleum	Water-soluble, good barrier	190 ÷ 210 °C	Biodegradable, recyclable, non-toxic	Expensive, deteriorates with moisture, special storage
PHA	Sugars with biosynthesis	Several copolymers, brittle and stiff	~160 °C	UV-stable, stiffness	Elasticity, brittle
HIPS	Petroleum	High impact resistance, soluble in limonene	190 ÷ 210 °C	Biodegradable, low cost, similar to ABS	Warping, heated printing bed
PET	Petroleum	Strong and Flexible	210 ÷ 230 °C	FDA approved, Recyclable	Absorbs moistness

PHA can be used as it is or as a mixture with PLA. PVA is a water-soluble and biodegradable material used for support structures. Duran et al. [13] printed PVA as support structure for ABS. They found that PVA is printable at dried condition until 45 min before it absorbs moisture from the air and it turns impossible to print.

Polyethylene terephthalate (PET) is a common plastic used for food containers and tools. It is fully recyclable and safe to use with foods. Recycled filament from PET is sold commercially by B-PET Company since 2015 [14].

High impact polystyrene (HIPS) is similar to ABS with good mechanical properties and extrusion temperature. It is used as a support material for ABS as it dissolves in limonene, but ABS does not.

In biocompatible and medical applications, 3D printing filaments are made of polymers with low melting temperatures. These materials can be used in FDM to create parts that integrate within the human tissues, such as scaffolds for example. Chia et al. [15] and Serra et al. [16] list some of these materials.

Biodegradable 3D printed hollow capsules made of hydroxypropyl cellulose (HPC) for drug delivery systems have been presented by Melocchi et al. [17] and Pietrzak et al. [18]. These capsules are orally consumed and the degrading of the capsule in stomach releases the drugs concealed inside.

2.2 Bio-composite FDM Filaments

Bio-composites filaments available on the market consist of a biodegradable polymeric matrix and bio-based fillers. Fillers can be fibres or particles. Filler contents starts from few % up to 40% in volume. The most used thermoplastic is PLA and the filler can be sawdust, cellulose fibres or other natural fibres. Filament manufacturers have developed many wood like filaments for FDM [9] using different types of fibres: Bamboo, Birch, Cedar, Cherry, Coconut, Cork, Ebony, Olive, Pine, Willow. These are used to provide a wood-like tactile feeling to aesthetic parts. Seppänen et al. [19] have printed thermoplastic cellulose derivatives on an FDM machine.

An example of filament modification is given by Kuo et al. [20] who prepared a biomass mixture of thermoplastic starches and ABS (TPS/ABS). Plastic pellets of ABS were first created and infiltrated with a compatibilizer for joining the two polymers together. These pellets were then molten and filtrated with TPS and TiO_2 particles into new pellets that were extruded into a new 3D printing filament.

At MIT self-assembly lab, David et al. [21] used a multi-material approach to apply wood-inspired design to 3D printing. By exploiting the hydrophilic nature of wood-based filaments as base layer and non-hydrophilic ABS or PA as reinforcement for top layers, they produced hydro-induced actuation. A similar work in fibre actuation was done by Duigou et al. [22], but with one deposited material only.

Xhang et al. [23] have created high conductive PLA filaments by mixing graphite flakes into filaments by melt extrusion. Graphene mixed well with PLA and conductive PLA filament was successfully printed in 2D and 3D.

A binder made of potato-starch was used by Marina Ceccolini, who graduated at the University of Design of San Marino, to fabricate AgriDust. This is a material composed of food waste for 64.5% and of the potato binder for the rest. AgriDust requires a cold technology and can be 3D printed by replacing the classic extruder with a syringe. Because of its constituents, the material is biodegradable and non-toxic [24].

A soy-based filament named FilaSoy was developed by students of the Purdue University in Indiana, USA [25]. The 20–25% soy additive improves PLA performances by providing anti-microbial properties and reducing the brittleness without inhibiting compostability and recycling.

3   Tensile Properties of 3D Printed Materials

A limited selection of materials is available for 3D printing and 3D printed parts usually have a worse mechanical behavior than the same materials processed by extrusion, compression or injection moulding. However, 3D printing empowers engineers and designer with more design freedom than other traditional processes [9, 22]. Tensile properties of FDM tensile test specimens manufactured from plastics, composites and recycled filaments are reported in Table 2 when available from the literature (NA = Not Available).

Table 2. Properties of tensile test specimens made from different filament materials

Material	Ultimate tensile strength (MPa)	Elongation (%)	Young modulus (MPa)
ABS [26]	19.9–29.1	1.5–8.9	1910–2050
PC [26]	29.5–36.9	3–6.7	1620–2000
PLA [27]	49.1–65.5	1.7–5.0	2800–3600
PLA recycled once [28]	51	1.88	3093 ± 194
PLA recycled 5 times [28]	48.8	1.68	3491 ± 98
PLA/PHA+10–20% fibre [22]	20–30	0.9–1.1	3500–4000
PLA/PHA+10–20% fibre water saturated [22]	15–20	0.5–0.7	3100–3600
PLA+5% pine lignin [29]	40.2–43.6	2.31–2.83	2160–2200
TPS/ABS biomass [20]	34.8–46.8	NA	NA
PLA+graphite 2% [23]	50	8.1	NA
PLA+graphite 8% [23]	62	6.1	NA
HDPE virgin [30]	25.5	16.1	463.4
HDPE recycled once [30]	25.6	16.1	428.4

Mechanical properties of 3D printed biodegradable plastics or composites are not as good as pure matrix materials. Fillers usually increase the melt viscosity, and therefore flowability problems arise. Lack of fusion between layers, porosity and swelling induced by natural fibres are also problematic [22]. Filler content in biodegradable filaments is usually below 40% in volume; layer adhesion and extrudability are reduced increasing this content, as well as mechanical properties [22].

Gkartzou et al. [29] investigated the processability and the mechanical properties of a PLA filament containing 5 wt.% of craft pine lignin. The addition of lignin makes PLA more brittle and reduces the elongation at break. Moreover, the filler causes a remarkable increase of the superficial roughness of the PLA filament.

Some strengths and weaknesses in bio-composites with different cellulose fibres are reported by Li et al. [9]. These are, for example, strength, hardness, flexibility and moisture sensitivity. Some of these issues might be altered through mixing differently refined cellulose fibres. Markstedt et al. have printed pure cellulose with a modified 3D printer, but the process was liquid-based instead of filament based [31].

4   Material Sustainability

Sustainability of 3D printing materials is of utmost importance for the future, since the 3D printing market is projected to grow at an annual rate of about 26% till 2020 [32]. Recycling the excess and unwanted material primarily into new feedstock or finding new methods for the material to degenerate or compost into harmless building blocks in nature is imperative. The same issues faced in general plastic recycling should be dealt with while recycling FDM wastes.

Recycling by transforming the waste material into new filaments is a good primary recycling method, especially for thermoplastics with similar resin properties and homogenous source. Thermoplastic materials have been recycled since 1970's and today there is more knowledge and expertise about the recycling process [33–35].

Nevertheless, material transformation implies also degradation. Degradation is an irreversible process leading to a significant change in the structure of the filament material resulting in loss of properties. Cruz et al. [28] investigated the degradation of a 3D printing PLA filament along 5 complete recycling cycles. They observed a trend of a slight reduction of mechanical performances through the cycles as in conventional recycling [36, 37]. Their results show a decrease of the polymer molecular weight of about 47% after 5 recycling cycles increasing crystallinity of the plastic. The loss in molecular weight caused a lower tensile strength, but did not affect the yield stress. In particular, the strain at break reduced from 1.88% to 1.68%. On the contrary, the reduction of molecular weight improved the material flowability and processability with a viscosity reduction of 80% after 5 recycling cycles.

Within the communities of makers, Kreiger [38] and Baecher [39] have created their own filament extruder and using a commercial manufacturing extruder Hamod [30] demonstrated that whoever owns a shredder and an extruder can directly turn waste plastics into new feedstock.

Kreiger et al. [38] have studied HDPE plastic collection in the USA. The study compared the environmental effect of centralized and decentralized filament making with recycled material available from the makers' community. Baechler et al. [39] calculated that the cost of producing recycled HDPE filament is between 2 and 3 USD per kilogram with homemade extruder, while the commercial filament is sold at an average price of 38 USD per kilogram. After calculating the environmental impact from both recycling cases of HDPE, Kreiger et al. [38] concluded that it would be better to centralize the plastic recycle into industrially made recycled filament as the scale would be larger and more viable. In their research, Baechler et al. [39] found out that more automation and product control for the extruder is needed to produce homogeneous filament. The properties of parts printed with recycled HDPE were lower than those of the same pieces made of virgin material. However, different results were reported by Hamod [30], whose study achieved similar mechanical properties for both virgin HDPE and recycled HDPE. The HDPE was only produced and recycled once and both virgin and recycled filaments were made using the same equipment. Further recycling and production cycles as well as impurities introduced into the material by recycling make the filament properties decay as described for PLA by Cruz et al. [28] or reported in the case of injection moulding [40].

To help identifying the thermoplastic resins after 3D printing, recycling material dependent coding is proposed and investigated by Hunt et al. [41]. Codes like those of conventional plastic industry could be added on the surface of 3D printed parts to make recycling straight forward by easy identification of the plastic blends. In industrial sorting, since plastics are sorted out by means of density and other properties, these markings are not relevant as in visual sorting.

5   Conclusions

Unlike injection moulding materials, not all thermoplastics are available or suitable for 3D printing. Moreover, 3D printed materials usually have worse properties than injection moulded ones. So far, the efforts of many researchers were directed to improve the 3D printing feasibility for new bio-based or compostable filaments, but there is a lack of performance testing for these filaments in the literature. The degradation of filaments during repeated recycling needs to be studied more in detail like Cruz et al. [28] did for PLA.

Contamination of extruded filament should be prevented. On one side, contamination can negatively affect material properties and mechanical performances. On the other side, contaminants can be the source for the development of UFA and VOC emissions and thus cause health hazards and risks.

Further development and testing of 3D printing filaments is needed to better characterize their properties in terms of mechanical behaviour and compostability. Because biodegradable plastics have lower mechanical properties and restricted life cycles, user should find short life-cycle applications. These applications for biodegradable filaments need to consider the environmental conditions (e.g. moistness, exposure to UV radiations, etc.) that can negatively affect the filament extrusion or the part swelling when part is in use. However, in some cases, swelling might also be a desired effect for part functionality as shown by David et al. [21]. On the contrary, the advantage of these biodegradable plastics is that printed components can be disposed at the end of their life in a more environmental-friendly manner by recycling them to make new filament or by composting. The characteristic of recyclability is fundamental for the sustainability of 3D printing, since the adoption and popularity of this AM process has been constantly increasing in recent years.

Filament recycling should be aimed at preserving interesting material properties from the engineering point of view allowing for a noble use of the recycled polymer. It might not be possible to use fully recycled filament as the mechanical properties decrease, but with a blend of virgin and recycled material may allow finding an acceptable trade-off. If additives are used for this purpose, they should not jeopardize the opportunity for further recycling cycles. Great losses in mechanical properties do not restrain the use of the recycled filament for models, visual prints, packaging and other short life-cycle products without structural demands or durability requirements. In addition to this, recycled filaments will need a retuning of 3D printing parameters to cope with the increase of flowability and reduced viscosity resulting from material recycling.

Up to date, research efforts about the development of bio-degradable or compostable 3D printing filaments have been primarily focused toward achieving material printability, whereas material recycling and the whole product life cycle has not been fully considered. In the circular economy framework, at least one or two recycling cycles for material re-use should be ensured as an alternative to direct eco-friendly disposal.

Since correct and successful filament recycling is a much more complex operation than 3D printing, the use of low-cost or crowdfunded extruders for makers appears restrictive. As a matter of fact, such kind of recycling devices are not optimized because of constrained development costs and production costs savings. Thus, their suitability is limited to the recycling of a few materials through control of a limited set of processing parameters.

As for the success of any other recycling process, the correct consumer behaviour and attitude towards waste management is not questioned, but specific policies or incentives might be proposed for material disposal in the 3D printing sector. Material traceability and the possibility of including specific marking into the design of 3D printed parts is still an open question. The best way to impose it could be by including an automatic marking function in the 3D printer software. The software should automatically add the mark to the part during the computation of the printing path, i.e. during the slicing operation. The selection of the area for marking should be left up to the user in order not to affect the aesthetics of the part significantly. Nonetheless, in open 3D printers the identification of the right material is not error free, because the software cannot automatically identify the filament material as with chipped filament spools.

In order to solve some of these issues, Chong et al. conceived a distributed recycling platform for 3D printed products to achieve the goal of zero waste production [42]. Beyond improvement of filament recyclability, recycled FDM material certification and other regulations about 3D printing waste management should be implemented to reach such a noble objective.

References

1. European Union action on circular economy. http://ec.europa.eu/environment/circular-economy/index_en.htm
2. Kohtala, C.: Addressing sustainability in research on distributed production: an integrated literature review. J. Clean. Prod. **106**, 654–668 (2015)
3. Hopewell, J., Drovak, R., Kosior, E.: Plastics recycling: challenges and opportunities. Philos. Trans. R. Soc. B, **364**(1526), 2115–2126 (2009)
4. Calignano, F., Manfredi, D., Ambrosio, E.P., Biamino, S., Lombardi, M., Atzeni, E., Salmi, A., Minetola, P., Iuliano, L., Fino, P.: Overview on additive manufacturing technologies. Proc. IEEE **105**(4), 593–612 (2017)
5. Jones, R., Haufe, P., Sells, E., Iravani, P., Olliver, V., Palmer, C., Bowyer, A.: RepRap – the replicating rapid prototyper. Robotica **29**(1), 177–191 (2011)
6. ISO/ASTM Standard, 52900:2015 – Additive manufacturing – General principles: Terminology

7. McAlister, C., Wood, J.: The potential of 3D printing to reduce the environmental impacts of production. Eceee Ind. Summer Study Proc. **2**(72), 213–221 (2014)
8. Gebler, M., Schoot Uiterkamp, A.J.M., Visser, C.: A global sustainability perspective on 3D printing technologies. Energy Policy **74**, 158–167 (2014)
9. Li, T., Aspler, J., Kingsland, A., Cormier, L.M., Zou, X.: 3d printing – a review of technologies, markets, and opportunities for the forest industry. J. Sci. Technol. For. Prod. Process. **5**(2), 30 (2016)
10. Enviro ABS. https://threedmaterials.com/products/enviro-abs-filament-1-75mm-blue
11. Stephens, B., Azimi, P., Orch, Z.E., Ramos, T.: Ultrafine particle emissions from desktop 3D printers. Atmos. Environ. **79**, 334–339 (2013)
12. Steinle, P.: Characterization of emissions from a desktop 3D printer and indoor air measurements in office settings. J. Occup. Environ. Hygiene **13**(2), 121–132 (2016)
13. Duran, C., Subbian, V., Giovanetti, M.T., Simkins, J.R., Beyette Jr., F.R.: Experimental desktop 3D printing using dual extrusion and water-soluble polyvinyl alcohol. Rapid Prototyp. J. **21**(5), 528–534 (2015)
14. B-PET Filament, http://bpetfilament.com/
15. Chia, H.N., Wu, B.M.: Recent advances in 3D printing of biomaterials. J. Biol. Eng. **4**(9), 1–14 (2015)
16. Serra, T., Planell, J.A., Navarro, M.: High-resolution PLA-based composite scaffolds via 3-D printing technology. Acta Biomater. **9**, 5521–5530 (2013)
17. Melocchi, A., Parietti, F., Loreti, G., Maroni, A., Gazzaniga, A., Zema, L.: 3D printing by fused deposition modeling (FDM) of a swellable/erodible capsular device for oral pulsatile release of drugs. J. Drug Delivery Sci. Technol. **30**, 360–367 (2015). Part B
18. Pietrzak, K., Isreb, A., Alhnan, M.A.: A flexible-dose dispenser for immediate and extended release 3D printed tablets. Eur. J. Pharm. Biopharm. **96**, 380–387 (2015)
19. Salminen, A., Seppälä, J.: 3D printing of thermoplastic cellulose derivatives. In: Design Driven Value Chains in the World of Cellulose project report 1, pp. 48–49 (2016)
20. Kuo, C.C., Liu, L.C., Teng, W.F., Chang, H.Y., Chien, F.M., Liao, S.J., Kuo, W.F., Chen, C.M.: Preparation of starch/acrylonitrile-butadiene-styrene copolymers (ABS) biomass alloys and their feasible evaluation for 3D printing applications. Compos. B **86**, 36–39 (2016)
21. David, C., Athina, P., Christophe, G., Nynika, J., Steffen, R., Achim, M., Skylar, T.: 3D-printed wood: programming hygroscopic material transformations. 3D Print. Addit. Manuf. **2**(3), 106–116 (2015)
22. Duigou, A.L., Castro, M., Bevanc, R., Martin, N.: 3D printing of wood fibre biocomposites: From mechanical to actuation functionality. Mater. Des. **96**, 106–114 (2016)
23. Zhang, D., Chi, B., Li, B., Gao, Z., Du, Y., Guo, J., Wei, J.: Fabrication of highly conductive graphene flexible circuits by 3D printing. Synth. Met. **217**, 79–86 (2016)
24. Ceccolini, M.: https://www.behance.net/gallery/24616719/agridust-biodegradable-material
25. S3D Innovations. http://s3dinnovations.wixsite.com/filasoy
26. Cantrell, J., Rohde, S., Damiani, D., Gurnani, R., DiSandro, L., Anton, J., Young, A., Jerez, A., Steinbach, D., Kroese, C., Ifju, P.: Experimental characterization of the mechanical properties of 3D-printed ABS and polycarbonate parts. Adv. Opt. Methods Exp. Mech. **3**, 89–105 (2016)
27. Letcher T.: Material Property Testing of 3D printed Specimen in PLA on an Entry level 3D printer, In: proceedings of the ASME 2014 International Mechanical Engineering Congress & Exposition (2014)
28. Cruz, F., Lanza, S., Boudaoud, H., Hoppe, S., Camargo, M.: Polymer recycling and additive manufacturing in an open source context: optimization of processes and methods. In: 2015 Annual International Solid Freeform Fabrication Symposium - An Additive Manufacturing Conference, Austin, Texas (USA), 10–12 August 2015

29. Gkartzou, E., Koumoulos, E.P., Charitidis, C.A.: Production and 3D printing processing of bio-based thermoplastic filament. Manuf. Rev. **4**(1), 14 (2017)
30. Hamod, H.: Suitability of recycled HDPE for 3D printing filament. B.Sc Thesis. Arcada University of Applied Science, Helsinki (2014)
31. Markstedt, K., Sundberg, J., Gatenholm, P.: 3D bioprinting of cellulose structures from an ionic liquid. 3D Print. Addit. Manuf. **1**(3), 115–121 (2014)
32. Business Wire. http://www.businesswire.com/news/home/20160325005103/en/Global-3D-Printing-Plastic-Market-Worth-USD
33. Al-Salem, S.M., Lettieri, P., Baeyens, J.: Recycling and recovery routes of plastic solid waste (PSW): a review. Waste Manage. **29**, 2625–2643 (2009)
34. Perugini, F., Mastellone, M., Arena, U.: A life cycle assessment of mechanical and feedstock recycling options for management of plastic packaging wastes. Environ. Progr. **24**, 137–154 (2005)
35. Hopewell, J., Dvorak, R., Kosior, E.: Plastics recycling: challenges and opportunities. Philos. Trans. R. Soc. B Biol. Sci. **364**(1526), 2115–2126 (2009)
36. Zenkiewicz, M., Richert, J., Rytlewski, P., Moraczewski, K., Stepczyńska, M., Karasiewicz, T.: Characterisation of multi-extruded poly(lactic acid). Polymer Test. **28**(4), 412–418 (2009)
37. Pillin, I., Montrelay, N., Bourmaud, A., Grohens, Y.: Effect of thermo-mechanical cycles on the physico-chemical properties of poly(lactic acid). Polymer Degrad. Stab. **93**(2), 321–328 (2008)
38. Kreiger, M.A., Mulder, M.L., Glover, A.G., Pearce, J.M.: Life cycle analysis of distributed recycling of post-consumer high density polyethylene for 3-D printing filament. J. Cleaner Prod. **70**, 90–96 (2014)
39. Baechler, C., DeVuono, M., Pearce, J.M.: Distributed recycling of waste polymer into RepRap feedstock. Rapid Prototyp. J. **19**(2), 118–125 (2013)
40. Torres, N., Robin, J.J., Boutevin, B.: Study of thermal and mechanical properties of virgin and recycled poly(ethylene terephthalate) before and after injection molding. Eur. Polymer J. **36**, 2075–2080 (2000)
41. Hunt, E.J., Zhang, C., Anzalone, N., Pearce, J.M.: Polymer recycling codes for distributed manufacturing with 3-D printers. Resour. Conserv. Recycl. **97**, 24–30 (2015)
42. Chong, S., Chiub, H., Liao, Y., Hung, S., Pan, G.: Cradle to cradle® design for 3D printing. Chem. Eng. Trans. **45**, 1669–1674 (2015)

Sustainable Mobility, Solar Vehicles and Alternative Solutions

Electric City Buses with Modular Platform: A Design Proposition for Sustainable Mobility

Cristiano Fragassa[✉]

Department of Industrial Engineering, Alma Mater Studiorum University of Bologna,
viale Risorgimento 2, 40136 Bologna, Italy
cristiano.fragassa@unibo.it

Abstract. The renewed interest for city buses and the constant growth of the urban population worldwide will contribute to inflate the global market size to more than 275,000 vehicles a year in 2020, up 60% compared to the current levels. Diesel buses are the most popular choice, even though their future market share will bear the brunt of shale gas discovery and advances in energy storage technology. This paper describes the dynamics of the city bus market and a business idea to enter the European public transportation market with sustainable vehicles featuring a new design platform, electric powertrains, and composite technology.

Keywords: Public transport · Bus market · Eco-sustainability · Pollution · New design methods · Green technology

1 Opportunity

Environmental sustainability has risen among the greatest concerns of mankind. Ambient air pollution has caused 3.7 million premature deaths worldwide in 2012 and about 50% of the urban global population is exposed to levels of particulate 2.5 times higher than the air-quality guideline values set by the World Health Organization [1]. Pollution scores high on the political agenda of many governments and the Doha Amendment of the Kyoto Protocol [2] further embitters the need for implementing unprecedented measures to reach the target of cutting carbon dioxide emission by 18% below 1990 levels from 2013 to 2020 [3]. Because of its 22% toll on global emissions, road transport has received much scorn in recent years. In several cities, where the problems of pollution and noise are more severe, a twofold pattern has arisen: from London to Stockholm to Milan, a progressive ban on private motorized vehicles from downtown is being followed by a political commitment to improve and modernize public transport.

After years of massive investments in the urban railway and trolley networks, however, the focus of attention has shifted towards city buses and their major advantages over alternative solutions, notably the subway (Fig. 1).

© Springer International Publishing AG 2017

G. Campana et al. (eds.), *Sustainable Design and Manufacturing 2017*, Smart Innovation, Systems and Technologies 68, DOI 10.1007/978-3-319-57078-5_74

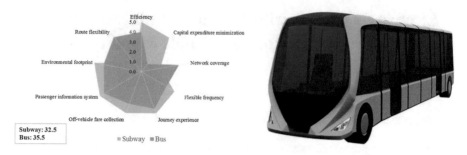

Fig. 1. Comparison subway vs bus [6] and current design proposition.

The issue is not trivial [4]. With around 30 billion passengers/year in Europe alone, buses represent 80% of total public transport [5]. More than 90% buses worldwide burn fossil fuels (Fig. 2). Despite technological advances, it appears as though vehicles equipped with internal-combustion engines only will struggle to keep up with the implementation of more and more stringent emission standards worldwide (Fig. 3), let alone their thermodynamic constraint to cut completely carbon dioxide emissions at the tailpipe.

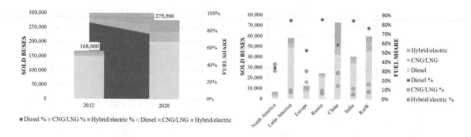

Fig. 2. Global and regional city bus market by fuel [6].

Tier	Date	CO [g/kWh]	HC [g/kWh]	NOₓ [g/kWh]	PM [g/kWh]	Smoke [m⁻¹]
Euro I	1992, < 85 kW	4.50	1.10	8.00	0.61	
	1992, > 85 kW	4.50	1.10	8.00	0.36	
Euro II	01/10/1996	4.00	1.10	7.00	0.25	
	01/01/1998	4.00	1.10	7.00	0.15	
Euro III	October 1999 EEVs only	1.00	0.25	2.00	0.02	0.15
	01/10/2000	2.10	0.66	5.00	0.10	0.80
Euro IV	01/10/2005	1.50	0.46	3.50	0.02	0.50
Euro V	01/10/2008	1.50	0.46	2.00	0.02	0.50
Euro VI	31/12/2013	1.50	0.13	0.40	0.01	

Fig. 3. EU (or equivalent) Emission standards for trucks and buses with timeline for implementation at the regional levels [7].

Any serious effort to address the issue of environmental sustainability should therefore entail, amongst the others, a requalification of bus fleets.

The political acknowledgement of this state of facts has resulted in the emergence of a demand for electric city buses that, although still embryonic, promises to explode within the next few years (Fig. 4).

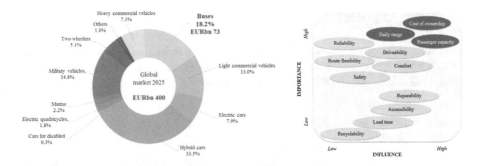

Fig. 4. Opportunities of electrification: global market electric vehicles 2025 and customers? requests with electric bus purchases (independent elaboration).

2   Customer Specifications

Purchases of city buses ensue from the planning of public and private bus operators, depending on the underlying business model adopted by the various municipalities. Despite a generalized agreement about the potential of electric buses to address the call for sustainable urban mobility, interviews with relevant stakeholders revealed that the acceptance of these vehicles is subordinate to the solutions of the following problems:

1. High cost of ownership: the sum of purchase and operating costs throughout the vehicle?s life cycle should equal the cost of comparable fossil-fuel-powered buses;
2. Limited daily service range: the bus should guarantee an uninterrupted service for the equivalent of 300 km/day;
3. Reduced passenger capacity: the sum of standing and seat passengers should be comparable to the capacity of fossil-fuel-powered buses; as a reference, 90+ passengers for 12 m bus.

In this respect, customers have similar expectations worldwide and the ability to meet such requirements represents a threshold capability to compete in the city bus industry. However, customers seek to fulfil additional needs (Fig. 4), which lack a basis of uniformity because city buses are operated within geographically restricted boundaries with their own idiosyncrasies. This translates into a heterogeneity of customer specifications that hampers regional cross-selling and gives a competitive hedge to those suppliers that best address local needs. Current attempts to satisfy the demand for electric city buses have almost exclusively relied on the conversion of traditional fossil-fuel-powered vehicles. Given the constraints of energy storage technology and the legal limits to buses? gross weight, however, not only have existing propositions failed to solve the aforementioned customers? problems, but they also have shown the architectural limits of current bus platforms.

3   Emerging Market

In general, it appears as though the electric bus market is still in its infancy and a dominant design has yet to arise. Despite a lack of standardization, however, current propositions fall within one the following three clusters:

– Fuel-cell powered
– Supercapacitor powered
– Battery powered

An overview on the full-electric bus technology and its positioning respects to the other traditional technologies are shown in Figs. 4, 5 and 6.

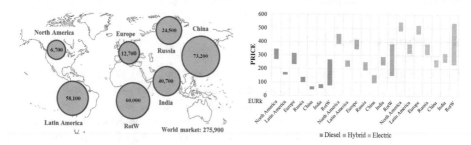

Fig. 5. Forecast global city bus market 2020 [9] and bus prices by region and fuel (independent elaboration from [6, 8] datasets).

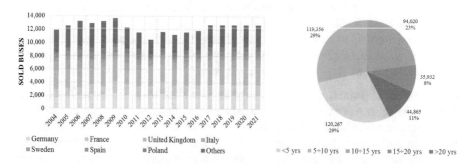

Fig. 6. Registrations, development and composition (as at 2012) of Western EU bus park [9].

The global city bus market is estimated at 275,000 new vehicles per year for 2020, mainly sold in developing countries with 7% between Europe and North America. Even in the case this investigation has to be limited to the European territory, since the larger consistency of data, the consequences are also interesting.

The current European bus fleet numbers in excess of 500,000 vehicles with an average age of more than eight years. Over the 20-year average life cycle of a bus, the typical usage pattern entails three changes of ownership, with the last one generally resulting in the export to developing countries before scrapping. The number of new registrations floats around 12,000 city buses/year, with an annual growth rate in the region of 0.4%. Germany, France, UK, Italy, Sweden, and Spain together account for about 70% of annual [6, 8, 9]. Large part of this demand could be covered by electrical buses with benefits for the environment and population.

In terms of market segments, the city bus market is traditionally clustered by vehicles? length. In Europe, the following clustering are commonly applied:

- *Midi:* between 8 and 10 m, two axles, single deck
- *Standard:* between 10 and 13.5 m, two axles, single or double deck
- *Articulated:* up to 25 m, three or four axles, single deck.

Standard buses have the largest market share, around 60%, and are operated almost uniformly around Europe (Fig. 7). Except a few specialized suppliers, all manufacturers offer standard buses. Heated competition tends to result in lower margins for these vehicles compared to Midi and Articulated buses, for which clear market preferences exists. For example, Midi buses find their way in Mediterranean countries with narrow streets, whereas Articulated buses are a rare sight in the UK. No European competitor in Europe offers a complete range of vehicles.

Category	2012	2015	2020	2025	2030
Midi	16%	16%	17%	16%	16%
Standard	60%	60%	59%	55%	56%
Articulated	23%	24%	24%	28%	29%

Fig. 7. Development of the European city bus market by category [10].

4 Customers

At the highest level of analysis, demand for city buses depends on the long-term vision of political rulers, who set broad targets for urban mobility and cater for necessary funds. In practical terms, demand turns into purchases according to the plans of bus operators. The predominant business model in Europe encompasses the presence of public bus operators, which set vehicles? specifications and then issue requests for quotation according to European public law. All qualified vendors can submit their offer, which generally undergoes a three-step revision process, until determining suppliers? so-called ?best and final? price. In the end, offers are evaluated quantitatively and ranked on the basis of pre-determined criteria. Price and consumption generally account for about 50% and 20%, respectively. Business is then awarded to the issuer of the so-called ? most economically convenient? offer. The typical terms of agreement entail full-sale of the vehicles, three-to-five year maintenance contract, and 30-day settlement with down-payments.

In the competing business model, which applies mainly in the UK and Scandinavia, the municipality outsources the service to private operators. In this case bus procurement formally encompasses a deal between two parties according to business law. Given the similarity of bus requirements worldwide, however, this business model does not result in significantly different outcomes compared to the public-tender arrangement, except for a preference for leasing agreements and slightly longer payment terms. Overall, the European city bus market is informed on principles of transparency and fairness aimed at favouring new entrants, but buyers leave little room to change the rules of the game. The heterogeneity of customer specifications may give a competitive hedge to those suppliers that best address local needs. This translates into a rather competitive European market, even though Mercedes, MAN, IVECO, and Volvo together command a 70% market share (Fig. 8). Furthermore, bus manufacturing as a stand-alone activity barely exists nowadays in Europe, but proxy analysis indicates that it tends to offer poor returns (Fig. 9).

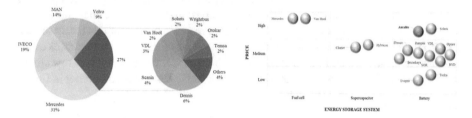

Fig. 8. Market share and positioning in Western Europe bus market (elaboration from [9]).

		Yutong	King Long	Anhui Ankai	Yaxing	New Flyer	Optare	Anadolu	Marcopolo
Fiscal year		31/12/13	31/12/13	31/12/13	31/12/13	31/12/13	31/03/14	31/12/13	31/12/13
Headquarters		China	China	China	China	Canada	UK	Turkey	Brazil
Employees		11,415	14,700	4,439	1,832	2,200	400	747	20,016
Exchange rate		0.1199	0.1199	0.1199	0.1199	0.6827	1.2099	0.3379	0.3073

INCOME STATEMENT		Yutong	King Long	Anhui Ankai	Yaxing	New Flyer	Optare	Anadolu	Marcopolo
SALES	[EURk]	2,649,150	2,495,434	424,339	144,844	818,587	68,963	217,945	1,124,580
GROSS PROFIT	[EURk]	515,585	311,510	24,460	15,827	82,610	7,259	35,479	224,652
Gross Margin	[% sales]	19.5%	12.5%	5.8%	10.9%	10.1%	10.5%	16.3%	20.0%
EBITDA	[EURk]	295,442	67,353	(31,390)	(2,409)	50,675	(2,289)	11,202	133,720
EBITDA Margin	[% sales]	11.2%	2.7%	-7.4%	-1.7%	6.2%	-3.3%	5.1%	11.9%
EBIT	[EURk]	225,898	44,707	(36,138)	(3,239)	34,887	(3,678)	8,627	121,359
EBIT Margin	[% sales]	8.5%	1.8%	-8.5%	-2.2%	4.3%	-5.3%	4.0%	10.8%
PRETAX INCOME	[EURk]	250,239	64,029	(5,396)	600	23,895	(4,913)	71,297	84,821
EBT Margin	[% sales]	9.4%	2.6%	-1.3%	0.4%	2.9%	-7.1%	32.7%	7.5%
NET INCOME	[EURk]	26,209	27,578	(4,197)	600	18,434	(4,840)	67,242	68,225
Return on sales	[% sales]	1.0%	1.1%	-1.0%	0.4%	2.3%	-7.0%	30.9%	6.1%

Fig. 9. Profitability **bus manufacturers** (elaboration from Morningstar & Financial Times)

When focusing on electric buses, however, it appears as though current market dynamics do not hold anymore, since only a handful of established manufacturers have stepped into the realm of alternative powertrains. Very often these players? foray into the electric bus business respond to a logic of opportunism or advertising rather than reflecting strategic intent. For instance, Mercedes and Van Hool have basically kept their fuel-cell buses to prototype stage and have now stepped down. At the other side of the spectrum, there are big manufacturers that have yet to embrace the electric credo and several startups blindly devoted to it. Such zealous commitment likely find its rationale in the generous government subsidies to promote sustainable mobility and in the lack of formal entry barriers, substantial scale economies, and retaliation. At this stage of development, it is difficult to depict the exact competitive landscape of the future. However, a review of the literature [11] suggests the following. First, as a dominant design arises, the market will sweep away non-compliant players. Second, as the market matures and margins erode, competitors will start to merge or exit. Finally, today?s market leaders may not be tomorrow?s leaders, especially when faced with a strong legacy of tradition and sunk costs [12?14].

5   Design Opportunity

To capture the opportunities of electrification and to fulfil a vision of affordable, efficient, enjoyable and safe urban mobility, this research highlights, as affordable market option, the business proposition of a deep range of low-floor electric city buses built on a modular platform in biocomposites.

   This gamma of public vehicles aims at offering a significant departure from the status quo, inside the Class I city bus market, thanks to several peculiarities in terms of flexible design, efficient manufacture and the large use of sustainable materials as:

- cost of ownership
- daily service range
- passenger capacity
- environmental friendliness
- corrosion resistance
- design supremacy
- full vehicle range coverage
- customizability
- accessibility
- driver?s safety
- lead time.

Market research and technical analysis have convinced of the need for stepping beyond the traditional idea of manufacturing buses by cutting and welding metals and then fitting the resulting vehicle with the desired motor [15]. This strategy has worked in the past with internal-combustion engines, but fails to accommodate the need for a light, cheap, and flexible framework necessary to offset the high cost and weight of electric power-trains.

   Unlike the akin passenger car and truck businesses, the bus industry has mainly a system-integrator nature. This feature makes the market somehow contestable and favours the entry of innovation solutions, which can also thrive on the right aggregation of already available technologies. It introduces a business proposition, whose source of competitive advantage stems from the combination of:

1. Lightweight ecosustainable materials
2. Uncommon treatments for alloys and polymers
3. Advanced bonding technology
4. Rapid and hybrid processes

These engineering characteristics are combined in a framed modular biocomposites platform, representing an advanced design solution for the sustainable urban mobility. In particular, it permits a higher flexibility in realisation of a large gamma of vehicles, spreading from 8 mt and 46 seats, up to 24.5 mt and 212 seats. This flexibility offers a real support to the business proposition and the return on the public/private investment.

6   Advanced Technology

Composites derive from the amalgamation of different materials to obtain compounds with specific technical features, especially in terms of strength and weight [16]. Originally confined to high-tech applications such as aircraft, boats, and race cars, composites have now made their successful appearance in the bus industry as well, with examples of both single components or entire vehicles with composite structure. Building on these experiences, several design solutions intend to fully capture the benefits of composite materials, but also to reduce the vehicles environmental footprint. Thus, in line with a deep action of material change toward a more sustainable design, it has been chased the methodical replacement, wherever possible, of common industrial reinforcements (as carbon or glass fibers) with natural fibers (as flax, jute, hemp...). This technical choice, if largely adopted as constructive solutions, will provide worldwide a different perspective of the sustainable mobility, also including additional aspects as the massive carbon dioxide storage by a systematic adoption of lignocellulosic materials [17].

It is also worth noting that the use of natural fibers in city buses would be, in same way, a simple redeployment of a proven technology in the automotive industry where several car manufactures (firstly Mercedes-Benz) already preferred eco-friendly materials for not structural parts.

In particular, current investigations envisage the use of a specific organic resin (a vinylest), characterized by a low content of styrene, coupled with flax fibers as an appropriate design solution for the bus semi-ring-shaped body. Given more stringent requirements in terms of torsional rigidity, a compound of carbon fiber and, even, wood seem to serve better for the chassis. But, actually, technology moves fast and can propose several solutions for improving the mechanical properties of naturally based composites.

In general, it is useful to remember that, according to the material technical datasheets, flax (and also other natural) reinforcements permit to improve the mechanical strengths of the constitutive resin between 30% and 300% in consideration of the direction of fibers [18]. These values, not negligible in general and suitable in several practical cases, is far away from carbon reinforced materials, aluminium or steel used in structural applications.

When a better performance is preferable, mixing fibers (?hybridization?) can represent a very interesting opportunity. In particular, basalt, with mechanical properties similar to carbon, seems to be the best solution for reinforcing flax and has be already investigated in ready-to-market applications [19?21].

In addition, the use of specific physical and chemical treatments for surface protection [22] or esoteric proposals, as the inclusion of PVDF nanofibers, in the matrix [23] can provide further improvements in material properties.

Finally, original design solutions can offer unexpected margins for enhancements. It is the case, for instance, of pre-stressed and curved composite laminate that exhibit higher resistance respect to static and impact loads. [24].

Beyond the specific expedients used to perfect the material, the choice of biocomposites inevitably imposes to carefully address the issues of reliability and safety. Without doubt, modern city buses have to guarantee, at least, the same level of reliability and safety offered by other vehicles, as private cars [25]. This target is not taken for granted considering the reduced number of buses (comparing to cars) in production line batches. As a consequence, a similar approach has to be adopted in their design and manufacturing.

In the former case, current legislation does not mandate neither crash nor crash and roll-over tests for Class I city buses, whereas fire-proof requirements can be meet with special coating. This surprising lack of regulation is quite similar to the situation characterizing the emerging minicars market [26], where, however, an intervention will be imminent, and design modification in vehicles necessary. In the case of city buses, since composites can be molded in the desired shape, it is planned to leverage this opportunity to engineer buses with a modular approach, in order to create self-contained modules to be eventually bonded in the desired length and fashion. Widely popularized by advanced sectors as airspace, motor-race, the idea has actually trailed efforts in the bus industry as well permitting to identify the business opportunity of a modular approach. However, composite molding requires very different skills from cutting and welding metals, thereby suggesting that established, vertically-integrated, and diversified players may lack the incentive to fully rethink bus manufacturing in this direction. On the one hand, the choice of using biocomposites find its rationale in the willingness to meet the targets of daily service range, passenger capacity, environmental friendliness and corrosion resistance. On the other, the modular bonded platform delivers much customer value on all other parameters. The simplification of the manufacturing process translates especially into lower operating costs, whereas the flexibility of molding and bonding allows, within the limits of homologation criteria, designers to propose bold and custom-made designs and engineers to accommodate easily and promptly requests of customization such as additional doors, tailored length, and separate driver?s cabin.

7   Competition-and-New-Entrants

In spite of potential issues of safety and recyclability, the on-board storage solution offers the best compromise between cost and flexibility. Competitors appear to share the same opinion, even more so that the price of best-in-class lithium-ion batteries should experience a significant drop in the foreseeable future. Moreover, evidence suggests that the greatest research effort is currently directed towards the development of improved energy storage systems, with such companies as IBM, Phinergy and Nanoflowcell already anticipating advances on metal-air and redox batteries to match fossil fuels? energy density within 10 years [27]. A synthesis of technical details for to the main competitors acting on the European market and the positioning of the presented vehicle is reported in Fig. 10.

	Auralite	Van Hool	Mercedes	Hybricon	Chariot / Higer	Evopro	Ebusco	Rampini
Model	e-bus	A330	Citaro	HAW 12E / 18E		Modulo	2.0	Ale El
Origin	Austria	Belgium	Germany	Sweden	Sweden	Hungary	Netherlands	Italy
Motor position	In-wheel	Rear mounted	In-wheel	In-wheel	Rear mounted	Rear mounted	Rear mounted	Rear mounted
Motor manufacturer	ZF	Siemens	N&A	ZA Wheel	Siemens	US Hybrid	N/A	Siemens
Cooling system	Water	N/A	N/A	Water	Water	N/A	N/A	N/A
Continuous power	70-140 kW	N/A	160 kW	226 kW	N/A	76 kW	N/A	85 kW
Peak power	140-280 kW	170 kW	240 kW	364 kW	N/A	150 kW	150 kW	150 kW
Energy brake recuperation	Yes	N/A	N/A	Yes	N/A	Yes	Yes	N/A
Energy storage system	Li-ion battery	Fuel cell	Fuel cell	Supercap	Supercap	Li-ion battery	Li-ion battery	Li-Fe battery
Energy capacity	176-621 kWh	120 kWh	250 kWh	N/A	21 kWh	150 kWh	311 kWh	180 kWh
Charging mode	Cable/plug-in	Refilling	Refilling	Wireless	Wireless	Cable/plug-in	Cable/plug-in	Cable/plug-in
Bus architecture	Modular molded	Traditional welded	Traditional welded	Traditional welded	Traditional welded	Modular molded	Traditional welded	Traditional welded
Material	Biocomposite	Steel	Steel	Aluminum/composite	Steel	Composite	Aluminum	Steel/aluminum
Length	8.0-24.5 m	13.2 m	12.0 m	12.0-18.0 m	12.0 m	8.5-12.0 m	12.0 m	7.7 m
Doors	1-7	3	3	3-4	3	2-3	2-3	2-3
Unladen weight	11.0t (12.5m)	N/A	13.0t	N/A	N/A	8.5-18.0t	N/A	11.8t
Driver Cabin	Open/Closed	Open	Open	Open	Open	Open	Open	Open
Unit price	EURk 425-1,222	N/A	N/A	N/A	EURk 475	EURk 300+	N/A	EURk 440
Range	300+ km	N/A	N/A	60 min driving	20 km	N/A	300 km	130-150 km
Recharging time	N/A	N/A	N/A	5 min	5 min	N/A	1.6 h	>2.0 h
Passenger capacity	46-212	105	77	N/A	90	45-147	90+	44

	Auralite	SOR	Solaris	VDL	Bozankaya	BYD	Optare	Trolza
Model	e-bus	EBN 10.5	Urbino Electric	Citea	Sileo	Ebus	Versa EV	52501
Origin	Austria	Czech Republic	Poland	Netherlands	Turkey / Germany	China	UK	Russia
Motor position	In-wheel	Rear mounted	In-wheel	In-wheel	In-wheel	In-wheel	Rear mounted	Rear mounted
Motor manufacturer	ZF	Pragoimex		ZA Wheel	ZF	BYD	Magtech P144	N/A
Cooling system	Water	Water	Water	Water	Water	Water	Water	Water
Continuous power	70-140 kW	N/A	N/A	226 kW	160 kW	N/A	N/A	N/A
Peak power	140-280 kW	120 kW	120-240 kW	364 kW	240 kW	180 kW	150 kW	125 kW
Energy brake recuperation	Yes	Yes	Yes	Yes	Yes	Yes	N/A	N/A
Energy storage system	Li-ion battery	Li-ion battery	Li-ion battery	Li-ion battery	Li-ion battery	Li-Fe-ion battery	Li-ion battery	Li-ion battery
Energy capacity	176-621 kWh	173 kWh	up to 210 kWh	260 kWh	200 kWh	324 kWh	92 kWh	N/A
Charging mode	Cable/plug-in	Cable/plug-in	Cable/plug-in	Cable/plug-in	Cable/plug-in	Cable/plug-in	Cable/plug-in	Cable/plug-in
Bus architecture	Modular molded	Traditional welded	Traditional welded	Traditional welded	Traditional welded	Traditional welded	Traditional welded	Traditional welded
Material	Biocomposite	Aluminum/composite	Steel/aluminum	Aluminum/composite	Aluminum/composite	Steel	Steel	Aluminum/composite
Length	8.0-24.5 m	10.5 m	8.9-18.0 m	12.0 m	10.7 m	12.0 m	9.0-11.5 m	11.7 m
Doors	1-7	3	3-4	3	3	2-3	1	3
Unladen weight	11.0t (12.5m)	10.2t	N/A	11.4t	12.0t	N/A	12.0t	N/A
Driver Cabin	Open/Closed	Open	Open	Open	Open	Open	Open	Open
Unit price	EURk 425-1,222	EURk 420	EURk 600 ca.	N/A	EURk 400	EURk 380	N/A	N/A
Range	300+ km	110-160 km	N/A	200 km	200 km	317 km	110-150 km	120km
Recharging time	N/A	1.0-8.0 h	N/A	7.0 h	2.0 h	5.5 h	6.0 h	3h
Passenger capacity	46-212	85	64-87	101	72	68	60	98

Fig. 10.   Main competitors European market as at 2016.

8   Conclusion

This paper proposed a study of feasibility regarding an uncommon electrical vehicle for public transportation in urban area. This city bus would be powered by several advances in technology, such as new materials and unconventional processes, permitting to amplify its environmental features. The engineering core is a structural design platform planned to permit light, but also modular, vehicles as a way to support the product competiveness. The draft estimation of costs, the plan of investment, the market study, together with the analysis of current and potential competitors suggest this business opportunity. The European public transportation market seems to lack of sustainable vehicles in some segments and a response can be maybe offered by this new design platform, electric powertrains and composite technology.

References

1. World Health Organization (WHO): Air quality deteriorating in many of the world?s cities (2014). http://www.who.int/mediacentre/news/releases/2014/air-quality/en/
2. United Nations (UN): Doha amendment of the Kyoto protocol (2015). http://unfccc.int/kyoto_protocol/items/2830.php
3. International Energy Agency (IEA): CO2 emissions from fuel combustion (2012). http://www.iea.org/publications/freepublications/publication/co2emissionfromfuelcombustionhighlights.pdf
4. Tegner, G.: Comparison of costs between bus, PRT, LRT and metro/rail (2003). http://www.jpods.com/JPods/004Studies/CostPerMileOperations_UWa.pdf

5. Community Research and Development Information Service (CORDIS): Final report summary - EBSF (European Bus System of the Future) (2014). http://cordis.europa.eu/result/rcn/140495_en.html

6. Frost and Sullivan (F&S): Strategic analysis of global hybrid and electric heavy-duty transit bus market (2013). http://www.slideshare.net/FrostandSullivan/strategic-analysis-of-global-hybrid-and-electric-heavy-duty-transit-bus-market

7. Transportpolicy: Global comparison: heavy-duty emissions (2015). http://transportpolicy.net/index.php?title=Global_Comparison:_Heavy-duty_Emissions#cite_note-0

8. Harrop, P.: Electric buses: big changes in market and technology. IDTechEx Webseminar, 13 February 2015

9. Analyse & Prognose (A&P): Bus study Europe: August 2014 consulting report, Wilhelmsfeld, Germany (2014)

10. Fuel Cell and Hydrogen Joint Undertaking (FCHJU): Urban buses: alternative powertrains for Europe (2012). http://www.fch-ju.eu/sites/default/files/20121029%20Urban%20buses,%20alternative%20powertrains%20for%20Europe%20-%20Final%20report_0_0.pdf

11. Porter, M.E.: How competitive forces shape strategy. Harv. Bus. Rev. **52**(2), 137?145 (1979)

12. Geroski, P.A.: The Evolution of New Markets. Oxford University Press, Oxford (2003)

13. Porter, M.E.: What is strategy. Harv. Bus. Rev. **74**(6), 61?78 (1996)

14. Festel, G., Wuermseher, M., Cattaneo, G.: Valuation of early stage high-tech start-up companies. Int. J. Bus. **18**(3), 216?231 (2013)

15. Proterra: Product & technology (2015). http://www.proterra.com/product-tech/gallery-2/photos/

16. Composite World: Built bus body (2015). http://www.compositesworld.com/articles/team-built-bus-body-bests-all

17. Food and Agriculture Organization of the United Nations (FAO): Natural fiber composites (2015). http://www.fao.org/docrep/007/ad416e/ad416e07.htm

18. Boria, S., Pavlovic, A., Fragassa, C., Santulli, C.: Modeling of falling weight impact behavior of hybrid basalt/flax vinylester composites. Procedia Eng. **167**, 223?230 (2016). doi:10.1016/j.proeng.2016.11.691

19. Zivkovic, I., Pavlovic, A., Fragassa, C., Brugo, T.: Influence of moisture absorption on the impact properties of flax, basalt and hybrid flax/basalt fiber reinforced green composites. Compos. B **111**, 148?164 (2017). doi:10.1016/j.compositesb.2016.12.018

20. Fragassa, C.: Effect of natural fibers and bio-resins on mechanical properties in hybrid and non-hybrid composites. In: Proceedings of the 8th Conference on Times of Polymers and Composites: From Aerospace to Nanotechnology, Ischia, Italy. AIP Conference Proceedings, vol. 1736, No. 4949693, 19?23 June 2016. doi:10.1063/1.4949693

21. De Paola, S., Minak, G., et al.: Green composites: a review of state of art.In: Proceedings of 30th Danubia Adria Symposium on Advanced Mechanics, Primosten, Croatia. Croatian Society of Mechanics Edition, pp. 77?78, 25-28 September 2013. ISBN:978-953-7539-17-7 (2013)

22. Zivkovic, I., Pavlovic, A., Fragassa, C.: Improvements in wood thermoplastic composite materials properties by physical and chemical treatments. Int. J. Qual. Res. **10**(1), 205?218 (2016)

23. Fotouhi, M., Saghafi, H., Brugo, T., et al.: Effect of PVDF nanofibers on the fracture behavior of composite laminates for high-speed woodworking machines. Proc. Inst. Mech. Eng. Part C J. Mech. Eng. Sci. **231**(1), 31?43 (2016). doi:10.1177/0954406216650711

24. Saghafi, H., Brugo, T., Zucchelli, A., et al.: Comparison the effect of pre-stress and curvature of composite laminate under impact loading. FME Trans. **44**(4), 353?357 (2016)

25. Fragassa, C., Pavlovic, A., Massimo, S.: Using a total quality strategy in a new practical approach for improving the product reliability in automotive industry. Int. J. Qual. Res. **8**(3), 297?310 (2014)
26. Pavlovic, A., Fragassa, C.: General considerations on regulations and safety requirements for quadricycles. Int. J. Qual. Res. **9**(4), 657?674 (2015)
27. Bernhart, W., Kruger, F.G.: Technology & market drivers for stationary and automotive battery systems (2012). http://www.rechargebatteries.org/wp-content/uploads/2013/04/Batteries-2012-Roland-Berger-Report1.pdf

Increasing the Energy Efficiency in Solar Vehicles by Using Composite Materials in the Front Suspension

Felipe Vannucchi de Camargo[1(✉)], Marco Giacometti[2], and Ana Pavlovic[3]

[1] Technological Institute of Aeronautics, São José dos Campos, SP, Brazil
fevannucchi@gmail.com
[2] Onda Solare Association, Castel San Pietro Terme, BO, Italy
webmillings@gmail.com
[3] Interdepartmental Industrial Research Centre Advanced Mechanics
and Materials, Alma Mater Studiorum University of Bologna, Bologna, Italy
ana.pavlovic@unibo.it

Abstract. The pursuance of energy efficiency is a constant endeavour in modern mobility. Accordingly, several institutions worldwide have been investing time and resources in developing solar powered vehicles, directing their efforts towards a continual search of technical solutions aiming at attaining the highest energy efficiency levels. This work investigates and compares the mechanical behaviour of a front suspension wheel hub, subjected to its operational forces, when made by three different materials: aluminium, carbon or basalt fiber reinforced composites. Despite of investigating the sole mechanical response of materials, the comparison focuses on the feasibility of applying light and stiff composites in structural parts in order to improve the energy efficiency in vehicles due to weight reduction whilst granting safety.

Keywords: Solar vehicles · Suspension system · Wheel hub · Material optimization

1 Introduction

According to the 21st century demand for novel technologies and products [1], the use of composite materials has undeniably shown to be a growing and perennial trend on industry due to their unique properties, allying remarkable resistance and lightweight. One might consider that their applications are somehow restricted, but recent developments have exhibited the wide range of fields in which extreme mechanical and environmental solicitations are demanded and where composite materials showed to withstand those efforts successfully, such as in aeronautic [2], industrial [3, 4] and naval [5] utmost applications.

Similarly, material developments have also embraced the automotive field where the finest properties of composites permit the achievement of the highest levels of energy efficiency and safety. The motorsport pioneer in adopting composites was John Barnard in 1981 with the McLaren MP4/1, a Formula 1 car assembled with a carbon

© Springer International Publishing AG 2017
G. Campana et al. (eds.), *Sustainable Design and Manufacturing 2017*, Smart Innovation, Systems and Technologies 68, DOI 10.1007/978-3-319-57078-5_75

fiber monocoque chassis [6]. From that moment, composite material innovations have spread to every automotive level, while, nowadays, many competition and commercial cars use composites aiming to lose weight, improve resistance, or both.

Additionally, light composites also permit to achieve remarkable improvements in energy efficiency of vehicles. This aspect is noteworthy, in particular, in the case of solar cars, where energy efficiency is not exclusively obtained by enhanced solar arrays or energy management techniques, but also from lighter vehicles, demanding less energy to overcome inertia [7] in alliance with a kinematics-focused design.

Composites certainly represent a large group of materials with a huge gamma of mechanical properties. Therefore, specific investigations on their mechanical behaviour are needed, especially when composites are used in structural components.

This study is focused on investigating a front wheel hub used in a solar racing vehicle, which connects the front wheels (with hub motors) to the rest of the suspension system. In particular, it is intended to explore the feasibility of manufacturing this critical structural part by the composites carbon fiber or basalt fiber reinforced polymers (CFRP and BFRP) instead of aluminium, traditionally preferred for structural components in racing cars when lightweight is required. A static finite element method (FEM) simulation is conducted to provide a comparison based on resistance, safety and weight of this part when made by such materials.

1.1   Solar Vehicles

Since the invention of solar car in 1955, when William G. Coob from General Motors exhibited his "Sun-mobile" vehicle in Chicago, USA, this segment of engineering never stopped to grow. Aiming to foment this progress, several solar races have been realized in places such as Australia, Chile, Belgium and Morocco, reuniting teams from universities and research centres of all around the world. As the years passed by, the competitiveness has increased, driven by continuous developments in materials, solar arrays, batteries, energy management and vehicle design.

Given the current outlook, renowned solar race teams, which use aluminium suspensions, presently affirm that it is better to use carbon fiber once it is lighter and stronger [8], reassuring that the present work is up to date on the research subject discussed.

1.2   Energy Efficiency

The energy efficiency of a solar car derives from the combination of several aspects such as aerodynamics, mechanics, energy management, material properties and so on. The aerodynamics design deserves a special attention for its impact on energy efficiency, once numerous variables must be taken into account (such as car body shape, fairing position, curvatures, canopy design, driver position and so on) with the aim to achieve the lowest aerodynamic drag [9]. However important, all those issues might not be enough to provide an optimal design.

Since a relevant quota of energy is dissipated through vibrations and motions [10], the challenge to minimize energy loss involves the suspension systems as key factor. Knowing that the main role of a suspension system is to absorb vibrations associated to the car motion to attenuate stresses on structures and passengers, its most characteristic feature becomes an important improvement aspect. Some uncommon technologies have been trying to solve this issue such as energy harvesting devices, which collect energy from mechanical vibrations, wind, electromagnetism and other sources [11, 12].

Besides the need to address an appropriate suspension design, it is also essential to validate the premise that lowering the weight of the vehicle is critical to save energy [13]. It is possible to quote that a 10% reduction in weight can generate up to 6–8% fuel economy in regular commercial cars, 5.1% on hybrid vehicles and can improve the electric range in electric vehicles by 13.7% [14].

In the case of solar vehicles for racing, where the demand for energy efficiency is a priority, each material choice has to be carefully considered: using simple and economic off-the-shelf parts might cause a car to not be able even to complete a competition circuit [15]. On the contrary, in [16] the design optimization of the wheel hub of a solar vehicle for the World Solar Challenge suggests an energy saving of 18.6 W for the battery supported by a 452 g reduction in weight. Hence, the study upon the appropriate material choice for the front wheel hub, as well as for all other mechanical parts of the car, is validated.

2   Materials and Methods

2.1   Aluminium

Commonly preferred as material choice when lightweight, resilience and affordable price are necessary, aluminium is widely used in structural parts, such as suspensions, in the case of non-solar vehicles [17]. Therefore, this material can represent a tangible base for investigating advantages and limits of using composites, being also sustainable due to its end-of-life recycling possibility. The alloy 6151-T6 is considered for this work.

2.2   Carbon Fiber Reinforced Polymer

One of the most widespread reinforcements for composites and known for its noteworthy mechanical properties, carbon fiber is actively present on modern manufacturing designs. Its advantages are mainly related to advanced properties such as tensile strength, elastic modulus, thermal stability, flame retardant, gas barrier [18] and even, sometimes, recyclability [19]; providing safety to vehicles, aspect that might be neglected in some cases despite being of utmost importance [20].

Carbon reinforcements, generally coupled with an epoxy resin matrix, are very common between solar car manufacturers, permitting to decrease the vehicle's weight significantly (down to 55 kg) [21]. The use of this material for realizing a suspension for a solar car was proposed in [7] highlighting that an adhesive bonding between CFRP and aluminium permitted not only to lighten the vehicle, but also to increase the safety factor of suspension by 18.7%.

2.3   Basalt Fiber Reinforced Polymer

Representing a relatively new and valid solution for reinforcing polymers, the use of basalt fiber in composites is encouraged for representing a natural material alternative, native from volcanic rocks originated by frozen lava, being the most abundant crustal rock in land and in the seabed [22].

Its mechanical properties are quite remarkable, being better than glass fiber [23] with an additional very high thermal resistance [24, 25]. When compared to carbon and aramid fibers, its qualities involve a wider temperature application range, higher radiation and oxidation resistances, plus comparable shear and compression strengths [26]. Its eco-friendly feature should also be emphasized once green fibers are an actual concern [27]: besides being an alternative to the carcinogenic asbestos fibers, it does not require pollutant and toxic additives on its production, unlike glass fiber [22].

2.4   Suspension Part Design

The main function of a vehicle's suspension system is to smooth the ride isolating the passengers from vibrations and shocks, to support the total weight of the car and to keep the tires in sturdy contact with the ground, providing safety and adherence for an improved propulsion.

Among the several technical specifications inherent to the suspension system, it is possible to highlight as traditional peculiarities often present in solar racing vehicles, either a MacPherson [28] or a wishbone [7, 8, 29] independent suspensions, with either a coil or leaf spring. These independent systems allow each wheel to perform vertically independent movements. In comparison, if MacPherson has a lighter and cheaper design even if its load carrying capacity is smaller, on the other hand, the wishbone systems offer more steering precision maintaining an adequate camber angle when in a curve or straight drive, while requiring more maintenance.

Other composite suspension parts from a leaf spring system were already studied for solar powered light vehicles, considering E-glass fiber impregnated by an epoxy resin [30], and all targets and requirements were successfully achieved by the composite [31].

2.5   Innovative Design Solution

The present investigation is focused, particularly, on a design solution proposed by Onda Solare Italian designers. Previous projects included a double wishbone suspension either in the whole car for a four-wheeled vehicle, or for the front suspension for a three-wheeled vehicle (which generally considers a trailing arm suspension for the rear wheel). Figure 1a shows one of the vehicles in which such technology is applied and Figs. 1b, c show the suspension set up of those cars, all of them featuring double wishbone and a transversal leaf spring.

Although a more recent evolution (showed by Fig. 2) still features a transversal leaf spring, the suspension system has been adapted to a sliding cart where the wheel hub, object of study of this work, translates.

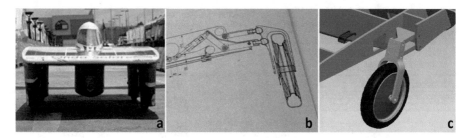

Fig. 1. Solar race vehicle (a) and double wishbone-leaf spring suspension layouts (b, c)

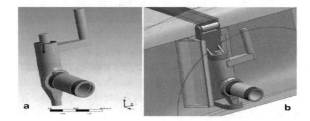

Fig. 2. Boundary conditions assumed for the simulation (a) and mounted wheel hub (b)

This improvement aims to reduce the weight of the mechanism as well as the friction caused by all the joints inherent of a wishbone. After all, the rubbing between moving parts of a car is absorbed by spending more energy to provide their relative movement, which is prevenient from the battery. Thus, this suspension design, despite proper for a vehicle that moves mainly in a straight trajectory, not only provides energy saving for its reduced weight, but as well as for its reduced friction.

The analysed part is a front wheel hub (Fig. 2b) from a suspension with transversal leaf spring. The front suspension is analysed because it is the most mechanically demanded due to cornering and breaking [7].

2.6   FEM Simulation

The mechanical response of the wheel hub was evaluated by a static analysis, which is already proven to represent a good simplification of the real behaviour for a suspension part [17]. Following design specifications, the part was modelled as a shell structure with 5 mm thickness.

As boundary condition, a 2 kN load was applied on the Z axis aiming to simulate the static load effect of the weights of car, battery pack and 4 passengers, when applied on a single wheel. The force intensity also contemplated the roughness of the road surface. Besides, a fixed face condition on the bearing mounting hubs was applied (Fig. 2a). These constraints represent a straight drive condition, without significant curves.

Table 1. Material properties

Material	Properties			
	Poisson coefficient	Young modulus [GPa]	Shear modulus [GPa]	Density [g/cm^3]
Aluminium	0.33	71	27	2.77
CFRP	$v_{12} = 0.20$	$E_1 = 112.5$; $E_2 = 49$	$G_{12} = 13$	1.54
BFRP	$v_{12} = 0.24$	$E_1 = 40$; $E_2 = 28$	$G_{12} = 3$	1.74

The material properties considered are exhibited in Table 1. Information regarding 6151-T6 aluminium alloy was extracted from [32]. As for the composite materials, an epoxy matrix was adopted for both types of reinforcement along with a 60% fiber volume. In order to establish the composites properties, it should be underlined that the approach used is the macromechanical (where the part is modelled as an equivalent homogeneous material) rather than the micromechanical (where it is treated as a multi-phase material) [33]. Thus, once this work concerns a preliminary study on the global behaviour of the component, the macromechanical approach is selected. This analyse is actually advised for materials which do not undergo beyond their elastic regime [34], once until the failure of the first ply, the assembly can be considered as a homogeneous part [35].

The designed component stack up sequence is built out of 20 pre-impregnated 0.25 mm thick plies assembled according to a [90/0$_3$/+45/−45/0$_3$/90]$_s$ orientation, resulting in a 5 mm thick part. This configuration is adequate to the effort the wheel hub is subjected, with the 0° fibers longitudinal to the cylindrical shapes of the part holding the biggest load share, while the ±45° plies contribute to torsional stiffness and the 90° plies grant a good mechanical resistance in all main directions making a quasi-isotropic material.

The carbon fiber reinforcement adopted is the unidirectional T800, by Toray, and the composite properties were calculated through different literature sources [35–37]. As for basalt fiber, a unidirectional composite made with Basfiber, by Kammeny Vek, is used as reference [22, 35]. The isotropic properties of aluminum and the orthotropic properties of the composites are shown in Table 1, where direction 1 stands for 0° and 2 for 90°. Given that a shell structure is analysed, only the in-plane properties are considered once the through-the-thickness stresses are negligible.

3   Results and Discussion

Initially, a mesh characterized by 1 mm element size was considered for the whole model showing a stress concentration present on the fillet (Fig. 3a). Even when assuming different mesh sizes, the maximum stress region was constant and provided of bilateral symmetry (Fig. 3b), but the stress values presented a major variation.

Therefore, aiming to obtain accurate results and define a suitable element size, a mesh convergence study was conducted in order to understand the relation between the

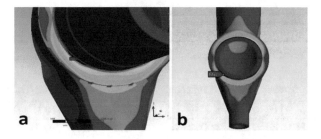

Fig. 3. Stress concentration region (a) and stress symmetry (b)

maximum stress and the element size on the fillet and its adjacent faces (Fig. 4a). The correspondent maximum stress results are displayed on Table 2. Once this mesh size analyse depends on geometry and not on material, only aluminium was considered to establish a suitable element dimension for all simulations.

The test was conducted with 5 iterations ranging from 1.0 mm to 0.2 mm with a 0.2 mm gradual decrease. On Table 2, the Variation and Convergence values represent the difference between the current and previous maximum stresses, in units of pressure and percentage. It is possible to notice a convergent trend among the results.

Table 2. Mesh optimization using aluminium as parameter

Iteration	Element size [mm]	von Mises Maximum Stress [MPa]	Variation [MPa]	Convergence [%]
1	1.0	49.219	–	–
2	0.8	54.103	4.884	9.92
3	0.6	56.095	1.992	3.68
4	0.4	57.993	1.898	3.38
5	0.2	59.552	1.559	2.68

Finally, a mesh with 0.2 mm element size was considered enough in terms of accuracy and used in all regions characterized by the highest level of stress concentrations (Fig. 4a). Other parts were discretized by 1 mm mesh elements (Fig. 4b). The mesh was realized by a hex dominant method, with a Quad/Tri mesh type, generating 149965 nodes and 151915 elements.

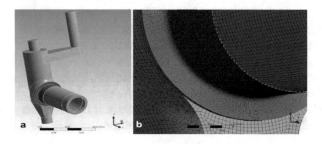

Fig. 4. Critical faces subjected to mesh refinement (a) and contrast between elements (b)

Table 3. Maximum and yield stresses [MPa] and safety factor for aluminium

	von Mises Max. Stress	Yield stress	Safety factor
Aluminium	60	280	4.7

Table 4. Maximum stress components and limits [MPa] and Hill-Tsai number

	σ_1	σ_2	τ_{12}	$\sigma_{1,rupture}$	$\sigma_{2,rupture}$	$\tau_{12,rupture}$	α^2	SF
CFRP	69.5	60.0	37.0	1133	467.5	241	0.04	4.0
BFRP	67.0	55.5	31.5	685.0	290.0	149	0.08	2.4

Maximum stress results are shown in Table 3 for aluminium (von Mises stress) and in Table 4 for the composites (maximum principal stress components), along with the respective safety factors. The gradients, as shown in Fig. 3, can be considered practically the same for all cases, demonstrating a very similar behaviour in stress distribution over the part. The maximum stress results also show to be very close to each other. This material independency is justified by the fact that in all cases the elastic limit is not reached.

As usual for metals, the safety factor of aluminum is calculated based on its yielding stress. For the composites, the appropriate Hill-Tsai [35] (Eq. 1) criterion is adopted for the laminate [38], which states that if the Hill-Tsai number (α^2) at the critical mesh element satisfies the condition [$0 < \alpha^2 < 1$], the design is safe. The conservative assumption of the tensile stress limits for the equation is made, once they are always smaller than the compressive ones (except when the 90° plies represent less than 10% of the structure and there are no ±45° plies, which is not the case) [35].

$$\alpha^2 = \left(\frac{\sigma_1}{\sigma_{1,rupture}}\right)^2 + \left(\frac{\sigma_2}{\sigma_{2,rupture}}\right)^2 - \frac{\sigma_1.\sigma_2}{\sigma_{1,rupture}^2} + \left(\frac{\tau_{12}}{\tau_{12,rupture}}\right)^2 \qquad (1)$$

where σ is the normal maximum stress in the principal directions 1 and 2 and τ is the maximum in-plane shear stress. In order to provide a comparison among all materials, the safety factor (SF) of the Hill-Tsai criterion [35] is calculated according to Eq. 2:

$$SF = \frac{1}{\alpha} - 1 \qquad (2)$$

Furthermore, a previous analyse on weight improvement of a solar vehicle [16] considering the World Solar Challenge race and its 3022 km course, showed that every kilogram of mass on the vehicle represents an energy consumption of approximately 0,8% of the total battery capacity. Therefore, for the wheel hub studied with volume of 257 cm^3, the usage of each material to build the part brings with it a consequent electric range spent due to weight, as stated in Table 5. W_1 states for the weight of the part in one wheel; W_4 is the total weight in the four wheels; and B is the battery capacity spent by the usage of each material. Calculations are made assuming the weight of the car and the pilot.

Table 5. Energy range spent by the usage of each material

Material	ρ [g/cm^3]	W_1 [kg]	W_4 [kg]	B [%]
Aluminium	2.77	0.71	2.85	2.28
CFRP	1.54	0.39	1.58	1.26
BFRP	1.74	0.45	1.78	1.42

4   Conclusion

Besides the fact that this wheel hub geometry is safer than other similar ones applied to solar cars from a modal analysis perspective [39], it showed to withstand safely to the static load applied regardless of the material. It is noticeable that in all alternatives a very similar response is reported in terms of maximum stress on the wheel hub since none of the materials reach the plastic regime.

Practically, if only these factors are considered, the enhanced safety factor of aluminium might lead to the precipitated conclusion that it is the most suitable alternative. However, the performance of the vehicle as a whole depends on many other factors that must be taken into account, such as the weight, which makes carbon fiber the most advantageous option for having a reasonably elevated safety level and adding a low share of weight in the vehicle.

If carbon fiber is selected for the part manufacturing, it is possible to reduce to half the energy that would be spent in the case of aluminium, with a 1.02% economy on battery capacity. Since solar cars are mainly built for racing, the energy savings prevenient from each aspect of the vehicle have to be considered as potentially crucial to determine its performance, making the difference between winning or losing.

Regarding basalt, it showed to be a suitable material with elevated resistance and reduced weight, being more advantageous than aluminium in terms of energy saving, but less than carbon (although a thickness increase in the wheel hub laminate would be advised to provide a higher safety factor). Furthermore, the use of basalt fibers as reinforcement is encouraged by its eco-friendly feature, in line with the culture of sustainability that characterizes solar vehicles.

Acknowledgements. This research was realized inside the Onda Solare project with the aim at developing an innovative solar vehicle. The authors acknowledge support of the European Union and the Region Emilia-Romagna (inside the POR-FESR 2014-2020, Axis 1, Research and Innovation).

References

1. Fragassa, C., Pavlovic, A., Massimo, S.: Using a total quality strategy in a new practical approach for improving the product reliability in automotive industry. Int. J. Qual. Res. **8**, 297–310 (2014)
2. Katsiropoulos, C., Chamos, A., Tserpes, K., Pantelakis, S.: Fracture toughness and shear behavior of composite bonded joints based on a novel aerospace adhesive. Compos. B Eng. **43**, 240–248 (2012)

3. Fragassa, C.: Investigations into the degradation of PTFE surface properties by accelerated aging tests. Tribol. Ind. **38**, 241–248 (2016)
4. Giorgini, L., Fragassa, C., Zattini, G., Pavlovic, A.: Acid aging effects on surfaces of PTFE gaskets investigated by Fourier Transform Infrared Spectroscopy. Tribol. Ind. **38**, 286–296 (2016)
5. Camargo, F.V., Guilherme, C.E.M., Fragassa, C., Pavlovic, A.: Cyclic stress analysis of polyester, aramid, polyethylene and liquid crystal polymer yarns. Acta Polytech. **56**, 402–408 (2016)
6. Tremayne, D.: The Science of Formula 1 Design: Expert Analysis of the Anatomy of the Modern Grand Prix Car. Haynes Publishing, Newbury Park (2011)
7. Hurter, W.S., van Rensburg, N.J., Madyira, D.M., Oosthuizen, G.A.: Static analysis of advanced composites for the optimal design of an experimental lightweight solar vehicle suspension system. In: ASME International Mechanical Engineering Congress and Exposition, Montreal (2014)
8. Mathijsen, D.: Redefining the motor car. Reinf. Plast. **60**, 154–159 (2016)
9. de Kock, J.P., van Rensburg, N.J., Kruger, S., Laubscher, R.F.: Aerodynamic optimization in a lightweight solar vehicle design. In: ASME International Mechanical Engineering Congress and Exposition, Montreal (2014)
10. Zuo, L., Scully, B., Shestani, J., Zhou, Y.: Design and characterization of an electromagnetic energy harvester for vehicle suspensions. Smart Mater. Struct. **19**, 045003 (2010)
11. Yusuf, S.T., Yatim, A.H.M., Samosir, A.S., Abdulkadir, M.: Mechanical energy harvesting devices for low frequency applications. J. Eng. Appl. Sci. **8**, 504–512 (2013)
12. Xie, X.D., Wang, Q.: Energy harvesting from a vehicle suspension system. Energy **86**, 385–392 (2015)
13. Taha, Z., Md Dawal, S.Z., Passarella, R., Kassim, Z., Md Sah, J.: Study of lightweight vehicle vibration characteristics and its effects on whole body vibration. In: 9th Asia Pacific Industrial Engineering and Management System Conference, Bali, pp. 629–633 (2008)
14. Joost, W.J.: Reducing vehicle weight and improving U.S. energy efficiency using integrated computational materials engineering. J. Miner. Metals. Mater. Soc. **64**, 1032–1038 (2012)
15. Sah, J.M., Passarella, R., Ghazilla, R., Ahmad, N.: A solar vehicle based on sustainable design concept. In: IASTED International Conference on Solar Energy, Phuket, pp. 38–43 (2009)
16. Betancur, E., Mejia-Gutierrez, R., Osorio-Gomez, G., Arbelaez, A.: Design of structural parts for a racing solar car. In: International Joint Conference on Mechanics, Design Engineering and Advanced Manufacturing, Catania, pp. 25–32 (2016)
17. Kakria, S., Singh, D.: CAE analysis, optimization and fabrication of formula SAE vehicle structure. In: 18th Asia Pacific Automotive Engineering Conference, Melbourne (2015)
18. Elmarakbi, A., Azoti, W.L.: Novel composite materials for automotive applications. In: 10th International Conference on Composite Science and Technology, Lisboa (2015)
19. Giorgini, L., Benelli, T., Mazzocchetti, L., et al.: Recovery of carbon fibers from cured and uncured carbon fiber reinforced composites wastes and their use as feedstock for a new composite production. Polym. Compos. **36**, 1084–1095 (2015)
20. Pavlovic, A., Fragassa, C.: General considerations on regulations and safety requirements for quadricycles. Int. J. Qual. Res. **9**, 657–674 (2015)
21. Tamura, S.: Teijin advanced carbon fiber technology used to build solar car for World Solar Challenge. Reinf. Plast. **60**, 160–163 (2016)
22. Lapena, M.H., Marinucci, G., de Carvalho, O.: Mechanical characterization of unidirectional basalt fiber epoxy composite. In: 16th European Conference on Composite Materials, Seville (2014)

23. Valentino, P., et al.: Mechanical characterization of basalt fiber reinforced plastic with different fabric reinforcements – Tensile tests and FE-calculations with representative volume elements (RVEs). In: 22nd Convegno Nazionale IGF, Rome, pp. 231–244 (2013)
24. Brandt, A.M., Olek, J., Glinicki, M.A., Leung, C.K.Y.: Brittle Matrix Composites. Woodhead Publishing Limited, Warsaw (2012)
25. Wittek, T., Tanimoto, T.: Mechanical properties and fire retardancy of bidirectional reinforced composite based on biodegradable startch resin and basalt fibres. Express Polym. Lett. **2**, 810–822 (2008)
26. Fragassa, C.: Effect of natural fibers and bio-resins on mechanical properties in hybrid and non-hybrid composites. In: VIII AIP International Conference on Times of Polymers and Composites, Ischia, pp. 020118-1–020118-4 (2016)
27. De Paola, S., Minak, G., Fragassa, C., Pavlovic, A.: Green composites a review of State of Art. In: 30th Danubia Adria Symposium on Advanced Mechanics, Zagreb, pp. 77–78 (2012)
28. Beres, J.: Sunrunner: the engineering report. Sol. Cells **31**, 425–442 (1991)
29. Burke, J., et al.: Viking XX - Western Washington University's solar race car. Sol. Cells **31**, 443–458 (1991)
30. Cranor, M., Bossert, J., Cloud, J., Cranor, N., et al.: The design and fabrication of "Texas Native Sun", the University of Texas entry in G.M. Sunrayce U.S.A., a solar powered vehicle race across the United States. In: Future Transportation Technology Conference and Exposition, San Diego (1990)
31. Sancaktar, E., Gratton, M.: Design, analysis, and optimization of composite leaf springs for light vehicle applications. Compos. Struct. **44**, 195–204 (1999)
32. Metallic Materials and Elements for Aerospace Vehicle Structures Military Handbook. Department of Defense of the United States of America (1998)
33. Ghafarizadeh, S., Chatelain, J.F., Lebrun, G.: Finite element analysis of surface milling of carbon fiber-reinforced composites. Int. J. Adv. Manuf. Technol. **87**, 399–409 (2016)
34. Dandekar, C.R., Shin, Y.C.: Modeling of machining of composite materials: a review. Int. J. Mach. Tools Manuf. **57**, 102–121 (2012)
35. Gay, D.: Materiaux Composites. Hermes, Paris (1997)
36. Siliotto, M.: Valutazione analitica delle aree di delaminazione in materiali compositi avanzati soggetti ad impatti a bassa velocità, Masters Thesis, School of Engineering and Architecture, Alma Mater Studiorum Università di Bologna, Forlì (2013)
37. T800H Technical Data Sheet CFA-007, Toray Carbon Fibers America Inc.
38. Palantera, M., Karjalainen, J.P., Saarela, O.: Laminate level failure criteria based on FPF analyses. J. Spacecraft Struct. Mater. Mech. Test. **428**, 365–370 (1999)
39. Camargo, F.V., Fragassa, C., Pavlovic, A., Martignani, M.: Analysis of the suspension design evolution in solar cars. FME Trans. **45**, 370–379 (2017)

History of Solar Car and Its Electric Components Advancement and Its Future

Hideki Jonokuchi[1(✉)] and Satoshi Maeda[2]

[1] IMRA Lab., Nagoya Institute of Technology, Room 202, Building No. 56, Gokiso, Showa, Nagoya, Aichi 466-8555, Japan
h.jonokuchi.894@nitech.jp
[2] Solar Car Archaeology Research Institute, Shiga, Japan

Abstract. The paper introduces a brief review about the history of solar cars, focusing on electronic aspects and with slight glances at the history of such vehicle in Japan, particularly. This emergent technology, by its conceptual nature, presents as limiting factor the availability of energy prevenient from sunlight, which induces challenging engineering endeavors in order to make such vehicle feasible, although the expectation around it noticeable. Hence, arising technologies have presented innovative solutions with the aim at minimizing running losses and improving energy efficiency in general. The present work explains in detail some design aspect inherent of solar vehicles, as well as the role it plays in a sustainable society.

Keywords: Solar car · In-wheel motor · Maximum Power Point Tracker · State-of-the-art

1 Introduction

Solar cars can be basically defined as vehicles using energy obtained from the sun for power. Within this context of generating driving power through solar energy, other examples of previously manufactured vehicles can be quoted such as airplanes, ships, and automobiles with two, three and four wheels. Historically, attempts of designing an electric vehicle with electric power derived from solar cells as the main power source have been made using Sterling engines and solar thermal steam engines [1, 3].

2 Solar Car Concept

2.1 History

The first solar car invented was the "Solar King", by IRF in "1958", shown in Fig. 1. The breakthrough encouraged later versions such as the "Solar Trek" developed in 1981 by Hans Tholstrup and Larry (Fig. 2). In 1982, these inventors became the first ones to cross a continent in a solar car, travelling autonomously from Perth to Sydney, Australia; powered by DC motors and a chain drive system.

© Springer International Publishing AG 2017
G. Campana et al. (eds.), *Sustainable Design and Manufacturing 2017*, Smart Innovation, Systems and Technologies 68, DOI 10.1007/978-3-319-57078-5_76

Fig. 1. Solar King

Fig. 2. Solar Trek

The importance of the visionary Hans Tholstrup went further, becoming the creator of the famous a competitive World Solar Challenge [4] that currently takes place in Australia every two years gathering solar teams from the finest and renowned universities and research centers worldwide. Following that, a lot of solar car races have been realized in other countries such as Chile, South Africa and Dubai to raise public awareness and spread this fully eco-friendly mobility approach leaded by solar cars.

2.2   Solar Car in Races

Energy efficiency is a critical factor in solar vehicles [5, 6] and all aspects of it require the finest materials and products available in industry [7], not only for efficiency, but also for safety [8]. Due to current cutting-edge technology limits, the application of solar cars to supply ordinary transportation needs is still unfeasible and certainly a barrier yet to overcome. Thus, aiming at fomenting research in such mobility category to develop novel technologies able to fulfill the energy efficiency demands required by these ideal eco-friendly vehicles, the focus spot is given to solar races, currently widely realized and spread all around the globe in competitive events.

The first official competition of the category took place in Australia in 1987, the World Solar Challenge (WSC), counting on 23 participants. After that, some other events followed such as the Europe Tour de Sol, the US Tour de Sol, and the Dream Cup Suzuka, in Japan [9], involving several great companies and universities. Also, General Motors have developed the "GM Sunrayce" in 1987 as a qualifying event for WSC. By then, the average speed of the best vehicles ranged about 68 km/h.

Company original solar cars have also being developed achieving remarkable results such as the one from GM and Ford at the WSC 1987 and the Honda dream at the WSC 1993.

3 Solar Car Drive System

The basic configuration of a solar drive system is based on an electric motor embedded into a wheel (in order to eliminate power transmission losses), a rechargeable battery and solar cells equipped with Maximum Power Point Trackers (MPPTs), as shown Fig. 3.

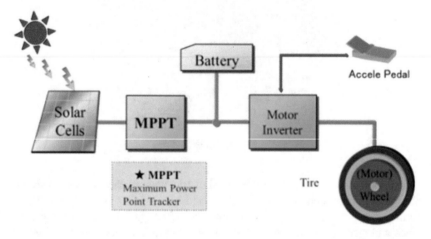

Fig. 3. Basic sketch of the solar car drive system

The battery functions as a power buffer between a solar cell that outputs a sunlight-dependent power and the motor that causes large load fluctuations according to speed and slope changes. It also helps to keep running a solar car that has entered the shade. An MPPT, by its turn, controls the charge that is inserted from the solar cells to the rechargeable battery; also a motor inverter is placed between the battery and the motor.

4 Advancement of Drive System

4.1 Motor Advancement

A solar car is a type of electric vehicle that drives a motor using electricity obtained exclusively from a solar cell. For example, in the Solar Trek vehicle, a 1 HP 24 V operating voltage DC motor from Bosch was used. Since the DC motor can be easily assembled by connecting it to the battery, it was often used for early solar cars. However, since the brushed DC motor has inferior durability, low efficiency, large dimensions and heavy weight; advances in power semiconductors enabled the replacement of those for more modern brushless DC motors (BLM) or AC motors.

Permanent Magnet (PM) DC brushless motors and AC synchronous motors differ in name, but the magnetic structure of the motor is exactly the same. While the first one has a rectangular drive waveform with rough position sensor (such as Hall Effect sensor), the AC synchronous motor has sinusoidal magnetization and driving waveform is sinusoidal with precise position sensor such as resolver. The aforementioned drive waveforms are illustrated in Fig. 4.

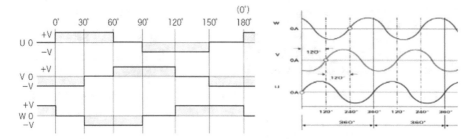

Fig. 4. Motor drive waveform of DC BLM and AC synchronous.

The early success of DC BLMs can be noticed by the fact that such a structure, developed by Seiko Epson, was applied to the solar car Honda Dream, famous for conquering the first place at the WSC 1993. Later, this motor was also supplied to the Golden Eagle car by the Kanazawa Institute of Technology.

Toshiba was also a developer of DC BLMs supplying it to Kansai Electric Esperanza in 1992 and Sky Ace I, from Ashiya University, in 1993. Figure 5 shows the solar car Esperanza I along with the Toshiba's motor specifications in Fig. 6. Sky Ace III (Fig. 7), so called TIGA in Malay, was completed in 2000 using the same Toshiba DC BLM, being the winner in 2000 and 2002, and second place in 2001 at the acclaimed Suzuka Dream cup.

For the WSC 1999, the CSIRO company developed a new concept direct drive in-wheel motor [10] for the Aurora team, applied until nowadays and considered as a major breakthrough (before that, the motor output was conventionally transmitted to the tire via chain, inferring considerable losses of power transmission). The motor was required to fit inside a wheel to reduce aerodynamic drag, and to have maximum efficiency. Furthermore, the usage of a "Halbach magnet array" configuration allowed cars to get a best performance; 10 W (20% loss decrease), but is not necessarily jointly applied by other teams that use in-wheel motors such as Nuna and Michigan.

The Mitsuba team has developed its own in-wheel motor for reassuring qualities such as low power loss and ease of handling (stability), encouraging other Japanese teams began to adopt the same system since 2003. Figure 8 shows the install of chain

Fig. 5. Esperanza I.

Fig. 7. Sky Ace III: TIGA in Suzuka.

Out Dia.	mm	158
Length	mm	155
Weight	kgf	8
Rot. Oot Dia.	mm	65.5
Rot. Length	mm	50
Core In Dia.	mm	67.5
Core Length	mm	45
Phase		3
poles		4
slots		27
Core Material		S9t0.35
Rated Voltage	V	160
Max. speed	rpm	5700
No Load current	A	0.6
Rated Torque	Nm	7.0
Rated current	A	28
Max Eff.	%	94

Fig. 6. Toshiba BLM data.

drive and direct drive systems [11], evidencing the decrease in space used and weight, which is also a fundamental factor for an efficient solar vehicle [5, 6].

In-wheel drive, although the transmission loss is very small, the motor rotation speed corresponds to the actual vehicle speed, so the reduction ratio is fixed. On the other hand, since the chain drive can easily adjust the speed reduction ratio by replacing the sprocket, there is a feature that makes it possible to travel more efficiently against bad weather conditions or uphill/downhill, however transmission losses occur. In the WSC 2001, this feature showed to be useful once the Ashiya University team actually exchanged speed reduction ratios between Alice Springs and Adelaide on climbing and between Darwin and Alice Springs descending.

Fig. 8. Drive system evolution: chain drive (left) to in-wheel motor (right).

These two conflicting functions may be solved with the so-called "magnetic field weakening technique". The method is realized by adjusting the opposing area with the magnet on the rotor surface and stator core, which is moved in axial direction by additional actuator. Meanwhile electrical control methods have been widely used with AC synchronous motor control by current phase control, these techniques are useful especially for uphill/downhill circuit races in order to improve the drive and regeneration efficiency.

4.2 Magnet Evolution and Motor Performance Improvement

The evolution of the motor is in great part due to progress of the magnet. In the 1970's, DC BLMs were developed and commercialized by the progress of ferrite magnets, especially for information devices such as FDD, VCR and so on. Furthermore, ferrite magnets evolved, and in the 1980's the application began to reach power motors such as air conditioner fan motors and compressors.

At that time, Sm-Co magnets and neodymium (Nd) magnets appeared as a novel resource. Nd magnets were used for HDD spindle motors and spread in large quantities and diverse applications. Finally, in 1986, GM developed the manufacturing process of the Nd magnet (Magnequench) and opened the way for power motor application. In WSC and GM SunRaycer of 1987, not by chance, the winning teams adopted the Nd bonded magnet motor, which surely played a part in the victory [12]. Then, Nd sintered magnets were developed by Sumitomo Material, a unique mobility was incorporated into solar car drive motors (DR 086) and commercialized, being adopted by many solar car teams in the early 1990s.

Interestingly, the performance of Nd magnet power motor had been verified with solar car firstly, and it evolved into a motor for EV/HEV. In 1997 Toshiba developed an EV drive DC BLM using an Nd motor together with a controller that had a boost circuit to reduce motor loss, shown at the 1997 Tokyo motor show. UQM, which developed a motor for solar cars, had also developed a motor system for EV drive using Nd magnet. Nowadays, the presence of these technologies is still important and is evolved into EV/HEV motor such as the ones from Prius or Leaf.

4.3 MPPT – Maximum Power Point Tracker

The significant role played by MPPTs in solar cell power generation systems may not be well known: the output of the solar cell varies depending on the irradiation intensity of the sun, the temperature, that is, season, weather and time, and the operating point also changes depending on the load stated on the output side. Therefore, if an external load such as a motor or a battery is directly connected to the solar cell, the electric power generated by the solar battery cannot be fully utilized. MPPT functions as an impedance matching device.

The first MPPT for solar cars is dated from early 1990's for the GM Sunrayce, presenting the performance of 96.6% of efficiency with a 65 kHz FET (Field Effect Transistor) chopper using an open-loop algorithm while acquiring an open-circuit

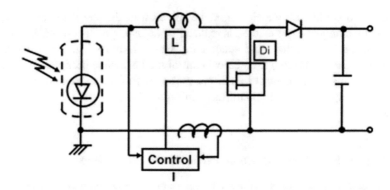

Fig. 9. Boost converter scheme.

voltage sample of 3 ms every 2 s and establishing optimal operating parameters [13]. When the MPPT output voltage is higher than the solar cell's voltage, that boost in the voltage by the MPPT is called a step-up boost type (scheme in Fig. 9). The first MPPT for solar car in Japan were developed for Esperanza for "solar car rally in Noto" held in 1992.

After that, the boost circuit technology was also applied to motor drive in electric vehicles [14], as shown in Fig. 10. By operating the booster only at high speeds, it is possible to increase the efficiency during low-speed driving and regenerative braking. This new system was released at the Tokyo Motor Show in 1997. Later this technology had been adopted in the hybrid car Prius, by Toyota, and Escape, by Ford.

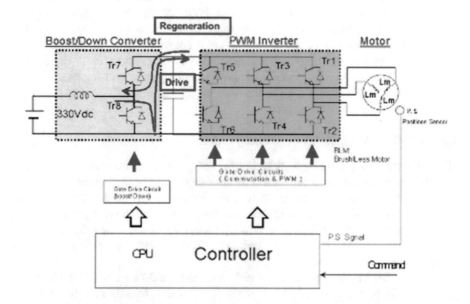

Fig. 10. Motor drive system with booster circuit for electric vehicles.

5 Summary

The development of the solar car has triggered the development of many power electronics technology branches; also giving a chance to verify different high efficiency systems. It is well known that the current technical reach on this area is not enough yet for supplying massive transportation demands, which are a significant problem to modern society in terms of cost and pollution. Nevertheless, given the outstanding evolutions in solar vehicles technology developed in the past three decades, the scenery is certainly auspicious.

Acknowledgments. We thank Sidd Bikkannvari (NASA) who gave me this opportunity. In addition, we must thank Dr. Howard Lavatto (CSIRO), Prof. Hatoh (Ashiya Univ.), Prof. Yamada (Shizuoka Solar Car Team) who cooperated in providing and organizing the materials.

References

1. New Zealand Geographic. https://www.nzgeo.com/stories/the-charge-of-the-light-brigade/. Accessed Jan 2017
2. The Steam Automobile: Modified vega demonstrates feasibility of solar power, vol. 23, no. 3, pp. 32–36 (1981)
3. Solar Car Archaeology Research Institute. http://sunlake.org/solar/archaeology/archaeology_j/japan3/japan_teams2.htm#aisin. Accessed Jan 2017
4. Automostory. http://www.automostory.com/first-solar-car.htm. Accessed Jan 2017
5. Minak, G., Fragassa, C., Camargo, F.V.: A brief review on determinant aspects in energy efficient solar car design and manufacturing. In: Campana, G., Howlet, R.J., Setchi, R., Cimatti, B. (eds.) Sustainable Design and Manufacturing 2017. Smart Innovation, Systems and Technologies Series. Springer International Publishing AG, Cham (2017)
6. Camargo, F.V., Giacometti, M., Pavlovic, A.: Increasing the energy efficiency in solar vehicles by using composite materials in the front suspension. In: Campana, G., Howlet, R.J., Setchi, R., Cimatti, B. (eds.) Sustainable Design and Manufacturing 2017. Smart Innovation, Systems and Technologies Series. Springer International Publishing AG, Cham (2017)
7. Fragassa, C., Pavlovic, A., Massimo, S.: Using a total quality strategy in a new practical approach for improving the product reliability in automotive industry. Int. J. Qual. Res. **8**(3), 297–310 (2014)
8. Pavlovic, A., Fragassa, C.: General considerations on regulations and safety requirements for quadricycles. Int. J. Qual. Res. **9**(4), 657–674 (2015)
9. Suzuka Circuit. http://www.suzukacircuit.jp/result_s/solar/2006/solar_dream_k.html
10. Lovatt, H.C., Ramsden, V.S., Mecrow, B.C.: Design of an in-wheel motor for a solar-powered electric vehicle. In: IEEE Proceedings of Electric Power Applications, vol. 145, no. 5, pp. 402–408 (1998)
11. Osaka University Solar Car Project. http://solarcar.osaka-sandai.ac.jp/. Accessed Jan 2017
12. Cambrier, C.S.: Brushless motors and controllers designed for GM Sunrayce. IEEE Aerosp. Electron. Syst. Mag. **5**(8), 13–15 (1990)
13. Kyle, C.R.: Racing with the Sun: The 1990 World Solar Challenge. Society of Automotive Engineers, London (1991)
14. Jonokuchi, H., Hirata, M., Hashimoto, S., Ishihara, H., Umeda, M., Itoh, H., Hashizume, S., Katoh, S.: EV drive system with voltage booster. In: Conference Rec. EVS-13, pp. 287–294 (1996)

Mg_2SiO_4:Er^{3+} Coating for Efficiency Increase of Silicon-Based Commercial Solar Cells

Rubia Young Sun Zampiva$^{(\boxtimes)}$, Annelise Kopp Alves, and Carlos Perez Bergmann

Department of Materials Engineering, Federal University of Rio Grande do Sul - UFRGS, Osvaldo Aranha 99, Porto Alegre, RS 90035-190, Brazil
rubiayoungsun@gmail.com, {annelise.alves,bergmann}@ufrgs.br

Abstract. Efficiency record commercial silicon solar panels convert about 25% of the sunlight into energy while the vast majority of conventional panels convert between 15% and 16%. The main factors of energy loss are the loss by light reflection on the cell surface and the loss by the energy emitted in the UV and IR band which is directly transmitted and/or converted to heat without being harnessed by the cell. To reduce these losses, it is proposed the use of rare earth doped Mg_2SiO_4 films. Preliminary absorption and emission tests for up and down conversion have indicated that the use of Mg_2SiO_4:Er^{3+}, as a cell coating generates the conversion of IR energy into VIS energy, allowing the solar cell to use this energy. The presented forsterite films antireflection property, together with the erbium upconversion properties, indicates the Mg_2SiO_4:Er^{3+} as promising to increase the commercial silicon-based solar cells efficiency.

Keywords: Antireflection · Forsterite · Rare-Earth · Solar cells · Upconversion

1 Introduction

Global installed capacity for solar-powered electricity has seen an exponential growth, reaching around 227 GW at the end of 2015. It produced 1% of all electricity used globally [1]. Silicon is the dominant material in the production of commercial solar cells. More than 80% of the world production is based on monocrystalline and/ or polycrystalline silicon [2].

According to data provided in 2016 by US National Renewable Energy Laboratory (NREL), commercial silicon solar panels, with a record of efficiency, convert about 25% of light (AM1.5 global spectrum, 1000 W/m^2 at 25 °C) while the vast majority of the panels convert between 15% and 16% [3].

The solar cells energy efficiency is affected by several factors such as recombination, bad contacts, etc. [4, 5] that occur at the solar device interfaces. Among these factors, the most relevant is the light capture conditions. The pure silicon in solar cells presents losses by surface reflection ranging from 31% to 51% between 1.1 μm and 0.40 μm respectively [6, 7]. A high amount of energy is lost by UV and IR solar emission that are transmitted directly without being harnessed by the cell [8] and also generating heat, which increases the losses due to thermalization process. Despite all technological

© Springer International Publishing AG 2017
G. Campana et al. (eds.), *Sustainable Design and Manufacturing 2017*, Smart Innovation, Systems and Technologies 68, DOI 10.1007/978-3-319-57078-5_77

innovation in the last years, there is still a lot to be done to make solar cell devices to reach their full potential [1, 2].

In order to reduce these losses, it is proposed the application of rare-earth (for this work, specifically erbium) doped Mg_2SiO_4 films. Forsterite exhibits chemical stability even at high temperatures and high mechanical properties [8–10]. In addition, studies show that forsterite thin films can be applied as anti-reflectivity coatings [11].

Forsterite can be a good host for rare-earth atoms due to the two non-equivalent crystallographic positions for Mg, the M1 site, with reverse symmetry and the M2 site, with mirror symmetry, present in its orthorhombic structure. Both sites can be replaced by various rare-earth and other ions [12–15].

Some of the rare-earth atoms have the photons upconversion property. This is a process in which the sequential absorption of two or more photons leads to light emission at a shorter wavelength than the excitation wavelength. Among the rare-earth atoms, ions such as erbium (Er^{3+}) stand out due their ability of upconversion of photons from near-IR (NIR) to VIS. The application of upconversion rare-earth doped hosts as solar cells coatings could allow the utilization of IR solar emission by the device. Silicon based commercial solar cells absorb wavelengths from 390 to 1100 nm, while the solar incidence at the sea level is from 300 to more than 1500 nm [8, 14]. A large solar emission spectral range is neglected.

Rare-earth atoms present different behavior depending on their host material. Optical activities are directly influenced by the host:dopant set [16]. The aim of this work is to validate forsterite as a rare-earth host in order to produce thin films that bind the anti-reflective property of forsterite with the upconversion property of rare earths (Er^{3+}), improving commercial solar cells overall efficiency.

2   Materials and Methods

The thin films production began with the synthesis of highly pure and homogeneous powders by reverse strike co-precipitation method. Magnesium Nitrate hexahydrate - $Mg(NO_3)_2 \cdot 6H_2O$, Tetraethyl orthosilicate - $Si(OC2H5)_4$ and erbium nitrate pentahydrate - $Er(NO_3)_3 \cdot 5H_2O$ were used as precursors. The precursors were dissolved in ethanol (analytical grade) and precipitated in 7 M aqueous ammonium hydroxide - $NH_4(OH)$. All the reactants were from Sigma-Aldrich. The forsterite samples were doped with X = 1, 3, 5, 7, 10 and 20% wt of Er^{3+}. The reaction follows the stoichiometry (Eq. 1):

$$2\,Mg(NO_3)_2 + Si(OHC_2H_5)_4 + X\,Er(NO_3)_3 \rightarrow Mg_2SiO:Er^{+3} \tag{1}$$

After chemical and optical characterization, the powder sample with better performance was compacted to produce a sputtering target. Forsterite:Er^{3+} thin films with different thicknesses were produced in order to evaluate the influence of the film thickness on the silicon wafer optical activity. They were deposited by magnetron sputtering on silicon wafer substrates with dimensions of 12 mm/4 mm. The thicknesses values chosen to produce the samples are based on ellipsometry analysis and are described in the results section.

2.1 Characterization

The sample chemical and structural composition were analyzed by Raman spectroscopy using a Renishaw in Via Spectrometer (532 nm). The crystallinity of the sample was evaluated by X-ray diffraction (XRD) using a PHILIPS diffractometer (model X'Pert MPD), 40 kV, 40 mA and Cu anode.

A preliminary optical characterization was carried out to investigate the applicability of the synthesized sample as host for rare earths. Absorption spectroscopy in the UV, VIS and near-NIR were performed in a UV-VIS. Ocean Optics covering a range of 200 to 1000 nm wavelength and in a InfraXact Foss NIR equipment that covers 570–1850 nm wavelength, respectively. The optical characterization was performed on the compacted powder. Photoluminescence measurements were performed using a Micro-PL system (HORIBA Jobin Yvon). An excitation source of 100 mW and a 532 nm solid state laser were used for analyze down-conversion emission and a 980 nm laser was applied for up-conversion analysis (Quantum Laser).

The ellipsometry tests were carried out in a SOPRA, GES - 5E spectral ellipsometer with angle variation of 10–90°. The thickness of the thin films where based on the Eq. 2. For the reflectance at normal incidence we define a series of parameters: r_1, r_2, and θ. The surrounding region has a refractive index of n_0 (1, for air), the thin-film coating has a refractive index of n_1 and a thickness of t_1, and the silicon has a refractive index of n_2 (around 3.5 for silicon wafer). For a single layer anti-reflective coating on a substrate the reflectivity is:

$$\mathbf{R} = \left\| r^2 \right\| = ((r_1)^2 + (r_2)^2 + 2\,r_1 r_2 \,\cos 2\theta)/(1 + (r_1)^2 \cdot (r_2)^2 + 2\,r_1 r_2\, \cos 2\theta) \tag{2}$$

$$\mathbf{r_1} = (n_0 - n_1)/(n_0 + n_1) \quad \mathbf{r_2} = (n_1 - n_2)/(n_1 + n_2) \quad \mathbf{\theta} = 2\pi n_1 t_1/\lambda$$

3 Results

The initial analyzes were performed for representative doped samples due the behavior linearity and then, emission analysis was performed for all produced samples. Figure 1a shows the diffractogram of the samples synthesized by reverse-strike coprecipitation with different doping concentrations, as well as the diffractogram of the pure forsterite and erbium oxide as reference.

No contamination or second phase formation were observed. It presented well defined peaks indicating high crystallinity, purity and homogeneity. Comparing the doped forsterite diffraction with the pure erbium oxide diffraction, it is observed no erbium oxide peaks present in the doped samples. All the Er^{3+} entered the forsterite crystalline lattice and did not form erbium oxide.

Raman spectra (Fig. 1b) presented completely different band positions to the doped samples compared to the pure forsterite and erbium oxide bands. It indicates that the doped samples optical behavior is not related to the erbium oxide or the forsterite separetedly. The Er^{3+} dopant is in the forsterite crystallite originating a new system $Mg_2SiO_4{:}Er^{3+}$.

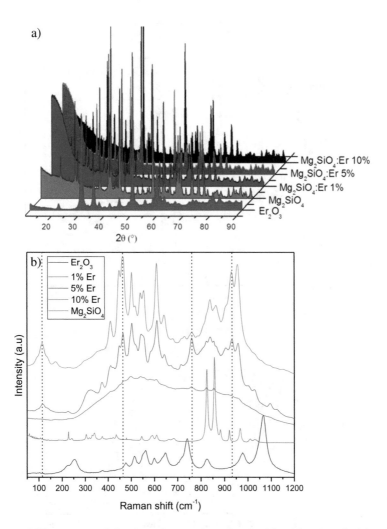

Fig. 1. (a) Diffractogram of the forsterite powders doped with erbium in 1% 5% and 10% concentration, as well the pure forsterite and erbium oxide diffractograms accordingly with PDF# 34-189 and #08-0050 respectively; (b) Raman spectra of doped forsterite samples with erbium in 1%, 5% and 10% concentration, as well the pure forsterite [17] and erbium oxide [18] spectra.

In order to carry out a more in-depth analysis of the properties presented by the Mg$_2$SiO$_4$:Er^{3+} system, some optical analyzes were performed. The absorbance spectra presented in the Fig. 2a shows pure forsterite, erbium oxide and the doped samples absorbance in the UV-VIS range. The Mg$_2$SiO$_4$ did not present optical activity in this range. That is an important property for crystal application as host in the optical area. The host should present elevated mechanical and physical-chemistry properties and a neutral optical behavior [13, 15]. Since the pure forsterite did not present optical activity, the doped samples absorbance bands are strongly related to erbium bands.

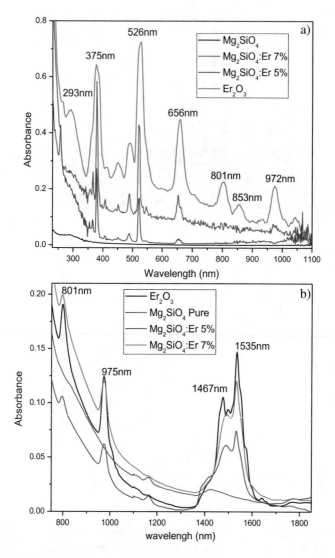

Fig. 2. (a) UV-VIS absorbance spectrum and (b) NIR absorbance spectrum for Erbium oxide, pure and 5%, 10% Er^{3+} doped forsterite

At the NIR range (Fig. 2b), pure forsterite did not shows expressive absorption and the doped samples presented high absorption band between 1400 nm and 1600 nm. An intense absorption in this range is essential for upconversion applications from the NIR to VIS. In all the UV-VIS-NIR analysed range, the doping concentration and the absorption intensity were directly related, bigger the dopant concentration the bigger the absorption intensity [19].

Based on $Mg_2SiO_4{:}Er^{3+}$ absorption analysis, it was possible to perform the emission analysis by exciting exactly the wavelength range that the system absorbs. Figure 3a

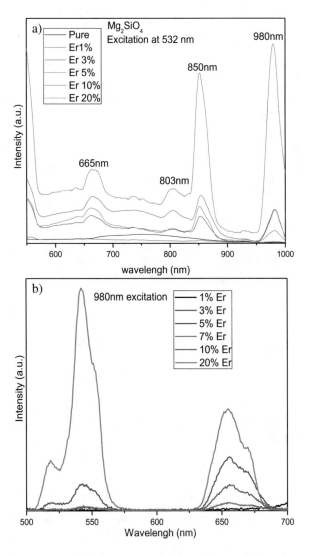

Fig. 3. Photoluminescence emission spectrum of the powders varying the dopant concentration (% wt) with (a) excitation at 532 nm and (b) with excitation at 980 nm, showing the upconversion phenomenon.

shows the luminescence spectrum for doped and pure forsterite samples with excitation at 532 nm wavelength laser.

The spectra presented the characteristic sharp emission peaks for Er^{+3} occupying voids in a crystalline host. The emission intensity was directly proportional to the dopant concentration in the forsterite structure. The 20% Erbium doped sample presented the highest emission. Pure forsterite did not present optical behaviour in the analysed UV-VIS range [16].

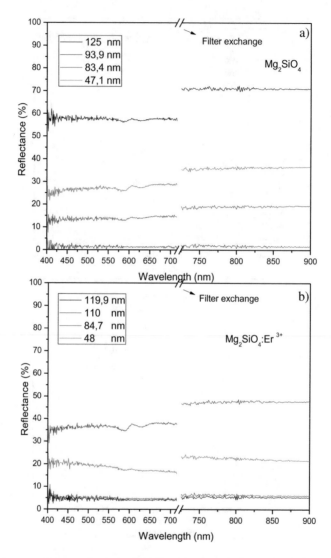

Fig. 4. Reflectance of (a) pure forsterite thin-films and (b) doped forsterite thin-films varying the thickness values.

Figure 3b presents the upconversion phenomenon for Mg_2SiO_4:Er^{3+} with excitation at 980 nm. Interestingly, the overall emission intensity increased following the dopant concentration up to the 7%. Above 7%, the emission decreased with the increasing of dopant concentration. This phenomenon is related to the concentration quenching effect. Quenching effect might have happened due the migration of excitation energy from one activator to another and eventually the migration to an imperfection that acts as an energy sink [20]. By increasing the concentration of emitter dopants in solid systems, the like hood of this phenomenon increases. In most cases, the doping concentration of the Er^{3+}

is typically confined below 2 mol% to avoid the luminescence concentration quenching effect [21]. Due to the forsterite crystalline structure, it was possible to reach high emission values with concentrations of 7% wt (16,1 mol%) and no concentration quenching effect.

Accordingly to the emission results, the 7% doped sample was employed to produce thin films by sputtering. Ellipsometry analysis indicated refractive index for the doped and non-doped samples ranged from 1.53 to 1.62 in the VIS range (300 to 800 nm). The thickness of the anti-reflection coating is chosen so that the wavelength in the dielectric material is one quarter the wavelength of the incoming wave [22].

Based on the range of the refraction index results, films with different thickness were produced in order to define the optimal thickness, avoiding destructive interferences. The doped thin film (Fig. 4b) which presented the lowest reflectance (5%) have 110 nm of thickness. This result was really close to the film with 119.9 nm. To the pure forsterite thin films (Fig. 4a), the lowest reflectance (2.5%) was presented by the 83.4 nm thickness film. This difference is explained by the Erbium light absorption in the doped films.

4 Conclusions

The Mg$_2$SiO$_4$:Er^{3+} samples synthesized by reverse strike co-precipitation presented high absorption area in the NIR range, manly between 1400 and 1600 nm and emission in the UV-VIS range, all with down-conversion and up-conversion excitation. The Mg$_2$SiO$_4$:Er^{3+} presented upconversion with 980 nm excitation as an indicative of the behaviour with excitation at around 1530 nm, since both excitation wavelengths will promote the upconversion phenomenon [23, 24]. Although the analyses presented in this paper still preliminary, forsterite presented the production of stable thin-films with high transparency in the VIS range. The use of Mg$_2$SiO$_4$:Er^{3+} as solar cell coating could allow the solar cell to utilize the energy from NIR range that is absorbed by the coating and re-emitted in the VIS range where the cell is able to absorb this energy (silicon-based solar cells absorb between 300 and 1100 nm). In addition to erbium, other rare-earth atoms can be added to this system to maximize up and down convertors. Forsterite proved to be an efficient host for optical applications. The Mg$_2$SiO$_4$:Er^{3+} coating showed to be promising to increase the efficiency of commercial solar cells.

References

1. World Energy Resources. World Energy Council, pp. 25–27 (2016)
2. EPIA: Global Market Outlook for Photovoltaics Until 2016, pp. 1–74 (2012)
3. Martin, A.G., Keith, E., Yoshihiro, H., et al.: Solar cell efficiency tables (version 47). Prog. Photovolt. Res. Appl. **24**, 3–11 (2016)
4. Ali, K., Kan, S.A., Matjafri, M.: 60Co γ-irradiation effects on electrical characteristics of monocrystalline silicon solar cell. Int. J. Electrochem. Sci. **8**(6), 7831 (2013)
5. Minak, G., Fragassa, C., De Camargo, F.V.: A brief review on determinant aspects in energy efficient solar car design and manufacturing. In: Smart Innovation, Systems and Technologies
6. Ali, K., Khan, S., Jafri, M.: Structural and optical properties of ITO/TiO$_2$ anti-reflective films for solar cell applications. Nanoscale Res. Lett. **9**(1), 1–6 (2014)

7. Ali, K., Khan, S.A., Jafri, M.Z.M.: Effect of double layer (SiO$_2$/TiO$_2$) anti-reflective coating on silicon solar cells. Int. J. Electrochem. Sci. **9**, 7865–7874 (2014)

8. Fischer, S., Goldschmidt, J.C., Loper, P., Bauer, G.H., et al.: Enhancement of silicon solar cell efficiency by upconversion. Optical and electrical characterization. J. Appl. Phys. **108**(4), 44912 (2010)

9. Fathi, M., Kharaziha, M.: Mechanically activated crystallization of phase pure nanocrystalline forsterite powders. Mater. Lett. **62**, 4306–4309 (2008)

10. Ringwood, A.E.: Significance of the terrestrial Mg/Si ratio. Earth Planet. Sci. Lett. **95**, 1–7 (1989)

11. Suleimanov, S., Dyskin, V.G., Settarova, Z.S., Dzhanklych, M.U., et al.: Antireflection coatings for solar cells based on an alloy of a mixture of MgO and SiO$_2$. Geliotekhnika **4**, 62–63 (2010)

12. Tsunooka, A., Androua, M., Higashidaa, Y., Sugiurab, H., et al.: Effects of TiO$_2$ on sinterability and dielectric properties of high-Q forsterite ceramics. J. Eur. Ceram. Soc. **23**, 2573–2578 (2003)

13. Abraham, E., Bordenave, E., Tsurumachi, N., et al.: Real-time two-dimensional imaging in scattering media by use of a femtosecond Cr^{4+}:forsterite laser. Opt. Lett. **25**, 929–931 (2000)

14. Haase, M., Schafer, H.: Upconverting Nanoparticles. Angew. Chem. Int. Ed. **50**, 5808–5829 (2011)

15. Chia, S., Liu, T., Ivanov, A., Fedotov, A.: A sub-100fs self-starting Cr:forsterite laser generating 1.4W output power. Opt. Exp. **18**, 23–25 (2010)

16. Niklaus, U.W., Alessandro, D.M., Izilda, M.R., et al.: Influence of excited-state-energy upconversion on pulse shape in quasi-continuous-wave diode-pumped Er:LiYF4 lasers. IEEE J. Quantum Electron. **46**(1), 99–104 (2010)

17. Zampiva, R.Y.S., Acauan, L., Alves, A.K., Bergmann, C.P.: Novel forsterite nanostructures with high aspect ratio via catalyst-free route. Mater. Res. Bull. **60**, 507–509 (2014)

18. Yan, D., Wu, P., Zhang, S.P., et al.: Assignments of the Raman modes of monoclinic erbium oxide. J. Appl. Phys. **114**, 193502 (2013)

19. Pal, M., Pal, U., et al.: Effects of crystallization and dopant concentration on the emission behavior of TiO2: Eu nanophosphors. Nanoscale Res. Lett. **7**(1), 6–12 (2012)

20. Raymond, F.C., Jay, K.R.: Mechanism of fluorescence concentration quenching of carboxyfluorescein in liposomes: energy transfer to nonfluorescent dimers. Anal. Biochem. **172**, 61–77 (1988)

21. Wu, X., Zhang, Y., Takle, K., et al.: Dye-sensitized core/active shell upconversion nanoparticles for optogenetics and bioimaging applications. ACS Nano **10**, 1060–1066 (2016)

22. Shiga, T.T., Tetsuya, S.: Anti-reflection optical article, United States Patent (4,904,525) (1987)

23. Quimby, R.S.: Upconversion and 980-nm excited-state absorption in erbium-doped glass. Fiber Laser Sources Amplifiers IV **1789**, 50–57 (1992)

24. Li, X., Zhou, S., Jiang, G., Wei, X., et al.: Blue upconversion of Tm^{3+} using Yb^{3+} as energy transfer bridge under 1532 nm excitation in Er^{3+}, Yb^{3+}, Tm^{3+} tri-doped CaMoO4. J. Rare Earths **33**(5), 475–479 (2015)

Experimental Temperature Modelization for Solar Racing Vehicle

Claudio Rossi[✉], Marco Bertoldi, Gabriele Fabbri, Davide Pontara,
and Gabriele Rizzoli

Department of Electrical Energy and Information Technology,
University of Bologna, Bologna, Italy
claudio.rossi@unibo.it

Abstract. This paper presents an experimental method to model the temperature response of the inverters of the solar car Emilia 3 which is the vehicle build by the Italian Onda Solare team for the World Solar Challenge 2013. Object of this paper is using the experimental data collected to create a thermal model of the inverters to predict dangerous situations and adding it to the energetic model of the overall vehicle powertrain to better simulate racing conditions and help the race strategy planning. The data is elaborated to identify the experimental transfer function between motor current and inverter temperature. The function is validated with the current measurements and its behavior was found to be consistent with the real temperature data.

Keywords: Solar car · Temperature transfer function · Temperature limitation · Vehicle model

1 Introduction

The Italian team Onda Solare built Emilia 3 to compete in the World Solar Challenge in Australia in 2013 in the Challenger Class. After this race, Emilia 3 has been used to compete in other events around the world, with different road conditions and race rules from the ones used for the initial car design. Figure 1 shows the Emilia 3 car during the 'Carrera Solar' race in the Atacama Desert (Chile) in 2016. The paper focuses on the identification of a thermal model of the inverter used during the development of the race strategy.

As per regulations, the vehicle has 4 wheels, 6 m^2 of solar panel and 21 kg of batteries. It is made entirely of carbon fiber, Nomex and Kevlar for a total weight of under 200 kg. The pilot has a minimum weight of 80 kg with ballast added to meet the limit. Aerodynamics and the mechanical characteristics of the vehicle are key to improve the efficiency of the solar car [1–4] as is the powertrain system.

The electric and electronic components have been developed by the University of Bologna and includes, among other smaller components, the BMSs, inverters, motors and MPPTs. The powertrain takes advantage of a direct drive configuration with one motor per back wheel, which sits inside the rim. Each motor is driven by its own inverter, which is commanded just by a torque reference, thus automatically

G. Campana et al. (eds.), *Sustainable Design and Manufacturing 2017*, Smart Innovation, Systems and Technologies 68, DOI 10.1007/978-3-319-57078-5_78

Fig. 1. Emilia 3 in Chilean Atacama Desert in 2016.

implementing a "virtual" differential. Section 2 lists the characteristics of the power-train installed on Emilia 3.

Given that the car is made for racing, having a good model that can predict certain key systems and subsystems is vital to be able to tune the perfect strategy. Such model already existed before this study, but it focused almost exclusively on the energy management and dynamic analysis.

Thermal analysis has been conducted, but only in laboratory setting on the single subsystems (specifically, only the motor and only the inverter) as part of the standard validation process of the hardware and software of the components. Of course, such measurements would require a lengthy thermodynamic study to be able to apply them on the real car configuration [5, 6].

In order to overcome the missing thermodynamic study of the inverter system, this paper focuses on the experimental analysis of real driving measurements. In this way, the temperature behavior of the system was identified by means of transfer functions.

Mathematical identification of the function and implementation of the results were carried out on Matlab Simulink [7] because the energy model was already set up and required only little modifications. This approach proved to be a viable way to get temperature predictions of the car during race conditions and confrontation with real data is presented in Sect. 5.

2   Powertrain Architecture

General description
The main physical components (or subsystems) of the powertrain are:

- electric motors
- inverters
- battery system
- photovoltaic panels
- PV panels to battery DC/DC converter.
- battery charger
- The main control functions are:

- motor control algorithm
- traction control algorithm
- power limiting functions
- battery monitoring and power limiting functions
- maximum power point tracking and charge control

Figure 2 shows the powertrain architecture, by emphasizing the role of the communication system.

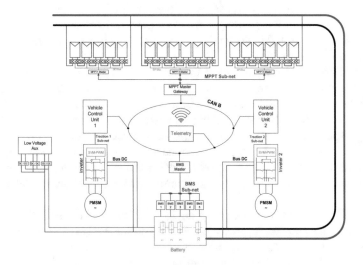

Fig. 2. Scheme of the powertrain of Emilia 3.

In Fig. 2 the red-black outer lines represents the DC power link that interconnects the following energy conversion systems:

- 391 PV cells subdivided in 17 independent PV sub-modules
- 17 DC/DC converters for supplying the energy generated in the 17 PV sub-modules into the DC power link. These DC/DC are grouped in three assemblies, 2 composed by 6 DC/DC and one by 5 DC/DC connected in series. These DC/DC implements local MPPT, regardless of the condition of neighbor modules.
- The battery pack, composed by 391 cells, assembled in 33 modules connected in series, each composed by 13 cells connected in parallel.
- The battery charger for recharging the batteries from the grid wall plug.
- Two inverters, one for each motor. The power stage is based on paralleled Mosfet, for the highest efficiency. All the inverter driver circuits and auxiliaries are on board.
- Three auxiliary DC/DC converters, for the highest reliable distribution of power to the control devices and vehicle auxiliaries.
- Communication system

- Figure 2 shows the smart communication network based on the CAN bus protocol. The network is subdivided into sub-nets, each of them connecting a limited number of devices and associated to a defined function:
- MPPT subnet
- battery system subnet
- traction 1 subnet
- traction 2 subnet.

Each subnet is connected to a secondary CAN bus called CAN B through a gateway unit. Each gateway allows bidirectional transferring of selected messages between the sub-net and CAN-B. In this way, transfer of information between any device is always possible. Furthermore, the gateway introduces electrical insulation between the two connected CAN-buses and in the unlikely event of a fault in a subnet, the fault does not propagate to the other sub-nets.

This architecture allows to reserve the CAN B for broadcasted messages that are exchanged among peripherals connected at the same sub-net. Messages used in a subnet only are not redirected on CAN B. In this way, it is possible to maximize the update rate of each information running on subnet mask, without affecting the communication speed on CAN B. On other hand, the traffic relying on CAN B is minimized and then the baud rate of this bus can be reduced for an increased reliability.

The functions associated to a subnet (MPPT, traction, battery management) operate in safe mode even in case of losing the communication with CAN B, and with other sub-nets.

CAN B is also used for connecting devices for data logging and PC for parameterization. For this purpose, a point-to point (P2P) protocol allows to query any variable or parameter from any device of the powertrain, with a scan rate up to 10 ms. For example, all battery cell voltage and temperature and all PV panel voltage and current are usually logged during a race with a sample rate of about 200 ms.

Hub wheel motors
The Emilia 3 powertrain is based on two hub wheel motor. These motors are SPMSM (Surface Permanent Magnet Synchronous Machine) in radial flux arrangement and

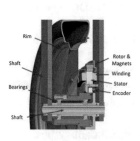

Fig. 3. Rendering of the hub-wheel motor

Fig. 4. Traction inverter-VCU

Fig. 5. Rendering of the MPPT board

Table 1. Electric motor data

Technology	PMSM
Rotor	outer
Pole pairs	10
Rated voltage [V_{RMS}]	65
Rated torque [Nm]	19
Rated speed [rpm]	750
Rated current [A_{RMS}]	15
Max. torque@50A [Nm]	49
Constant power at @15A [W]	1700
Speed of constant power@15A rpm	700–1200
Efficiency in rated cond.	97%

external rotor. The design of this motor, shown in Fig. 3 was shared with the design of the carbon fiber hub-wheel. In this way, high torque density, low weight and high efficiency is reached. Magnetic position sensor is embedded in the motor assembly. One motor only is able to drive the vehicle up to a speed of 70 km/h. Main data of the electric motor are given in Table 1.

Traction inverter

Traction inverter is designed for this special traction application. Table 2 gives the main data of the traction inverter and Fig. 4 shows a picture of the inverter. The power stage is based on the use of paralleled Mosfet. A powerful microcontroller was used on the inverter control stage. Motor control algorithm is based on an advanced field oriented control of the motor [8], that maximizes torque all over speed and torque range, with particular emphasis on the flux weakening region. Torque generation from the motor is obtained by searching the minimum losses operating point.

The traction control algorithm is implemented in the same microcontroller of the motor control algorithm. This algorithm generates the torque reference from the driver command by applying all the required mapping and computation (ramping, smoothing, power limiting, torque limiting, current limiting…). Three maps can be selected directly by the driver for obtaining different vehicle characteristics.

Table 2. Inverter data

Power stage technology	Mosfets
Mosfet in parallel per switch	2
Rated output current [A_{RMS}]	15
Rated DC voltage [V]	100
Switching frequency [kHz]	7
Rated efficiency	98%
Microcontroller	320F2812

The inverter hardware also holds the VCU microcontroller, that realize the interfacing between the vehicle command (gas pedal, levers) and the traction control functions.

In each drive, VCU and traction inverter shares the same sub-net. The VCU microcontroller also operates as gateway between traction sub-net and CAN B.

Acquisition of vehicle signals for the two VCU are completely separated, allowing reduced power operation of the vehicle, with one traction drive only.

Battery Management System

The battery management system is based on a modular solution, composed by 5 slave modules and one master unit.

Each slave module acquires voltage and temperature across up to 8 cells connected in series and sends the acquired data to the BMS master unit, through the BMS dedicated sub-net. This unit calculates on line the fundamental variables representing the battery pack status and send it to the CAN B for the need of traction systems and MPPT systems. In this way the BMS master operates as gateway between BMS sub-net and CAN B. Main data of the BMS system are given in Table 3.

Table 3. Battery Management System

Configuration	Modular
Number of slave modules	5
Cell voltage meas. accuracy [mV]	10
Cell voltage meas. resolution [mV]	5
Cell voltage scan rate [ms]	9
Number of temperature meas.	2 per module
Battery current meas. accuracy [A]	0.1
Equalization current [A]	0.5
SOC estimation	Current integration
On-line calculation by master unit	• high cell voltage • low cell voltage • average cell voltage • total pack voltage • max. cell temperature
Refresh rate of calc. quantities [ms]	10

Low and high cell voltages, calculated by BMS master unit, are mainly used by the traction control system for limiting the power, either absorbed or generated by the traction drives, in case of cell under-voltage or cell over-voltage respectively.

High cell voltage, calculated by BMS master unit, is used by the MPPT control system for limiting the power generated by the PV panels in case of cell overvoltage.

Maximum battery temperatures are also used in traction and MPPT controls for limiting power exchange with the battery pack.

BMS master unit manages the automatic passive equalization procedure, during charging, in case of battery misalignment of cell voltage. The BMS directly controls the battery charger by setting the reference current for the charger. The BMS master unit, on the basis of available batteries quantities, calculates the State of Charge (SOC) of the batteries.

MPPT control system

The Maximum Power Point Tracking system is based on a modular solution, composed by three slave modules and one master unit. Two slave modules controls six DC/DC converter and one module controls five DC/DC converters. The input of a single DC/DC converter, of each slave module, is connected to a string of PV cells. The output of the DC/DC converters, of each module, are connected in series. DC/DC converter are all boost topology. All the output ports of the three modules are connected to the main DC-link of the powertrain.

Any single DC/DC converters is controlled by an independently operated MPPT algorithm, yielding to splitting the PV panel of the car in 17 independently controlled sub-modules. In other words, these DC/DC implements local MPPT, regardless of the condition of neighbor modules. This is a very important feature of the conversion system, in order to maximize energy harvesting in case of partial shadowing or not uniform exposure to the sun's rays.

Figure 5 shows the MPPT slave module with six DC/DC conversion stages installed on the same board. Table 4 gives main data of the MPPT converters.

Table 4. Maximum Power Point Tracking MPPT

Configuration	Modular
Nr. of slave modules	3
Nr. of DC/DC per module	2×6 and 1×5
DC/DC topology	Boost
DC/DC configuration	4 interleaved
DC/DC converter model	ST -SPV1020
Switching frequency [kHz]	100
Max. output current [A]	4

The MPPT slave module is equipped with a microcontroller for controlling the operation of the connected DC/DC. This internal communication is based on SPI protocol, and is mainly used by setting DC/DC operating mode and acquiring DC/DC operating state and variables.

Each MPPT slave module sends the acquired data through a dedicated sub-net to the MPPT master unit. This unit makes all the necessary computation and send the result of energy production on the CAN B. In this way the MPPT master operates as gateway between MPPT sub-net and CAN B.

In Tables 5 and 6 are summarized the specifications for the solar panel and the battery pack.

Table 5. Solar panel data

Solar cell type	SunPower C60
Cell efficiency	22.5%
Voc [V]	0.687
Isc [A]	6.28
Temperature coeff. power [/°C]	−0.32%
Number of cells	391
Sub-modules	17
Active surface [m^2]	6
Panel peak power [W]	1350

Table 6. Battery pack data

Cell manufacture	Panasonic
Cell type	NCR18650B
Nominal cell capacity [mAh]	3200
Cell weight [g]	48
Series connection	13
Parallel connection	33
Rated pack voltage [V]	130
Pack energy [Wh]	5020
Gravimetric energy density [Wh/kg]	250

3 Model Development

After the good result of WSC 2013, the car has competed in other international races around the world. In particular, from April 21st to April 26th 2016, it participated in the Carrera Solar Atacama in Chile where it finished 2nd in the "Evolucion" class.

During the preparation for the race, the team brought the car on the road that leads to the La Silla Observatory located at the outskirts of the Chilean Atacama Desert, 600 km north of Santiago de Chile and at an altitude of 2400 m. Figure 6 shows its location.

The road steepness was outside of the design limits of the car, thus stressing the powertrain and rising the temperatures of both the motors and the inverters. So much, in fact, as to activate the safety limitations of the control system. In particular, the inverter temperatures reached 70 °C on both of them.

The team anticipated the possibilities of dangerous temperatures during these extreme conditions and set up the logging of the data. It is one of the few experimental data collection of limit intervention outside of the lab tests on single components.

Therefore, it was a perfect opportunity to conduct thermal analysis on real road conditions and modeling the car even for high temperatures.

The car was driven only for about 15 km before the inverter temperatures caused the limitation to intervene and the car was not able to win the steepness of the road with the torque available (around 14%).

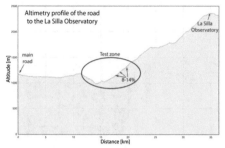

Fig. 6. Location *of the La Silla Observatory in Chile.*

Fig. 7. Topographic profile of the road and test zone.

In Fig. 7 the topographic profile of the complete road is presented. With a red circle it is shown the segment of road where the data was collected. At the beginning, there is a downhill part, but then the road is steady climbing with grades between 8% and 14%.

Figure 8 shows the data collected during the test. In particular, the upper graph relates to the power used by the left and right motors; the middle one presents the inverter currents for the left and right one; in the last one the temperatures recorded can be seen, specifically the inverter and motor temperatures for both left and right drive.

Therefore, during the first part the temperature do not rise more than expected, mainly because the road was downhill. Power is high and currents low because the vehicle was driving at relatively high speeds.

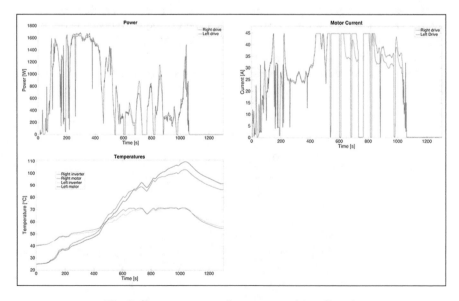

Fig. 8. Power, current and temperature data collected.

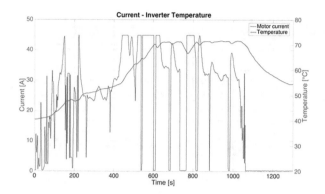

Fig. 9. Current and temperature relation.

The environment temperature during the test was around 20 °C.

After about 400 s, the road started to climb, the vehicle slowed down and the temperature started to rise significantly. It is important to notice that the temperature limits activated were the ones on the inverter temperature that are set at 70 °C. Motor temperatures reached 110 °C, but the limit is set at 120 °C.

At around 600 s, only the right temperature limitation is activated and, therefore, in the current graph, only the right motor current (in blue) is reduced as it should.

In Fig. 9 it is clear to see how the motor current directly influences the inverter temperature: it rises when the current is applied and it falls when there is no current flowing through the inverter. It is, therefore, between these two values that the relation is investigated.

For the identification of the correlation it was used Matlab's Identification Toolbox, with the application of the Instrumental Variable and IV initialization method, for it is well known. Moreover, this method allows to identify the system even when external noise and disturbance are not white.

Figure 10 shows the toolbox interface with both sets of data for the right and left inverter.

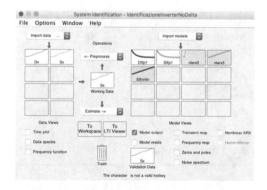

Fig. 10. Matlab identification toolbox interface.

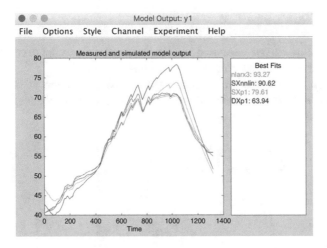

Fig. 11. Estimation results. Cyan curve is the best.

Multiple refinements of the estimations were needed to reach a good approximation of the data. In Fig. 11 some of these results can be seen. The cyan one, named "nlarx3", is the best one found at 93.27% confidence.

Such data estimation was calculated for both inverters and the results of the Matlab toolbox are the zeroes and poles of the two transfer functions.

Since the usual transfer functions do not allow to set an initial condition, but they always have to start from 0, the Matlab function tf2ss was used to calculate the relative state spaces.

The obtained coefficient for the transfer function and state space representation are given in Eqs. (1) to (4) for both inverters.

Right side inverter:

$$G_{right}(s) = \frac{0.004179}{s + 0.001557} \tag{1}$$

$$\begin{cases} \frac{dx}{dt} = -0.001558x + 0.0041789u \\ \quad\quad y = 1x + 0u \end{cases} \tag{2}$$

Left side inverter:

$$G_{left}(s) = \frac{0.003582}{s + 0.001412} \tag{3}$$

$$\begin{cases} \frac{dx}{dt} = -0.001412x + 0.0035822u \\ \quad\quad y = 1x + 0u \end{cases} \tag{4}$$

This way, in the simulation application, it was possible to set an initial condition of 40 °C and reduce the time of the simulation by skipping the "warming up" of the transfer function. Details will be explained in the next section.

4   Model Validation

The results of the analysis were used in a simple Simulink program to compare them to the real data in the environment used for the modelization.

In Fig. 12 is presented the diagram used for the verification of the results. The Simulink blocks Transfer Fcn and State Space were used to implement the following function for the right and left drive:

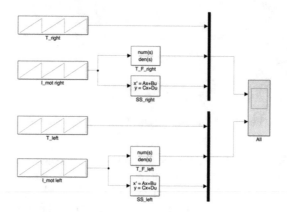

Fig. 12. Simulink model for verification.

In the state space function, the A coefficient represents the thermal time constant, while the B coefficient is the thermal resistance. It is interesting to note that the first one results very similar between left and right, while the other differs more significantly. This can justify the fact that the right drive reached first the temperature limitation.

Figure 13 shows the results of the simulation. In red the measured temperature data is compared with both the transfer function (blue) and the state space (green) outputs.

It is easy to spot how the transfer function starts from 0, while the initial condition of the state spaces was set to 40 °C. Therefore, 2 whole cycles of the measured data were fed to the model in order to verify the equivalence of the outputs sources.

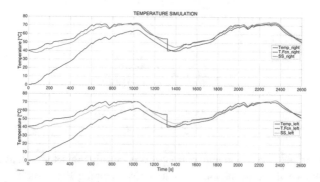

Fig. 13. Simulation comparison of transfer function and state space to the measured data.

This way it can be seen how, towards the end, the functions outputs correctly behave the exact same way. The transfer function just needs more time to compensate the initial offset. From Fig. 13, the state space blocks were validated and chosen to be used on a more complete vehicle dynamic model in which they are useful to predict possible critical condition during a complete race simulation.

5   Complete Vehicle Model

The Simulink model used for simulation of the Emilia 3 model is the one presented in Fig. 14. In particular, the red block on the left is where the vehicle parameters such ad mass, efficiency, aerodynamic coefficient and rolling coefficient are set. The gray blocks represent the dynamic simulation and converts the torque input in the correct vehicle speed according to the vehicle parameters and the road grade.

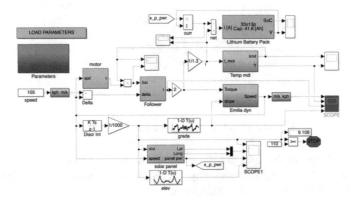

Fig. 14. Simulink model of Emilia 3.

The green blocks simulate the various subsystems of the car: battery, motor, solar panel and driver. Finally, the orange block is the one that implements the temperature behavior. It has as output the maximum temperature between left and right electric drive and a limit port that limits the torque if the temperature exceeds 70 °C, like on the real vehicle.

The panel block is presented in Fig. 15.

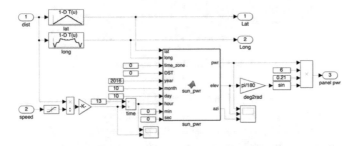

Fig. 15. Solar panel simulation block.

The key element is the function block that calculates the sun position and the solar irradiation anywhere on the planet and anytime for the next 100 years. As inputs it needs the latitude and longitude in degrees and the date and hour of the race. Then the time zone and DST (Daylight Savings Time) can be set to adjust to the local time of the event.

The function has been confronted to the correct calculation that can be found on the NOAA site [9] and the error on the position is less than 0.1°.

The irradiation is then multiplied for the area, the efficiency and the inclination of the panel which is parallel to the ground while the vehicle is moving.

The solar power is then converted in current to correctly simulate the recharging effect on the battery. It is therefore subtracted to the battery current.

The battery block is shown in Fig. 16.

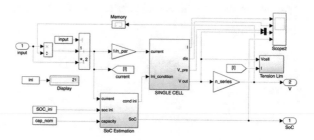

Fig. 16. Simulink *battery pack simulation model.*

From a mixture of nominal and experimental data, this block considers the battery pack as a single cell. The inputs and outputs are scaled accordingly to provide the right current and voltage range of the entire pack.

The block estimates the state of charge (SoC) by integrating the current drained and it has the possibility to evaluate a safety limitation on the cell voltage, but, on this model, it is not used.

The motors are simulated implementing a lookup table for just one characteristic and then multiplying the torque request by 2. Inside the block looks like Fig. 17. The number 1.3 used in the model is the torque coefficient of the motors than links torque and current.

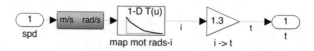

Fig. 17. Simulink motor block.

The "Follower" block that simulates the driver is a simple PI regulator with a dynamic saturation to the maximum torque of the motors.

The dynamic block in gray implements Newton's second law of motion $\Delta F = ma$ and the necessary integration to g from acceleration to speed. It looks like Fig. 18.

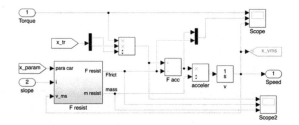

Fig. 18. Simulink dynamic block calculations.

The force applied is the difference between the traction force from the motors and the friction force from the outside. The latter is calculated by accounting the aerodynamic drag of the vehicle, the rolling friction of the tires and the slope of the road.

The last main subsystem is the one under study and its implementation is quite straightforward: it receives the current used by the vehicle, divides it by 2 (the number of electric drives) and delivers the inverter temperatures based on the transfer functions identified before on the paper. See Fig. 19 for the inside of the block.

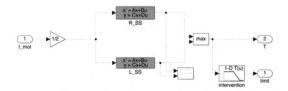

Fig. 19. Simulink inverter temperature model.

The state space implementation has been used instead of the transfer function to set the initial condition and have better results even at the model start.

As stated before, a simple limitation map is implemented to simulate the safety intervention on high temperatures. It is set to intervene at 70 °C as the one the real vehicle.

One of the first use of this model was in the strategy planning for a solar race held in Morocco in October 2016. The race was held on public roads for a total distance of around 110 km with some elevation changes, but not very significant. Because of the shortness of the race and a regulation change that allowed the start with full battery, it meant the speed would be higher than usual at around 90 km/h.

After calculating that the energy would have been enough even with little sun irradiation, the problem was the verification that the high speed and high power application would not put in danger the electronics because of temperature.

Tests on the model were conducted at 90 km/h and they did not show any dangerous temperatures. During the race, however, the conditions were different: more sun than expected because the starting time was moved closer to midday, more speed and a higher starting temperature of the inverters.

After the race, therefore, the simulation was re-run with:

- Starting time of 13:00 local time.
- 35 °C as starting temperature of the inverters.
- 105 km/h as reference speed.

Unfortunately, some data acquired from the car was corrupted during the download and full graphs of the measurements of interest cannot be displayed. However, the temperature recorded from the halfway point to the finish line was almost constant at 44 °C. Moreover, the final state of charge of the battery was 37.11% on the inboard display.

The first graph, presented in Fig. 20, shows the speed of the simulated vehicle in respect to the reference of 105 km/h and the grade of the road during the race which never exceeds 2° (that translate to 3.5%).

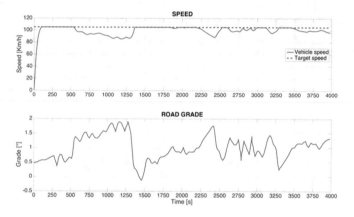

Fig. 20. Race simulation - speed and grade.

The second graph is Fig. 21 and shows the torque used by one motor and two currents: in blue the one absorbed by the electric drive and in red the one provided by the solar panel.

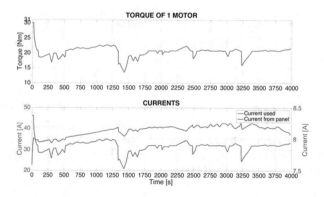

Fig. 21. Race simulation - torque and currents.

The simulation finished with a battery SoC of 37.86% which means a difference less than 3% from the actual value.

Finally, the maximum inverter temperature is presented in Fig. 22.

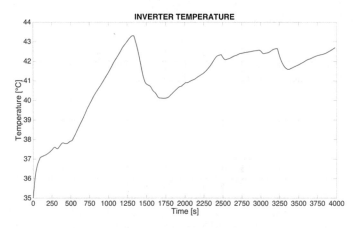

Fig. 22. Race simulation – Inverter temperature.

It can be seen how the temperature reaches 43 °C in the first third of the simulation and then it slowly ripples around 42 °C. During the actual race, it stayed at around 44 °C at least from mid-race until the end.

6 Conclusions

The paper deals with the identification from experimental data of the transfer function of the inverter thermal system in a racing solar vehicle, Emilia 3. This is done to predict possible dangerous temperatures during a race.

The inputs used were the data collected during a photo shoot drive in Chile where the safety limitations, always present onboard the vehicle, were activated and effectively caused the car to stop. Thanks to the Matlab Identification toolbox, two transfer functions (one for the right inverter and one for the left one) were calculated with a 93% confidence level.

These functions were first validated with the original data and then deployed on a more complex Simulink model of the complete vehicle.

They were then used to simulate an actual race both to predict if dangerous temperatures would occur and, as post-analysis, to validate again the model.

Both objectives were met inside the approximations introduced.

First, the prediction of the model showed that temperatures over the limit would not have been a problem in such a race, even if the speed, starting temperature and starting time were not correct at the time.

Second, the temperature simulated after the fact, with more accurate data, differed only by 2 °C from the real one. This can be safely considered inside the approximation of the model especially given that the transfer function assumed equal the environment temperature and it was tuned for higher values.

For sure, better results would have been possible knowing the outside temperature and, therefore, deriving the model from the over-temperature of the system. Even so, the results establish confidence in the overall method and more research will be spent in this direction with specific tests.

Acknowledgement. This investigation has been realized inside the TIME collaborative project. It is co-financed by the European Regional Development Fund POR-FESR 2014–2020, Axis 1, Research and Innovation of the Region Emilia-Romagna.

References

1. Camargo, F.V., Fragassa, C., Pavlovic, A., Martignani, M.: Analysis of the suspension design evolution in solar cars. FME Trans. **45**, 370–379 (2017)
2. Camargo, F.V., Giacometti, M., Pavlovic, A.: Increasing the energy efficiency in solar vehicles by using composite materials in the front suspension. In: Campana, G., Howlet, R.J., Setchi, R., Cimatti, B. (eds.) Sustainable Design and Manufacturing 2017. Smart Innovation, Systems and Technologies series. Springer International Publishing AG, Cham (2017)
3. Minak, G., Fragassa, C., Camargo, F.V.: A brief review on determinant aspects in energy efficient solar car design and manufacturing. In: Campana, G., Howlet, R.J., Setchi, R., Cimatti, B. (eds.) Sustainable Design and Manufacturing 2017. Smart Innovation, Systems and Technologies series. Springer International Publishing AG, Cham (2017)
4. Betancur, E., Fragassa, C., Coy, J., Hincapie, S., Osorio-Gómez, G.: Aerodynamic effects of manufacturing tolerances on a solar car. In: Campana, G., Howlet, R.J., Setchi, R., Cimatti, B. (eds.) Sustainable Design and Manufacturing 2017. Smart Innovation, Systems and Technologies series. Springer International Publishing AG, Cham (2017)
5. Koschan, A., Govindasamy, P., Sukumar, S., Page, D., Abidi, M., Gorsich, D.: Thermal modeling and imaging of as-built vehicle components. In: Military Vehicles, SAE 2006 World Congress, SAE SP-2040, Detroit, MI, SAE Technical Paper 2006-01-1167, April 2006
6. Pesaran, A.A.: Battery thermal management in EVs and HEVs: issues and solutions. In: Advance Automotive Battery Conference, Las Vegas, Nevada, USA, February 2001
7. Jimenez, M.J., Madsen, J., Andersen, K.K.: Identification of the main thermal characteristics of building components using MATLAB. Build. Environ. **43**(2), 170–180 (2008). Science Direct
8. Rossi, C., Pilati, A., Casadei, D.: Unified model and field oriented control algorithm for three-phase ac machines. In: Proceedings of 15th European Conference on Power Electronics and Applications (EPE), Lille, France, 2–6 September 2013, pp. 1–13. ISBN: 9781479901166. doi:10.1109/EPE.2013.6634406
9. NOAA solar calculator. https://www.esrl.noaa.gov/gmd/grad/solcalc/

A Brief Review on Determinant Aspects in Energy Efficient Solar Car Design and Manufacturing

Giangiacomo Minak[✉], Cristiano Fragassa, and Felipe Vannucchi de Camargo

Interdepartmental Industrial Research Centre Advanced Mechanics and Materials, Alma Mater Studiorum University of Bologna, Bologna, Italy
{giangiacomo.minak,cristiano.fragassa,felipe.vannucchi}@unibo.it

Abstract. In the past decades, sustainable means of transportation have become an important issue once they are potentially able to supply modern transport needs whilst not harming the environment. Accordingly to this general interest, solar vehicles have been developed by several institutions worldwide to participate in international class races, promoting this research field. As the competitiveness increased, solar technologies evolved toward noteworthy solutions for a modern and sustainable mobility. Hence, this work intends to provide a general overview on solar vehicles, particularly regarding the main design and manufacturing features that allowed to increase energy efficiency, considering the relevance of this factor for solar cars. Due to the huge amount of information available, a limited number of aspects was selected for further analysis, mainly related to design and engineering, such as: weight reduction, aerodynamics and kinematics, mechanics and advanced materials.

Keywords: Solar cars · State-of-the-art · Energy efficiency · Sustainable design

1 Introduction

The demand for more efficient means of transportation arose as an important segment, once fossil fuel powered vehicles are dependent on a progressively scarce energy source while being responsible for producing massive toxic pollutant substances, aggravating the green-house effect [1]. Addressing to supply the need for environmentally harmless means of transportation, solar cars emerged as a remarkable option, receiving a crescent interest from renowned research centers worldwide [2–5].

Although powered by a renewable energy source, a solar vehicle is only viable if embedded with the highest level of energy efficiency, which depends on various design and manufacturing facets. Aiming to provide a broad review on those factors showing their characteristics and evolutions, this work provides a state-of-the-art background on the variables related to mechanical aspects, e.g. design, materials and aerodynamics. Most of the researches are performed on racing solar cars, the most propitious area for innovations, considering the attention given by universities and research centers in developing novel technologies, focused on competitive race events realized throughout the globe.

© Springer International Publishing AG 2017
G. Campana et al. (eds.), *Sustainable Design and Manufacturing 2017*, Smart Innovation, Systems and Technologies 68, DOI 10.1007/978-3-319-57078-5_79

2   Energy Optimization

The main performance difference that apart conventional motor sport races and solar racing, is the limitation on the total power available for generating movement: the power output of the solar panels is limited [6]. Then, it is fundamental to include solutions for the optimal use of energy. Several design factors influence the performance of the vehicle in terms of energy, and weight is a major concern [7]: the lower the weight, the higher the energy efficiency.

Physically, the power needed to provide movement to a body is dependent on its weight. Not differently for solar cars, their design should be focused on a lightweight goal approach. However, there is a limit even for decreasing the weight of the car if one desires to keep the stability of the vehicle at a safe level: the center of gravity should be kept close to the road, and the weight should be high enough for the car to remain stable when subjected to side wind gusts, which is a potentially threatening influence [5], especially on solar races held in desert regions [3].

The energy efficiency improved by weight optimization can be noticed on electric vehicles, once a 10% reduction results in a 13.7% electric range increase [8]; or specifically in the case of a solar car where a reduction as small as 452 g is responsible for an energy saving of 18.6 Wh [4]. In another solar racing vehicle, classified as 2nd in the Cruiser class of the most competitive solar race, the World Solar Challenge (WSC), the innovation of manufacturing a carbon fiber monocoque chassis provided a weight reduction as high as 55 kg [3]; which leads to the importance of the materials for performance.

Significant improvements that relate the impact on energy efficiency by design, aerodynamics and materials have been thus realized in solar vehicles.

3   Design

3.1   Number of Wheels

The decision on the number of wheels for solar races might either be indicated on the competition categories regulations, or depend on the designer decision. Up to now, the basic parameters to decide which layout to adopt can be defined by the advantages and disadvantages of each one [6]:

- Three wheels: The major concern of this layout is related to instability. After all, if the center of gravity moves towards the axle with two wheels, the rolling tendency is decreased. But, if it moves towards the single wheel, the vehicle will tend to tip over. On the other hand, the less number of parts infers in decreased weight and cost, while the single wheel, most likely configured with a steering arm suspension [2], brings with it an ideal steering and conserves energy on bumpy roads once it allows only vertical movements [9].
- Four wheels: Despite provided of more movable parts, hence with more weight, this system grants a higher stability and improved aerodynamics, once the frontal area is considerably smaller.

At first sight, aiming at achieving the most energy efficient design, the three wheels layout seems adequate for offering a weight reduction on this aspect. However, it is important to highlight that the aerodynamics improve caused by the reduction of frontal area in the case of a four-wheeled-vehicle is important on its performance. The drag caused by the air resistance can be optimized at a higher level, and it might not be a coincidence that solar vehicles provided with such design had achieved outstanding results, even though in different categories, at the most recent edition of the World Solar Challenge 2015 [2, 3].

3.2   In-Wheel Motors

Electric propulsion has been used for many years in vehicles [10], and it is a crucial topic of study if an efficient design is required, especially in solar vehicles where driving force is fundamental to acceleration and power performance [10]. Therefore, in-wheel motors emerged as an efficient solution for solar cars, with early applications in solar competitions dated of 1990 [11].

What makes this alternative so attractive to designers respects to the facts that a direct drive eliminate drivetrain power losses [12], generally excluding the need for a gearbox and its associated power dissipation [2]; backed up by a simple in-wheel packaging where the individual wheel control improves handling and safety [13]. Also, an important consideration for the design of a car is the unsprung weight reduction, which is covered by a motor entirely housed by the wheel (as an integral part of it) [13].

Further studies [14] on how to enhance its efficiency have been performed, comparing Axial Flux Permanent Magnet (AFPM) to Radial Flux Permanent Magnet (RFPM) motors. It was realized that even though AFPM motors are a bit more expensive, they are essential in improving the performance of solar vehicles at the WSC; being a good alternative at low torque and low-to-medium speed applications. The explanation is based on the fact that ironless AFPM motors have a high torque-to-mass ratio compared with RFPM, and lack of iron eddy currents.

Some in-wheel motor application examples for the WSC show that the brushless DC motors are more efficient (converting approximately 97% of their electric energy) and lighter (about 8 kg) than other alternatives (with efficiency ranging between 92–95% and weight 12–16 kg) [12]. To exemplify their power, the 2nd placed competitor in Cruiser category at the WSC 2015 has a maximum power of 42 kW, i.e. 57 BHP [2].

3.3   Suspension

Responsible for absorbing external and internal vibrations of a car, the suspension is certainly one of the most vital systems in any terrestrial vehicle, softening shocks, preserving all mechanical components, and granting comfort to the passengers. In solar cars, the suspensions are designed to be as stiff as possible without abandoning the aforementioned characteristics, in order to decrease energy loss in driving.

An energy efficient-focused optimization of a suspension can either involve diverse factors due to its kinematic complexity, or be as simple as in a weight reduction process [15]. For example [4], the shape of a wheel knuckle being redesigned in order to save

452 g of overall mass in the vehicle can generate an energy saving of approximately 18.6 Wh; which is especially important once this change was done in a solar vehicle competing in the competitive Challenger category at the WSC.

Basically, the suspension systems can be subdivided into two main groups [16]: dependents and independents. The first ones have an axle connecting both right and left wheels in front and in the rear, in a way that every vertical displacement on one side, affects the performance of the opposite. When used in solar vehicles, are generally in form of a transversal leaf spring; aiming at reducing the total weight of the structure [17]. The independent systems, as the name states, are free to realize vertically autonomous movements, without affecting other wheels. Nowadays, the vast majority of solar cars adopt independent systems. The principal types of independent suspensions, and their contributions to improve energy efficiency are:

- MacPherson: consists of an A-arm or a compression link stabilized by a second link that offers a bottom mounting point for the axle (or hub) of the wheel; carrying a coil spring and a shock absorber. It is low cost, lightweight and compact [18].
- Wishbone (Fig. 1b): generally in double wishbone configuration, it is the most spread configuration [19, 20], composed by two A-arms that provide great steering precision and an adequate camber angle. Allows the vehicle height to be set lower, hence lowering the center of gravity and increasing stability. As for energy efficiency, this system allows the suspension to be set stiffer (decreasing energy loss) and keep the wheels always in contact with the road (even in hard cornering), providing good propulsion.
- Trailing arm (Fig. 1a): It is generally used in three-wheeled solar cars in alliance with two wishbones in the other wheels. Its contribution to energy efficiency is given by the fact that it is a lightweight alternative that does not perform sideways movements, hence saving energy on bumpy roads [9].

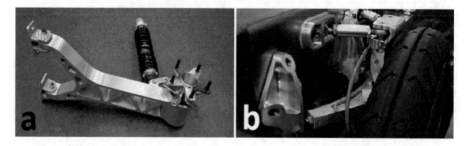

Fig. 1. (a) Trailing arm [2] and (b) double wishbone suspensions applied to solar cars

4   Aerodynamics

Every body that moves in space has to overcome the resistance imposed by the environment around it. In solar cars, this challenge is particularly crucial because, given the limited source of energy available and the dependency on the area of the solar arrays, the more efficient the design of the vehicle, the less energy it will spend to move (i.e. more efficient it will be). To define the ideal shape of a car in these terms, simulations are required using Finite Element Analysis (FEA) techniques performed through the use of computational software.

In this context, the two parameters to be defined to make a design feasible are the coefficient of drag (C_D) and lift coefficient (C_L), determined with the use of FEA considering a certain speed. In short, C_D is a positive value that quantifies how much the vehicle is being held by the resistance of the air to move. The closer to zero the C_D value, the more aerodynamic the body. For example, a reduction of 31% in C_D can provide an energy saving as high as 442.6 W considering a 100 km/h speed [21].

As for C_L, it defines the following behavior of the vehicle in motion: if the value is negative, it means that the car continues with a vertical downwards force resultant; if the value is positive, the car suffers the possibility of lifting from the ground. Therefore, this value must always be kept negative for maneuverability and safety of the driver [5].

For optimizing the shape of the vehicle and achieving a good efficiency, numerous factors must be taken in account [21], such as the fairing position, body-and-fairing fillet blend, fairing leading edge curvature, driver position and canopy design. It is possible to notice a few aspects in common from the first three classified cars in the Challenger category of the WSC 2015, in other words, the fastest cars of the most competitive category and event: for example, the drivers were placed sideways in all cases, between one front and one rear wheel. One can infer that the reason to justify this pattern in the fact that the frontal area of the vehicle is reduced, increasing the efficiency. This case is illustrated on Fig. 2.

Fig. 2. Solar vehicles designed by the universities of (a) Delft, (b) Twente and (c) Tokai [3]

The teardrop shape is known to provide, ideally, a perfect shape design in order to maximize the aerodynamic efficiency [22], which is actually a trend that can be noted in all competitive solar cars (Fig. 2). Nevertheless, there was a case regarding an attempt to realize a new design [23] inspired on a boxfish; but this attempt showed to be inefficient producing a car with drag and lift coefficient respectively equal to 0.35 and −0.19; while a common teardrop design given as good [5] grants more efficiency, backed up by drag and lift coefficients of approximately 0.11 and −0.11, respectively.

The shape of the car, however, is not exclusively focused to decrease the drag influence, but also to enhance the heat removal from the solar array, thus increasing the power generated by photovoltaic modules [24]. The heat removal is an important issue on solar car design, and as having an active cooling system is unfeasible due to the weight it would bring to the car composition, convection is the only source of cooling [24].

5   Materials, Manufacturing and Safety

According to the 21st century demand for innovative products and technologies [25], the selection of materials for building efficient solar cars have become an important task due to the alliance of the importance of having a safe and efficient design, while novel materials have been gradually developed. After all, the solar races realized nowadays present such high technological standards and extensive courses, that if one might build a car with off-the-shelf parts regardless of the material properties, the competitor probably won't even be able cross the finish line [26].

Essentially, mechanical systems are designed to maintain both weight and friction to a low level, and at the same time keeping strength and stiffness. In most vehicle architectures, the preferred materials to attain this goal are aluminum, titanium and composites [27]. Even though all of them might be strong and stiff while being fairly light; the application of metals in solar cars, such as aluminum chassis [26], present a massive weight share to the detriment of energy efficiency. Thus, composites are the most indicated solution as structural material in this case.

Therefore, although polymeric materials present particular deterioration characteristics highly dependent on environmental conditions [28–30], this is not relevant to motorsport parameters once each vehicle is basically built for a reduced number of races, as the regulations suffer constant changes and new improved technologies arise, not exposing the vehicles to noxious environments for long periods.

The application of composite materials in structural parts is a reality in modern mobility industry aiming at providing more energy efficiency, ranging from aviation [31] to high tech Formula 1 competition cars; which nowadays are composed of over 75% of various forms of composites [32]. Particularly in solar cars, these materials are applied in mechanical components such as suspension springs [33], or even linked to aluminum suspension parts by adhesive bonding, not only decreasing the overall weight, as well as increasing their safety factors [19].

The evolution of chassis in solar cars has also shown to be a determinant factor to a good performance, relying on its material to maintain a strong and safe chassis and decreasing its weight. While past designs used to involve aluminum assemblies [26], the latest trend recognized as the most advantageous is to build a carbon fiber monocoque chassis; proved to grant a competitive design [2, 3]. It's preferred manufacturing technique, vacuum infusion, consists of assembling the composite layers on a solid mold, cover the material with plastic, and make a vacuum bagging (Fig. 3) which is succeeded by an autoclave curing process at controlled temperature and pressure.

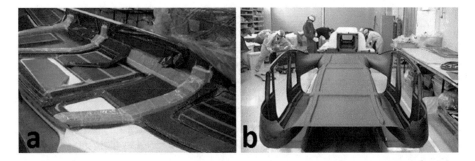

Fig. 3. Vacuum bagging (a) and post cured (b) monocoque chassis for a solar car [3].

Safety is a mandatory characteristic in all means of transportation, and should be emphasized as priority in all projects and classes of vehicles, even though it might be neglected in some occasions despite its utmost importance [34]. Thus, the materials selected for the arrangement of the vehicle must present great resistance, which, in the case of solar cars, should be necessarily allied to low weight to provide an energy efficient design.

Fulfilling this premise, the application of carbon fiber as reinforcement for composite materials is a concern of study for energy-efficient and safe vehicles (EESVs) [35], presenting interesting properties in terms of elastic modulus, tensile strength, thermal and electrical conductivity, thermal stability, gas barrier and flame retardant [35]; although the last characteristic can also be noticed in basalt composites [36].

6   Sustainability

As aforementioned, regarding the application of composites in solar cars, carbon fiber reinforced composites are undoubtedly the most widely used material due to its advantageous properties. However, the intensification of usage of such material is noticeable in the last years, with an expected global demand of 208 000 tons per year in 2020 [37]. This results in an increase of waste, inferred by both end of life products and production processes [38]. Therefore, accordingly to the sustainable products premise of being easy recycling, easy reuse and easy degradation [39]; studies involving recycling methods for this material have been developed, boosted by pyrolysis/oxidative recovering processes [38].

Besides recycling, another sustainable solution in composites is the use of biodegradable fibers, if possible allied with eco-friendly resins, such as vinylester [40]. Basalt, even though not biodegradable, is a natural and relatively new alternative in reinforcing polymers and structural composites [40], not only being made with by a toxic-free process (unlike glass fiber), as well as featuring enhanced mechanical properties, which have been studied and quantified [41]. In comparison with carbon fiber and aramid, for instance, it presents a wider temperature resistance [36, 42], higher oxidation and radiation resistance, and higher compressive and shear strengths [40]. Also, comparing to glass fiber, the mechanical properties of basalt are either similar or better [43–45].

7 Conclusions

This work provided a brief review of some important topics on design of solar cars crucial to provide an energy efficient design. The topics analyzed were weight optimization; materials; design; aerodynamics and sustainability.

Attaining an optimal overall vehicle weight is proved to be determinant to its performance. Also, in-wheel motors have shown to be a common trend among all solar car teams for decades, once eliminating the energy losses caused by extensive power transmission chains is a necessity for solar vehicles considering their high efficiency demand.

The materials used in solar cars vary from metals to composites in diverse systems among the solar cars nowadays, mainly restricted to budget. After all, the crescent usage of carbon fiber is a noticeable tendency, powered by its lightweight, safety and sustainability with recent studies regarding recyclability techniques for its composite. Yet speaking of sustainability, basalt is a material with various advantageous structural properties, having also a low toxicity production process. Its usage is encouraged, even if in non-structural parts, once it would be a great eco-friendly accomplishment to design a solar car in the less environmentally harmful way possible.

Suspension, as well as some aerodynamic factors and the decision on adopting either a three or a four-wheeled design, are subjective settlements particular to each design team. However, some characteristics commonly found on the current world top solar cars include a four-wheeled design, with sideways driver position and wishbone suspension; so it is advisable to take these in account as well.

The study on improving solar cars energy efficiency is an ongoing research topic, presenting significant evolutions in the most diversified aspects lately. Therefore, it is advised to constantly seek for updating the state-of-the-art on this topic, keeping up with all recent developments.

Acknowledgements. This research was realized inside the *Onda Solare* collaborative project, an action with the aim at developing an innovative solar vehicle. The authors acknowledge support of the European Union and the Emilia-Romagna Region (inside the POR-FESR 2014–2020, Axis 1, Research and Innovation).

References

1. Rizzo, G.: Automotive applications of solar energy. In: 6th IFAC Symposium Advances in Automotive Control, Munich, pp. 174–185 (2010)
2. Mathijsen, D.: Redefining the motor car. Reinf. Plast. **60**, 154–159 (2016)
3. Tamura, S.: Teijin advanced carbon fiber technology used to build solar car for world solar challenge. Reinf. Plast. **60**, 160–163 (2016)
4. Betancur, E., Mejia-Gutierrez, R., Osorio-Gomez, G., Arbelaez, A.: Design of Structural Parts for a Racing Solar Car. In: Eynard, B., Nigrelli, V., Oliveri, S.M., Peris-Fajarnes, G., Rizzuti, S. (eds.). LNME, pp. 25–32. Springer, Catania (2016)

5. Kin, W.D., Kruger, S., van Rensburg, N.J., Pretorius, L.: Numerical assessment of aerodynamic properties of a solar vehicle. In: ASME International Mechanical Engineering Congress and Exposition, San Diego (2013)
6. Ersoz, E.: Development of a racing strategy for a solar car. MSc thesis, Ancara, Turkey (2006)
7. Taha, Z., Md. Dawal, S.Z., Passarella, R., Kassim, Z., Md. Sah, J.: Study of lightweight vehicle vibration characteristics and its effects on whole body vibration. In: 9th Asia Pacific Industrial Engineering and Management System Conference, Bali, pp. 629–633 (2008)
8. Joost, W.: Reducing vehicle weight and improving U.S. energy efficiency using integrated computational materials engineering. JOM **64**, 1032–1038 (2012)
9. Thosar, A.: Design, analysis and fabrication of rear suspension system for an all terrain vehicle. Int. J. Sci. Eng. Res. **5**, 258–263 (2014)
10. Taha, Z., Passarella, R., Rahim, N.A., Sah, J.M.: Driving force characteristic and power consumption of a 4.7 kW motor. ARPN J. Eng. Appl. Sci. **5**, 26–31 (2010)
11. Cambier, C.S.: Brushless motors and controllers designed for GM Sunrayce. IEEE AES Mag. **5**(8), 13–15 (1990)
12. Lovatt, H.C., Ramsden, V.S., Mecrow, B.C.: Design of an in-wheel motor for a solar-powered electric vehicle. IEEE Proc. Online Power Appl. **145**, 402–408 (1998)
13. Lovatt, H.C., et al.: Design procedure for low cost, low mass, direct drive. In: Wheel Motor Drivetrains for Electric Vehicles, pp. 4558–4562. IEEE (2011)
14. Al Zaher, R., de Groot, S., Polinder, H., Wieringa, P.: Comparison of an axial flux and a radial flux permanent magnet motor for solar race cars. In: XIX International Conference on Electrical Machines, Rome (2010)
15. Camargo, F.V., Fragassa, C., Pavlovic, A., Martignani, M.: Analysis of the suspension design evolution in solar cars. FME Trans. **45**, 394–404 (2017)
16. Gadade, B., Todkar, R.G.: Design, analysis of A-type front lower suspension arm in commercial vehicle. Int. Res. J. Eng. Technol. **2**, 759–766 (2015)
17. Vivekanandan, N., Gunaki, A., Acharya, C., Gilbert, S., Bodake, R.: Design, analysis and simulation of double wishbone suspension system. Int. J. Mech. Eng. **2**, 1–7 (2014)
18. Purushotham, A.: Comparative simulation studies on MacPherson suspension system. Int. J. Mod. Eng. Res. **3**, 1377–1381 (2013)
19. Hurter, W.S., van Rensburg, N.J., Madyira, D.M., Oosthuizen, G.A.: Static analysis of advanced composites for the optimal design of an experimental lightweight solar vehicle suspension system. In: ASME International Mechanical Engineering Congress and Exposition, Montreal (2014)
20. Paterson, G., Vijayaratnam, P., Perera, C., Doig, G.: Design and development of the Sunswift eVe solar vehicle: a record-breaking electric car. J. Automobile Eng. **230**, 1972–1986 (2016)
21. de Kock, J.P., van Rensburg, N.J., Kruger, S., Laubscher, R.F.: Aerodynamic optimization in a lightweight solar vehicle design. In: ASME International Mechanical Engineering Congress and Exposition, Montreal, pp. 1–8 (2014)
22. Beres, J.: Sunrunner: the engineering report. Solar Cells **31**, 425–442 (1991)
23. Passarella, R., Rahim, N.A., Sah, J.M., Ahmad-Yazid, A.: CFD for a solar vehicle. J. Comput. Theoret. Nanosci. **4**, 2807–2811 (2011)
24. Vinnichenko, N.A., Uvarov, A.V., Znamenskaya, I.A., Ay, H., Wang, T.H.: Solar car aerodynamic design for optimal cooling and high efficiency. Sol. Energy **103**, 183–190 (2014)
25. Fragassa, C., Pavlovic, A., Massimo, S.: Using a total quality strategy in a new practical approach for improving the product reliability in automotive industry. Int. J. Qual. Res. **8**, 297–310 (2014)
26. Sah, J.M., Passarella, R., Ghazilla, R., Ahmad, N.: A solar vehicle based on sustainable design concept. In: IASTED International Conference on Solar Energy, Phuket (2009)

27. Mohsan, M.D.B.M.: Front and rear suspension for a solar car. Bachelor thesis, Pahang, Malaysia (2010)

28. Camargo, F.V., Guilherme, C.E.M., Fragassa, C., Pavlovic, A.: Cyclic stress analysis of polyester, aramid, polyethylene and liquid crystal polymer yarns. Acta Polytech. **56**, 402–408 (2016)

29. Fragassa, C.: Investigations into the degradation of PTFE surface properties by accelerated aging tests. Tribol. Indus. **38**, 241–248 (2016)

30. Giorgini, L., Fragassa, C., Zattini, G., Pavlovic, A.: Acid aging effects on surfaces of PTFE gaskets investigated by Fourier transform infrared spectroscopy. Tribol. Ind. **38**, 286–296 (2016)

31. Katsiropoulos, C., Chamos, A., Tserpes, K., Pantelakis, S.: Fracture toughness and shear behavior of composite bonded joints based on a novel aerospace adhesive. Compos. B Eng. **43**, 240–248 (2012)

32. Tremayne, D.: The Science of Formula 1 Design: Expert Analysis of the Anatomy of the Modern Grand Prix Car. Haynes Publishing, California (2011)

33. Sancraktar, E., Gratton, M.: Design, analysis, and optimization of composite leaf springs for light vehicle applications. Compos. Struct. **44**, 195–204 (1999)

34. Pavlovic, A., Fragassa, C.: General considerations on regulations and safety requirements for quadricycles. Int. J. Qual. Res. **9**, 657–674 (2015)

35. Elmarakbi, A., Azoti, W.L.: Novel composite materials for automotive applications. In: 10th International Conference on Composite Science and Technology, Lisbon (2015)

36. Wittek, T., Tanimoto, T.: Mechanical properties and fire retardancy of bidirectional reinforced composite based on biodegradable startch resin and basalt fibres. Express Polym. Lett. **2**, 810–822 (2008)

37. Jahn, B., Witten, E.: Composites Market Report. AVK – Industrievereinigung Verstarkte Kunststoffe e.V. (Federation of Reinforced Plastics) (2013)

38. Giorgini, L., Benelli, T., Mazzocchetti, et al.: Recovery of carbon fibers from cured and uncured carbon fiber reinforced composites wastes and their use as feedstock for a new composite production. Polym. Compos. **36**, 1084–1095 (2015)

39. Zhou, C.C., Yin, G.F., Hu, X.B.: Multi-objective optimization of material selection for sustainable products: artificial neural networks and genetic algorithm approach. Mater. Design. **30**, 1209–1215 (2009)

40. Fragassa, C.: Effect of natural fibers and bio-resins on mechanical properties in hybrid and non-hybrid composites. In: VIII AIP International Conference on "Times of Polymers and Composites", Ischia, vol. 1736 (2016). Article No. 4949693. doi:10.1063/1.4949693

41. Lapena, M.H., Marinucci, G., Carvalho, O.: Mechanical characterization of unidirectional basalt fiber epoxy composite. In: 16th European Conference on Composite Materials, Seville (2014)

42. Brandt, A.M., Olek, J., Glinicki, M.A., Leung, C.K.Y.: Brittle Matrix Composites. Woodhead Publishing Limited, Warsaw (2012)

43. Valentino, P., et al.: Mechanical characterization of basalt fiber reinforced plastic with different fabric reinforcements – Tensile tests and FE-calculations with representative volume elements (RVEs). In: XXII Convegno Nazionale IGF, pp. 231–244, Rome (2013)

44. Zivkovic, I., Pavlovic, A. et al.: Influence of moisture absorption onthe impact properties of flax, basalt and hybrid flax/ basalt fiberreinforced green composites. Compos. Part B **111**, 148–164 (2017)

45. Boria, S., Pavlovic, A., et al.: Modeling of Falling Weight ImpactBehavior of Hybrid Basalt/ Flax Vinylester Composites. ProcediaEng. **167**, 223–230 (2016)

Market Growth and Perspective for Solar Mobility: The Case of India

Vikas Badiger[1], Riccardo Paterni[2(✉)], and Cristiano Fragassa[3]

[1] International Master School in Business Administration, University of Pisa, Pisa, Italy
vrb007@gmail.com
[2] Business Development Entrepreneurial Division, Synergy Pathways, Lucca, Italy
riccardo@synergypathways.net
[3] Department of Industrial Engineering, Alma Mater Studiorum University of Bologna,
Bologna, Italy
cristiano.fragassa@unibo.it

Abstract. This research aims at exploring and tapping the emerging market of solar powered technology as a way to fuel the transport industry in a vision of modern and sustainable mobility. It is based on three main aspects: an overview of solar powered vehicles and their recent developments; the current market research and analysis, including industry analysis, country analysis and market trends in the specific case of fast developing country as India; the assessment on potential investment opportunities and strategies toward an active role on the developing solar car market.

Keywords: Sustainable mobility · Solar powered vehicles · Market research · Developing countries · India

1 Introduction

1.1 Need for Research

With a vision of non-renewable resources rapidly fading out, this study is carried out to research for alternative ways of fuelling the day-to-day mobility. Oil and gas industry already realised that oil extinction is at its brink. This paper investigates the market potentiality of Solar Energy for mobility. The flow will include both an extensive industry analysis and a country analysis which will be taken from secondary data since the market for Solar Energy is still not yet a mature market.

1.2 How Solar Cells Work

A solar cell is defined as a device able to capture part of the energy of a photon transforming it in electricity [1]. It is also commonly called photovoltaic cell since this energy conversion uses to be realized by a photoelectric phenomenon. But, not all solar energy has photovoltaic origin: some solar technologies collect, in fact, the heat of absorbed

© Springer International Publishing AG 2017
G. Campana et al. (eds.), *Sustainable Design and Manufacturing 2017*, Smart Innovation,
Systems and Technologies 68, DOI 10.1007/978-3-319-57078-5_80

photons, rather than energy. Thus, with such a general definition, the term photovoltaics encompasses a wide variety of different technologies [1].

However, all solar cells have one thing in common: they use the energy of a photon to excite electrons in the cell's semi-conducting material from a nonconductive energy level to a conductive one. What makes this complex is that not all photons are created equal. Light arrives as an unhelpful amalgamation of wavelengths and energy levels, and no one semi-conducting material is capable of properly absorbing all of them. This means that to increase the efficiency of capture of solar radiation, hybrid ("multi-junction") cells are preferred, created using more than one absorbing material. Each semi-conducting material has a characteristic "band gap" or a spectrum of electron energies which the material simply cannot abide. This gap is present between the electron's excited and unexcited states. An electron in its rest state cannot be excited into usefulness unless it receives enough excess energy to jump right over this band gap. Silicon has a proper, achievable band gap that can be bridged by a single photon's-worth of extra energy. This allows silicon to be either on (conducting) or off (not) as defined by the position of its potentially conductive electrons [1, 3].

A new material as graphene could, in a sense, be a far better basis for a photovoltaic cell than silicon due an incredible electrical efficiency and the potential to be packed far more densely on the panels themselves [4]. Problems come back to the band gap, and graphene's inability to be properly excited by the power of an incoming photon. Even if some complex devices, like dual gate bilayer graphene transistors, already exist, their manufacturing limits represents a real challenge.

While waiting for suitably super-materials, interim solutions permitted to greatly enlarge operability and efficiency of traditional solar panels [5]. Anti-reflective coatings increase the overall amount of light absorbed, while chemical "doping" of the transistors improves silicon's optical abilities. Moreover, some solar setups use fields of mirrors to concentrate as much solar radiation as possible on just a few highcapacity cells at the centre. Many are now even designed cells' protections as lightcapture devices, so light that enters gets bounced around internally, till absorbed. Each technical expedient can represent an important step in rolling out the green solar power on a practical scale. Unless fusion energy makes huge leaps forward or fission nuclear power solves its criticalities in terms of safety and eco-compatibility, the Sun will certainly assume a central role in supplying us our future energy.

1.3   History of Solar Cars

According to the Guinness Book of World Records, the first completely powered solar car was developed in 1984 by Greg Johansson and Joel Davidson. But the true history started some decades before, in 1955, with a fully solar based car debuting at the General Motors Powerama auto show in Chicago. This car, named as Sunmobile, introduced photovoltaic systems to the automotive sector, showing the energy benefit of installing 12 photoelectric cells (in selenium).

Since that moment competition on solar mobility started and solar vehicles constantly improved their technology level, especially in terms of energy efficiency [6] and eco-friendship. However, one of the most curious facts about solar vehicles is that they have been mainly developed inside universities and research centers [7], as noteworthy prototypes, with car manufacturers and large OEMs in second lines.

For years, manufacturers have tried to produce solar powered cars and launch them on the global market with marginal effects. Toyota Motor Corp originally declared that a first solar car would be positioned on the market within 2009. Even if this prototype has never been developed, investments moved in the direction of a massive electrification of vehicles, up of transforming a Prius in a solar hybrid car.

A general overview of hybrid vehicles and technologies is available in [8]. The public has had an on again/off again love affair with solar cars. The idea of using the sun as fuel is inherently appealing. However, many people have also raised valid doubts about the usefulness of solar car as a reliable means of transportation.

Solar cars adopt photovoltaic cells to convert the sun's rays into electricity [9]. Energy is used to power the motor in the way of providing motion, but also to be stored in a battery so that the vehicle is able to work on cloudy days, at night, and in dark spaces like parking garages.

Photovoltaic cells can cost from \$10 to \$400 per unit: the more expensive the cells, the more efficient they are at gathering and converting the sun's energy [2]. This means that constructing an efficient solar car may be way beyond set budgets. Additionally, solar powered transportation cannot be considered comfortable yet: solar cars, made with lightweight frames and uncommon materials [10], can generally only accommodate one person, the driver. Furthermore, these vehicles can get very warm on the inside and can be uncomfortable to operate. Nevertheless, solar powered cars are more and more popular nowadays. Many people are deeply interested in these types of vehicles, as they can be really economical, as well as eco-friendly.

2   Methods and Tools

2.1   Market Research and Country Analysis

India represents the 7th largest and the 2nd most populous country in the world with more than 1.2 billion people. Independent since 1947, for the past decade or so, India has been experiencing a constant growth in its GDP, along with a continuous growth of liberalisation. Since 1991, the country has also been opening its doors to attract investors and foreign companies to further promote growth.

2.2   Pestle Analysis

To get a better understanding of the business environment in India, the PESTLE analysis was considered in this research [11]. It is basically a framework used for scanning and analysing an organization's external macro environment by considering factors which include political, economic, socio-cultural, technological, legal and environmental details. it allows to identify the proper framework in which innovative projects can find fertile ground of development and develop awareness on the barriers to overcome. Specifically, those insights can provide useful information about the most significant factors able to affect the external macro environment, impacting in the mobility industry and, therefore, on the potentiality of the solar powered mobility.

2.3 Political Factors

Being one of the largest democracies in the world, India runs on a federal form of government. The political environment is greatly influenced by factors such as government policies, politician interests, and the ideologies of several political parties. As a result, the business environment in India is affected by multivariate political factors. The taxation system is well-developed and several taxes, such as income tax, services tax and sales tax are imposed by the Union Government. Other taxes, such as *octroi* and utilities, are taken care of by local bodies. Privatisation is also influenced and the government encourages free business through a variety of programs.

2.4 Economic Factors

The economy of India has been significantly stable since the introduction of industrial reforms in 1991. As per the policy, reductions in industrial licensing, liberalisation of foreign capital, formation of FIBP and so on, has resulted in a constant improvement of India's economic environment. The country registered a GDP of $5.07 trillion in 2013 following a further improved GDP growth rate of 5% in 2014 as compared to 4.35% in 2013 [11].

2.5 Social Factors

The social factors refer to any changes in trends which would impact a business environment. In particular, the rise in India's ageing population is resulting in a considerable rise in pension costs and increase in the employment of older workers. In India about 70% of its 1.2 billion of people has an age between 15 and 65. Therefore, there are structures with percentages according to age. These structures contain varying flexibility, in education, work attitudes, income distribution, and so on.

2.6 Technological Factors

Technology significantly influences product development and also introduces fresh cost-cutting processes. The country possesses one of the strongest IT sectors in the world, promoting constant IT development, software upgrades and other technological advancements. Recently, India has also attempted to launch satellites into space.

2.7 Legal and Environmental Factors

In the recent past, a number of legal changes have been implemented in India, such as recycling, minimum wage increase and disability discrimination, which has directly affected businesses there. However, when it comes to environment, the quality of air in

India has been adversely affected by industrialisation and urbanisation, also resulting in health problems. As a result, there have been establishments of environmental pressure groups, noise controls, and regulations on waste control and disposal or energy saving and optimisation.

3   Automotive Industry Analysis

India is native to a diverse auto industry of more than 40 million units. It has been one of the few worldwide which saw growing passenger car sales during the recession of the past two years. 2009–2010 was told to be record of highest volume of sales ever. This is due to a strong domestic market and increased push on exports. The Indian economy has grown at an average rate of around 9% over the past five years and is expected to continue this growth in the medium term. This is predicted to drive an increase in the percentage of the Indian population able to afford vehicles. India's car per capita ratio (expressed in cars per 1,000 population) is currently among the lowest in the world's top 10 auto markets. The twin phenomena of low car penetration and rising incomes, when combined with increasing affordability of cars, are expected to contribute to an increase in India's demand for automobile [12].

This segment of the business has been a major driver of India's economy. Automobile industry accounts to 4% of the country's GDP.

Considering the current scenario with FDI allowed in India for automobile sector is 100%. Further, the "Make in India" campaign by the PM of India Narendra Modi and the team is helping in terms of cost reductions while manufacturing automobiles [12, 13]. Some of the car manufacturing companies that had initially installed solar panels in their cars were Ford, Mazda and Cadillac. The 2006 Ford Reflex installed solar panels in the headlights, and the 2005 Mazda Senku featured solar panels on its roof to help charge its battery. The 2008 Cadillac Provoq uses solar panels to power accessories, such as interior lights and the audio system [12].

While the Reflex, Provoq and Senku are merely concept models, cars outfitted with solar panels may be tiptoeing their way into the consumer automotive industry as companies try to find innovative methods for dodging gasoline dependence. French car company Venturi has made one of the most publicised efforts with its unveiling of the Eclectic model prototype at the 2006 Paris Auto Show. The Eclectic combines solar, wind and battery power to run a three-passenger car specifically for city driving. Solar panels cover its roof, and a wind turbine can also catch energy on blustery days. The Venturi Eclectic isn't cut for highway travel, however, since it only goes up to 30 mph (48 kph) [12].

The automotive regulatory context is witnessing increased activity and some bold decisions by the Government and judiciary to battle the high pollution levels in Delhi and other metros. In the medium term, we expect heightened regulatory activity in order to make transport cleaner and safer [12].

3.1   Cleaner Transport - Advancement of BS VI Norms

The government recently decided to skip the BS V norms altogether, and adopt the BS VI norms from April 2020. Automakers and the oil companies will need to make significant investments to meet these deadlines. We expect a continued thrust toward cleaner fuels in the policy formation and some possible announcements in the 2016 Union budget [12, 13].

3.2   Indian Scheme and "End of Life" Policy

The scheme, launched in April 2015, provides incentives for the purchase of green vehicles. While the scheme has pushed electric vehicle (EV) sales, efforts need to be undertaken in areas such as setting up of charging infrastructure, launching compelling EV models, and reducing battery costs. The Government is likely to introduce a policy promoting the destruction of old vehicles. This is a welcome move and is likely to reduce pollution and drive new vehicle sales [13].

3.3   Safer Transport

In Compliance with international safety standard, the Government has announced October 2017 as the deadline for automakers to ensure international standards in terms of vehicle quality and safety [14, 15]. Mandatory crash tests will be implemented from October 2017 for new models (Oct 2019 for existing models). To conduct the tests, India is likely to have seven world-class automotive design and testing centres (being set up by NAT Rip) by the end of 2016. New road transport and safety bill: The bill seeks to drive faster clearances, stricter road and vehicle safety norms and define the recall policy. It focuses on transparency and computerisation, heavy penalties for traffic violations and incorporates global best practices for issues related to vehicle regulation and road safety.

3.4   Regulations for New Mobility Initiatives in Advanced Mobility

The Indian market has recently witnessed a slew of advanced mobility offerings and new mobility initiatives such as technology-based cab aggregation and ride sharing. A regulatory environment conducive to promote innovation for these new mobility initiatives is needed. The road transport ministry has issued some guidelines, which could be the starting point for states to establish regulations [16].

4   Solar Energy Market Analysis

4.1   Political and Economic Perspective

As recently as a few months ago, the alternative energy targets announced by Indian Prime Minister Narendra Modi's government in 2014 seemed truly overambitious even to solar electricity enthusiasts in the industry. But a surge of investment in solar power stations now makes them look merely optimistic. Azure Power, one of the country's biggest producers, pointed out that in 2009, many considered impossible to have in India two to three GWs of solar energy by 2016; despite of that currently 5 solar plants in the country produced around 5 GWs. In India huge opportunities exist for solar energy since the shortage of electricity and the need to replace costly diesel generators with more sustainable energy sources [17].

Azure Power, part-owned by the World Bank's International Finance Corporation as well as Foundation Capital and Helion Ventures, is one of dozens of domestic and international investors putting money into building or supplying equipment to largescale solar photovoltaic power stations across India.

The Government has championed solar power and helped launch a global solar alliance at the Paris climate summit to mobilise an attention-grabbing $1tn of funds worldwide by 2030. With current electricity grid capacity of less than 300 GWs, India aims to increase its solar installations to 100 GW by 2022, more than double the present solar capacity of China and Germany, the two biggest solar nations. The latest contract awards suggest that, in sunny India at least, solar power can compete head-on with coal in terms of price, albeit not in terms of 24-hour availability [17].

SB Energy, a joint venture between Japan's SoftBank (which has promised 20 GW or $20bn of solar investment in India), Bharti Enterprises of India and Foxconn of Taiwan, won a reverse auction for a 25-year, 350 MW project in Andhra Pradesh. The price of the electricity to be sold, equaled the record-low winning bid of SunEdison, the struggling US group, for a 500 MW project auctioned earlier in the same state. That price is cheaper than for the power to be supplied by recently auctioned projects using imported coal. This situation transformed the Indian conglomerates with interests in coal are among the active bidders for solar plants [17]. According to the USbased Institute for Energy Economics and Financial Analysis, which promotes green power, investments of over $100bn have been announced for Indian renewable energy by companies and lenders from East Asia, Europe and India itself.

4.2 Market Growth and Peculiarities

According to IEEFA's estimation, India is executing one of the most radical transformations ever undertaken in the energy sector and the flow of finance is matching this ambition [17]. Several 9 GW of solar projects are under development, portending a massive jump in India's total installed capacity within 2017. BMI Research, part of Fitch, has raised its forecasts for solar projects in the coming years, projecting solar capacity to reach nearly 45 GW by 2024, short of the government targets but a ninefold increase from current installed capacity levels [17].

For all the optimism among investors and suppliers, the Sweden engineering group ABB sees India's massive aspiration on renewable energy as an opportunity for newest infrastructures. The first problem is the grid, which needs to be extended and modernised. The current grid can handle 272 GW in all, of which less than 30 GW is from renewable sources, with issues of evacuation of power from the plants.

When solved, and organisations such as the Asian Development Bank are investing heavily in this, investors face the question of whether bankrupt Indian state electricity distributors or even central institutions can be relied on to pay the agreed cost for a quarter of a century when prices are in constant decline [16].

Such issues are not unique to India: Spain enraged solar investors by retroactively cutting subsidised prices five years ago and investors hope the reverse auction system applied to most contracts since 2010 is sufficiently robust and transparent to protect their interests.

The Council on Energy, Environment & Water stated India is trying to do what Germany did to develop alternative energy over two decades, but in half the time.

For starters, India simply needs more power. Between April 2014 and March 2015, for instance (Fig. 1a), the country had to deal with a 3.6% deficit in peak hour energy supply (the period when demand for power is significantly higher than the average supply level). Power requirement in India is estimated to grow at an average of 5.2% during the 10 years between 2014 and 2024 (Fig. 1b). Currently, India requires 1,068,923 million units of electricity annually but the supply falls short by 3.6%.

Fig. 1. India's power deficit (a) and requirement (b) [17].

All of this demand is not likely to be met by traditional energy sources. India's coal fed power plants, which contribute to nearly 60% of the total production, have been grappling with periodic fuel shortages. Domestic production of coal has not quite kept pace with demand, which alongside expensive imported coal, has made things difficult [16]. Simultaneously, there has been a steady decline in solar power prices, on the back of cheaper solar panel costs and lower financing costs, that has made the sector increasingly attractive to investors, also in the case of unconventional projects [18]. By 2019, according to some estimates, India could achieve grid parity between solar and conventional energy sources. It means that solar power will cost less than or equal to power from conventional sources, pushing up further interests.

4.3   Future Investments in Solar Energy

Japan's Softbank Corp is going to invest additional $20 billion in solar projects in one of the biggest investment in the country's renewable energy sector. Softbank aims at a minimum commitment of generating 20 GWs of energy. India has a sun irradiation two times higher than in Japan. Adding, the cost of construction of the solar park is half of Japan. It means that efficiency of the investment is four times higher [17].

The rapidly falling cost of solar power, expected to reach parity with conventional energy in 2017, has ignited interest in its potential in India, as the country steps up its own efforts to encourage investment in renewable energy. Despite more than 300 days of sunshine a year, India relies on coal for three-fifths of its energy needs, while solar supplies less than 1%. Indian Government has looked to industry for help in funding what government advisers hope will be a $160 billion push into renewable energy over five years. It aims to make India one of the world's largest renewable energy markets, targeting 100,000 MW of output by 2022 from just 3,000 MW currently [13].

5 Conclusion

Solar powered cars have come a long way since the creation of the first solar cell in 1883. Nevertheless, the solar mobility is far from representing a consolidated reality. Universities and technology centers have invested billions of dollars all over the world with the aim at realizing solar vehicles, especially for using them in solar races. This competitive atmosphere has positively fostered the research on the field, so that technical solutions used as solar powered vehicles are now probably ready to move from the stage of prototypes to market-orient products (Fig. 2a). At the same time, during these long years, a similar interest about solar technologies has unfortunately missed between car manufactures and OEMs.

Fig. 2. Solar powered vehicles by Bochum University (a) and Toyota (b)

The next Toyota Prius (Fig. 2b), perhaps available in 2017, will be the mean of mass transportation on the market closest to a solar vehicle. It will feature a hybrid electric-gasoline engine and solar panels to power its air-conditioning system. Even if the use of solar cells will be very limited, it will make Toyota the first major automaker to use solar power for a vehicle.

At the same time, there is an unquestionable need to rapidly achieve energy sustainability as we are running out of non-renewable resources. Sun can represent a relevant source of energy for the future transport and automobile industry. Huge markets and huge opportunities of investment in sustainable energy can maybe modify the global perception on solar mobility, also attracting automakers and stakeholders [19].

In this perspective, India probably represents an ideal place for companies to invest and manufacture solar powered cars. In fact, almost certainly, India is shortly going to be the largest country powered by solar energy. This emerging country is also characterized by a very low labour cost and an extremely large internal market. Beyond the fact that more than

1 billion of potential customers of automotive exist in India and move throughout its narrow and crowded streets, the current vehicle fleet is pour, obsolete, polluting and with the largest consistency of light cars (as quadricycles) in the world: a perfect social and economic environmental for testing a new vision of sustainable mobility, made by little and efficient microcars.

Acknowledgments. This preliminary investigation on solar mobility has been realised inside the Onda Solare project. The authors acknowledge support of the European Union and the Region Emilia-Romagna (inside the POR-FESR 2014–2020, Axis 1, Research and Innovation).

References

1. Olabi, A.G.: State of the art on renewable and sustainable energy. Energy **61**, 2–5 (2013)
2. Green, M.A., Emery, K., Hishikawa, Y., Warta, W., Dunlop, E.D.: Solar cell efficiency tables. Prog. Photovoltaics Res. Appl. **23**(1), 1–9 (2015)
3. Brabec, C.J., Dyakonov, V., Parisi, J., Sariciftci, N.S. (eds.): Organic Photovoltaics: Concepts and Realization. Springer Series in Material Science, vol. 60. Springer, Berlin (2013)
4. Yin, Z., Zhu, J., He, Q., Cao, X., Tan, C., et al.: Graphene-based materials for solar cell applications. Adv. Energy Mater. **4**, 1–19 (2014)
5. Taguchi, M., Yano, A., Tohoda, S., Matsuyama, K., et al.: 24.7% record efficiency HIT
6. Solar cell on thin silicon wafer. IEEE J. Photovoltaics **4**(1), 96–99 (2014)
7. Minak, G., Fragassa, C., Vannucchi De Carmago, F.: A brief review on determinant aspects in energy efficient solar car design and manufacturing. In: Campana, G., Howlett, R.J., Setchi, R., Cimatti, B. (eds.) Sustainable Design and Manufacturing 2017. Smart Innovation, Systems and Technologies, vol. 68, pp. 847–856. Springer, Cham (2017). doi:10.1007/978-3-319-57078-5_76
8. Arkesteijn, G.C.M., de Jong, E.C.W., Polinder, H.: Loss modeling and analysis of the nuna solar car drive system. In: International Conference on Ecologic Vehicles and Renewable Energies. MC2D & MITI Press (2007)
9. Prajapati, K.C., Patel, R., Sagar, R.: Hybrid vehicle: a study on technology. Power (kW) **2**, 10–20 (2014)
10. Urabe, S., Kimura, K., Kudo, Y., Sato, A. Effectiveness and issues of automotive electric power generating system using solar modules. SAE Tech. Paper No. 01-1266 (2016)
11. Zivkovic, I., Pavlovic, A., Fragassa, C.: Improvements in wood thermoplastic composite materials properties by physical and chemical treatments. Int. J.Qual. Res. **10**(1), 205–218 (2016)
12. Rosy, Ms.: Make in India: prospects and challenges. Int. J. Bus. Quant. Econ. Appl. Manag. Res. **2**(8), 60–75 (2016)
13. Veracruz, C.: Solar vehicles design for urban use. In: International Solar Energy Society: Solar World Congress. ISES Press, Cancun (2013)
14. Booz&Co and PWC: Indian Automotive Market 2020. 1–21, printed in USA (2011)
15. Fragassa, C., Pavlovic, A., Massimo, S.: Using a total quality strategy in a new practical approach for improving the product reliability in automotive industry. Int. J. Qual. Res. **8**(3), 297–310 (2014)
16. Pavlovic, A., Fragassa, C.: General considerations on regulations and safety requirements for quadricycles. Int. J. Qual. Res. **9**(4), 657–674 (2015)
17. Becker, D., Nagporewalla, Y.: The Indian Automotive Industry. Evolving Dynamics KPMG India (2010)

18. ElecRama: Why India's solar sector has turned into a $100 billion investment magnet. The World Electricity Forum, Bengaluru (2016)
19. Fragassa, C.: Electric City Buses with Modular Platform: A Design Proposition for Sustainable Mobility. In: Campana, G., Howlett, R.J., Setchi, R., Cimatti, B. (eds.) Sustainable Design and Manufacturing 2017. Smart Innovation, Systems and Technologies, vol. 68, pp. 789–800. Springer, Cham (2017). doi:10.1007/978-3-319-57078-5_71

Aerodynamic Effects of Manufacturing Tolerances on a Solar Car

Esteban Betancur[1(✉)], Cristiano Fragassa[2], Jairo Coy[1], Sebastian Hincapie[1], and Gilberto Osorio-Gómez[1]

[1] Universidad EAFIT, Medellin, Colombia
{ebetanc2,jcoycoy,jhinca15,gosoriog}@eafit.edu.co
[2] Università di Bologna, Bologna, Italy
cristiano.fragassa@unibo.it

Abstract. In the case of solar vehicles, since the primary necessity is to optimise the energy efficiency during motion, many efforts are addressed by designers in searching the perfect aerodynamics. It means, in particular, the minimization of the drag force at cruising speeds and an elaborated vehicle's Computer-Aided Design (CAD) are the principal result of this activity. Despite, these efforts can be nullified by geometrical tolerances emerging from manufacturing. In this paper, the effects of tolerances introduced by composite manufacturing processes are investigated combining 3D scanning technology and Computational Fluid Dynamics (CFD). After the solar car manufacturing, a reverse engineering process is executed with the aim to scan the vehicle's body and compare it to the initial theoretical design. Geometric deviations are found and their aerodynamic consequences are evaluated in terms of aerodynamic losses.

Keywords: Solar car · Composite manufacturing · Reverse engineering · CFD simulation · Aerodynamics

1 Introduction

Even at their current stage of prototypes, solar vehicles represent a valid contribution to a sustainable mobility. Zero-emissions transport is achieved with a fully autonomous electric car, in terms of power, able to collect energy with its body-embedded solar panels. As ultra-efficient vehicle, both in energy collection and consumption, its purpose is to travel with the same energy that is collected by the panels. Then, cutting-edge developments are implemented in order to achieve the highest efficiency level [1].

The aerodynamic performance of solar vehicles represents, in particular, an essential issue for reducing the energy consumption and, therefore, improving the energy efficiency. Even for conventional cars aerodynamics is a strategical aspect to be considered in the body design. At a cruising speed between 80 and 120 km/h the aerodynamic drag usually contributes on around 70% of the total energy consumption [2]. It means that the aerodynamic represents the main

© Springer International Publishing AG 2017
G. Campana et al. (eds.), *Sustainable Design and Manufacturing 2017*, Smart Innovation, Systems and Technologies 68, DOI 10.1007/978-3-319-57078-5_81

cause of energy expenses. The residual part can be related to a large number of additional (and less controllable) factors, as tire roll resistance [3], vehicle dynamics [4], material lightness [5], technical solutions for improving reliability [6] or safety of vehicles [7] among the others.

The instantaneous power required for a racing solar car to travel at 100 km/h is around 1.5 kW [8], near 10 times lower than an average conventional car. This value is achieved with an extreme aerodynamic and lightweight design. The aerodynamic drag force on a vehicle can be estimated as defined on Eq. 1 where ρ stands for the air density, v for the vehicle velocity, and $C_d A$ for the drag area (Drag coefficient multiplied by the frontal area) that is a property of the car shape.

$$F_d = \frac{1}{2} C_d A \rho v^2 \tag{1}$$

The $C_d A$ value of a conventional car e.g. the Tesla Model S is near $0.58\,\mathrm{m}^2$ [9]. Therefore, at a constant speed of 100 km/h, the drag force is 265 N and it needs around 7.4 kW of power only to overcome this aerodynamic drag. For solar cars, in order to be energy autonomous, the $C_d A$ must be less than the half of these conventional vehicles on the market [10].

Then, the purpose of the aerodynamic design is to minimize the $C_d A$ of the vehicle. This result can be achieved with a streamlined body design, frontal area reduction, high quality (class A) surfaces, among other aspects. After an aerodynamic shape is obtained on the design stage, the manufacturing process should guarantee the conservation of this designed shape. Computer Numerical Control (CNC) milling machines are commonly used for the mock-up and molding fabrication, then the body parts are produced. This manufacturing stage brings shape changes, with respect to the original design, that are converted into changes of the vehicle aerodynamic properties. These shape variations can also affect the driver ergonomics and visibility, solar panel integration, mechanical system mounting among other issues.

A comparison between CFD analysis and wind tunnel tests was made by Yang and Liou [11]. They consider that differences in results can be related to CFD meshing and ground clearance assumption and deem that the designed and real shapes are equal. On the contrary, the manufacturing errors are specifically taken into account for the final aerodynamic study of the car. As detailed below, the realization of the Onda Solare Emilia 3 vehicle began with the design based on CFD analyses, then a viable manufacturing process was implemented and the vehicle was built.

During the following on-the-road tests, the vehicle showed an energy consumption higher than what expected. With the aim at investigating and solving this discrepancy, the manufacturing errors were measured by a complete 3D laser scanning of the real shape of the vehicle and a new CFD simulation was carried out.

2   Aerodynamic Design

The solar vehicle Emilia 3 was designed by the Italian Onda Solare team for competing in the World Solar Challenge (WSC) 2013, the most recognized solar car race (see Fig. 1). The objective of this competition is to travel 3022 Km only using solar energy and the respecting main regulations (for challenger class) as listed below:

- 1 Occupant
- 4 Wheels
- Maximum 6 m^2 of Si solar panel
- Maximum 21 kg of lithium battery cells
- Maximum 4.5 m long and 1.8 m wide

Fig. 1. Solar vehicles during the World Solar Challenge (Australia, 2015) [12]

According to the WSC conditions, a fully composite monocoque vehicle was designed and built. The body shape was obtained based on the land vehicles aerodynamic literature theory and innovative design proposals. The low consumption necessity of these kind of vehicles demands a great effort on the aerodynamics area. Since the 90's, important academic disclosures about solar car aerodynamics were published. "The Leading Edge" [13] is a full book about solar car aerodynamic design. The book "Speed of Light" [14] dedicates one complete chapter to the solar car aerodynamics, the article [15] explains the shape design process of the "Honda Dream" solar car, winner on the 1996 WSC and the article by Doig and Beves [2] discloses the design of the Sunswift IV solar car that competed on WSC 2009 and WSC 2011 on challenger class, also breaking, with a speed of 88.74 km/h the solar car high speed Guinness record on 2011. In particular, this last paper details the CFD analysis of all the body parts and crosswind conditions. More recently, Paterson et al. [10] exposed the main developments of the Sunswift eVe solar car showing how their aerodynamics design process was based on CFD iterations and satisfactory results were obtained.

On the body design process, the main issue is to safeguard the solar panel area, minimize the drag force and guarantee neutral or slightly negative lift force (downforce), all these for a vehicle running very close to the ground.

3 CFD Analysis

As part of the design process and final optimization, several CFD analyses were executed using ANSYS CFX and OpenFOAM software packages. No real wind tunnel tests were done due to availability limitations. To improve the simulation accuracy, important indications for the analysis given by Damjanovic et al. [16], Lanfrit [17] and Gaylard et al. [18] were taken into account. The computational tunnel, namely fluid domain, was defined in order to diminish the boundary influence on the results; the tunnel dimensions were defined as Damjanovic et al. [16]. The meshing considered 15 structured layers around the car body in order to capture the boundary layer phenomena such as laminar flow. A size refinement in the proximity of the vehicle was included, specially downstream to simulate the small scale vortices. The final mesh consisted in around 70×10^6 elements. The Shear Stress Transport (SST) turbulence model was included; this model guaranteed a proper combination between the k-epsilon on the free flow regions and the k-omega turbulence model close to the vehicle body. The condition for solver convergence stopping was defined as the achievement of a value of 1×10^{-5} or lower residuals on the momentum and mass conservation equations.

According to an optimal racing strategy for the WSC, a cruising speed between 70 and 100 km/h is expected. This velocity has to be as constant as possible for optimal energy usage. It is possible to estimate that less than the 5% of the race time is characterized by changes in speed, mainly related to road limits, traffic, crossing roads, overtaking, etc. Therefore, the aerodynamic analyses were done around cruising speeds. After the Emilia 3 design stage based on CFD iterations, a final model was obtained. Figure 2 depicts the body shape.

4 Body Manufacture

The vehicle structural body was built as a carbon fiber monocoque. More than 10 main pieces were generated independently and bonded together. Every piece was realized starting from a specific model depicting the final shape of this part. The large models and mock-ups were made with metal sheets, cut and assembled

(a) Front view   (b) Rear view

Fig. 2. Emilia 3 CAD model

as an arched structure, then plastered and refined to optimize the surface finish. The small ones were made in aluminum, composites (as glass fiber) or even wood, shaped by CNC machining. Starting from these models, molds in Carbon Fiber Reinforced Plastics (CFRP) were realized by lamination (instead of other cheaper solutions) to guarantee a dimensional stability of the final parts, to be also realized in CFRP, under positive pressure and high temperature solidification of the autoclave.

Using the molds, the laminate process was manually done with pre-impregnated carbon fiber and Nomex cores of different thickness according to the structural necessities. Finally, the cure process was executed using vacuum bag, temperatures around 100 °C and positive pressures between 1.5 and 6 Bar on autoclave.

After the separate manufacture of the pieces, final parts were aligned and assembled with fixation pins. Then, the complete body was polished and bonded together using special glues. Although a visually continuous and smooth surface was obtained, several variations on the geometry with respect to the design were raised. These changes were related to the accuracy in the manufacturing and assembly processes and were investigated by a Reverse Engineering (RE) technology.

To guarantee driver's ergonomics and visibility, the canopy had to be bigger than the designed one and a truncation was done to keep the solar panel area on the top of the vehicle. Then, major changes were done in this part of the vehicle.

5   Reverse Engineering

A FARO® portable measurement arm with over 0.1 mm accuracy was used to scan the vehicle body (See Fig. 3). Around 800,000 representative geometrical coordinates were obtained and the deviations with respect to the original geometry was calculated. Figure 4 illustrates the deviation magnitude: the largest errors were evident in the canopy and wheel fairings, up to ±50 mm differences. The principal airfoil body (top and belly) of the vehicle was found with geometric differences of less than 10 mm with respect to the original design.

Fig. 3. Vehicle reverse engineering by 3D laser scanner

Fig. 4. Difference between original and manufactured geometries (units: mm)

6 Results

With the 3D scan data, a new CAD solid was generated and used as input for a further CFD analysis. The same turbulence model and boundary conditions were used comparing these results with the ones obtained in the case of the original geometry while different velocities were simulated. No crosswind, yaw or pitch conditions were considered. The total drag force (pressure and viscous drag) was calculated and the $C_d A$ was obtained from Eq. 1.

The total surface, i.e. wet area of the original CAD was $24.15\,\mathrm{m}^2$ against a scanned one of $24.92\,\mathrm{m}^2$ ($+3.2\%$). Figure 5 shows the pressure distribution and flow streamlines for both geometries. In particular, no counterflow regions or large vortices were observed and the pressure magnitudes and distributions were similar. The main difference was observed on the canopy region, the truncation on the real geometry generates small vortices downstream that can be appreciated on Fig. 5b.

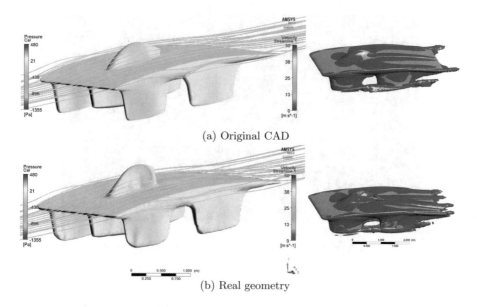

(a) Original CAD

(b) Real geometry

Fig. 5. CFD results at 100 km/h: streamlines and pressure distribution (left) and 600 m/s² helicity density isosurface (right)

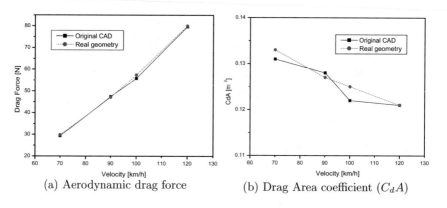

(a) Aerodynamic drag force   (b) Drag Area coefficient (C_dA)

Fig. 6. CFD results for the original CAD and real geometry at different speeds

The resulting drag force for the two bodies at different cruising speeds is displayed on Fig. 6a. Figure 6b plots the resulting C_dA for both geometries. A marginal difference can be appreciated, the scanned geometry C_dA is, on average, 0.9% higher than the original CAD.

The impact of this difference in terms of energy consumption of the vehicle can be calculated according to aerodynamic results and taking into account a car weight of 280 kg, a tire roll coefficient of 0.005, a drivetrain efficiency of 90% and a flat road. Table 1 depicts these estimations for several speeds.

Table 1. Energy consumption of Emilia3 vehicle at different speeds. units: Wh/km

Velocity [km/h]	70	90	100	120
Original CAD	13.31	18.85	21.47	28.91
Real geometry	13.42	18.78	21.93	28.92
Difference	0.83%	−0.37%	2.14%	0.03%

7 Discussion and Conclusion

The manufacturing process of carbon fiber solar cars includes a series of steps that clearly generate changes in geometry and dimensions.

From the ideal CAD geometry, geometric differences up to 50 mm were measured. On a 4.5 m long car, these deviations are visually unnoticeable but they can negatively impact the aerodynamic performance.

A CFD analysis was used to find this aerodynamic losses. In the case under investigation, in particular, the energy consumption is increased around 1% on the real geometry with respect to the original CAD, this marginal difference has not statistical significance because is under the estimated error for CFD analysis (±5%).

The main differences on the two geometries were on the canopy and fairings regions, while the principal airfoil body remained constant. Therefore, no main variations were found on the flow behaviour and pressure distribution. This leads to the high correlation on the results.

The helicity plot on the original CAD shows a vortex generation in the frontal part of the canopy, producing early flow separation and, therefore, turbulent boundary layer. Because of this, the small shape variations are not having major effects on the flow behaviour.

In addition to the aerodynamic performance, special care should be taken with the shape variations resulting on the manufacturing process for all the systems integration, such as solar panel, driver ergonomics and suspension.

Acknowledgement. This investigation has been realized inside the Onda Solare collaborative project. The authors acknowledge support of the European Union and the Region Emilia-Romagna (inside the POR-FESR 2014–2020, Axis 1, Research and Innovation). Special thanks to Massimo Mele, Aurea Servizi, for 3D laser scanning and geometrical modeling.

The CFD simulations were executed on the Scientific Computing Center "APOLO" at Universidad EAFIT with the support of Eng. Juan David Pineda C.

References

1. Minak, G., Fragassa, C., Camargo, F.V.: A brief review on determinant aspects in energy efficient solar car design and manufacturing. In: Campana, G., Howlet, R.J., Setchi, R., Cimatti, B. (eds.) Sustainable Design and Manufacturing 2017. Smart Innovation Systems and Technologies series. Springer, Cham (2017)

2. Doig, G., Beves, C.: Aerodynamic design and development of the sunswift iv solar racing car. Int. J. Veh. Des. **66**(2), 143–167 (2014)
3. Betancur, E., Mejía-Gutiérrez, R., Osorio-Gómez, G., Arbelaez, A.: Design of structural parts for a racing solar car. In: Eynard, B., Nigrelli, V., Oliveri, S.M., Peris-Fajarnes, G., Rizzuti, S. (eds.) Advances on Mechanics, Design Engineering and Manufacturing, pp. 25–32. Springer, Cham (2017)
4. Camargo, F.V., Giacometti, M., Pavlovic, A.: Increasing the energy efficiency in solar vehicles by using composite materials in the front suspension. In: Campana, G., Howlett, R.J., Setchi, R., Cimatti, B. (eds.) Sustainable Design and Manufacturing 2017, Smart Innovation, Systems and Technologies, vol. 68, pp. 801–811. Springer, Cham (2017). doi:10.1007/978-3-319-57078-5_72
5. Camargo, F.V., Fragassa, C., Pavlovic, A., Martignani, M.: Analysis of the suspension design evolution in solar cars. FME Trans. **45**(3), 394–404 (2017)
6. Fragassa, C., Pavlovic, A., Massimo, S.: Using a total quality strategy in a new practical approach for improving the product reliability in automotive industry. Int. J. Qual. Res. **8**(3), 297–310 (2014)
7. Pavlovic, A., Fragassa, C.: General considerations on regulations and safety requirements for quadricycles. Int. J. Qual. Res. **9**(4), 657–674 (2015)
8. Mocking, C.: Optimal design and strategy for the SolUTra. Master's thesis, Masters Thesis: University of Twente (2006)
9. Palin, R., Johnston, V., Johnson, S., D'Hooge, A., Duncan, B., Gargoloff, J.I.: The aerodynamic development of the tesla model s-part 1: Overview. Technical report, SAE Technical Paper (2012)
10. Paterson, S., Vijayaratnam, P., Perera, C., g, G.: Design and development of the sunswift eve solar vehicle: a record-breaking electric car. Proc. Inst. Mech. Eng. Part D: J. Automobile Eng., 1972–1986 (2016)
11. Yang, Y., Liou, W.W.: Comparison of computational and experimental aerodynamics results for a wmu solar car model. Technical report, SAE Technical Paper (2005)
12. Schubert, S.: ABC News Austraila website. http://www.abc.net.au/news/2015-10-18/collection-of-solar-vehicles-in-solar-challenge-race/6863982. Accessed 4 Jan 2017
13. Tamai, G.: The Leading Edge: Aerodynamic Design of Ultra-streamlined Land Vehicles. Engineering and Performance. Robert Bentley, Cambridge (1999)
14. Roche, D.: Speed of Light: The 1996 World Solar Challenge, vol. 1. University of New South Wales Press, Sydney (1997)
15. Ozawa, H., Nishikawa, S., Higashida, D.: Development of aerodynamics for a solar race car. JSAE Rev. **19**(4), 343–349 (1998)
16. Damjanović, D., Kozak, D., Živić, M., Ivandić, Ž., Baškarić, T.: Cfd analysis of concept car in order to improve aerodynamics. A Jovo Jarmuve, A Magyar Jármûipar Tudományos Lapja **1**(2), 63–70 (2011)
17. Lanfrit, M.: Best practice guidelines for handling automotive external aerodynamics with fluent (2005)
18. Gaylard, A., Baxendale, A.J., Howell, J.: The use of CFD to predict the aerodynamic characteristics of simple automotive shapes. Technical report, SAE Technical Paper (1998)

Eco Designed Through Systematic Innovation

How to Build Guidelines for Eco-Improvement

Davide Russo[✉], Caterina Rizzi, and Christian Spreafico

Department of Management, Information and Production Engineering,
University of Bergamo, Viale G. Marconi, n.5, 24044 Dalmine, BG, Italy
{davide.russo,caterina.rizzi,christian.spreafico}@unibg.it

Abstract. Over the last 30 years the number of methods for Eco-design increased dramatically. LCA in Eco-assessment has established itself as a reference methodology and with it some tools that reached an international resonance. On the contrary, in the Eco-improvement world, the growth of methods has not been accompanied by a method or a tool better than other ones. One of the main reasons is the different type of users; there are people skilled in problem solving and those who have no experience. In addition, in order to be universal, the methods based on guidelines often do not go into too much detail, thus limiting their effectiveness. The balance between completeness and simplicity is the key issue around which the authors have attempted in recent years.

In such a context, this paper aims at solving this contradiction and proposes an ontological framework to build guidelines for eco-improvements. Their content has been structured into five parts, according to well-known conceptual design frameworks, such as Function-Behaviour-Structure (FBS) methods and similar.

The result is a set of over than two hundreds suggestions that can be comfortably used through a web portal following a recommended step-by-step methodological path.

Keywords: Eco-design · Guidelines · Ontology · FBS · Web portal

1 Introduction

In the last three decades the number of articles dealing with Eco-design has always increased. From the few sporadic articles indexed by Scopus at the beginning of 90's we now count more than 3500 articles in 2015, containing the pool of keywords built according to the taxonomy proposed by Maxwell et al. (2006), such as green and eco design in title and abstract. The eco-design mainly aims at designing products with special consideration for the environmental impacts during its whole lifecycle. If we extend this horizon towards "sustainable design", including buildings and product service, shifting the focus also to the social, economic and ethical topics (Knight and Jenkins 2009), the trend is similar but even more important in absolute values (4850 papers only in 2015).

This is mainly due to the boost of magazines about Engineering, Environmental Science, Social Sciences and Energy, encompassing a vast number of tools, procedures,

© Springer International Publishing AG 2017

G. Campana et al. (eds.), *Sustainable Design and Manufacturing 2017*, Smart Innovation, Systems and Technologies 68, DOI 10.1007/978-3-319-57078-5_82

and standards, developed by private and public initiatives, aimed at reducing the environmental footprint of products, processes and services.

Already in 2000, a survey presented by Baumann et al. (2002) identified more than 150 existing methods and tools for green product development, highlighting the lack of practical relevance or testing, and the tendency to develop new tools rather than to evaluate and improve existing ones. Thereafter the situation has not changed and the proliferation of methods has led to the appearance of many papers whose intention was only to make order and pursue a higher comprehension, by developing different classifications and comparisons (Russo et al. 2015).

Among them, Byggeth and Hochschorner (2006) provided a classification according to three criteria: (1) Methods and tools for environmental impact assessment (e.g., LCA-Life Cycle Assessment); (2) Methods and tools for the comparison of environmental design strategies and product solutions (e.g., spider web diagrams); (3) Methods and tools for active eco-improvement (e.g., Eco guidelines). Other classifications of the eco-design tools consider also the nature of the provided results and the stages of the conceptual design process where the tools can be applied (Bovea and Pérez-Belis 2012), and the ideation mechanisms provided by tools, in terms of originality and environmental impact (Tyl et al. 2014).

In this paper, we focus the attention on this last category. They consist of a collection of guidelines or step-by-step approaches (e.g., checklist) that help the designers to adopt green solutions and reduce product environmental impacts.

Another useful classification was proposed by (Le Pochat et al. 2007). He identified only two main categories: "Analysis and evaluation" and "Improvement". Therefore, eco-design methods are divided in Eco-Assessment methods, that allow the analysis and evaluation of products environmental impacts, and Eco-Improvement methods, i.e. approaches that suggest possible solutions to overcome environmental criticalities by means of design guidelines.

Eco-guidelines work at different level of detail according to the different topics they face to (Russo et al. 2015). There are numerous and comprehensive guidelines addressing topics like material or energy consumption, reuse/recyclability, use and functionality, while on the contrary, maintenance, emission, reparability and logistic are less considered.

More generally, there is a preponderance of qualitative approaches. The lack of quantitative and objective data is noticeable. There is a need for consistent methods for the assessment and selection of available Eco-design methodologies, in order to help designers in choosing the suitable tool for their needs.

The aim of this paper is to fill this gap. Section 2 presents an ontological framework developed to structure a guideline, and in the following sections a new list of Eco guidelines on which the authors have been working for years. They are the result of various revisions, updated after tests with companies and tests with students.

2   How to Build a Guideline

Several studies have addressed the problem of well-written guidelines. Starting from the work of Newell and Simon (1972), Jonassen (2000) and Anderson (2009), the authors propose an ontological framework constituted by the four main key elements:

Description of problem type (Main goal): The first part of the guideline suggests the problem to be faced by providing the information about the initial state of the problem (e.g., "Presence of a harmful action"), or the description of the present situation that the guideline wants to change, and the goal of the guideline (e.g., "Reduce the harmful action"), or in what manner the guideline can change the initial situation.

Description of the sub-goal: This description clarifies the declared main goal, by explaining, for instance, the conceptual solution the guideline can provide. If the guidelines are a lot, the selection of the proper one can be based on the information provided by the declared goal and sub-goal.

Suggestion: This is the part of the guideline that explains how to manipulate the current state for achieving the declared goal and sub-goal. Differently to the sub-goal, the suggestion works with a more practical and operative point of view. Let's consider for instance the Newton's formula ($F = m * a$). If we want to increase the acceleration of a body, we can use the Newton's formula to define two sub-goals: "Reducing the mass of the body without changing the force" and "Increasing the force without changing the mass of the body". These sub-goals are directly related to the main goal "Increase the acceleration of the body", while a possible suggestion can be "Reduce the volume of the body" or "Change the temperature of the air".

Suggestions contain not only strategies but also details tools and introduce knowledge details about of the specific situation.

Examples: An example presents the possible solution for the considered guideline, derived from its application in other problems.

3   ITree Guidelines

Since 2009, the authors have been working on the implementation of a set of more than 300 guidelines to support green design, based on TRIZ, the theory for inventive problem solving (Altshuller 1984; Chechurin and Borgianni 2016). The main novelty of that work, compared to already existing methods for eco-design, was on the operative level. Adapting classical TRIZ fundamentals, originally conceived only for problem solving, into suggestions focused on green design, the authors tried to increase the level of detail at which other well-known guidelines usually work. Laws of Technical System Evolution (LTSE), Resources and Functionality methods, inventive principles and other classical TRIZ tools constituted the content of these new guidelines since the first version. Therefore, several other versions have been proposed and continuously updated and enlarged in number according to the results of tests conducted in industrial contexts,

European projects and scientific collaborations with academic groups working on environmental area. Step by step, they gradually acquired an easier and pragmatic aspect.

In the current version, the ITree guidelines are organized into four main groups and satisfy 28 main goals (Fig. 1). Each goals contains sub-goals (61) in turn divided into suggestions (207). The main groups are the following:

- Pre-manufacturing: in this group we take into account the choice of the product material, the redesign of its shape and the kind of supply in order to optimize transport and packaging before entering in the gate to be transformed.
- Manufacturing: in this group we consider all industrial operations of transforming the components into a product, i.e. by acting on the choice and optimization of the machineries and the semi-assembled flows, optimizing internal logistic, minimizing waste, reducing packaging and auxiliaries flows, designing a less impact product maintenance, etc.
- Product use: in this group we consider the impact of the product design during the use phase. Here all directions for minimization of mass, volume, inertia, energy consumption are included. It also includes hazardous materials exploitation, air emission reduction, maintenance management for increasing lifetime.
- End of life: in this group we consider the design of the product and its packaging disassembly, recycling and disposal. There also suggestions dealing with logistic and transport.

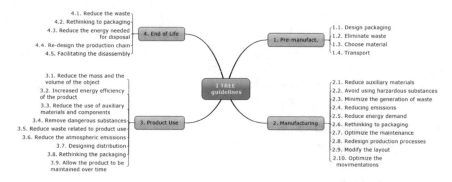

Fig. 1. ITree guidelines: groups and main goals.

Figure 1 portrays the main topics addressed by ITREE guidelines.

Each branch of the Fig. 1 represents an eco-design goal, at highest level of detail, that can be split in sub-goals as presented in Fig. 2.

Fig. 2. ITree guidelines: goal 2.6. and its sub-goals.

In order to fulfil a sub-goal the user has to follow precise design suggestions. To be really efficient, each guidelines has to propose tools and/or a strategy to exploit not obvious solving directions, and all resources already present in the system but not fully exploited, introduce new substances into the system without complicating it, and optimize the synergy among parts. Providing a so complex set of data with a user-friendly approach is a critical task. In the following we describe a short evolution of the guidelines.

1st version (Russo et al. 2011): The first set of guidelines was built starting from the LTSE and TRIZ tools. They were not organised as "well-written" guidelines according to above mentioned structure. Specific suggestions required a big effort for interpreting how to apply them to each specific problem. Let's consider for instance the following goals dealing with packaging. The first version offered different ways of intervention: reduction of packaging mass, reduction of packaging volume, reduction of energy linked to the packaging, use reusable/recyclable packaging. For each direction a specific suggestion was given as shown in Table 1.

Table 1. An example from the 2011 version of the guidelines.

Guideline "Reduction of packaging mass"
Suggestion: *Simplify the packaging (or use simplified packaging) trying to eliminate useless components. To do this perform a trimming activity associated with a traditional TRIZ functional analysis (to perform a functional analysis it's necessary to write the components of a system and show the functions with which they interact among them. Trimming activity consists to eliminate one or more elements and to assign their positive and essential functions to other components of the system). Explorer other technologies for the deposition of the packaging only where it is necessary.*

From the experimentation in small and medium size enterprise carried out in six countries within the European project REMAKE emerged that only people skilled in TRIZ were able to effectively solve problems adopting them and part of those guidelines were misinterpreted if not preceded by a small introductory course.

2nd version (Russo et al. 2014, 2015): on the basis of the feedback provided by REMAKE project, a second version of the guidelines has been developed. The content of each specific suggestion was reformulated in a more practical way by explicitly defining the goal and sub-goals, reformulating the specific suggestion in more synthetic way, specifying the interested phase during product lifecycle, and improving the examples. In addition, the guidelines have been specifically customised to the product lifecycle phases. An example is the packaging simplification for the product end of life as shown in Table 2.

Table 2. An example from the 2015 version of the guidelines.

Guideline reduction of packaging mass	Example	
Goal: *reduce mass flow of packaging in the use phase.*	*Procter & Gamble introduced a stand-alone rigid tube. So it was been possible eliminate the box.*	
Suggestion: *Eliminate useless components.*		

3rd version (2016): This version does not contain any reference to TRIZ labels and language. It is conceived for people that never heard about conceptual design model or problem solving tools. New triggers were added to cover logistic, design for disassembly, and other topics. Figure 3 shows an example of the new guideline for packaging redesign.

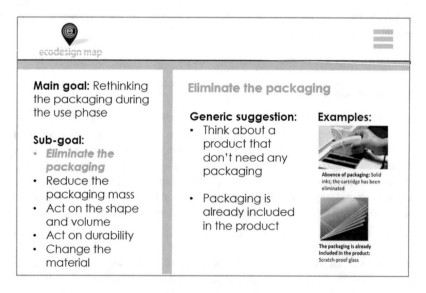

Fig. 3. An example from the 2016 version of the guidelines.

4   A Web Portal for Exploiting ITREE Guidelines

A web-based portal for exploiting ITREE Guidelines has been developed to provide a practical support to the user, and manage the addressed knowledge (e.g., databases).

The web portal guides the users following a predefined step-by-step procedure. First the user is required to fill a questionnaire; then, the system automatically proposes the set of pertinent guidelines structured as mentioned above. In the following a short description of the user interface is presented.

4.1   Interactive Questionnaire

The interactive questionnaire permits to collect all the main data about the product and the context of application. Figure 4 shows the structure and its main components. Required data are:

(1) Information about the designer in order to create her/his personal account and search history;
(2) Generic information about the product to be improved (name and field of application) and the name of the project;
(3) Specific data about the product through two checklists. The first one (Subsect. 3.1) investigates the possibility of intervention on product lifecycle (e.g. the choice of the suppliers, the decision making during conceptual design, the managing of the logistic, etc.). The second one (Subsect. 3.2) requires instead information about the product use (e.g., the energy consumption, the modalities of use, the user's interactions and manipulations, etc.).

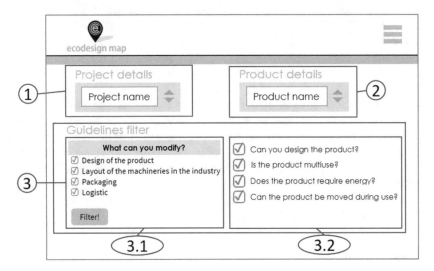

Fig. 4. Structure of window 1: "Interactive questionnaire".

4.2   Guidelines Interface

Once the user filled the questionnaire, s/he can access to a narrow set of guidelines dealing with one or more life cycle phases. Figure 5 shows the guidelines interface. The menu (1) shows the product life cycle phases: in grey the unselected ones and in green those active. According to the selected phase, menu (2) lists the corresponding goals ordered on the basis of a specific problem solving path. Once the user selected one goal, the system automatically displays the related sub-goals (3). Each sub-goal includes a short explanation of the objective, advantages and disadvantages (3.1) and the links to the generic suggestions (3.2). This allows the user to start from the most promising one. On the right side (4) the system displays the content of the generic suggestion (4.1) together with one or more descriptive examples (4.2).

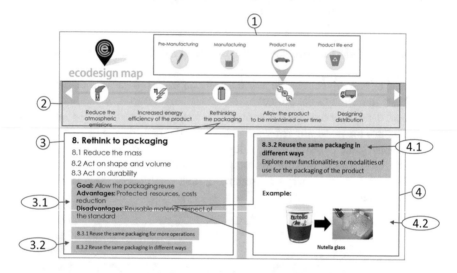

Fig. 5.   Structure of window 4: "Product use".

This procedure can be repeated until the final satisfying solution is reached so far.

In the Fig. 5 the menu (1) shows an example dealing with the disposal of the packaging during the product use. The user selected the first goal from the menu 2, concerning "Rethink to packaging". Menu 3 presents 2 different sub-goals, "8.1. Reduce the mass", "8.2. Act on shape and volume" and "8.3. Act on durability" (selected). In particular, two suggestions to realize sub-goal 8.3. are available: 8.3.1. "Reuse the same packaging for more operations" and 8.3.2. "Reuse the same packaging in different ways". In frame 4 the suggestion 8.3.2. is explained together with an example.

5   Conclusions

This paper presents an ontological structure for the construction of guidelines for Eco-design. In the last years, a set of guidelines for eco-design has been updated and revised according with the development of the ontological framework. A survey on different historical versions of the guidelines highlights how critical it is to succeed in transforming guidelines for problem solving in a form accessible also to user with no specific skills in problem solving. The presented ontology divides each guideline in goal, sub goal, general and specific suggestions, case study and examples. A web-based portal has been conceiving to exploit guidelines during product innovation as a light problem-solving path. The balance between the depth of information provided and simplicity of its content is the object of the analysis addressed during our research works and guidelines implementation.

References

Altshuller, G.S.: Creativity as an exact science: The Theory of the Solution of Inventive Problems. Gordon & Breach Science Publishing, New York (1984)

Anderson, J.R.: Cognitive psychology and its implications. WH Freeman/Times Books/Henry Holt & Co, New York (1990)

Baumann, H., Boons, F., Bragd, A.: Mapping the green product devel-opment field: engineering, policy and business perspectives. J. Cleaner Prod. **10**(5), 409–425 (2002). doi:10.1016/S0959-6526(02)00015-X

Bovea, M., Pérez-Belis, V.: A taxonomy of ecodesign tools for integrating environmental requirements into the product design process. J. Cleaner Prod. **20**(1), 61–71 (2012)

Byggeth, S., Hochschorner, E.: Handling trade-offs in Ecodesign tools for sustainable product development and procurement. J. Cleaner Prod. **14**(15–16), 1420–1430 (2006). doi:10.1016/j.jclepro.2005.03.024

Chechurin, L., Borgianni, Y.: Understanding TRIZ through the review of top cited publications. Comput. Ind. **82**, 119–134 (2016)

Jonassen, D.H.: Toward a design theory of problem solving. Educ. Technol. Res. Dev. **48**(4), 63–85 (2000)

Knight, P., Jenkins, J.O.: Adopting and applying eco-design techniques: a practitioner's perspective. J. Cleaner Prod. **17**(5), 549–558 (2009). doi:10.1016/j.jclepro.2008.10.002

Le Pochat, S., Bertoluci, G., Froelich, D.: Integrating ecodesign by conducting changes in SMEs. J. Cleaner Prod. **15**(7), 671–680 (2007). doi:10.1016/j.jclepro.2006.01.004

Maxwell, D., Sheate, W., van der Vorst, R.: Functional and systems aspects of the sustainable product and service development approach for industry. J. Cleaner Prod. **14**(17), 1466–1479 (2006). doi:10.1016/j.jclepro.2006.01.028

Newell, A., Simon, H.A.: Human Problem Solving, vol. 104, No. 9. Prentice-Hall, Englewood Cliffs (1972)

Russo, D., Regazzoni, D., Montecchi, T.: Eco-design with TRIZ laws of evolution. Procedia Eng. **9**, 311–322 (2011)

Russo, D., Rizzi, C., Montelisciani, G.: Inventive guidelines for a TRIZ-based eco-design matrix. J. Cleaner Prod. **76**, 95–105 (2014)

Russo, D., Serafini, M., Rizzi, C.: Comparison and classification of eco improvement methods. In: DS 80-1 Proceedings of the 20th International Conference on Engineering Design (ICED 2015) vol. 1: Design for Life, Milan, Italy, 27–30 July 2015 (2015)

Tyl, B., Legardeur, J., Millet, D., Vallet, F.: A comparative study of ideation mechanisms used in eco-innovation tools. J. Eng. Des. **25**(10–12), 325–345 (2014)

Sustainability as a Value-Adding Concept in the Early Design Phases? Insights from Stimulated Ideation Sessions

Lorenzo Maccioni[1], Yuri Borgianni[1(✉)], and Federico Rotini[2]

[1] Faculty of Science and Technology, Free University of Bozen-Bolzano, Bolzano, Italy
lorenzo.maccioni@natec.unibz.it, yuri.borgianni@unibz.it
[2] Department of Industrial Engineering, Università Degli Studi Di Firenze, Florence, Italy
federico.rotini@unifi.it

Abstract. As creativity is increasingly important in order to achieve differentiation and competitiveness in industry, designers face the challenge of conceiving and rating large numbers of new product development options. The authors' recent studies show the effectiveness of ideation procedures guided by stimuli that are submitted to designers in the form of abstract benefits. A rich collection of said benefits has been created to this scope; more specifically, the authors have performed a detailed clustering of the categories described in TRIZ ideality, i.e. useful functions, attenuation of undesired effects and reduction of consumed resources. Aspects related to sustainability and environmental friendliness manifestly appear in the list of stimuli and these issues are reflected in several ideas emerged in initial experiments. However, many promising product development objectives conflict with sustainability or, at least, their adherence to eco-design is arguable. The paper assesses the share of ideas that are supposed to comply with sustainability in experiments described in recent literature. Subsequently, it intends to stimulate a discussion about the introduction of measures to attract attention of designers on sustainability in the critical early product development stages also when green aspects do not represent the fundamental driver to achieve greater customer value. As well, it discusses which sustainability aspects are worth being considered adequately during the very early design phases and which ones could result as exceedingly constraining.

Keywords: Product value · Sustainability · Idea generation · Very early design phases · TRIZ

1 Introduction

The ability to discover consumers' valuable attributes has become a key factor for products/services' success or failure. In turn, product value manifests when users perceive that such a value enables the satisfaction of (some of) their needs. Many designers are aware of the necessity to individuate and meet human needs today and in the future through an enduring process of value delivery. Taking care of the time dimension, thus of future generations, is the baseline for sustainable development according to the definition that follows. The World Commission on Environment and Development [1] has indeed defined *sustainable* the development that meets the needs of the present without

© Springer International Publishing AG 2017
G. Campana et al. (eds.), *Sustainable Design and Manufacturing 2017*, Smart Innovation, Systems and Technologies 68, DOI 10.1007/978-3-319-57078-5_83

compromising the ability of future generations to meet their own needs too. This requires a multidisciplinary overview, which takes into consideration the interaction between environmental, economic and social issues [2].

In light of the above considerations, readers could figure out that the objectives of innovative value-oriented design and sustainability are completely aligned, if not overlapping. On the contrary, Sect. 2 documents potential conflicts between perceived value and sustainability and strives to frame the role of design in enabling the development of products that comply with both the fulfilment of human needs and sustainability issues, with a particular reference to environmental aspects.

Before multiple nuances of this potential conflict are illustrated, it is worth clarifying which design phases and activities are discussed in the paper. According to the conventional structure of the New Product Development (NPD) provided in the literature [3], design progresses by completing two fundamental milestones. The former, namely the front end or Fuzzy Front End (FFE) in light of its fairly unpredictable nature, articulates the development of products in terms of design goals and abstract solutions that should fulfil these goals. The latter, the back end, brings NPD further by making envisioned solutions more and more concrete as far as detailed design and the effective transformation of virtual prototypes into physical objects through the manufacturing process. It is intuitive that the creation of successful products requires that both macro-phases be carried out accurately.

However, literature tends to assign the FFE a predominant role [4] and the definition of design objectives, clearly shaped also by market inputs, is to be considered the most critical activity to future product success within the FFE [5]. Such an activity leading to the definition of new product attributes can be supported by proper techniques. Diffused methodologies include specific forms of customer surveys, brainstorming and stimulated idea generation, which stands for a more proactive approach to the identification of latent human needs [6]. The individuation of new benefits a novel product should display is clearly connected with the delivery of value the paper has introduced at the very beginning.

Despite the increasing importance of sustainability and the difficulties in implementing new (sustainable) requirements as NPD progresses, it emerges that the effectiveness of eco-design methods is arguable in this sense. The latter strive to satisfy sustainable principles through new solutions with the additional aim of fulfilling new requirements and environmental regulations [7]. Many of these tools are reviewed, summarized and clustered in [8, 9]. Most of them contribute in finding technical solutions in light of previously defined design specifications and objectives. A possible explanation of this issue is the circumstance that a large number of these tools have been developed in the past to facilitate problem solving and have been adapted to sustainability goals. Among them, a considerable number of instruments originate from the Theory of Inventive Problem Solving (TRIZ), which is considered a viable set of techniques and design principles to ease the conceptualization of sustainable solutions [10, 11]. As a result, many eco-design methods are unsuitable for addressing the ideation of benefits product design should deliver.

At the same time, general-purpose idea generation tools do not usually show any specific preference to sustainable aspects, as their overall purpose is product success and the identification of unexplored market opportunities. Therefore, the attention to sustainability is random, not taken for granted and presumably dependent on designers' sensibility towards environmental and human problems. Given the shifted shortcomings of eco-design and idea generation approaches, a valuable methodological goal would be represented by the development of tools that are capable of both focusing on environmental issues and stimulating value-adding creativity. In essence, this stance is supported by [8], which suggests sustainability and other critical performances be proposed simultaneously.

The objective of the paper is more limited, but attempts to provide an original contribution to the above goal. The authors claim that the degree of sustainability emerging from ideation activities should be evaluated in order to understand the magnitude and outreach of initiatives making value-oriented very early design activities greener and closer to social aspects. A preliminary assessment of such a kind has been carried out by exploiting a large number of ideas generated by engineering students, who have deployed a general-purpose tool developed by the authors. Such an instrument has been selected because of its claimed adaptability to variegated industrial domains, its supposed capability to explore NPD opportunities thanks to a large variety of stimuli and, especially, its demonstrated performance in improving idea generation creativity (especially in terms of quantity). The tool and experiment are presented in Sect. 3. The subsequent evaluation of the share of sustainable ideas is illustrated in Sect. 4. Discussions and concluding remarks can be found in Sects. 5 and 6, respectively.

2   The Argued Relationship Between Sustainability and Value: Relevant Background

In the sustainability field, product value is the ability to meet needs resulting from environmental, economic and social issues. A tentative list of said needs is described in terms of a set of principles, approaches and strategies in [11], which will be elucidated in Sect. 4 for the sake of the assessment. With a particular reference to environmental features, said principles well mirror the possible objectives individuated by Vezzoli and Manzini [12] in order to diminish the ecological footprint. Still in [12], existing eco-design methods are classified according to their capability to support the achievement of these objectives. Additional and complementary criteria have been introduced to categorize these methods. The taxonomy proposed by Bovea and Pérez-Belis [8] clusters available eco-design tools according to, among the others, the different stages of product development that best fit their application. This categorization supports designers in the individuation of the right tool to apply for achieving the improvement of certain environmental performances in a given activity of the NPD cycle. The capability of the discussed instruments to support different design stages suggests that their outreach should have expanded well beyond the optimization of product architectures in order to minimize the ecological footprint. [13] points out how the scope of eco-design tools rapidly turned in the 1990s from isolated environmental-friendly measures to complex business strategies that provide benefits to multiple stakeholders. [14] highlights how their goal has moved from the introduction of environmental issues in companies to a proactive engagement aiming at strategic advantage.

However, the profitability of eco-designed products is often hypothesized, but not fully demonstrated [15]. Especially in certain sectors, sustainability and profitability still represent a trade-off to overcome [16].

The relationship between the capability to deliver value and sustainability, which involves very early design phases more specifically, can be considered as a specific aspect of the conflict concerning environmental friendliness and profitability. As achievable e.g. from a case study presented in [17], strictly following the principles of sustainability can lead to a worsening of some general performance and to the decrease of value perception enjoyed by some stakeholders. According to [18], trying to develop a product that overlooks customer requirements for increasing environmental benefits is counterproductive.

These claims stress the relevance of the goal of the present study, which has been outlined in Sect. 1.

3   Description of the Experiment

The experiment presented in this paper has been carried out through the employment of a computer-aided system, namely iDea, specifically dedicated to support designers in the identification of new product features [19]. The aim of the tool is to enhance the creative process through providing users with textual stimuli that act as triggers around which ideas might enucleate and come out. These stimuli:

- are administered in an abstract form in order to fit multiple industrial domains and limit design fixation;
- originate from a wide analysis of previously analyzed NPD successes; the authors have striven to provide an all-encompassing set of viable design directions;
- are articulated according to a logic that compels the designer to explore the following main dimensions:
  - Benefits: typologies of human needs new products might fulfil by displaying new or improved characteristics; these benefits may refer and vary according to the three categories that follow. TRIZ terms of ideality (useful functions, harmful effects, and resources) have resulted critical to gather and cluster the kinds of benefits potentially meaningful for the design process.
  - Stakeholders (SHs): actors that are influenced by the product in the lifecycle phases that follow its market launch, i.e. users, beneficiaries, service recipients and outsiders.
  - Lifecycle phases (LCs): circumstances that may occur along the different stages of product existence mirroring the horizontal axis of TRIZ System Operator. Consistently with the observation raised for the previous dimension, the relevant domain to categorize product attributes starts with the market introduction, thus ranging from this moment to the end of product functioning.
  - Hierarchies of system (SYSs): hierarchical levels of the systems, as indicated by classical TRIZ System Operator (especially in the nine-screen version).

Therefore, they range from the external environment, which strongly characterizes design requirements, to parts and components.

In this sense, it emerges that the deployment of TRIZ does not refer to its distinguished problem solving capabilities, but to its capability to structure design knowledge and ideation sources. The exploitation of its articulated body of knowledge in the design domain is diffused according to a recent review of the most popular publications dealing with TRIZ [20].

With reference to the above dimensions, the designer can choose to use the bundle of stimuli belonging to a specific dimension or ask for more articulated stimuli through a selected combination of different prompts belonging to different dimensions. Therefore, the described computer-aided system allows to customize the search; both a quick and superficial exploration of the design space and an exhaustive, deepened and/or specialized investigation are feasible. Readers can experience performances and limitations of the system by trying the free version of iDea computer application, which can be downloaded from the link http://goo.gl/AwzZHF, while further information about the system and its use are available in [19].

The same publication documents an extensive testing campaign, which spotlights the impact of iDea on the proliferation of ideas generated by students in Mechanical and Energy Engineering attending the course in *Product Development and Engineering* at the *Università degli Studi di Firenze*. While a first test in which participants received no stimuli was carried out in order to establish outcomes from a control group, the second step of the experiment involved the employment of iDea. Two separate groups of twelve students were asked to produce as many ideas as possible in terms of potential novel benefits that a camera or a coffee machine could exhibit. Extremely similar quantities of new ideas were produced for the two kinds of products under the same test conditions: 248 for cameras and 237 for coffee machines. Given the reference to the very early design phases, ideas were expected in the form of abstract benefits and product characteristics and features without any reference to feasibility or potential embodied structure and/or technical solution.

4   Analysis of Sustainable Outcomes

The complete set of generated ideas was post-processed in order to identify the ones that clearly cope with sustainability aspects. All the three domains of sustainability, i.e. environmental, social and economic, were considered carefully. To this scope, an acknowledged sample of sustainable aspects was selected, namely the set of design principles and strategies proposed by [11]. This cluster resulted the best option available in literature, not only due to the abundance of sustainable nuances that are illustrated. In fact, the explanations that are provided by the scholars (*objectives* in Table 1) enable a considerably straightforward association between said sustainable design principles and students' ideas. Table 1 provides a selected set of sustainable aspects, their reasons of existence in terms of the objectives they aim to fulfil and illustrative ideas that match these indications.

Table 1. Illustrative sustainable aspects that have been recognized with the set of ideas generated in the experiment

Sustainable aspect in terms of design principles	Objective	Illustrative generated idea matching the design principle
Source Reduction	Reduce the quantity of materials entering a waste stream	Edible coffee pods
Renewable Resources	Do not use limited resources	Using the heat of the hands to recharge camera battery
Reuse	Reuse the product for uses different from initial ones with the aim of extending the product life	Using the package as a backpack
Product Service System (PSS) strategy	Fulfil consumers' needs through the provision of more dematerialized services	Operating and maintenance guides displayed through video available on the camera instead of on printed paper
Minimization of Resource Usage	Preserve natural resources	Making coffee without hot water

Beyond the shown sustainability aspects, other design principles were found to which ideated benefits can be associated: Remanufacturing, Repair, Recycling, Degradation, End-of-pipe and Social purpose are to be mentioned here. It can be noted that, while environmental issues are clearly predominant, economic aspects did not emerge

Table 2. Number of ideas concerning new product attributes for coffee machines and cameras that exhibit sustainable concepts

Sustainability aspect	Coffee machine	Camera	Total
Source Reduction	11	6	17
Remanufacturing	4	0	4
End-of-pipe	3	0	3
Repair	5	1	6
Recycling	2	1	3
Reuse	7	4	11
Renewable Resources	4	8	12
Minimization of resource Usage	5	2	7
Product Service System (PSS) strategy	3	7	10
Recovery	4	0	4
Degradation	0	3	3
Incentive to Recycling	0	1	1
Social purpose	0	5	5
Total	48	38	86

in the set of generated ideas; social principles seldom emerged as well. Table 2 documents the number of ideas associated to the detected sustainable aspects – redundancies were overlooked also in this case in order to compare data with the global generation of ideas.

5 Discussion of the Results

Table 2 depicts the ideas that have been generated by students and underlines the frequency of each sustainability concept. At the end of the categorization of ideas in terms of their consistence with sustainability issues, 86 out of 485 (roughly 18%) new product attributes cope with one of the principles described in eco-design literature.

The results lend themselves to two kinds of discussion, namely the centrality of sustainability in idea generation and, from a methodological perspective, the potential effects of sustainability prescriptions on the design process.

With regard to the former, it is arguable to affirm whether this number of ideas is small or large, because of the empirical nature of the study and due to lack of previous investigations of this kind. Authors too did not converge to any kind of expectation beforehand. Furthermore, due to the unbalance of new sustainable benefits for coffee machines and cameras (whose complete sets showed very similar quantities), it can be claimed that treated topics may affect the frequency of sustainable aspects during idea generation. As a result, any conclusions descending from the above percentage are preliminary and qualitative in essence.

On the one hand, given the structure of iDea, the share of stimuli that directly suggest environmental or sustainability benefits is lower than the indicated percentage. This means that, either sustainability-related prompts are overall more productive than other stimuli, or devising sustainable ideas is likely to happen with the exploitation of several kinds of hints. Such an issue will be verified in a future experiment in which participants will be asked to indicate which specific stimulus has triggered their ideas, so that the prolificacy of prompts will be compared. It is worth noting that this specific information could not be extracted in the first testing campaign because students were asked to generate as many ideas as possible in a given amount of time and any issue limiting productivity was avoided. Otherwise said, the capability of the employed tool to support the exploration of the design space and boost creativity represented the primary scopes of the documented test. Moreover, it can be stated that ideas seldom emerged that are clearly incompatible with sustainability principles in order to prioritize other performances. Thus, the compatibility with sustainability of ideas that are not highlighted in the above Table should be checked in the subsequent design phases at a less abstract level. Contradictions between new benefits, which are supposed to be the primary carriers of increased customer value, and sustainability features are likely to emerge during conceptual design or later stages. In these circumstances, TRIZ, or any of its adaptations to comply with eco-design better [9, 10] is supposed to be a valuable instrument for both fulfilling new value-oriented product characteristics and reducing (or, at least, not increasing) harmful effects in terms of sustainability.

On the other hand, readers of the present paper might argue that the number of unambiguously sustainable ideas is low, especially if it is taken into account that a share of students followed a program in Energy Engineering and should have developed heightened sensibility to issues such as resources' safeguard and reduction of the ecological footprint. According to this potential interpretation, the background of the participants in the experiment could have biased the outcomes, which would have displayed less sustainable ideas if testers had been picked up randomly. Hence, the share of selected ideas, i.e. 18%, likely represents an overestimation of the general perception of sustainability as a value-adding driver. In addition, as all highlighted ideas refer to unique sustainability aspects, contradictions with other principles might emerge in the subsequent design phases as well.

From a methodological viewpoint, it is worth highlighting that many sustainability aspects were not displayed in the set of new ideas. At the same time, some specific environmental aspects outweighed the residual sustainability features, e.g. diminishing the channeling of resources, use of renewable forms of energy, opportunities to reuse products, servitization strategies. The limited share of ideas embodying social sustainability is restricted to qualities that can be referred to the Quality of Life concept [17]. Conversely, economic aspects did not arise at all in the set of generated new ideas. The asymmetric distribution of these aspects between coffee machines and cameras lowers the robustness of the results' analysis with regard to the identification of the most diffused sustainability principles. Therefore, the absence or low diffusion of particular sustainability aspects can be due to case. However, the authors tend to provide a different interpretation of this issue, also in consideration of the high variety of ideas throughout the whole set of proposed unprecedented benefits [21]. Many of the discussed sustainability principles appear as forms of constraints for the New Product Development process, with a particular reference to social and economic domains. According to this vision, they cannot constitute any sort of superior value apart from additional benefit-increasing cues. Consequently, they can be overlooked during the ideation of new product attributes, but they have to be carefully considered when decisions are made concerning the choice of technical working principles or the definition of business models in order to cope with sustainable development.

6 Conclusions

The present article attempts to provide a contribution in the literature discussion about the argued essence of sustainability to represent a carrier of value. Eco-design methods and other techniques that focus on the reduction of undesired effects for the environment supposedly tend to limit the space for potential solutions. According to this reading, they tend to underestimate the need for introducing other performances that are critical for product acceptability. It is therefore crucial to determine whether a sustainability-oriented product development can take place from the very early design stages and if this can lead to valuable and successful outputs. In this perspective, the declared problem is undertaken in an original way, namely by providing a preliminary understanding about the role of sustainability within the whole domain of value-adding strategies.

The paper has firstly outlined an experiment about the use of a tool developed by the authors, whose objective consisted in supporting the generation of innovative ideas through textual stimuli, so that the design space is better explored in the very early product development phases. Said stimuli include all gathered hints that are supposed to provide customer value, regardless their reference to the sustainability sphere. The result of the test is a set of potential new product attributes for future versions of coffee machines and cameras, which stand for candidate directions to guide the design of products with superior value. As the exploration of stimuli was left free to participants, the outcomes of the test are supposed to represent a good baseline for investigating the diffusion of sustainability issues as drivers for innovating artefacts and services, despite the limitations recalled in the previous section.

Sections 4 and 5 remark that sustainable characteristics, among which environmental-friendly ones are the most diffused, can be clearly identified in the set of outcomes and that their number is not negligible. However, they surely represent a minority within the whole set of ideas. In this sense, it is very likely that sustainability issues have to be tackled in subsequent product development phases, where they might give rise to design conflicts. In case dichotomies are not overcome and trade-off solutions are chosen, product development faces situations in which design objectives are just partially fulfilled and the performances that should be spotlighted get blurred. In this sense, new value profiles should comprise sustainability issues from the very beginning of product design, including them in coherent propositions and lists of design specifications. This demands new methodological roadmaps, whose development represents the core of authors' future work.

Other planned activities are still concerned with the assessment of the contribution of sustainability features to product success and customer satisfaction. As outlined in the discussion section, the authors commit to monitoring the capability to provide value of prompts broadly expressing enhanced quality of life and environmental friendliness, which have emerged as the best sustainability-related candidates according to the performed analysis. Another possible means to investigate the perception of sustainable principles is collecting and analyzing previous studies that have assessed the role played by large samples of customer requirements in terms of affecting satisfaction or appreciation. This survey and examination could be complemented by other typologies of studies that aim to investigate users' emotional reactions up against products featured by particular properties – the focus would be clearly on sustainable characteristics in this case. This kind of analysis might exploit the capabilities of biometric measurement devices, whose deployment is increasingly diffused in the design field.

References

1. Brundtland, G.H.: World Commission on Environment and Development. Our common future: Report of the World Commission on Environment and Development (1987)
2. Elkington, J.: Cannibals with Forks. The Triple Bottom Line of 21st Century. New Society Publishers, Gabriola Island (1997)
3. Pahl, G., Beitz, W., Feldhusen, J., Grote, K.H.: Engineering Design: A Systematic Approach, vol. 157. Springer Science & Business Media, London (2007)

4. Achiche, S., Appio, F.P., McAloone, T.C., Di Minin, A.: Fuzzy decision support for tools selection in the core front end activities of new product development. Res. Eng. Design **24**(1), 1–18 (2013)
5. Bacciotti, D., Borgianni, Y., Cascini, G., Rotini, F.: Product Planning techniques: investigating the differences between research trajectories and industry expectations. Res. Eng. Des. **27**(4), 367–389 (2016)
6. Schaffhausen, C.R., Kowalewski, T.M.: Large-scale needfinding: methods of increasing user-generated needs from large populations. J. Mech. Des. **137**(7), 071403 (2015)
7. Bey, N., Hauschild, M.Z., McAloone, T.C.: Drivers and barriers for implementation of environmental strategies in manufacturing companies. CIRP Ann. Manufact. Technol. **62**(1), 43–46 (2013)
8. Bovea, M., Pérez-Belis, V.: A taxonomy of ecodesign tools for integrating environmental requirements into the product design process. J. Cleaner Prod. **20**(1), 61–71 (2012)
9. Tyl, B., Legardeur, J., Millet, D., Vallet, F.: A comparative study of ideation mechanisms used in eco-innovation tools. J. Eng. Des. **25**(10–12), 325–345 (2014)
10. Russo, D., Serafini, M., Rizzi, C.: Is TRIZ an ecodesign method? In: Setchi, R., Howlett, R., Liu, Y., Theobald, P. (eds.) Sustainable Design and Manufacturing, vol. 52, pp. 525–535. Springer International Publishing, Cham (2016)
11. Glavič, P., Lukman, R.: Review of sustainability terms and their definitions. J. Cleaner Prod. **15**(18), 1875–1885 (2007)
12. Vezzoli, C.A., Manzini, E.: Design for Environmental Sustainability. Springer Science & Business Media, London (2008)
13. Stevels, A.: Application of ecodesign: ten years of dynamic development. In: Proceedings EcoDesign 2001: Second International Symposium on IEEE Environmentally Conscious Design and Inverse Manufacturing, pp. 905–915 (2001)
14. Fiksel, J.: Design for Environment. A Guide to Sustainable Product Development, 2nd edn. McGraw-Hill, New York (2009)
15. Plouffe, S., Lanoie, P., Berneman, C., Vernier, M.F.: Economic benefits tied to ecodesign. J. Cleaner Prod. **19**(6), 573–579 (2011)
16. Figge, F., Hahn, T.: Is green and profitable sustainable? Assessing the trade-off between economic and environmental aspects. Int. J. Prod. Econ. **140**(1), 92–102 (2012)
17. D'Anna, W., Cascini, G.: Adding quality of life to design for Eco-Efficiency. J. Cleaner Prod. **112**, 3211–3221 (2016)
18. Cluzel, F., Yannou, B., Millet, D., Leroy, Y.: Eco-ideation and eco-selection of R&D projects portfolio in complex systems industries. J. Cleaner Prod. **112**, 4329–4343 (2016)
19. Bacciotti, D., Borgianni, Y., Rotini, F.: An original design approach for stimulating the ideation of new product features. Comput. Ind. **75**, 80–100 (2016)
20. Chechurin, L., Borgianni, Y.: Understanding TRIZ through the review of top cited publications. Comput. Ind. **82**, 119–134 (2016)
21. Bacciotti, D., Borgianni, Y., Rotini, F.: A CAD tool to support idea generation in the product planning phase. Comput. Aided Des. Appl. **13**(4), 490–502 (2016)

QFD and TRIZ to Sustain the Design of Direct Open Moulds

Gianni Caligiana[✉], Alfredo Liverani, Daniela Francia,
Leonardo Frizziero, and Giampiero Donnici

ALMA MATER STUDIORUM University of Bologna, v.le Risorgimento, 2, Bologna, Italy
{gianni.caligiana,alfredo.liverani,d.francia,leonardo.frizziero,
giampiero.donnici}@unibo.it

Abstract. Sustainable design aims at the creation of physical objects, environment and services that complies to optimize social, economic, and ecological impact. QFD is able to assess the product design by the choice and definition of parameters that can be qualitatively discussed. The purpose of design is to meet a need in new ways and in innovative ways. In this context, the QFD aims at evaluating the quality of a design process. TRIZ is a design method that aim at defining and overcome some critical issue that can affect the development of a product, by means of potential innovative solutions. In this paper QDF and TRIZ analysis have been adopted in order to validate a design method for direct open moulds, by a new strategy: hybrid manufacturing can reduce the production time, the use of material, the energy and the waste consumption, employing subtractive and addictive techniques efficiently combined.

Keywords: Hybrid manufacturing · Additive · Subtractive · Direct open mould · QFD · TRIZ

1 Introduction and State of the Art

Sustainability is a challenge that requires an integrated approach to link the community's economy, environment and society. Many attempts to explain the concept of sustainability has led to different meanings of the same concept, depending on the field of application as human, social, ecological, biological, industrial etc. In 1987 the Brundtland Commission defined first sustainability as "meets the needs of the present without compromising the ability of future generations to meet their own needs" [1, 2].

In the engineering context, the sustainable development refers to advances in technology, economics, environment, health and welfare. Traditionally, good strategies for manufacturing were considered increasing the volume of production, reducing the time and costs [3]. Nowadays, concerns as environmental implication and use of natural resources strongly influence the manufacturing strategies choice, also in the preliminary design phases.

Emerging design strategies are the Design for Environment and Life Cycle Assessment, the Resource and Energy Sustainability, the Design for Sustainability, the Design for Disassembly [4], whose focus is to conceive a product by taking care of all the effect

© Springer International Publishing AG 2017
G. Campana et al. (eds.), *Sustainable Design and Manufacturing 2017*, Smart Innovation, Systems and Technologies 68, DOI 10.1007/978-3-319-57078-5_84

that its use can cause to the economy, to the society and to the environment, also at its disposal.

In order to support sustainability in manufacturing, this paper presents an enhancing automatic process for direct open moulding manufacturing. Open moulding is a process suitable for parts that require wide range of size part, large and complex shapes, low-volume and rate of few thousand parts per year. In this process, raw materials (resins and fibre reinforcements) are deposited on a mould through different processes, including hand lay-up, spray-up, casting, and filament winding and then they are exposed to air as they cure or harden.

This technique allows for a rapid product development cycle because the tooling fabrication process is simple and relatively low cost because of low cost tooling option.

On the other hand, all the open moulding processes are manual, slow and labour consuming and require the mould preparation by means of a model. The mould should be as accurate as possible and it reproduce the final part as the design project describes. Furthermore, the entire process reveals not time-cost efficient and hazardous in terms of harmful emission because in the open mould processes volatile organic compound (VOC) is pretty high [5]. This is the reason why, in order to improve air quality, the open mould process has been converted to closed mould process [5, 6] that allows the fabrication of parts with complex geometry [7] and accounts for environmental protection, but requires more expensive tooling.

Direct open moulding does not require the model preparation for the mould and starts processing the mould directly, catching information about the piece to reproduce only by a digital model. Commonly, the CAD model includes all the information necessary to give instruction to the CAM. Thus, the CNC processing can be elaborated in order to directly machining a block of material, up to the final mould shape. This technique allows reducing the material use, wastes and energy consumption.

The goal of this paper is to suit a solution that can integrate the advantages of direct open moulding with sustainable design, in order to find the optimal trade-off between time and cost-effective manufacturing and the respect of the environment and of the human labours.

Traditional design strategies may be inadequate to meet these two requirements, thus a customer-driven approach has been evaluated in order to develop product and design quality in the manufacturing industry [8–10]. A QFD (Quality Function Deployment) plan has been developed in this paper in order to improve the design of products/services according to the customer requirements. Relationships between the technical process requirements have been evaluated through the morphological matrix. Then the fundamental requirements have been determined and further analysed by means of the dependence/independence matrix. QFD made it possible to translate the process requirements into design attributes, but some contradiction arises from the QFD evaluation. Thus, in order to enhance the attributes arose from QFD, a systematic analysis based on the Theory of Creative Problem Solving (TRIZ) has been developed in order to propose innovative solutions that meet also the sustainable design principles. In literature, studies about the integration between QFD and TRIZ can be found. In particular M. Mayda and H.R. Borklu [11] implemented a composed method using TRIZ to identify innovative concepts at first, then QFD to meet customers' needs. Moreover, also C.H. Yeh, C.Y. Huang Jay and C.K. Yu [12] developed a case study for the integration of QFD and TRIZ. Differently from the previous authors, they developed the method using first a four phase QFD plan, followed by TRIZ application to enable the development of breakthrough products.

In this paper a new method is presented that is inspired to the direct moulding and that eliminates manual operations by the introduction of an automatic process that combines additive and subtractive techniques to enhance the fabrication of mould for direct open moulding.

2   QFD Approach to Solve the Open Mould Production

2.1   QFD

Quality Function Deployment (Q.F.D.) is a methodology to structuralise the data stream, which accompanies each design development. QFD is a practice to translate clients' desires into appropriate businesses' necessities at every phase, from research through production design and development, to manufacture, distribution, installation and marketing, sales and services. It was developed to bring personal characters to modern manufacturing and business processes. It helps designers looking for both spoken and unspoken requirements, translating these into actions and designs, and focusing various business functions toward achieving this common goal.

QFD can be exploited as follows:

1. Considerate the desires of the customers;
2. Enlightening quality systems thinking + psychology + knowledge/epistemology;
3. Maximizing positive quality to add value;
4. Realizing comprehensive quality system for customers' satisfaction;
5. Creating strategy to stay ahead of the competitive game

This methodology starts with the explanation of the task, composed by the following important steps:

(1) Analysis of the market environment; (2) Analysis of the competitors' production; (3) The six questions to characterize the products; (4) The evaluation and comparative matrixes.

Then, when the definition of the technical requirements is reached, it will be assumed to design the product. After the explanation of the task, the problem is completely defined. The analysis of the environment and the six questions are part of the explanation of the task.

In particular, the six questions serve immediately to characterize those features that the object must necessarily embody to be designed. The questions are:

(1) Who: who uses our product? (2) What: what is the use of the product? (3) Where: where is it used? (4) When: when is it used? (5) Why: why is it used? (6) How: how is it used?

For estimating the relative importance among the requirements defined above, the interrelation matrix is used. The interrelation matrix is an instrument evaluating the relationships of dependency (first use) and/or of relative importance (second use) among various necessities or ideas; the instrument is also used to define the priorities and to establish the optimal sequence of actions. In the first use, the variables on the columns must be considered as the causes and the same ones, on the rows, as the effects. The dependence between the requirements can be *weak, medium* or *strong*: it can be amountable, for example, with the following conventional values: 0, 1, 3, 9.

Instead, in the second use, the matrix is employed to evaluate relative importance among all the variables. For the relative importance analysis, the following conventional votes are used:

1 if the row element has the same importance of the column one;

0 if the row element is most important than the column one;

2 if the column element is most important than the row one.

All the votes can be obtained by market analysis upon a large sample of people.

Using this instrument, the highest values of the sums per rows indicate which is the most important variable among all of them (i.e. those which have more influence on the others).

Finally, after explanation of the task, the conceptual design starts at first, followed by the constructive one. In this paper the investigation has been made among five researchers of the Design and Methods Research Group of the University of Bologna. The design study is obtained through the Morphological Matrix [13] and the conceptual CAD drawings.

2.2 QFD vs. Open Mould Production

2.2.1 Six Questions

Applying the six questions to the case of the Open Mould Production, the team were able to give the following answers, in order to find out the requirements to be analysed:

(1) *Who: who uses the open mould? who produced the open mould?* The open mould is used by company producing shell component. It is produced through industry processes.

(2) *What: what is the use of the open mould?* The open mould needs to produce industrial components.

(3) *Where: where is the open mould produced/used?* It is produced inside industrial mould production departments and it is used in industrial production.

(4) *When: when is the open mould produced?* It is produced on customer request; it is usually available not before than six months.

(5) *Why: why is the open mould used?* It is used for giving a shape to products; often to give a complex shape.

(6) *How: how is the open mould used/produced?* It can be composed by several parts; it can be used matched with a machine.

From these answers, nine most important characteristics, which the open mould has to own, have been identified:

1. Industriabilty - 2. Structural Strength - 3. Customization - 4. Production Speed - 5. Workability - 6. Precision - 7. Thermal Resistance - 8. Complex Shaping - 9. Reliability.

2.2.2   The Evaluation Matrixes

Below (Figs. 1 and 2) are shown the interrelation matrixes, used in both kinds of employment for the open mould where the above mentioned requirements, obtained by the six-answers analysis, have been inserted.

Relative importance matrix	Industriability	Structural Strenght	Customization	Production Speed	Workability	Precision	Thermal Resistance	Complex Shaping	Reliability
Industriability	1	1	2	1	1	1	0	1	1
Structural Strenght	1	1	1	0	1	1	0	1	1
Customization	0	1	1	1	1	1	0	1	0
Production Speed	1	2	1	1	2	1	0	1	0
Workability	1	1	1	0	1	2	1	1	1
Precision	1	1	1	1	0	1	0	1	1
Thermal Resistance	2	2	2	2	1	2	1	2	2
Complex Shaping	1	1	1	1	1	1	0	1	1
Reliability	1	1	2	2	1	1	0	1	1
Importance	9	11	12	9	9	11	2	10	8

Fig. 1.   The relative importance evaluation matrix

Indipendence/Dependence Matrix	Industriability	Structural Strenght	Customization	Production Speed	Workability	Precision	Thermal Resistance	Complex Shaping	Reliability	DEPENDENCE
Industriability		1	9	9	1	3	1	3	9	36
Structural Strenght	0		1	1	3	3	1	3	3	15
Customization	3	9		3	9	3	1	9	1	38
Production Speed	9	1	9		9	9	3	9	3	52
Workability	3	1	3	1		1	9	1	1	20
Precision	9	3	3	9	3		1	3	1	32
Thermal Resistance	0	3	1	3	1	1		0	1	10
Complex Shaping	9	9	3	3	3	3	1		3	34
Reliability	9	9	3	3	1	3	3	3		34
INDIPENDENCE	42	36	32	32	30	26	20	31	22	

Fig. 2.   The independence/dependence evaluation matrix

By the analysis of the matrixes, QFD parameters were defined as:

(1) Industriability – (2) Structural Strength – (3) Customization – (4) Precision

These parameters can be linked in order to define the final project proposal. The link is obtained through the morphological matrix (Fig. 3) in which, for each of the requirements evidenced by the previous analysis, four possible technical solutions have been proposed. The combinations of requirements traced a feasible quality assisted project path.

MORPHOLOGICAL MATRIX	TECHNICAL SOLUTIONS 1	TECHNICAL SOLUTIONS 2	TECHNICAL SOLUTIONS 3	TECHNICAL SOLUTIONS 4
Industriability	CNC	Additive Manufacturing	Manual Work	Other Technologies (FOR EX Electric Erosion)
Customization	Complex Shape	Advanced Material	Dedicated Technology	Specific Architecture
Precision	Material characteristics	Bi-layer Architecture	Workabilty	Structural and thermal strenght
Structural Strenght	Structure/ Frame	Architecture/ Assembly	Material	Shape

Fig. 3. The morphological matrix

The technical solutions path has been suggested by the application of other relative importance evaluation matrixes, for each line (requirement vs. technical solution).

So, switching from the conceptual analysis to the technical design study, the proposed concept of open mould can be outlined, as depicted in Fig. 4 that follows.

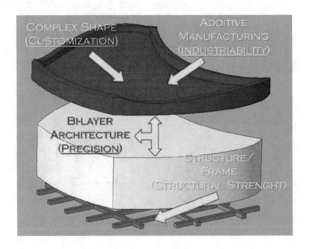

Fig. 4. QFD concept of open mould

In order to refine the conceptual design for direct open moulding, this technical solution, suggested by QFD, can be submitted to TRIZ analysis for a further optimization, in order to enhance innovative aspects [14].

3 TRIZ Approach to Find the Innovative Solution

TRIZ is a Russian acronym for "Theory of Inventive Problem Solving" and consists, at the same time, in theory, operational procedures and a set of tools made since 1946 by Genrich Altshuller Saulovich, with the goal of capturing the creative process in technical and technological fields, encode it and make it repeatable and applicable, in short a real theory of the invention. We used TRIZ for the analysis and optimization of the solution that has been indicated in the previous step by QFD. A dedicated software for TRIZ problem analysis [15], TechOptimizer®, has been employed for the analysis of the open mould. TechOptimizer includes all of the analysis tools to optimize the inventive solutions, such as the trimming tool that we're going to use to find the optimal solution.

The first task of this method is the formulation of the "Ideal Final Result", the final objective of the work, the best one among the possible solutions. Then, from this ideal starting point, it is possible to deploy a functional analysis and moving backwards towards less ideal but more workable solutions. The analysis began first by the assumption of a mould with three layers: a <u>structural strength frame</u>, the <u>intermediate support</u> in a light and easily workable material, a <u>complex shape additive layer</u> easy to be shaped and to be finished. This is a feasible "Ideal Final Result".

In the environment "Product Analysis" are provided in input the objectives to be achieved and the problem limitations (Fig. 5). These data have been captured from previous QFD analysis. For each requirement, the relative importance has been evaluated on a scale from 1 to 10.

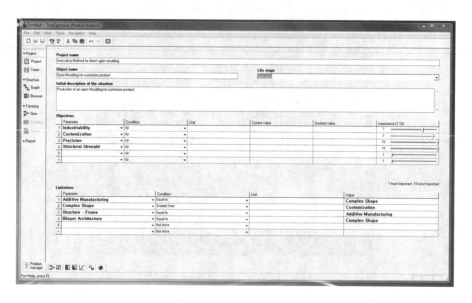

Fig. 5. The product analysis

The program allows to build a chart in which are placed all the relationships between the different parts that compose the mould and all operators that are involved. It was built a functional diagram (Fig. 6) in which each relation can be described as good or harmful; then, it was necessary to set each relation to make it interacting with parameters of the mould, project parameters and so on.

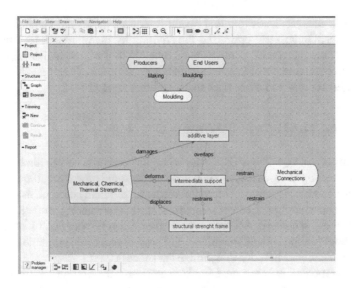

Fig. 6. TRIZ analysis

The following step consists in defining the same relationships and the data can be inserted both in qualitative and quantitative manner. Additional information to be included are about the interaction with the parameters entered at the beginning of the analysis. Finally, the software process consists in the launch of the command "Trimming". The "Trimming" is a tool that allows to optimize the problem in order to choose the optimal path and architecture of the project, eliminating negative components and actions (Fig. 7).

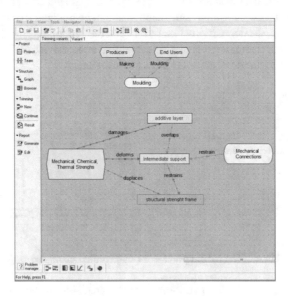

Fig. 7. Results of trimming

The TRIZ analysis leads to a first result: it reduces the mould layers from three to two ones. TechOptimizer® optimised the solution by the elimination of the strength frame and by the transfer of its mechanical function to the intermediate support. Thus, the innovative solution is a bi-layered architecture composed of a self-sustaining (former intermediate) support and a complex shape additive layer [16–18].

4   Results: The Hybrid 3D Manufacturing

The analysis of both QFD and TRIZ suggested the need for a new method of manufacturing for direct open moulds that could account for rapid processing time, accuracy of dimensional tolerances and roughness of surfaces, automatization of the process and low material consumption, in order to optimise the manufacturing process not affecting the environment safeguard. All these requirements can be satisfied by a new way of thinking about the open mould manufacturing process that could combine addictive and subtractive techniques in order to optimize the process.

This new method is based on a fundamental consideration on the functionality of open mould. In open moulding, only few surfaces of the entire mould are useful to the part lamination and these have to be subjected to machining in order to reach specific roughness ad dimensional tolerances.

The remaining volume of the mould does not need particular accuracy and it can be roughly shaped. The idea is to reproduce the mould in two parts: an inner support of rough material and, upon it, the deposition of a thin bed of plastic material that has to follow the shape of the mould. Figure 8 shows an open mould and its relative shaped part (a) and a sequence of phases in which the processes of hybrid manufacturing for the open mould preparation is realised. In the figure are shown, consecutively: a support (b), optimized in terms of volume and shape, that can be realized from a rough block of material by milling, the deposition of material upon the support (c), that has to be layered as close as possible to the target shape, and eventual conclusive CNC machining (d), in order to conform to dimensional tolerances and roughness.

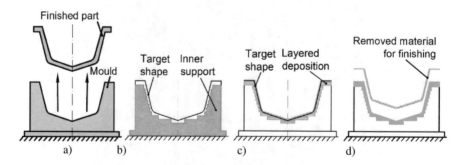

Fig. 8.   The hybrid process for open mould manufacturing

In order to improve the industriability and the precision of the process, it could be advisable that all these additive/subtractive operations can be automatized. In the laboratories of the Department of Industrial Engineering, University of Bologna is available an hybrid 3D 5 axis printer, shown in Fig. 9, able to work as addictive and sub-tractive manufacturing system at the same time by a head/nozzle replacement for both milling and addictive manufacturing. The system spans over a huge volume ($5 \times 3 \times 2$ m) and may be equipped by a nozzle in order to spray a film coat on the surface. Through a very user-friendly interface, the user can choose a process, can simulate it and then can make the system working [19].

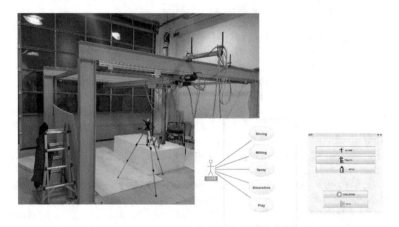

Fig. 9. The hybrid 3D printer own in the laboratories of the University of Bologna

5 Conclusions

In this paper, QDF and TRIZ analysis have been investigated in order to validate a design method for direct open moulds, by a new strategy that intends to combine additive and subtractive manufacturing in order to obtain an innovative product. The work was developed through the following steps: 1. QFD analysis composed by Six-questions analysis, evaluation matrixes and morphological matrix analysis; 2. Output of QFD analysis as Product Requirements and Conceptual Product Architecture; 3. TRIZ analysis using as Input the above mentioned QFD Output; 4. final optimized solution QFD&TRIZ achieved. Finally, from the integration of QFD and TRIZ innovative

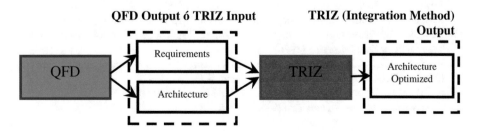

Fig. 10. QFD-TRIZ integration method

solutions, arose a proposal of a new design method for the manufacturing of direct open moulds, as shown in the following picture (Fig. 10), considering that QFD (and Integration Method) Input are "Customers' Needs".

References

1. World Commission on Environment and Development (WCED). Our Common Future. Oxford University Press, Oxford, New York (1987)
2. Rosen, M.A., Hossam, A.: Kishawy sustainable manufacturing and design: concepts. Practices Needs Sustain. **4**, 154–174 (2012)
3. Hayes, R.H., Wheelwright, S.C.: The dynamics of product-process life cycles. Harv. Bus. Rev. **57**, 127–136 (1979)
4. Francia, D., Caligiana, G., Liverani, A.: DFD evaluation for not automated products. Res. Interact. Des. **4**, 439–445 (2016)
5. U.S. Environmental Protection Agency: 1987–1993 Toxics Release Inventory; EPA-749/ C-95-004 (NTIS PB95-503793); U.S. Environmental Protection Agency, Office of Pollution Prevention and Toxics: Washington, DC (1995)
6. U.S. Department of Labor: Occupational Safety and Health Administration (OSHA). Air Contaminants, Subpart Z: Toxic and Hazardous Substances, 29 CFR 1910.1000 (1993)
7. Felix, L., Merritt, R., Williamson, A.: Evaluation of Styrene Emissions from a Shower Stall/ Bathtub Manufacturing Facility EPA-600/R-96-138 (NTIS PB97-1254); U.S. Environmental Protection Agency, Office of Pollution Prevention and Toxics: Washington, DC (1996)
8. Chan, L.K., Wu, M.L.: Quality function deployment: a comprehensive review of its concepts and methods. Qual. Eng. **15**(1), 23–35 (2002)
9. Akao, Y., Mazur, G.H.: The leading edge in QFD: past, present, and future. Int. J. Qual. Reliab. Manag. **20**(1), 20–35 (2003)
10. Vinodh, S., Chintha, S.K.: Application of fuzzy QFD for enabling leanness in a manufacturing organization. Int. J. Prod. Res. **49**(6), 1627–1644 (2011)
11. Mayda, M., Borklu, H.R.: Development of an innovative conceptual design process by using Pahl and Beitz's systematic design, TRIZ and QFD. J. Adv. Mech. Des. Syst. Manuf. **8**(3), 1–12 (2014)
12. Yeh, C.H., Huang Jay, C.Y., Yu, C.K.: Integration of four-phase QFD and TRIZ in product R&D: a notebook case study. Res. Eng. Des. **22**(3), 125–141 (2011)
13. Freddi, A.: Imparare a progettare. Pitagora, Bologna (2002)
14. Shingley, J.E., Mische, C.E., Budynas, R.G.: Progetto e costruzione di macchine. McGraw-Hill, Milano (2005)
15. Altshuller, H.: The art of inventing (And Suddenly the Inventor Appeared). Translated by Lev Shulyak, Technical Innovation Center, Worcester (1994)
16. Frizziero, L., Ricci Curbastro, F.: QFD and TRIZ methods in the mechanical design. In: YSESM– 5th Youth Symposium on Experimental Solid Mechanics, Puchov (Slovakia) (2006)
17. Terninko, J., Zusman, A., Zlotin, B.: Innovazione Sistematica, un'introduzione a TRIZ. Translated by Sergio Lorenzi (1996)
18. Ko, H.K., Kim, K.H.: Design evolution of an actuation system for manipulator upper arm based on TRIZ. J. Adv. Mech. Des. Syst. Manuf. **6**(1), 131–139 (2012)
19. Caligiana, G., Francia, D., Liverani, A.: CAD-CAM integration for 3D hybrid manufacturing. Adv. Mech. Des. Eng. Manuf. **2**, 329–337 (2016)

An Industrial Application of a TRIZ Based Eco-Design Approach

Davide Russo[✉], Caterina Rizzi, and Pierre-Emmanuel Fayemi

Department of Management, Information and Production Engineering, University of Bergamo,
Viale G. Marconi, n.5, 24044 Dalmine, BG, Italy
{davide.russo,caterina.rizzi}@unibg.it,
p.fayemi@aim-innovation.com

Abstract. ITree, a step by step procedure for supporting eco-assessment and eco-design is presented. The assessment phase is carried out combining life cycle assessment, for calculating the environmental impacts, with an innovative technique, called "IFR index", for selecting the main LCA criticalities. IFR index is inspired by Ideal Final Result tool from TRIZ, the Theory of inventive problem solving. Also part of the design phase is based on the use of TRIZ: a set of Eco-guidelines, have been conceived introducing TRIZ fundamentals onto green design. An industrial case study dealing with the production of a chemical product for the agricultural market illustrates how the method has been applied.

Keywords: Eco-design · Eco-assessment · Eco-improvement · Guidelines · TRIZ

1 Introduction

During the last years, several methods and tools for supporting eco-design have been developed. According to Byggeth and Hochschorner (2006) and Le Pochat et al. (2007), the approaches can be brought back to the following main categories:

- Methods and tools for environmental impact assessment. They study and assess environmental impacts associated with all the stages of a product life cycle, in order to highlight environmental criticalities and focus the designer attention on the most impacting aspects of the product. Such methods can be used both for the evaluation of a single product and for the comparison with a best in class.
- Methods and tools for active eco-improvement. They consist of guidelines that help designers to conceive solutions for reducing product environmental impacts. Some of them work as a step-by-step guide (checklist) that can be followed in order to maximize the product environmental performance (Bovea 2012; Russo 2015a, b; Spreafico 2016).

However, only few approaches integrate both eco-assessment and eco-improvement, and those who do, they offer only a partial integration. For this reason, companies often use different methods for assessment and improvement, which implies some limitations:

© Springer International Publishing AG 2017

G. Campana et al. (eds.), *Sustainable Design and Manufacturing 2017*, Smart Innovation,
Systems and Technologies 68, DOI 10.1007/978-3-319-57078-5_85

- The suggestions provided for improving the system can be not sufficient to cover all the shortcomings highlighted during the assessment.
- The identified limitations and the guidelines are organized with different classifications, forcing the designer to identify the most suitable suggestions according to personal experience and trial and error.

In order to overcome this limitations, in these paper, we propose a procedure integrating eco-assessment and eco-improvement, both based on the same strict classification of the phases of product lifecycle and able to suggest the proper guidelines for each identified problem (Russo 2015a, b). The approach integrates a quantitative analysis of the material flows based on abridged LCA software, called eVerEE (Masoni et al. 2004; Buttol et al. 2012), an evaluation of the critical parameters with Ideal Final Result (IFR) index, from TRIZ (Altshuller 1984) and a series of guidelines also derived from TRIZ (Russo 2016).

An extensive application of the proposed approach to improve a chemical product is here presented and discussed.

2 Proposal

In synthesis, the proposed approach is constituted by the following phases: first, a data collection campaign about the environmental impact during product lifecycle is carried out. Then, the achieved data are analyzed through a simplified LCA methodology based on the software tool and database "eVerdEE". The main highlighted criticalities of each phase of product lifecycle (pre-manufacturing, manufacturing, product use and end of life) from LCA are then evaluated through Ideal Final Result (IFR) index. On the basis of the combination between the impacts and IFR index, the more appropriate flows for intervention are identified. The most suitable suggestions within the proposed Eco TRIZ guidelines, which are structures like eVerdEE, are then applied on these flows. Finally, a second assessment with eVerdEE is executed again in order to evaluate the modifications suggested by the guidelines.

The proposed step divided procedure is summarized in the following scheme (Fig. 1):

Fig. 1. The proposed ITree procedure.

In the following the single phases are explained in detail:

2.1 PHASE 1: Data Collection About Product Lifecycle

The main data about the product lifecycle are collected:

- Data about the design: the choice of the material, the choice of the production processes and the machineries, etc.
- Data about the production: the duration of the processes, the energy required, the mass of the raw materials and of the finished product, etc.
- Data about packaging and transportation: the used pallets and packaging, the list of auxiliary materials, the distances of transportations and the logistic, etc.

2.2 PHASE 2: LCA Assessment with eVerdEE

LCA assessment approach proposed by eVerdEE SW allows us to show clearly the AS-IS situation about quantitative analysis of the criticalities of each of the main phases of product lifecycle. The environmental indicators and the percentage impact of every flow are provided by the output of eVerdEE (Fig. 2).

View Results » First Level of Detail

BASE CASE REMOTE OPEN VERTICAL CHILLED MULTI DECK

Indicator	Total	Pre-manufacture
Consumption of mineral resources (kg antimony eq)	0.0738	0.0738
Consumption of biomass (kg)	62.4	62.4
Consumption of fresh water (m^3)	5.78	5.78
Consumption of non-renewable energy (MJ)	$5.67 \ 10^5$	$2.81 \ 10^5$
Consumption of renewable energy (MJ)	70200	36500
Climate change (kg CO_2 eq)	36400	18000
Acidification (kg SO_2 eq)	307	126

Fig. 2. A partial view of results from eVerdEE SW

2.3 PHASE 3: Evaluation with IFR-Ideal Final Result

The novelty of the proposed approach is to introduce an index in order to identify critical areas on which to perform redesign actions. The index is called Ideal Final Result (IFR), as inspired by the homonymous TRIZ tool that forces the analyst to look for the ideal, magical solution to a problem (it works only when needed, only where needed, and it uses existing or very few resources). IFR indicates what could be the maximal theoretical reduction of each flow. In order to estimate this index, it is necessary to think of the ideality of the system as suggested by TRIZ. In case of energy flow for example IFR is calculated on the base of the following definition:

The "technical system" should not only be a suitable power conductor but should also operate with minimal energy losses (such as losses incurred by transformation, production of useless wastes, and withdrawal of energy with ready-made artifact).

The "IFR" system should use energy and materials only to provide the main useful function in according to TRIZ definition of IFR.

Adopting this index, it is possible to associate each flow of energy or substance in input with the maximum potential reduction that can be achieved theoretically. In this way a sensitivity analysis on all flows based on realistic design criteria is done. Using this new index, the assessment is then made not only on actual criticality of existing flows but also on possible future theoretical improvement. Percentage impact rates are calculated by simplified LCA software and the hotspots are found after the calculation of the IFR potential reduction index and the application of that index to every flow. According to our experience, the application of the IFR index can overturn the initial ranking of the percentage impact rate of the considered flows.

2.4   PHASE 4: Eco TRIZ Guidelines

Since 2009, the authors have been working on the implementation of a set of more than 300 guidelines to support green design (Russo et al. 2014), based on TRIZ. The main novelty of that work, compared to already existing methods for eco-design, was on the operative level. Adapting classical TRIZ fundamentals, that was originally conceived only for problem solving, into suggestions focused to green design, the authors tried to increase the level of detail at which other well-known guidelines usually work. Laws of Technical System Evolution (LTSE), Resources and Functionality methods, inventive principles and other classical TRIZ tools constituted the content of these new guidelines since the initial version. Therefore, several other versions have been proposed and continuously updated and enlarged in number according to experimentation conducted in industrial contexts, European projects and scientific collaborations with academic

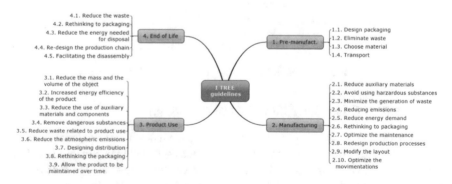

Fig. 3.   Organization of the guidelines: main groups and goals are shown.

groups working on environmental area. Step by step, they gradually acquired an easier and pragmatic aspect.

In the current version, the guidelines are organized into 4 main groups (pre-manufacturing, manufacturing, product use and end of life) and they satisfy 28 main goal. Each goals contains sub-goals (61) in turn divided into suggestions (207). The following figure summarizes the organization of the guidelines (groups and main goals) (Fig. 3).

We consider for instance the following guideline:

Description of problem type (Main goal): Allow the packaging reuse to not waste it.
Description of the sub-goal: Increase the durability of the packaging.
General suggestion: Use a packaging completely recyclable: convert raw materials in packaging, use the same packaging to return the product, exploit the packaging for other uses.
Examples: (Fig. 4)

Nutella glass

Fig. 4. An example from the last version of the Eco TRIZ guidelines.

3   Case Study

The objective of this case study is to illustrate the application of the suggested method on a chemical product, part of the family of anionic powder wetting agents for the agricultural market.

The functional unit of the analysis of the system is fixed in a pallet of 400 kg of powdered product (finished and packaged).

3.1   PHASE 1: Data Collection About Product Lifecycle

Production: The product is obtained by means of a preliminary reaction of components as sulfuric acid, sodium hydroxide, isopropyl alcohol. It is performed within a Batch reactor, with the aim of producing the semi-finished liquid. Then, the step of pulverization (i.e. the removal of water by evaporation) is carried out through an industrial atomization process.

The production of a batch of product requires about 28 h of machine time, and allows you to produce about 9,300 kg of finished product; for every kg of semi-finished liquid, 0,63 kg of exhausted acid will be generated.

The atomization plant is a continuous process that allows achieving the final product with a theoretical productivity of 130 kg/h.

For the LCA all the data related to procurement, packaging and transport of each material listed above and all the auxiliary materials required for the process that are omitted for brevity are included.

Packaging and transport: the finished product is packed within bags of white wrapping paper (weight: 100 g), coupled internally with a layer of polyethylene (50 g for the paper and 50 g for PET). The bags are laid on a wooden pallet (cm 100 × 120) of type CP1 HT; each pallet can accommodate 20 bags; each bag contains 20 kg of finished product.

LCA is evaluated in its path "from cradle to market", i.e. from pre-manufacturing (supply of materials), up to the stage of marketing and distribution. Thus, the use phase and end of life phase are not dealt with. This choice is mainly due to the poor availability of information about the methods of use of this product by the final customer.

3.2   PHASE 2: Eco Assessment with eVerdEE

The assessment was carried out using the software eVerdEE (Fig. 5). Several indicators used for the Life Cycle Assessment provide a relatively high concentration of bad values in the pre-manufacturing phase. This is probably due to the high level of complexity to the manufacturing process and the high consumption of resources (mainly energy and water) caused by the reagents involved in the production of product, upstream of the production chain.

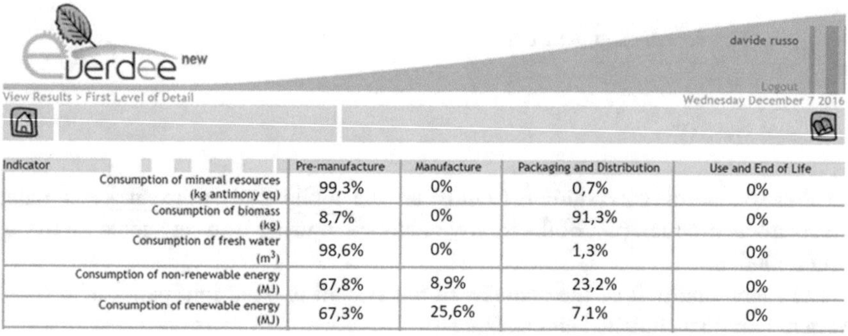

Indicator	Pre-manufacture	Manufacture	Packaging and Distribution	Use and End of Life
Consumption of mineral resources (kg antimony eq)	99,3%	0%	0,7%	0%
Consumption of biomass (kg)	8,7%	0%	91,3%	0%
Consumption of fresh water (m³)	98,6%	0%	1,3%	0%
Consumption of non-renewable energy (MJ)	67,8%	8,9%	23,2%	0%
Consumption of renewable energy (MJ)	67,3%	25,6%	7,1%	0%

Fig. 5.   Global impact of the process (eVerdEE screenshot).

As a conclusion, the phase of pre-manufacturing has the greatest environmental impact. It depends on the ingredients of the composition and there is not an element that predominates in terms of environmental impact on others.

Analyzing the manufacturing stage, as expected in this phase, there is a high contribution in terms of the generation of hazardous waste (exhausted acid) and a significant impact from the point of view of the particulate emissions (arising from the process of atomization).

The phase of packaging and distribution appears to be marginal in most indicators except for the consumption of biomass. It is mainly due to the use of wooden pallets for storage and handling of materials and paper bags for the packaging of the powder.

3.3   PHASE 3: Evaluation with Ideal Final Result

Analyzing IFR, since the quantities of the individual components are designed according to appropriate stoichiometric proportions, it is particularly difficult to think of an improvement involving the mass of the product. The production of hazardous waste exhausted acid itself is impossible to eliminate, unless considering the re-industrialization of the process. These considerations are taken into account thanks to the index of IFR, very low in all the processes of pre-manufacturing (below 10%).

From the combination of impacts and the index IFR, the flows that are most appropriate for intervention are the stage of manufacturing and distribution of the product (which was considered marginal).

One of the hypotheses concerning the improvement on the manufacturing phase considers the aspect of energy saving.

Table 1.   Eco TRIZ guidelines extract for energy saving.

SUGGESTED GUIDELINES	EXAMPLES
#3.1 IFR Choose a machine tools and service plants that use only the energy needed to transform the object in product, reduce energy conversions	
#3.3 EXPLORE OTHER TECHNOLOGIES • Choose machine tools and service plants that interact with the object more efficiently only inside the right zone to be worked (operative zone) and exclusively at the right time (i.e. light up only working zone, cool down only where heat is excessive, light up only when someone passes) • From macro to micro interaction: a good way to reduce the operative zone is to change the level of detail of the interaction • About controllability of machine tools and service plants: shift from poor controllable fields to more controllable fields following the next fields list: gravitational, mechanical, thermal, magnetic, electric and electromagnetic. • Benchmarking analysis on Web and Patent DBs	• High efficiency machine tools

3.4   PHASE 4: Eco TRIZ Guidelines

Among the various guidelines, ideality suggested to consider only the essential steps of the process, assuming the elimination of all ancillary costs and suggesting the elimination of the process of transformation from liquid to powder (Table 1). Indeed, it may be considered not necessary since the product is then used in liquid form by the client.

The advantage of the distribution of the product in a liquid form rather than powder is obvious. It would be possible to eliminate a whole phase of the production process (atomization), reducing both emissions in the air caused by the process itself and the marginal production costs (essentially related to the consumption of electric energy and methane consumption). On the other hand, there would be several problems for transport. For this reason, before creating the new scenario, the list of guidelines related to packaging and transport are also considered (Table 2). Among them there are some tips on how to optimize the loads to decrease the number of transports.

Table 2.   Eco TRIZ guidelines extract for transportation.

SUGGESTED TOOLS	EXAMPLES
#Use packaging that dissolving after the fulfillmenbf its function	• Biodegradable bag 
#Use packaging made of low energy r elate material	
# Use packaging made of recycled material	• Recyclable wine packaging • Recycled paper packaging 

3.5 PHASE 5: 2nd Eco Assessment with eVerdEE

The new scenario LCA is arranged with eVerdEE. It takes into account transportation by tanker instead of pallets, where the total number of loads is reduced from 14 (in the case of powders) to 8, despite the volume of the liquid is about 2 times wider than the volume of the powder.

Table 3 shows the outcomes of this solution.

Table 3. Screenshot of comparisons of process variants.

View Results > First Level of Detail

Indicator	Reference study	Comparison study
Consumption of mineral resources (kg antimony eq)	0,0196	0,0196
Consumption of biomass (kg)	15400	15100
Consumption of fresh water (m³)	3520	3520
Consumption of non-renewable energy (MJ)	1,82 10^7	1,51 10^7
Consumption of renewable energy (MJ)	1,29 10^6	92200

Transporting the liquid product in tankers brings substantial improvements especially in terms of eutrophication (increase of the concentration of nitrogen and phosphorus in the water with a consequent proliferation of microorganisms etc.) and acidification (increased acidity of precipitation, due to emissions of sulphur oxides in the atmosphere).

Moreover, the changes introduced for packaging and distribution appear to be in some way beneficial (although they are of less impact compared to the interventions on the manufacturing stage). From a practical point of view there is a substantial reduction in the consumption of non-renewable energy and a reduction of the impact on climate change (due both to reduction of mileage and to the increase in lots of dispatch of finished product), and an elimination of emissions of CFC (thanks to the elimination of the bags used for the packaging of the powder, in particular of their PET coating).

4 Conclusions

In this paper, a systematic procedure for Eco-design based on Eco-assessment and Eco-improvement is proposed and an application is presented.

The proposed approach has been applied to the improvement of the production of a chemical product for agricultural market.

References

Altshuller, G.S.: Creativity as an Exact Science: The Theory of the Solution of Inventive Problems. Gordon & Breach Science Publishing, New York (1984)

Bovea, M., Pérez-Belis, V.: A taxonomy of ecodesign tools for integrating environmental requirements into the product design process. J. Cleaner Prod. **20**(1), 61–71 (2012)

Byggeth, S., Hochschorner, E.: Handling trade-offs in Ecodesign tools for sustainable product development and procurement. J. Cleaner Prod. **14**(15–16), 1420–1430 (2006). doi:10.1016/j.jclepro.2005.03.024

Buttol, P., Buonamici, R., Naldesi, L., Rinaldi, C., Zamagni, A., Masoni, P.: Integrating services and tools in an ICT platform to support eco-innovation in SMEs. Clean Technol. Environ. Policy **14**, 211–221 (2012)

Le Pochat, S., Bertoluci, G., Froelich, D.: Integrating ecodesign by conducting changes in SMEs. J. Cleaner Prod. **15**(7), 671–680 (2007). doi:10.1016/j.jclepro.2006.01.004

Masoni, P., Sara, B., Scimia, E., Raggi, A.: eVerdEE: a tool for adoption of life cycle assessment in small and medium sized enterprises in Italy. Prog. Ind. Ecol. Int. J. **1**, 203–228 (2004)

Russo, D., Schöfer, M., Bersano, G.: Supporting ECO-innovation in SMEs by TRIZ Eco-guidelines. Procedia Eng. **131**, 831–839 (2015a)

Russo, D., Serafini, M., Rizzi, C.: Comparison and classification of eco improvement methods. In: DS 80-1 Proceedings of the 20th International Conference on Engineering Design (ICED 2015) vol. 1: Design for Life, Milan, Italy, 27–30 July 2015 (2015b)

Russo, D., Rizzi, C., Montelisciani, G.: Inventive guidelines for a TRIZ-based eco-design matrix. J. Cleaner Prod. **76**, 95–105 (2014)

Russo, D., Serafini, M., Rizzi, C.: Is TRIZ an Ecodesign Method? In: Setchi, R., Howlett, R., Liu, Y., Theobald, P. (eds.) Sustainable Design and Manufacturing 2016, pp. 525–535. Springer International Publishing, Cham (2016)

Spreafico, C., et al.: TRIZ industrial case studies: a critical survey. Procedia CIRP **39**, 51–56 (2016)

An Eco-Design Methodology Based on a-LCA and TRIZ

Giacomo Bersano[1(✉)], Pierre-Emmanuel Fayemi[1],
Malte Schoefer[2], and Christian Spreafico[3]

[1] Levallois-Perret, France
{g.bersano,p.fayemi}@aim-innovation.com
[2] phi Engineering Services AG, Sonnenbergstrasse 41, 8603 Schwerzenbach, Switzerland
malte.schoefer@gmx.net
[3] Department of Management, information and Production Engineering, University of Bergamo,
Viale G. Marconi, n.5, 24044 Dalmine, Bergamo, Italy
christian.spreafico@unibg.it

Abstract. An Eco-design methodology based on two abridged Life Cycle Assessment (aLCA) tools and TRIZ Eco guidelines is presented. This method is one of the outputs of the European project REMake, which developed and tested new approaches for eco-innovation and optimization of energy and materials for 250 manufacturing SMEs in six countries. Unlike other Eco-design methods, this method couples a simplified but solid assessment phase, realized with an abridged LCA, to an advanced and structured product improvement phase (that normally consists of basic design suggestions). A set of over 300 Eco-design guidelines, coming from problem solving techniques as TRIZ and conceptual design are selectively introduced to develop design variants to the given system with the aim of providing a lower global environmental impact. The advantages and limits of the method have been evaluated versus other methods inside European project REMake, and in this article are presented two case study realized in an independent way by two research groups that have tested it in two industrial case studies.

Keywords: Eco-assessment · Eco-design · Guidelines · TRIZ

1 Introduction

Nowadays, the importance of sustainable development no longer needs highlighting. The scarcity of resources and higher levels of pollution are progressively orienting consumers and therefore industry towards cleaner production.

This topic is currently at the heart of reflections in European Community, as testified by various initiatives as European Project CIP REMake (Resources and Material efficiency in manufacturing industries, 2009–2012), which had the following goals:

- to help small and medium sized enterprises (SMEs) to grow along the green development in the manufacturing industry by focusing on energy and resource efficiency and

G. Campana et al. (eds.), *Sustainable Design and Manufacturing 2017*, Smart Innovation,
Systems and Technologies 68, DOI 10.1007/978-3-319-57078-5_86

- to provide SMEs advice and support in the fields of innovation and change management, Life Cycle Assessment, Eco-design and eco-innovation as well as project management.

Active Innovation Management, as project member, has been strongly involved in the setting-up, testing and training of new eco-innovation approaches adapted to SMEs.

Based on a state of the art-analysis on current methods for eco-design, Life Cycle Thinking (LCT) and LCA are highly recommended for industries, but their penetration is still weak. Amongst the causes of this poor penetration, some authors indicate the complexity of the said methods (Dutch Ministry on Environment 2010). This fact is strongly limiting the adoption of LCA in SMEs.

Moreover, in Eco-design it is common practice to under-evaluate the role of resources; actually, most methods focus only on materials and energy and this quite superficially. For instance, the "companies' guidelines" for the choice of material are limited to a simple classification that ranges from good materials, the usage of which is unrestricted, to awful materials, which should not to be taken into account (Dutch Ministry on Environment 2010; Knight and Jenkins 2009). More generally, all methods are very effective either during the assessment phase or the improvement phase, but not in both.

Another mapping of eco design methods/tools tested in the REMake project (Russo 2011) is presented in the following Fig. 1; the methods are organized in three sequential phases (pre-diagnosis, audit, and implementation) and three major areas:

- Potentiality analysis (methods/tools: REMake Self Assessment Tool–SAT, INNO-WATER Water management Audit, Material Flow Analysis STAN, Standard & regulation compliance REMake Guidebook, Environmental Alternative Assessment EAMA)

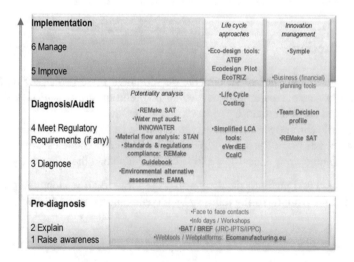

Fig. 1. Synthetic table of eco-innovation methods according to EU project REMake.

- Life cycle approaches (methods/tools: ATEP, EcoDesign Pilot, EcoTRIZ guidelines, Life Cycle Cost, eVerdee, CCalc)
- Innovation management (methods/tools: Symple, various Business planning tools, Team Decision Profile, REMake SAT).

Further information on these methods/tools is available on REMake web site.

The major findings of the analysis of state of the art are the following:

1. The eco-innovation activity is mainly composed of two phases: one is the assessment of the existing product and its impact on the environment, the other one is the improvement of the product by adopting some de-sign guidelines
2. A plethora of LCA methods exist but most of these methods are too complex and long to meet the requirements of Eco-design teams in SMEs.
3. Once a complex LCA has been performed, the support provided by existing Eco-design methods to designers is very weak, not going beyond obvious and generic suggestions.
4. The eco-innovation activity is mainly composed of two phases: one is the assessment of the existing product and its impact on the environment, the other one is the improvement of the product by adopting some de-sign guidelines
5. A plethora of LCA methods exist but most of these methods are too complex and long to meet the requirements of Eco-design teams in SMEs.
6. Once a complex LCA has been performed, the support provided by existing Eco-design methods to designers is very weak, not going beyond obvious and generic suggestions.

The scope of this chapter is to present a methodology developed in order to: (1) support eco innovation in SMEs, being user friendly, even for designers with limited knowledge in LCA and Eco-design; (2) provide to designers easy-to-understand outcomes and data; (3) provide problem-based improvement heuristics and examples for communicating strategies and results in a simple and powerful way.

Limitation of the method: precaution is necessary when environmental certifications are needed, the assessment phase being by choice simplified.

The result of this work is provided in an exemplary case study, dealing with machine tools.

This methodology has been finalized after a benchmarking process among several other LCA and Eco-design methods: five LCA and Eco-design methods have been tested in five case studies. In order to be able to compare the methods under neutral conditions, all five methods have finally been applied on the same case study (redesign of a mechanical component). The following tools have been tested: French Standard NF E 01-005 (also known as ATEP), EcoTRIZ-Guidelines & eVerdEE, Ecodesign Pilot, CcaLC, Simapro.

After the evaluation of the application of the methods on the case studies and based on the results which were produced, the REMake partners performed an evaluation of the respective methods according to two performance criteria. The first criterion is related to the method's potential for Eco-assessment and the second criterion is associated to the potential for eco-improvement. The score for the two criteria ranged from

"0" for "no performance capacity" to "5" for "tool of reference". Figure 2 presents the result of this evaluation process.

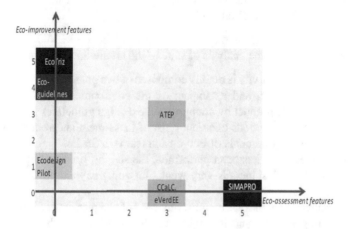

Fig. 2. Mapping various Eco design tools (Source: REMake project).

From Fig. 2, it can be inferred that TRIZ (Altshuller 1984) based eco design has the best performance in terms of Eco-improvement/eco-innovation whereas eVerdEE is at a medium advanced level in terms of Eco-assessment. Considering SMEs typical budget constraints, for the development of the method the freeware eVerdEE has been then preferred as an input.

2 Assessment Phase - An Introduction to Simplified LCA

LCA is the most established and well-developed tool to evaluate the environmental impacts of a product or a service throughout its life cycle. The Swiss Agency of the Environment and the Dutch Ministry of the Environment state: "Although LCA is a good tool to assess the environmental performance of a product, and although it is widely adopted by designer, it is time consuming and costly and results need to be interpreted and weighted" (Consultants 2000; Hur and Lee 2005). Other barriers to a wider LCA diffusion (mainly in SMEs) are:

- complexity of data collection;
- complexity of interpretation of results;
- expensive Software and databases;
- high LCA required knowledge;
- no support provided to designers to improve situation AS-IS.

Therefore, there is a need for simplified methods that require less cost, time and effort but yet provide similar results (Hur and Lee 2005).

Specific simplified (or abridged or streamlined) LCA methods have thus been developed (Hur and Lee 2005; Hochschorner and Finnveden 2003) and different depth levels of LCA analysis have been defined (Wenzel 1998).

In order to improve the LCA approach, some specific development projects have been supported by the EC (European community) such as the E-LCA (Sara et al. 2002) and E-LCA2 (ENEA 2002) projects. The goal of these projects was to develop a simplified LCA methodology with a related software tool and a database called Abridged LCA that:

- has to maintain the life cycle approach, simplifying the approach of ISO 14040;
- minimizes time and resource investments and doesn't require high LCA knowledge;
- has to be clear end easy to use;
- has to contain a high quantity of supporting "background" information such as databases of substances and processes.

Even if these improvements have increased the ease of use, nevertheless the interpretation of results requires yet expert analysts. Therefore, databases of materials and processes are often inadequate.

In this method, abridged LCA is not exploited as a tool for environmental certification but it is integrated into an Eco-design procedure as a strategic tool for the identification of the hotspots to work on.

Always having in mind the SME constraints, considering a rough time scale of the design activity, a detailed quantified LCA analysis could need some months, a simplified LCA could demand a few weeks. This is still too much for most of SMEs as seen in the case studies. This large amount of time and complexity is an important reason to prevent SME to adopt this approach. In order to manage these problems and provide companies with a simplified but nevertheless effective LCA method, the French Centre Technique des Industries Mécaniques (CETIM), in 2010, has developed a simplified approach of eco-design, which has been converted later into a French Environmental Standard, named NF E 01-005 - Eco-conception des produits mécaniques (ANFOR).

This process is formed of a multiple choice questionnaire, capable of arriving to the definition of a product environmental profile, indicating in a qualitative way of a scale from 0 to 4 the following phases of a product life cycle: Raw material extraction, Manufacturing, Use, End of life, Hazardous substances, Transport, Packaging.

According to his answers, the analyst is guided through an algorithm which leads to the products environmental profile. This profile allows comparing in a qualitative manner the different impacts of the product's environmental aspects. Based on these results the analyst decides on which aspects he or she wants to concentrate his/her design efforts. The typical length for the application of this preliminary qualitative analysis is only a few hours.

After the realization of this audit, the following steps are suggested:

- Identification of most critical phases/aspects/adapted improvement indicators
- Choice of suggested strategies for eco design within a brainstorming
- Calculation of improvements derived by implementation of these strategies.

3 The Integrated Approach

Starting from these assessment methods, the eco-guidelines based on TRIZ theory have been developed and integrated in order to jointly provide a quantitative assessment of ecological impact for a product or process, and to provide to designer relevant improvement strategies.

Firstly, it is necessary to define the goal of the activity, indicating what are the expected outcomes, the timing and the nature of the topic (process or product).

In order to map a given process, previously to the use of LCA software, a system modeling is done, using IDEF0 language. The IDEF0 language is an updated version of the Structured Analysis and Design Technique (SADT) (IDEF0 1993). The aim of the modeling phase is to chart all the data of process and products, keeping track of the quality and others metadata and additional information, needed for the use of chosen LCA software, eVerdEE. Employing such a model allows us to show clearly the AS-IS situation; in particular, it is easy to define all flows as well as their loops, with the values really used into eVerdEE during the quantitative analysis.

Then IDEF0 modeling is used again after the eVerdEE calculation and it is enriched of specific indicators associated to the flows and the operations inside each phase.

Particularly, in order to highlight the hotspots to improve, a diagram for every considered environmental indicator has to be made, and every flow is mapped with its percentage impact rate on the considered indicator. The environmental indicators and the percentage impact of every flow are provided by the output of eVerdEE.

To amplify the simplicity and the immediately of reading, in addition to a label with the flow rate, higher is the percentage rate, higher is the size of the arrow referring to that flow.

Finally, with the aim to identify the hotspot, that is the flow with the greater potential improvement, the IFR index is applied to every flow. In order to reduce the environmental impact of the identified hotspot, the guidelines are then applied.

A specific path is chosen on material flow analysis for a process (a useful tool is the software STAN) or on product environmental profile using French Standard NF E 01 005 (i.e. Mapeco). Then, if there is the need of quantified results and a detailed LCA, the Everdee tool is applied.

Once this activity is realized, the inventive activity can be launched, using eco TRIZ guidelines. It is to be noted that the Life cycle approach has been chosen as a foundation system in order to integrate inventing TRIZ capabilities to provide a complete approach; then, eco TRIZ guidelines have been structured following the life phases decomposition, as in eVerdEE, that is a software for abridged LCA developed by ENEA (Italy). The phases are, respectively, as follows:

- Pre-manufacturing, i.e. the identification of all elements bought and that will be transformed inside the company later on.
- Manufacturing, i.e. the industrial transformation of components into a product.
- Operation (product use), i.e. the time during which the product operates, including maintenance activity and consumptions.
- End of life, i.e. the recycling part of the product.

At the end, the LCA analysis can be used to quantify the improvement. The overall Eco-design process is shown in Fig. 3.

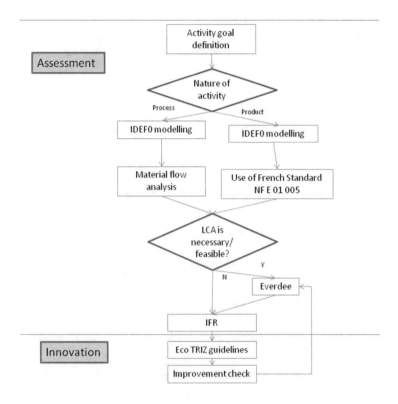

Fig. 3. Proposed Eco-design approach.

4   Illustrative Case Studies

The company is a developer and manufacturer of precision mechanics for the automotive sector. During its 80-year history the company specialized in the manufacture of CNC gear shaving and gear honing machines. The system under study is a specific honing machine that according to the company represents the latest frontier of post-hardening fine finishing. The company wanted to improve the preliminary design of the first machine, in order to make it more industrial, profitable and in order to reduce its environmental impact by significantly reducing its energy and material consumption.

The main characteristics of the machine are the following: Honing wheel with external tooth finishing design, easy accessibility for maintenance and operations, Vertical work piece axis for an easier link to automation, main machining axes operated by direct-drive motors, work piece/tool synchronization, axis operated by electro-spindles, integrated pre-process measuring station.

At the beginning of the study the company defined the timeline of the analysis as one month. Because of this short delay and the machine complexity (more than 5000 electronic, hydraulic, pneumatic and metallic parts), the chosen approach has been then French Standard NF E 01 005 and the IFR.

The analyst filled the related multiple choice questionnaire jointly with company Technical Manager, arriving to the definition of the environmental profile of the cogwheel machine (please refer to Fig. 4).

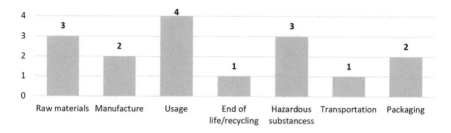

Fig. 4. Environmental profile of the honing machine.

The analysis showed that the machines highest environmental impacts are, respectively:

- the raw material necessary for its production,
- the consumption of energy and resources during the operation phase,
- the toxic materials used (as lubricants and rare materials in electronics).

Thus, it was jointly decided to focus the Eco-design activities on these to aspects.

For this case study, a Pareto diagram for major consumers in terms of material and energy was created and used in working sessions on these topics. In these working sessions the R&D team started generating some ideas on how to reduce the impact, then was driven by dedicated eco-TRIZ guidelines selected according to the focus.

Analyzing the use phase of the machine, the methodologist, together with the mechanical designers, applied the Eco-Guidelines aimed at reducing the amount of auxiliary materials and components. These guidelines suggest as example to:

- modify the system accordingly to the previously identified Idea Final Result,
- explore other technologies dealing with the same problem and bench-mark them,
- reduce the energy conversion in the process in order to make auxiliary materials for the purpose of energy conversion obsolete,
- make the process or specific actions resonant, i.e. replace a continuous action by a periodic or pulsating one;
- dynamize the system, i.e. make the system auto-adaptable to different process conditions and requirements;
- shift to auxiliary materials that are available at low or no added energetic and material cost in the environment, i.e. use or reuse material which has been used for the same other purposes (Fig. 8);
- shift from mechanic actions to actions fulfilled by physical or chemical effects.

During a two 2 h workshop, the group developed 60 ideas to reduce the amount of material for manufacturing the cogwheel machine and to reduce the amount of auxiliary materials necessary for the use of the machine.

In a third workshop five of these ideas have finally been chosen to enter into a detailed study.

According to participants, those five ideas would probably lead to reduction of motors power consumption and mass by 20% and to a reduction of work consumables by 50%. These major savings in terms of material and energy consumption go along with substantial cost reductions.

According to the CTO, "the approach has provided ideas for the second machine we are developing right now with an immediate impact; moreover, we also learned a very efficient method for innovation with a concrete demonstration"

5 Conclusions

The approach mentioned above has been developed to support eco design in SMEs. It has been tested in real case studies in manufacturing SMEs in France and Italy with very good results. Strengths of the method are:

- the reduced time for the assessment (from few hours to few days), to be compared to long LCA assessment,
- the IDEF0 language is user friendly and identifies quickly the product/process spot to improve,
- eco-TRIZ guidelines high efficiency in robust solutions generation.

It is also important to notice that the use of IFR as "direction" for the design orients the solutions through a radical reduction of resources, in a more global sustainable development direction. In one case, the design group, together with the methodologist, was able to reduce the material consumption of a production machine by somewhat 50%. Considering the success factors for the implementation in SMEs, the commitment of top management is very important to provide the good impulse on design teams. Another key element for choosing to use this approach was the timely presence of a product or a process needing a redesign or an optimization.

The method is then currently taught in some universities, in order to enlarge the base of application.

References

Altshuller, G.S.: Creativity as an Exact Science: The Theory of the Solution of Inventive Problems. Gordon and Breach Science Publishers, New York (1984). ISBN 9780677212302

Consultants, P.R.: Eco-indicator 99 Manual for designers. Ministry of Housing, Spatial Planning and the Environment, The Hague (2000)

Dutch Ministry on Environment: Eco-indicator 99: Manual for Designers (2010). http://www.pre.nl/

ENEA: Web site and database to help SMEs adopt Integrated Product Policy 2010 (2002). http://www.elca.enea.it/

Finnveden, G.: On the limitations of life cycle assessment and environmental systems analysis tools in general. Intl. J. Life Cycle Assess. **5**(4), 229–238 (2000)

Finnveden, G., Moberg, Å.: Environmental systems analysis tools - an overview. J. Clean. Prod. **13**(12), 1165–1173 (2005)

Finnveden, G., Hauschild, M.Z., et al.: Recent developments in life cycle assessment. J. Environ. Manage. **91**(1), 1–21 (2009)

Hochschorner, E., Finnveden, G.: Evaluation of two simplified life cycle assessment methods. Intl. J. Life Cycle Assess. **8**(3), 119–128 (2003)

Hur, T., Lee, J., et al.: Simplified LCA and matrix methods in identifying the environmental aspects of a product system. J. Environ. Manage. **75**(3), 229–237 (2005)

IDEF0: Integration Definition for Function Modelling. FIPS Publication 183. National Institute of Standards (1993)

Knight, P., Jenkins, J.O.: Adopting and applying eco-design techniques: a practitioners perspective. J. Clean. Prod. **17**(5), 549–558 (2009)

Luttropp, C., Lagerstedt, J.: EcoDesign and The Ten Golden Rules: generic advice for merging environmental aspects into product development. J. Clean. Prod. **14**(15–16), 1396–1408 (2006)

Ness, B., Urbel-Piirsalu, E., et al.: Categorising tools for sustainability assessment. Ecol. Econ. **60**(3), 498–508 (2007)

Russo, D., Bersano, G., et al.: European testing of the efficiency of TRIZ in eco-innovation projects for manufacturing SMEs. Procedia Eng. **9**, 157–171 (2011)

Sára, B., Buonamici, R., et al.: Methodological framework and prototype realisation of a WEB-based LCA tool for Small and Medium sized Enterprises (2002)

Wenzel, H.: Application dependency of LCA methodology: key variables and their mode of influencing the method. Intl. J. Life Cycle Assess. **3**(5), 281–288 (1998)

Author Index

Printed in the United States
By Bookmasters